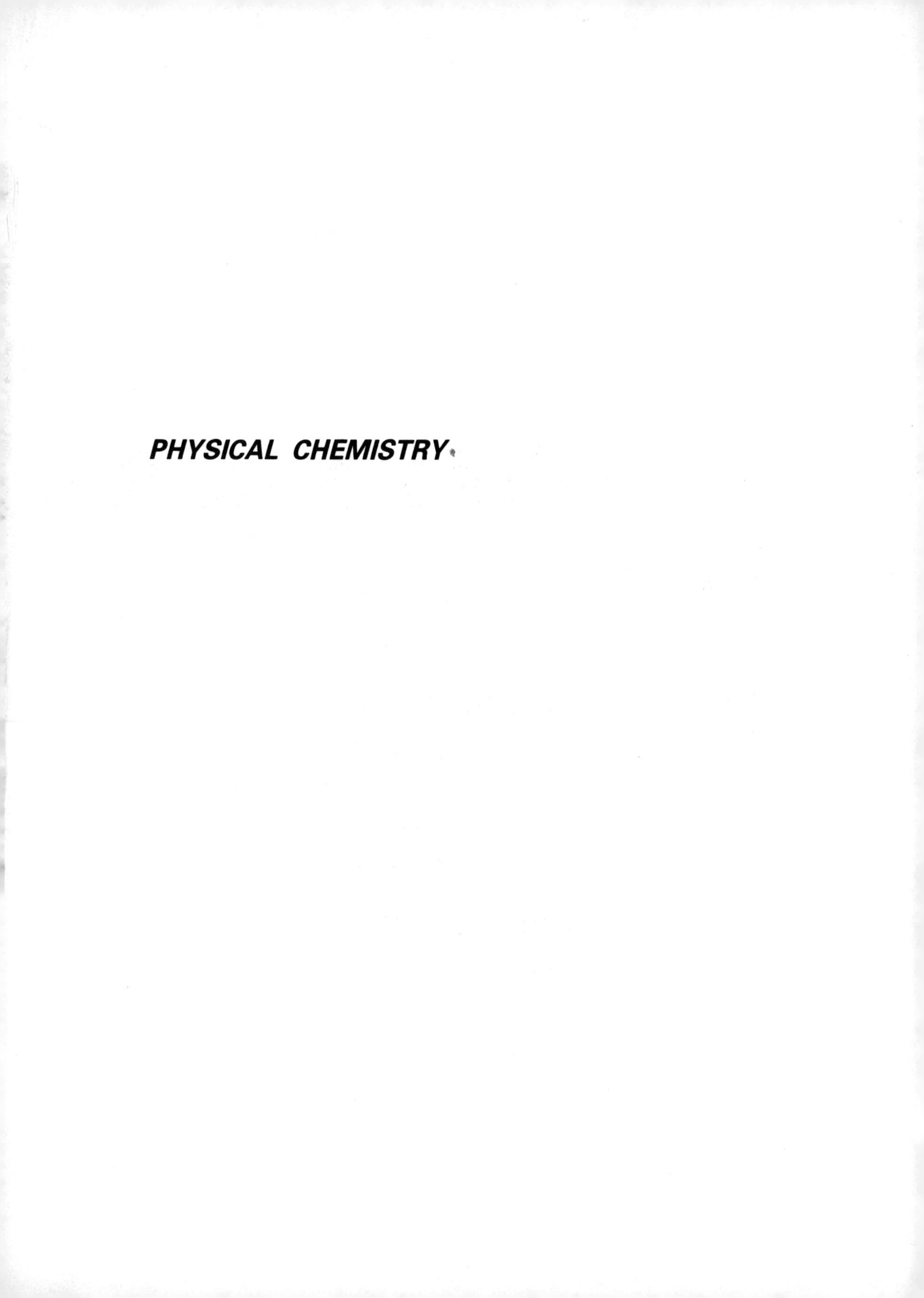

PHYSICAL CHEMISTRY

PHYSICAL

Ignacio Tinoco, Jr.
University of California, Berkeley
Kenneth Sauer
University of California, Berkeley
James C. Wang
Harvard University

CHEMISTRY

Principles and Applications in Biological Sciences

PRENTICE-HALL, INC., ENGLEWOOD CLIFFS, NEW JERSEY 07632

Library of Congress Cataloging in Publication Data

Tinoco, Ignacio.
PHYSICAL CHEMISTRY.

Includes bibliographies and index.
1. Biological chemistry. 2. Chemistry, Physical and theoretical. I. Sauer, Kenneth, (date) joint author. II. Wang, James C., joint author. III. Title.

QH345.T56 541'.02'4574 77-25417
ISBN 0-13-665901-2

PHYSICAL CHEMISTRY
Principles and Applications in Biological Sciences

Ignacio Tinoco, Jr., Kenneth Sauer, and James C. Wang

Printed in the United States of America

10 9 8 7 6 5 4 3

PRENTICE-HALL INTERNATIONAL, INC., *London*
PRENTICE-HALL OF AUSTRALIA PTY. LIMITED, *Sydney*
PRENTICE-HALL OF CANADA, LTD., *Toronto*
PRENTICE-HALL OF INDIA PRIVATE LIMITED, *New Delhi*
PRENTICE-HALL OF JAPAN, INC., *Tokyo*
PRENTICE-HALL OF SOUTHEAST ASIA PTE. LTD., *Singapore*
WHITEHALL BOOKS LIMITED, *Wellington, New Zealand*

Contents

9 QUANTUM MECHANICS 366

10 SPECTROSCOPY 421

Preface

We have been teaching a course in physical chemistry for the biological science majors at Berkeley for the past fifteen years. Most of the students are undergraduates from biochemistry, physiology, bacteriology, botany, and zoology departments. They include most of the premedical students. In addition a few graduate students in the biological sciences are always represented. We have found it difficult to find a suitable text for these students. Standard physical chemistry texts have too much material irrelevant to the biochemist, and they mainly ignore macromolecules and biological systems. Life science physical chemistry texts often omit too much fundamental physical chemistry, and although they discuss biological problems, they do so only at an elementary level. Modern physical methods used by molecular biologists and biochemists, such as density gradient sedimentation, fluorescence energy transfer, and electron microscopy, are usually not included.

We have written a book to provide a fundamental understanding of the physical chemistry important to biochemists and biologists. Each physical chemical principle is related and applied to biological problems. We thus try to teach basic physical chemistry, while motivating the student by stressing biochemical applications. We also discuss most of the physical methods and ideas that modern biological scientists use.

The book contains an introductory chapter describing the types of problems in biology or medicine to which physical chemistry can be applied. The next four chapters cover thermodynamics. The emphasis is on solution thermodynamics. One complete chapter deals with the equilibrium constant, activities, and free energy. A chapter on physical equilibria includes equilibrium dialysis, active transport, surfaces, and the use of colligative properties in measuring molecular weights.

Transport properties are treated in Chapter 6. A short discussion on kinetic theory of gases is followed by a thorough description of diffusion, viscosity, sedimentation, and electrophoresis. The application of each of these properties to the determination of size and shape of macromolecules is stressed. Molecular weights, axial ratios, hydration, and flexibility of nucleic acids and proteins are described.

There are two chapters on kinetics. The kinetics and mechanisms of reactions in solution are emphasized. The chapter on enzyme kinetics includes a discussion of temperature-jump kinetics.

The first eight chapters require only a minimum of mathematical knowledge beyond algebra. Although calculus is used in these chapters, only differentiation and integration of powers of x are required. Partial derivatives are introduced, but the text and problems mainly can be understood without them. The next four chapters assume more mathematical confidence on the part of the students. The mathematical steps are explained in detail (as before), but the student will feel more confident with a course in calculus for background.

Chapter 9 introduces quantum mechanics to provide a vocabulary for spectroscopy and chemical structure. A particle in a box is treated and applied to the free electron model for conjugated molecules. Qualitative discussions of molecular orbitals, and molecular structure and reactivity are given. The spectroscopy chapter deals mainly with electronic spectra of organic molecules. Ultraviolet and visible absorption, fluorescence, circular dichroism, and optical rotatory dispersion are discussed. Application of each of these methods to the study of proteins and nucleic acids is included. A short section on nuclear magnetic resonance is presented.

The application of statistical methods to macromolecules is emphasized in Chapter 11. Cooperative binding of small molecules to a macromolecule is described. The random walk is applied to diffusion and polymer dimensions. Helix-coil transitions in polypeptides and polynucleotides are discussed. The statistical thermodynamic definitions of energy, work, and entropy are described.

The last chapter gives the basic fundamentals of x-ray diffraction. Applications to structures of macromolecules are discussed. Qualitative descriptions of neutron scattering, electron microscopy, and their application to biological samples are given.

The Appendices contain tables of thermodynamic data, conversion factors, abbreviations, and structures of biological molecules mentioned in the text. The reader is encouraged to keep looking through the Appendix while using the book.

At Berkeley we have found it possible to cover most of the book in two quarters. This keeps the students and the instructor working hard. Two semesters or three quarters might be a more reasonable time to spend on the material covered in the book. The first five chapters form a complete treatment of the thermodynamics most useful for the biological scientists. The remaining chapters

can essentially stand by themselves. The instructor can teach them in any order or omit any of them. For example, the applications in the spectroscopy chapter do not require knowledge of quantum mechanics.

We have tried to make this book useful to students and instructors with a wide range of backgrounds. There are many worked examples in the text; there are many problems at the end of each chapter. Each chapter has a summary and also a review of the mathematics needed for the chapter. The first half of the book uses less mathematics (and describes it more fully) than the last half. There are references to the current literature for most chapters. We hope that the readers of this book will tell us about aspects of the book which need improving.

We are happy to acknowledge the many people who contributed to this project. In the past several years, the students of Chemistry 109A,B at Berkeley and the teaching assistants made many helpful suggestions. Our faculty colleagues who also taught the course were often properly critical. Dr. Leonard Peller, University of California, San Francisco, read essentially the entire book and made detailed criticisms. Dr. Helen Berman carefully reviewed Chapter 12. Reviewers chosen by Prentice-Hall gave useful suggestions. Drs. Paul Hartig, Che-Hung Lee, and Esther Yang kindly contributed original data. Suzanne Pfeffer and Robert Sauer solved all the problems from several chapters and helped to clarify their presentation. We wish to thank all of these people.

We are grateful to Marshall Tuttle who typed and retyped the various editions of the manuscript. His cheery disposition encouraged us to continue.

Berkeley, California

Cambridge, Massachusetts

IGNACIO TINOCO, JR.
KENNETH SAUER
JAMES C. WANG

1

Introduction

Physical chemistry is a group of principles and methods that are helpful in solving many different types of problems. Just as the methods of algebra can be applied to many problems, so can the methods of thermodynamics. In the following chapters we shall present the principles of thermodynamics, kinetics, quantum mechanics, and statistical thermodynamics. We shall also discuss various experimentally measurable properties, such as viscosity, light absorption, and x-ray diffraction. All these experimental and theoretical methods can give us useful information about the part of the universe we are interested in. We will stress biochemical and biological applications in this book, but it is up to the reader to see how the methods presented can be applied to other specific problems of interest. For applications of physical chemistry to other areas, the reader is directed to standard physical chemistry texts. Biochemistry and molecular biology texts can provide specific information about such areas as enzyme mechanisms, metabolic paths, and structure of membranes. Finally, a good physics textbook is useful for learning or reviewing the fundamentals of forces, charges, photons, and energy. A list of such books is given at the end of the chapter.

In this introductory chapter we shall give a few examples of the types of problems that physical chemistry can solve.

Energy Efficiency

The minimum amount of food that an average, adult human being at rest needs every day to survive corresponds to about 1500 kcal of energy. This 1500 kcal day^{-1} is the basal metabolic rate of a 60-kg human; it is equal to 70 W of power. Light daily activity, such as walking, studying, and eating, increases the human energy requirement to about 2000 kcal day^{-1}, or roughly 100 W. The energy we eat is food energy converted from solar energy by photosynthesis. We convert this food energy directly to heat or to some form of work outside the body. There is obviously a long chain of energy-converting steps involved in this solar-energy to human-energy process. It is to our advantage to understand these steps and to be able to improve their efficiency. The efficient conversion of energy from one form to another is the subject of thermodynamics.

The amount of solar energy reaching the earth's atmosphere is 1.4 kW m^{-2} (2 cal min^{-1} cm^{-2}), but only about one half of this reaches the surface of the earth. This is the total energy input available to us for food productivity. Naturally occurring photosynthesis in plants is only about 1% efficient, so about 100 kcal of food can be grown per day per square meter of the earth's surface. This means that a minimum of 20 m^2 is needed to raise food for one person. However, in the United States we spend about 10 cal of energy to produce and prepare 1 cal of food (Steinhart and Steinhart, 1974). Figure 1.1 shows how this added energy has increased over the years. The energy input occurs about one fourth at the farm (fertilizer, machinery, irrigation), one third in refrigeration and cooking, and the rest in the processing industry. How can knowledge of physical chemistry and thermodynamics help? Just knowing the amounts of energy involved may make us more conscious of the real cost of eating highly processed foods. But more important is the fact that thermodynamics can tell us the maximum possible efficiency for performing a certain task. The first and second laws of thermodynamics tell us the maximum amount of heat or work that can be transferred in a particular process for a given energy input. Analysis of U.S. energy uses shows an average efficiency of only 10% of the maximum efficiency possible (*Physics Today*, 1975). For example, refrigeration efficiency is only 4% of theoretical thermodynamic efficiency, and truck transportation is 10% of theoretical efficiency. Improving refrigeration and transport efficiency would allow us to have food variety all year round yet not spend 10 times its food-energy value.

How about human efficiency? The energy cost of transportation can be defined as the amount of power used divided by the weight and velocity of the moving object. The energy cost is unitless; it is energy input per time (power) divided by the weight (a force) times distance moved per time. The minimum cost of walking occurs at a velocity of 3.8 mph for a 150-lb human. The metabolic rate for this fast walker is about 450 W and the cost of transport is 0.38. The minimum cost of transport for a bicyclist is 0.1 and is the most efficient transportation by an animal; only large fish or whales may be as efficient. Freight trains and freight steamers are the most efficient form of any kind of transportation; their minimum cost is about 0.01 (Tucker, 1975).

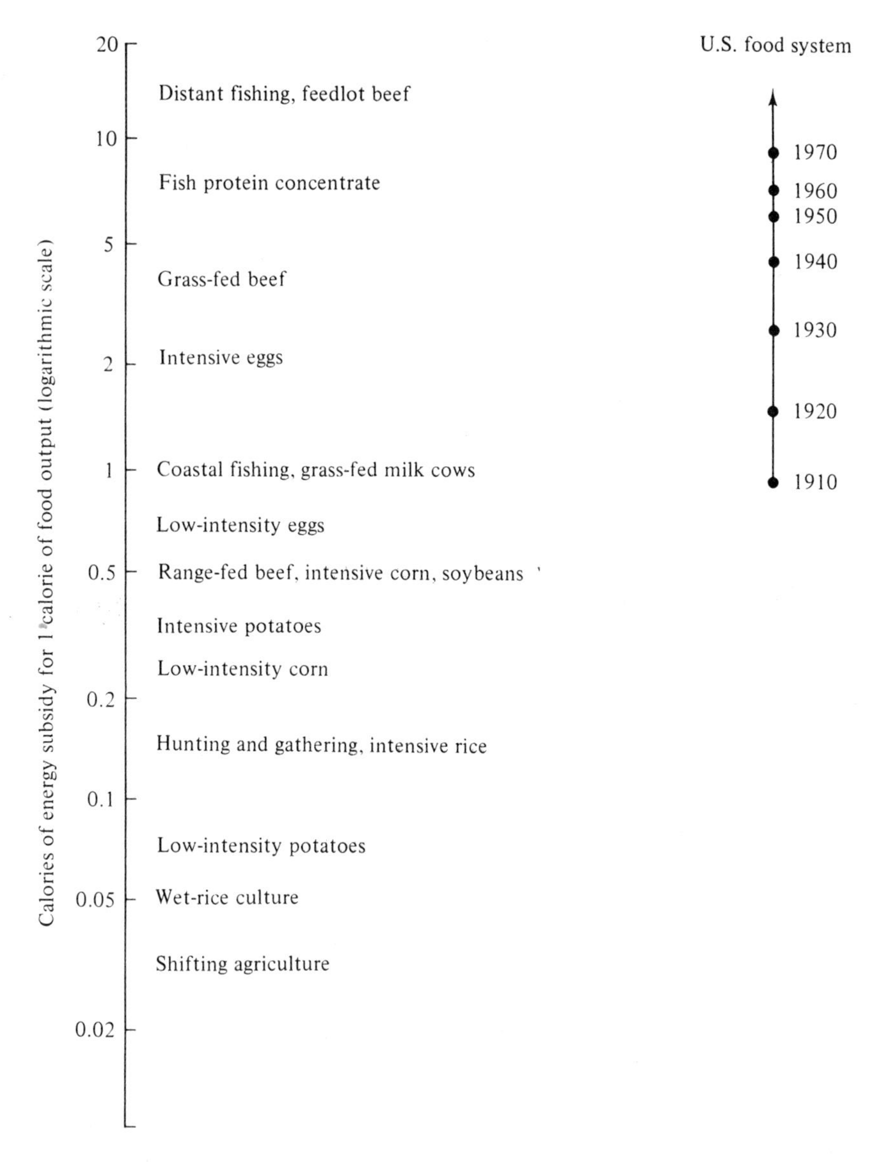

Fig. 1.1 Number of calories used to produce 1 calorie of food by various methods. The average number of calories used in the United States for all food production is also given. [Data from J. S. Steinhart and C. E. Steinhart, *Science 184*, 307 (1974), Fig. 5.] Originally published in *Energy: Sources, Use, and Role in Human Affairs* by C. E. Steinhart and J. S. Steinhart, © 1974 by Wadsworth Publishing Co., Belmont, Calif., 94002. Used by permission of the publisher, Duxbury Press.

Human Age

The human life span is of interest to many of us, so reports of populations that normally live to be 120 years old or more attract attention. The reasons proposed for their longevity range from their lack of the type of tension associated with life in a modern society to the presence of unique nutrients in their diets. However, before trying to understand their reported longevity, we must be sure that it exists. Systematic record of births and deaths are very scarce in these populations. An independent method of determining the age of these old people would be very useful.

A kinetic method has been developed for determing the age of human beings (Helfman and Boda, 1976). It is based on the finding that L-amino acids tend to racemize (to be converted to D-amino acids) with time. The D-residues accumulate in the metabolically inert protein in tooth dentine. For aspartic acid in tooth dentine, the following relation was found:

$$\ln [1 + (\mathrm{D}/\mathrm{L})] = 7.87 \times 10^{-4} t + 0.014$$

where ln = natural logarithm
D/L = ratio of D- to L-aspartic acid measured
t = age of tooth, yr

The authors claim that this equation will give the age of a person within 10%. The uncertainty depends partly on experimental error in determining (D/L) and partly on the body temperature of the individual (which affects the rate of racemization).

Detection of Polarization

We can all see that the sky is blue and we should all know why the sky is blue—the short-wavelength blue light is scattered more than the long-wavelength red light. However, insects can see that the sky is polarized while we need Polaroids or other polarizing devices to detect this. It has been proposed that some insects, such as bees and ants, use the polarization of the sky for navigation (Wehner, 1976). Insects and human beings use the same molecule (rhodopsin) for detection of light. The absorbing part of rhodopsin is retinal, which contains a conjugated system of double bonds. This π system of electrons will absorb only light of certain wavelengths and with a polarization that is parallel to the axis of the conjugated system. The molecular photoreceptor in vertebrates and invertebrates is therefore sensitive to both color and polarization of light. Spectroscopists know that the absorption of all molecules depends on wavelength and polarization.

The difference between insects and human beings is that the insects have the rhodopsin rigidly oriented in the eye, whereas human rhodopsin is free to rotate in the eye. The fixed orientation of retinal in the insects may provide the necessary polarization-dependent receptor for navigation.

Hemoglobin

Hemoglobin is a protein of molecular weight about 66,000 which carries oxygen in many animals from the lungs to all cells. Its complete three-dimensional structure has been determined by x-ray diffraction. This means that we know the position of each of the nearly 600 amino acid residues in the molecule. The molecule contains four polypeptide chains as two pairs of identical subunits; it is designated as $(\alpha)_2(\beta)_2$. Figure 1.2 shows a schematic of the three-dimensional

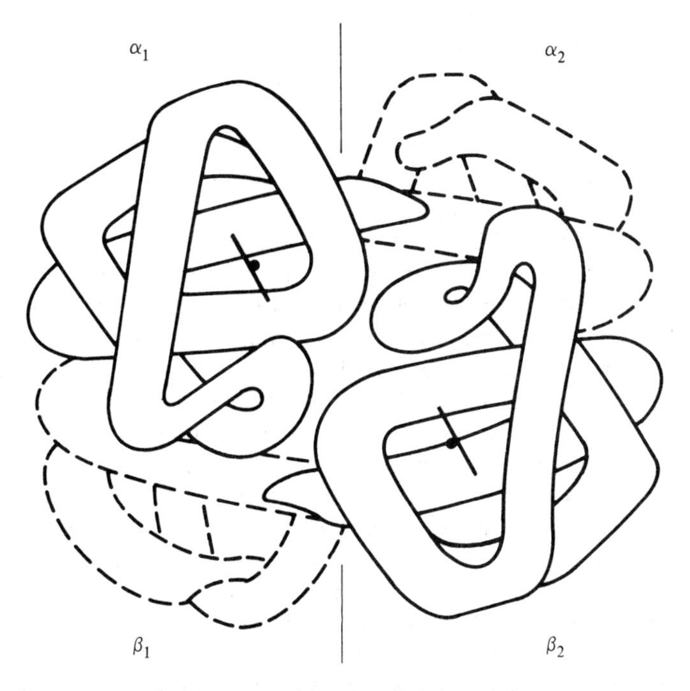

Fig. 1.2 Schematic of hemoglobin, showing the folding of the four polypeptide chains. The molecule has a twofold symmetry axis, shown as the vertical line. Rotation about this axis by 180° transforms α_1 into α_2 and β_2 into β_1. The two heme groups are seen from the side of the α_1 and β_2 subunits. [From G. Fermi, *J. Mol. Biol. 97*, 237 (1975).]

arrangement of the four chains. Hemoglobin has a twofold symmetry axis; if the molecule is rotated by 180° around the twofold axis, it will look the same. Each polypeptide chain carries one heme group, which can bind an O_2 molecule. In fact, each of the four subunits in hemoglobin is similar to myoglobin, an O_2-storing molecule present in muscle. How can this knowledge help us to understand the structure and function of hemoglobin?

Sickle-cell anemia is a disease in which the red blood cells are different from normal. They tend to become trapped in small blood vessels and they are more fragile than normal cells. The molecular difference that causes this is the change of a glutamic acid residue to a valine residue at the sixth residue from the amino terminal end of the β chain. This change occurs on the outside surface of the hemoglobin; it removes two negative charges from the hemoglobin and causes aggregation of hemoglobin molecules. The effect is oxygen-dependent and is consistent with the increased sickling that occurs with reduced oxygen concentration. More than 200 abnormal human hemoglobins have now been detected. The clinical symptoms of blood diseases can be correlated with the positions of the amino acid changes in the hemoglobins (Perutz and Lehmann, 1968; Perutz, 1976). Changes that occur on the inside of the hemoglobin are usually more damaging than those on the surface of the molecule. Amino acid changes that occur in the regions of contact between chains affect the stability of the molecule. Amino acid changes near the heme group decrease the oxygen-binding ability. The usual clinical symptom is cyanosis, a bluish discoloration of the skin caused by a deficiency of oxygen.

How can physical chemistry help? The first step is to determine what changes in structure occur for each amino acid replacement. The sickle-cell hemoglobin was easy to detect because the change in charge caused a change in electrophoretic mobility (velocity in an electric field). The next step is to study the stability of individual hemoglobin molecules and to study their aggregation. Physical methods are effective here. Sedimentation and viscosity measurements tell us about the interaction of the four subunits to form hemoglobin and the interaction of hemoglobin molecules to form large aggregates. Absorbance studies will quickly tell us if the colored heme group is present. Next we measure the kinetics and thermodynamics of O_2 binding. The hemoglobin can be stable and nonaggregating but still be unable to bind O_2 because an amino acid change has favored the ferric form of iron rather than the normal ferrous form in heme.

Once we know the changes in structure and properties of hemoglobin caused by an amino acid change, what can we do about treating the symptoms? Chemicals that reduce protein aggregation in general have been prescribed for sickle-cell anemia. A molecule that is specific for hemoglobin aggregation would obviously be better. Antioxidants to keep the heme iron reduced are being tried for some of the abnormal hemoglobin diseases.

There are many other areas of medicine, biochemistry, and biology that could also benefit from application of physical chemistry. We hope that the reader will provide some of these applications.

REFERENCES

The following textbooks can be useful for the entire course.

Physical Chemistry

BARROW, G. M., 1973. *Physical Chemistry*, 3rd ed., McGraw-Hill, New York.

DANIELS, F., and R. A. ALBERTY, 1975. *Physical Chemistry*, 4th ed., John Wiley, New York.

MOORE, W. J., 1972. *Physical Chemistry*, 4th ed., Prentice-Hall, Englewood Cliffs, N.J.

Biochemistry

LEHNINGER, A. L., 1975. *Biochemistry*, 2nd ed., Worth Publishers, New York.

MAHLER, H. R., and E. H. CORDES, 1971. *Biological Chemistry*, 2nd ed., Harper & Row, New York.

STRYER, L., 1975. Biochemistry, W. H. Freeman, San Francisco.

Physics

HALLIDAY, D., and R. RESNICK, 1970. *Fundamentals of Physics*, John Wiley, New York.

Molecular Biology

WATSON, J. D., 1976. *Molecular Biology of the Gene*, 3rd ed., W. A. Benjamin, Menlo Park, Calif.

Many applications of physical chemistry to biological systems are given in

Biophysical Chemistry, 1975. Readings from Scientific American, Introductions by V. A. Bloomfield and R. E. Harrington, W. H. Freeman, San Francisco.

For useful compilations of data, see

Handbook of Biochemistry and Molecular Biology, 1976. 3rd ed., G. D. Fasman (ed.), CRC Press, Cleveland.

SUGGESTED READINGS

HELFMAN, P. M. and J. L. BODA, 1976. Aspartic Acid Racemisation in Dentine as a Measure of Ageing, *Nature 262*, 279.

PERUTZ, M. F., 1976. Fundamental Research in Molecular Biology: Relevance to Medicine, *Nature 262*, 449.

PERUTZ, M. F., and H. LEHMANN, 1968. Molecular Pathology of Human Hemoglobin, *Nature 219*, 902.

Physics Today, 1975. Efficient Use of Energy, August, p. 23.

Scientific American, 1971. Energy and Power, September.

STEINHART, J. S., and C. E. STEINHART, 1974. Energy Use in the U.S. Food Systems, *Science 184*, 307.

TUCKER, V. A., 1975. The Energetic Cost of Moving About, *American Scientist 63*, 413.

WEHNER, R. 1976. Polarized-Light Navigation by Insects, *Sci. Amer. 235* (July), 106.

2

Conservation of Energy

A *scientific law* is an attempt to describe, in a few words, one aspect of nature. Therefore, in a sense all scientific laws will usually be "wrong": they are incomplete, approximate, or in error. In fact, the only useful scientific laws are those that in principle can be proved to be wrong. Many scientists spend their time testing theories in an attempt to disprove them or to discover their limitations. Other scientists spend their time trying to formulate more and more general laws—laws that always apply to all things. The cooperation and competition between scientists with these different approaches leads to progress in science.

Thermodynamics deals with interchanges among different forms of energy. The laws of thermodynamics are excellent examples of both the generality and the limitations of scientific laws. The *first law* states that energy is conserved; different forms of energy can interconvert, but the sum remains constant. The law was originally (about 1800) based mainly on experiments in which mechanical energy was converted into heat. A falling weight turned some paddles in a bucket of water; the water got hot. Other forms of energy were later recognized and included in the first law. The form of energy most difficult to believe in was matter. In 1923, eighteen years after Einstein postulated that $E = mc^2$, thermodynamicists were still not sure whether thermodynamics applied to radioactive materials. Now we think that the laws of conservation of mass and conservation of energy are each incomplete, but that the law of conservation of mass-energy is correct. Therefore, to make the first law correct, we must in principle consider mass itself as a form of energy.

One can see that the first law evolved from a simple description of a few experiments to a general statement about *all* forms of energy. Any new forms of energy that may be discovered can presumably be incorporated into the first law.

The *second law*, which states that the entropy of an isolated system always increases, has had a different history. It started (about 1820) as a description of experiments on the efficiency of heat engines. The law was later generalized as the first law was, but it was eventually found that the second law did not always apply to *very* small systems. Entropy was shown to be closely linked to probability; thus the second law could only be applied to a sample which contained enough molecules so that statistical predictions would work. Statistics can tell us that if we flip a coin 1000 times, we can be reasonably sure of obtaining around 500 heads, but it cannot predict whether one flip will be heads or tails. The second law is now known to be slightly limited in its application, but we think we know when and how to correct it.

The *third law* is the most recent addition to the principles of thermodynamics. It was clearly stated in the 1920s and essentially it has not been modified since. The third law states that the entropy of any pure, perfect crystal can be chosen as equal to zero at a temperature of 0 K (absolute zero). The law can be tested experimentally, and it is found to be correct for most chemicals. Whenever there has been a discrepancy, the problem has been traced to the lack of a perfectly ordered crystal at 0 K. The carbon monoxide molecule (CO), for example, forms what looks like a perfect crystal, but the molecule can fit into its crystal structure in two orientations, because the C and the O are similar in size. This disorder remains at absolute zero and therefore the entropy is not zero. The relation of entropy and probability allows us to calculate the expected entropy, and we conclude that the entropy would be zero at 0 K if the CO crystal were perfectly ordered. Helium is an example of a substance that does not form a perfect crystal at absolute zero, but one whose entropy, nevertheless, becomes zero. Helium becomes a perfect superfluid with zero entropy at 0 K.

The additions, exceptions, and corrections to the third law, and to all other scientific laws, provide some of the reasons for our continuing study of science. We assume that new ideas will lead to new experiments and eventually to new laws. Presumably the next two hundred years of science will be filled with as many new discoveries as were the past two hundred years.

THE MECHANISM OF ENERGY CONSERVATION

Many experiments, done over a period of many years, have shown that energy is easily converted from one form to another but that the amount of energy remains constant. We shall eventually discuss this quantitatively, but a few examples will make the idea clearer.

Consider a brick on the window ledge of the fifth floor of an apartment building. Owing to its height above the sidewalk, it possesses gravitational potential energy. If the brick falls, most of the potential energy will become kinetic energy of motion (a small amount becomes heat due to air friction). What happens when the brick hits bottom? The kinetic energy is converted into many new forms of energy. Much heat will be produced. If the brick hits the sidewalk, there might be some light energy in the form of sparks. Some energy is used to make and break chemical bonds in the brick and sidewalk fragments. Some sound is produced. Although complicated changes in different forms of energy are involved, the first law tells us that the total energy will remain constant. We can also calculate how the potential energy of the brick will depend on its height above the sidewalk and the weight of the brick. This will tell us the maximum amount of damage to expect. It might also make us more careful of bricks (or flowerpots) on fifth-story ledges.

A more practical and important example is to consider the total amount of energy available from the sun, and to consider what forms of energy it is presently transformed into. Sunlight hitting a desert is mainly transformed into heat. However, sunlight striking a green leaf is partly transformed into useful chemical energy by photosynthesis. Sunlight absorbed by a solar energy cell generates electrical energy. It is vitally important to us to know and understand the various kinds of energy that are available.

Systems and Surroundings

To be able to treat energy and its conversion quantitatively, we must define some new terms. Actually, we shall take common terms and give them specific meanings. We define *system* as the part of the universe that we are interested in. As we want to consider energy changes, we must specify the object within which the energy changes. The system we consider might be the sun, the earth, a person, the liver, a single cell, or a mole of liquid water at 15°C and 1 atm pressure. That is, we think of the part of the universe we are interested in, draw an enclosure around it, and label it the *system*. Everything else we call the *surroundings* (Fig. 2.1).

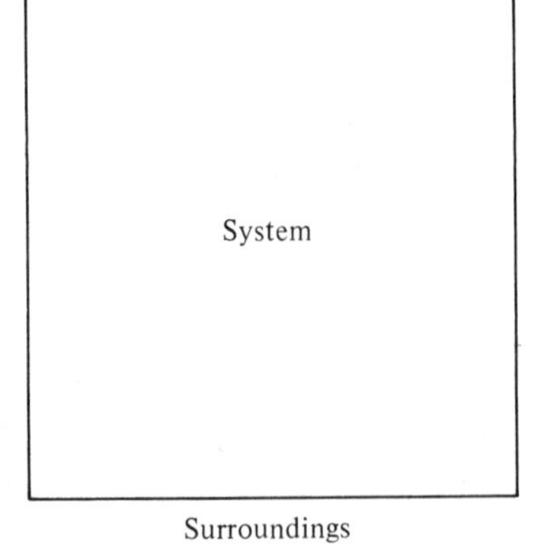

Fig. 2.1 In thermodynamics the *system* is what we focus our attention on. The *surroundings* is everything else in the universe, but we need to consider only the part that interacts with the system.

First Law of Thermodynamics

Energy can be transferred between the system and surroundings, but the total energy of the system plus surroundings is constant. This is a statement of the *first law of thermodynamics.* The first law can be stated more usefully as: the change in energy of a system equals the amount of energy that entered from the surroundings minus the energy that went out into the surroundings. All of this sounds obvious and even trivial, but it is not. For example, whenever we notice energy coming out of a system into the surroundings, we know that the system is losing energy, and we should think about the source of the energy. Nineteenth-century thermodynamicists who thought about the sun were worried. The sun shines about 2 cal of energy per square centimeter per minute on the upper atmosphere of the earth. This amounts to about 10^{15} kcal min^{-1} for the entire earth. The sun radiates its energy in all directions and we catch only a very small fraction. But the total amount of energy radiated by the sun per minute can be calculated. Not much energy comes into the sun, so the sun must be losing energy at a high rate. What is producing its energy?

The mass of the sun and its composition of roughly half hydrogen and half helium has long been known. Therefore, the early thermodynamicists could estimate how much energy was available from the sun for any chemical reaction that could occur. The answer was very discouraging. Either the sun should have burned out after a few million years, or thermodynamics did not work, or there was an unknown energy source. We are all happy to learn about the discovery of nuclear energy as the source of solar energy—which saved us, and thermodynamics, from an early death.

Energy Exchanges

We shall now concentrate on the system and how its energy can be changed. First, a few more definitions are useful. The simplest system to consider is one in which the system is defined to have no exchange of any kind with the surroundings; this is an *isolated system.* It is difficult, if not impossible, to construct such a system, but it is useful to think of one. Thermodynamics, like mathematics, defines ideal situations which can be obtained only approximately. The energy of an isolated system does not change, according to the first law. A *closed system* is defined to have no exchange of matter with the surroundings; it can exchange energy. A closed system can be made by actually putting a physical box around the system. Many chemical reactions are performed in a closed system, a stoppered flask being one example. The chemicals stay in the flask, but heat can come in or out. The most difficult type of system to consider (and of course also the most interesting and useful) is the open system. An *open system* can exchange both matter and energy with the surroundings. A fertilized egg being hatched by a hen is a good example of an open system. Oxygen comes in and carbon dioxide goes out of the egg. Heat is also exchanged between the egg and its surroundings.

We should emphasize that it is up to us to draw whatever box we wish to separate our system from its surroundings. For example, if a solution containing the enzyme catalase is added to an open beaker containing a hydrogen peroxide solution, the enzyme will accelerate the reaction $H_2O_2 \rightarrow H_2O + \frac{1}{2}O_2$, and oxygen gas will come out of the beaker. If we choose the liquid content in the beaker as our system, we have an open system. But if we choose the liquid content plus the oxygen evolved as our system, we have a closed system. We usually make our choices based on convenience.

It is also important to realize that it is the *change* of energy which characterizes what has happened to the system. We are less concerned about the absolute value of energy for the system. We can add energy to a system in various ways; for example, to an open system, we can add matter. Adding matter to a system increases its chemical energy, because the matter can undergo various chemical reactions. But we do not have to think about the large amount of energy potentially available from nuclear reactions if we are only considering chemical reactions. That is, we do not have to include the $E = mc^2$ energy term, because it does not change significantly in an ordinary chemical reaction.

Work

It is convenient to divide energy exchange between system and surroundings into two types: heat and work. *Work* is defined as the product of a force times a distance. If a force is applied which causes a movement, one multiplies the force applied times the displacement, that is, the distance moved in the direction of the force. This applied force is often called the *external force*:

$$\text{Work} \equiv \text{external force} \cdot \text{displacement} \tag{2.1}$$

We must be careful about the sign of the work, as we shall be combining heat and work in the application of the first law of thermodynamics. If the direction of the external force is the same as the direction of the displacement, the work is positive; if the direction of the external force is opposite to the direction of the displacement, the work is negative. For example, if a spring is compressed by an external force, the directions of the external force and the displacement are the same, and the work done on the spring by the external force is positive. If a compressed spring expands against an external force, the directions of the external force and the displacement are opposite to each other, and the work done on the spring by the external force is negative.

Note also that if work done by an external force is positive, it means that work is done by the surroundings on the system; if work done by an external force is negative, it means that work is done by the system on the surroundings. To summarize:

Sign of work	Convention
+	Work done on system
−	Work done by system

Work of compressing or stretching a spring

Consider a system that consists of a spring. The force that must be applied to stretch or compress a spring is characterized by *Hooke's law*, which states that the force is directly proportional to the change in length of the spring. Since the force changes as the length changes, we must integrate to obtain the work done. We choose our x axis as the direction along the spring, and consider one end of the spring labeled A in Fig. 2.2a, as *always* fixed at the position $x = 0$. The position

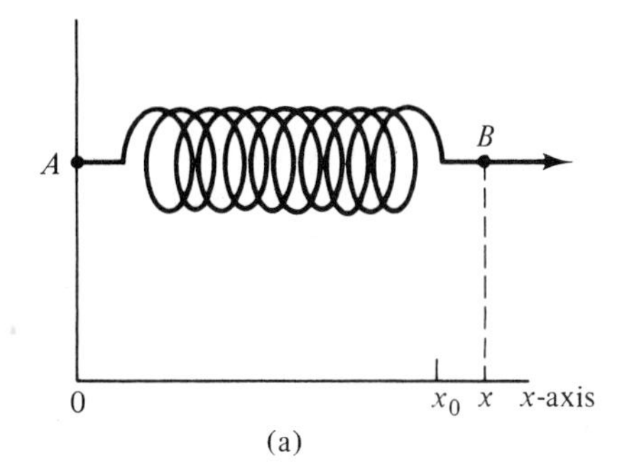

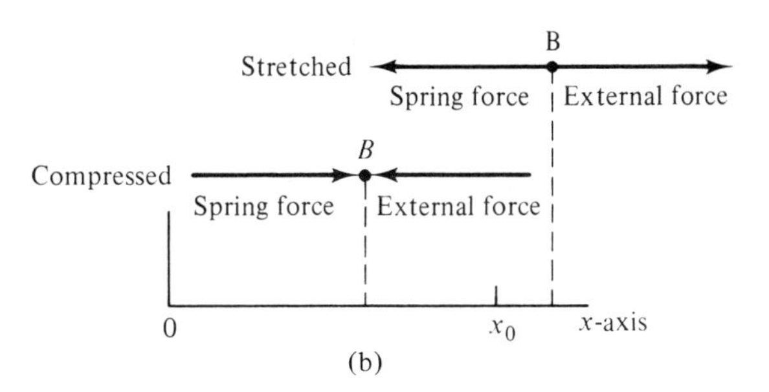

Fig. 2.2 Direction of forces when spring is stretched or compressed. Newton's law requires that the external force and the spring force be equal in magnitude when point B is not being accelerated. At x_0 the external force on the spring is zero.

of the other end, labeled B in the figure, is at $x = x$. Note that x is also the length of the spring. Let $x = x_0$ be the position of B when no force is acting on the spring.

According to Hooke's law,

$$\text{spring force} = -k \cdot (x - x_0) \tag{2.2}$$

where k is a constant for a given spring (if we do not stretch or compress it too much). The magnitude of k will be different for different springs. The negative sign states that the direction of the spring force is opposite to the direction of displacement from the equilibrium position of B.

As illustrated in Fig. 2.2b, the external force is the negative of the spring force, as the two are equal in magnitude but opposite in direction:

$$\text{external force} \equiv f = k \cdot (x - x_0) \tag{2.3}$$

The work done on the spring when its length is changed from x_1 to x_2 is

$$
\begin{aligned}
w &= \int_{x_1}^{x_2} f\,dx \\
&= \int_{x_1}^{x_2} k \cdot (x - x_0)\,dx \\
&= \tfrac{1}{2}k \cdot (x_2^2 - x_1^2) - kx_0 \cdot (x_2 - x_1) \\
&= k \cdot (x_2 - x_1)\left(\frac{x_2 + x_1}{2} - x_0\right) \qquad (2.4)
\end{aligned}
$$

We have thus quantitatively calculated the work done on the system by compressing or expanding the spring. If the spring was originally compressed or expanded, the system would do work on the surroundings when the spring returned to its equilibrium position. A muscle fiber works the same way. It can do work by stretching or contracting.

The units of work are in ergs if k is in dyn cm^{-1} and x is in centimeters: 1 dyn cm $\equiv$ 1 erg. One dyne is the magnitude of the force that will cause an acceleration of 1 cm s^{-2} when applied to a body of mass 1 g: 1 dyn $\equiv$ 1 g cm s^{-2}. The unit calorie (cal) is also frequently used: 1 cal = 4.184 $\times$ 10^7 erg. The trend now is to use *Standard International*, or *SI, units*. The basic units of length, mass, and time in the SI system are meter (m), kilogram (kg), and second (s). The unit of force in this system is the newton: 1 N = 1 kg m s^{-2}. The unit of work or other forms of energy is the joule (J): 1 J = 1 N m = 1 kg m^2 s^{-2}. In this book we shall use erg, cal, or kcal (kilocalorie) as our energy units. The SI energy unit is related to these units as follows:

$$
\begin{aligned}
1\text{ J} &= 10^7\text{ erg} \\
4.184\text{ J} &= 1\text{ cal} \\
&= 10^{-3}\text{ kcal}
\end{aligned}
$$

Additional energy-conversion tables will be found in the Appendix.

Work of increasing or decreasing a volume

Consider a system enclosed in a container with a movable wall, such as a piston (Fig. 2.3). We can do expansion or compression work on the system by moving the piston. This example is very similar to the spring example that we discussed previously. The force per unit area, f/A, applied to the system is the external pressure, P_{ex}. This pressure may depend on the volume of the system, so we generally integrate to obtain the work.

If we move the piston by a distance dx in the direction of the external pressure, the work done on the system by the external force is $f\,dx$, which is identical to $(f/A) \cdot A\,dx$ or $P_{ex} \cdot A\,dx$, with A being the cross-sectional area of the piston. The volume change dV of the system corresponding to moving the piston by dx is

$$dV = -A\,dx$$

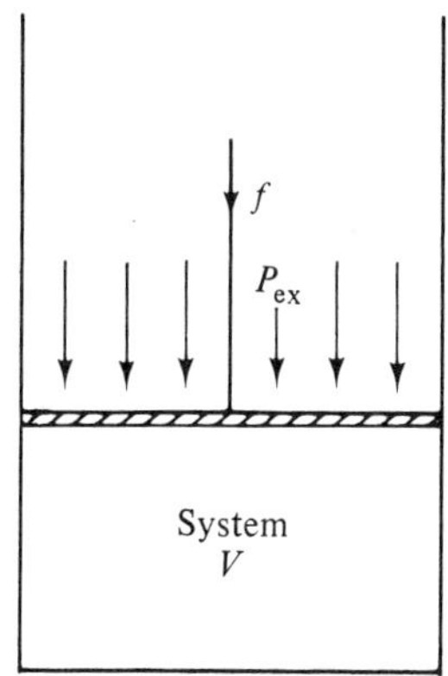

Fig. 2.3 Work done on a system by an external force, *f*. The work is positive when the volume, *V*, decreases.

The negative sign is needed because we have defined dx to be in the same direction as the external pressure. Thus, if the system is being compressed, dx is positive by definition but dV is negative, since the volume of the system is decreasing. The work done by the external force when the volume of the system changes from V_1 to V_2 is therefore

$$w = -\int_{V_1}^{V_2} P_{ex}\, dV \tag{2.5}$$

We put a subscript "ex" on the pressure to remind us that the pressure which must be used here is the *external* pressure applied to the system. To obtain work in ergs, we must use pressure in dyn cm^{-2} and volume in cm^3. If pressure is in atmospheres and volume in liters, the resulting units, ℓ atm, must be converted to more common energy units, such as ergs or calories.

Equation (2.5) tells us that we can calculate the work of expansion or compression of a system if we know the pressure we are applying and the volume change of the system. Often the pressure will change as the volume changes, so we write the equation as an integral. If the pressure is kept constant, however, the integration can be done easily. For *constant pressure*,

$$w_P = -P_{ex} \cdot (V_2 - V_1) \tag{2.6}$$

Expansion work done by a system containing only gases has been very important in thermodynamics, because that system is a model for heat engines. Engineers (and others) need to know how much work can be done by the expanding gas inside the cylinder of a car, for example. Biologists may be interested in the work done by the lungs on the air we breathe or the work done in systems containing liquids. The work done when liquid freezes may be important to them. The density of ice is less than that of liquid water, so work can be done by the system when water expands on freezing. The amount of work (and the freezing temperature) will depend on the external pressure. In a complete vacuum no pressure-volume work is done in an expansion, because P_{ex} in Eq. (2.6) is exactly zero.

Example 2.1 Calculate the pressure-volume work done when a system containing a gas expands from 1.0 ℓ to 2.0 ℓ against a constant external pressure of 10 atm. Express the answer in calories and ergs.

Solution

$$w_P = -P_{ex} \cdot (V_2 - V_1) \tag{2.6}$$

$$= -(10 \text{ atm})(2\,\ell - 1\,\ell)$$

$$= -10\,\ell \text{ atm}$$

$$w = -(10\,\ell \text{ atm})\left(\frac{1.987 \text{ cal}}{0.08205\,\ell \text{ atm}}\right)$$

$$w = -242 \text{ cal}$$

$$w = (-242 \text{ cal})\left(4.184 \times 10^7 \frac{\text{erg}}{\text{cal}}\right)$$

$$w = -1.01 \times 10^{10} \text{ erg}$$

The system does work; the sign of w is negative. Note that in converting ℓ atm to cal, we used the values of the gas constant R in these units ($R = 1.987$ cal deg^{-1} mol^{-1} = 0.08205 ℓ atm deg^{-1} mol^{-1}). We can, of course, also do the conversion by looking up in the Appendix the conversion factor 1 ℓ atm = 24.22 cal.

Example 2.2 Calculate the pressure-volume work done in calories and ergs when a sphere of water 1.00 micrometer (1 μm = 10^{-4} cm) in diameter freezes to ice at 0°C and 1 atm pressure.

Solution Find the volume of the sphere of water, then use the density of ice and liquid water at 0°C to calculate the volume of the frozen sphere.

The volume of the liquid water is

$$V_l = \frac{1}{6}\pi D^3 = \left(\frac{3.142}{6}\right)(10^{-4} \text{ cm})^3$$

$$= 5.237 \times 10^{-13} \text{ cm}^3$$

The density of liquid water at 0°C is 1.000 g cm^{-3}. The density of ice at 0°C is 0.915 g cm^{-3}. The volume of the solid water is

$$V_s = (5.237 \times 10^{-13} \text{ cm}^3)\left(\frac{1.000}{0.915}\right)$$

$$= 5.723 \times 10^{-13} \text{ cm}^3$$

The work done by the system is

$$w_p = -P_{ex} \cdot (V_2 - V_1) \tag{2.6}$$
$$= -(1 \text{ atm})(5.723 \times 10^{-13} \text{ cm}^3 - 5.237 \times 10^{-13} \text{ cm}^3)$$
$$= -0.486 \times 10^{-13} \text{ cm}^3 \text{ atm}$$
$$w = (-0.486 \times 10^{-13} \text{ cm}^3 \text{ atm})(2.422 \times 10^{-2} \text{ cal cm}^{-3} \text{ atm}^{-1})$$
$$= -1.18 \times 10^{-15} \text{ cal}$$
$$= (-1.18 \times 10^{-15} \text{ cal})(4.184 \times 10^7 \text{ erg cal}^{-1})$$
$$= -4.94 \times 10^{-8} \text{ erg}$$

The system expands: therefore, work is done by the system and the sign of w is negative.

It should be clear that all the examples of work given so far have been very similar. In each case, work can be done on the system by changing it from its equilibrium state, or the system can do work on the surroundings by tending to return to its equilibrium state.

Friction

The force of friction causes an energy change whenever two surfaces in contact move relative to each other. The frictional force is opposite in direction to the force causing the motion. We will not discuss frictional effects in this chapter, but one should realize that it may be an important source of energy loss in real systems. Engineers, of course, try to maximize expansion work and minimize friction in engines. Frictional losses are also important in biological engines. As arteries get rougher and narrower with aging, the energy needed to circulate blood increases. The blood pressure must then also increase, and the heart must do more expansion and compression work. Friction is not all bad, however. Driving on icy roads is treacherous enough; think of what would happen if there were no friction at all!

Work in a gravitational field

All processes that occur on the earth are affected by the earth's gravitational field (and to a much lesser extent by the moon's and other astronomical bodies' gravitational fields). This means that all systems on earth can do gravitational work. A simple example is illustrated in Fig. 2.4. If an object of mass m is lowered at a constant velocity from a height h_1 to a height h_2 above the earth's surface, the work done by the external force on the system is

$$w = mg \cdot (h_2 - h_1) \tag{2.7}$$

The work done by the external force on the system is negative ($h_2 - h_1$ is negative), as the direction of displacement is opposite to the direction of the external force. In other words, the system is doing work on the surroundings.

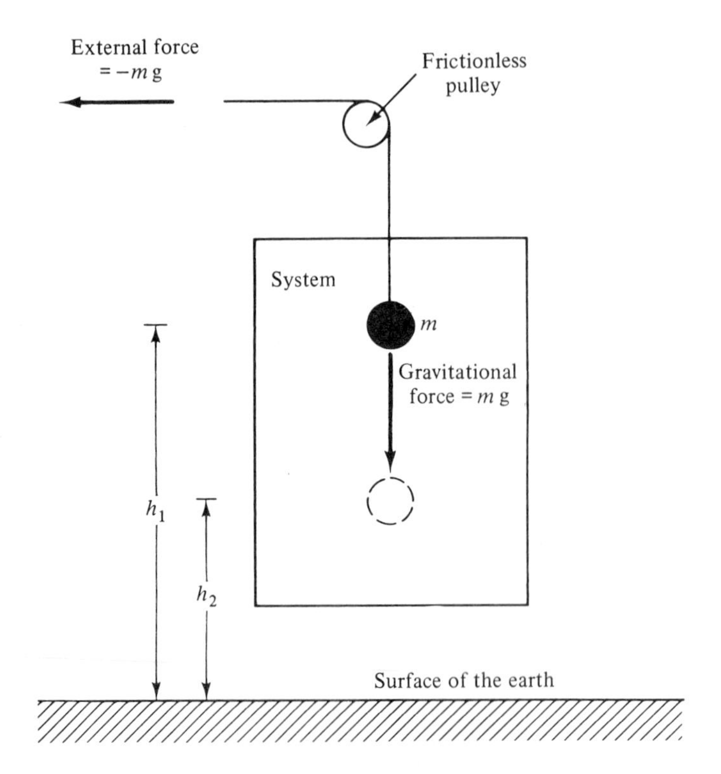

Fig. 2.4 System doing gravitational work.

For mass m, in grams, height h, in centimeters, and g, in cm s^{-2} (the standard acceleration of gravity = 980.7 cm s^{-2}), the work is obtained in ergs. With SI units of kg, meter, and acceleration of gravity = 9.807 m s^{-2}, the work is in joules.

Work in an electric field

If a system contains electrical charges, an electric field will produce a force on the charges which will cause them to move, and thus a current will flow. The work done on the system can be shown to be

$$w \text{ (electrical)} = -EIt \tag{2.8}$$

where E is the voltage ≡ potential difference ≡ electromotive force, I is the current, and t is the time. For E in volts, I in amperes, and t in seconds, the units of work obtained are joules ≡ 10^7 ergs. The cost of electricity is for electrical work, and it is usually calculated in kilowatt-hours. Since a watt-second is a joule, 1 kWh is 3.6×10^6 J.

Example 2.3 Calculate the electrical work done in joules, calories, and ergs by a 12.0-V storage battery which discharges 0.1 A for 1.00 h.

Solution The electrical work done on the battery is

$$w = -EIt \tag{2.8}$$

$$= -(12\text{ V})(0.1\text{ A})(1\text{ h})$$

$$= -1.2\text{ VA h} = -1.2\text{ W h}$$

$$= -(1.2\text{ W h})(3600\text{ s h}^{-1})$$

$$= -4.32 \times 10^3\text{ W s}$$

$$= -4.32 \times 10^3\text{ J}$$

$$= (-4.32 \times 10^3\text{ J})(10^7\text{ erg J}^{-1})$$

$$= -4.32 \times 10^{10}\text{ erg}$$

$$= (-4.32 \times 10^3\text{ J})(0.2389\text{ cal J}^{-1})$$

$$= -1.03 \times 10^3\text{ cal}$$

The minus sign tells us that the battery is doing work on the surroundings.

Heat

When two bodies are in contact with each other, their temperatures tend to become equal. Energy is being exchanged. The hot body will lose energy and cool; the cold body will gain energy and warm. The energy exchange is said to occur by *heat transfer.* Heat is defined in terms of temperature changes, but it is recognized in many processes. Chemical reactions, electrical currents, friction, and absorption of radiation all involve heat transfer.

For a closed system, the amount of heat transferred, q, is proportional to the difference in temperature of the system before (T_1) and after (T_2) the heat exchange. The proportionality constant depends on the system and is called the *heat capacity* (C) of the system; in general, it will vary with temperature. We must therefore write the equation for the heat gained as an integral.

$$q = \int_{T_1}^{T_2} C\,dT \tag{2.9}$$

For a hot body in contact with a cold body, the heat capacity of each body must be known and Eq. (2.9) must be applied to each body separately.

Example 2.4 Calculate the heat in calories necessary to change the temperature of 100.0 g of liquid water by 50°C at constant pressure. The heat capacity of liquid water at constant pressure is 1.00 cal g^{-1} deg^{-1} and is nearly independent of temperature.

Solution The heat absorbed by the system is

$$q = \int_{T_1}^{T_2} C\,dT$$

$$= C \cdot (T_2 - T_1)$$

$$= (100.0\text{ g})\left(1.00\frac{\text{cal}}{\text{g deg}}\right)(50\text{ deg})$$

$$= 5000\text{ cal}$$

Heat capacity is an experimental quantity for every material which characterizes how much heat is necessary to raise its temperature by 1 degree (Celsius or Kelvin). The units are often cal deg^{-1} mol^{-1} or J deg^{-1} mol^{-1}. It is easy to remember that the heat capacity of liquid water is about 1 cal deg^{-1} g^{-1}. The heat capacity will, in general, depend on temperature and on whether P or V is held constant during the heating. The symbol C_P means heat capacity at constant pressure, and C_V is heat capacity at constant volume. For solids or liquids C_P and C_V are nearly equal. For gases we will show later that the molar heat capacities differ approximately by R, and that ($\bar{C}_P \cong \bar{C}_V + R$). Table 2.1 gives some representative values.

Table 2.1 Heat capacities at constant pressure of various substances near 25°C

Substance	$\bar{C}_P$	Substance	$\bar{C}_P$	Substance	$\bar{C}_P$
Gases	molar heat capacities, cal deg^{-1} mol^{-1}	Liquids	molar heat capacities, cal deg^{-1} mol^{-1}	Solids	specific heat capacities, cal deg^{-1} g^{-1}
He	5.0	Hg	6.7	Au	0.0308
O_2	7.0	H_2O	18.0	Fe	0.106
N_2	7.0	Ethanol	27.0	C (diamond)	0.124
H_2O	7.9	Benzene	32.4	Glass (pyrex)	0.2
CH_4	8.6	*n*-Heptane	53.7	Brick	~0.2
CO_2	9.0			Al	0.215
				Glucose	0.30
				Urea	0.50
				H_2O (0°C)	0.50
				Wood	~0.5

There are many practical applications of Eq. (2.9). We often want to know how much heat can be transferred from one system to another and what the final temperature will be. One of the main problems in solar heating is how to store the energy for use at night. It is simple to raise the temperature of a storage system such as water or rocks during the day and to transfer the heat at night to the cold air in your house. The amount of heat transferred depends on the temperature difference and the heat capacity of each system.

Example 2.5 Cold air at 0°C is passed through 100 kg of hot crushed rock that has been heated to 110°C. The air is heated to 20°C by the time it leaves the rock and is admitted into a house for heating. Calculate the total volume of 20°C air that can be obtained by this process. The heat capacity at constant pressure of the rock is 0.20 cal g^{-1} deg^{-1} and of the air is 0.25 cal g^{-1} deg^{-1}. The density of air at 1 atm and 20°C is 1.20×10^{-3} g cm^{-3}; the density of the rock is 2.5 g cm^{-3}.

Solution We need to calculate the heat transferred from the rock in cooling to 20°C. This is equal to the heat absorbed by the air. We can then calculate the weight of the air heated and therefore the volume of the air heated.

The heat lost by the rock is

$$\begin{aligned} q &= C \cdot (T_2 - T_1) \\ &= (100 \times 10^3 \text{ g})(0.20 \text{ cal g}^{-1} \text{ deg}^{-1})(90 \text{ deg}) \\ &= 1.8 \times 10^6 \text{ cal} \end{aligned}$$

The weight of the air heated to 20°C by this amount of heat is

$$\begin{aligned} \text{air wt} &= \frac{1.8 \times 10^6 \text{ cal}}{(0.25 \text{ cal g}^{-1} \text{ deg}^{-1})(20 \text{ deg})} \\ &= 3.6 \times 10^5 \text{ g} = 360 \text{ kg} \end{aligned}$$

The volume of the air heated to 20°C is

$$\begin{aligned} \text{air vol} &= \frac{3.6 \times 10^5 \text{ g}}{1.2 \times 10^{-3} \text{ g cm}^{-3}} \\ &= 3 \times 10^8 \text{ cm}^3 \\ &= 3 \times 10^2 \text{ m}^3 \end{aligned}$$

This corresponds to the volume of a medium-sized room. The volume of crushed rock necessary is only 100×10^3 g/2.5 g cm^{-3}, or 4×10^4 cm^3 = 0.04 m^3.

Radiation

A very important method of energy exchange is that of radiation. Most of the earth's useful energy comes from the sun's radiation. It is important to be able to calculate the energy present in a given number (N) of photons of frequency ν, given in sec^{-1}:

$$\text{radiation energy} = Nh\nu \tag{2.10}$$

where $h \equiv$ Planck's constant $= 6.6256 \times 10^{-27}$ erg s.

Another useful equation, the Stefan–Boltzmann equation, describes how much energy is radiated by a body as a function of its temperature. The equation is for an ideal body that radiates and absorbs all wavelengths, a *black body*:

$$\text{radiation energy cm}^{-2}\,\text{s}^{-1} = \sigma T^4 \tag{2.11}$$

where $\sigma \equiv$ Stefan–Boltzmann constant $= 5.96 \times 10^{-5}$ erg cm^{-2} s^{-1} deg^{-4}
T = absolute temperature.

Example 2.6 The average surface temperature of the sun is about 6000 K; its diameter is about 1.4×10^{11} cm. Estimate the total energy radiated by the sun in erg s^{-1}.

Solution The Stefan–Boltzmann equation provides the radiation rate per cm^2. From the diameter of the sun we can obtain the surface area and thus the total energy radiated per second.

$$\begin{aligned}\text{radiation} &= \sigma T^4 \qquad (2.11)\\ &= (5.69 \times 10^{-5}\ \text{erg cm}^{-2}\,\text{s}^{-1}\,\text{deg}^{-4})(6000)^4\\ &= 7.4 \times 10^{10}\ \text{erg cm}^{-2}\,\text{s}^{-1}\end{aligned}$$

$$\begin{aligned}\text{area} &= \pi D^2\\ &= (\pi)(1.4 \times 10^{11}\ \text{cm})^2\\ &= 6.2 \times 10^{22}\ \text{cm}^2\end{aligned}$$

$$\begin{aligned}\text{energy radiation per second} &\equiv \text{luminosity}\\ &= (7.4 \times 10^{10}\ \text{erg cm}^{-2}\,\text{s}^{-1})(6.2 \times 10^{22}\ \text{cm}^2)\\ \text{luminosity} &= 4.6 \times 10^{33}\ \text{erg s}^{-1}\end{aligned}$$

This number agrees well with the measured luminosity of the sun.

Example 2.7 Radiation can cause chemical reactions to occur. For some reactions each photon produces one molecule of product. How many photons are there per erg of red light? The wavelength is 700 nm; the frequency is 4.29×10^{14} s^{-1}.

Solution

$$N = \frac{\text{energy}}{h\nu} \tag{2.10}$$

$$\begin{aligned}&= \frac{1\ \text{erg}}{(6.63 \times 10^{-27}\ \text{erg s})(4.29 \times 10^{14}\ \text{s}^{-1})}\\ &= 3.5 \times 10^{11}\ \text{photons}\end{aligned}$$

VARIABLES OF STATE

We have been using the terms pressure, volume, and temperature without defining them further, because they are familiar to us. We must be careful about units, however. For pressure we will use atm, or torr ≡ mm Hg, or dyn cm^{-2}.

$$1 \text{ atm} \equiv 760 \text{ torr} \equiv 1.013 \times 10^6 \text{ dyn cm}^{-2}$$

For volume we use cm^3 = milliliter, or liter. For temperature we use degrees absolute ≡ degrees Kelvin, K; deg means K. In the SI system, the unit of pressure is the pascal. 1 pascal ≡ 1 N m^{-2} and 1.013×10^5 pascals = 1 atm. Whenever T occurs in thermodynamic equations, degrees Kelvin is indicated. To convert from other temperature scales, use

$$\text{K} = {}^\circ\text{C} + 273.1$$

$$\text{K} = \frac{{}^\circ\text{F} - 32}{1.8} + 273.1$$

°C = degrees Celsius, or centigrade
°F = degrees Fahrenheit

We now want to describe P, V, and T as variables that help specify the state of a system. Consider a closed system in the absence of all external fields. This statement describes a system that we can only approximate on earth. We cannot turn off gravity, for example. However, for many practical purposes we can ignore the effects of gravity. An essential part of learning a science is learning which approximations are useful and which are not.

If the system consists of a pure liquid, specifying the pressure (P), volume (V), and temperature (T) of the liquid is sufficient to specify many other properties of the liquid. These other properties, and P, V, and T, are called *variables of state*. Such variables depend only on the state of the system, not on how the system arrived at its state. The useful characteristic of variables of state is that when a few (in general, more than P, V, and T) are used to specify a system, all other variables of state are determined implicitly. It is thus possible and practical to measure and tabulate certain properties of the system: those that are variables of state. The discovery of the first law of thermodynamics showed that *energy* ≡ *internal energy*, E, is a variable of state. Heat and work are *not* variables of state; they depend specifically on the method used to change from one state to another. In other words, they depend on the *path* between states. It is sometimes convenient for thermodynamicists to define new variables of state by combining previously defined ones. *Enthalpy*, H, is a variable of state defined as

$$H \equiv E + PV \tag{2.12}$$

H can be thought of as being no more than a shorthand notation for $E + PV$. The units must be the same as E, so the PV units must be converted to calories or ergs or joules before adding to E.

Variables of state are divided into two classes: extensive and intensive. An *extensive* variable of state is directly proportional to the mass of the system. If you double the mass of the system you double the magnitude of an extensive variable. An *intensive* variable of state is independent of the mass of the system. Of the variables that we have considered so far, P and T are intensive, whereas V, E, H, and the heat capacity C in Eq. (2.9) are extensive. However, we can always change an extensive variable to an intensive one by expressing it per-unit amount of material. Energy is extensive, but energy mol^{-1} or energy g^{-1} is intensive.

Equations of State

An *equation of state* is an equation that relates variables of state. A few variables of state are usually sufficient to specify all the others. This means that equations exist that can relate the variables of state. The simplest and most frequently used equations of state link P, V, and T.

Solids or liquids

The volume of a solid or a liquid does not change very much with either pressure or temperature. We are not considering changes from solid to liquid or any other change except P, V, and T. Therefore, a first approximation for the *equation of state of a solid or liquid is* $V \cong$ *constant.* This means that to calculate the volume of the solid or liquid, one just finds the density or specific volume at one temperature and uses that value for any temperature and pressure. The volume is related to the density by

$$V = \frac{\text{mass}}{\text{density}}$$

Actually, the volume of a solid or liquid does change somewhat with T and P. Experimental data for 1 mol of liquid water are shown in Fig. 2.5. The molar volume is plotted versus temperature at constant pressure in the upper half and the volume is plotted versus pressure at constant temperature in the lower half. Equations can be obtained for V as a function of P and T, $V(P, T)$, for liquid water from these data. However, for most of our applications we shall use the first approximation, that V is constant independent of T and P for a solid or liquid.

Gases

For gases the volume varies greatly with T and P, but the variation is nearly independent of the type of gas. Thus there is a simple approximate equation of state for gases. The *ideal gas equation* is

$$PV = nRT \tag{2.13}$$

where n = number of moles
$R \equiv$ universal gas constant $= 0.08205\ \ell\ \text{atm deg}^{-1}\ \text{mol}^{-1}$

A plot of how the volume of 1 mol of an ideal gas depends on pressure is shown in Fig. 2.6. Note that, although 1000 atm was necessary to change the volume of

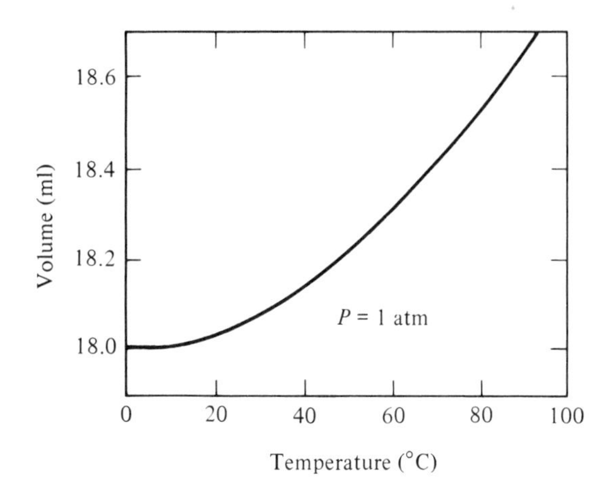

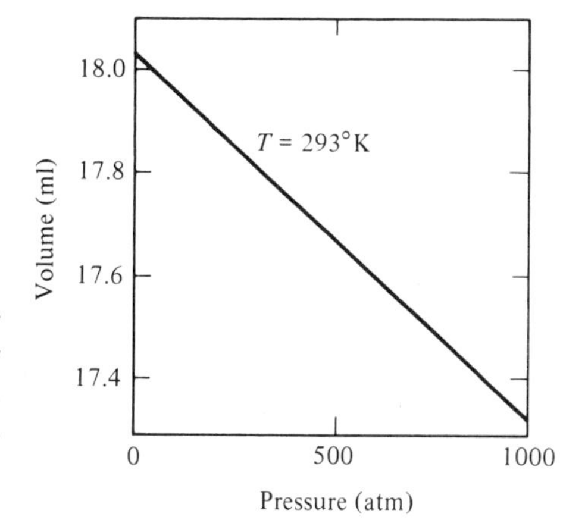

Fig. 2.5 Volume of 1 mol of liquid water as a function of temperature and pressure. Note that the volume changes by less than 5% over the temperature and pressure ranges shown.

liquid water by 3%, a change from 1 atm to 1000 atm for an ideal gas will cause a change in volume by a factor of 1000. The ideal gas equation has the great advantage that it contains no constants applying to individual gases; it applies to all gases if the pressure is low enough. It is an exact limiting equation for all gases as P approaches zero. For higher pressures it is an approximation. The answers obtained using Eq. (2.13) are usually correct to within $\pm$ 10% for most gases near room temperature and atmospheric pressure. Of course, if the pressure causes the gas to liquify, the ideal gas equation cannot be used to calculate the volume of the liquid.

Many other more accurate equations of state for gases have been developed. They are corrections to the ideal gas equation which contain parameters relating to individual gases. One example is the *van der Waals gas equation*:

$$\left(P + \frac{n^2 a}{V^2}\right)(V - nb) = nRT \tag{2.14}$$

where a and b are constants that characterize each gas.

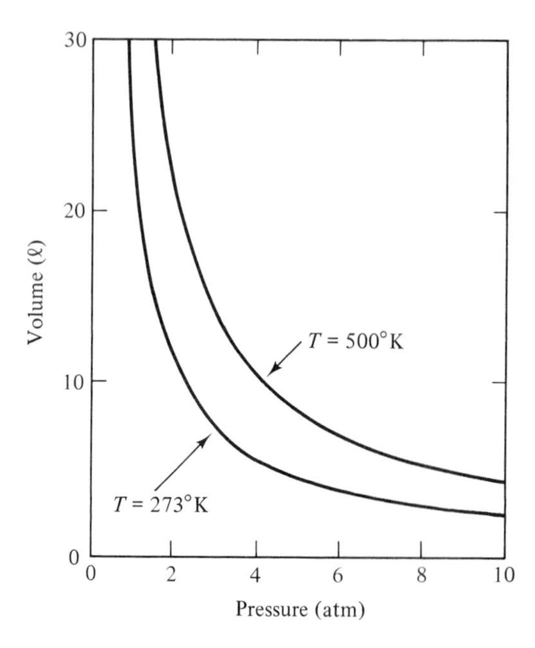

Fig. 2.6 Volume of 1 mol of an ideal gas as a function of temperature and pressure.

The equations of state that we have been discussing have all been applied to systems containing only one component. For mixtures, the number of grams or moles of each component must be specified, and the equation of state will depend on the concentrations. For gases that can be approximated by the ideal gas equation, the results are particularly simple. The ideal gas equation can be applied to each gas in the mixture as if the others were not there. The partial pressure of each gas can be calculated as follows:

$$P_i = \frac{n_i RT}{V} \tag{2.15}$$

where P_i = partial pressure of component i in the ideal gas mixture
n_i = number of moles of component i
V = total volume of gases

The total pressure is just the sum of the partial pressures:

$$P_{\text{total}} = \sum_i P_i \tag{2.16}$$

$$P_{\text{total}} = \frac{n_{\text{total}} RT}{V} \tag{2.17}$$

The partial pressures can also be obtained from the total pressure and the mole fractions X_i of each component:

$$\begin{aligned} P_i &= X_i P_{\text{total}} \\ X_i &= \frac{n_i}{\sum_i n_i} \end{aligned} \tag{2.18}$$

The equations for partial pressures are useful, because often we are only interested in one of the components of a gas mixture. For example, in the air we breathe, the partial pressure of oxygen or carbon dioxide (or even sulfur dioxide) is much more important than the total pressure.

Energy and Enthalpy Changes

It is important to be able to calculate the energy and enthalpy changes that occur in a system when the system changes from one state to another, or to be able to calculate the amount of energy necessary to cause a change in the state of the system.

We shall discuss various methods of calculating energy (E) and enthalpy (H) changes. The most important fact to remember is that E and H are variables of state; they depend only on the state of the system. Therefore, the changes in E and H depend only on the initial and final states; they do not depend on the path that we find convenient to bring about the changes.

Heat and Work Changes

When heat is transferred to or from a system and work is done by or on a system, the energy of the system will change. For a closed system, if heat and work are the only forms of energy that the system exchanges with the surroundings, we can write

$$E_2 - E_1 = q + w \tag{2.19}$$

where $E_2 - E_1$ = change in energy from the initial state 1 to the final state 2

q = net heat transferred *to* the system (the heat *in*)

w = net work done *on* system (the work *in*)

To use this equation we pick a convenient process, or path, to get from state 1 to state 2 and find the heat and work changes along the path. The heat and work will each depend on the path we choose, but their sum will not. We use the convention that both heat, q, and work, w, are positive when they increase the energy of the system. Some other books have work done *by* the system as positive; their equation corresponding to Eq. (2.19) would have $q - w$ in it, with w being the work done *by* the system.

Reversible Path

There are an infinite number of paths to get from one state to another, but some are so convenient that they have been given names. The most important path is the reversible path. In a *reversible path* the system always remains very near equilibrium.

As an example, let us consider the expansion of an ideal gas in a cylinder with a frictionless and weightless piston, as illustrated in Fig. 2.7a. The cylinder

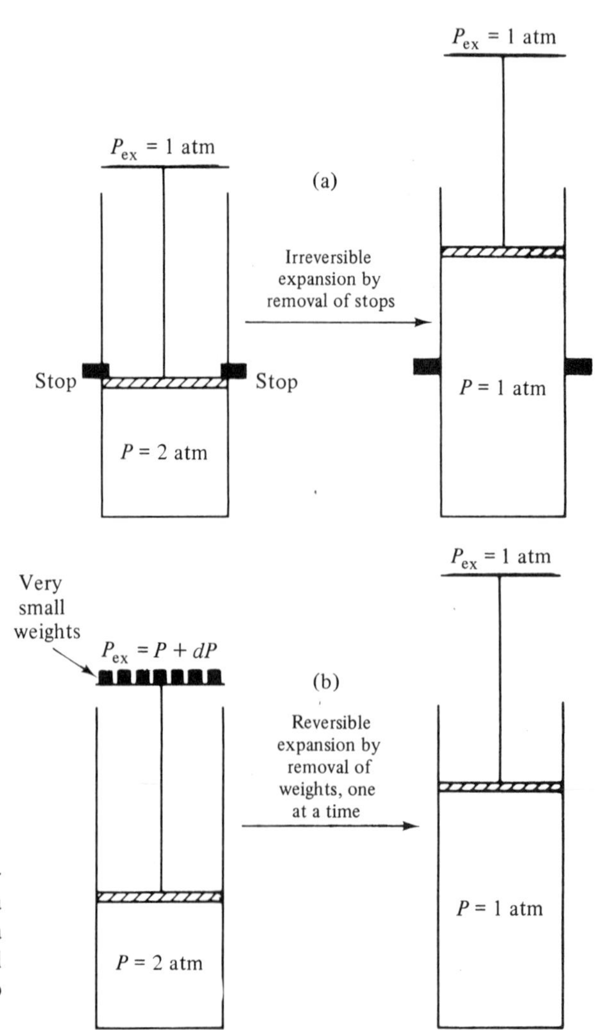

Fig. 2.7 Comparison of an irreversible expansion (a) and a reversible expansion (b). In a reversible expansion the internal pressure is always nearly equal to the external pressure.

conducts heat well, so the temperature of the system (the gas in the cylinder) is always the same as the temperature of the surroundings, which we specify to be a constant. Suppose that the pressure outside the cylinder is always at 1 atm and the pressure inside the cylinder is initially at 2 atm. If we remove the stops that hold the piston in position, the gas will expand irreversibly until a final state is reached at which its pressure becomes 1 atm. During the course of expansion, the pressure of the system is always greater than that of the surroundings (that is why the expansion is called *irreversible*). The two become the same only at the end of the expansion.

We can carry out the expansion in a different way. Instead of holding the piston in position with stops at the beginning, we put many small weights on top of the piston to make up for the pressure difference (Fig. 2.7b). The expansion is then carried out in a stepwise manner by removing one weight at a time. If the number of weights is very large (and the weight of each very small),the pressure of the system is always almost the same as that of the surroundings during the course of expansion. Furthermore, if we add rather than remove a weight, the process will be reversed.

A reversible path is one in which the process can be reversed at any instant by an infinitesimal change of a certain parameter of the system. The path illustrated in Fig. 2.7a is not a reversible one; the path illustrated in Fig. 2.7b becomes a reversible one when the weights are very small and their number approaches infinity.

As a second example, let us consider the transition from liquid water to gaseous water (steam). The transition is reversible at 100°C and 1 atm. If we maintain the temperature at 100°C, liquid water will evaporate to steam if the pressure is lowered to slightly below 1 atm, and steam will condense to liquid water if the pressure is increased to slightly above 1 atm. Similarly, if the pressure is maintained at 1 atm, the direction of change can be reversed by a small change in temperature.

There are many reversible paths for a process. For example, if we want to use a reversible path to calculate the energy needed to evaporate 1 mol of liquid water at 25°C and 1 atm,* we can choose either of the reversible paths shown in Fig. 2.8. In the first reversible path (1) the temperature is raised (assumed reversibly) to 100°C, the normal boiling point, where the evaporation can be done reversibly. The system is always near equilibrium; the process is reversible. In reversible path (2) the pressure is reduced until water will evaporate reversibly at 25°C. This reversible, equilibrium pressure is the vapor pressure. The vapor pressure is the pressure of a gas in equilibrium with a solid or liquid; the vapor pressure is temperature dependent. A few values for water are given in Table 2.2. The direct irreversible path is shown in Fig. 2.8. Whichever path we use we would calculate the same $E_2 - E_1$. Reversible paths are stressed for two reasons. It is often easier to calculate q and w along reversible paths; therefore, $E_2 - E_1$ can be easily calculated. More important, it can be shown that the work a system can do on the surroundings for a given change in energy is a maximum along a reversible path. Paths (1) and (2) also illustrate two other named paths. Path (1) is at constant pressure; a *constant-pressure path* is called *isobaric.* Path (2) is at constant temperature; a *constant-temperature path* is called *isothermal.* A *constant-volume path* is called *isochoric.* A path in which *q is zero* is called *adiabatic.* A *cyclic path* is one in which the system comes back to the state it started from. It is obviously very easy to calculate the change in any variable of state for a cyclic path; there is *no change in any variable of state for a cyclic path.*

Some numerical examples should make these ideas clearer. The examples will be simple ones, but they should illustrate the methods.

* The reader may wonder about the practicability of calculating the energy of a process that is difficult to carry out. Supercooled gaseous water with a temperature of 25°C and a pressure of 1 atm is exceedingly difficult to obtain. However, there are many valid reasons for wanting to know the difference in energy between liquid and gaseous water at 25°C. For example, the properties of gases at 1 atm pressure and 25°C are the standard values given in tables. One of the important points to remember is that thermodynamics does allow you to determine the energy changes for processes in which direct measurement is very difficult. The reader may prefer to think about processes closer to equilibrium, such as the evaporation of superheated liquid water at 110°C and 1 atm, or the condensation of supercooled gaseous water at 25°C and 0.05 atm.

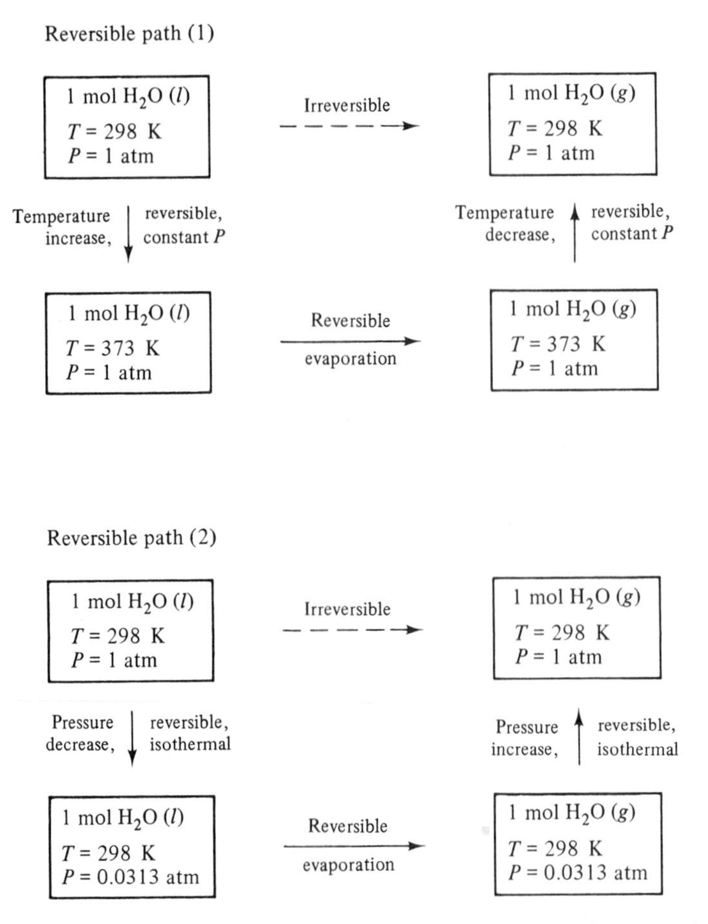

Fig. 2.8 Two possible reversible paths that can be taken between the same initial and final states. The letters in parentheses, (*l*) and (*g*), denote liquid and gas.

Temperature and Pressure Changes for a Liquid or Solid

Consider a system of one component which undergoes a change in P, V, and T only. External fields, friction, and surface effects are ignored. Only two of the three variables need be specified, because the equation of state allows calculation of the third. The number of moles, or grams, of the component does have to be specified. The goal is to calculate the energy change and enthalpy change for a variety of processes. One can get from the initial state to the final state by many paths. It is usually easiest to use a combination of reversible steps, where either P, or V, or T is held constant, and then add the energy or enthalpy contributions from each step. For the calculation we will use one of the most important biological molecules there is, water. Some properties of water are given in Table 2.2.

Table 2.2 Physical properties of water, H_2O (mol wt = 18.016), at 1 atm*

Solid H_2O ≡ ice

(at 0°C)

Density = 0.915 g cm^{-3}; specific volume = 1.093 cm^3 g^{-1}
Vapor pressure = 4.579 torr
Heat of melting = 79.7 cal g^{-1}
Absolute entropy = 9.8 cal mol^{-1} deg^{-1}
Heat capacity = 0.505 cal g^{-1} deg^{-1}

Liquid H_2O

Temperature (°C)	Density (g cm^{-3})	Surface tension (erg cm^{-2})	Vapor pressure (torr)	Heat of vaporization (cal g^{-1})	Viscosity (poise)
0	0.9999	75.64	4.579	595.9	1.7921 × 10^{-2}
20	0.9982	72.75	17.535	584.9	1.0050 × 10^{-2}
40	0.9922	69.56	55.324	574.0	0.6560 × 10^{-2}
60	0.9832	66.18	149.38	563.2	0.4688 × 10^{-2}
80	0.9718	62.61	355.1	551.3	0.3565 × 10^{-2}
100	0.9584	58.85	760.00	540.0	0.2838 × 10^{-2}

Absolute entropy = 15.1 cal mol^{-1} deg^{-1} at 0°C
= 20.8 cal mol^{-1} deg^{-1} at 100°C
Heat capacity = 1.00 cal g^{-1} deg^{-1} between 0°C and 100°C
Heat of freezing = − 79.7 cal g^{-1} at 0°C

Gaseous H_2O ≡ steam

(at 100°C)

Density = 5.880 × 10^{-4} g cm^{-3}; volume = 1701 cm^3 g^{-1}
Absolute entropy = 46.9 cal mol^{-1} deg^{-1}
Heat capacity at constant pressure = 0.448 cal g^{-1} deg^{-1}
Heat of condensation = − 540.0 cal g^{-1}

* Some of the properties listed will be defined and discussed in later chapters.

First, let us consider heating or cooling some liquid water in an open flask at a constant pressure provided by the atmosphere. The reaction is

$$n \text{ mol } H_2O(l) \text{ at } T_1, P_1, V_1 \longrightarrow n \text{ mol } H_2O(l) \text{ at } T_2, P_1, V_2$$

The heat is calculated from Eq. (2.9); at constant P,

$$q_P = \int_{T_1}^{T_2} C_P \, dT \tag{2.20}$$

The subscript P on the heat q and the heat capacity C remind us that the heat effects depend on the path. C_P is the heat capacity at constant pressure. In general, C_P will depend on the pressure and the temperature. However, we can usually neglect the effects of P and often neglect the effect of temperature. That is, C_P for 1 g of liquid water is close to 1 cal deg^{-1} from 0°C to 100°C and for pressures less than a few hundred atm. Therefore, for a temperature change at constant P, if C_P is independent of T,

$$q_P = C_P \cdot (T_2 - T_1) \tag{2.21}$$

Because values of heat capacity are often given per mole, we can write

$$q_P = n\bar{C}_P \cdot (T_2 - T_1) \tag{2.22}$$

where $\bar{C}_P$ = molar heat capacity at constant P (the bar over the C means per mole)
n = number of moles

If T_2 is greater than T_1, we know that heat is absorbed, which is consistent with the positive sign of q_P.

The work is calculated from Eq. (2.6); at constant P

$$w_P = -P_1 \cdot (V_2 - V_1) \tag{2.6}$$

But if we assume that the volume change of the liquid water is negligible, then $V_2 - V_1 \cong 0$ and $w_P \cong 0$.

If the water is heated in a closed and very strong container which keeps the volume constant, the heat and work are, at constant V,

$$q_V = n\bar{C}_V \cdot (T_2 - T_1) \tag{2.23}$$

$$w_V = 0 \tag{2.24}$$

C_V is the heat capacity at constant volume; for a solid or a liquid, it is not very different in magnitude from C_P. The pressure-volume work for a constant volume process is obviously zero.

If the water is kept at constant temperature while the pressure is changed, there is only a negligible volume change, and there is no appreciable work done or heat transferred; for an isothermal process involving liquid or solid

$$q_T \cong 0 \tag{2.25}$$

$$w_T \cong 0 \tag{2.26}$$

Calculation of $E_2 - E_1$ and $H_2 - H_1$ for a solid or liquid

To obtain $E_2 - E_1$ for the changes discussed above, we just add q and w. To obtain $H_2 - H_1$, we use the definition of enthalpy:

$$H_2 - H_1 \equiv E_2 - E_1 + (P_2V_2 - P_1V_1) \tag{2.12}$$

For any change of P_1, V_1, T_1 to P_2, V_2, T_2 we can obtain $E_2 - E_1$ and $H_2 - H_1$ by choosing a path and then combining the q's and w's. For example, we could use an isothermal plus a constant-pressure path, or an isothermal plus a constant-volume path. Because the volume of a solid or a liquid does not change much with temperature or pressure, we find that, for a solid or a liquid,

$$E_2 - E_1 \cong H_2 - H_1 \tag{2.27}$$

$$C_P \cong C_V \tag{2.28}$$

The exact relations between C_P and C_V or $E_2 - E_1$ and $H_2 - H_1$ depend on the equation of state.

Example 2.8 Calculate $E_2 - E_1$ and $H_2 - H_1$ in calories for heating 1 mol of liquid water from 0°C and 1 atm to 100°C and 10 atm. The volume per gram of the water is essentially independent of pressure; it can be calculated from the average density of water given in Table 2.2, 0.98 g cm^{-3}.

Solution Choose a path such as an isothermal path plus a constant-pressure path. First, the pressure is raised from 1 atm to 10 atm at 0°C. Then, the temperature is raised from 0°C to 100°C at 10 atm:

$$\begin{aligned} E_2 - E_1 &= q_T + w_T + q_P + w_P \\ &= 0 + 0 + n\bar{C}_P \cdot (T_2 - T_1) + 0 \\ &= (1\ \text{mol})\left(18 \frac{\text{cal}}{\text{mol deg}}\right)(100\ \text{deg}) \\ &= 1800\ \text{cal} \end{aligned}$$

$$\begin{aligned} H_2 - H_1 &= E_2 - E_1 + P_2V_2 - P_1V_1 \\ &= E_2 - E_1 + (P_2 - P_1)V_1 \\ &= 1800\ \text{cal} + (10\ \text{atm} - 1\ \text{atm})\left(\frac{18\ \text{g}}{\text{mol}}\right)\left(\frac{\text{cm}^3}{0.98\ \text{g}}\right)(1\ \text{mol}) \\ &= 1800\ \text{cal} + 165\ \text{cm}^3\ \text{atm} \\ &= 1800\ \text{cal} + (165\ \text{cm}^3\ \text{atm})\left(\frac{1.987\ \text{cal}}{82.05\ \text{cm}^3\ \text{atm}}\right) \\ &= 1800\ \text{cal} + 4.0\ \text{cal} \\ &= 1804\ \text{cal} \end{aligned}$$

Temperature and Pressure Changes for a Gas

Let us now calculate $E_2 - E_1$ and $H_2 - H_1$ when H_2O as a gas changes from P_1, V_1, T_1 to P_2, V_2, T_2. The heat transferred at constant P or V has the same form as for a liquid:

$$q_P = n\bar{C}_P \cdot (T_2 - T_1) \tag{2.29}$$

$$q_V = n\bar{C}_V \cdot (T_2 - T_1)$$

Of course, the heat capacities for gaseous H_2O must be used here instead of those for the liquid.

The constant-pressure work deserves more discussion. For liquids, we can ignore the expansion work, but for gases it is important. The equation for calculating constant-pressure work is the usual one:

$$w_P = -P_{ex} \cdot (V_2 - V_1) \tag{2.6}$$

Figure 2.9 illustrates the process. Note that for an expansion, the external pressure, P_{ex}, can be any pressure smaller than P. The volumes V_1 and V_2 are fixed by stops that hold the cylinder at these chosen values. The temperatures T_1 and T_2 are controlled by thermostats, and the pressure of the gas P depends on

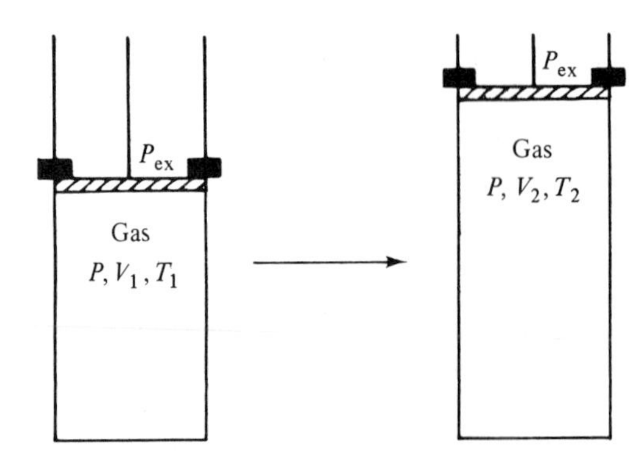

Fig. 2.9 Expansion of a gas at constant pressure, P_{ex}. For a reversible expansion, $P_{ex} = P$; for an irreversible expansion, $P_{ex} < P$.

the number of moles of the gas and the equation of state. For the ideal gas $PV_1 = nRT_1$ and $PV_2 = nRT_2$. The maximum constant-pressure work would be done by the gas if the gas could be heated reversibly with the external pressure just slightly less than the gas pressure P at all points during the expansion. For such a reversible expansion or compression,

$$P_{ex} \cong P \tag{2.30}$$

We usually express pressure as the absolute value of force per unit area, and therefore P_{ex} and P are chosen positive.

Combining Eqs. (2.6) and (2.30), we obtain

$$w_P = -P \cdot (V_2 - V_1) = -nR \cdot (T_2 - T_1) \tag{2.31}$$

Figure 2.10 illustrates that constant-pressure work can be thought of as the area of a rectangle of height P_{ex} and width $(V_2 - V_1)$ in a P versus V plot. If P_{ex} equals P for the gas, the constant-pressure work done by the gas is a maximum for a given change of volume, $V_2 - V_1$. If P_{ex} equals zero, the work is zero. For any P_{ex} between these extremes, the work done is intermediate. In practical engines the external pressure must be significantly less than the gas pressure

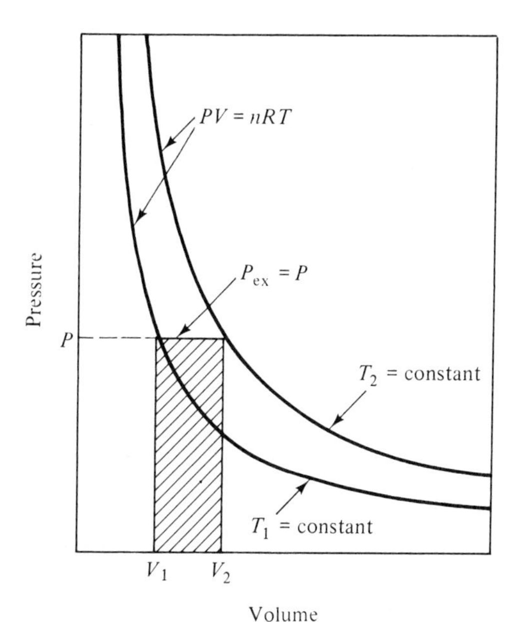

Fig. 2.10 Constant-pressure work is the area of rectangle of length P_{ex} and width $V_2 - V_1$ on a P versus V plot. The cross-hatched area shown is the work that is done by the system when $P_{ex} = P$.

inside the piston, however. Equation (2.31) thus represents the maximum work that would be available from an ideal engine of this type.

If a gas is expanded reversibly and isothermally from an initial volume V_1 to a final volume V_2, the work is

$$w = -\int_{V_1}^{V_2} P_{ex}\, dV \tag{2.5}$$

$$P_{ex} \cong P \qquad \text{(reversible)}$$

$$w = -\int_{V_1}^{V_2} P\, dV$$

$$= -\int_{V_1}^{V_2} \frac{nRT}{V}\, dV$$

$$= -nRT\int_{V_1}^{V_2} \frac{dV}{V}$$

(T = constant for an isothermal process, n = constant for a fixed number of moles of a gas)

$$= -nRT \ln \frac{V_2}{V_1}$$

This work is represented by the crosshatched area in Fig. 2.11. For an isothermal, reversible, ideal gas,

$$w_T = -nRT \ln \frac{V_2}{V_1} \tag{2.32}$$

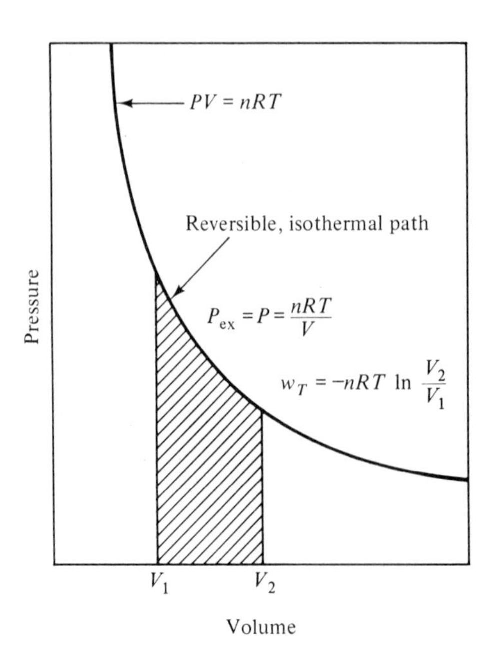

Fig. 2.11 Reversible isothermal expansion for an ideal gas. The crosshatched area is equal to the work done.

For an ideal gas it can be shown that the energy is independent of the volume and depends only on temperature. Thus the heat absorbed during the isothermal expansion of an ideal gas is equal to minus the work done: for an isothermal, ideal gas,

$$q_T = -w_T$$

For an isothermal, reversible, ideal gas,

$$q_T = nRT \ln \frac{V_2}{V_1} \tag{2.33}$$

Calculation of $E_2 - E_1$ and $H_2 - H_1$ for an ideal gas

For a change of P_1, V_1, T_1 to P_2, V_2, T_2 for an ideal gas, we can use any convenient path. For an isothermal path plus a constant-pressure path, we have

$$E_2 - E_1 = q_T + w_T + q_P + w_P$$

$$= nRT \ln \frac{V_2}{V_1} - nRT \ln \frac{V_2}{V_1} + C_P \cdot (T_2 - T_1) - nR \cdot (T_2 - T_1)$$

$$= (C_P - nR)(T_2 - T_1) \tag{2.34}$$

$$H_2 - H_1 = E_2 - E_1 + P_2 V_2 - P_1 V_1$$

$$= E_2 - E_1 + nR \cdot (T_2 - T_1)$$

$$= C_P \cdot (T_2 - T_1) \tag{2.35}$$

For an isothermal path plus a constant-volume path,

$$\begin{aligned} E_2 - E_1 &= q_T + w_T + q_V + w_V \\ &= nRT \ln \frac{V_2}{V_1} - nRT \ln \frac{V_2}{V_1} + C_V \cdot (T_2 - T_1) + 0 \\ &= C_V \cdot (T_2 - T_1) \end{aligned} \tag{2.36}$$

$$\begin{aligned} H_2 - H_1 &= E_2 - E_1 + P_2V_2 - P_1V_1 \\ &= C_V \cdot (T_2 - T_1) + nR \cdot (T_2 - T_1) \\ &= (C_V + nR) \cdot (T_2 - T_1) \end{aligned} \tag{2.37}$$

Using the fact that $E_2 - E_1$ and $H_2 - H_1$ must be independent of path, we see that for an ideal gas,

$$\begin{aligned} C_P &= C_V + nR \\ \bar{C}_P &= \bar{C}_V + R \end{aligned} \tag{2.38}$$

We also notice that *for an ideal gas*, $E_2 - E_1$ *and* $H_2 - H_1$ *depend only on* T. For *any* change of P, V, T for an ideal gas,

$$E_2 - E_1 = n\bar{C}_V \cdot (T_2 - T_1) \tag{2.39}$$

$$H_2 - H_1 = n\bar{C}_P \cdot (T_2 - T_1) \tag{2.40}$$

We have assumed for simplicity that $\bar{C}_P$ and $\bar{C}_V$ are independent of temperature in Eqs. (2.34) through (2.40). In general, for an ideal gas,

$$\begin{aligned} E_2 - E_1 &= n \int_{T_1}^{T_2} \bar{C}_V \, dT \\ H_2 - H_1 &= n \int_{T_1}^{T_2} \bar{C}_P \, dT \end{aligned} \tag{2.41}$$

Example 2.9 Calculate $E_2 - E_1$ and $H_2 - H_1$ in calories for heating 1 mol of gaseous water from 0°C and 0.001 atm to 100°C and 1 atm. The gaseous water can be approximated as an ideal gas with a constant heat capacity.

Solution For an ideal gas, E and H are independent of P and $\bar{C}_V = \bar{C}_P - R$; therefore, we use

$$E_2 - E_1 = n\bar{C}_V \cdot (T_2 - T_1) \tag{2.39}$$

$$\begin{aligned} &= (1 \text{ mol})[(18 \text{ g mol}^{-1})(0.448 \text{ cal g}^{-1} \text{ deg}^{-1}) \\ &\quad - 1.987 \text{ cal mol}^{-1} \text{ deg}^{-1}](100 \text{ deg}) \\ &= 608 \text{ cal} \end{aligned}$$

$$H_2 - H_1 = n\bar{C}_P \cdot (T_2 - T_1) \tag{2.40}$$

$$\begin{aligned} &= (1 \text{ mol})(8.06 \text{ cal mol}^{-1} \text{ deg}^{-1})(100 \text{ deg}) \\ &= 806 \text{ cal} \end{aligned}$$

All of these equations for ideal gases will apply approximately to real gases. However, the reader must be careful not to confuse equations that apply only to gases with those that apply only to liquids or solids.

Properties of $E_2 - E_1$ and $H_2 - H_1$ Independent of Equation of State

Of course, the first law [Eq. (2.19)] and the definition of H [Eq. (2.12)] are independent of whether the system is a gas, a liquid, a solid, or a mixture of these. But there are also some special cases that are independent of the equation of state.

For a closed system at constant volume, the heat absorbed is equal to the energy change, because the work at constant volume is zero:

$$E_2 - E_1 = q_V \tag{2.42}$$

$$E_2 - E_1 = \int_{T_1}^{T_2} C_V\,dT = C_V \cdot (T_2 - T_1) \tag{2.43}$$

Note that this conclusion is obtained if the only type of work is the pressure-volume work. If the system can do other types of work (electrical, for example), the conclusion will not apply.

For a closed system at constant pressure, the heat absorbed is equal to the enthalpy change. The energy change is the heat absorbed plus the reversible expansion work done on the system:

$$E_2 - E_1 = q_P - P \cdot (V_2 - V_1) \tag{2.44}$$

At constant pressure,

$$H_2 - H_1 = E_2 - E_1 + P \cdot (V_2 - V_1)$$

Thus, at constant pressure,

$$H_2 - H_1 = q_P \tag{2.45}$$

$$= \int_{T_1}^{T_2} C_P\,dT = C_P \cdot (T_2 - T_1)$$

Note again that in obtaining Eq. (2.44), and therefore (2.45), it is assumed that the only type of work is the pressure-volume type. The enthalpy H is also known as the *heat content*, because $H_2 - H_1$ is equal to the heat absorbed at constant pressure. It should be clear that E and H are variables of state; only for certain special processes are $E_2 - E_1$ or $H_2 - H_1$ equal to the heat transferred to the system.

PHASE CHANGES

The previous section has dealt with changes of P, V, and T only. Now we want to consider changes in phase; that is, the system changes from a solid to a liquid, for example. Names of phase changes are as follows:

Phase change	Name
Gas → liquid or solid	Condensation
Solid → liquid	Fusion, melting
Liquid → solid	Freezing
Liquid → gas	Vaporization
Solid → gas	Sublimation

There can also be phase changes between different solid phases and between different liquid phases. We are usually interested in the thermodynamics of the reversible change that occurs at constant T and P. A phase change is usually a good way to store energy. It takes only 1 cal to heat 1 g of water from 99°C to 100°C, but it takes 540 cal to change 1 g of water from liquid to gas at 100°C.

Consider a reversible phase change from phase a to phase b at constant T and P. The work done on the system is

$$w_P = -P \cdot (\Delta V)$$

where ΔV = volume change ≡ V(phase b) − V(phase a). The heat absorbed by the system at constant P is q_P. It is equal to $\Delta H \equiv H$(phase b) − H(phase a), according to Eq. (2.45).

The reversible phase change at constant P is given as

$$\Delta H = q_P \tag{2.45}$$

We have introduced the symbol Δ to represent a difference. To avoid confusion, we shall reserve this symbol for phase changes or chemical reactions. For changes caused solely by T or P, we will continue to use $V_2 - V_1$, $E_2 - E_1$, and so on. Thus, we can use $\Delta H(T_2) - \Delta H(T_1)$ to represent the change in the enthalpy of a reaction with temperature.

Values of ΔH have been tabulated for various reversible phase changes. Because the change is at constant pressure and the only work involved is the pressure-volume type, the heat and enthalpy are equal. Thus one can speak about a heat of vaporization or a heat of fusion equally as well as about an enthalpy of vaporization or fusion. The energy change of a phase change is

$$\Delta E = \Delta H - P \cdot (\Delta V) \tag{2.46}$$

Often a ΔH or a ΔE value is known at one T and P, but is needed at another. We know how to calculate the change of E and H with T and P so the calculation is easy. Suppose that we want to calculate the amount of heat removed when liquid water evaporates at human skin temperature (around 35°C). We want to know how effectively vaporization of sweat can cool us. If we know only ΔH for vaporization of water at 100°C, we can use the lower path to calculate the ΔH for the desired reaction at 35°C:

$$\begin{array}{ccc} H_2O(l),\ T_1 = 35°C,\ P_1 = 1\ \text{atm} & \xrightarrow{\Delta H(35°C)} & H_2O(g),\ T_1 = 35°C,\ P_1 = 1\ \text{atm} \\ \downarrow \text{Constant } P & & \uparrow \text{Constant } P \\ H_2O(l),\ T_2 = 100°C,\ P_1 = 1\ \text{atm} & \xrightarrow{\Delta H(100°C)} & H_2O(g),\ T_2 = 100°C,\ P_1 = 1\ \text{atm} \end{array}$$

We can calculate ΔH for each step in the chosen path and add the ΔH's to obtain the overall ΔH:

$$\Delta H(35°C) = C_P(l) \cdot (373\text{–}308) + \Delta H(100°C) + C_P(g) \cdot (308\text{–}373)$$

$$= \Delta H(100°C) + n[\bar{C}_P(g) - \bar{C}_P(l)]\,(-65) \tag{2.47}$$

The generalization for the temperature dependence of enthalpy of a phase change is

$$\Delta\bar{H}(T_2) = \Delta\bar{H}(T_1) + \Delta\bar{C}_P \cdot (T_2 - T_1) \tag{2.48}$$

where $\Delta\bar{H} = \bar{H}(\text{phase b}) - \bar{H}(\text{phase a})$
$\Delta\bar{C}_P = \bar{C}_P(\text{phase b}) - \bar{C}_P(\text{phase a})$

Example 2.10 Check the value for ΔH of vaporization of water at 20°C given in Table 2.2 by using Eq. (2.48) and $\Delta H(100°C)$ from the table.

Solution

$$\Delta\bar{H}(20°C) = \Delta\bar{H}(100°C) + \Delta\bar{C}_P \cdot (T_2 - T_1) \tag{2.48}$$

Equation (2.48) is written for 1 mol of substance, but it applies equally if all extensive quantities are for 1 g of substance.

$$\Delta H(20°C) = 540.0\ \text{cal g}^{-1} + (0.448\ \text{cal g}^{-1}\ \text{deg}^{-1} - 1.00\ \text{cal g}^{-1}\ \text{deg}^{-1})(-80\ \text{deg})$$

$$= 540.0\ \text{cal g}^{-1} + 44.2\ \text{cal g}^{-1}$$

$$= 584.2\ \text{cal g}^{-1}$$

The table gives 584.9 cal g^{-1}. The temperature dependence of the heat capacities accounts for the small discrepancy.

The energy of a phase change at constant P can be obtained from Eq. (2.46). If one of the phases in the phase change is a gas (vaporization or sublimation), the volume of the solid or liquid is so much smaller than the volume of the gas that it can be ignored. Furthermore, the gas phase can be approximated as an ideal gas:

$$\Delta\bar{E} = \Delta\bar{H} - P\bar{V}(\text{gas}) \tag{2.49}$$

$$\Delta\bar{E} \cong \Delta\bar{H} - RT \tag{2.50}$$

Example 2.11 Calculate (a) the energy of freezing 1.00 g of liquid water at 0°C and 1 atm, and (b) the energy of vaporizing 1.00 g of liquid water at 0°C and 1 atm.

Solution

$$\text{(a) } \Delta E = \Delta H - P \cdot (\Delta V) \tag{2.46}$$

$$= -79.7 \text{ cal g}^{-1} - (1 \text{ atm})(1.093 \text{ cm}^3 \text{ g}^{-1} - 1.000 \text{ cm}^3 \text{ g}^{-1})$$

$$= -79.7 \text{ cal g}^{-1} - 0.093 \text{ cm}^3 \text{ atm g}^{-1}$$

$$= -79.7 \text{ cal g}^{-1} - (0.093 \text{ cm}^3 \text{ atm g}^{-1})\left(\frac{1.987 \text{ cal}}{82.05 \text{ cm}^3 \text{ atm}}\right)$$

$$= -79.7 \text{ cal g}^{-1} - 0.002 \text{ cal g}^{-1}$$

$$= -79.7 \text{ cal g}^{-1}$$

$$\text{(b) } \Delta \bar{E} = \Delta \bar{H} - RT \tag{2.50}$$

$$= 595.9 \text{ cal g}^{-1} - (1.987 \text{ cal mol}^{-1} \text{ deg}^{-1})(273.1 \text{ K})\left(\frac{1 \text{ mol}}{18.016 \text{ g}}\right)$$

$$= 595.9 \text{ cal g}^{-1} - 30.1 \text{ cal g}^{-1}$$

$$= 565.8 \text{ cal g}^{-1}$$

Note that there is a significant difference between ΔH and ΔE when one of the phases is a gas [part (b)], but when neither phase is a gas [part (a)], the difference is insignificant.

CHEMICAL REACTIONS

We come next to the most important way the energy and enthalpy of a system can be changed: a chemical reaction can occur. Here the properties of variables of state are most useful. The fact that the enthalpy change when 1 teaspoon of sugar is burned to CO_2 and H_2O is the same whether the reaction occurs in a human being or in a calorimeter is very convenient. Of course, we do have to ensure that the initial and final states for the reaction are the same inside the human being and inside the calorimeter. Once this is done, however, it does not matter whether the path involves 10 enzyme-catalyzed steps or a direct reaction (combustion) with O_2.

The change in the system that we are concerned with is represented by the general chemical reaction

$$n_A A + n_B B \longrightarrow n_C C + n_D D$$

That is, n_A moles of A react with n_B moles of B to give n_C moles of C and n_D moles of D. The conditions of P, V, and T must be specified for both the products and the reactants. The change in the variables of state, such as ΔE and ΔH, are desired. To repeat, it does not matter how the reaction actually takes place; only the initial and final states are important.

Heat Effects of Chemical Reactions

If the reaction takes place at constant P, the heat given off is equal to the decrease in enthalpy when there is no work other than the pressure-volume type:

$$q_P = \Delta H \tag{2.51}$$

A negative ΔH and a negative q_P means that heat is released (exothermic); a positive ΔH and a positive q_P means that heat is absorbed (endothermic). If the reaction takes place at constant V in a sealed container such as a bomb calorimeter, the heat given off is equal to the decrease in energy:

$$q_V = \Delta E \tag{2.52}$$

By measurement of the heat effects of chemical reactions at constant P or V, values of ΔH and ΔE for the reactions are obtained. The heat effect is usually measured by surrounding the reaction with a known amount of water and measuring the temperature rise of the water. If either ΔH or ΔE is measured, the other can be obtained from the definition of H:

$$\Delta H = \Delta E + \Delta(PV) \tag{2.53}$$

where $\Delta(PV) = PV(\text{products}) - PV(\text{reactants})$. If gases are involved in the reaction, we can ignore the volumes of the solids or liquids and use the ideal gas equation for the gases.

The amount of heat that can be obtained from a chemical reaction is obviously of great practical importance. We may be interested in obtaining the maximum amount of heat for a given weight of fuel. For example, if cost is unimportant, is hexane, methanol, benzene, or polyethylene the best fuel per unit weight? Or it may be that we want to minimize the heat. In a battery we want to convert chemical energy into electrical work and not waste the chemical energy in the form of heat released.

The biochemical reactions necessary to sustain life in a person produce about 1500 kcal day^{-1} of heat at constant pressure. This is the basal metabolic rate. Each person thus continually produces about 70 W (1 W $\equiv$ 1 J s^{-1}). If one does more than lie around in bed, further energy is produced. The 2000 to 3000 kcal day^{-1} needed is normally replenished in the person in the form of food (remember that a nutritionist's Cal $\equiv$ a chemist's kcal). Each gram of protein or carbohydrate provides about 4 kcal, and fat provides about 9 kcal g^{-1}.

The heat at constant pressure, and thus the enthalpy changes for many reactions, have been measured. These reactions and their enthalpies can be combined to calculate the enthalpies for many other reactions. For example, suppose that we know ΔH_1 for the oxidation of solid glycine at 25°C to form CO_2, ammonia, and liquid water:

(1) $3\,O_2(g, 1\text{ atm}) + 2\,\underset{\text{Glycine}}{NH_2CH_2COOH(s)} \longrightarrow$

$4\,CO_2(g, 1\text{ atm}) + 2\,H_2O(l) + 2\,NH_3(g, 1\text{ atm}) \qquad \Delta H_1 = -787.58\text{ kcal}$

The ΔH_2 for the hydrolysis of solid urea is also known.

(2) $H_2O(l) + \underset{\text{Urea}}{H_2NCONH_2(s)} \longrightarrow$

$$CO_2(g, 1\text{ atm}) + 2\,NH_3(g, 1\text{ atm}) \qquad \Delta H_2 = 31.82\text{ kcal}$$

If we subtract these two reactions, treating the chemicals and their ΔH's as algebraic quantities, we get

$$3\,O_2(g, 1\text{ atm}) + 2\text{ glycine}(s) - \text{urea}(s) - H_2O(l) \longrightarrow$$
$$4\,CO_2(g, 1\text{ atm}) + 2\,H_2O(l) + 2\,NH_3(g, 1\text{ atm}) - CO_2(g, 1\text{ atm}) - 2\,NH_3(g, 1\text{ atm})$$

Rearranging and canceling, we get

(3) $3\,O_2(g, 1\text{ atm}) + 2\text{ glycine}(s) \longrightarrow 1\text{ urea}(s) + 3\,CO_2(g, 1\text{ atm}) + 3\,H_2O(l)$

$$\Delta H_3 = \Delta H_1 - \Delta H_2$$
$$= -819.40\text{ kcal}$$

This equation is of more biochemical interest, because urea rather than ammonia is the main oxidative metabolic product of amino acids. However, the biological reaction does not involve solid glycine and solid urea, but rather aqueous solutions. Therefore, we use the reactions and heats for the dissolution of 1 mol of urea and of glycine.

(4) $\text{glycine}(s) + \infty\,H_2O(l) \longrightarrow \text{glycine}(aq) \qquad \Delta H_4 = 3.75\text{ kcal}$

(5) $\text{urea}(s) + \infty\,H_2O(l) \longrightarrow \text{urea}(aq) \qquad \Delta H_5 = 3.33\text{ kcal}$

The enthalpies of solution will depend on concentration; here we will use enthalpies for very dilute solutions (designated *aq*) and assume that they do not depend on concentration. That means that we choose the number of moles of $H_2O(l)$ to be a very large number, infinite ($\equiv\infty$), in the preceding equations. From reaction (3) we now subtract two times reaction (4) and add reaction (5):

(6) $3\,O_2(g, 1\text{ atm}) + 2\text{ glycine}(s) - 2\text{ glycine}(s)$

$$- \infty\,H_2O(l) + \text{urea}(s) + \infty\,H_2O(l) \longrightarrow$$

$$\text{urea}(s) + 3\,CO_2(g, 1\text{ atm}) + 3\,H_2O(l) - 2\text{ glycine}(aq) + \text{urea}(aq)$$

(6) $3\,O_2(g, 1\text{ atm}) + 2\text{ glycine}(aq) \longrightarrow \text{urea}(aq) + 3\,CO_2(g, 1\text{ atm})$

$$+ 3\,H_2O(l)$$

$$\Delta H_6 = \Delta H_3 - 2\,\Delta H_4 + \Delta H_5$$
$$= -823.57\text{ kcal}$$

We now know the enthalpy for the reaction of a dilute aqueous solution of glycine with O_2 gas to form a dilute aqueous solution of urea plus CO_2 gas plus 3 mol of liquid H_2O. We cannot ignore the synthesized H_2O in the reaction; the 3 mol cannot be added to the "infinities" introduced from the dissolution reaction and thus made negligible. The enthalpy for reaction (6) may be quite close to that of the naturally occurring reaction in the human body. However, it should be clear that by adding or subtracting the enthalpies of other reactions we can calculate

ΔH for *any* reaction we like. That is, if we think it is important, we can find the heat of solution of glycine in a defined buffer solution instead of pure water. We can specify that instead of $CO_2(g, 1\ \text{atm})$ as a product, we have a carbonic acid solution of a certain pH. It may be very difficult to directly measure the ΔH of the reaction we want, but we can always find the ΔH by using a convenient alternative path. In other words, if we want the ΔH for reaction $A \rightarrow B$, we may use a path that is a sum of many other reactions.

$$\begin{array}{ccccccccc} A & & & \xrightarrow{\Delta H} & & & & & B \\ \Delta H_1 \downarrow & & & & & & & & \uparrow \Delta H_6 \\ C & \xrightarrow{\Delta H_2} & D & \xrightarrow{\Delta H_3} & E & \xrightarrow{\Delta H_4} & F & \xrightarrow{\Delta H_5} & G \end{array}$$

$$\Delta H = \Delta H_1 + \Delta H_2 + \Delta H_3 + \Delta H_4 + \Delta H_5 + \Delta H_6$$

It is convenient to remember that if

$$A \longrightarrow C \quad \text{has} \quad \Delta H_1$$

then

$$C \longrightarrow A \quad \text{has} \quad -\Delta H_1$$

and

$$nA \longrightarrow nC \quad \text{has} \quad n\,\Delta H_1$$

Temperature Dependence of ΔH

By the same reasoning used for phase changes, if ΔH is known at one temperature, it can be calculated at other temperatures.

$$\begin{array}{ccc} A(T_2) & \xrightarrow{\Delta H(T_2)} & B(T_2) \\ -C_P^A \cdot (T_2 - T_1) \downarrow & & \uparrow C_P^B \cdot (T_2 - T_1) \\ A(T_1) & \xrightarrow{\Delta H(T_1)} & B(T_1) \end{array}$$

$$\Delta H(T_2) = \Delta H(T_1) + \Delta C_P \cdot (T_2 - T_1) \tag{2.54}$$

where $\Delta H = H(\text{products}) - H(\text{reactants})$

$\Delta C_P = C_P(\text{products}) - C_P(\text{reactants})$

ΔE for a Reaction

For a reaction at constant pressure, ΔE can be calculated from the definition of H:

$$\Delta E = \Delta H - P\,\Delta V \tag{2.55}$$

If gases are involved, we ignore the volumes of solids and liquids, and approximate the gases as ideal gases.

$$\Delta E \cong \Delta H - \Delta nRT \tag{2.56}$$

where Δn is the number of moles of *gaseous* products minus number of moles of *gaseous* reactants.

Standard Enthalpies of Formation

In combining chemical reactions and their enthalpies, we treated the products and reactants as algebraic quantities. For the reaction

$$n_A A + n_B B \longrightarrow n_C C + n_D D \tag{2.57}$$

the enthalpy change can be written

$$\Delta H = H(\text{products}) - H(\text{reactants})$$

$$\Delta H = n_C \bar{H}_C + n_D \bar{H}_D - n_B \bar{H}_B - n_A \bar{H}_A \tag{2.58}$$

where $\bar{H}$ = enthalpy mol^{-1}. However, enthalpies are all relative; only differences of enthalpy are defined. We can talk of the volume mol^{-1}, $\bar{V}$, but we can only specify an enthalpy difference. Therefore, we can arbitrarily choose a zero of enthalpy. For example, in the reaction

$$H_2(g, 1\text{ atm}) + \tfrac{1}{2} O_2(g, 1\text{ atm}) \xrightarrow{\Delta H} H_2O(g, 1\text{ atm})$$

$$\Delta H = 1\bar{H}(H_2O, g, 1\text{ atm}) - \tfrac{1}{2}\bar{H}(O_2, g, 1\text{ atm}) - 1\bar{H}(H_2, g, 1\text{ atm})$$

If we choose $\bar{H}$ for O_2 and H_2 equal to zero, the enthalpy of H_2O is equal to the enthalpy of the reaction.

$$\Delta H = 1\bar{H}(H_2O, g, 1\text{ atm}) - \tfrac{1}{2}\cdot 0 - 1\cdot 0$$

$$\bar{H}(H_2O, g, 1\text{ atm}) = \Delta H$$

Thermodynamicists have assigned zero enthalpy to all elements in their most stable states at 1 atm pressure. These are called *standard states* and are designated by a superscript zero. *The standard enthalpy mol*$^{-1}$ *of a molecule is defined to be equal to the enthalpy of formation of* 1 *mol of the molecule at* 1 *atm pressure from its elements in their standard states.*

$$\bar{H}^0(\text{molecule}) \equiv \Delta \bar{H}_f^0 \tag{2.59}$$

where $\Delta \bar{H}_f^0$ = enthalpy of formation of 1 mol from elements under standard conditions. The superscript zero means 1 atm pressure and the most stable form of the element. It should be clear that the standard enthalpy is a defined quantity that depends on the choice of the standard state. Assigning the elements *in their most stable state* at 1 atm pressure to have zero enthalpy is purely arbitrary.

The standard enthalpies of thousands of substances have been determined; the Appendix gives a few of them at 25°C. The enthalpies at 25°C and 1 atm pressure for very many reactions can be calculated from this table. One atmosphere pressure is part of the definition of the standard state, but the temperature is not specified in the definition. However, nearly all the tables available are for 25°C. The standard chemical reaction at 25°C is given as

$$\Delta H^0(298) = n_C \bar{H}_C^0 + n_C \bar{H}_D^0 - n_A \bar{H}_A^0 - n_B \bar{H}_B^0$$

where $\bar{H}^0$ = standard enthalpy mol^{-1} at 25°C = $\Delta \bar{H}_f^0$

$\Delta \bar{H}_f^0$ = 0 for all elements in their standard states

Example 2.12 Use the Appendix tables to calculate the value of ΔH for reacting 1 g of solid glycylglycine with oxygen to form solid urea, CO_2, and liquid H_2O at 25°C, 1 atm.

Solution The reaction is

$$3\,O_2(g) + \underset{\text{Glycylglycine}}{C_4H_8N_2O_3(s)} \longrightarrow \underset{\text{Urea}}{CH_4N_2O(s)} + 3\,CO_2(g) + 2\,H_2O(l)$$

$$\begin{aligned}\Delta H^0 &= \bar{H}^0(\text{urea}) + 3\bar{H}^0(CO_2) + 2\bar{H}^0(H_2O, l)\\ &\quad - \bar{H}^0(\text{glycylglycine}) - 3\bar{H}^0(O_2)\\ &= -79.6 + 3(-94.05) + 2(-68.32) - (-178.12) - 3(0)\\ &= -320.3 \text{ kcal}\end{aligned}$$

The enthalpy change calculated is for 1 mol of glycylglycine. To find ΔH per gram, we must divide by the molecular weight, 132.12 g mol^{-1}:

$$\Delta H = -2.42 \text{ kcal g}^{-1}$$

The negative sign means that heat is given off in the reaction.

What about other temperatures and pressures? The enthalpy at any temperature can be obtained from Eq. (2.54). The temperature dependence of the standard chemical reaction

$$\Delta H^0(T) = \Delta H^0(298) + \Delta C_P^0 \cdot (T - 298) \tag{2.60}$$

where $\Delta C_P^0 = n_C\bar{C}_P^0(C) + n_D\bar{C}_P^0(D) - n_A\bar{C}_P^0(A) - n_B\bar{C}_P^0(B)$. It is necessary to remember that ΔC_P^0 here must include all products and reactants. The value of $\bar{H}^0$ for elements is chosen as zero, but $\bar{C}_P^0$ for an element is not zero.

The pressure dependence of ΔH is not large and we will ignore it. For ideal gases, H is independent of P; for solids or liquids, H is not very dependent on P. We shall also generally ignore the effect of concentration on ΔH.

It should be clear that a heat of reaction and a ΔH can be measured for any condition of concentration, solvent, pH, pressure, and so on. We make these approximations because they are accurate enough for most purposes and because they simplify the calculations. For a discussion of the pressure dependence of ΔH and of heats of dilution, the reader is referred to standard thermodynamics textbooks.

Bond Energies

Although many values of enthalpies of formation have been determined, there are many more compounds whose ΔH_f^0 is not known. However, we can approximate ΔH_f^0 and heats of reaction by using bond dissociation energies, D. The *bond dissociation energy* is the enthalpy at 25°C and 1 atm for the reaction

$$\text{A—B} \longrightarrow \text{A}(g) + \text{B}(g)$$

The bond dissociation energy should be called a bond dissociation enthalpy, but the tradition for energy is strong. It has been found that the amount of energy necessary to break a particular type of bond in a molecule is not too dependent on the molecule. For example, the average energy necessary to break a C—H bond is 99 kcal mol^{-1} ± 10% in a wide range of organic compounds.

Some average bond dissociation energies are given in Table 2.3. We expect the value for ΔH estimated from the table to be reasonable. One notable exception is for molecules which are stabilized greatly by electron delocalization. For example, the data that were used to obtain values for C—C and C=C bonds in Table 2.3 came from molecules with single bonds, and isolated double bonds. Molecules with conjugated double bonds (C=C bonds separated by only one C—C bond) are more stable than those containing the same number of isolated single and double bonds. This difference in energy is called the *resonance energy*. For such molecules the difference in heat of formation calculated from bond energies and that experimentally measured is an estimate of the resonance energy.

Table 2.3. Average bond dissociation energies at 25°C

Bond	D (kcal mol^{-1})
C—C	83
C=C	146
C≡C	200
C—H	99
C—N	70
C—O	86
C=O	178
N—H	93
O—H	111
H_2	104.18
N_2	226.8
O_2	118.86
C(graphite)	171.70

The easiest way to show how Table 2.3 can be used is by some examples.

Example 2.13 Calculate the heat of formation for gaseous cyclohexane using Table 2.3 and compare with the measured values in the Appendix.

Solution The reaction for the formation of cyclohexane is

$$6\,\text{C(graphite)} + 6\,H_2(g) \longrightarrow C_6H_{12}(g)$$

We can write it as a sum of bond-breaking and bond-forming reactions:

(1) $6\,\text{C(graphite)} \longrightarrow 6\,\text{C}(g)$

$$\Delta H_1 = 6D(\text{graphite}) = (6)(171.70) = 1030 \text{ kcal}$$

(2) $6\,H_2(g) \longrightarrow 12\,H(g)$

$$\Delta H_2 = 6D(H_2) = (6)(104.18) = 625 \text{ kcal}$$

(3) $6\,C(g) + 12\,H(g) \longrightarrow (6\,C—C + 12\,C—H) = C_6H_{12}(g)$

$$\Delta H_3 = -6D(C—C) - 12D(C—H) = -(6)(83) - (12)(99)$$

$$= -1686 \text{ kcal}$$

$$\Delta H_f^0(C_6H_{12}) = \Delta H_1 + \Delta H_2 + \Delta H_3$$

$$= -31 \text{ kcal}$$

The value in the Appendix is -29.43 kcal for cyclohexane, which is in excellent agreement.

Example 2.14 Calculate the heat of formation for gaseous benzene using Table 2.3 and compare with the measured value in the Appendix.

Solution Because of resonance effects, we expect the measured value for benzene to differ greatly from that calculated for the classical structure of benzene. An explanation of the large thermodynamic stability of benzene was an important goal for chemists interested in chemical bonding. The reaction for benzene is

$$6\,C(\text{graphite}) + 3\,H_2(g) \longrightarrow C_6H_6(g)$$

(1) $6\,C(\text{graphite}) \longrightarrow 6\,C(g)$

$$\Delta H_1 = (6)(171.70) = 1030 \text{ kcal}$$

(2) $3\,H_2(g) \longrightarrow 6\,H(g)$

$$\Delta H_2 = (3)(104.18) = 313 \text{ kcal}$$

(3) $6\,C(g) + 6\,H(g) \longrightarrow (3\,C—C + 3\,C{=}C + 6\,C—H) = C_6H_6(g)$

$$\Delta H_3 = -(3)(83) - (3)(146) - (6)(99)$$

$$= -1281 \text{ kcal}$$

$$\Delta H_f^0(C_6H_6) = \Delta H_1 + \Delta H_2 + \Delta H_3$$

$$= 62 \text{ kcal}$$

The value in the Appendix is 19.82 kcal for $C_6H_6(g)$! Benzene is about 42 kcal lower in enthalpy than would be expected for a molecule made up of 3 C—C single bonds, 3 C=C double bonds, and 6 C—H single bonds. This energy is what we call the resonance energy.

SUMMARY

State Variables

Name	Symbol	Units	Definition
Volume	V	liters (ℓ), mℓ, cm^3	(length)^3
Pressure	P	atm, torr, mm Hg, dyn cm^{-2}, pascals, N m^{-2}	force area^{-1}
Temperature	T	K, °C, °F	
Energy	E	cal, J, ergs	
Enthalpy	H	cal, J, ergs	$H \equiv E + PV$

Unit Conversions

Volume:

$$1\ \ell \equiv 1000\ \text{m}\ell;\ 1\ \text{m}\ell = 1\ \text{cm}^3$$

Pressure:

$$1\ \text{atm} = 760\ \text{torr} = 1.0132 \times 10^6\ \text{dyn cm}^{-2} = 1.013 \times 10^5\ \text{pascals}$$

$$1\ \text{torr} \equiv 1\ \text{mm Hg}$$

$$1\ \text{dyn cm}^{-2} \equiv 1\ \text{g cm}^{-1}\ \text{s}^{-2}$$

$$1\ \text{pascal} \equiv 1\ \text{N m}^{-2}$$

Temperature:

$$\text{K} = {}^\circ\text{C} + 273.16$$

$${}^\circ\text{C} = \frac{{}^\circ\text{F} - 32}{1.8}$$

Energy and enthalpy:

$$1\ \text{cal} = 4.184\ \text{J} = 4.184 \times 10^7\ \text{erg}$$

$$1\ \text{J} \equiv 1\ \text{kg m}^2\ \text{s}^{-2}$$

$$1\ \text{erg} = 1\ \text{g cm}^2\ \text{s}^{-2}$$

$$1\ \ell\ \text{atm} = 24.22\ \text{cal}$$

GENERAL EQUATIONS

Energy, E: closed system, heat and work are the only forms of energy the system exchanges with surroundings.

$$E_2 - E_1 = q + w \tag{2.19}$$

Enthalpy, $H \equiv E + PV$:

$$H_2 - H_1 = E_2 - E_1 + P_2V_2 - P_1V_1$$

Heat, q (heat absorbed by system is positive):

$$q = \int_{T_1}^{T_2} C\,dT \tag{2.9}$$

$$C = \text{heat capacity} = \frac{dq}{dT}$$

Work, w (work done on system is positive)
Stretching or compressing a spring

$$w = k \cdot (x_2 - x_1)\left(\frac{x_2 + x_1}{2} - x_0\right) \tag{2.4}$$

k = Hooke's law constant
x_0 = length of spring in the absence of a force
x_1, x_2 = initial and final lengths of the spring, respectively

Expansion or compression of a gas:

$$w = -\int_{V_1}^{V_2} P_{ex}\,dV \tag{2.5}$$

P_{ex} = external pressure

$$w_P = -P \cdot (V_2 - V_1) \qquad \text{(constant pressure)} \tag{2.6}$$

Electrical work done by a system:

$$w = -EIt \tag{2.8}$$

E = voltage
I = current
t = time

Closed System; Pressure-Volume Work Only

$$E_2 - E_1 = q_V = \int_{T_1}^{T_2} C_V\,dT \qquad \text{(constant volume)}$$

C_V = heat capacity at constant volume

$$H_2 - H_1 = q_P = \int_{T_1}^{T_2} C_P\, dT \qquad \text{(constant pressure)}$$

C_P = heat capacity at constant pressure

Solids and Liquids

We assume in these equations that the volume of a solid or liquid is independent of T and P, that $C_P = C_V = C$, that they do not depend on T and P.

$$E_2 - E_1 = n\bar{C} \cdot (T_2 - T_1) \qquad \text{(any change of } P, T\text{)}$$

$$H_2 - H_1 = n\bar{C} \cdot (T_2 - T_1) + (P_2 - P_1) \cdot V \qquad \text{(any change of } P, T\text{)}$$

n = number of moles
$\bar{C}$ = heat capacity per mole

Gases

We assume that gas properties can be approximated by the ideal gas equation and that C_P and C_V are independent of T. $PV = nRT$. $\bar{C}_P = \bar{C}_V + R$.

$$E_2 - E_1 = n\bar{C}_V \cdot (T_2 - T_1) \qquad \text{(any change of } P, V, T\text{)} \qquad (2.39)$$

$$H_2 - H_1 = n\bar{C}_P \cdot (T_2 - T_1) \qquad \text{(any change of } P, V, T\text{)} \qquad (2.40)$$

n = number of moles
$\bar{C}_P$ = heat capacity per mole at constant P
$\bar{C}_V$ = heat capacity per mole at constant V

$$w_P = -nR \cdot (T_2 - T_1) \qquad \text{(reversible, constant } P\text{)} \qquad (2.31)$$

$$w_T = -nRT \ln \frac{V_2}{V_1} \qquad \text{(reversible, constant } T\text{)} \qquad (2.32)$$

$$q_T = nRT \ln \frac{V_2}{V_1} \qquad \text{(reversible, constant } T\text{)} \qquad (2.33)$$

$\ln x = 2.303 \log x$
$R = 1.987 \text{ cal deg}^{-1} \text{ mol}^{-1}$

Phase Changes

For a phase change, phase a → phase b, which occurs at constant T and P,

$$\Delta H = q_P \qquad (2.45)$$

$\Delta H = H(\text{phase b}) - \text{H}(\text{phase a})$

$$\Delta E = \Delta H - P \cdot (\Delta V) \quad (2.46)$$

$\Delta E = E(\text{phase b}) - E(\text{phase a})$
$\Delta V = V(\text{phase b}) - V(\text{phase a})$

$$\Delta H(T_2) = \Delta H(T_1) + n\,\Delta\bar{C}_P \cdot (T_2 - T_1) \quad (2.48)$$

n = number of moles
$\Delta\bar{C}_P$ = heat capacity per mole at constant P of phase b − heat capacity per mole at constant P of phase a

$$\Delta H(P_2) \cong \Delta H(P_1)$$
$$w_P = -P \cdot \Delta V$$

Chemical Reactions

For a chemical reaction

$$n_A\text{A} + n_B\text{B} \longrightarrow n_C\text{C} + n_D\text{D}$$

which occurs at constant T and P,

$$\Delta H = n_C\bar{H}_C + n_C\bar{H}_D - n_A\bar{H}_A - n_B\bar{H}_B = q_P$$

$\bar{H}^0_{298}(\text{A}) \equiv \Delta\bar{H}^0_{f,298}(\text{A})$ = heat of formation of A per mole from the elements in their most stable states at standard conditions (1 atm) and 25°C

$$\Delta E = \Delta H - \Delta(PV) \quad (2.53)$$

$\Delta(PV) = PV(\text{products}) - PV(\text{reactants})$
$1\ \ell\ \text{atm} = 24.22\ \text{cal}$

$$\Delta E = \Delta H - \Delta nRT \quad (2.56)$$

Δn = number of moles of *gaseous* products − number of moles of *gaseous* reactants
$R = 1.987\ \text{cal deg}^{-1}\ \text{mol}^{-1}$

$$\Delta H(T_2) = \Delta H(T_1) + \Delta C_P \cdot (T_2 - T_1) \quad (2.54)$$
$$\Delta C_P = n_C\bar{C}_P(\text{C}) + n_D\bar{C}_P(\text{D}) - n_A\bar{C}_P(\text{A}) - n_B\bar{C}_P(\text{B})$$
$$\Delta H(P_2) \cong \Delta H(P_1)$$

MATHEMATICS NEEDED FOR CHAPTER 2

Students should be able to integrate simple powers of x.

Indefinite integral of ax^n:

$$\int ax^n\,dx = \frac{ax^{n+1}}{n+1} \qquad (n \neq -1) \tag{1}$$

$$\int ax^{-1}\,dx = a\int \frac{dx}{x} = a \ln x \tag{2}$$

Here a is a constant independent of x.

Definite integral of ax^n:

$$\int_{x_1}^{x_2} ax^n\,dx = \frac{a \cdot (x_2^{n+1} - x_1^{n+1})}{n+1} \qquad (n \neq -1) \tag{3}$$

$$\int_{x_1}^{x_2} a\frac{dx}{x} = a \cdot (ln\, x_2 - \ln x_1) = a \ln \frac{x_2}{x_1} \tag{4}$$

Remember that $\ln ab = \ln a + \ln b$; $\ln (a/b) = \ln a - \ln b$; $\ln a = 2.303 \log a$.

Example $P = aV + bV^2$, with a and b constant. Using Eqs. (1) and (3), we obtain

$$\int P\,dV = \int (aV + bV^2)\,dV = a\frac{V^2}{2} + b\frac{V^3}{3}$$

$$\int_{V_1}^{V_2} P\,dV = \frac{a}{2}\cdot(V_2^2 - V_1^2) + \frac{b}{3}\cdot(V_2^3 - V_1^3)$$

REFERENCES

The following textbooks on thermodynamics can be useful as supplements to Chapters 2–5.

DICKERSON, R. E., 1969. *Molecular Thermodynamics*, W. A. Benjamin, Menlo Park, Calif.

KLOTZ, J. M., and R. M. ROSENBERT, 1972. *Introduction to Chemical Thermodynamics*, 2nd ed., W. A. Benjamin, Menlo Park, Calif.

LEWIS, G. N., and M. RANDALL, 1961. (Revised by K. S. Pitzer and L. Brewer.) *Thermodynamics*, 2nd ed., McGraw-Hill, New York.

For articles on applications of calorimetry to biological molecules, see

BROWN, H. D., ed., 1969. *Biochemical Microcalorimetry*, Academic Press, New York.

PROBLEMS

1. Many remedies for acid indigestion contain sodium bicarbonate, $NaHCO_3$. Assuming that this reacts in the stomach according to the equation

$$NaHCO_3 + HCl \longrightarrow NaCl + H_2O + CO_2(g)$$

calculate the volume of gaseous CO_2 from 0.5 g of sodium bicarbonate at a pressure of 1 atm and a temperature of 37°C.

2. Photosynthesis by land plants leads to the fixation each year of about 1 kg of carbon on the average for each square meter of an actively growing forest. The atmosphere is approximately 20% O_2 and 80% N_2 but contains 0.046% CO_2 by weight.
   (a) What volume of air (25°C, 1 atm) is needed to provide this 1 kg of carbon?
   (b) At this rate, how long would it take to use all the CO_2 in the entire atmosphere directly above the forest?

   (This assumes that atmospheric circulation and replenishment from the oceans, rocks, combustion of fuels, respiration of animals, and decay of biological materials are cut off.) *Hint*: Atmospheric pressure measures the weight of the atmosphere; 1 atm is equivalent to 1.033×10^4 kg m^{-2}.

3. Calculate the work (in calories) done on the system for each of the following examples. Specify the sign of the work.
   (a) A mass of 80 kg (about 176 lb) is lifted 310 m (about 1000 ft). The system is the mass.
   (b) The volume of the system changes from 3 ℓ to 1 ℓ under a constant pressure of 1 atm.

4. Calculate the heat (in calories) absorbed by the system for each of the following examples. Specify the sign of the heat.
   (a) 100 mℓ of liquid water is heated from 0°C to 100°C at 1 atm.
   (b) 100 mℓ of liquid water is frozen to ice at 0°C at 0.01 atm.
   (c) 100 mℓ of liquid water is evaporated to steam at 100°C at 1 atm.

5. One mole of an ideal gas initially at 27°C and 1 atm pressure is heated and allowed to expand reversibly at constant pressure until the final temperature is 327°C. For this gas, $\bar{C}_V = 5.0$ cal deg^{-1} mol^{-1} and is constant over the temperature range.
   (a) Calculate the work, w, done on the gas in this expansion.
   (b) What are ΔE and ΔH for the process?
   (c) What is the amount of heat, q, absorbed by the gas?

6. For the following processes, state whether each of the thermodynamic quantities q, w, ΔE, and ΔH is greater than, equal to, or less than zero for the *system* described.
   (a) An ideal gas expands adiabatically against an external pressure of 1 atm.
   (b) An ideal gas expands isothermally against an external pressure of 1 atm.
   (c) An ideal gas expands adiabatically into a vacuum.
   (d) A liquid at its boiling point is converted reversibly into its vapor, at constant temperature and 1 atm pressure.
   (e) H_2 gas and O_2 gas are caused to react in a closed bomb at 25°C and the product water is brought back to 25°C.

7. One mole of liquid water at 100°C is heated until the liquid is converted entirely to vapor at 100°C and 1 atm pressure. Calculate q, w, ΔE, and ΔH for each of the following paths for the process:

(a) The vaporization is carried out in a cylinder where the external pressure on the piston is maintained at 1 atm throughout.
(b) The cylinder is first expanded against a vacuum ($P_{ex} = 0$) to the same volume as in part (a), and then sufficient heat is added to vaporize the liquid completely to 1 atm pressure.

8. Calculate the work (in calories) done on the system for each of the following examples. Specify the sign of the work.
(a) The volume of the system changes from 1 ℓ to 3 ℓ at an initial temperature of 25°C and a constant pressure of 1 atm.
(b) The volume of the system changes from 1 ℓ to 3 ℓ at an initial temperature of 25°C and a constant pressure of 10^{-6} atm.
(c) The volume of the system changes from 1 ℓ to 3 ℓ at a constant temperature of 25°C and the expansion is done reversibly for an ideal gas with an initial pressure of 1 atm.

9. For the following processes, state whether each of the four thermodynamic quantities q, w, ΔE, and ΔH is greater than, equal to, or less than zero for the *system* described. Consider all gases to behave ideally. Each system is indicated by italic type. State explicitly any reasonable assumptions that you may need to make.
(a) *Two copper bars*, one initially at 80°C and the other initially at 20°C, are brought into contact with one another in a thermally insulated compartment and then allowed to come to equilibrium.
(b) A *sample of liquid* in a thermally insulated container (a calorimeter) is stirred for 1 h by a mechanical linkage to a motor in the surroundings.
(c) A *sample of H_2 gas* is mixed with *an equimolar amount of N_2 gas* at the same temperature and pressure under conditions where no chemical reaction occurs between them.

10. If you set out to explore the surface of the moon, you would want to wear a space suit with thermal insulation. In such activity you might expect to generate roughly 1 kcal of heat per kilogram of mass per hour. If all of this heat is retained by your body, by how much would your body temperature increase per hour owing to this rate of heat production? (Assume that your heat capacity is roughly that of water.) What time limit would you recommend for a moon walk under these conditions?

11. If a breath of air, in a volume of 1 ℓ, is drawn into the lungs and comes to thermal equilibrium with the body at 37°C while the pressure remains constant, calculate the increase in enthalpy of the air if the initial air temperature is 20°C and the pressure is 1 atm. At a breathing rate of 30 per minute, how much heat is lost in this fashion in 1 day? Compare your answer (and that of Problem 10, assuming a body weight of 80 kg) with a typical daily intake of 3000 kcal of food energy. What problems might you foresee in arctic climates, where the air temperature can reach −40°C and below? The heat capacity of air is about 7 cal mol^{-1} deg^{-1}.

12. A reaction that is representative of those in the glycolytic pathway is the catabolism of glucose by complete oxidation to carbon dioxide and water:

$$C_6H_{12}O_6(s) + 6\,O_2(g) \longrightarrow 6\,CO_2(g) + 6\,H_2O(l)$$

Calculate ΔH^0_{298} for the glucose oxidation.

13. One mol of liquid H_2O at 15°C is mixed with 1 mol of liquid H_2O at 65°C. The pressure is kept constant and no heat is allowed to leave the system. The process is adiabatic. Calculate the final temperature for the system.

14. Alcoholic fermentation by microorganisms involves the breakdown of glucose into ethanol and carbon dioxide by the reaction

$$\text{glucose}(s) \longrightarrow 2\,\text{ethanol}(l) + 2\,CO_2(g)$$

(a) Calculate the amount of heat liberated in a yeast brew upon fermentation of 1 mol of glucose at 25°C, 1 atm.

(b) What fraction is the heat calculated in part (a) of the amount of heat liberated by the complete combustion (reaction with O_2) of glucose to carbon dioxide and liquid water at 298K?

15. The enzyme catalase catalyses the decomposition of hydrogen peroxide by the exothermic reaction

$$H_2O_2(aq) \xrightarrow{\text{catalase}} H_2O(\ell) + \tfrac{1}{2}O_2(g)$$

Estimate the minimum detectable concentration of H_2O_2 if a small amount of catalase (solid) is added to a hydrogen peroxide solution in a calorimeter. Assume that a temperature rise of 0.02°C can be distinguished. You can use a heat capacity of 1 cal $m\ell^{-1}$ deg^{-1} for the hydrogen peroxide solution.

16. For each of the following processes, calculate ΔE and ΔH, if possible. If values cannot be calculated, state the sign of ΔE and ΔH.

(a) Twenty-four grams of molybdenum oxide at −204°C, P = 15.7 atm, is converted to the difluoride at a temperature of 1053°C and a pressure of 0.1 mm Hg. The difluoride is transformed to the pure metal, which is then reacted with $O_2(g)$ to form 24 g of molybdenum oxide at −204°C, P = 15.7 atm.

(b) An unknown substance is heated 5°C at constant pressure.

(c) An unideal gas expands at constant temperature.

(d) A substance absorbs 10 cal of heat and does 7 cal of work, irreversibly, at constant T and P.

(e) A substance absorbs 10 cal of heat and does 7 cal of work reversibly at constant T and P.

17. Consider the reaction

$$CH_3OH(l) \longrightarrow CH_4(g) + \tfrac{1}{2}O_2(g)$$

(a) Calculate ΔH^0_{298}.

(b) Estimate ΔE^0_{298}.

(c) Write an equation that would allow you to obtain ΔH at 500°C and 1 atm.

18. Yeasts and other organisms can convert glucose ($C_6H_{12}O_6$) to ethanol or acetic acid. Calculate the change in enthalpy, ΔH, when 1 g of glucose is oxidized to (a) ethanol, or (b) acetic acid by the following path at 298°K:

glucose ⟶ glucose-6-phosphate ⟶ fructose-6-phosphate ⟶ glyceraldehyde-3-phosphate ⟶ ethanol ⟶ acetaldehyde ⟶ acetic acid

You can ignore all heats of solution of products or reactants. The overall reactions are:

(a) $C_6H_{12}O_6(s) \longrightarrow 2\,CH_3CH_2OH(l) + 2\,CO_2(g)$

(b) $2\,O_2(g) + C_6H_{12}O_6(s) \longrightarrow 2\,CH_3COOH(l) + 2\,CO_2(g) + 2\,H_2O(l)$

(c) Calculate the ΔH for the complete combustion of glucose to $CO_2(g)$ and $H_2O(l)$.

19. Estimate the change in ΔH if each reaction in Problem 18(a) and 18(b) is carried out by a thermophilic bacterium at 80°C. $\bar{C}_P$ for ethanol(l) = 26.64 cal mol^{-1} deg^{-1}, for acetic acid(l) = 29.5 cal mol^{-1} deg^{-1}, for glucose(aq) ≅ 50 cal mol^{-1} deg^{-1}.

20. (a) Calculate the enthalpy change on burning 1 g of $H_2(g)$ to $H_2O(l)$ at 25°C and 1 atm.

(b) Calculate the enthalpy change on burning 1 g of n-octane to $CO_2(g)$ and $H_2O(l)$ at 25°C and 1 atm.

(c) Compare $H_2(g)$ and n-octane(g) in terms of calories of heat available per gram.

21. Consider the reaction in which 1 mol of aspartic acid(s) is converted to alanine(s) and $CO_2(g)$ at 25°C and 1 atm pressure. The balanced reaction is

$$H_2NCH(CH_2COOH)COOH \rightleftharpoons H_2NCH(CH_3)COOH + CO_2$$

(a) How much heat (in kcal) is evolved or absorbed?

(b) Write down a cycle you could use to calculate the heat effect for the reaction at 50°C. State what properties of molecules you would need to know and what equations you would use to calculate the answer.

22. One mole of ice at 0°C is mixed with 1 mol of liquid water at 100°C in an insulated container. What is the final temperature of the mixture?

23. What experiments would you have to do to measure the (a) energy, and (b) enthalpy change for the following reaction at 25°C and 1 atm:

$$ATP^{4-} + H_2O \rightleftharpoons ADP^{3-} + HPO_4^{2-} + H^+$$

The reaction takes place in aqueous solution using the sodium salts of each compound. Give as much detail as possible and show the equations you would need to use. ATP^{4-} is adenosine triphosphate; ADP^{3-} is adenosine diphosphate.

24. A household uses 22 kW h day^{-1} of electricity.

(a) How many kcal day^{-1} are used?

(b) About 1 cal cm^{-2} min^{-1} of energy from the sun hits the surface of the earth. With a 10% efficient solar battery, what area (in m^2) of solar battery would be needed to provide sufficient solar energy to supply the household? Assume an average of 5 h day^{-1} of sunshine. (A suitable energy-storage device would, of course, be needed to have electricity at night.)

25. Formic acid has the structure H—C(=O)—OH

(a) Calculate the heat of combustion of gaseous formic acid (forming liquid water) at 25°C.

(b) Calculate the bond energy of the carbon-to-oxygen double bond in formic acid.

(c) Give the most likely reason for any discrepancy between your answer to part (b) and the average value in Table 2.3.

26. The enzyme catalase efficiently catalyzes the decomposition of hydrogen peroxide to give water and oxygen. At room temperature the reaction goes essentially to completion.

(a) Using heats of formation, estimate ΔH^0_{298} for the reaction

$$2\,H_2O_2(g) \longrightarrow 2\,H_2O(g) + O_2(g)$$

$\Delta H^0_f(298)$ for gaseous H_2O_2 is -31.83 kcal mol^{-1}.

(b) Calculate the bond dissociation energy for the O—O single bond.

(c) The enzyme normally acts on an aqueous solution of hydrogen peroxide, for which the equation is

$$2\,H_2O_2(aq) \longrightarrow 2\,H_2O(l) + O_2(g)$$

What is ΔH^0_{298} for this process?

(d) A solution initially 0.01 M in H_2O_2 and at 25.00°C is treated with a small amount of the enzyme. If all the heat liberated in the reaction is retained by the solution, what would be the final temperature? (Take the heat capacity of the solution to be 1 cal deg^{-1} g^{-1}.)

27. Use bond-energy values to estimate the heat of formation of n-heptane(g). Compare your answer with the value given in the Appendix.

3

Spontaneous Reactions, Entropy, and Free Energy

Everybody would like to be able to predict the future. Chemists, at least in their professional work, are interested mainly in predicting the future of chemical reactions. That is, they want to know (1) what reactions are impossible under given conditions, and (2) how the conditions can be changed so that impossible reactions become probable. A knowledge of thermodynamics will allow us to answer these questions. However, we can decide only *which* reactions are possible, not *when* they will actually occur. We can thus learn when it is useful to search for a catalyst (such as an enzyme) to make the reaction go, or when we should try a different reaction. Thermodynamics applies to other processes besides chemical reactions, and so we will consider very general changes and see whether they can occur.

An industrious inventor might ask the following questions during the course of a busy day. (1) Can diamonds be made out of pencil lead (graphite)? (2) Is it possible to make a heat shield that allows heat to go through one way but not the other? There would be many useful applications of this shield. (3) Can an engine be constructed that can convert heat quantitatively into work?

We can answer the questions asked by the inventor (who is obviously not well educated in thermodynamics) once we understand entropy and the second law of thermodynamics. We do not even have to do the experiments suggested by the questions to determine the answers. Experiments alone might not convince the inventor. Just because a particular catalyst added to the pencil lead did not produce diamonds does not mean that some other catalyst will also not work. We shall jump ahead of our quantitive thermodynamic discussion and give the answers to question (1) as yes and to questions (2) and (3) as no.

At high pressures, graphite has been converted to diamonds. High temperatures are used to speed up the reaction, and certain transition metals act as catalysts. A one-way heat shield can never be found, and no engine can be constructed that can convert heat quantitatively into work.

HISTORICAL DEVELOPMENT OF THE SECOND LAW: THE CARNOT CYCLE

An important milestone in the development of the second law of thermodynamics was Carnot's analysis of the efficiency of a heat engine in the early nineteenth century.

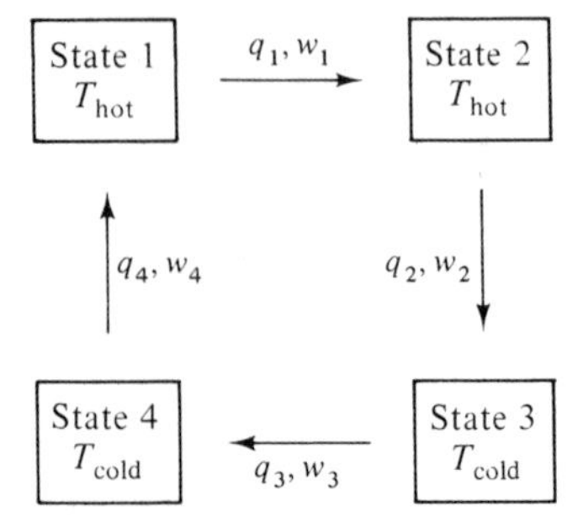

Fig. 3.1 Cycle of a heat engine. Each step is carried out reversibly.

Consider the cycle shown in Fig. 3.1. The system goes through a series of reversible steps and returns to its original state. At each step some heat or work is exchanged with the surroundings. This cycle is an idealized version of a real engine such as a steam engine or an automobile engine. In general, a hot gas at T_{hot} expands in a cylinder and does work $(w_1 + w_2)$. As it does work it cools to T_{cold}. The gas is recompressed $(w_3 + w_4)$ and returned to its original condition. For one complete cycle, the total work is $w = (w_1 + w_2 + w_3 + w_4)$ and the heat absorbed is $q = (q_1 + q_2 + q_3 + q_4)$.

For simplicity, Carnot considered an ideal gas in the engine, which underwent a cycle of four successive reversible steps to return to its original state. These four steps are illustrated in Fig. 3.2.

From the first law and the ideal gas law, $PV = nRT$, the q's and w's of each step can be calculated. For the first step,

$$w_1 = -\int_{V_1}^{V_2} P\,dV = -nRT_{hot} \ln \frac{V_2}{V_1}$$

$$E_2 - E_1 = 0 \qquad (E \text{ for an ideal gas is dependent only on temperature})$$

$$q_1 + w_1 = E_2 - E_1 = 0 \qquad (\text{first law})$$

$$q_1 = -w_1 = nRT_{hot} \ln \frac{V_2}{V_1}$$

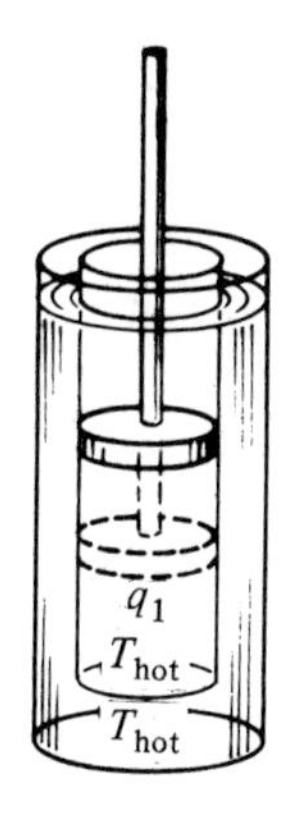

1. Isothermal reversible expansion
$P_1, V_1 \rightarrow P_2, V_2$
q_1 is positive
w_1 is negative

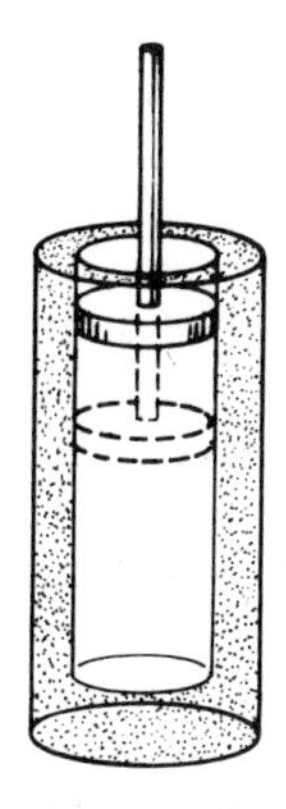

2. Adiabatic reversible expansion
$P_2, V_2 \rightarrow P_3, V_3$
$q_2 = 0$
w_2 is negative

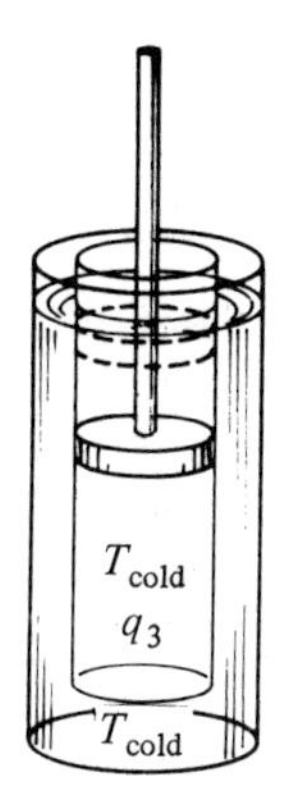

3. Isothermal reversible compression
$P_3, V_3 \rightarrow P_4, V_4$
q_3 is negative
w_3 is positive

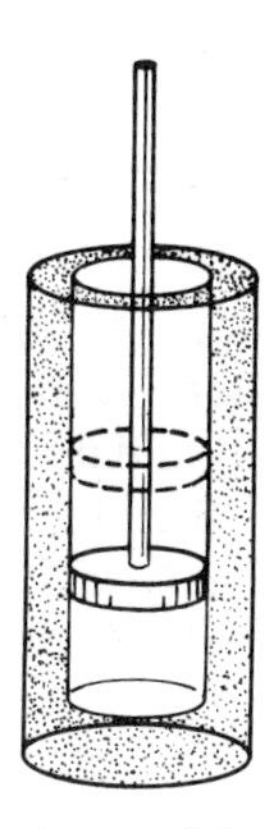

4. Adiabatic reversible compression
$P_4, V_4 \rightarrow P_1, V_1$
$q_4 = 0$
w_4 is positive

Fig. 3.2 Expansion and compression steps of a Carnot-cycle heat engine. The engine is in thermal contact with either a hot or a cold heat reservoir.

For the second step,

$$q_2 = 0$$

$$w_2 = E_2 - E_1$$

$$= C_V \cdot (T_{\text{cold}} - T_{\text{hot}})$$

The third and fourth steps are similar to the first and second, respectively:

$$q_3 = -w_3 = nRT_{\text{cold}} \ln \frac{V_4}{V_3}$$

$$q_4 = 0 \qquad w_4 = C_V \cdot (T_{\text{hot}} - T_{\text{cold}})$$

The total heat absorbed is

$$q = q_1 + q_2 + q_3 + q_4 = nRT_{\text{hot}} \ln \frac{V_2}{V_1} + 0 + nRT_{\text{cold}} \ln \frac{V_4}{V_3} + 0$$

The total work done by the engine is

$$-w = -(w_1 + w_2 + w_3 + w_4) = nRT_{\text{hot}} \ln \frac{V_2}{V_1} + nRT_{\text{cold}} \ln \frac{V_4}{V_3}$$

because w_2 and w_4 cancel. Therefore, the total work done by the engine is just

$$-w = q$$

We could also have reached this conclusion by using the fact that $E_2 - E_1 = 0$ for any cyclic change. Therefore, from the first law, $q = -w$ for any cyclic change. Carnot noticed that the sum of the quantities (q_i/T_i) for all the steps of the cycle had an interesting property, that the sum is zero. We can derive this easily:

$$\frac{q_1}{T_{\text{hot}}} + 0 + \frac{q_3}{T_{\text{cold}}} + 0 = \text{nR} \ln \frac{V_2}{V_1} + nR \ln \frac{\text{V}_4}{\text{V}_3}$$

$$\frac{q_1}{T_{\text{hot}}} + \frac{q_3}{T_{\text{cold}}} = nR \ln \frac{V_2 V_4}{V_1 V_3}$$

We need a relation among the volumes of the cycle. For a *reversible*, adiabatic step, such as steps 2 and 4 of the Carnot cycle, there is a relation between the initial and the final values of V and T. Consider a small change dV in volume. Let the corresponding change in temperature be dT. The energy change dE is $C_V\,dT$. Since the process is adiabatic, the energy change is equal to the work, $-P\,dV$:

$$C_V\,dT = -P\,dV$$

$$= -\frac{nRT}{V}dV$$

Dividing both sides by T and integrating, we have, for step 2,

$$C_V \int_{T_{\text{hot}}}^{T_{\text{cold}}} \frac{dT}{T} = -nR \int_{V_2}^{V_3} \frac{dV}{V}$$

$$C_V \ln \frac{T_{\text{cold}}}{T_{\text{hot}}} = -nR \ln \frac{V_3}{V_2} = nR \ln \frac{V_2}{V_3} \tag{3.1a}$$

Similarly, for step 4,

$$-C_V \ln \frac{T_{\text{cold}}}{T_{\text{hot}}} = nR \ln \frac{V_4}{V_1} \tag{3.1b}$$

Adding Eqs. (3.1a) and (3.1b), we see that

$$nR \ln \frac{V_2 V_4}{V_3 V_1} = 0$$

which is what we wanted to prove. Therefore,

$$\frac{q_1}{T_{\text{hot}}} + \frac{q_3}{T_{\text{cold}}} = 0 \tag{3.2}$$

Why is this conclusion important? Our system (gas in cylinder) has gone through a cycle and returned to its original state. The sum of the quantity q_{rev}/T for this cyclic path is zero (we use the subscript "rev" on q here to emphasize that all steps are carried out reversibly). This is what we learned about state functions such as E and H. For a cyclic path, state functions exhibit no change. Could it be that by dividing q_{rev}, a path-dependent quantity, by T, we have generated a path-independent function? Carnot showed that q_{rev}/T indeed represents a state function, now called *entropy*. The symbol S will be used for entropy:

$$\Delta S = q_{\text{rev}}/T \tag{3.3}$$

Equation (3.3) can be considered a definition of entropy.

ENTROPY IS A STATE FUNCTION

The logic which led to the realization that entropy is a state function is as follows. The efficiency of a heat engine is expressed as the total work done by the engine divided by the heat absorbed at the high temperature:

$$\text{efficiency} = \frac{-w}{q_1} \tag{3.4}$$

Now it just takes algebra to solve for efficiency in terms of temperature. From Eq. (3.2),

$$\frac{q_1}{T_{\text{hot}}} = -\frac{q_3}{T_{\text{cold}}}$$

or

$$\frac{q_1}{q_3} = -\frac{T_{\text{hot}}}{T_{\text{cold}}} \tag{3.5}$$

From the first law, the net work is

$$-w = q_1 + q_3$$

Then

$$\text{efficiency} = \frac{q_1 + q_3}{q_1} = 1 + \frac{q_3}{q_1}$$

From Eq. (3.5),

$$\text{efficiency} = 1 - \frac{T_{\text{cold}}}{T_{\text{hot}}}$$

or

$$\text{efficiency} = \frac{T_{\text{hot}} - T_{\text{cold}}}{T_{\text{hot}}} \tag{3.6}$$

Since all steps are reversible, the Carnot engine can also be operated in the reverse as a refrigerator or heat pump. In the forward direction as a heat engine, there is a net transfer of heat from the hot reservoir to the cold one, and the system does work on the surroundings. In the reverse direction, as a heat pump, the surroundings perform work on the system, and there is a net flow of heat from the cold reservoir to the hot one.

From our experience heat flows spontaneously only from a hot body to a cooler one. A consequence of this is that all heat engines operating with reversible cycles between T_{hot} and T_{cold} must have the same efficiency. Suppose this were not true. We can always operate the less efficient heat engine in reverse as a heat pump, and use the more efficient one as the heat engine. The work output of the engine is used to drive the heat pump. The combined effect of the two reversible engines operated in the fashion described is to cause a heat flow from the cold reservoir to the hot reservoir with no change in the surroundings (see Problem 1). Heat would be transferred spontaneously from a low temperature to a high temperature, which we do not believe can happen. Therefore, we must conclude that all reversible engines have the same efficiency.

Since all reversible engines have the same efficiency, it is straightforward to show that Eq. (3.5), and therefore Eq. (3.2), cannot be a consequence of using an ideal gas in our engine or of the particular paths chosen. In other words, we have defined a new state function, *entropy*:

$$\Delta S \equiv \frac{q_{rev}}{T} \tag{3.3}$$

Entropy is an extensive variable of state that depends only on the initial and final states of the system.

We have "proven" that entropy is a state function only because we accepted as true that heat flows spontaneously from a hot body to a cooler one. We can also consider Eq. (3.3) as *defining* a thermodynamic temperature scale. The temperature scale T is the one that makes q_{rev}/T a state function. The fact that Eq. (3.3) can also be derived for an ideal gas, as we have done, means that the thermodynamic temperature scale defined by Eq. (3.3) is identical to the temperature scale defined by the ideal gas law, $PV = nRT$.

THE SECOND LAW OF THERMODYNAMICS: ENTROPY IS NOT CONSERVED

When we calculate the entropy changes of both the system and the surroundings for simple processes that can occur spontaneously, such as the flow of heat from a hot body to a cooler one, the expansion of an ideal gas into a vacuum, or the flow of water down a hill, we find that the sum of the entropy changes is not zero. In other words, unlike energy, entropy is not conserved. The generalization of a great deal of experience, which is the second law of thermodynamics, is that the sum of the entropy changes of the system and the surroundings is always positive.

Even zero values can be approached only as a limit, and negative values are never found. If there is a decrease in entropy in a system, there must be an equal or larger increase in entropy in the surroundings. The first two laws of thermodynamics can now be stated briefly as: the total energy of the system plus the surroundings remains constant; the total entropy of the system plus surroundings never decreases. Equivalent statements of the second law are: (1) heat spontaneously flows from a hot body to a cold body, but work must be done to transfer heat from a cold body to a hot body; and (2) it is impossible by a *cyclic* process to remove heat from a hot body and convert it solely into work; some of the heat removed must be transferred to a cold body. The *second law* can be stated:

$$\Delta S(\text{system}) + \Delta S(\text{surroundings}) \geq 0 \tag{3.7}$$

For an *isolated system*, since there is no energy or material exchange between such a system and the surroundings, there is no change in the surroundings. Therefore,

$$\Delta S(\text{isolated system}) \geq 0 \tag{3.8}$$

The $\geq$ sign means greater than or equal to; the former applies for irreversible processes and the latter for reversible ones.

We express the second law by Eqs. (3.7) and (3.8), because these definitions lead to general conclusions which are consistent with experiments. This is illustrated in the following example.

Example 3.1 One mole of an ideal gas initially at $P_1 = 2$ atm, T, and V_1 expands to $P_2 = 1$ atm, T, and $2V_1$. Consider two different paths: (a) the expansion occurs irreversibly into a vacuum as shown below, and (b) the expansion is reversible. Calculate $q_{\text{irreversible}}$, ΔS(system), and ΔS(surroundings) for (a) and q_{rev}, ΔS(system), and ΔS(surroundings) for (b). We know immediately that ΔS(system) is independent of path and is the same for (a) and (b). However, q and ΔS(surroundings) will be different for (a) and (b); q depends on path, and the states of the surroundings are different for (a) and (b).

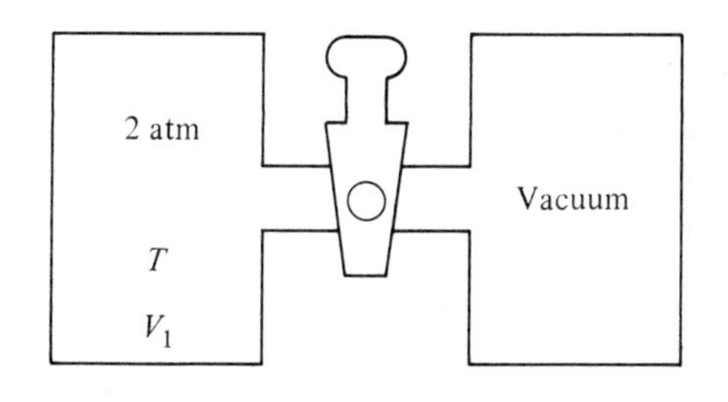

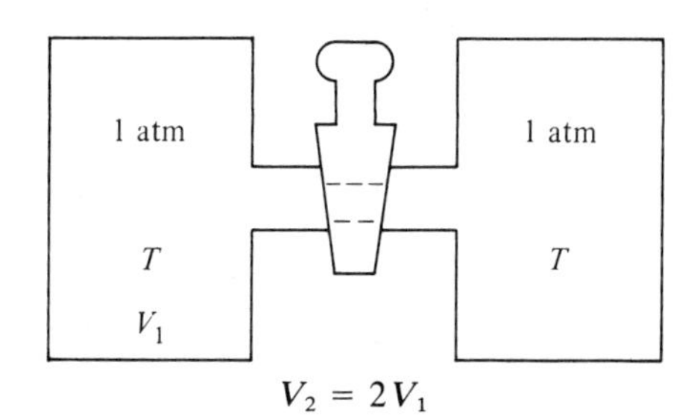

Solution

(a)

$$w = 0 \text{ (no work done against surroundings)}$$

$$\Delta E = 0 \text{ (}E\text{ for an ideal gas is independent of volume)}$$

$$q_{\text{irreversible}} = \Delta E - w = 0$$

$$\Delta S(\text{surroundings}) = 0 \text{ (no work or heat exchange between system and surroundings; surroundings not affected in any way by this process)}$$

To calculate ΔS(system), we must obtain q_{rev}. We consider a different path in which the gas expands isothermally and reversibly to the same final state,

$$w_{rev} = -\int_{V_1}^{V_2} P\,dV = -RT\int_{V_1}^{V_2} \frac{dV}{V} = -RT\ln\frac{V_2}{V_1}$$

$$= -RT\ln 2$$

$$\Delta E = 0$$

Therefore,

$$q_{rev} = \Delta E - w_{rev} = RT\ln 2$$

and

$$\Delta S(\text{system}) = \frac{q_{rev}}{T} = R\ln 2$$

Note that for this spontaneous, irreversible process,

$$\Delta S(\text{system}) + \Delta S(\text{surroundings}) = R\ln 2 + 0$$

$$> 0$$

consistent with Eqs. (3.7) and (3.8).

(b) We have already calculated that

$$q_{rev} = RT\ln 2$$

$$\Delta S(\text{system}) = R\ln 2$$

An amount q_{rev} of heat is transferred into the system. One way for this transfer to occur reversibly is to choose the surrounding temperature higher than T by only an infinitesimal amount. In other words, the surrounding temperature is also T. For the surroundings, the heat input is $-q_{rev}$ and

$$\Delta S(\text{surroundings}) = \frac{-q_{rev}}{T} = -R\ln 2$$

Note that for this reversible process,

$$\Delta S(\text{system}) + \Delta S(\text{surroundings}) = 0$$

consistent with Eq. (3.7).

Measurement of Entropy

The more modern interpretation of entropy is that it is a measure of disorder (see Chapter 11). The more disorder, the higher the entropy. Therefore, the second law tells us that the universe is continually becoming more disordered. This does not mean that a small part of the universe (the system) cannot become more ordered. Everytime we freeze an ice tray full of water in the refrigerator, we are increasing the order inside the ice tray. However, the disorder caused by the heat released outside the freezer compartment more than balances the order created in the ice. The total disorder in the universe has increased; the total entropy of the universe has increased. Many biological processes involve decreases of entropy for the organism itself but they are always coupled to other processes which increase the entropy, so that the sum is always positive. It is a pessimistic, but accurate, view that anything we do will always add to the disorder of the universe.

Feynman (1963) defines disorder as the number of ways the insides can be arranged so that from the outside the system looks the same. For example, 1 mol of ice is more ordered than 1 mol of water at the same temperature, because the water molecules in the liquid may have many different arrangements but still have the properties of liquid water. The water molecules in the solid ice can only have the arrangement corresponding to the crystal structure of ice to have the properties of ice. The molar entropies at 0°C and 1 atm are 9.8 cal deg^{-1} for ice, 15.1 cal deg^{-1} for liquid water, and 45.0 cal deg^{-1} for water vapor.

All entropies increase as the temperature is raised, because increasing molecular motion increases disorder and thus increases the entropy. For simple systems the disorder and the entropy can be calculated quantitatively (see Chapter 11). However, an entropy change can always be determined by measuring the *reversible* heat transfer for the process and dividing it by the absolute temperature [Eq. (3.3)]. If the temperature changes during the process, the reversible heat absorbed must be divided by the temperature at which it was absorbed to find the entropy change at each temperature. The total entropy change is just the sum of the individual entropy changes. It is clear, then, that for any process we can write the entropy change as an integral*:

$$S_2 - S_1 = \int_1^2 \frac{dq_{\text{rev}}}{T} \tag{3.9}$$

We must be sure to understand entropy and its relation to the reversible heat given by Eq. (3.9). Remember that entropy only depends on the initial state (1) and the final state (2). It does not matter how we get from state 1 to state 2. However, the entropy difference can be measured by finding a reversible path between states 1 and 2, measuring the heat change, and using Eq. (3.9). For an irreversible path the entropy change is *not* equal to the heat absorbed divided by T; furthermore, the entropy change is always greater than the irreversible heat divided by the temperature.

$$S_2 - S_1 > \int \frac{dq_{\text{irreversible}}}{T} \tag{3.10}$$

* Although we use the symbol dq_{rev} to express a differential, reversible heat change, we should keep in mind that there is an important difference between dq_{rev} and the differentials of state functions, such as dE and dH. For any cyclic path from state 1 back to state 1, the integrals $\int_1^1 dE$ and $\int_1^1 dH$ are always 0. But $\int_1^1 dq_{\text{rev}}$ is dependent on the particular path and is generally not zero. Mathematically, we say that dE and dH are *exact* differentials, but dq_{rev} is not.

Exercise Show that Example 3.1(a) is consistent with Eq. (3.10).

From Eq. (3.9), one sees that the dimensions of entropy are energy/temperature; we shall use units of cal $\deg^{-1}$. A calorie per degree is also called an *entropy unit*, eu. The unit for entropy in the SI system is J $\deg^{-1}$,

$$4.184 \text{ J deg}^{-1} = 1 \text{ cal deg}^{-1}$$

One-way Heat Shield

We can now consider if it is possible to place a one-way heat shield between two identical pieces of metal so that one piece gets hotter while the other gets cooler. In Fig. 3.3 the transparent side of the shield is facing left so that heat can only move from left to right. An equivalent but more general question is whether there is any method by which the change shown in the figure can occur without any change in the surroundings. Let us calculate the entropy change of the system and see. The entropy change of the surroundings must be zero, because there is no change in the surroundings. We must specify the initial and final states of the system so that we can calculate the entropy change. We shall choose the pressure to be constant, and from Fig. 3.3 we see that for half the system (left side) the

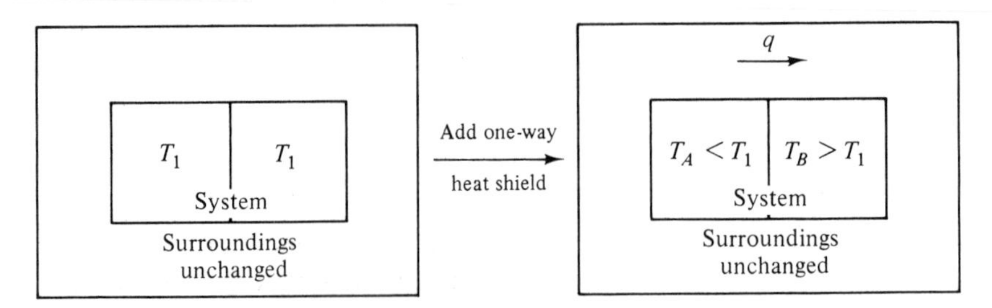

Fig. 3.3 One-way heat shield that will not work. The inventor, however, argues that by analogy with one-way mirrors, only a few more months of experiments are necessary to develop the right material. The second law says never.

temperature decreases from T_1 to T_A; for the other identical half (right side), the temperature increases from T_1 to T_B. From Eq. (3.9),

$$\begin{aligned} S_2 - S_1 &= \Delta S(\text{left side}) + \Delta S(\text{right side}) \\ &= \int_1^A \frac{dq_{\text{rev}}}{T}(\text{left}) + \int_1^B \frac{dq_{\text{rev}}}{T}(\text{right}) \end{aligned} \tag{3.11}$$

Heat is being lost from the left side, so dq_{rev} (left) is negative and the integral from state 1 to state A is therefore negative. As heat is being gained by the right side, the integral from state 1 to state B is positive. However, from the first law the magnitudes of the heats are identical; the heat lost by the left side must equal the heat gained by the right side:

$$|dq_{\rm rev}(\text{left})| = |dq_{\rm rev}(\text{right})|$$

The vertical lines, | |, indicate absolute magnitude; the magnitudes are equal, but the signs are different. As the heat flows from left to right, the temperature on the left decreases and the temperature on the right increases:

$$T_{\rm left} < T_{\rm right}$$

This means that

$$\left|\frac{dq_{\rm rev}}{T}(\text{left})\right| > \left|\frac{dq_{\rm rev}}{T}(\text{right})\right|$$

Remembering that $dq_{\rm rev}$ (left) is negative, we see that

$$\int_1^A \frac{dq_{\rm rev}}{T}(\text{left})$$

is negative and greater in magnitude than

$$\int_1^A \frac{dq_{\rm rev}}{T}(\text{right})$$

We conclude that $S_2 - S_1$ would be negative for the process shown in Fig. 3.3! This is contrary to the second law, so we say that it cannot actually take place.

Exercise For the problem above, show that if C_P, the heat capacity of the system, is constant, independent of T:
(a) $T_1 = (T_A + T_B)/2$ because of the first law.
(b) $S_2 - S_1 = C_P \ln (T_A T_B / T_1^2) = C_P \ln [4T_A T_B/(T_A + T_B)^2]$.
(c) $S_2 - S_1$, given by the expression in part (b), is always negative.

Hints: For part (b), substitute

$$dq_{\rm rev} = C_P\, dT \tag{2.9}$$

into Eq. (3.11). For part (c), the ln of a quantity is negative if the quantity is less than 1. Therefore, we want to show that

$$(T_A + T_B)^2 > 4T_A T_B$$

This is so because $(T_A + T_B)^2 - 4T_A T_B \equiv (T_A - T_B)^2 > 0$ if $T_A \neq T_B$ (the square of a quantity is positive).

What about the reverse of the process in Fig. 3.3? Can we start with a system in which the right side is hot (T_B), the left side is cool (T_A), and with no change in the surroundings reach a uniform temperature (T_1)? The answer from the second law and from experience is "yes." The only change in the previous discussion is that the left integral is now positive and greater in magnitude than the right integral, because heat is flowing to the cool side, so $S_2 - S_1$ is positive. The second law, therefore, tells us that a system will spontaneously tend to reach uniform temperature.

Fluctuations

As we have already shown (Example 3.1), if we assume that a system will spontaneously reach uniform pressure, we will arrive at the conclusion that the sum of entropy changes of the system and surroundings is positive (the second law). Or if we start with the second law, we can predict that a system will spontaneously tend to reach uniform pressure if the temperature of the system is uniform (Fig. 3.4a). Similarly, starting with the second law, we predict that a system such as the one shown in Fig. 3.4b will tend to reach uniform composition

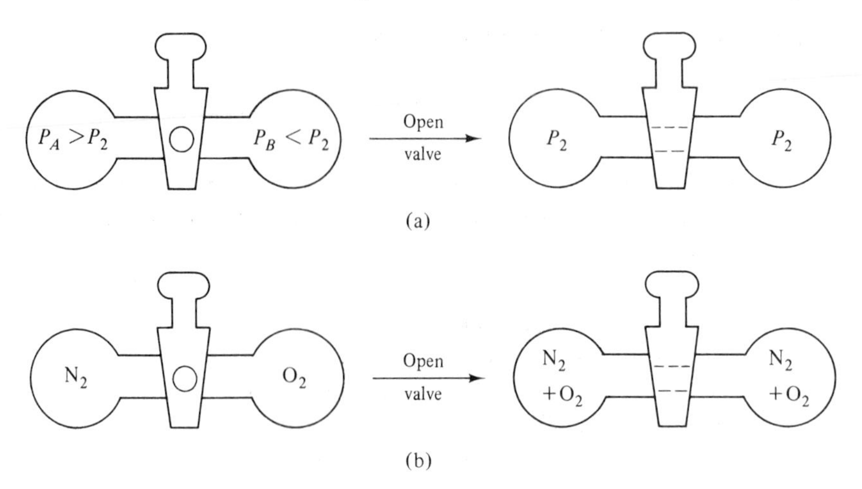

Fig. 3.4 The second law states that (a) pressures tend to become uniform, and (b) composition tends to become uniform.

spontaneously. That is, at constant pressure, gases originally separated in two halves of a system will tend to mix. Both of these processes can take place without changing the surroundings. The reverse of these reactions—unmixing gases, or going from a uniform pressure to unequal pressures—can only be done if the surroundings are also changed. They will not occur spontaneously. The quantitative statement of the second law was made to be consistent with experiment, so it does not surprise us to be able to draw experimentally correct conclusions from Eqs. (3.7) and (3.8).

But the second law is not quite correct for very small systems and very short times. In a system of uniform pressure, there will be very slight increases of pressure on one side and decreases on the other side owing to the random motion of the molecules. These fluctuations in pressure are too small to be measured by a pressure gauge, but they can be measured indirectly by other methods, such as light scattering. Similarly, a system of uniform composition will have fluctuations in composition in any volume of the system. If, for a very short time, we watched a system containing 10 N_2 molecules and 10 O_2 molecules, we might notice that molecules segregated spontaneously so that all 10 molecules of N_2 were on one side of the system and all 10 molecules of O_2 were on the other side. We could then quickly say that this observation disproved the second law. We would have to say it very quickly because if we watched the system for a reasonable time, we would find that *on the average* the 20 molecules are randomly distributed.

The second law therefore does not apply to fluctuations. Usually, we cannot detect the fluctuations, but if we can, the second law cannot be used to interpret the results. Fluctuations will be important only for very small systems and very short times. For large numbers of molecules the second law always applies. This means that thermodynamics is a macroscopic theory; it applies to matter in bulk. Even if molecules were shown not to exist, the laws of thermodynamics would remain unchanged. However, we do believe in the existence of molecules and we can estimate the probability that a process will occur which is not consistent with the second law. From the relation between entropy and disorder (Chapter 11), one can show that the probability of observing a violation of the second law depends exponentially on the number of molecules, N, in the system. The probability that you will observe a change contrary to the second law in a system containing N molecules is e^{-N}. We thus find that for 5 molecules, the probability of contradicting the second law is about 1/100. For 10 molecules it is $1/10^4$; for 20 molecules, it is $1/10^8$; and for 100 molecules, it is $1/10^{43}$. Furthermore, observing 5 molecules for a long time is equivalent to observing a system of many molecules. Therefore, the second law is actually a very reliable and useful generalization, in spite of its slight limitations.

CHEMICAL REACTIONS

So far we have been discussing processes and changes that have not involved chemical reactions. One way to obtain the entropy for a chemical reaction is to know the entropies of the reactants and products. For a general reaction

$$n_A\text{A} + n_B\text{B} \longrightarrow n_C\text{C} + n_D\text{D}$$

at a chosen T and P, the entropy change is

$$\Delta S = n_C\bar{S}_C + n_D\bar{S}_D - n_A\bar{S}_A - n_B\bar{S}_B$$

where $\bar{S}_A$ is the entropy mol^{-1} of compound A at T and P. The standard molar entropies of compounds can be obtained from tables (see Appendix), which give the entropy mol^{-1} for some compounds at 25°C and 1 atm pressure. Note that, by contrast with the standard enthalpies, the entropies of the elements are not equal to zero at 25°C and 1 atm. We remember that by convention the enthalpies of the elements are taken as zero at 25°C and 1 atm. The values of the entropy given in the Appendix for both elements and compounds were obtained by using the third law of thermodynamics.

Example 3.2 Calculate the entropy change at 25°C and 1 atm for the decomposition of 1 mol of liquid water to form H_2 and O_2 gas.

Solution The reaction for 1 mol of H_2O is

$$H_2O(l) \longrightarrow H_2(g) + \tfrac{1}{2} O_2(g)$$

The entropy change at 25°C and 1 atm is

$$\begin{aligned}\Delta S^0(25°C) &= \tfrac{1}{2}\bar{S}^0_{O_2(g)} + \bar{S}^0_{H_2(g)} - \bar{S}^0_{H_2O(l)} \\ &= \tfrac{1}{2}(49.00) + 31.21 - 16.72 \\ &= 39.99 \text{ eu}\end{aligned}$$

The entropy change is positive, which is consistent with our expectations for a reaction that involves the formation of two gases from one liquid. There is an increase in disorder.

THIRD LAW OF THERMODYNAMICS

The *third law* states that the entropy of all pure, perfect crystals is zero at a temperature of 0K (absolute zero):

$$S_A(0K) \equiv 0$$

where A is any pure, perfect crystal. The third law is an experimental one, as the first two laws are, but we can understand it in terms of the relation of entropy and disorder. At 0K the disorder of a substance can become zero. For this to happen, the substance must be crystalline; a liquid or a gas is still disordered at 0K*. Also, the substance must be pure; in a mixture the entropy could be reduced by separating the mixture. So, for a perfect crystal of any pure compound, we know the entropy at absolute zero. We can obtain the entropy at any other temperature if we know how entropy changes with temperature.

* One notable exception is liquid helium (F. London, *Superfluids*, Vol. II, Dover, New York, 1964).

Temperature Dependence of Entropy

The entropy change for heating or cooling a system is easy to calculate. The heating or cooling can be done essentially reversibly so that the entropy change is

$$S_2 - S_1 = \int \frac{dq_{\text{rev}}}{T} = \int_{T_1}^{T_2} \frac{C\,dT}{T}$$

At constant P,

$$S_2 - S_1 = \int_{T_1}^{T_2} \frac{C_P\,dT}{T} \tag{3.12}$$

$$S_2 - S_1 = C_P \ln \frac{T_2}{T_1} \qquad \text{(if } C_P \text{ is independent of } T\text{)} \tag{3.13}$$

where C_P is the heat capacity at constant P. At constant V,

$$S_2 - S_1 = \int_{T_1}^{T_2} \frac{C_V\,dT}{T} \tag{3.14}$$

$$S_2 - S_1 = C_V \ln \frac{T_2}{T_1} \qquad \text{(if } C_V \text{ is independent of } T\text{)} \tag{3.15}$$

where C_V = heat capacity at constant V. Because C_P and C_V are always positive, Eqs. (3.13) and (3.15) show that raising the temperature will always increase the entropy; this is expected from the increase in disorder.

Example 3.3 Calculate the change in entropy at constant P when 1 mol of liquid water at 100°C is brought in contact with 1 mol of liquid water at 0°C. Assume that the heat capacity of liquid water is independent of temperature and is equal to 18 cal mol^{-1} deg^{-1}. No heat is lost to the surroundings.

Solution As we bring equal amounts of the water in contact, C_P is constant, and no heat is lost; the first law tells us that the final temperature of the mixture will be the average of the two temperatures, 50°C. We use Eq. (3.13) to calculate the change in entropy of the hot water and of the cold water and add the results:

Hot water:

$$S(50°\text{C}) - S(100°\text{C}) = \bar{C}_P \ln \frac{323}{373} = 2.303\bar{C}_P \log \frac{323}{373}$$

$$= -2.59 \text{ eu}$$

Cold water:

$$S(50°\text{C}) - S(0°\text{C}) = C_P \ln \frac{323}{273}$$

$$= 3.03 \text{ eu}$$

Sum for $H_2O(100°C) + H_2O(0°C) \longrightarrow 2\,H_2O(50°C)$:

$$S_2 - S_1 = 0.44 \text{ eu}$$

The entropy change is positive, as it must be for a spontaneous process in an isolated system.

Temperature Dependence of the Entropy Change for a Chemical Reaction

To calculate the entropy change for a chemical reaction at 1 atm and some temperature other than 25°C, we can use Eq. (3.13). We consider a cycle in which products and reactants are heated or cooled to the new temperature and then add the entropy change for the heating or cooling to $\Delta S^0(25°C)$:

$$\begin{array}{ccc} A(T_2) & \xrightarrow{\Delta S^0(T_2)} & B(T_2) \\ \downarrow & & \uparrow \\ A(25°C) & \xrightarrow{\Delta S^0(25°C)} & B(25°C) \end{array}$$

$$\Delta S^0(T_2) = \Delta S^0(25°C) + \int_{T_2}^{298} C_P(A)\frac{dT}{T} + \int_{298}^{T_2} C_P(B)\frac{dT}{T}$$

We can generalize the result and rewrite the equation in a more compact form by using the identity

$$\int_a^b = -\int_b^a$$

$$\Delta S^0(T_2) = \Delta S^0(T_1) + \int_{T_1}^{T_2} \Delta C_P \frac{dT}{T} \tag{3.16}$$

where $\Delta C_P = C_P(\text{products}) - C_P(\text{reactants})$.

Example 3.4 If a spark is applied to a mixture of $H_2(g)$ and $O_2(g)$, an explosion occurs and water is formed. Calculate the entropy change when 2 mol of gaseous H_2O is formed at 100°C and 1 atm from $H_2(g)$ and $O_2(g)$ at the same temperature and each at a partial pressure of 1 atm.

Solution

The reaction for 2 mol of H_2O is

$$2\,H_2(g) + O_2(g) \longrightarrow 2\,H_2O(g)$$

The entropy change at 25°C, 1 atm is

$$\Delta S^0(25°\text{C}) = 2\bar{S}^0_{H_2O(g)} - \bar{S}^0_{O_2(g)} - 2\bar{S}^0_{H_2(g)}$$

$$= 2(45.11) - 49.00 - 2(31.21)$$

$$= -21.2 \text{ eu}$$

To find $\Delta S^0(100°\text{C})$, we need to use Eq. (3.16); therefore, we need to know the heat capacities of $H_2(g)$, $O_2(g)$, and $H_2O(g)$. They can be taken as constants over the temperature range 25°C to 100°C with the following values:

	$\bar{C}^0_P$ (cal mol^{-1} deg^{-1})
$H_2(g)$	6.9
$O_2(g)$	7.0
$H_2O(g)$	8.0

$$\Delta C^0_P = 2\bar{C}^0_{PH_2O(g)} - \bar{C}^0_{PO_2(g)} - 2\bar{C}^0_{PH_2(g)}$$

$$= 2(8.0) - 7.0 - 2(6.9)$$

$$= -4.8 \text{ cal mol}^{-1}\text{ deg}^{-1} = -4.8 \text{ eu}$$

From Eq. (3.16)

$$\Delta S^0(100°\text{C}) = \Delta S^0(25°\text{C}) + \int_{298}^{373} \Delta C_P \frac{dT}{T}$$

$$= \Delta S^0(25°C) + \Delta C_P \ln\frac{373}{298}$$

$$= -21.2 - (4.8)(0.224)$$

$$= -22.3 \text{ eu}$$

The entropy of the system decreased; the second law thus requires that the entropy of the surroundings must have increased. The reaction is exothermic and heat is lost to the surroundings at constant temperature.

Entropy Change for a Phase Transition

On heating many compounds from 0K to room temperature, various phase transitions, such as melting and boiling, may occur. The reversible heat absorbed divided by the *equilibrium* temperature of the transition gives the entropy change for the phase transition. It is important to stress equilibrium transition temperature. Otherwise, the transition is not reversible and the heat absorbed is not the reversible heat. For a phase transition, at constant P and T,

$$q_{\text{rev}} = \Delta H_{\text{tr}} \tag{2.45}$$

$$\Delta S_{\text{tr}} = \frac{\Delta H_{\text{tr}}}{T_{\text{tr}}} \tag{3.17}$$

where ΔS_{tr} = entropy of phase transition
ΔH_{tr} = enthalpy of phase transition at equilibrium temperature T_{tr}

To obtain the third law entropy of a liquid compound at 25°C and 1 atm, we would use the equation

$$\bar{S}(25°\text{C}, 1 \text{ atm}) = \int_0^{T_m} \bar{C}_P(s)\frac{dT}{T} + \frac{\Delta H_m}{T_m} + \int_{T_m}^{298} \bar{C}_P(l)\frac{dT}{T}$$

where T_m is the melting temperature at 1 atm and we have assumed that there are no solid-solid transitions. If there were, we would add $(\Delta H_{\text{tr}}/T_{\text{tr}})$ for each transition and use an appropriate $\bar{C}_p$ for each solid phase.

Pressure Dependence of Entropy

The pressure dependence of entropy will not be so important for us as the temperature dependence. Raising the pressure usually lowers the entropy. For solids and liquids we will ignore the direct effect of pressure on entropy:

$$S(P_2) - S(P_1) \cong 0 \tag{3.18}$$

For gases we will approximate the effect by the effect of pressure on an ideal gas:

$$S(P_2) - S(P_1) = -nR \ln \frac{P_2}{P_1} \tag{3.19}$$

(The origin of this equation is given at the end of the chapter.) We notice in Eq. (3.19) that if either P_2 or P_1 is zero, the logarithm and the entropy become infinite. This should not surprise us because zero pressure means infinite volume, and infinite volume implies infinite disorder of the molecules in the volume. The point to remember is that as the pressure of a gas decreases, the entropy will increase.

Equations (3.18) and (3.19) could be used to calculate the entropy change for a chemical reaction at some pressure other than 1 atm.

Spontaneous Chemical Reactions

We began this chapter by wondering if pencil lead could be converted into diamonds. One way to answer this question is to learn if the entropy of the universe increases when the reaction occurs. We can easily learn from the

Appendix whether the entropy of the system increases, but it is not so easy to learn about the change in entropy of the surroundings. *Entropy is a useful criterion for spontaneity in isolated systems.* Then the surroundings do not change, and we can limit our attention to the system. If neither the energy nor the volume of the system change during a reaction, the sign of the entropy change tells whether the reaction can occur. Very few reactions occur at constant energy and volume. Racemization of an optically active molecule to a racemic mixture is an example. The D form and the L form have identical energies, volumes, and entropies, but the mixture has a larger entropy and nearly the same energy and volume. The reaction is spontaneous.

GIBBS FREE ENERGY

For most reactions there are large changes in energy and enthalpy, and we are interested in whether the reaction will occur at some constant T and P. We need a criterion of spontaneity that applies to the system for these conditions. A new thermodynamic variable of state, the Gibbs free energy, is useful in this situation. The *Gibbs free energy*, G, is an extensive variable of state defined as a combination of enthalpy, temperature, and entropy:

$$G \equiv H - TS \qquad (3.20)$$

The Gibbs free energy has the same units as enthalpy or energy; it also depends only on initial and final states. Its definition in Eq. (3.20) was chosen because it can thus characterize whether a process will occur spontaneously at constant temperature and pressure. If ΔG for a reaction at constant T and P is negative, the reaction can occur spontaneously; if ΔG at constant T and P is positive, the reaction will not occur spontaneously; if ΔG at constant T and P is zero, the reaction is at equilibrium.

Spontaneous Reactions at Constant *T* and *P*

Let us derive the Gibbs free energy criterion for a spontaneous reaction at constant T and P. Combining Eqs. (3.9) and (3.10), we see that for any reaction at constant T,

$$S_2 - S_1 \geq \frac{q}{T} \qquad (3.21)$$

The equal sign applies to a reversible reaction; that is, one at equilibrium. The "greater than" sign applies to an irreversible reaction; that is, a spontaneous reaction. We can replace q by using the first law:

$$S_2 - S_1 \geq \frac{E_2 - E_1 - w}{T} \qquad \text{(constant } T\text{)} \qquad (3.22)$$

If we consider only expansion and compression work, at constant P we have

$$S_2 - S_1 \geq \frac{E_2 - E_1 + P \cdot (V_2 - V_1)}{T} \qquad \text{(constant } T, P\text{)} \tag{3.23}$$

But the numerator of the right-hand side of Eq. (3.23) is $H_2 - H_1$ at constant P:

$$S_2 - S_1 \geq \frac{H_2 - H_1}{T} \qquad \text{(constant } T, P\text{)} \tag{3.24}$$

Rearranging, we obtain

$$T \cdot (S_2 - S_1) - (H_2 - H_1) \geq 0$$

$$(H_2 - H_1) - T \cdot (S_2 - S_1) \leq 0$$

The left-hand side of the equations is just the difference in Gibbs free energy between final and initial states at the same temperature, as seen from Eq. (3.20). We thus obtain the desired criterion for a spontaneous reaction at constant temperature and pressure:

$$\Delta G < 0 \qquad \text{(a spontaneous reaction)} \tag{3.25}$$

$$\Delta G = 0 \qquad \text{(an equilibrium reaction)} \tag{3.26}$$

$$\Delta G > 0 \qquad \text{(no spontaneous reaction)} \tag{3.27}$$

Calculation of Gibbs Free Energy

The Gibbs free energy change for a reaction at constant temperature can be obtained from the enthalpy and entropy changes. At constant temperature,

$$\Delta G = \Delta H - T\,\Delta S \tag{3.28}$$

The Appendix gives values of $\bar{H}^0 \equiv \Delta H_f^0$ and $\bar{S}^0$ for various substances at 25°C and 1 atm. These values can therefore be used to calculate ΔH^0 and ΔS^0, and hence ΔG^0, at 25°C and 1 atm.

Values of ΔG_f^0, the standard free energy of formation, of various substances are also tabulated in the Appendix. The molar standard free energy of formation is defined as the free energy of formation of 1 mol of any compound at 1 atm pressure from its elements in their standard states at 1 atm. In a manner completely analogous to our discussion on standard enthalpy of formation, we arbitrarily assign the elements in their most stable state at 1 atm to have zero free energy.

Example 3.5 Calculate the Gibbs free energy for the following reaction at 25°C and 1 atm. Will the reaction occur spontaneously?

$$H_2O(l) \longrightarrow H_2(g) + \tfrac{1}{2}O_2(g)$$

Solution

$$\Delta G^0(25°C) = \bar{G}^0_{H_2(g)} + \tfrac{1}{2}\bar{G}^0_{O_2(g)} - \bar{G}^0_{H_2O(l)}$$
$$= 0 + 0 - (-56.69)$$
$$= 56.69 \text{ kcal mol}^{-1}$$

The Gibbs free energy change is just the negative of the Gibbs free energy of formation of liquid water.

Alternatively, we can calculate ΔH^0 and ΔS^0 from the Appendix:

$$\Delta H^0 = \bar{H}^0_{H_2(g)} + \tfrac{1}{2}\bar{H}^0_{O_2(g)} - \bar{H}^0_{H_2O(l)}$$
$$= 0 + 0 - (-68.32)$$
$$= 68.32 \text{ kcal mol}^{-1}$$
$$\Delta S^0 = \bar{S}^0_{H_2(g)} + \tfrac{1}{2}\bar{S}^0_{O_2(g)} - \bar{S}^0_{H_2O(l)}$$
$$= 31.21 + \tfrac{1}{2}\cdot 49.00 - 16.72$$
$$= 38.99 \text{ cal deg}^{-1}\text{ mol}^{-1}$$

ΔG^0 can then be obtained:

$$\Delta G^0 = \Delta H^0 - T\,\Delta S^0$$
$$= 68.32 \text{ kcal mol}^{-1} - 298\cdot 38.99\cdot 10^{-3}\text{ kcal mol}^{-1}$$
$$= 56.70 \text{ kcal mol}^{-1}$$

The same answer is obtained. The reaction will *not* occur spontaneously, because ΔG is positive.

A proposed method of storing solar energy is to use sunlight to decompose water. The sunlight provides the driving force to overcome the large positive free energy. The hydrogen and oxygen gas produced make an excellent fuel.

Example 3.6 We wish to know whether proteins in aqueous solution are unstable with respect to their constituent amino acids. As an example, let us calculate the standard free energy of hydrolysis for the dipeptide glycylglycine at 25°C and 1 atm in dilute aqueous solution.

Solution The reaction is

$$\underset{\text{Glycylglycine}}{{}^+H_3NCH_2CONHCH_2COO^-(aq)} + H_2O(l) \longrightarrow 2\,\underset{\text{Glycine}}{{}^+H_3NCH_2COO^-(aq)}$$

The standard free energy change when solid glycine dissolves has been measured and is small. We shall assume that this is also true for solid glycylglycine. Therefore, we use the free energy values in the Appendix for solid glycine and glycylglycine in the following calculation.

$$\Delta G^0(25°\text{C}) = 2\bar{G}^0(\text{glycine}, s) - \bar{G}^0\ (\text{glycylglycine}, s) - \bar{G}^0(\text{H}_2\text{O}, l)$$
$$= 2\,(-90.27) - (-117.25) - (-56.69)$$
$$= -6.60\,\text{kcal mol}^{-1}$$

The reaction is spontaneous, but luckily it normally occurs slowly. However, appropriate catalysts such as proteolytic enzymes can cause the reaction to occur rapidly. If this catalyst finds its way into your bloodstream, the effects would be very unpleasant.

It is clear from Eq. (3.28) that ΔG depends explicitly on T. Both ΔH and ΔS may be independent of T, but ΔG will depend on T with slope equal to $-\Delta S$. Often, as an approximation for temperatures not too different from 25°C, Eq. (3.28) can be used to calculate ΔG at other temperatures:

$$\Delta G(T) \cong \Delta H(25°\text{C}) - T \cdot \Delta S(25°\text{C}) \tag{3.29}$$

For example, values of ΔH and ΔS at 25°C from the Appendix can be used to calculate the approximate free energy of the reaction at physiological temperatures. A useful equation that is easily derived from Eq. (3.29) is

$$\Delta G(T) - \Delta G(25°\text{C}) \cong -(T - 298) \cdot \Delta S(25°\text{C}) \tag{3.30}$$

This equation shows that the sign of ΔS for a reaction indicates how ΔG will change with temperature. If ΔS is negative, ΔG increases with increasing temperature. Another useful equation is

$$\frac{\Delta G(T)}{T} - \frac{\Delta G(25°\text{C})}{298} \cong \left(\frac{1}{T} - \frac{1}{298}\right) \cdot \Delta H(25°\text{C}) \tag{3.31}$$

which can be derived from Eqs. (3.28) and (3.30). These equations can be used to calculate the temperature dependence of ΔG, or they can obviously be used to calculate ΔH and ΔS if the temperature dependence of ΔG is known. It is clear that Eqs. (3.30) and (3.31) apply to any two temperatures; the approximation involves the assumption that ΔH and ΔS are independent of temperature. If ΔH or ΔS depend greatly on temperature, Eqs. (3.30) and (3.31) must be replaced by integrals. The most useful one to use is

$$\frac{\Delta G(T_2)}{T_2} - \frac{\Delta G(T_1)}{T_1} = -\int_{T_1}^{T_2} \frac{\Delta H(T)}{T^2}\,dT \tag{3.32}$$

called the *Gibbs-Helmholtz equation.*

Example 3.7 What is the free energy of hydrolysis of glycylglycine at 37°C and 1 atm?

Solution We can use Eq. (3.30) to find $\Delta G^0(37°C)$ if we assume that ΔS is independent of temperature. Or we can use Eq. (3.31) if we assume that ΔH is independent of temperature. We shall use both equations and thus show that the two equations are indeed equivalent. Because the temperature range is small, the constancy of ΔH and ΔS is a good approximation. To calculate ΔS and ΔH, we use the data for the solid compounds.

$$\begin{aligned}\Delta S^0(25°C) &= 2\bar{S}^0(\text{glycine}) - \bar{S}^0(\text{glycylglycine}) - \bar{S}^0(H_2O, l)\\ &= 2(24.74) - 45.4 - 16.72\\ &= -12.64 \text{ eu mol}^{-1} = -12.64 \text{ cal deg}^{-1}\text{ mol}^{-1}\\ \Delta H^0(25°C) &= 2(-128.4) - (-178.12) - (-68.32)\\ &= -10.36 \text{ kcal mol}^{-1}\end{aligned}$$

With Eq. (3.30)

$$\begin{aligned}\Delta G^0(37°C) &= \Delta G^0(25°C) - 12 \cdot \Delta S^0(25°C)\\ &= -6.60 \text{kcal mol}^{-1} - (12 \text{ deg})(-0.01264 \text{ kcal deg}^{-1}\text{ mol}^{-1})\\ &= -6.45 \text{ kcal mol}^{-1}\end{aligned}$$

With Eq. (3.31),

$$\begin{aligned}\frac{\Delta G^0(37°C)}{310} - \frac{\Delta G^0(25°C)}{298} &= \left(\frac{1}{310} - \frac{1}{298}\right) \cdot \Delta H^0(25°C)\\ &= \frac{-6.60}{298} + (-1.299 \times 10^{-4})(-10.36)\\ &= -0.0208\end{aligned}$$

$$\Delta G^0(37°C) = -6.45 \text{ kcal mol}^{-1}$$

Both methods lead to a decrease in the free energy of hydrolysis and agree pretty well.

Pressure Dependence of Gibbs Free Energy

From values in the Appendix we can calculate the Gibbs free energy change, ΔG^0, exactly at 25°C and approximately at any temperature. However, the values of ΔG^0 are all at 1 atm pressure. To calculate free energy at some other pressure we must know its dependence on pressure. The change of Gibbs free energy with pressure at constant temperature is directly proportional to the volume:

$$G(P_2) - G(P_1) = \int_{P_1}^{P_2} V\,dP \tag{3.33}$$

(We will discover the origins of this equation later in this chapter.) If the volume is independent of pressure, it can be taken out of the integral. This is usually a good approximation for a solid or liquid:

$$G(P_2) - G(P_1) = V \cdot (P_2 - P_1) \tag{3.34}$$

For a gas we use the equation of state to write V as a function of P. For an ideal gas, and approximately for any gas, we can substitute the ideal gas equation in Eq. (3.33):

$$G(P_2) - G(P_1) = \int_{P_1}^{P_2} \frac{nRT}{P} dP = nRT \ln \frac{P_2}{P_1} \tag{3.35}$$

To calculate the effect of pressure on the free energy of a chemical reaction, we just apply Eqs. (3.34) and (3.35) to each product and reactant. If all products and reactants are solids or liquids, we use Eq. (3.34):

$$\Delta G(P_2) - \Delta G(P_1) = \Delta V \cdot (P_2 - P_1) \tag{3.36}$$

where $\Delta V = V(\text{products}) - V(\text{reactants})$. Equation (3.36) shows that if the volume of products is greater than the volume of reactants (ΔV is positive), increasing the pressure will increase the free energy of the reaction. If at least one gaseous product or reactant is involved, we can ignore the volume of the solids or liquids compared to that of the gases and use Eq. (3.35).

$$\Delta G(P_2) - \Delta G(P_1) = \Delta nRT \ln \frac{P_2}{P_1} \tag{3.37}$$

where Δn is the number of moles of *gaseous* products minus the number of moles of *gaseous* reactants.

Example 3.8 Can graphite (pencil lead) be converted spontaneously into diamond at 25°C and 1 atm?

Solution The reaction at 25°C and 1 atm

$$\text{C}(s, \text{graphite}) \longrightarrow \text{C}(s, \text{diamond})$$

has a Gibbs free energy of + 0.68 kcal from the Appendix. Therefore, the answer is "no"; graphite will not spontaneously convert into diamond at 25°C and 1 atm. The result does mean that diamond will spontaneously convert into graphite. Experimentally, the reaction has been found to be very slow, but a catalyst may eventually be found which can speed it up. There has not been much economic incentive to pursue this particular research.

Example 3.9 Will increasing the pressure favor the conversion of graphite to diamond? If so, what is the minimum pressure necessary to make this reaction spontaneous at 25°C? Will increasing the temperature favor the reaction?

Solution We can use Eq. (3.36) to answer the first two questions. We need to know the volume change for the reaction. We will use the densities of graphite and diamond at 25°C to calculate the volumes and assume that the volumes are independent of pressure. Just knowing that diamond is denser than graphite allows us to conclude that ΔV for the reaction is negative ($\bar{V}$ of diamond is less than $\bar{V}$ of graphite). This means that *increasing the pressure* will decrease the free energy and *favor* the reaction graphite to diamond.

To calculate the minimum pressure necessary to allow the spontaneous reaction, we calculate the pressure that produces $\Delta G = 0$. The densities at 25°C and 1 atm are as follows:

	Density (g cm^{-3})
C (graphite)	2.25
C (diamond)	3.51

The molar volumes at 25°C and 1 atm are obtained by dividing the atomic weight of carbon by the densities:

	$\bar{V}$ (cm^3 mol^{-1})
C (graphite)	5.33
C (diamond)	3.42

For the reaction

$$\text{C (graphite)} \longrightarrow \text{C (diamond)}$$

the Gibbs free energy at 25°C and P atm is, from Eq. (3.36),

$$\Delta G(P) = \Delta G(1 \text{ atm}) + \Delta V \cdot (P - 1)$$

$$= 0.68 \text{ kcal mol}^{-1} + (3.42 - 5.33) \text{ cm}^3 \text{ mol}^{-1} \cdot (P - 1) \text{ atm}$$

We must use the same units throughout our equation, so we convert cm^3 atm to kcal by multiplying by a ratio of gas constants, $R/R = 1.99 \times 10^{-3}$ kcal/82.05 cm^3 atm $= 2.43 \times 10^{-5}$ kcal cm^{-3} atm^{-1}:

$$\Delta G(P) = 0.68 - 4.63 \times 10^{-5} \cdot (P - 1) \text{ kcal mol}^{-1}$$

We want to find the pressure that makes $\Delta G(P) = 0$.

$$0 = 0.68 - 4.63 \times 10^{-5} \cdot (P - 1)$$

$$P - 1 = \frac{0.68}{4.63 \times 10^{-5}}$$

$$P = 15{,}000 \text{ atm}$$

The assumption that $\Delta \bar{V}$ is constant is not valid over this large pressure range, and the actual pressure needed is higher than 15,000 atm. Nevertheless, the effect is real, and small diamonds have been made in this way for many years for industrial uses.

To decide whether increasing temperature will favor the reaction, we have to know the sign of ΔS [see Eq. (3.30)]. From the Appendix we find that $\Delta S = 0.58 - 1.37 = -0.79$ eu; therefore, *increasing the temperature* will *not favor* the formation of diamond. We do not use Eq. (3.30) to calculate at what low temperature ΔG becomes zero, because ΔS will change significantly with temperature over the wide temperature change. The industrial process does use high temperatures, but for kinetic reasons. The increase in pressure provides the necessary free energy change.

Phase Changes

For a phase change that takes place at its equilibrium temperature and pressure, the change in Gibbs free energy is zero:

$$\Delta G = 0 \text{ (at equilibrium)}$$

To calculate the Gibbs free energy at other temperatures and pressures, we can use

$$\Delta G = \Delta H - T\,\Delta S \tag{3.28}$$

if ΔH and ΔS are known for the phase change at arbitrary T and P. Otherwise, we can use Eq. (3.30), (3.31), (3.35), or (3.36) for the change of ΔG with T and P.

HELMHOLTZ FREE ENERGY

For a process that takes place at constant T and V, the sign of the Gibbs free energy is not a criterion for equilibrium. A new thermodynamic variable of state is needed; it is called the *Helmholtz free energy*, A. The Helmholtz free energy is defined as

$$A \equiv E - TS \tag{3.38}$$

and the criterion for equilibrium is, at constant T and V,

$$\Delta A < 0 \quad \text{(spontaneous reaction)} \tag{3.39}$$

$$\Delta A = 0 \quad \text{(reaction at equilibrium)} \tag{3.40}$$

$$\Delta A > 0 \quad \text{(no spontaneous reaction)} \tag{3.41}$$

We will not use the Helmholtz free energy very much in this book, although for reactions at constant volume it is as useful as the Gibbs free energy is for reactions at constant pressure. The Helmholtz free energy is widely used for geochemical problems, where the pressure may vary widely and the constant-volume restriction is more appropriate.

SOME THERMODYNAMIC DATA OF NONCOVALENT REACTIONS

The examples that we have given so far of the changes in H, S, and G for chemical reactions involve the breaking and formation of covalent bonds. Many important biochemical processes involve reactions in which no covalent bonds are made or broken; only weaker bonds and interactions are involved. Examples include the reaction of an antigen with its antibody, the binding of many hormones and drugs to nucleic acids and proteins, the reading of the genetic message (codon-anticodon recognition), denaturation of proteins and nucleic acids, and so on. Thermodynamics can help us to understand these reactions and to decide which proposed mechanisms are reasonable and which are not. Table 3.1 gives some measured enthalpies for simple reactions which illustrate the forces involved.

Charged species have ionic interactions. In a NaCl crystal, for example, very strong ionic interactions exist between the positively charged Na^+ ions and the negatively charged Cl^- ions. This is reflected in the large positive enthalpy change when solid NaCl is separated into Na^+ and Cl^- ions:

$$\mathrm{NaCl}(s) \longrightarrow \mathrm{Na}^+(g) + \mathrm{Cl}^-(g) \qquad \Delta H^0_{298} = 180\ \mathrm{kcal\ mol^{-1}}$$

In spite of such strong ionic forces, solid NaCl nevertheless dissolves in water easily, with a small ΔH^0_{298} of only 1 kcal mol^{-1}:

$$\mathrm{NaCl}(s) + \infty\mathrm{H_2O}(l) \longrightarrow \mathrm{Na}^+(aq) + \mathrm{Cl}^-(aq) \qquad \Delta H^0_{298} = 1\ \mathrm{kcal\ mol^{-1}}$$

The reason is that there are strong interactions between the charged ions and the water molecules, so the net ΔH change is small.

The electronic distribution in an uncharged water molecule

```
H     H
 \   /
   O
```

is such that the bonding electrons are localized more on the oxygen atom than on the hydrogen atoms. Therefore, the oxygen atom is slightly negative and the hydrogen atoms are slightly positive. Since the molecule is not linear, the geometric centers of the positive charges and the negative charges do not coincide. We say that H_2O has an electric dipole. The extent of charge separation is expressed in terms of the dipole moment. If two point charges $+e$ and $-e$ are separated by a distance r, the dipole moment is er. Experimentally, the dipole moment of H_2O is found to be equivalent to an electron and a unit positive charge separated by 0.38 angstrom ($1\ \text{Å} = 10^{-8}$ cm). Interactions between a charged ion and a neutral molecule with a dipole moment are called *ion-dipole interactions*. Ions can also interact with neutral molecules with zero dipole

Table 3.1 Some enthalpies of noncovalent bonds and interactions*

Reaction	Characteristic interaction	ΔH° (kcal mol^{-1})
$Na^+(g) + Cl^-(g) \rightarrow NaCl(s)$	Ionic	−180
$NaCl(s) + \infty\ H_2O(l) \rightarrow Na^+(aq) + Cl^-(aq)$	Ionic and ion-dipole	1
Argon $(g) \rightarrow$ argon (s)	London	−2
n-Butane $(g) \rightarrow n$-butane (l)	London–van der Waals	−5
Acetone $(g) \rightarrow$ acetone (l)	London–van der Waals	−7
$2\,[CH_3\text{–O–H}]_{(g)}$ (Methanol) $\rightarrow [CH_3(H)O\cdots H\text{–O–}CH_3]_{(g)}$	Hydrogen bond	−5
$2\,[NH_3]_{(g)}$ (Ammonia) $\rightarrow [H_3N\cdots H\text{–}NH_2]_{(g)}$	Hydrogen bond	−4
$2\,[H\text{–C(=O)–N(H)–}CH_3]$ (benzene) (N-Methyl formamide) $\rightarrow [HC(=O)\text{–N(}CH_3\text{)–H}\cdots O\text{=C(H)–N(H)–}CH_3]$ (benzene)	Hydrogen bond	− 4

moment. Take a molecule of CCl_4, for example. Though the bonding electrons are localized more on the chlorine atoms, the *permanent dipole moment* of the molecule is zero because the four Cl atoms are symmetrically located at the four corners of a tetrahedron, with the C atom occupying the center of the tetrahedron. However, if a charge is placed near a CCl_4 molecule, the charge will distort the electronic distribution. We say that the CCl_4 molecule becomes *polarized.* The centers of the positive and negative charges in the polarized CCl_4 molecule no longer coincide, and the molecule has now an *induced dipole moment.* Interactions between a charged ion and polarized molecules are called *charge-induced dipole interactions.*

Induced dipole-induced dipole interactions exist even between neutral molecules with no permanent dipole moments. This is because of fluctuations in the electronic distributions in the molecules. A molecule with no permanent dipole moment may acquire an instantaneous dipole moment because of a

Table 3.1 Some enthalpies of noncovalent bonds and interactions (*cont.*)

Reaction	Characteristic interaction	ΔH° (kcal mol^{-1})
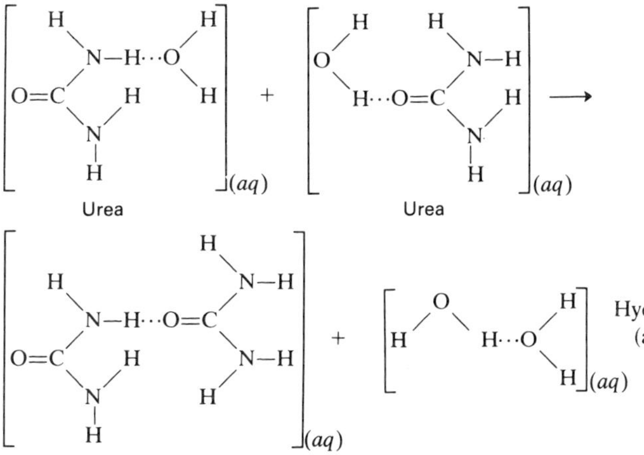	Hydrogen bond (aqueous)	−1
$C_3H_6(l) + \infty\ H_2O(l) \rightarrow C_3H_6(aq)$	Hydrophobic	−2
Benzene $(l) + \infty\ H_2O(l) \rightarrow$ benzene(aq)	Hydrophobic	0

* The enthalpies were obtained near room temperature except for the vaporization of solid argon; the standard state is dilute solution extrapolated to 1 M. Data are from various sources and are rounded to the nearest integer. For the precise values, see: G. C. Pimentel and A. L. McClellan, *The Hydrogen Bond*, W. H. Freeman, San Francisco, 1960; W. Kauzmann, *Adv. Protein Chem. 14*, 1 (1959); and J. A. A. Ketelaar, *Chemical Constitution*, Elsevier, Amsterdam, 1958.

fluctuation. This instantaneous dipole can induce a dipole in a neighboring molecule. Interactions between such fluctuation dipole-induced dipoles are called *London interactions*. London interaction is always present and is always an attractive force between molecules. It is the only force acting between identical rare gas atoms. It is responsible for the 2 kcal mol^{-1} enthalpy change necessary to vaporize solid argon to gaseous argon, for example. The force depends on the *polarizability* of the interacting molecules, which is a measure of how easy it is to distort the electron clouds of the molecule. In addition to the London interaction, uncharged molecules have van der Waals interactions. *Van der Waals interactions* include permanent dipole–permanent dipole interactions, permanent dipole-induced dipole attractions and steric repulsions. The London–van der Waals interactions are usually nonspecific forces which contribute to the energies of all reactions. For example, the heat of vaporization of liquid n-butane is 5 kcal mol^{-1}. The London–van der Waals forces can become specific when careful fitting of molecules is required. Binding of a particular substrate to an enzyme, antigen-antibody binding, and the function of specific membrane lipids may be dominated by London–van der Waals interactions.

The *hydrogen bond* is one of the important bonds that determines the three-dimensional structures of proteins and nucleic acids. A hydrogen atom covalently attached to one oxygen or nitrogen can form a weak bond to another oxygen or nitrogen. For O—H···O, N—H···N, and N—H···O hydrogen bonds, the bond enthalpy of 4 to 5 kcal mol^{-1} can be compared with values of about 100 kcal mol^{-1} for covalent bonds. This weak bond becomes even weaker in aqueous solution, *because of competition between solute-solute hydrogen bonds and solute-water and water-water hydrogen bonds.* Between urea molecules in water the hydrogen-bond strength is just 1 kcal mol^{-1}. Urea can be considered as a model for a peptide bond, so this is the magnitude of the heat that is expected for breaking the hydrogen bond involving the peptide links in a protein. Therefore, the free energy for breaking a peptide hydrogen bond in aqueous solution at 25°C is expected to be close to zero.

One type of interaction that is important in aqueous solutions is the *hydrophobic* (fear-of-water) interaction. Water molecules have a strong attraction for each other, primarily as a consequence of hydrogen-bond formation. The oxygen atom of most molecules of liquid water is hydrogen-bonded to two hydrogen atoms of two other H_2O molecules, and the hydrogen atoms of most molecules are hydrogen-bonded to the oxygen atoms of two other water molecules. Therefore, the molecules of liquid water form a mobile network: most water molecules are primarily interacting, through hydrogen bonds, with four tetrahedrally oriented neighbors. The network is not a rigid one, and change of neighbors occurs rapidly because of thermal motions.

Let us consider what happens if a molecule such as propane (C_3H_8) is introduced into this network. A hole is created; some hydrogen bonds in the original network are broken. The C_3H_8 does not interact with water strongly; it does not form hydrogen bonds. The water molecules around the C_3H_8 molecule must orient themselves in a way which reforms the hydrogen bonds that were disrupted by the hydrocarbon molecule. The net result is that water molecules around the C_3H_8 actually become more ordered. Since there is little change in the number of hydrogen bonds, the enthalpy change is small. The ordering of water molecules around the hydrocarbon molecule, however, is associated with a negative entropy change. These interpretations are consistent with experimental data:

$$C_3H_8(l) \longrightarrow C_3H_8(aq)$$

$$\Delta H^0_{298} \approx -2 \text{ kcal mol}^{-1}$$

$$\Delta S^0_{298} \approx -20 \text{ eu mol}^{-1}$$

$$\Delta G^0_{298} \approx +4 \text{ kcal mol}^{-1}$$

Now let us consider two such hydrocarbon groups, R. Each separate group when exposed to liquid water will, from the data above, cause an unfavorable free-energy change. If the two groups cluster together, the disruptive effect on the solvent network will be less than the combined effects of two separate groups. Therefore, the association of the groups will be thermodynamically favored:

$$\underset{\text{Separate}}{R(aq) + R(aq)} \longrightarrow \underset{\text{Clustered}}{R{-}R(aq)}$$

The clustering of the groups is not because they like each other, but because they are both disliked by water. Such hydrophobic interactions are important in many biological systems. For example, hydrocarbon groups in a water-soluble protein are usually found to cluster in the interior of the protein. Similarly, typical lipid molecules form bilayer sheets or membranes in water, in which the hydrocarbon portions are buried inside and the polar or charged portions are on the surface, exposed to the water. Such molecules are called *amphiphilic*. Hydrophobic interactions are characterized by low enthalpy changes and are entropy-driven. We should note that hydrophobic interaction is a term that we use to describe the combined effects of London, van der Waals, and hydrogen-bonding interactions in certain processes in aqueous solutions; it is not a "force" different from the others we have discussed. In particular, there is no such thing as a hydrophobic bond.

Table 3.2 Thermodynamics of transitions for biochemical conformational changes

Transition	ΔH^0 (kcal)	ΔS^0 (cal deg^{-1})
Poly-L-glutamate helix to coil transition (0.1 *M* KCl, 30°C)	1.1/amide	—
Poly-γ-benzyl-L-glutamate helix to coil transition (ethylene dichloride-dichloroacetic acid, 19:81 wt %, 39°C)	− 0.95/amide	− 3/amide
Polyadenylic acid unstacking (0.1 *M* KCl, 25°C)	8.5/nucleotide	27/nucleotide
Polyadenylic acid polyuridylic acid double strand to single strand (0.1 *M* NaCl, 24°C)	6.0/base pair	18/base pair
Calf thymus DNA double strand to single strand (0.15 *M* NaCl, 72°C)	7.0/base pair	20/base pair
A_7U_7 double strand to single strand* (1 *M* NaCl, 37°C)	107/mol	324/mol

SOURCE: *Handbook of Biochemistry and Molecular Biology*, 2nd ed., Chemical Rubber Co., Cleveland, Ohio, 1970, and references cited there.

*A_7U_7 represents an oligonucleotide consisting of 7 adenylic acid residues attached to 7 uridylic acid residues. Since A can pair with U, two A_7U_7 can pair in an antiparallel manner to form 14 base pairs. Data from P. N. Borer, B. Dengler, I. Tinoco, Jr., and O. C. Uhlenbeck, *J. Mol. Biol. 86,* 843 (1974).

Table 3.2 lists some measured enthalpies for biochemical reactions involving changes in shape (conformation) of molecules. By changes in conformation we mean changes in the secondary structure of the molecule. (The primary structure involves the covalent bonds.) Examples include the change in a polypeptide from a rigid helix to a flexible coil. Denaturation of proteins involves this type of change. The corresponding change in nucleic acids and polynucleotides is the change from a two-strand helix to two single strands. Hydrogen bonds between the amides, the nucleic acid bases, and the solvent are important, but so are London–van der Waals types of interactions. The magnitudes of ΔH^0 and ΔS^0 can help us to understand the various interactions involved.

The ΔH^0 for breaking a hydrogen bond between two urea molecules in water and forming urea-water hydrogen bonds is about 1 kcal mol^{-1}. From Table 3.2 we see that this is consistent with ΔH^0 for the helix-coil transition of poly-L-glutamate in aqueous solution. For calf thymus DNA and polyadenylic acid · polyuridylic acid double-strand to single-strand transitions, we need to break two hydrogen bonds per A · U base pair and three hydrogen bonds per G · C base pair. Therefore, if hydrogen bonds were dominant, we would expect about 2 to 3 kcal per base pair, instead of 6 to 7 kcal per base pair. Polyadenylic acid unstacking refers to the transition between an ordered helical molecule with the adenine bases stacked on top of each other and a much more disordered molecule with the adenine bases not oriented relative to each other. This change in stacking does not involve hydrogen bonds directly, but it does have a ΔH^0 of 8.5 kcal per nucleotide. This ΔH^0 must be mainly London–van der Waals interactions among the bases and between the bases and the solvent. The entropy change for the double-strand to single-strand transition in nucleic acids is also consistent with our conclusions above. There is an entropy increase due to the freedom of motion of the two single strands relative to the rigidity of the double strand.

USE OF PARTIAL DERIVATIVES

So far we have considered eight thermodynamic variables of state: E, H, S, G, A, P, V, and T. We can choose a few parameters as independent variables and express the others as functions of these independent variables. For example, if our system is a fixed amount of a pure substance, and if we choose P and T as our independent variables, we can express V as a function of P and T:

$$V = V(P, T)$$

The specific form of the function $V(P, T)$ depends on the nature of the substance. If it is an ideal gas,

$$V = V(P, T) = \frac{nRT}{P}$$

When we are interested in how V changes with T at constant P, we can take the derivative of V with respect to T, keeping P constant:

$$\frac{dV}{dT} = \frac{nR}{P} \tag{3.42}$$

Instead of stating explicitly that $P = \text{constant}$, we can use the notation of partial derivatives:

$$\left(\frac{\partial V}{\partial T}\right)_P = \frac{nR}{P} \tag{3.43}$$

$(\partial V/\partial T)_P$ means simply the derivative of V with respect to T at constant P. Similarly, instead of writing

$$\frac{dV}{dP} = nRT\frac{d}{dP}\left(\frac{1}{P}\right) \quad \text{at } T = \text{constant}$$

$$= -\frac{nRT}{P^2} \tag{3.44}$$

we can write

$$\left(\frac{\partial V}{\partial P}\right)_T = -\frac{nRT}{P^2} \tag{3.45}$$

Although in this book the usage of partial derivatives will be limited, we should point out that many thermodynamic relationships can be thus derived. For the state functions used in thermodynamics, it is generally true that if

$$V = V(P, T)$$

the total differential of V is

$$dV = \left(\frac{\partial V}{\partial P}\right)_T dP + \left(\frac{\partial V}{\partial T}\right)_P dT \tag{3.46}$$

This equation states, essentially, that for small changes, dV can be treated as the sum of how V changes with P (at constant T) and how V changes with T (at constant P). Also, the order of integration is not important.

$$\left[\frac{\partial}{\partial T}\left(\frac{\partial V}{\partial P}\right)_T\right]_P = \left[\frac{\partial}{\partial P}\left(\frac{\partial V}{\partial T}\right)_P\right]_T \tag{3.47}$$

An example of the use of these equations follows.

Example 3.10 Show that:

$$\text{(a)} \left(\frac{\partial G}{\partial P}\right)_T = V \qquad \text{(b)} \left(\frac{\partial V}{\partial T}\right)_P = -\left(\frac{\partial S}{\partial P}\right)_T$$

Solution (a) Since we want an expression for $(\partial G/\partial P)_T$, our first task is to express G as a function of T and P. Start with the definition of G.

$$G = H - TS$$
$$= E + PV - TS$$

We apply the standard rules of differentiation to obtain

$$dG = dE + d(PV) - d(TS)$$
$$= dE + P\,dV + V\,dP - T\,dS - S\,dT$$

Since G is a state function, if we can derive an expression $G = G(T, P)$ for a specific path, the expression will hold for any other path. For simplicity, we pick a reversible path with pressure-volume work only. Applying the first law to such a path,

$$dE = dq_{\text{rev}} + dw_{\text{rev}}$$
$$= dq_{\text{rev}} - P\,dV$$

but

$$dq_{\text{rev}} = T\,dS$$

Thus,

$$dE = T\,dS - P\,dV$$

and, by substitution,

$$dG = T\,dS - P\,dV + P\,dV + V\,dP - T\,dS - S\,dT$$
$$= V\,dP - S\,dT \tag{3.48}$$

Comparing (3.48) with the relationship according to (3.46),

$$dG = \left(\frac{\partial G}{\partial P}\right)_T dP + \left(\frac{\partial G}{\partial T}\right)_P dT$$

we obtain

$$\left(\frac{\partial G}{\partial P}\right)_T = V \tag{3.49}$$

and

$$\left(\frac{\partial G}{\partial T}\right)_P = -S \tag{3.50}$$

Equation (3.49) has been given previously in this chapter [in the integral form Eq. (3.33)] without proof. Similarly, Eq. (3.50) can be integrated to provide an exact replacement for Eq. (3.30).

(b) Since, according to (3.47),

$$\left[\frac{\partial}{\partial T}\left(\frac{\partial G}{\partial P}\right)_T\right]_P = \left[\frac{\partial}{\partial P}\left(\frac{\partial G}{\partial T}\right)_P\right]_T \tag{3.51}$$

from Eqs. (3.49) and (3.50) we obtain

$$\left(\frac{\partial V}{\partial T}\right)_P = -\left(\frac{\partial S}{\partial P}\right)_T$$

Note that in the special case of an ideal gas, since

$$\left(\frac{\partial V}{\partial T}\right)_P = \frac{nR}{P}$$

we have

$$\left(\frac{\partial S}{\partial P}\right)_T = -\frac{nR}{P}$$

Integrating at constant temperature gives

$$\int_{S_1}^{S_2} dS = \int_{P_1}^{P_2} -\frac{nR}{P}\,dP = -nR\ln\frac{P_2}{P_1}$$

This is Eq. (3.19).

Many useful thermodynamic relations can be derived by the use of partial derivatives, including the Gibbs-Helmholtz equation (3.32), which we gave without proof. Examples can be found in standard texts on thermodynamics.

SUMMARY

New State Variables

Name	Symbol	Units	Definition
Entropy	S	cal deg^{-1}, eu, J deg^{-1}	$dS \equiv dq_{rev}/T$
Gibbs free energy	G	cal, J, ergs	$G \equiv H - TS$
Helmholtz free energy	A	cal, J, ergs	$A \equiv E - TS$

Unit Conversions

Entropy:

1 cal deg^{-1} $\equiv$ 1 eu = 4.184 J deg^{-1}

Free energy:

1 cal = 4.184 J = 4.184 × 10^7 erg

EQUATIONS FOR CHAPTER 3

General Equations

Entropy, S:

$$S_2 - S_1 = \int_1^2 \frac{dq_{\text{rev}}}{T} \quad \text{(reversible path)} \tag{3.9}$$

$$S(T_2) - S(T_1) = \int_{T_1}^{T_2} \frac{C_P\,dT}{T} \quad \text{(constant pressure)} \tag{3.12}$$

$$S(T_2) - S(T_1) = \int_{T_1}^{T_2} \frac{C_V\,dT}{T} \quad \text{(constant volume)} \tag{3.14}$$

Gibbs free energy, G:

$$G \equiv H - TS \tag{3.20}$$

$$G_2 - G_1 = H_2 - H_1 - T\cdot(S_2 - S_1) \quad \text{(constant temperature)}$$

$$G(P_2) - G(P_1) = \int_{P_1}^{P_2} V\,dP \tag{3.33}$$

1 ℓ atm = 24.22 cal

Carnot-cycle efficiencies, for a heat engine:

$$\text{efficiency} = \frac{-w}{q_{\text{hot}}} = \frac{T_{\text{hot}} - T_{\text{cold}}}{T_{\text{hot}}} \tag{3.6}$$

Second law of thermodynamics and criterion for a spontaneous process:

In the following equations, the equal sign applies for reversible processes and the other sign applies for spontaneous (irreversible) processes:

$$\Delta S\ \text{(system)} + \Delta S\ \text{(surroundings)} \geq 0 \tag{3.7}$$

$$\Delta S\ \text{(isolated system)} \geq 0 \tag{3.8}$$

$$\Delta S\ \text{(system)} \geq \frac{q}{T} \qquad (3.3), (3.10)$$

$$\Delta G \leq 0 \quad \text{(constant } T,\ P) \qquad (3.25), (3.26)$$

$$\Delta A \leq 0 \quad \text{(constant } T,\ V) \qquad (3.39), (3.40)$$

Third law of thermodynamics

$$S_A(0\ \text{K}) \equiv 0$$

A = any pure, perfect crystal

Solids and Liquids

We assume in these equations that the volume of a solid or liquid is independent of T and P; that $C_P = C_V$; and that they do not depend on T and P.

$$S(P_2, T_2) - S(P_1, T_1) = n\bar{C}_P \ln\frac{T_2}{T_1} = n\bar{C}_V \ln\frac{T_2}{T_1}$$

n = number of moles
$\bar{C}_P$ = heat capacity per mole at constant P
$\bar{C}_V$ = heat capacity per mole at constant V

Gases

We assume that gases can be approximated by ideal gas equations and that C_P and C_V are independent of T. $PV = nRT$. $\bar{C}_P = \bar{C}_V + R$.

$$S_2 - S_1 = -nR\ln\frac{P_2}{P_1} = nR\ln\frac{V_2}{V_1} \qquad \text{(constant temperature)} \qquad (3.19)$$

$$G_2 - G_1 = nRT\ln\frac{P_2}{P_1} = -nRT\ln\frac{V_2}{V_1} \qquad \text{(constant temperature)} \qquad (3.35)$$

$$S(T_2) - S(T_1) = n\bar{C}_P\ln\frac{T_2}{T_1} \qquad \text{(constant pressure)} \qquad (3.13)$$

$R = 1.987\ \text{cal mol}^{-1}\ \text{deg}^{-1}$

Phase Changes at Constant T and P

$$\Delta G = 0 \qquad \text{(at equilibrium)}$$

$$\Delta S = \frac{\Delta H_{\text{transition}}}{T_{\text{transition}}} \qquad \text{(at equilibrium)} \qquad (3.17)$$

Chemical Reactions

For a chemical reaction

$$n_A A + n_B B \longrightarrow n_C C + n_D D$$

which occurs at constant T and P:

$$\Delta S^0 = n_C\bar{S}_C^0 + n_D\bar{S}_D^0 - n_A\bar{S}_A^0 - n_B\bar{S}_B^0$$

$$\Delta G^0 = n_C\bar{G}_C^0 + n_D\bar{G}_D^0 - n_A\bar{G}_A^0 - n_B\bar{G}_B^0$$

the superscript ^{0} means standard conditions, which here is 1 atm.

$$\Delta S(T_2) = \Delta S(T_1) + \int_{T_1}^{T_2} \Delta C_P \frac{dT}{T} \qquad (3.16)$$

$$\Delta C_P = n_C\bar{C}_{P,C} + n_D\bar{C}_{P,D} - n_A\bar{C}_{P,A} - n_B\bar{C}_{P,B}$$

$$\Delta G(T_2) = \Delta G(T_1) - (T_2 - T_1)\cdot\Delta S \qquad \text{(if } \Delta S \text{ is independent of } T)$$

$$\frac{\Delta G(T_2)}{T_2} = \frac{\Delta G(T_1)}{T_1} + \left(\frac{1}{T_2} - \frac{1}{T_1}\right)\cdot\Delta H \qquad \text{(if } \Delta H \text{ is independent of } T)$$

$$\frac{\Delta G(T_2)}{T_2} - \frac{\Delta G(T_1)}{T_1} = -\int_{T_1}^{T_2} \frac{\Delta H(T)}{T^2} dT \qquad \text{(if } \Delta H \text{ depends on } T) \quad (3.32)$$

$$\Delta G(P_2) - \Delta G(P_1) = \Delta V\cdot(P_2 - P_1) \qquad \text{(for solids and liquids only, if } \Delta V \text{ is independent of } T)$$

$$\Delta V = n_C\bar{V}_C + n_D\bar{V}_D - n_A\bar{V}_A - n_B\bar{V}_B \qquad (3.36)$$

$$\Delta G(P_2) - \Delta G(P_1) = \Delta nRT \ln\frac{P_2}{P_1} \qquad \text{(at least one gaseous product or reactant)} \quad (3.37)$$

Δn = number of moles of gaseous products − number of moles of gaseous reactants.

REFERENCES

In addition to the textbooks listed in Chapter 2, see

BENT, H. A., 1965. *The Second Law*, Oxford University Press, New York.

SUGGESTED READINGS

FEYNMAN, R. P., R. B. LEIGHTON, and M. SANDS, 1963.*The Feynman Lectures on Physics*, Addison-Wesley, Reading, Mass.

PROBLEMS

1. The temperature of the heat reservoirs for a Carnot cycle (reversible) engine are T_{hot} = 1200 K and T_{cold} = 300 K. The efficiency of the engine is calculated to be 0.75, from Eq. (3.6).

(a) If $w = -100$ kcal, calculate q_1 and q_3. Explain the meaning of the signs of w, q_1, and q_3.
(b) The same engine can be operated in the reverse order. If $w = +100$ kcal, calculate q_1 and q_3. Explain the meaning of the signs.
(c) Suppose that it were possible to have an engine with a higher efficiency, say 0.80. If $w' = -100$ kcal, calculate q_1' and q_3'. The superscript denotes quantities for this engine.
(d) If we use all the work done by the engine in part (c) to drive the heat pump in part (b), calculate $(q_1 + q_1')$ and $(q_3 + q_3')$, the amount of heat transferred. Explain the signs. Note that the net effect of the combination of the two engines is to allow a "spontaneous" transfer of heat from a cooler reservoir to a hotter reservoir, which should not happen.
(e) Show that if it were possible to have a reversible engine with a lower efficiency operating between the same two temperatures, heat could also be transferred spontaneously from the cooler reservoir to the hotter reservoir. (*Hint*: Operate the engine with lower efficiency as a heat pump.)

2. One suggestion for solving the fuel-shortage problem is to use electric power (obtained from solar energy) to electrolyze water to form $H_2(g)$ and $O_2(g)$. The hydrogen could be used as a pollution-free fuel.
(a) Calculate the enthalpy, entropy, and free-energy change on burning 1 g of $H_2(g)$ to $H_2O(l)$ at 25°C and 1 atm.
(b) Calculate the enthalpy, entropy, and free-energy change on burning 1 g of *n*-octane(g) to $H_2O(l)$ and $CO_2(g)$ at 25°C and 1 atm.

3. Use the Appendix to answer the following questions.
(a) A friend wants to sell you a catalyst that allows benzene to be formed by passing $H_2(g)$ over carbon (graphite) at 25°C and 1 atm. Should you buy? Why?
(b) Is the reaction $2\,Fe(s) + \frac{3}{2}\,O_2(g) \rightarrow Fe_2O_3(s)$ spontaneous at 25°C and 1 atm?
(c) Is the reaction $2\,CH_4(g) + NH_3(g) + \frac{5}{2}\,O_2(g) \rightarrow H_2NCH_2CO_2H$ (solid glycine) + $3\,H_2O(l)$ spontaneous at 25°C and 1 atm?

4. For each of the following processes, state whether ΔS and ΔG increase, decrease, do not change, or are impossible to determine. The changes in E and H were considered in Problem 16, Chapter 2.
(a) Twenty-four grams of molybdenum oxide at −204°C, $P = 15.7$ atm, is converted to the difluoride at a temperature of 1053°C and a pressure of 0.1 mm Hg. The difluoride is transformed to the pure metal, which is then reacted with $O_2(g)$ to form 24 g of molybdenum oxide at −204°C, $P = 15.7$ atm.
(b) An unknown substance is heated 5°C at constant pressure.
(c) A nonideal gas expands at constant temperature.
(d) A substance absorbs 10 cal of heat and does 7 cal of work, irreversibly, at constant T and P.
(e) A substance absorbs 10 cal of heat and does 7 cal of work, reversibly, at constant T and P.

5. Consider the reaction

$$CH_3OH(l) \longrightarrow CH_4(g) + \tfrac{1}{2}\,O_2(g)$$

Calculate ΔH^0_{298}, ΔS^0_{298}, and ΔG^0_{298}. Estimate ΔE^0_{298}. State what further data would be needed to obtain ΔH^0 at 500°C and 1 atm.

6. Consider the reversible, isothermal, constant-pressure freezing of 1 mol of water at 0°C and 1 atm.
   (a) Calculate ΔE in kcal.
   (b) Calculate ΔH in kcal.
   (c) Calculate ΔS in cal deg^{-1}.
   (d) Calculate ΔG in kcal.
   (e) Calculate q in cal. Is heat absorbed or evolved?
   (f) Calculate w in cal. Is work done by the system or on the system?

7. For the statements below, choose the word or words inside the parentheses that serve(s) to make a correct statement. More than one answer may be correct. At 100°C, the equilibrium vapor pressure of water is 1 atm. Consider the process where 1 mol of water vapor at 1 atm pressure is reversibly condensed to liquid water at 100°C by slowly removing heat into the surroundings.
   (a) In this process the entropy of the system will (increase, remain unchanged, decrease).
   (b) The entropy of the universe will (increase, remain unchanged, decrease).
   (c) Since the condensation process occurs at constant temperature and pressure, the accompanying free-energy change for the system will be (positive, zero, negative).
   (d) In practice, this process cannot be carried out reversibly. For the real process, compared with the ideal reversible one, different values will be observed for the (entropy change of the system, entropy change of the surroundings, entropy change of the universe, free-energy change of the system).

   At a lower temperature, of 90°C, the equilibrium vapor pressure of water is only 0.692 atm.

   (e) If 1 mol of water is condensed reversibly at this temperature and pressure, the entropy of the system will (increase, remain unchanged, decrease).
   (f) The entropy of the universe will (increase, remain unchanged, decrease).
   (g) The free energy of the system will (increase, remain unchanged, decrease).
   (h) Since the molar heat capacity of water vapor is only about one-half that of the liquid, the entropy change upon condensation at 90°C will be (more negative than, the same as, more positive than) that at 100°C.

8. Calculate the entropy change when:
   (a) Three moles of $H_2O(g)$ is cooled irreversibly at constant P from 110°C to 100°C.
   (b) One mol of $H_2O(g)$ is expanded at constant pressure of 2 atm from an original volume of 20 ℓ to a final volume of 25 ℓ. You can consider the gas to be ideal.
   (c) One gram of $H_2O(s)$ at −10°C and 1 atm is heated to $H_2O(l)$ at +10°C and 1 atm.

9. You are asked to evaluate critically the following situations. Some of the proposals or interpretations are reasonable and others violate very basic principles of thermodynamics.
   (a) It is commonly known that one can supercool water and maintain it as a liquid at temperatures as low as −10°C. If a sample of supercooled liquid water is isolated in a closed, thermally insulated container, after a time it spontaneously changes to a mixture of ice and water at 0°C. Thus it increases its temperature spontaneously with no addition of heat from outside, and furthermore, some low-entropy ice is produced.

(b) An inventor proposed a new scheme for heating buildings in the winter in arctic climates. Since freezing temperatures reach only a few feet down into the soil, she proposes digging a well and immersing a coil of copper tubing in the water. She will then connect the ends of the coil to the radiators in the building and use a heat pump to transfer heat into the building. What would be your advice to an attorney who is assigned to evaluate this patent application? Base your evaluation on the relevant thermodynamics.

(c) On a hot summer's day your laboratory partner proposes opening the door of the lab refrigerator to cool off the room.

(d) A sample of air is separated from a large evacuated chamber, and the entire system is isolated from the surroundings. A small pinhole between the two chambers is opened and roughly half the gas is allowed to effuse into the second chamber before the pinhole is closed. Because nitrogen effuses faster than oxygen, the gas in the second chamber is richer in nitrogen and the gas in the first chamber is richer in oxygen than the original air. The gases have, thus, spontaneously unmixed (at least partially), and this is held to be a violation of the second law of thermodynamics.

(e) Supercooled water at −10°C has a higher entropy than does an equal amount of ice at −10°C. Therefore, supercooled water cannot go spontaneously to ice at the same temperature in an isolated system.

(f) A volume of an aqueous solution of hydrogen peroxide is placed in a cylinder and covered with a tight-fitting piston. A small amount of the enzyme catalase is placed on a probe and inserted through an opening in the base of the cylinder. The catalase catalyses the decomposition of the hydrogen peroxide, and the oxygen gas formed serves to raise the piston. The catalase is then withdrawn and the hydrogen peroxide is re-formed, causing the piston to return to its initial position. The piston is connected to an engine and net work is obtained indefinitely by cycles of simply inserting and withdrawing the catalase.

(g) High temperatures can be achieved in practice by focusing the rays of the sun on a small object using a large parabolic reflector, such as is used in astronomical telescopes. Since the energy gathered increases as the square of the diameter of the reflector, it should be possible to produce temperatures higher than those in the sun by using a large-enough reflector.

(h) Hydroelectric plants generate electric power (work) by using water falling in a gravitational potential. The water running down a mountain passes through a power plant; flows out to the ocean, where it subsequently evaporates; and gets reprecipitated in the mountains. This seems to be a perpetual-motion cycle and must, therefore, lie outside the realm of the second law of thermodynamics.

(i) The maximum efficiency of a steam engine can be calculated using the second law of thermodynamics. If it operates between the boiling point of water and room temperature, the maximum efficiency is about

$$\frac{373 - 293}{373} = \frac{80}{373} = 0.215 \quad \text{or} \quad 21.5\%$$

Photosynthesis by green plants occurs almost entirely at ambient temperatures, yet recent publications report theoretical limits as high as 85% of the fraction of the light energy absorbed that can be converted into chemical energy (free energy or net work). Clearly, such estimates must be wrong, or else the second law cannot apply.

10. For the statements below, choose the word or words inside the parentheses that serve to make a correct statement. Each statement has at least one, and may have more than one, correct answer.

(a) According to the second law of thermodynamics, a spontaneous process, such as a balloon filled with hot gas cooling to the surroundings at constant pressure, will always occur (adiabatically, reversibly, irreversibly, without work done).

(b) Associated with such a process there is always an increase in entropy of (the system, the surroundings, the system plus the surroundings, none of these).

(c) For the example given, the heat gained by the surroundings is just equal to the negative of the (internal energy change, enthalpy change, entropy change, Gibbs free energy change) of the system.

(d) To return the system to its initial state requires from the surroundings an expenditure of entropy whose magnitude is (greater than, equal to, less than) that which it gained during the spontaneous process.

11. (a) Consider 1 mol of liquid water to be frozen reversibly to ice at 0°C by slowly removing heat to the surroundings. In this process the entropy of the system will (increase, remain unchanged, decrease).

(b) If the freezing process occurs at 0°C and a pressure of 1 atm, the accompanying change in the Gibbs free energy for the system will be (positive, zero, negative).

(c) For this process the enthalpy of the system will (increase, remain unchanged, decrease).

12. For the statements below, choose the word or words inside the parentheses that serve to make a correct statement. Each statement has at least one, and may have more than one, correct answer.

(a) For a sample of an ideal gas, the product PV remains constant as long as the (temperature, pressure, volume, internal energy) is held constant.

(b) The internal energy of an ideal gas is a function of only the (volume, pressure, temperature).

(c) The second law of thermodynamics states that the entropy of an isolated system always (increases, remains constant, decreases) during a spontaneous process.

(d) If the system is not isolated from the surroundings, then $T\,\Delta S$ for a system undergoing an isothermal, reversible process is (less than, equal to, greater than) the heat absorbed.

(e) $T\,\Delta S$ for an isothermal, spontaneous irreversible process is (less than, equal to, greater than) the heat absorbed by the system.

(f) When a sample of liquid is converted reversibly to its vapor at its normal boiling point, (q, w, ΔP, ΔV, ΔT, ΔE, ΔH, ΔS, ΔG, none of these) is equal to zero for the system.

(g) If the liquid is permitted to vaporize isothermally and completely into a previously evacuated chamber that is just large enough to hold the vapor at 1 atm pressure, then (q, w, ΔE, ΔH, ΔS, ΔG) will be smaller than for the reversible vaporization.

13. For the following processes, determine whether each of the thermodynamic quantities listed is greater than, equal to, or less than zero for the *system* described. Consider all gases to behave ideally. The system is shown in italic type in each case. State explicitly any reasonable assumptions that you need to make.

(a) An *ideal gas* expands adiabatically and reversibly to twice its initial volume: q, w, ΔP, ΔT, ΔE, ΔH, ΔS, ΔG. Example: $\Delta V > 0$.

(b) A *sample of liquid water* supercooled to $-10°C$ placed in a thermally insulated compartment undergoes a spontaneous crystallization to form a mixture of ice and liquid water: q, w, ΔT, ΔE, ΔH, ΔS, ΔG.

(c) Separate samples of *two strongly interacting liquids* (e.g., H_2SO_4 and water) are mixed together under open conditions. Heat is evolved. Both the initial and final states are in thermal equilibrium with the surroundings: q, w, ΔT, ΔE, ΔH, ΔS, ΔG.

(d) A dilute *aqueous solution of hydrogen peroxide*, H_2O_2, in a closed thermally insulated container is suddenly exposed to a small amount of the enzyme catalase, which catalyzes its decomposition according to the reaction

$$H_2O_2(aq) \xrightarrow{\text{catalase}} H_2O(l) + \tfrac{1}{2}O_2(g)$$

(you may assume that the amount of enzyme involved is negligible for thermodynamic purposes): q, w, ΔP, ΔT, ΔE, ΔH, ΔS, ΔG.

Data for $H_2O_2(aq)$:

$$\Delta H^0_{f,298} = -45.68 \text{ kcal mol}^{-1}$$

$$\Delta G^0_{f,298} = -31.57 \text{ kcal mol}^{-1}$$

4

The Concentration Dependence of Free Energy

We have been discussing free energy and criteria for chemical equilibria and spontaneity without mentioning concentration. Yet we know that the concentrations of products and reactants have an important effect on whether a reaction will occur. The transport of O_2 by hemoglobin depends on the fact that the high partial pressure of O_2 in the lungs causes O_2 molecules to be bound to the hemoglobin, while in other parts of the body the lower pressure of O_2 favors the dissociation of the complex. The O_2 binding efficiency of the hemoglobin depends on the O_2 pressure in a complex way, and only recently have detailed molecular mechanisms been proposed. However, the thermodynamics is straightforward.

The equations relating free energy and equilibrium constants to concentrations are quantitative descriptions of *Le Châtelier's principle*. Le Châtelier observed that when a reaction at equilibrium was perturbed, the concentrations changed so as to minimize the effect of the perturbation. Adding reactant to a reaction at equilibrium produces more product and uses up some of the reactant added. Adding product decreases the extent of reaction and converts some of the added product back to reactant. The concentrations of products and reactants control the driving force for the reaction. This driving force is the negative free energy at constant T and P. Therefore, we say that adding reactants decreases the free energy for the forward reaction; adding products increases the free energy for the forward reaction.

IDEAL GASES

Free energy Changes

We shall derive the equation for the concentration dependence of free energy for a reaction involving only ideal gases. For concentrations we use partial pressures of the gases. Consider a reaction chamber where we can control the pressure of each gaseous reactant and product. This can usually be done in a flow reactor such as the one shown in Fig. 4.1. For example, let us use our favorite reaction. We flow into the reactor $H_2(g)$ and $O_2(g)$ at pressures P_{H_2} and P_{O_2}. These pressures are completely under our control. The gases are both introduced at temperature T, which we choose. A reaction occurs in the reaction chamber to produce some $H_2O(g)$. The $H_2O(g)$ is cooled, if necessary, to temperature T, and its partial pressure, P_{H_2O}, is determined. We can control P_{H_2O} by adjusting flow rates and experimental conditions in the reactor.

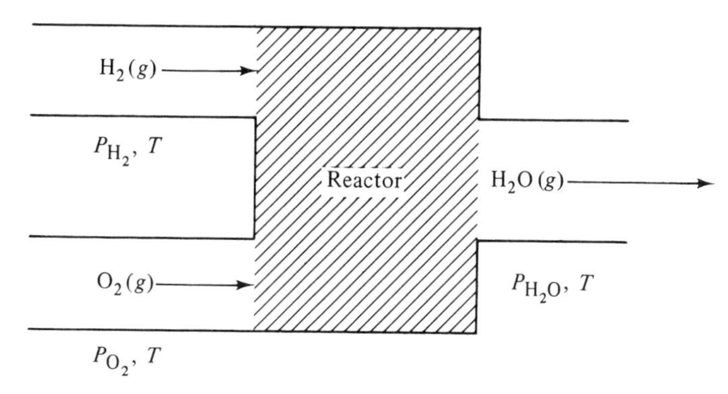

Fig. 4.1 Flow reactor in which reactants $H_2(g)$ at P_{H_2} and T and $O_2(g)$ at P_{O_2} and T are introduced and the product $H_2O(g)$ is removed at P_{H_2O} and T.

We want to calculate the Gibbs free energy change for the reaction. We have specified the initial state (P_{O_2}, P_{H_2}, T) and the final state (P_{H_2O}, T), so the free energy is determined once we specify how many moles have reacted. There are various points to note:

1. It does not matter what happens within the reactor. The temperature may be higher than T; it will surely be nonuniform. The partial pressures of reactants and products will change rapidly in an unknown way in the reactor. There may or may not be catalysts in the reactor. None of this matters as long as we know P_{O_2}, P_{H_2}, P_{H_2O}, and T.
2. We need not have stoichiometric amounts of H_2 and O_2. Any unreacted H_2 or O_2 is unchanged and therefore does not contribute directly to the change of free energy. Excess reactants may affect the partial pressure of H_2O and P_{H_2O}, but we determine P_{H_2O}.

3. We emphasize once more that we can choose any partial pressures of products or reactants.

If all the partial pressures are 1 atm and the temperature is 25°C, then we can calculate the free energy change, ΔG^0_{298}, from the Appendix. For any other temperature T we can use Eqs. (3.29) through (3.32) to find ΔG^0_T. What about other partial pressures? We must use an alternative path with individual steps which we can calculate to obtain a general ΔG_T. For formation of 1 mol of $H_2O(g)$, the two paths are

$$\begin{array}{ccc} \underbrace{H_2(g, P_{H_2}) + \tfrac{1}{2}O_2(g, P_{O_2})} & \xrightarrow{\Delta G_T} & H_2O(g, P_{H_2O}) \\ \downarrow \Delta G_1 & & \uparrow \Delta G_2 \\ H_2(g, 1\text{ atm}) + \tfrac{1}{2}O_2(g, 1\text{ atm}) & \xrightarrow{\Delta G^0_T} & H_2O(g, 1\text{ atm}) \end{array}$$

ΔG_1 and ΔG_2 are free-energy changes due to changing the pressure of (assumed) ideal gases. From Eq. (3.35),

$$G(P_2) - G(P_1) = nRT\ln\frac{P_2}{P_1} \tag{3.35}$$

$$\Delta G_1 = n_{H_2}RT\ln\frac{1}{P_{H_2}} + n_{O_2}RT\ln\frac{1}{P_{O_2}}$$

$$\Delta\bar{G}_1 = RT\ln\frac{1}{P_{H_2}} + \frac{1}{2}RT\ln\frac{1}{P_{O_2}} \qquad \text{(per mol of } H_2\text{)}$$

$$= RT\ln\frac{1}{P_{H_2}} + RT\ln\frac{1}{P_{O_2}^{1/2}} \tag{4.1}$$

We have used the algebraic identity $a \cdot \ln x \equiv \ln x^a$. Similarly, we find

$$\Delta\bar{G}_2 = RT\ln P_{H_2O} \qquad \text{(per mol of } H_2O\text{)} \tag{4.2}$$

Equating the free-energy changes along both paths linking initial and final states,

$$\Delta\bar{G}_T = \Delta\bar{G}^0_T + \Delta\bar{G}_1 + \Delta\bar{G}_2$$

$$= \Delta\bar{G}^0_T + RT\ln\frac{1}{P_{H_2}} + RT\ln\frac{1}{P_{O_2}^{1/2}} + RT\ln P_{H_2O}$$

$$= \Delta\bar{G}^0_T + RT\ln\frac{P_{H_2O}}{(P_{H_2})(P_{O_2})^{1/2}} \tag{4.3}$$

Equation (4.3) is the equation we want. It relates the free energy change for any partial pressures, $\Delta\bar{G}_T$, to the standard free-energy change at 1 atm partial pressures, $\Delta\bar{G}^0_T$. It should be clear now how important it is to note the presence or absence of the superscript "0". Its presence means standard conditions of 1 atm; its absence means different conditions, which must be specified. The superscript "0" is less important for ΔH, because enthalpy does not change very much with pressure, even for gases. However, for ΔG it is vital.

It is also important to remember that the partial pressures in Eq. (4.3) must be in atmospheres. That is, the P's that occur are actually the ratio of P atm to 1 atm. It may help to remember that algebraically one cannot take the logarithm of a unit; therefore, one must be sure that the correct units (here atm) which cancel those in the standard conditions are used.

Example 4.1 What will be the free energy of forming 1 mol of H_2O at 1000 K if: (a) 10 atm of O_2 and 10 atm of H_2 are reacted to give 0.01 atm of H_2O; (b) 0.01 atm of O_2 and 0.01 atm of H_2 are reacted to give 10 atm of H_2O? For part (b), $H_2O(g)$ could be introduced with the reactants to give a high enough product partial pressure.

Solution

(a)

$$\Delta\bar{G}_T - \Delta\bar{G}_T^0 = RT \ln \frac{P_{H_2O}}{(P_{H_2})(P_{O_2})^{1/2}} \tag{4.3}$$

$$= (1.99 \text{ cal mol}^{-1} \text{ deg}^{-1})(1000 \text{ deg})(2.303) \log \frac{0.01}{(10)(3.162)}$$

$$= -1.60 \times 10^4 \text{ cal mol}^{-1}$$

The free energy is decreased (relative to the standard free energy) by having high pressures of reactants and low pressures of products; the reaction is favored relative to standard conditions.

(b)

$$\Delta\bar{G}_T - \Delta\bar{G}_T^0 = (1.99)(1000) \ln \frac{10}{(0.01)(0.1)}$$

$$= 1.83 \times 10^4 \text{ cal mol}^{-1}$$

The free energy is increased by having high pressures of products and low pressures of reactants; the reaction is not favored relative to standard conditions.

We can generalize Eq. (4.3) for any reaction involving only (ideal) gases. For a reaction of ideal gases at temperature T,

$$n_A A + n_B B \longrightarrow n_C C + n_D D$$

$$\Delta G_T = \Delta G_T^0 + RT \ln Q \tag{4.4}$$

$$Q = \frac{(P_C)^{n_C}(P_D)^{n_D}}{(P_A)^{n_A}(P_B)^{n_B}} \tag{4.5}$$

where P_A, P_B, P_C, and P_D = partial pressure (in atm) of each reactant or product. The quotient, Q, is the ratio of the arbitrary partial pressures (in atm) of reactants and products, each raised to the power of its coefficient in the chemical equation.

If we double the number of moles involved in the reaction, each of the coefficients is multiplied by 2, and Q is squared. We see that Q will be large if product pressures are large or reactant pressures are small. A large Q means a positive (unfavorable) contribution to the free energy. Q is small if product pressures are small or reactant pressures are large. A small Q means a negative (favorable) contribution to the free energy. The table below gives quantitative values for the change of free energy with Q at 25°C. The reader can easily calculate values at any other temperature, T, by multiplying by $(T/298)$.

Q	$\Delta G_{298} - \Delta G^0_{298}$ (cal)
100	+2730
10	+1365
1	0
0.1	−1365
0.01	−2730

We have emphasized here the distinction between standard free-energy changes, ΔG^0_T, for defined concentrations, and actual free-energy changes, ΔG_T, for any other concentrations specified by Q.

Equilibrium Constant

For every chemical reaction at any temperature there are partial pressures of products and reactants for which the system is at equilibrium. We can find these pressures by letting the reaction attain equilibrium. That is, we wait for equilibrium to be attained at a chosen T, then we measure the equilibrium partial pressures of products and reactants. When these equilibrium partial pressures are substituted into the expression for Q, we obtain the equilibrium constant, K. For the reaction

$$n_A\text{A} + n_B\text{B} \longrightarrow n_C\text{C} + n_D\text{D}$$

the equilibrium constant is

$$K = \frac{(P_C^{eq})^{n_C}(P_D^{eq})^{n_D}}{(P_A^{eq})^{n_A}(P_B^{eq})^{n_B}} \tag{4.6}$$

where P_A^{eq}, P_B^{eq}, and so on, are partial pressures (in atm) at equilibrium. Equation (4.4) of course still applies, but at equilibrium the free energy change for the reaction is zero. Therefore,

$$0 = \Delta G^0_T + RT \ln K \qquad \text{(at equilibrium)}$$

$$\Delta G^0_T = -RT \ln K \tag{4.7}$$

Note that determination of the equilibrium constant experimentally permits calculation of the *standard* free energy change. Combining Eqs. (4.4) and (4.7), we have

$$\Delta G_T = -RT \ln K + RT \ln Q$$

$$= RT \ln \frac{Q}{K} \tag{4.8}$$

Equations (4.4) through (4.8) may be the most useful thermodynamic equations a biochemist learns. They relate the free energy change for a reaction to experimentally measurable quantities. Furthermore, Eq. (4.6), which defines the equilibrium constant, presents a ratio of equilibrium partial pressures that is constant for a chemical reaction at a given temperature. At this point the reader may not be convinced of the importance of these equations, because they were all derived for ideal gases and the biochemist reader is presumably more interested in reactions in solution. The introduction of *activity* in the next section provides the necessary connection between the real world and Eqs. (4.4) through (4.8).

Anticipating applications to reactions other than those involving ideal gases, we shall write Eq. (4.7) in different but equivalent forms. Solving for K, we obtain

$$K = e^{-\Delta G^0/RT}$$

or

$$K = 10^{-\Delta G^0/2.303RT} \tag{4.9}$$

Substituting for ΔG^0 gives

$$K = e^{\Delta S^0/R} e^{-\Delta H^0/RT}$$

$$= 10^{\Delta S^0/2.303R} 10^{-\Delta H^0/2.303RT} \tag{4.10}$$

(Remember that $e^a e^b = e^{a+b}$.) The superscript is vital on ΔG and ΔS, which depend markedly on concentration; but it is not so necessary for ΔH, which is essentially independent of concentration.

Example 4.2 Calculate the equilibrium constant at 25°C for the decarboxylation of liquid pyruvic acid to form gaseous acetaldehyde and CO_2.

Solution The reaction is

$$CH_3\overset{\overset{\displaystyle O}{\|}}{C}COOH(l) \longrightarrow CH_3\overset{\overset{\displaystyle O}{\|}}{C}H(g) + CO_2(g)$$

From the Appendix,

$$\Delta G^0 = \Delta G_f^0(\text{acetaldehyde}) + \Delta G_f^0(CO_2) - \Delta G_f^0(\text{pyruvic acid})$$

$$= -31.86 + (-94.26) - (-110.75)$$

$$= -15.37 \text{ kcal mol}^{-1}$$

From Eq. (4.9),

$$K = 10^{-\Delta G^0/2.303RT} \tag{4.9}$$

The factor $2.303RT = 1365 \text{ cal mol}^{-1}$ for 25°C; it will occur often and it is useful to remember this value for 25°C.

$$\begin{aligned} K &= 10^{-\Delta G^0/1.365} \qquad (\Delta G^0 \text{ in kcal}) \\ &= 10^{15.37/1.365} \\ &= 10^{11.25} \\ &= 1.79 \times 10^{11} \end{aligned}$$

Pyruvic acid is very unstable with respect to CO_2 and acetaldehyde, but a suitable catalyst is needed to cause a rapid reaction leading to equilibrium.

SOLUTIONS

Activity and Chemical Potential

It was clear to thermodynamicists that Eqs. (4.4) through (4.8) were simple and easy to use; therefore, it would be convenient if they also applied to real gases, liquids, solids, and solutions. The thermodynamicists decided that the only way to ensure this was to define Eqs. (4.4), (4.7), and (4.8) to be generally true, and to change Eqs. (4.5) and (4.6) for Q and K so as to agree. This is done by introducing a new quantity, the activity, a. For our usual reaction of $n_A A + n_B B \rightarrow n_C C + n_D D$,

$$Q \equiv \frac{(a_C)^{n_C}(a_D)^{n_D}}{(a_A)^{n_A}(a_B)^{n_B}} \tag{4.11}$$

the equilibrium constant

$$K \equiv \frac{(a_C^{eq})^{n_C}(a_D^{eq})^{n_D}}{(a_A^{eq})^{n_A}(a_B^{eq})^{n_B}} \tag{4.12}$$

where a_A, a_B, etc. = activities of each component
a_A^{eq}, a_B^{eq}, etc. = activities of each component at equilibrium

Instead of writing a_A, one can use brackets around A to designate activity:

$$a_A \equiv [A]$$

and the equilibrium constant

$$K = \frac{[C^{eq}]^{n_C}[D^{eq}]^{n_D}}{[A^{eq}]^{n_A}[B^{eq}]^{n_B}} \tag{4.13}$$

We often omit the superscript "eq" when writing the expression for K. However, *equilibrium* activities are understood. The activities are unitless numbers, so we do not have to worry about taking logarithms. Equation (4.11), together with Eq. (4.4), define activities. By substituting $n_C\bar{G}_C + n_D\bar{G}_D - n_A\bar{G}_A - n_B\bar{G}_B = \Delta G$ and a similar expression for ΔG^0 in Eq. (4.4) and using Eq. (4.11) for Q, we obtain

$$\bar{G}_A \equiv \bar{G}_A^0 + RT \ln a_A \tag{4.14}$$

where $\bar{G}_A$ = free energy per mole of A in the mixture of molecules, A, B, C, D
$\bar{G}_A^0$ = standard free energy per mole of A
a_A = activity of A in the mixture (unitless)

Identical expressions hold for B, C, and D, so Eq. (4.14) defines the activity of any molecule. We repeat that activities were introduced so that the Eqs. (4.4) through (4.8), which relate free energy to measurable quantities, would keep a simple form. Both $\bar{G}_A$ and $\bar{G}_A^0$ depend on temperature, and $\bar{G}_A$ depends not only on the concentration of A but, in general, to some extent on the concentrations of B, C, and D as well.

These molar free energies are also called *chemical potentials*, μ. The definition of chemical potential is

$$\mu_A \equiv \bar{G}_A \tag{4.15}$$

$\bar{G}_A$ is also called a *partial molal free energy*, because it can be defined in terms of a partial derivative. We shall return to partial molal quantities at the end of this chapter.

An important aspect of Eq. (4.14) is that activity is defined with respect to a standard state. The difference in free energy between two states, $\bar{G}_A - \bar{G}_A^0$, is a measurable quantity. But $\bar{G}_A^0$ can only be obtained after the standard conditions (standard state) are defined. Once $\bar{G}_A$ and $\bar{G}_A^0$ are determined, the activity, a, can be obtained. However, until the standard state is specified, the activity cannot be determined. The trouble is that various standard states are useful (and used), so before the activity can be determined it is necessary to know the particular standard state that is involved.

Standard States

Equation (4.14) tells us how to determine the activity for the substance A in a system. We determine the difference in free energy per mole between A in the system and A in a specified standard state. This difference allows us to calculate a_A.

$$\ln a_A = \frac{\bar{G}_A - \bar{G}_A^0}{RT} \tag{4.16}$$

It follows immediately that the *activity of a substance in its standard state is equal to unity*. In the following paragraphs we will give the standard states that are commonly used in specifying activities. However, we shall reserve until later, discussion of how to measure free energies of molecules in solution.

Ideal gases

The standard state for an ideal gas is the gas with a partial pressure equal to 1 *atm.* The activity of an ideal gas is defined as its actual partial pressure divided by 1 atm, its partial pressure in its standard state:

$$a_A \equiv \frac{P_A\,(\text{atm})}{1\ \text{atm}} = P_A \tag{4.17}$$

where P_A = partial pressure of ideal gas (in atm). The standard conditions for a reaction involving only ideal gases are that each product and reactant has a partial pressure equal to 1 atm, as discussed in Chapter 3.

Real gases

The activity of a real gas is a function of pressure, which we write

$$a_A = \gamma_A P_A \tag{4.18}$$

where γ_A = activity coefficient
P_A = partial pressure, atm

We know that all gases become ideal at low-enough pressures, so γ_A must become 1 as the total pressure in the system approaches zero. Near atmospheric pressure the activity coefficient of gases is close to 1, so we will approximate real gases by ideal gases.

Pure solids or liquids

The standard state for a solid or liquid is the pure substance (*solid or liquid*) *at* 1 *atm pressure.* Therefore, the activity is equal to 1 for a pure solid or liquid at 1 atm. The free energy of a solid or liquid changes so slightly with pressure [see Eq. (3.33)] that we can usually neglect the change and use 1 for the activity of a solid or liquid at any pressure:

$$a_A = 1 \tag{4.19}$$

This convention is familiar from elementary chemistry courses.

Solutions

By solution we mean a homogeneous mixture of two or more substances. We can have solid solutions, liquid solutions, or very complicated mixtures of components as might be found in a biological cell. The activity of each substance will depend upon its concentration and the concentrations of everything else in the mixture. We write this dependence in a deceptively simple equation:

$$\text{activity} = (\text{activity coefficient}) \cdot (\text{concentration}) \tag{4.20}$$

The activity coefficient is not a constant; it incorporates all the complicated dependence of the activity of A on the concentrations of A, B, C, and so on. Another complication is that different concentration units are routinely used in Eq. (4.20). We shall discuss each of the concentrations units that are commonly used and discuss the standard state connected with each unit.

Mole fraction and solvent standard state

The standard state that uses mole fraction as a concentration unit is often called the solvent standard state. *The solvent standard state for a component of a solution defines the pure component as the standard state.* Let us choose the solvent standard state for component A in a solution. A useful concentration unit is the *mole fraction X*, the number of moles of species A divided by the total number of moles of all components present in the solution:

$$X_A = \frac{n_A}{n_T} \tag{4.21}$$

where n_A = number of moles of A
n_T = total number of moles of all components

The activity of A is, then, from Eq. (4.20),

$$a_A = \gamma X_A \tag{4.22}$$

where a_A = activity of species A
γ = activity coefficient of A on the mole fraction scale
X_A = mole fraction of A

We choose the standard state so that the activity, a_A, becomes equal to the mole fraction, X_A, as the mole fraction of A approaches unity. Mathematically, we write this as

$$\lim_{X_A \to 1} a_A = X_A \tag{4.23}$$

which is read "in the limit as X_A approaches 1, $a_A = X_A$." Because $\gamma = a_A/X_A$, we can express the same idea as

$$\lim_{X_A \to 1} \gamma = 1 \tag{4.24}$$

In the limit as X_A approaches 1, $\gamma = 1$. Choosing $a_A \to X_A \to 1$ defines the standard state, the state where the activity is 1. For a liquid solution, the standard state for the solvent is defined as the pure liquid. The best example is dilute aqueous solutions, for which $a_{H_2O} = 1$ is used in introductory chemistry courses. That is, we ignore the water in writing equilibrium constants in aqueous solutions. The logic is that we are using an activity on the mole fraction scale for the H_2O and that the solution is dilute, so $X_{H_2O} \approx 1$ and $a_{H_2O} \approx 1$. The mole fraction scale for activities is traditionally used for the *solvent* in a solution.

We now summarize the useful equations for a solvent in a solution:

Very dilute solution: $a_{\text{solvent}} = 1$ (4.25)

Dilute solution or ideal solution: $a_{\text{solvent}} = X_{\text{solvent}}$ (4.26)

Real solution: $a_{\text{solvent}} = \gamma X_{\text{solvent}}$ (4.27)

Equation (4.26) defines an ideal solution: for any concentration the solvent has the properties of the solvent in a dilute solution. To determine the activity coefficient and therefore the activity in Eq. (4.27), we have to measure the free energy of the solvent in the solution. This can be done by methods involving measurements of the vapor pressure of the solvent in a solution, the freezing-point depression, the boiling-point elevation, or the osmotic pressure. These will be discussed in Chapter 5.

Solute standard states

For solutions in which certain components never become very concentrated, such as dilute aqueous salt solutions, we define a solute standard state whose free energy can be obtained from measurements on the dilute solution. *The solute standard state for a component is defined as the extrapolated state where the concentration is equal to* 1 *molar* (M) *or* 1 *molal* (m), *but the properties are those extrapolated from very dilute solution.* The solute standard state is a hypothetical state in the sense that it corresponds to a 1 M or 1 m *ideal* solution. For real solutions the ideal behavior is not approached except for concentrations that are much less than 1 M or 1 m. The definition of the solute standard state should become clearer after reading the following paragraphs.

Molarity

Concentrations are commonly measured in *molarity*, with units of mol ℓ^{-1}. We shall use the symbol c for molarities. The activity of solute B on the molarity scale and with a solute standard state is, from Eq. (4.20),

$$a_B = \gamma c_B \tag{4.28}$$

where a_B = activity of species B
γ = activity coefficient of B on the molarity scale
c_B = concentration of B, mol ℓ^{-1}, M

One should immediately note that the activity of B, a_B, for a molecule B in a solution will be different if one uses Eq. (4.28) rather than Eq. (4.22). The free energy per mole of B in a specified solution has a definite value, but different choices of standard states produce different values of a_B. The standard state for the molarity scale is chosen so that the activity becomes equal to the concentration as the concentration approaches zero.

$$\lim_{c_B \to 0} a_B = c_B \tag{4.29}$$

and

$$\lim_{c_B \to 0} \gamma = 1 \tag{4.30}$$

The molarity scale for activities is used mainly for *solutes*, whose molarities often cannot become very large because of limited solubility. The molarity scale can be used, however, for any component in solution. For example, in an aqueous solution containing NaCl, sugar, and H_2O, the molarity scale of activities could be used for the NaCl and the sugar, whereas the mole fraction scale could be used for the H_2O. However, for very concentrated solutions of sugar, one might prefer to use the mole fraction scale for the activity of the sugar and the molarity scale for the H_2O. Any convenient activity scale can be chosen for any component, as long as it is specified.

For the molarity scale we have:

Dilute solution or ideal solution: $$a_B = c_B \tag{4.31}$$

Real solution: $$a_B = \gamma c_B \tag{4.32}$$

To determine the activity coefficient and therefore the activity in Eq. (4.32), we have to measure the free energy of component B in the solution. This can be done indirectly through its effect on the vapor pressure of the solvent in solution. Depending on the properties of B, various methods exist for measuring the free energy directly. If component B is volatile, its vapor pressure can be measured. Electrolytic cells provide easy methods for measuring certain activities.

The free energy in the standard state must also be measured to obtain a_B, so we must understand the definition of the standard state for the molarity scale. This is most easily seen in Fig. 4.2, in which a_B is plotted against c_B. This shows that the standard state for the molarity scale (solute standard state) is an extrapolated point. The dual requirement that $a_B = c_B$ in the limit as c_B approaches zero and that $a_B = 1$ in the standard state defines the extrapolated standard state. To obtain the free energy in the standard state, we measure $\bar{G}_B$ as a function of $\ln c_B$ in dilute solution and extrapolate linearly to $\ln c_B = 0$

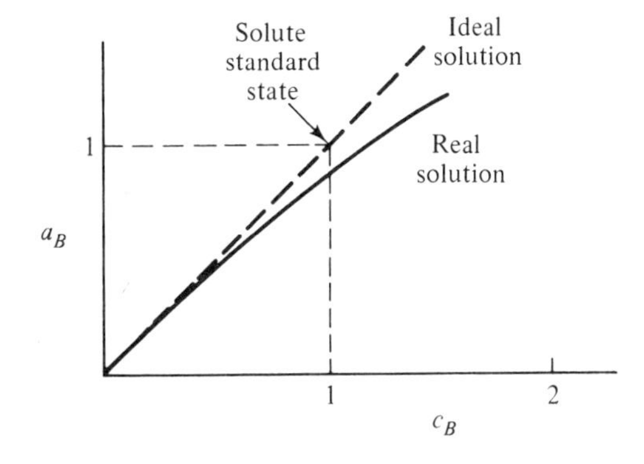

Fig. 4.2 Activity of a solute as a function of molarity for a real solution (solid curve) compared with that of an ideal solution (dashed line) extrapolated from very dilute conditions. The solute standard state lies on this extrapolated line, as shown.

($c_B = 1$); we thus obtain $\bar{G}^0_B$, as shown in Fig. 4.3. The solute standard state is thus the state that has the properties of a very dilute solution extrapolated to a concentration of 1 *M*. It is a hypothetical state rather than an actual solution that can be prepared.

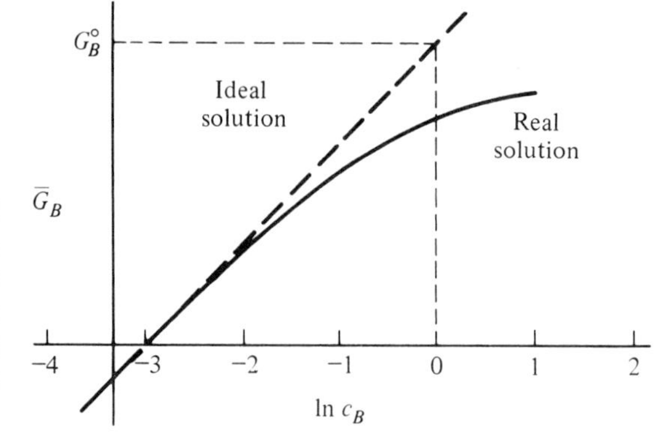

Fig. 4.3 Free energy per mole of a solute plotted against logarithm of molarity for a real solution (solid curve), compared with that of an ideal solution (dashed line) extrapolated from very dilute conditions.

Molality

Another concentration unit that is frequently used is the *molality*, *m*, with units of moles of solute per kilogram of solvent. The activity of B on the molality scale is

$$a_B = \gamma m_B \tag{4.33}$$

The discussion about molarity applies identically to molality. We have

$$\text{Dilute solution or ideal solution:} \quad a_B = m_B \tag{4.34}$$

$$\text{Real solution:} \quad a_B = \gamma m_B \tag{4.35}$$

The standard state is an extrapolated state; $\bar{G}^0$ is obtained by linearly extrapolating $\bar{G}$ measured in dilute solution to $\ln m_B = 0$ ($m_B = 1$). Molality is used instead of molarity for the most accurate thermodynamic measurements. Because molalities are defined by weight, not volume, they can be measured quite accurately, and they do not depend on temperature.

Biochemist's Standard State

We have assumed in our discussion of activities and concentrations that we knew the concentration of each species involved in a system. For a molecule that dissociates in solution, this may be very difficult to determine. For example, a reaction may involve $H_2PO_4^-$; however, the species in solution may include H_3PO_4, $H_2PO_4^-$, HPO_4^{2-}, and PO_4^{3-}. The distribution actually present will depend markedly on pH, so the concentration of $H_2PO_4^-$ will be difficult to specify. To simplify this situation, biochemists have chosen pH 7.0, which is near physiological pH, as their standard condition for a_{H+} and set the activity of a molecule equal to the *total concentration* of all species of that molecule at pH 7.0.

$$\text{Dilute solution:} \quad a_A = \sum_i^{\text{species}} c_i^A \qquad (\text{at pH } 7.0) \tag{4.36}$$

In our example, this sum over all species is the total concentration of phosphate added, that is, the concentration determined analytically. Knowledge of ionization constants is thus not needed, nor is it necessary to specify the concentration of the actual species involved in the reaction. When this biochemical standard state is used, the standard free energy is designated $\Delta G^{0\prime}$. $\Delta G^{0\prime}$ is the free energy change for a reaction at pH 7 when each product and reactant (except H^+ ion) has a total concentration of 1 M but the solution is ideal. The equilibrium is actually measured in dilute solution and the free energy is obtained by extrapolation to 1 M concentrations. The important difference between the biochemist's standard state and all the others discussed is that the equilibrium constant applies only at pH 7. For a reaction involving, for example, the hydrolysis of adenosine triphosphate, ATP, pH can be considered a variable analogous to temperature. The biochemist's standard state is then a very practical and useful choice. The standard free energy and the equilibrium constant can be used directly from the tables at, or near, pH 7. For other pH's, one needs either to repeat the experiments to determined the new equilibrium concentrations, or to use known pK's to calculate how the concentrations of reactive species depend on pH.

STANDARD FREE ENERGY AND THE EQUILIBRIUM CONSTANT

The easiest way to measure free energies of reactions, particularly complicated biochemical reactions, is to measure the equilibrium constant for the reaction. Once standard free energies are tabulated, it is easy to calculate equilibrium constants and it is straightforward to calculate equilibrium concentrations of reactants and products. We will repeat the appropriate equations (because they are so important) and give examples of how the various standard states are actually used.

$$\Delta G^0 = -RT \ln K \tag{4.7}$$

Experimental determination of the equilibrium constant allows us to calculate the *standard* free-energy change for the reaction, with all products and reactants in their standard states. The free-energy change, ΔG, for the reaction at equilibrium is of course zero. The free-energy change at arbitrary concentrations specified by Q is

$$\Delta G = \Delta G^0 + RT \ln Q \tag{4.4}$$

We shall consider several reactions and see how the equilibrium constant, the standard free energy, and the actual free energy depend on the choice of standard states. We begin with the dissociation of acetic acid, HOAc, into hydrogen ions, H^+, and acetate ions, OAc^-, in aqueous solutions:

$$HOAc(aq) \rightleftharpoons H^+(aq) + OAc^-(aq)$$

The equilibrium constant is

$$K = \frac{[a_{H^+}][a_{OAc^-}]}{[a_{HOAc}]} \tag{4.37}$$

If we choose the molarity scale for all species, $a = \gamma c$ and

$$K = \frac{(c_{H^+})(c_{OAc^-})}{(c_{HOAc})} \frac{\gamma_+\gamma_-}{\gamma_{HOAc}} = K_c \frac{\gamma_+\gamma_-}{\gamma_{HOAc}} \tag{4.38}$$

The activity coefficients γ_{HOAc}, γ_+, and γ_- depend on concentration and approach 1 as the solution becomes very dilute. Therefore, from the thermodynamic equilibrium constant we can calculate concentration ratios directly for dilute solutions. For more concentrated solutions we need to know activity coefficients if we want to calculate concentration ratios. In practice, it is not possible to determine γ_+ and γ_- individually. Consequently, a mean ionic activity coefficient is used instead. This is the geometric mean value, and for a 1–1 electrolyte such as HOAc, it is defined as

$$\gamma_\pm = (\gamma_+\gamma_-)^{1/2}$$

The concentrations that require activity coefficients depend on the accuracy of the desired result and on the nature of the species in solution. Ions are very unideal; for example, in 0.1 M NaCl, the mean ionic activity coefficient, $\gamma_\pm$, is 0.78, and in 0.01 M NaCl, it is 0.90. Therefore, 0.01 M solutions might be the upper limit for "dilute" ionic solutions.

From the thermodynamic equilibrium constant we can calculate the standard free energy, ΔG^0 [Eq. (4.7)]. This is the free energy change for 1 mol of HOAc at 1 M concentration in water (but with the properties of a very dilute solution) to dissociate to H^+ and OAc^- (each at 1 M concentration but with dilute solution properties):

$$HOAc(1\,M, aq) \xrightarrow{\Delta G^0} H^+(1\,M, aq) + OAc^-(1\,M, aq)$$

$$\Delta G^0 = \bar{G}^0(H^+, a = 1, aq) + \bar{G}^0(OAc^-, a = 1, aq) - \bar{G}^0(HOAc, a = 1, aq)$$

For the free-energy change at any arbitrary concentration, we use Eq. (4.4). For example, for the dissociation of 1 mol of HOAc at 10^{-4} M concentration in water, the free-energy change is

$$HOAc(10^{-4} M, aq) \xrightarrow{\Delta G} H^+(10^{-4}\,M, aq) + OAc^-(10^{-4}\,M, aq)$$

$$\Delta G = \Delta G^0 + RT \ln Q = \Delta G^0 + RT \ln \frac{(10^{-4})(10^{-4})}{10^{-4}}$$

At 25°C this is

$$\Delta G = \Delta G^0 + (2.303)(1.99)(298) \log (10^{-4})$$

$$= \Delta G^0 - 4(1365) \text{ cal}$$

From Le Châtelier's principle, we know that the lower the concentration, the better acetic acid dissociates. The equation tells us that every factor of 10 dilution decreases the free energy by 1365 cal at 25°C.

A hydrolysis reaction illustrates the use of other standard states. Consider the hydrolysis of ethyl acetate:

$$H_2O + CH_3COOCH_2CH_3 \rightleftharpoons CH_3COOH + CH_3CH_2OH$$

$$H_2O + EtOAc \rightleftharpoons HOAc + EtOH$$

$$K = \frac{[a_{HOAc}][a_{EtOH}]}{[a_{H_2O}][a_{EtOAc}]}$$

We could choose the molarity standard state for all molecules, but the trouble with this choice is that in dilute solution, the concentration of water would be near 55.5 M (1000/18) and the activity coefficient of the water would not be 1. A better choice would be the mole fraction for the water:

$$K = \frac{(c_{HOAc})(c_{EtOH})}{(X_{H_2O})(c_{EtOAc})} \cdot \frac{\gamma_{HOAc}\gamma_{EtOH}}{\gamma_{H_2O}\gamma_{EtOAc}}$$

In dilute solution the mole fraction of H_2O approaches 1 and all the activity coefficients approach 1; hence, only concentrations appear in K. It should be clear that the numerical value of K will be very different, depending on our choice of standard state. The corresponding standard free energies will also be different; they refer to different reaction conditions. The free energy change, ΔG, for any particular concentration of reactants and products is a definite measurable quantity, however.

Because one of the products, acetic acid, is ionizable, we might wish to choose the biochemist's standard state for HOAc. When we study the hydrolysis of ethyl acetate, we find that the equilibrium ratio of concentrations,

$$\frac{(c_{HOAc})(c_{EtOH})}{(c_{EtOAc})}$$

is very dependent on the concentration of HOAc. The reason for this is the ionization of HOAc to H^+ and OAc^-. However, if we are mainly interested in the reaction near physiological conditions, we can simplify our problem by choosing the biochemist's standard state. We study the hydrolysis in a pH 7 buffer. We now find that the equilibrium ratio of concentrations is nearly constant in dilute solution. The pH 7 buffer ensures that the acetic acid is essentially all in the form of acetate, and the HOAc concentration will be directly proportional to the acetate over a wide range of total concentrations.

The decision among the possible standard states and equilibrium constants is simply to choose the most convenient for the system being studied.

Most of the time we will assume that the solutions are dilute enough so that we can set all activity coefficients equal to unity. However, we should realize that in concentrated solutions our calculations may be in error by factors of 2 or more, because of activity coefficients. Later we shall discuss methods of measuring activity coefficients. Here it suffices to repeat that the activity coefficient of a molecule depends on the concentrations of *all* species in solution. If one is studying the ionization constant of dilute acetic acid in water, activity coefficients of the ions are nearly 1; however, the addition of NaCl to 1 *M* concentration will have a strong affect on the activity coefficients of the dilute acetic acid. This is true even though NaCl does not play a direct role in the dissociation equilibrium.

Calculation of Equilibrium Concentrations: Ideal Solutions

One reason for studying equilibrium constants is to enable us to calculate concentrations at equilibrium. We want to know how to make buffer solutions of any pH, how much product is obtained from a reaction, how much metal ion is bound in a complex, and so on. Some of these problems are very simple, but many of them require a computer for complete analysis. We shall discuss here the general method, which in principle allows a solution for any equilibrium problem. We assume that we know the equilibrium constants for all the equilibria involved and we assume that all solutions are ideal; that is, we ignore all activity coefficients.

The way most scientists solve equilibrium problems is to make intuitive approximations to simplify the arithmetic operations. The "intuition" comes either from having solved many similar problems or from a good memory of what we learned in introductory chemistry. The method we present here is a general one; it applies when intuition fails on simple problems or when the problems become more difficult.

The usual requirement is to learn the concentrations of all species present in a mixture. The problem is essentially an algebraic one of finding simultaneous solutions for a number of equations. So the first requirement is to have as many equations as unknowns (species). The equations are equilibrium expressions, plus two types of conservation equations. The conservation equations are the ones we tend to use intuitively, but they are always included. They are *conservation of mass*—the total mass of each element is not altered by any chemical reaction—and *conservation of charge*—the number of positive charges must always equal the number of negative charges in the mixture. Once the number of equations equals the number of unknowns, we look for approximations that allow a simple solution. A solution, not necessarily simple, can always be obtained, however.

Let us consider a simple problem and see how the method works. The first equilibrium problem one usually encounters is the ionization of a weak acid; then one proceeds to buffers and to hydrolysis. Each problem is often treated as a separate case and one memorizes the appropriate simple method for each. Here let us do the general problem and explicitly see what approximations are made to simplify the arithmetic. Suppose that a solution is prepared by adding c_A moles of acetic acid and c_S moles of sodium acetate to form 1 ℓ of aqueous solution. The solution will contain HOAc, OAc^-, Na^+, H^+, and OH^-. Even for this simple equilibrium we have five species in addition to the solvent water. We therefore need five equations. They are:

Mass balance:

$$[Na^+] = c_S = \text{constant}$$

$$[HOAc] + [OAc^-] = c_A + c_S = \text{constant}$$

Charge balance:

$$[Na^+] + [H^+] = [OAc^-] + [OH^-]$$

Equilibria:

$$K_{HOAc} = \frac{[H^+][OAc^-]}{[HOAc]} = 1.8 \times 10^{-5}$$

$$K_{H_2O} = [H^+][OH^-] = 1.0 \times 10^{-14}$$

We are using the solute (1 M) standard state for H^+, OAc^-, HOAc, OH^-, and Na^+, and the mole fraction standard state for H_2O. These five equations allow us to solve any problem involving only these five species.

Example 4.3 What are the concentrations of all species in pure water?

Solution

$$\text{Mass balance:} \quad [Na^+] = 0, [HOAc] = 0, [OAc^-] = 0$$

$$\text{Charge balance:} \quad [H^+] = [OH^-]$$

$$\text{Equilibrium:} \quad [H^+][OH^-] = 1.0 \times 10^{-14}$$

Therefore,

$$[H^+]^2 = 1.0 \times 10^{-14}$$

and

$$[H^+] = 1.0 \times 10^{-7}$$

$$[OH^-] = 1.0 \times 10^{-7}$$

Example 4.4 What are the concentrations of all species in a 0.1 M HOAc solution?

Solution

$$\text{Mass balance:} \quad [Na^+] = 0$$

$$[HOAc] + [OAc^-] = 0.1$$

$$\text{Charge balance:} \quad [H^+] = [OAc^-] + [OH^-]$$

$$\text{Equilibria:} \quad K_{HOAc} = \frac{[H^+][OAc^-]}{[HOAc]} = 1.8 \times 10^{-5}$$

$$K_{H_2O} = [H^+][OH^-] = 1.0 \times 10^{-14}$$

Except for pure water or solutions close to pH 7, either $[H^+]$ or $[OH^-]$ will be negligible in the charge-balance equation. The solutions will be acidic or basic, so either $[OH^-]$ or $[H^+]$ can be ignored. Even for a solution at pH 6.5, the $[H^+]$ concentration is 10 times the $[OH^-]$. We know that an acetic acid solution is acidic, so we ignore $[OH^-]$ in the charge-balance equation. Assuming that $[OAc^-] \gg [OH^-]$,

$$\text{Charge balance:} \quad [H^+] \cong [OAc^-]$$

Remember that we can sometimes ignore species in sums, but we can never ignore them in products. We can never ignore $[H^+]$ in K_{H_2O}.

If we set $[H^+] = x$, we also have $[OAc^-] = x$ and $[HOAc] = 0.1 - x$:

$$K_A = \frac{(x)(x)}{0.1 - x} = 1.8 \times 10^{-5}$$

We can solve for x in the quadratic equation

$$x^2 + 1.8 \times 10^{-5}x - 1.8 \times 10^{-6} = 0$$

$$x = \frac{-1.8 \times 10^{-5} + \sqrt{3.24 \times 10^{-10} + 7.2 \times 10^{-6}}}{2}$$

$$= \frac{-1.8 \times 10^{-5} + 2.68 \times 10^{-3}}{2}$$

$$= 1.33 \times 10^{-3}$$

$$[H^+] = [OAc^-] = 1.33 \times 10^{-3}$$

$$[HOAc] = 9.87 \times 10^{-2}$$

$$[OH^-] = 7.52 \times 10^{-12}$$

Note that our assumption that $[OAc^-] \gg [OH^-]$ is well verified.

Alternatively, we can solve the problem using successive approximations. The secret in solving by approximations comes in setting x equal to concentrations of species that are *not* the largest involved in the equilibrium. In this case, we have set $x = [H^+]$ or $[OAc^-]$. We do not set $x = [HOAc]$, because it is the species that we expect would have the largest concentration in 0.1 M acetic acid. Even before knowing the exact answer, we can guess that $[H^+]$ and $[OAc^-]$ will have to be much smaller than 0.1 M to satisfy the equilibrium constant $K_{HOAc} = 1.8 \times 10^{-5}$.

As a first approximation when $x \ll 0.1$,

$$K_A = \frac{x^2}{0.1 - x} \cong \frac{x^2}{0.1} = 1.8 \times 10^{-5}$$

$$x \cong 1.34 \times 10^{-3}$$

Indeed, we see that our initial approximation that $x \ll 0.1$ is justified. To obtain a more precise answer, we can use this first solution to generate a second approximation:

$$\frac{x^2}{0.1 - 0.0013} = 1.8 \times 10^{-5}$$

$$x = 1.33 \times 10^{-3}$$

Clearly, no further approximations are needed.

Example 4.5 What are the concentrations of all species in a 0.20 M NaOAc solution?

Solution This is a hydrolysis problem, because NaOAc is the salt of a weak acid and we might be tempted to write the hydrolysis reaction and derive a value for the hydrolysis equilibrium constant, $K_h = K_{H_2O}/K_{HOAc}$. However, this is not necessary or particularly useful. Instead, our general method gives

$$\text{Mass balance:} \quad [Na^+] = 0.20$$

$$[HOAc] + [OAc^-] = 0.20$$

$$\text{Charge balance:} \quad [H^+] + [Na^+] = [OH^-] + [OAc^-]$$

$$\text{Equilibria:} \quad K_{HOAc} = \frac{[H^+][OAc^-]}{[HOAc]} = 1.8 \times 10^{-5}$$

$$K_{H_2O} = [H^+][OH^-] = 1.0 \times 10^{-14}$$

When the salt of a weak acid hydrolyzes, the reaction

$$OAc^- + H_2O \rightleftharpoons HOAc + OH^-$$

occurs, and the resulting solution is basic; therefore, we can ignore $[H^+]$ in the charge balance. We already know that $[Na^+] = 0.20$; therefore, assuming that $[Na^+] \gg [H^+]$,

$$\text{Charge balance:} \quad 0.20 = [OH^-] + [OAc^-]$$

Comparing this with the mass-balance equation for $[HOAc] + [OAc^-]$, we see that

$$[HOAc] = [OH^-]$$

Now we set $x = [HOAc] = [OH^-]$ to facilitate making approximations. (The choice for x is between $[OAc^-]$ on the one hand and $[HOAc] = [OH^-]$ on the other. The equilibrium expression for K_{HOAc} tells us that $[OAc^-]/[HOAc] \gg 1$ because $[H^+] \ll 10^{-7}$ for a basic solution.) Thus

$$[OAc^-] = 0.20 - [HOAc] = 0.20 - x$$

$$[H^+] = \frac{10^{-14}}{x}$$

and

$$K_{\mathrm{HOAc}} = \frac{(10^{-14}/x)(0.20 - x)}{x} = \frac{10^{-14}(0.20 - x)}{x^2} = 1.8 \times 10^{-5}$$

$$\frac{0.20 - x}{x^2} = 1.8 \times 10^9$$

We can solve the quadratic, but it is easier to use successive approximations.

As our first approximation, when $x \ll 0.20$,

$$\frac{0.20}{x^2} = 1.8 \times 10^9$$

$$x = 1.05 \times 10^{-5} \qquad \text{(consistent with assumption)}$$

No further approximations are needed.

$$[\mathrm{OH^-}] = [\mathrm{HOAc}] = 1.05 \times 10^{-5}$$

$$[\mathrm{H^+}] = \frac{1.0 \times 10^{-14}}{1.05 \times 10^{-5}} = 9.5 \times 10^{-10}$$

$$[\mathrm{OAc^-}] = 0.20$$

$$[\mathrm{Na^+}] = 0.20$$

Note that $[\mathrm{Na^+}] \gg [\mathrm{H^+}]$, which justifies our early approximation in the charge-balance equation.

This general method works for any equilibrium. For more complicated examples we must be sure that the equilibrium expressions are independent. In the example above, instead of the $K_A = K_{\mathrm{HOAc}}$ and $K_{\mathrm{H_2O}}$, we could have used K_B:

$$K_B = \frac{[\mathrm{HOAc}][\mathrm{OH^-}]}{[\mathrm{OAc^-}]}$$

and $K_{\mathrm{H_2O}}$, or K_B and K_A. However, we cannot count K_A, K_B, and $K_{\mathrm{H_2O}}$ as three independent equations; only two are independent, because $K_A K_B$ always equals $K_{\mathrm{H_2O}}$.

For buffer problems the computations are usually simpler, because both the acid and the salt (or the base and its corresponding salt) are present at an appreciable concentration. Furthermore, these concentrations are large compared with either $[\mathrm{H^+}]$ or $[\mathrm{OH^-}]$. In the case of a buffer involving acetic acid and sodium acetate, for example, we have:

$$\text{Mass balance:} \quad [\mathrm{Na^+}] = c_S = \text{constant}$$

$$[\mathrm{HOAc}] + [\mathrm{OAc^-}] = c_A + c_S = \text{constant}$$

$$\text{Charge balance:} \quad [\mathrm{Na^+}] + [\mathrm{H^+}] = [\mathrm{OAc^-}] + [\mathrm{OH^-}]$$

which becomes

$$[Na^+] = [OAc^-] = c_S$$

because $[H^+]$ and $[OH^-] \ll [Na^+]$ or $[OAc^-]$. Thus

$$[HOAc] = c_A$$

and the equilibrium constant is

$$K_{HOAc} = \frac{[H^+][OAc^-]}{[HOAc]} = \frac{[H^+]c_S}{c_A}$$

This equation is often written in the form

$$pH = pK_A + \log\frac{c_S}{c_A} \tag{4.39}$$

where $pH = -\log[H^+]$

$pK_A = -\log K_A$

and is sometimes called the *Henderson-Hasselbalch equation.*

Example 4.6 What amount of solid sodium acetate is needed to prepare a buffer at pH 5.0 from 1 ℓ of 0.10 M acetic acid?

Solution

$$pH = pK_{HOAc} + \log\frac{c_S}{c_A}$$

$$5.0 = 4.75 + \log\frac{c_S}{0.10} = 5.75 + \log c_S$$

$$\log c_S = -0.75$$

$$c_S = 0.18 \text{ mol } \ell^{-1}$$

$$\text{wt of NaOAc} = (0.18 \text{ mol})(82.0 \text{ g mol}^{-1}) = 14.8 \text{ g}$$

Of course, this answer assumes that the solutions are ideal and that the addition of sodium acetate has a negligible effect on the volume. In practice, one always makes the final adjustment of pH using a calibrated pH meter or other electrode system that measures a_{H^+} directly.

Temperature Dependence of the Equilibrium Constant

Once we know an equilibrium constant at one temperature, what can we say about the equilibrium constant at another temperature? Le Châtelier can help us if we think of heat as a product of the reaction. If heat is given off (an exothermic reaction, a negative ΔH), the products have a lower enthalpy than the reactants, and raising the temperature will favor reactants. The equilibrium constant will become smaller. If heat is absorbed (an endothermic reaction, a positive ΔH), raising the temperature will favor products. The equilibrium constant will become larger.

The quantitative relation is easily obtained if ΔH is independent of temperature. We use

$$\Delta G^0 = \Delta H^0 - T\,\Delta S^0 = -RT \ln K \tag{4.40}$$

at two temperatures T_1 and T_2:

$$\ln K_2 = -\frac{\Delta H^0}{RT_2} + \frac{\Delta S^0}{R} \tag{4.41}$$

$$\ln K_1 = -\frac{\Delta H^0}{RT_1} + \frac{\Delta S^0}{R} \tag{4.42}$$

We have assumed that ΔH^0 and ΔS^0 are independent of T. Subtracting Eq. (4.42) from (4.41), we obtain the desired equation:

$$\ln \frac{K_2}{K_1} = -\frac{\Delta H^0}{R}\left(\frac{1}{T_2} - \frac{1}{T_1}\right) \tag{4.43}$$

This equation can be used to calculate an equilibrium constant, K_2, at T_2 when K_1 and the standard enthalpy of the reaction, ΔH^0, are known. Alternatively, from equilibrium constants measured at two different temperatures, one can calculate the enthalpy. The usual way of obtaining ΔH^0 from K as a function of T is to plot $\log K$ versus $1/T$. The slope is $-\Delta H^0/2.303R$:

$$\log K(T) = -\frac{\Delta H^0}{2.303R}\frac{1}{T} + \text{constant} \tag{4.44}$$

We now know how to get the thermodynamic variables ΔG^0, ΔH^0, and ΔS^0 for a reaction simply by measuring equilibrium concentrations (in dilute solutions where activity coefficients are equal to 1). A more rigorous derivation shows that Eq. (4.44) is correct even if ΔH^0 does change with temperature. For such a system the plot of $\log K$ versus $1/T$ will show some curvature, however.

Example 4.7 The equilibrium constant for ionization of 4-aminopyridine is 1.35×10^{-10} at 0°C and 3.33×10^{-9} at 50°C. Calculate ΔG^0 at 0°C and 50°C as well as ΔH^0 and ΔS^0.

Solution We assume that ΔH^0 is independent of temperature and use Eq. (4.43) with $K_2 = 1.35 \times 10^{-10}$, $K_1 = 3.33 \times 10^{-9}$, $T_2 = 273$, and $T_1 = 323$:

$$\ln \frac{K_2}{K_1} = -\frac{\Delta H^0}{R}\left(\frac{1}{T_2} - \frac{1}{T_1}\right) \quad (4.43)$$

$$\frac{1}{T_2} - \frac{1}{T_1} = \frac{1}{273} - \frac{1}{323} = (3.663 \times 10^{-3} - 3.096 \times 10^{-3})$$

$$= 5.67 \times 10^{-4} \text{ deg}^{-1}$$

$$\ln \frac{K_2}{K_1} = \ln \frac{1.35}{33.3} = -3.205$$

$$\Delta H^0 = -\frac{(1.99 \text{ cal deg}^{-1} \text{ mol}^{-1})(-3.205)}{5.670 \times 10^{-4} \text{ deg}^{-1}}$$

$$= 11{,}250 \text{ cal mol}^{-1} = 11.25 \text{ kcal mol}^{-1}$$

This value of ΔH^0 is an average value over the temperature range from 0°C to 50°C.

From Eq. (4.40), we obtain ΔG^0 at each temperature:

$$\Delta G^0 = -RT \ln K \quad (4.40)$$

$$\Delta G^0(0°\text{C}) = 12.35 \text{ kcal mol}^{-1}$$

$$\Delta G^0(50°\text{C}) = 12.55 \text{ kcal mol}^{-1}$$

The value of ΔS^0, assumed to be independent of temperature between 0°C and 50°C, is

$$\Delta S^0 = \frac{\Delta H^0 - \Delta G^0}{T}$$

$$= \frac{11{,}250 - 12{,}350}{273}$$

$$= -4.03 \text{ cal deg}^{-1} \text{ mol}^{-1}$$

or, as a check,

$$\Delta S^0 = \frac{11{,}250 - 12{,}550}{323}$$

$$= -4.02 \text{ cal deg}^{-1} \text{ mol}^{-1}$$

For a small temperature range (less than 25°) Eq. (4.43) should be adequate. However, in general we must consider the temperature dependence of ΔH.

Knowledge of equilibrium constants, free energies, and their dependence on concentration and temperature is very important in biochemistry and biology. Here we shall discuss a few simple examples. Table 4.1 lists equilibrium constants and heats of ionization for various acids. It summarizes a great deal of information about the acids in a small space. For example, some biochemical systems (such as human beings) work best at 37°C (98.5°F). Table 4.1 allows you to calculate the pK's at 37°C. We see that acetic acid is the only acid listed whose pK is the same at 37°C as at 25°C, because its $\Delta H = 0$. We usually consider that a neutral aqueous solution has a pH of 7, but this is only true at 25°C.

Example 4.8 What is the pH of pure water at 37°C?

Solution The necessary equation is

$$\ln \frac{K_2}{K_1} = -\frac{\Delta H^0}{R}\left(\frac{1}{T_2} - \frac{1}{T_1}\right) \tag{4.43}$$

The ionization constant of water is 10^{-14} at 25°C, its $\Delta H^0 = 13.5$ kcal (see Table 4.1), $K_1 = 10^{-14}$, $\Delta H^0 = 13.5$ kcal, $T_1 = 298$, and $T_2 = 310$. Then

$$\ln \frac{K_2}{10^{-14}} = \frac{-13{,}500}{1.99}\left(\frac{1}{310} - \frac{1}{298}\right)$$

$$= +0.88$$

$$\frac{K_2}{10^{-14}} = e^{+0.88} = 2.4$$

$$K_2 = 2.4 \times 10^{-14}$$

This is the ionization constant of water at 37°C. The H^+ and OH^- in pure water is surely low enough so that the activities of the ions equal concentrations: $[H^+] = [OH^-]$. At 37°C,

$$[H^+][OH^-] = 2.4 \times 10^{-14}$$

$$[H^+] = 1.55 \times 10^{-7}$$

$$\text{pH} = -\log [H^+]$$

$$= 6.81$$

From the pK's given in Table 4.1 or those that we calculate at other temperatures, we can calculate the activities and concentrations of various ionized species present in solution. For example, consider histidine in solution at 25°C.

Table 4.1 Ionization constants and heats of ionization at 25°C

Compound	Ionizing species	pK*	ΔH^0 (kcal mol^{-1})
Acetic acid	CH_3COOH	4.76	0
Adenosine	$^+HN_1$ of adenine	3.55	3.8
Adenosine triphosphate (ATP)	pK_1, pK_2, pK_3	<2	—
	$^+HN_1$ of adenine (pK_4)	4.0	3.7
	$[RO(PO_3)_3H]^{3-}$ (pK_5)	7.0	−1.2
5′-Adenylic acid (adenosine monophosphate)	$^+HN_1$ of adenine (pK_1)	3.7	4.2
	$ROPO_3H_2$ (pK_2)	6.4	−1.8
	$[ROPO_3H]^-$ (pK_3)	13.06	10.9
Alanine	$H_3^+NRCOOH$ (pK_1)	2.35	0.7
	$H_3^+NRCO_2^-$ (pK_2)	9.83	10.8
Ammonia	NH_4^+	9.24	12.5
Aspartic acid	$HO_2CR(NH_3)^+COOH$ (pK_1)	2.05	1.8
	$HO_2CR(NH_3)^+CO_2^-$ (pK_2)	3.87	1.0
	$^-O_2CR(NH_3)^+CO_2^-$ (pK_3)	9.0	10.0
Carbonic acid	H_2CO_3 (pK_1)	6.36	2.1
	HCO_3^- (pK_2)	10.24	3.5
Fumaric acid	$R(COOH)_2$ (pK_1)	3.10	0.1
	$^-O_2CRCOOH$ (pK_2)	4.60	−0.7
Histidine	$H_2^+NR(NH_3)^+COOH$ (pK_1)	1.82	—
	$H_2^+NR(NH_3)^+CO_2^-$ (pK_2)	6.00	7.1
	$HNR(NH_3)^+CO_2^-$ (pK_3)	9.16	10.4
Hydrocyanic acid	HCN	9.21	10.4
Phenol	ϕOH	9.98	5.6
Phosphoric acid	H_3PO_4 (pK_1)	2.12	−1.9
	$H_2PO_4^-$ (pK_2)	7.20	0.9
	HPO_4^{2-} (pK_3)	12.40	4.2
Pyruvic acid	RCOOH	2.49	2.9
Tyrosine	$HOR(NH_3)^+COOH$ (pK_1)	2.20	—
	$HOR(NH_3)^+COO^-$ (pK_2)	9.11	—
	$HOR(NH_2)COO^-$ (pK_3)	10.05	6.0
Water	H_2O	14.00	13.5

* p$K \equiv -\log K$. The K values are thermodynamic equilibrium constants on the molarity scale. $\Delta G^0 = -RT \ln K = 2.303RT$ pK.

SOURCE: *Handbook of Biochemistry and Molecular Biology*, 2nd ed., Chemical Rubber Co., Cleveland, Ohio, 1970.

The equilibria are

$$\underset{\text{HisH}_3^{2+}}{\text{(HN—NH}^+\text{ ring)—CH}_2\text{—C(NH}_3^+\text{)(H)—COOH}} \xrightleftharpoons{pK_1 = 1.82} \underset{\text{HisH}_2^+}{\text{(HN—NH}^+\text{ ring)—CH}_2\text{—C(NH}_3^+\text{)(H)—COO}^-} + H^+$$

$$\underset{\text{HisH}_2^+}{\text{(HN—NH}^+\text{ ring)—CH}_2\text{—C(NH}_3^+\text{)(H)—COO}^-} \xrightleftharpoons{pK_2 = 6.00} \underset{\text{HisH}}{\text{(HN—N ring)—CH}_2\text{—C(NH}_3^+\text{)(H)—COO}^-} + H^+$$

$$\underset{\text{HisH}}{\text{(HN—N ring)—CH}_2\text{—C(NH}_3^+\text{)(H)—COO}^-} \xrightleftharpoons{pK_3 = 9.16} \underset{\text{His}^-}{\text{(HN—N ring)—CH}_2\text{—C(NH}_2\text{)(H)—COO}^-} + H^+$$

We often need to know the concentration of each species at a given pH. Each ionic species has different chemical reactivity and physical properties; therefore, the properties of the solution will depend on the concentration of each species.

Example 4.9 Calculate the concentration of each species in a $0.10M$ solution of histidine at pH 7.

Solution The three equilibrium expressions are

$$K_1 = 1.51 \times 10^{-2} = \frac{[H^+][HisH_2^+]}{[HisH_3^{2+}]}$$

$$K_2 = 1.00 \times 10^{-6} = \frac{[H^+][HisH]}{[HisH_2^+]}$$

$$K_3 = 6.92 \times 10^{-10} = \frac{[H^+][His^-]}{[HisH]}$$

We assume that activities are equal to concentrations. If necessary, we could measure or estimate activity coefficients to obtain more accurate concentrations. The mass-balance equation is

$$[HisH_3^{2+}] + [HisH_2^+] + [HisH] + [His^-] = 0.100$$

We now have four equations and four unknowns, so we can solve for the unknowns. As often happens, it will be convenient to solve by successive approximations. First, we find the ratios of species from the three equilibrium constants and the value of $[H^+] = 10^{-7}$.

$$\frac{[HisH_2^+]}{[HisH_3^{2+}]} = 1.51 \times 10^5$$

$$\frac{[HisH]}{[HisH_2^+]} = 10.0$$

$$\frac{[His^-]}{[HisH]} = 6.92 \times 10^{-3}$$

From the ratios of species we see that a good first approximation will be to ignore $[HisH_3^{2+}]$ and $[His^-]$ in the mass-balance equation. That is, we need consider only the two species involved in the equilibrium whose pK is closest to the pH of the solution.

The first-approximation mass balance is, assuming that $[HisH_3^{2+}] + [His^-] \ll [HisH] + [HisH_2^+]$,

$$[HisH] + [HisH_2^+] = 0.100$$

but

$$[HisH] = [HisH_2^+] \cdot 10.0$$

$$[HisH_2^+] \cdot 10.0 + [HisH_2^+] = 0.100$$

$$[HisH_2^+] = \frac{0.100}{11.0} = 9.1 \times 10^{-3}$$

$$[HisH] = 10[HisH_2^+] = 9.1 \times 10^{-2}$$

$$[His^-] = (6.92 \times 10^{-3})(9.1 \times 10^{-2}) = 6.3 \times 10^{-4}$$

$$[HisH_3^{2+}] = \frac{9.1 \times 10^{-3}}{1.51 \times 10^5} = 6.0 \times 10^{-8}$$

We have ignored activity coefficients, so the concentrations are at best accurate to ±10%. If the second approximation does not change values by more than 10%, we can stop at the first approximation.

The second-approximation mass balance is

$$[HisH] + [HisH_2^+] = 0.100 - [His^-] - [HisH_3^{2+}]$$
$$= 0.0994$$

The first approximation is good enough; the concentrations are

$$[HisH_2^+] = 9.1 \times 10^{-3}$$

$$[HisH] = 9.1 \times 10^{-2}$$

$$[His^-] = 6.3 \times 10^{-4}$$

$$[HisH_3^{2+}] = 6.0 \times 10^{-8}$$

Note that the sum of the last two species is much less than the first two.

These methods can, of course, be used to calculate the concentrations of species needed to produce a buffer of a given pH and salt concentration. For common buffers the algebra has already been done, and buffer tables can be found in such handbooks as the *Handbook of Biochemistry and Molecular Biology* (Chemical Rubber Co., Cleveland, Ohio). For new buffers or special requirements of salt concentrations you will have to be able to do the calculations yourself.

Thermodynamics of Metabolism

Table 4.2 gives free energies and enthalpies for hydrolysis of various biochemical substances. The data can be used to calculate equilibrium constants and ratios of concentrations at any temperature. Tables 4.3 and 4.4 give free energies at 25°C and pH 7 for steps in the photosynthetic synthesis and metabolic degradation of glucose. The biochemical standard state is used in these three tables. This means that the pH is 7, the activity of water is equal to 1, and the activity of all other chemicals is replaced by their total concentration in mol ℓ^{-1}. For example, adenosine triphosphate, ATP, in the reaction refers to the sum of all ionized species of ATP present at pH 7. Many useful insights can be gained from studying these tables; we shall mention only two.

Table 4.2 Standard free energies and enthalpies of hydrolysis at 25°C and pH 7

Reaction	$\Delta G^{0\prime}$ (kcal mol^{-1})	ΔH^0 (kcal mol^{-1})
Ethylacetate + $H_2O \rightarrow$ acetate + ethanol	−4.7	—
Acetylcholine + $H_2O \rightarrow$ acetate + choline	−6.0	—
Sucrose + $H_2O \rightarrow$ fructose + glucose	−7.0	−1.1
Asparagine + $H_2O \rightarrow$ aspartate + ammonia	−3.6	−5.7
3-Phosphoglycerate + $H_2O \rightarrow$ glycerol + phosphate	−2.4	−8.2
Glucose-6-phosphate + $H_2O \rightarrow$ glucose + phosphate	−4.0	−8.4
Fructose-1,6-diphosphate + $H_2O \rightarrow$ fructose-6-phosphate + phosphate	−3.4	—
ATP + $H_2O \rightarrow$ ADP + phosphate	−7.4	−5.8
Acetylphosphate + $H_2O \rightarrow$ acetate + phosphate	−10.3	—
Phosphoenolpyruvate + $H_2O \rightarrow$ pyruvate + phosphate	−14.8	—

SOURCE: Free-energy data from *Handbook of Biochemistry and Molecular Biology*, 2nd ed., Chemical Rubber Co., Cleveland, Ohio, 1970. Enthalpy data from H. D. Brown, *Biochemical Microcalorimetry*, Academic Press, New York, 1969.

Table 4.3 Standard free energies of reaction at 25°C, pH 7 for steps in the photosynthetic formation of glucose by plants

	$\Delta G^{0\prime}$ (kcal mol^{-1})
CO_2 + ribulose-1,5-diphosphate + H_2O → 3-phosphoglycerate + $2H^+$	−8.4
H^+ + 3-phosphoglycerate + ATP + NADPH → ADP + glyceraldehyde-3-phosphate + $NADP^+$ + phosphate	+4.3
Glyceraldehyde-3-phosphate → dihydroxyacetone phosphate	−1.8
Glyceraldehyde-3-phosphate + dihydroxyacetone phosphate → fructose-1,6-diphosphate	−5.2
Fructose-1,6-diphosphate + H_2O → fructose-6-phosphate + phosphate	−3.4
Fructose-6-phosphate + glyceraldehyde-3-phosphate → erythrose-4-phosphate + xylulose-5-phosphate	+1.5
Erythrose-4-phosphate + dihydroxyacetone phosphate → sedulose-1,7-diphosphate	−5.6
Sedulose-1,7-diphosphate + H_2O → sedulose-7-phosphate + phosphate	−3.4
Sedulose-7-phosphate + glyceraldehyde-3-phosphate → ribose-5-phosphate + xylulose-5-phosphate	+0.1
Ribose-5-phosphate → ribulose-5-phosphate	+0.5
Xylulose-5-phosphate → ribulose-5-phosphate	+0.2
Ribulose-5-phospate + ATP → ribulose-1,5-diphosphate + ADP + H^+	−5.2
Fructose-6-phosphate → glucose-6-phosphate	−0.5
Glucose-6-phosphate + H_2O → D-glucose + phosphate	−4.0

SOURCE: J. A. Bassham and G. H. Krause, *Biochem. Biophys. Acta 189*, 207 (1969). The biochemist's standard state is used.

The first step in the metabolism of glucose (Table 4.4) is the formation of glucose-6-phosphate. From Table 4.2 we see that the direct reaction of glucose with phosphate has a positive standard free energy and will not occur significantly at standard conditions.

(1) glucose + phosphate ⟶ glucose-6-phosphate + H_2O

$$\Delta G^{0\prime} = +4.0 \text{ kcal mol}^{-1}$$

But if we couple this reaction to the hydrolysis of ATP to form adenosine

Table 4.4 Standard free energies of reaction at 25°C, pH 7 for steps in the metabolism of glucose in animals

	$\Delta G^{0\prime}$ (kcal mol^{-1})
D-glucose + ATP → D-glucose-6-phosphate + ADP	−3.4
D-glucose-6-phosphate → D-fructose-6-phosphate	+0.5
D-fructose-6-phosphate + ATP → D-fructose-1,6-diphosphate	−3.4
Fructose-1,6-diphosphate → dihydroxyacetone phosphate + glyceraldehyde-3-phosphate	+5.8
Dihydroxyacetone phosphate → glyceraldehyde-3-phosphate	+1.8
Glyceraldehyde-3-phosphate + phosphate + NAD^+ → 1,3-diphosphoglycerate + NADH + H^+	+1.5
1,3-diphosphoglycerate + ADP + H^+ → 3-phosphoglycerate + ATP	−6.8
3-Phosphoglycerate → 2-phosphoglycerate	+1.1
2-Phosphoglycerate → phosphoenolpyruvate	+0.4
2-Phosphoenolpyruvate + ADP + H^+ → pyruvate + ATP	−5.7
Pyruvate + NADH + H^+ → lactate + NAD^+	−6.0

SOURCE: J. A. Bassham and G. H. Krause, *Biochem. Biophys. Acta 189*, 207 (1969). The biochemist's standard state is used.

diphosphate, ADP, which is very favorable, the sum of the reactions has a negative standard free energy and will occur.

$$\text{(2)}\quad \text{ATP} + H_2O \longrightarrow \text{ADP} + \text{phosphate} \qquad \Delta G^{0\prime} = -7.4 \text{ kcal mol}^{-1}$$

Adding (1) and (2), we obtain

$$\text{(3)}\quad \text{ATP} + \text{glucose} \longrightarrow \text{ADP} + \text{glucose-6-phosphate} \qquad \Delta G^{0\prime} = -3.4 \text{ kcal mol}^{-1}$$

as shown in Table 4.4. In this manner the strongly spontaneous hydrolysis of ATP is coupled to the otherwise nonspontaneous glucose phosphorylation. This reaction is typical of the role played by ATP in metabolism.

Another condition that allows a reaction with a positive standard free energy to take place occurs because of the concentration dependence. Having high concentrations of reactants and low concentrations of products will favor the occurrence of a reaction.

Example 4.10 Under standard conditions one of the steps in the photosynthetic production of glucose does not occur spontaneously.

$$\text{Fructose-6-phosphate} + \text{glyceraldehyde-3-phosphate} \longrightarrow$$
$$\text{erythrose-4-phosphate} + \text{xylulose-5-phosphate}$$
$$\Delta G^{0\prime} = +1.5 \text{ kcal mol}^{-1}$$

Can the reaction take place in a chloroplast where the concentrations are as follows?

$$\text{F} = \text{fructose-6-phosphate} = 53 \times 10^{-5}\,M$$
$$\text{G} = \text{glyceraldehyde-3-phosphate} = 3.2 \times 10^{-5}\,M$$
$$\text{E} = \text{erythrose-4-phosphate} = 2 \times 10^{-5}\,M$$
$$\text{X} = \text{xylulose-5-phosphate} = 2.1 \times 10^{-5}\,M$$

Solution We use Eq. (4.4):

$$\Delta G = \Delta G^0 + RT \ln \frac{[\text{E}][\text{X}]}{[\text{F}][\text{G}]} \tag{4.4}$$

$$= 1500 + (1.99)(298)(2.303) \log \frac{(2 \times 10^{-5})(2.1 \times 10^{-5})}{(53 \times 10^{-5})(3.2 \times 10^{-5})}$$

$$= 1500 + 1365 \log (2.48 \times 10^{-2})$$

$$= -690 \text{ cal mol}^{-1} = -0.69 \text{ kcal mol}^{-1}$$

Under the actual conditions in the chloroplast, the free energy of the reaction is negative and the reaction will occur.

The reaction of fructose-1,6-diphosphate to give dihydroxyacetone phosphate + glyceraldehyde-3-phosphate has a large positive standard free energy ($\Delta G^{0\prime} = +5.8$ kcal mol^{-1}). However, because 1 mol of reactants goes to 2 mol of products, the concentration dependence of the free energy is large enough to allow the reaction to proceed in dilute solution. For equal numbers of moles of products and reactants, dilution does not affect the free energy; for an increase in the number of moles of products, dilution favors the reaction. Inspection of Eq. (4.4) will convince the reader of this conclusion, which can also be reached intuitively by using Le Châtelier's principle.

GALVANIC CELLS

We have used free energies and presented tables of standard free energies without discussing the easiest and most accurate method of measuring free energies. If a chemical reaction can be made to occur in a galvanic cell, the voltage of the cell is a direct measure of the free energy of the reaction. Alternatively, from the known free energy change of a reaction, we can calculate the maximum voltage that can be obtained when the reaction takes place in an electrochemical cell. Clearly, this method can be applied to any reaction that involves oxidation and reduction, but it is even more general than that. As we shall see, it can be applied to equilibrium processes such as the solubility of a salt, the dissociation of a weak acid or base, the formation of complex ions, or concentration differences across osmotically active membranes. It is sufficient to carry out the thermodynamic studies in the presence of electrodes that are sensitive to the chemical or ionic species involved.

For a reversible process at constant T and P, one can show that the free energy change is equal to the maximum useful work (useful work is defined here as work other than pressure-volume work):

$$-\Delta G = -w_{\text{reversible}} - P\,\Delta V = -w_{\text{max, useful}} \tag{4.45}$$

In a galvanic cell the maximum useful work is the reversible electric work:

$$-\Delta G = -w_{\text{reversible electrical}} \tag{4.46}$$

We use minus signs in the equation to emphasize that the *decrease* in free energy is equal to the reversible electrical work done *by* the system. Electrical work is equal to the voltage times the current multiplied by the time [Eq. (2.8)]. Because current multiplied by time is charge, the electrical work done by a chemical reaction in a galvanic cell is the voltage of the cell multiplied by the amount of charge transferred in the reaction. For n moles of electrons transferred, the free energy is

$$\Delta G(\text{electron volts}) = -n\mathscr{E} \tag{4.47}$$

where n = number of moles of electrons involved in the reaction
$\mathscr{E}$ = reversible voltage of cell (the sign convention for a cell will be discussed shortly)

An electron volt is a unit of energy. One mol of electrons moving 1 cm in a field of 1 V cm^{-1} acquires an energy of 23.06 kcal mol^{-1}.

$$\Delta G(\text{kcal}) = -23.06n\mathscr{E} \tag{4.48}$$

The free energy is also often written with units of joules:

$$\Delta G(J) = -n\mathscr{E}F \tag{4.49}$$

where F = Faraday = 96,487 coulombs mol^{-1} of electrons. Because the reversible voltage is proportional to free energy, it follows from $(\partial G/\partial T)_P = -S$ that the entropy change for the reaction in the cell depends on the temperature dependence of the voltage.

$$\Delta S(\text{cal deg}^{-1}) = 23{,}060n\left(\frac{\partial \mathscr{E}}{\partial T}\right)_P \tag{4.50}$$

Also, from the definition of $G \equiv H - TS$,

$$\Delta H(\text{kcal}) = 23.06n\left[-\mathscr{E} + T\left(\frac{\partial \mathscr{E}}{\partial T}\right)_P\right] \tag{4.51}$$

Therefore, the thermodynamic variables of the cell reaction can be obtained from measurements of the reversible voltage of a galvanic cell at several temperatures.

When a reaction occurs spontaneously in a galvanic cell, the free energy change must be negative, and the voltage, by convention, is positive. By convention we write the oxidation reaction as occurring at the left electrode and reduction at the right electrode (an easy way to remember this is to associate the two r's: reduction, right). For example, in the reaction between Zn metal and Cu^{2+} ions, the equation is

$$\text{Zn}(s) + \text{Cu}^{2+}(aq) \longrightarrow \text{Zn}^{2+}(aq) + \text{Cu}(s)$$

If a Zn rod is put into a solution of Cu^{2+} ions, the Zn will dissolve and Cu metal will precipitate. The free energy change for the reaction will depend on T and P and the concentrations of Cu^{2+} and Zn^{2+} ions. A galvanic cell in which this reation occurs can be made by placing a Zn rod into a $Zn(NO_3)_2$ solution, placing a Cu rod into a $Cu(NO_3)_2$ solution, and having the two solutions separated by a porous barrier or a salt bridge. Such a cell is represented in Fig. 4.4. The

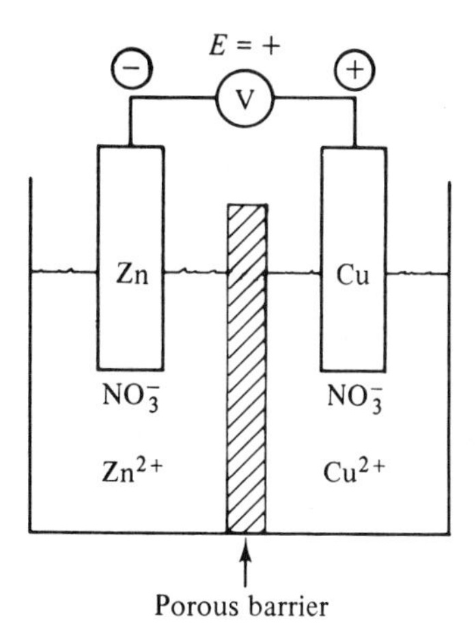

Fig. 4.4 Electrochemical cell with two different metal electrodes in solutions of the corresponding ions.

oxidation takes place at the left electrode:

$$\text{Oxidation:} \quad \text{Zn}(s) \longrightarrow \text{Zn}^{2+}(c_{\text{Zn}^{2+}}) + 2e^-$$

Thus, the left terminal is negative; electrons are being produced. The reduction takes place at the right electrode:

$$\text{Reduction:} \quad 2e^- + \text{Cu}^{2+}(c_{\text{Cu}^{2+}}) \longrightarrow \text{Cu}(s)$$

Thus, the right terminal is positive; electrons are being used up. The purpose of the porous barrier is to keep the two electrode solutions from mixing while still maintaining a path for electrical conduction (by ion diffusion). Such cells are often described using an abbreviated notation.

$$\mathrm{Zn}(s)|\mathrm{Zn}^{2+}(c_{\mathrm{Zn}^{2+}})|\,|\mathrm{Cu}^{2+}(c_{\mathrm{Cu}^{2+}})|\mathrm{Cu}(s)$$

To determine the free energy of the reaction for the concentrations of ions present in the cell, $c_{\mathrm{Zn}^{2+}}$, $c_{\mathrm{Cu}^{2+}}$, we use the measured reversible voltage of the cell in Eqs. (4.47) through (4.49) with $n = 2$. It is easy to measure the reversible voltage of the cell quite accurately with a potentiometer (represented by Ⓥ in the diagram). The concentration dependence of the reversible voltage can be measured, and by extrapolation the reversible voltage for standard conditions can be determined. This provides the standard free energy change for the reaction.

Standard Electrode Potentials

Reversible voltages have been determined for many reactions at standard conditions. Each reaction of course includes both an oxidation and a reduction half-cell reaction. If we *define* the voltage of one half-cell reaction as zero, all others can be obtained relative to this arbitrary choice. Chemists have chosen the reduction of H^+ to H_2 gas under standard conditions to have a standard electrode potential of zero:

$$\mathrm{H}^+(a = 1) + e^- \longrightarrow \tfrac{1}{2}\mathrm{H}_2(g, P = 1\text{ atm}) \qquad \mathscr{E}^0 = 0.000\text{ V}$$

With this choice the standard reduction potentials given in Table 4.5 were obtained.

To calculate the standard potential (and therefore the standard free energy) for a chemical reaction from Table 4.5, it is necessary to add the appropriate potentials for the half-cell reactions. The value of $\mathscr{E}^0_{\text{cell}}$ is $\mathscr{E}^0$ for the reduction half-cell reaction minus $\mathscr{E}^0$ for the oxidation half-cell reaction. By convention for galvanic cells, this is equal to $\mathscr{E}^0$ for the right-hand side of the cell minus $\mathscr{E}^0$ for the left-hand side of the cell. For example, let us reconsider the reaction

$$\mathrm{Zn}(s) + \mathrm{Cu}^{2+}(a = 1) \longrightarrow \mathrm{Zn}^{2+}(a = 1) + \mathrm{Cu}(s)$$

The half-cell reactions and their potentials are

$$\begin{array}{ll} \mathrm{Cu}^{2+}(a = 1) + 2e^- \longrightarrow \mathrm{Cu}(s) & \mathscr{E}^0 = +0.337\text{ V} \\ \mathrm{Zn}(s) \longrightarrow \mathrm{Zn}^{2+}(a = 1) + 2e^- & \mathscr{E}^0 = -(-0.763\text{ V}) \\ \hline & \mathscr{E}^0_{\text{cell}} = +1.100\text{ V} \end{array}$$

The standard reversible voltage is +1.100 V and the standard free energy is obtained from Eqs. (4.47) through (4.49).

$$\Delta G^0 = -2.20\text{ eV} = -50.73\text{ kcal} = -212 \times 10^3\text{ J}$$

Because Table 4.5 gives reduction potentials, the sign of the potential must be changed for the oxidation half-cell reaction. However, the voltages must *not*

Table 4.5 Standard reduction electrode potentials at 25°C

Electrode	Electrode reaction	$\mathscr{E}^0$ (V)*	$\mathscr{E}^{0\prime}$ (V)† (pH 7)
Li^+/Li	$Li^+ + e^- \rightarrow Li$	−3.045	
K^+/K	$K^+ + e^- \rightarrow K$	−2.925	
Cs^+/Cs	$Cs^+ + e^- \rightarrow Cs$	−2.923	
Ca^{2+}/Ca	$Ca^{2+} + 2e^- \rightarrow Ca$	−2.866	
Na^+/Na	$Na^+ + e^- \rightarrow Na$	−2.714	
Mg^{2+}/Mg	$Mg^{2+} + 2e^- \rightarrow Mg$	−2.363	
$OH^-/H_2/Pt$	$2H_2O + 2e^- \rightarrow H_2 + 2OH^-$	−0.8281	
Zn^{2+}/Zn	$Zn^{2+} + 2e^- \rightarrow Zn$	−0.7628	
Acetate/acetaldehyde	$OAc^- + 3H^+ + 2e^- \rightarrow CH_3CHO + H_2O$		−0.58
Fe^{2+}/Fe	$Fe^{2+} + 2e^- \rightarrow Fe$	−0.4402	
Gluconate/glucose	$C_6H_{11}O_7^- + 3H^+ + 2e^- \rightarrow C_6H_{12}O_6 + H_2O$		−0.44
Spinach ferredoxin	$Fe(III) + e^- \rightarrow Fe(II)$		−0.43
CO_2/formate	$CO_2 + 2H^+ + 2e^- \rightarrow HCO_2^- + H^+$	−0.20	−0.42
NAD^+/NADH‡	$NAD^+ + H^+ + 2e^- \rightarrow NADH$	−0.105	−0.32
Fe^{3+}/Fe	$Fe^{3+} + 3e^- \rightarrow Fe$	−0.036	
$H^+/H_2/Pt$	$2H^+ + 2e^- \rightarrow H_2$	0	−0.421
Mn hematoporphyrin IX	$Mn(III) + e^- \rightarrow Mn(II)$		−0.342
Acetoacetate/β-hydroxybutyrate	$CH_3COCH_2CO_2^- + 2H^+ + 2e^- \rightarrow CH_3CHOHCH_2CO_2^-$		−0.27
Horseradish peroxidase	$Fe(III) + e^- \rightarrow Fe(II)$		−0.27
Cytochrome c	$Fe(III) + e^- \rightarrow Fe(II)$		−0.25
$FAD/FADH_2$§	$FAD + 2H^+ + 2e^- \rightarrow FADH_2$		−0.22
Acetaldehyde/ethanol	$CH_3CHO + 2H^+ + 2e^- \rightarrow CH_3CH_2OH$		−0.20
Pyruvate/lactate	$CH_3COCO_2^- + 2H^+ + 2e^- \rightarrow CH_3CHOHCO_2^-$		−0.19
Oxaloacetate/malate	$^-O_2CCOCH_2CO_2^- + 2H^+ + 2e^- \rightarrow {}^-O_2CCHOHCH_2CO_2^-$		−0.17
Fumarate/succinate	$^-O_2CCH{=}CHCO_2^- + 2H^+ + 2e^- \rightarrow {}^-O_2CCH_2CH_2CO_2^-$		+0.031
Myoglobin	$Fe(III) + e^- \rightarrow Fe(II)$		+0.046
Cu^{2+}/Cu	$Cu^{2+} + 2e^- \rightarrow Cu$	+0.337	
$I_2/I^-/Pt$	$I_2 + 2e^- \rightarrow 2I^-$	+0.5355	
$O_2/H_2O_2/Pt$	$O_2 + 2H^+ + 2e^- \rightarrow H_2O_2$	+0.69	+0.295
$Fe^{3+}/Fe^{2+}/Pt$	$Fe^{3+} + e^- \rightarrow Fe^{2+}$	+0.771	

Table 4.5 Standard reduction electrode potentials at 25°C (*Cont.*)

Electrode	Electrode reaction	$\mathscr{E}^0$ (V)*	$\mathscr{E}^{0\prime}$ (V)† (pH 7)
Ag^+/Ag	$Ag^+ + e^- \rightarrow Ag$	+0.799	
$NO_3^-/NO_2^-/Pt$	$NO_3^- + 2H^+ + 2e^- \rightarrow NO_2^- + H_2O$	+0.94	+0.421
$Br_2/Br^-/Pt$	$Br^2 + 2e^- \rightarrow 2Br^-$	+1.087	
$O_2/H_2O/Pt$	$O_2 + 4H^+ + 4e^- \rightarrow 2H_2O$	+1.229	+0.816
$Cl_2/Cl^-/Pt$	$Cl_2 + 2e^- \rightarrow 2Cl^-$	+1.359	
$Mn^{3+}/Mn^{2+}/Pt$	$Mn^{3+} + e^- \rightarrow Mn^{2+}$	+1.4	
$Ce^{4+}/Ce^{3+}/Pt$	$Ce^{4+} + e^- \rightarrow Ce^{3+}$	+1.61	
$F_2/F^-/Pt$	$F_2 + 2e^- \rightarrow 2F^-$	+2.87	

* $\mathscr{E}^0$ refers to the solute standard state with unit activity for all species.
† $\mathscr{E}^{0\prime}$ refers to the biochemist's standard state with pH 7.
‡ NAD^+ is nicotinamide adenine dinucleotide.
§ FAD is flavin adenine dinucleotide.

be multiplied by 2 because there are two electrons involved. The voltage of the cell is independent of the number of electrons involved. The free energy does depend on the number of electrons, and therefore n appears explicitly in Eqs. (4.47) through (4.49). Free energy is extensive, but voltage is intensive.

If the reaction as written does not occur spontaneously, the calculated voltage is found to be negative and the standard free energy is positive.

Concentration Dependence of $\mathscr{E}$

Table 4.5 is analogous to a table of standard free energies. The standard free energies of a great many chemical reactions can be obtained from the table. What if we are interested in the free energies at arbitrary concentrations? We know the relation between standard free energy and free energy

$$\Delta G = \Delta G^0 + RT \ln Q \tag{4.4}$$

Substituting Eq. (4.49) for ΔG and ΔG^0, we obtain the *Nernst equation*:

$$\mathscr{E} = \mathscr{E}^0 - \frac{RT}{nF} \ln Q \tag{4.52}$$

where $\mathscr{E}$ = cell voltage for arbitrary concentrations specified by Q
$\mathscr{E}^0$ = standard cell voltage
R = gas constant in joules = 8.314 J deg^{-1} mol^{-1}
n = number of moles of electrons involved in the reaction

For the reaction,

$$aA + bB \longrightarrow cC + dD$$

$$Q = \frac{[a_C]^c[a_D]^d}{[a_A]^a[a_B]^b}$$

At 25°C the Nernst equation can be written explicitly as

$$\mathscr{E} = \mathscr{E}^0 - \frac{0.0591}{n} \log Q \tag{4.53}$$

The Nernst equation can be used to calculate the voltage and free energy for any concentrations of reactants and products in a cell. At equilibrium $\Delta G = 0$; therefore, $\mathscr{E} = 0$, also. Thus the standard potential gives the equilibrium constant for the reaction in the cell:

$$\mathscr{E}^0 = \frac{RT}{nF} \ln K \tag{4.54}$$

We can solve for K, and at 25°C write

$$K = 10^{n\mathscr{E}^0/0.0591} \tag{4.55}$$

Applications of Galvanic Cells

The types of problems that we can solve using galvanic cells are many and varied. The two most important for thermodynamics are the determination of accurate free energies for chemical reactions in aqueous solutions and the determination of activity coefficients.

A very important ionic reaction in aqueous solutions is the ionization of water. Suppose that we wish to determine the ionization of D_2O using a galvanic cell to provide the thermodynamic parameters for the reaction. The reaction is

$$D_2O \longrightarrow D^+(c_{D^+}) + OD^-(c_{OD^-})$$

First, we must think of a cell in which the reaction can occur as two half-cell reactions. Possible reactions are

$$\text{Oxidation:} \quad \tfrac{1}{2} D_2(g) \longrightarrow D^+(c_{D^+}) + e^-$$
$$\text{Reduction:} \quad e^- + D_2O \longrightarrow OD^-(c_{OD^-}) + \tfrac{1}{2} D_2(g)$$

Note that although the reaction of interest does not involve oxidation or reduction, we can devise two half-cell reactions that do. The cell could therefore have deuterium gas bubbling into a DCl solution on one side of the cell and deuterium gas bubbling into an NaOD solution on the other side. The solutions are separated by a porous barrier which allows electrical contact but prevents mixing of the solutions. The reversible voltage of the cell is measured by two inert platinum electrodes connected to a potentiometer, as shown in Fig. 4.5. The

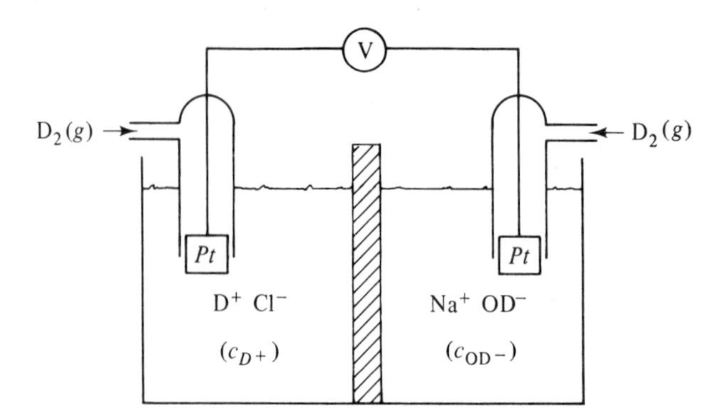

Fig. 4.5 Electrochemical cell designed for the measurement of the dissociation constant of D_2O.

reversible voltage of the cell which we measure is related to concentrations by the Nernst equation [Eq. (4.52)]. If the cell is at 25°C, we have

$$\mathscr{E} = \mathscr{E}^0 - \frac{0.0591}{1} \log \frac{(a_{D^+})(a_{OD^-})}{a_{D_2O}}$$

We use the solvent standard state for the D_2O and use sufficiently dilute solutions so that $a_{D_2O} = X_{D_2O} = 1$. We use the solute standard state on the molarity scale for the D^+ and OD^-. The pressure of deuterium gas does not appear in the expression as long as the pressure is identical on both sides of the cell. Therefore, at 25°C,

$$\begin{aligned} \mathscr{E} &= \mathscr{E}^0 - 0.0591 \ \log (\gamma_+ c_{D^+})(\gamma_- c_{OD^-}) \\ &= \mathscr{E}^0 - 0.0591 \ [\log (c_{D^+})(c_{OD^-}) + \log \gamma_\pm^2] \end{aligned} \tag{4.56}$$

To obtain the thermodynamics of the reaction, we measure voltage and the concentrations of the DCl and NaOD solutions in the cell. At low enough concentrations, $\log \gamma_\pm^2$ is zero, because the activity coefficients are equal to 1. The $\mathscr{E}^0$ can be obtained from the measured voltage and concentrations:

$$\mathscr{E}^0 = \mathscr{E} + 0.0591 \ \log (c_{D^+})(c_{OD^-})$$

This gives ΔG^0 from Eq. (4.48) and K from Eq. (4.55). The experiment can be repeated at any temperature T (we remember to correct 0.0591 by multiplying by $T/298$) to obtain the temperature dependence of $\mathscr{E}^0$. This gives ΔH^0 and ΔS^0 from Eqs. (4.50) and (4.51).

We can also obtain activity coefficients for D^+ and OD^- in the DCl and NaOD solutions. Once we have $\mathscr{E}^0$, we see from Eq. (4.56) that $\log \gamma_\pm^2$ is obtainable from the measured $\mathscr{E}$ and concentrations.

Activity Coefficients of Ions

We have usually set activity coefficients equal to 1 in our thermodynamic calculations. For small uncharged molecules and for univalent ions in aqueous solutions, this approximation may not be too bad, but for polymers or multivalent ions such as Mg^{2+} or PO_4^{3-}, activity coefficients may be very different from 1, even at millimolar concentrations. Figure 4.6 illustrates some measured activity coefficients for a 1–1 electrolyte (HCl), a 1–2 electrolyte (H_2SO_4), and a 2–2 electrolyte ($ZnSO_4$). The activity coefficients are much less than 1 even at 0.1 *M* solutions; at concentrations above 1 *M* the activity coefficients for some electrolytes increase and even become greater than 1.

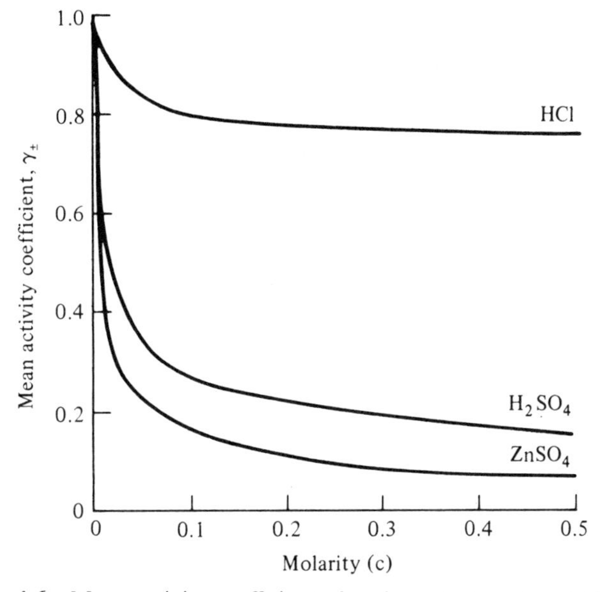

Fig. 4.6 Mean activity coefficient of various electrolytes at 25°C.

Previously in this chapter we mentioned the necessity of using mean ionic activity coefficients. The definitions for different electrolytes is best illustrated by examples.

For HCl, $ZnSO_4$ or any 1—1 electrolyte, the mean ionic activity coefficient is

$$\gamma_\pm = (\gamma_+\gamma_-)^{1/2} \qquad \text{1-1 electrolyte} \tag{4.57}$$

For H_2SO_4 or any 1–2 electrolyte,

$$\gamma_\pm = (\gamma_+^2\gamma_-)^{1/3} \qquad \text{1-2 electrolyte} \tag{4.58}$$

For $LaCl_3$ or any 1—3 electrolyte,

$$\gamma_\pm = (\gamma_+\gamma_-^3)^{1/4} \qquad \text{1-3 electrolyte} \tag{4.59}$$

The reader can generalize these equations to any type of salt. To calculate activity coefficients theoretically, we need to know the interaction between the solute particles; attraction leads to activity coefficients less than 1; repulsion leads to activity coefficients greater than 1. Debye and Hückel were able to calculate the attraction between ions in dilute solution to obtain the following quantitative result. In dilute (less than $0.01M$) aqueous solution at 25°C:

$$\log \gamma_{\pm} = -0.509|Z^{+}Z^{-}|\sqrt{I} \tag{4.60}$$

where $\gamma_{\pm}$ = mean activity coefficient of ion with solute standard state
$|Z^{+}Z^{-}|$ = absolute value of the product of the charges on the ions ($Z = 1, 2, 3, \ldots$)
I = ionic strength
0.509 = numerical value of a collection of constants appropriate to aqueous solutions at 25°C

The ionic strength is a charge–weighted measure of the total concentration of all ions in the solution. It is defined as follows:

$$I \equiv \frac{1}{2}\sum_{i} c_i Z_i^2 \tag{4.61}$$

where c_i = concentration of ion i, mol ℓ^{-1}
Z_i = charge on ion i ($Z_i = 1, 2, 3, \ldots$)

Ionic strength has been found to be a good measure of how "ionic" an environment is. That is, general effects of different salts on equilibria and reaction rates depend on the ionic strength rather than the concentration of the salt. Note that 1 M NaCl, 0.33 M $MgCl_2$, and 0.25 M $MgSO_4$ all have an ionic strength equal to 1.

PARTIAL MOLAL QUANTITIES

We introduced partial derivatives in Chapter 3 to discuss changes in G, H, E, S, A with variables T, P, and V. In this chapter we have considered the change in concentration of a chemical and its effect on G, H, and S. This change can be represented as a partial derivative, the partial molal quantity. The partial molal free energy, or chemical potential, for substance A in a system containing other chemicals is

$$\mu_A \equiv \bar{G}_A \equiv \left(\frac{\partial G}{\partial n_A}\right)_{T,P,n_{j\neq A}} \tag{4.62}$$

The T, P, and n_j outside the parentheses tells us that T, P, and the number of moles of all *other* chemicals present in the system are constant; so the partial molal free energy, $\bar{G}_A$, describes the way in which the free energy of an open system changes when the number of moles of A changes, but all other variables are held constant. We see that this becomes equal to the free energy per mole of A if the system consists of pure A. In general, however, the free energy change will depend on what else is in the system, and we must specify what the other components are by specifying the values of n_j. We should remember that $\bar{G}_A$ is related to the activity; $\bar{G}_A = \bar{G}_A^0 + RT \ln a_A$. Some examples should help clarify the ideas. Consider three systems: one is an empty beaker, one is a liter of H_2O, and one is a liter of 1 M KNO_3. We add a small pinch of NaCl to each system and ask how the free energy of each system changes. The free energy change will surely be different for each system.

The free energy change for the empty beaker will depend on the free energy per mole of pure sodium chloride:

$$\left(\frac{\partial G}{\partial n_{\text{NaCl}}}\right)_{25^\circ\text{C, 1atm}} = \Delta\bar{G}_f^0(\text{NaCl}, s) = -91.78 \text{ kcal mol}^{-1}$$

For the liter of H_2O the free energy will depend on the final concentration, c, of NaCl and the standard free energy per mole corresponding to an infinitely dilute aqueous solution extrapolated to $1\,M$ [$\Delta G_f^0(\text{NaCl}, aq) = -93.94$ kcal mol^{-1}]:

$$\left(\frac{\partial G}{\partial n_{\text{NaCl}}}\right)_{25^\circ\text{C, 1 atm, dilute aq. soln.}} = -93.94 \text{ kcal mol}^{-1} + RT \ln c$$

For the $1\,M$ KNO_3 solution, although the NaCl may be dilute, the presence of the KNO_3 will affect the activity of the NaCl. We can no longer calculate the partial molal free energy simply from tables; we must measure the NaCl activity in the KNO_3 solution:

$$\left(\frac{\partial G}{\partial n_{\text{NaCl}}}\right)_{25^\circ\text{C, 1 atm, } 1\,M\,\text{KNO}_3} = -93.94 \text{ kcal mol}^{-1} + RT \ln a$$

The mean ionic activity coefficient for K^+ and NO_3^- in $1\,M$ KNO_3 is 0.443. If we assume that the value applies also to small added amounts of Na^+ and Cl^-, then

$$\begin{aligned}\left(\frac{\partial G}{\partial n_{\text{NaCl}}}\right)_{25^\circ\text{, 1 atm, } 1\,M\,\text{KNO}_3} &= -93.94 \text{ kcal mol}^{-1} + RT \ln a \\ &= -93.94 \text{ kcal mol}^{-1} + RT \ln (\gamma_\pm)^2 + RT \ln c \\ &= -93.94 \text{ kcal mol}^{-1} - 0.97 \text{ kcal mol}^{-1} + RT \ln c \\ &= -94.91 \text{ kcal mol}^{-1} + RT \ln c\end{aligned}$$

Whenever we talk about the free energy per mole of a chemical in a mixture, we mean the partial molal free energy. For accurate work we must determine the free energy of the chemical in the system at the appropriate concentration. Often, however, we approximate the standard free energy in solution by the molar free energy of the pure substance. In the Appendix values are listed for some pure compounds and for the infinitely dilute aqueous solution extrapolated to $1\,M$ concentration (labeled $1\,M$ activity); one can see that the values differ by less than 5%.

Most of what we have said about partial molal free energy applies equally to the other extensive thermodynamic variables E, H, S, V, and A. Partial molal volumes are interesting to discuss because they illustrate the large difference that may occur between the volume and the partial molal volume. The volume per mole of solid $MgCl_2$ is about 40 mℓ mol^{-1}. However, the partial molal volume of $MgCl_2$ in dilute aqueous solution is negative! Adding a small amount of $MgCl_2$ to water causes the volume of the solution to decrease, because the Mg^{2+} ion strongly binds water molecules to it.

This decrease in solution volume upon adding ionized solutes is known as *electrostriction*. It results from the electrostatic charge on small ions which can organize the surrounding water molecules into a more compact structure than they have in the absence of the ion. This is seen in the case of such molecules as the amino acid glycine, which occurs as a zwitterion (double ion) in dilute aqueous solution. Its partial specific volume is

$$\bar{V}(H_3N^+CH_2COO^-) = 43.5 \text{ m}\ell \text{ mol}^{-1}$$

The molecule glycolamide, $HOCH_2CONH_2$, has the same empirical formula but it is not ionic in aqueous solutions.

$$\bar{V}(HOCH_2CONH_2) = 56.3 \text{ m}\ell \text{ mol}^{-1}$$

The partial molal volume is significantly smaller for the zwitterionic molecule. For polyelectrolytes, such as the sodium salts of nucleic acids, there is pronounced electrostriction upon forming aqueous solutions. Such information is useful in understanding solute-solvent interactions at the molecular level.

A thermodynamic property of any mixture is defined as the sum of the number of moles of each component times the partial molal property of the component. For volume the equation is

$$V \equiv n_1\left(\frac{\partial V}{\partial n_1}\right)_{T,P,n_j} + n_2\left(\frac{\partial V}{\partial n_2}\right)_{T,P,n_j} + \cdots \tag{4.63}$$

Similarly, for free energy,

$$G = n_1\mu_1 + n_2\mu_2 + \cdots \tag{4.64}$$

SUMMARY

Chemical Reactions

$$a\text{A} + b\text{B} \longrightarrow c\text{C} + d\text{D}$$

$$\Delta G = \Delta G^0 + RT \ln Q \tag{4.4}$$

ΔG = free energy change for the reaction at temperature T and concentrations specified by Q

$\Delta G^0 = c\bar{G}^0_C + d\bar{G}^0_D - a\bar{G}^0_A - b\bar{G}^0_B$

= standard free energy change for the reaction at 1 atm pressure and standard conditions

$$Q = \frac{[a_C]^c[a_D]^d}{[a_B]^b[a_A]^a}$$

a_A, a_B, a_C, a_D = activities of reactants and products at the concentrations specified

$$\Delta G^0 = -RT \ln K \tag{4.7}$$

K = equilibrium constant

$= \dfrac{[a_C^{eq}]^c [a_D^{eq}]^d}{[a_B^{eq}]^b [a_A^{eq}]^a}$

a_A^{eq}, etc. = activities of reactants and products at equilibrium

R = 1.987 cal deg^{-1} mol^{-1}

$2.303RT$ = 1365 cal at 25°C

$$K = e^{-\Delta G^0/RT} = 10^{-\Delta G^0/2.303RT} \tag{4.9}$$

$$K = e^{\Delta S^0/R} e^{-\Delta H^0/RT} \tag{4.10}$$

Activities

$$\bar{G}_A \equiv \bar{G}_A^0 + RT \ln a_A \tag{4.14}$$

$\bar{G}_A$ = free energy per mole of A at temperature T and specified concentration
$\bar{G}_A^0$ = standard free energy per mole of A at temperature T, 1 atm pressure, and standard concentration
a_A = activity of A at temperature T and specified concentration

$$\bar{G}_A \equiv \mu_A \equiv \left(\frac{\partial G}{\partial n_A}\right)_{T,P,n_{j \neq A}} \tag{4.62}$$

μ_A = chemical potential

$\left(\dfrac{\partial G}{\partial n_A}\right)_{T,P,n_{j \neq A}}$ = partial molal free energy

$$a_A = \gamma_A \cdot (\text{concentration of A}) \tag{4.20}$$

γ_A = activity coefficient of A

Standard States

Ideal gases: Standard state is gas at 1 *atm pressure:*

$$a_A = P_A \qquad \text{in atm} \tag{4.17}$$

Pure solids or liquids: Standard state is pure solid or liquid at 1 *atm pressure:*

$$a_A = 1 \qquad \text{at 1 atm pressure} \tag{4.19}$$

Solvent in solution: Standard state is pure liquid at 1 *atm pressure:*

$$a_A = \gamma_A X_A \qquad (\gamma_A \longrightarrow 1 \quad \text{as} \quad X_A \longrightarrow 1) \tag{4.27}$$

Solutes in solution: Standard state is solute in solution with the properties of an infinitely dilute solution but at a concentration of 1 [*either molar* (c) *or molal* (m) *concentrations can be used*]:

$$a_A = \gamma_A c_A \qquad (\gamma_A \longrightarrow 1 \quad \text{as} \quad c_A \longrightarrow 0) \tag{4.28}$$

$$a_A = \gamma_A m_A \qquad (\gamma_A \longrightarrow 1 \quad \text{as} \quad m_A \longrightarrow 0) \tag{4.33}$$

Temperature dependence of equilibrium constant:

$$\ln \frac{K_2}{K_1} = -\frac{\Delta H^0}{R}\left(\frac{1}{T_2} - \frac{1}{T_1}\right) \tag{4.43}$$

Galvanic Cells

$$\Delta G = -n\mathscr{E} \text{ (in eV)} = -23.06n\mathscr{E} \text{ (in kcal)} = -n\mathscr{E}F \text{ (in J)} \tag{4.47)–(4.49}$$

$F = 96{,}487$ coulombs mol^{-1}

$$\Delta H(\text{kcal}) = 23.06n\left[-\mathscr{E} + T\left(\frac{\partial \mathscr{E}}{\partial T}\right)_P\right] \tag{4.51}$$

$$\Delta S(\text{cal deg}^{-1}) = 23{,}060n\left(\frac{\partial \mathscr{E}}{\partial T}\right)_P \tag{4.50}$$

Nernst equation:

$$\mathscr{E} = \mathscr{E}^0 - \frac{RT}{nF}\ln Q \tag{4.52}$$

$$\mathscr{E}^0 = \frac{RT}{nF}\ln K \tag{4.54}$$

Debye-Hückel equation at 25°C in H_2O *solutions:*

$$\log \gamma_\pm = -0.509|Z^+Z^-|\sqrt{I} \tag{4.60}$$

$\gamma_\pm$ = mean activity coefficient of ions of charge Z^+ and Z^-

Ionic strength:

$$I = \tfrac{1}{2}\sum_i c_i Z_i^2$$

c_i = molarity of ion with charge Z_i

REFERENCES

In addition to the biochemistry texts listed in Chapter 1, see

KLOTZ, I. M., 1967. *Energy Changes in Biochemical Reactions*, Academic Press, New York.

LEHNINGER, A. L., 1971. *Bioenergetics*, 2nd ed., W. A. Benjamin, Menlo Park, Calif.

PROBLEMS

1. A key step in the biosynthesis of triglycerides (fats) is the conversion of glycerol to glycerol-1-phosphate by ATP:

$$\text{glycerol} + \text{ATP} \xrightarrow[\alpha\text{-glycerol kinase}]{Mg^{2+}} \text{glycerol-1-phosphate} + \text{ADP}$$

At a steady state in the living cell, (ATP) = $10^{-3}\,M$ and (ADP) = $10^{-4}\,M$. The maximum (equilibrium) value of the ratio (glycerol-1-P)/(glycerol) is observed to be 770 at 25°C and pH 7.

(a) Calculate K and $\Delta G^{0\prime}_{298}$ for the reaction.

(b) Using the value of $\Delta G^{0\prime}_{298}$ for the reaction ADP + P → ATP + H_2O, together with the answer to part (a), calculate $\Delta G^{0\prime}_{298}$ and K for the reaction

$$\text{glycerol} + \text{phosphate} \longrightarrow \text{glycerol-1-phosphate} + H_2O$$

2. (a) In the frog muscle rectus abdominis, the concentrations of ATP, ADP, and phosphate are $1.25 \times 10^{-3}\,M$, $0.50 \times 10^{-3}\,M$, and $2.5 \times 10^{-3}\,M$, respectively. Calculate the free-energy change $\Delta G'$ for the hydrolysis of ATP in this muscle. Take the temperature and pH of the muscle as 25°C and 7, respectively.

(b) For the muscle described, what is the maximum amount of mechanical work it can do per mole of ATP hydrolyzed?

(c) In muscle, an enzyme creatine phosphokinase catalyzes the following reaction:

$$\text{phosphocreatine} + \text{ADP} \longrightarrow \text{creatine} + \text{ATP}$$

The standard free energy of hydrolysis of phosphocreatine at 25°C and pH 7 is $-10.3\ \text{kcal mol}^{-1}$:

$$\text{phosphocreatine} + H_2O \longrightarrow \text{creatine} + \text{phosphate}$$

Calculate the equilibrium constant of the creatine phosphokinase reaction.

3. An important step in the glycolytic pathway is the phosphorylation of glucose by ATP, catalyzed by the enzyme hexokinase and Mg^{2+}:

$$\text{glucose} + \text{ATP} \xrightarrow[\text{hexokinase}]{Mg^{2+}} \text{glucose-6-P} + \text{ADP}$$

In the absence of ATP, glucose-6-P is unstable at pH 7, and in presence of the enzyme glucose-6-phosphatase, it hydrolyzes to give glucose:

$$\text{glucose-6-P} + H_2O \xrightarrow[\text{G-6-phosphatase}]{} \text{glucose} + \text{phosphate}$$

(a) Using data from Table 4.2, calculate $\Delta G^{0\prime}$ at pH 7 for the phosphorylation of glucose by ATP at 298 K.
(b) If the reaction of part (a) is allowed to proceed to equilibrium in the presence of equal concentrations of ADP and ATP, what is the ratio (glucose-6-P)/(glucose) at equilibrium? Assume a large excess of ATP and ADP; that is, (ATP) = (ADP) $\gg$ [(glucose) + (glucose-6-P)].
(c) In the absence of ATP (and ADP), calculate the ratio (glucose-6-P)/(glucose) at pH 7, if phosphate = 10^{-2} *M*.

4. The equilibrium constant at 400°C for the reaction

$$\tfrac{3}{2}H_2(g) + \tfrac{1}{2}N_2(g) \longrightarrow NH_3(g)$$

is 0.0129. Assume that the gases behave ideally.
(a) Calculate ΔG^0_{673} for this reaction.
(b) If the total pressure of the gas mixture is 1 atm, what is the partial pressure of NH_3 at equilibrium at 400°C? Assume that pure NH_3 was present initially.

5. The following reactions can be coupled to give alanine and oxalacetate:

$$\text{glutamate} + \text{pyruvate} \rightleftharpoons \text{ketoglutarate} + \text{alanine}$$

$$\Delta G^{0\prime}_{303} = -240 \text{ cal mol}^{-1}$$

$$\text{glutamate} + \text{oxalacetate} \rightleftharpoons \text{ketoglutarate} + \text{aspartate}$$

$$\Delta G^{0\prime}_{303} = -1150 \text{ cal mol}^{-1}$$

(a) Write the form of the equilibrium constant for the reaction

$$\text{pyruvate} + \text{aspartate} \rightleftharpoons \text{alanine} + \text{oxalacetate}$$

and calculate the numerical value of the equilibrium constant at 30°C.
(b) In the cytoplasm of a certain cell, the components are at the following concentrations: pyruvate = 10^{-2} *M*, aspartate = 10^{-2} *M*, alanine = 10^{-4} *M*, and oxalacetate = 10^{-5} *M*. Calculate the free-energy change for the reaction of part (a) under these conditions. What conclusion can you reach about the direction of this reaction under cytoplasmic conditions?

6. The biosynthesis of glutamine from glutamate and ammonium ion is a "coupled" reaction, in the sense that one molecule of ATP is required for each glutamine molecule formed, and follows the scheme

$$GLU^- + NH_4^+ + ATP \underset{\text{glutamine synthetase}}{\rightleftharpoons} GLN + ADP + \text{phosphate}$$

The equilibrium constant for the entire system in the presence of the enzyme has been measured to be $K' = 1200$ at pH 7 and 37°C.
(a) What is the standard free-energy change, $\Delta G^{0\prime}$, for the reaction at pH 7 and 37°C?
(b) In the absence of ATP, ADP, phosphate, and the enzyme, the equilibrium constant for the reaction

$$GLU^- + NH_4^+ \rightleftharpoons GLN + H_2O$$

is $K'_b = 0.0035$ at pH 7 and 37°C. Use this observation and the information above to calculate $\Delta G^{0\prime}_{310}$ for the reaction

$$ATP + H_2O \rightleftharpoons ADP + \text{phosphate}$$

(c) Explain briefly the function of the ATP in the biosynthetic pathway. What is the nature of the difference between the role of ATP and the role of the enzyme in the biochemical process?

7. An important metabolic step is the conversion of fumarate to malate. In aqueous solution an enzyme (fumarase) allows equilibrium to be attained.

$$\text{fumarate} + H_2O \rightleftharpoons \text{malate}$$

At 25°C the equilibrium constant $K = [a_M]/[a_F] = 4$. The activity of malate is a_M and the activity of fumarate is a_F, defined on the molarity concentration scale ($a = c$ in dilute solution).

(a) What is the standard free-energy change for the reaction at 25°C?
(b) What is the free-energy change for the reaction at equilibrium?
(c) What is the free-energy change when 1 mol of 0.1 M fumarate is converted to 1 mol of 0.1 M malate?
(d) What is the free-energy change when 2 mol of 0.1 M fumarate are converted to 2 mol of 0.1 M malate?
(e) If $K = 8$ at 35°C, calculate the standard enthalpy change for the reaction. Assume that the enthalpy is independent of temperature.
(f) Calculate the standard entropy change for the reaction.

8. Consider the hydrolysis of ATP to ADP (ATP → ADP + phosphate).
(a) Calculate the equilibrium constant for the reaction at 25°C, 1 atm.
(b) Calculate the equilibrium constant for the reaction at 37°C, 1 atm.
(c) Calculate the equilibrium constant for the reaction at 0°C, 1 atm.

9. Use the free energy of hydrolysis of ATP under standard conditions at 25°C, 1 atm, to answer the following questions.
(a) Calculate the ΔG for the reaction when ATP = 10^{-2}, ADP = 10^{-4}, and phosphate = 10^{-1}.
(b) Calculate the maximum available work under the conditions of part (a) when 1 mol of ATP is hydrolyzed. This work could be used, for example, to contract a muscle and raise a weight.
(c) Calculate ΔG and the maximum available work if ATP = 10^{-7}, ADP = 10^{-1}, and phosphate = 2.5×10^{-1}.

10. What is the equilibrium pressure of CO(g) in equilibrium with the $CO_2(g)$ and $O_2(g)$ in the atmosphere at 25°C? The partial pressure of $O_2(g)$ is 0.2 atm and the partial pressure of $CO_2(g)$ is 3×10^{-4} atm. CO is extremely poisonous because it forms a very strong complex with hemoglobin. Should you worry?

11. In a certain solvent a polypeptide is found to change from a stable coil at low temperature to a stable helix at high temperature. The equilibrium constant can be written approximately as

$$K = \frac{[\text{helix}]}{[\text{coil}]}$$

At 50°C, K = 1; at 60°C, $K = 10$.

(a) Calculate ΔH^0 for the reaction. Is heat evolved or absorbed?
(b) Calculate ΔS^0 for the reaction at 50°C.
(c) The helix is thought to be a rigid structure held together by hydrogen bonds. The coil is thought to be a flexible structure with the hydrogen bonds broken. Is this model consistent with the signs and magnitudes of ΔH^0 and ΔS^0? Explain your answer.

12. (a) From the ionization constant, calculate the standard free-energy change for the ionization of acetic acid in water at 25°C.
(b) What is the free-energy change at equilibrium for the ionization of acetic acid in water at 25°C?
(c) What is the free-energy change for the following reaction in water at 25°C? (Assume that activity coefficients are all = 1.)

$$H^+ \text{ (conc. } = 10^{-4}\,M) + OAc^- \text{ (conc. } = 10^{-2}\,M) \longrightarrow HOAc\,(1\,M)$$

(d) What is the free-energy change for the following reaction in water at 25°C? (Assume that activity coefficients are all = 1.)

$$H^+ \text{ (conc. } = 10^{-4}\,M) + OAc^- \text{ (conc. } = 10^{-2}\,M) \longrightarrow HOAc\,(10^{-5}\,M)$$

(e) What is the free-energy change for transferring 1 mol of acetic acid from an aqueous solution of 1 M concentration to an aqueous solution of $10^{-5}\,M$? (Assume that activity coefficients are all equal to 1.)

13. You want to make a pH 7 buffer using NaOH and phosphoric acid. The sum of the concentrations of all phosphoric acid species is 0.1 M. The equilibrium constants for concentrations given in mole ℓ^{-1} for the following equilibria are

$$H_3PO_4 \rightleftharpoons H^+ + H_2PO_4^- \qquad K_1 = 7.5 \times 10^{-3}$$

$$H_2PO_4^- \rightleftharpoons H^+ + HPO_4^{2-} \qquad K_2 = 6.2 \times 10^{-8}$$

$$HPO_4^{2-} \rightleftharpoons H^+ + PO_4^{3-} \qquad K_3 = 2.2 \times 10^{-13}$$

(a) Write the equation which specifies that the solution is electrically neutral. Use (PO_4^{3-}), (HPO_4^{2-}), etc., for the concentrations.
(b) Calculate the concentrations of all species in the buffer.
(c) If about 10 kcal of heat is absorbed when 1 mol of $H_2PO_4^-$ ionizes, calculate K_2 at 37°C.

14. Consider the reaction

$$A \longrightarrow 2\,B$$

in aqueous solution at 25°C and 1 atm pressure. The equilibrium constant is 10 for this reaction and the standard enthalpy change is 50 kcal.
(a) Calculate the standard free-energy change at 25°C.
(b) Calculate the standard entropy change at 25°C.
(c) Will the equilibrium constant increase or decrease with increasing temperature?
(d) Is heat evolved or absorbed by this reaction if it takes place as written? How much heat?
(e) If the concentration of B is doubled, will the free-energy change increase or decrease? Calculate the change in ΔG caused by doubling the activity of B.
(f) Is the reaction spontaneous when A (a = 1) reacts to form B (a = 1)?
(g) If the activity of B is held equal to 1, what is the minimum activity of A that is sufficient to make the reaction spontaneous?

15. Consider the reaction

$$2\,Fe^{3+} + 3\,Zn(s) \longrightarrow 3\,Zn^{2+} + 2\,Fe(s)$$

at 25°C.

(a) Calculate the thermodynamic equilibrium constant from Table 4.5.

(b) Substitute molarities and activity coefficients for activities ($a = \gamma m$) in K_a. It should now be clear that the equilibrium constant in terms of molarities (K_m) will be changed if the activity coefficients are changed, but that K_a does not change.

(c) Use the Debye–Hückel theory for activity coefficients to calculate how $\log K_m$ depends on $\sqrt{I}$.

16. Cytochromes are iron-heme proteins in which a porphyrin ring is coordinated through its central nitrogens to an iron atom that can undergo a one-electron oxidation-reduction reaction. Cytochrome f is an example of this class of molecules, and it operates as a redox agent in chloroplast photosynthesis. The standard reduction potential, $\mathscr{E}^{0\prime}$, of cytochrome f at pH 7 can be determined by coupling it to an agent of known $\mathscr{E}^{0\prime}$, such as ferricyanide/ferrocyanide:

$$Fe(CN)_6^{3-} + e^- \longrightarrow Fe(CN)_6^{4-} \qquad \mathscr{E}^{0\prime} = 0.440\ \text{V}$$

In a typical experiment, carried out spectrophotometrically, a solution at 25°C and pH 7 containing a ratio

$$[Fe(CN)_6^{4-}]/[Fe(CN)_6^{3-}] = 2.0$$

is found to have a ratio $[\text{Cyt f}_{\text{reduced}}]/[\text{Cyt f}_{\text{oxidized}}] = 0.10$ at equilibrium.

(a) Calculate $\mathscr{E}^{0\prime}$ (reduction) for cytochrome f.

(b) On the basis of a standard reduction potential, $\mathscr{E}^{0\prime}$, for a reduction of O_2 to H_2O at pH 7 and 25°C, is oxidized cytochrome f a good enough oxidant to cause the formation of O_2 from H_2O at pH 7?

17. Consider the following reaction, in which two electrons are transferred:

$$2\ \text{cytochrome c (ferrous)} + \text{pyruvate} + 2\,H^+ \longrightarrow$$

$$2\ \text{cytochrome c (ferric)} + \text{lactate}$$

(a) What is $\mathscr{E}^{0\prime}$ for this reaction at pH 7 and 25°C?

(b) Calculate the equilibrium constant for the reaction at pH 7 and 25°C.

(c) Calculate the standard free-energy change for the reaction at pH 7 and 25°C.

(d) Calculate the free-energy change (at pH 7 and 25°C) if the lactate concentration is 5 times the pyruvate concentration and the cytochrome c (ferric) is 10 times the cytochrome c (ferrous).

18. The cell Ag(s), AgI(s) $|$KI($10^{-2}\,M$)$||$KCl($10^{-3}\,M$)$|$Cl$_2$(g, 1 atm), Pt(s) is galvanic and has the voltage 1.5702 V at 298 K.

(a) Write the cell reaction.

(b) What is ΔG_{298}?

(c) What is ΔG^0_{298}?

(d) Calculate the standard reduction potential for the half-cell on the left.

(e) Calculate the solubility product of AgI; $K_{AgI} = [a_{Ag^+}][a_{I^-}]$.

(f) The cell has a potential of 1.5797 V at 288 K. Estimate ΔS^0_{298} for the reaction.

19. Ferredoxins (Fd) are iron- and sulfur-containing proteins that undergo redox reactions in a variety of microorganisms. A particular ferredoxin is oxidized in a one-electron reaction, independent of pH, according to the equation

$$\mathrm{Fd_{red}} \rightleftharpoons \mathrm{Fd_{ox}} + e^-$$

To determine the standard emf of Fd_{red}/Fd_{ox}, a known amount was placed in a buffer at pH 7.0 and bubbled with H_2 at 1 atm pressure. (Finely divided platinum catalyst was present to ensure reversibility.) At equilibrium, the ferredoxin was found spectrophotometrically to be exactly $\frac{1}{3}$ in the reduced form and $\frac{2}{3}$ in the oxidized form.
(a) Calculate K, the equilibrium constant, for the system

$$\tfrac{1}{2}\mathrm{H_2} + \mathrm{Fd_{ox}} \rightleftharpoons \mathrm{H^+} + \mathrm{Fd_{red}}$$

(b) Calculate $\mathscr{E}^{0\prime}$ for the Fd_{red}/Fd_{ox} system at 25°C.

20. The conversion of β-hydroxybutyrate, β-HB$^-$, to acetoacetate, AAC$^-$, is an important biochemical redox reaction that uses molecular oxygen as the ultimate oxidizing agent:

$$\beta\text{-HB}^- + \tfrac{1}{2}\mathrm{O_2} \longrightarrow \mathrm{AAC^-} + \mathrm{H_2O}$$

(a) Using the standard reduction potentials given in Table 4.5, calculate $\Delta G^{0\prime}$ and the equilibrium constant for this system at pH 7 and 25°C.
(b) In a solution saturated at 1 atm with respect to dissolved air (which is 20% oxygen), what is the ratio of [AAC$^-$] to [β-HB$^-$] at equilibrium?

21. Consider the reaction

$$\underset{\text{ethanol}}{\mathrm{CH_3CH_2OH}(aq)} + \tfrac{1}{2}\mathrm{O_2}(g) \longrightarrow \underset{\text{acetaldehyde}}{\mathrm{CH_3\overset{\overset{\displaystyle O}{\|}}{C}H}(aq)} + \mathrm{H_2O}$$

(a) Calculate $\mathscr{E}^0$ for this reaction at 25°C.
(b) Calculate the standard free energy (in kcal) for the reaction at 25°C.
(c) Calculate the equilibrium constant at 25°C for the reaction.
(d) Calculate $\mathscr{E}$ for the reaction at 25°C when: a(ethanol) = 0.1, P_{O_2} = 4 atm, a(acetaldehyde) = 1, and $a(H_2O)$ = 1.
(e) Calculate ΔG for the reaction in part (d).

5

Physical Equilibria

A very important type of reaction is the transfer of a chemical from one phase to another. The evaporation of liquid water into the vapor phase is a familiar example. We know that the heat removed from our bodies by water evaporation allows us to survive in the radiation from the sun. Another example is the transport of ions from inside a cell to outside a cell, which is vital to nerve conduction. We can consider the inside of the cell and the outside of the cell as two different phases.

PHASE EQUILIBRIA

In this chapter we shall consider the equilibrium of a substance between two phases. The thermodynamic equations we shall develop lead to some surprising conclusions. For example, we will show that at equilibrium the chemical potential of each species must be the same in all phases. This means that water which has been in contact with mercury long enough to reach equilibrium has the same vapor pressure of mercury as pure mercury! The difference is that as the mercury evaporates from the aqueous solution, the vapor pressure of mercury quickly decreases, while the vapor pressure of mercury remains constant above the pure mercury, or above the water that is kept saturated with mercury. The mercury that has reached equilibrium with water has the same water vapor pressure as pure water. Of course, we worry less about slightly wet mercury than we do about slightly contaminated water.

Consider a system containing mercury and water in equilibrium. If we shake a closed container containing mercury and water, we obtain the result illustrated in Fig. 5.1. The system has three phases: the bottom one is mercury saturated

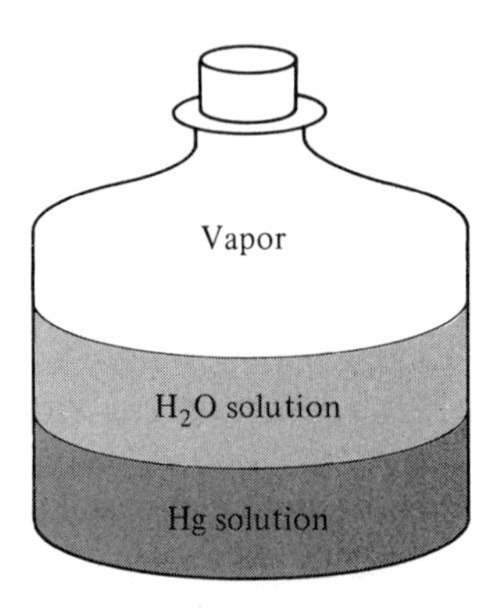

Fig. 5.1 Distinct phases occupy separate regions of space. The two liquid phases and one vapor phase are all in equilibrium with one another.

with water; the second one is water saturated with mercury; the top one contains water and mercury gas. The reason that we have distinct layers is because of gravity; however, we can ignore the effect of gravity. It is much smaller than the chemical forces. We would obtain essentially the same results if the system were in a zero-gravity spaceship. There would still be three phases, but they would consist of intermingled bubbles. The conditions for equilibrium require that there is (1) no net heat flow between phases, (2) no net work done, and (3) no net matter flow. Conditions (1) and (2) mean that the temperature and pressure are equal for all three phases, T and P are constant in the system. For a system at equilibrium at constant T and P, the free-energy change must be zero. Consider the process of moving a small amount (n moles) of mercury from the mercury solution to the water solution.

$$n\ \mathrm{Hg}\ (\mathrm{Hg\ solution}) \longrightarrow n\ \mathrm{Hg}\ (\mathrm{H_2O\ solution})$$

$$\Delta G(\mathrm{eq}) = 0 = n\bar{G}_{\mathrm{Hg}}(\mathrm{H_2O\ soln}) - n\bar{G}_{\mathrm{Hg}}(\mathrm{Hg\ soln})$$

Therefore, at equilibrium,

$$\bar{G}_{\mathrm{Hg}}(\mathrm{Hg\ soln}) = \bar{G}_{\mathrm{Hg}}(\mathrm{H_2O\ soln})$$

It follows that this is also true for the water in the system and for the components in the vapor phase. At equilibrium,

$$\bar{G}_{\mathrm{Hg}}(\mathrm{Hg\ soln}) = \bar{G}_{\mathrm{Hg}}(\mathrm{H_2O\ soln}) = \bar{G}_{\mathrm{Hg}}(\mathrm{vapor})$$

$$\bar{G}_{\mathrm{H_2O}}(\mathrm{Hg\ soln}) = \bar{G}_{\mathrm{H_2O}}(\mathrm{H_2O\ soln}) = \bar{G}_{\mathrm{H_2O}}(\mathrm{vapor})$$

The $\bar{G}$ is the partial molal free energy or chemical potential, μ:

$$\bar{G}_{\mathrm{A}} \equiv \left(\frac{\partial G}{\partial n_{\mathrm{A}}}\right)_{T,P,n_{j\neq \mathrm{A}}} \equiv \mu_{\mathrm{A}} \tag{4.62}$$

We have presented criteria for a special case. We can generalize. For phase equilibrium, with T and P constant:

$$\left.\begin{array}{ll} \text{Species A:} & \\ & \mu_A(\text{phase 1}) = \mu_A(\text{phase 2}) = \mu_A(\text{phase 3}) = \cdots \\ \text{Species B:} & \\ & \mu_B(\text{phase 1}) = \mu_B(\text{phase 2}) = \mu_B(\text{phase 3}) = \cdots \end{array}\right\} \tag{5.1}$$

The chemical potential depends on the activity of each species in each phase.

$$\mu_A = \mu_A^0 + RT \ln a_A = \mu_A^0 + RT \ln \gamma_A c_A \tag{4.14}$$

Therefore, if the same standard state is used for a component in two different phases, its activity in the two phases will be the same at equilibrium.

For phase equilibrium (same standard state in both phases),

$$a_A(\text{phase 1}) = a_A(\text{phase 2}) \tag{5.2}$$

It seems to worry students that the equality of activities depends on our choice of standard states. Remember that the free energies and chemical potentials are the measurable quantities (analogous to T and P), which are equal at equilibrium. Activities, by contrast, are defined quantities that depend on our choices of standard states. If the standard states are the same and the activity coefficients are equal in both phases, concentrations in the two phases will be equal at equilibrium.

For phase equilibrium (activity coefficients equal in both phases),

$$c_A(\text{phase 1}) = c_A(\text{phase 2}) \tag{5.3}$$

In summary, if equilibrium is attained for a chemical component among two or more phases at the same T and P, (1) its chemical potential is the same in all phases; (2) if the same standard state is used, its activity is the same in all phases; and (3) if activity coefficients are equal, its concentration is the same in all phases.

Free Energies of Transfer Between Phases

If T and P are constant, but the concentrations in the different phases do not correspond to equilibrium, the free energy per mole for transferring A from one phase to another is

$$\begin{aligned} \Delta\bar{G} &= \mu_A(\text{phase 2}) - \mu_A(\text{phase 1}) \\ &= RT \ln \frac{a_A(\text{phase 2})}{a_A(\text{phase 1})} \end{aligned} \tag{5.4}$$

If the activity coefficients for A are the same in the two phases, the free energy of transfer per mole is

$$\Delta\bar{G} = RT \ln \frac{c_A(\text{phase 2})}{c_A(\text{phase 1})} \tag{5.5}$$

These equations apply to uncharged molecules in the absence of an electric field. If there is a voltage difference between the two phases, an additional term is required for the free energy of transferring 1 mol of an ion:

$$\Delta\bar{G} = RT \ln \frac{a_A(\text{phase 2})}{a_A(\text{phase 1})} + ZFV \tag{5.6}$$

where Z = charge on ion, ± 1, ± 2, $\pm 3, \ldots$
F = Faraday = 23,060 cal $(\text{eV})^{-1}$
V = potential difference between the two phases, V

The sign of V is positive when phase 2 is electrically positive relative to phase 1.

These equations will be useful in understanding active transport in biological cells. The transport of ions and metabolites across membranes is characterized by Eqs. (5.4) through (5.6).

Equilibrium Dialysis and Scatchard Plots

A common application of Eqs. (5.1) through (5.3) in biochemistry is the study of binding of small molecules to macromolecules by equilibrium dialysis. The binding of drugs, hormones, inhibitors, and so on to proteins and nucleic acids has been widely studied by this method, which is illustrated in Fig. 5.2. A

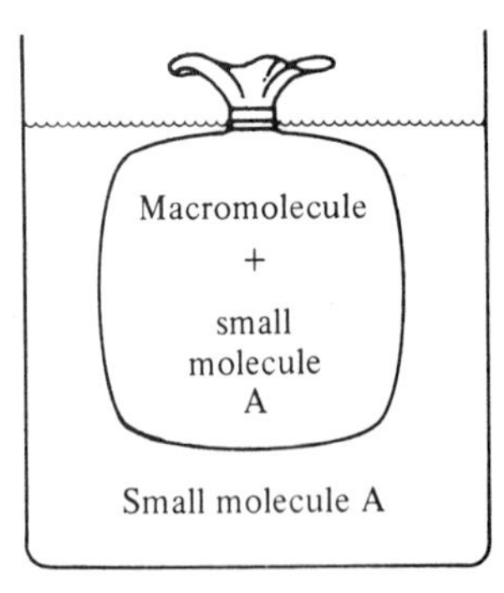

Fig. 5.2 Two liquid phases can be separated by a dialysis membrane that permits partial equilibration of the solution components. In this case, small molecules such as the solvent and the small molecule A can pass freely through the membrane, but large molecules cannot.

macromolecular species is placed on the inside of a dialysis membrane. The semipermeable dialysis membrane allows water and other small molecules to pass through freely but prevents passage of the macromolecule. At equilibrium,

$$\mu_A(\text{inside}) = \mu_A(\text{outside})$$

It is convenient to use the same standard state for small molecule A in both phases; therefore,

$$a_A(\text{inside}) = a_A(\text{outside}) \tag{5.2}$$

We can measure the concentrations of the small molecule in both phases; this gives us directly the ratio of activity coefficients for A in the two phases:

$$\frac{\gamma_A(\text{inside})}{\gamma_A(\text{outside})} = \frac{c_A(\text{outside})}{c_A(\text{inside})}$$

For simplicity, let us assume that we have made the concentrations of all molecules on the outside small enough so that the activity coefficient of A outside is equal to 1. We have

$$\gamma_A(\text{inside}) = \frac{c_A(\text{outside})}{c_A(\text{inside})}$$

The activity coefficient of A inside will usually be smaller than 1; consequently, the concentration of A outside will be less than that inside. Thermodynamically, we can say only that the activity of A is lowered in the presence of the macromolecule, but a reasonable interpretation of this is that A is bound to the macromolecule. To calculate the amount bound, we assume that the *free* molecule A *that is not bound to the macromolecule* has activity coefficient equal to 1. The only difference between inside and outside is the presence of the macromolecule inside; it binds some A molecules but has no effect on the free A molecules. Then we can write at equilibrium

$$c_A(\text{free inside}) = c_A(\text{outside})$$

But the total concentration of A inside is just the sum of the free and bound A:

$$c_A(\text{inside}) = c_A(\text{free inside}) + c_A(\text{bound inside})$$

Therefore, from equilibrium dialysis,

$$c_A(\text{bound inside}) = c_A(\text{inside}) - c_A(\text{outside}) \qquad (5.7)$$

or

$$c_A(\text{bound to macromolecule}) = c_A(\text{macromolecule side}) - c_A(\text{solvent side})$$

We have obtained the concentration of A bound to a macromolecule in terms of measurable quantities: the total concentration of A on each side of the dialysis membrane. We are usually interested in the number, ν, of small molecules bound per macromolecule. This is

$$\nu = \frac{c_A(\text{bound})}{c_M} \qquad (5.8)$$

where ν = average number of A's bound to macromolecule
$c_A(\text{bound})$ = concentration of A bound
c_M = concentration of macromolecule

The number of A's bound will depend on the equilibrium constant for binding, the number of binding sites on the macromolecule, and the concentrations of macromolecules and of A. When there is only one binding site, ν varies from 0 to

1 and is identical to θ, the *fraction of sites bound.* If there are N binding sites, the fraction of sites bound is

$$\theta = \frac{\nu}{N} \tag{5.9}$$

If the binding sites on the macromolecule are identical and independent, the equilibrium expression for θ does not depend further on the number of sites per macromolecule. The terms "identical" and "independent" mean that binding at one site does not change the binding at another site, and all of the sites have the same binding equilibrium constant. We can rather easily derive an expression for θ for one site and apply it to any number of sites. Consider the equilibrium

$$\mathrm{P + A \overset{K}{\rightleftharpoons} P \cdot A}$$

$$K = \frac{[\mathrm{P \cdot A}]}{[\mathrm{P}][\mathrm{A}]} = \text{equilibrium constant for binding}$$

where $[\mathrm{P}]$ = concentration of free polymer at equilibrium
$[\mathrm{A}]$ = concentration of free A at equilibrium
$[\mathrm{P \cdot A}]$ = concentration of complex at equilibrium

The fraction of sites with A bound is

$$\theta = \frac{[\mathrm{P \cdot A}]}{[\mathrm{P}] + [\mathrm{P \cdot A}]}$$

We are ignoring activity coefficients. Using the equilibrium constant, we substitute for $[\mathrm{P \cdot A}]$:

$$[\mathrm{P \cdot A}] = K[\mathrm{P}][\mathrm{A}]$$

$$\theta = \frac{K[\mathrm{P}][\mathrm{A}]}{[\mathrm{P}] + K[\mathrm{P}][\mathrm{A}]}$$

$$= \frac{K[\mathrm{A}]}{1 + K[\mathrm{A}]} \tag{5.10}$$

This equation is true for all N, if the sites are identical and independent. Solving for $K[\mathrm{A}]$, we obtain

$$\frac{\theta}{1 - \theta} = K[\mathrm{A}] \tag{5.11}$$

Substituting ν/N for θ and simplifying, we have

$$\frac{\nu}{N - \nu} = K[\mathrm{A}]$$

By rearranging, we obtain the *Scatchard equation*:

$$\frac{\nu}{[\mathrm{A}]} = K \cdot (N - \nu) \tag{5.12}$$

The Scatchard equation (5.12) is very often used to study binding to a macromolecule. The binding can be measured by any method, but equilibrium dialysis is often convenient. The value of [A] which we designated as c_A(free) is the concentration on the solvent side of the dialysis membrane at equilibrium. The value of ν comes from the concentration of A on the macromolecule side at equilibrium [see Eqs. (5.7) and (5.8)]. A plot of $\nu/[A]$ versus ν should give a straight line with slope of minus K, y intercept of NK, and x intercept of N, as shown in Fig. 5.3. If a Scatchard plot does not give a straight line, this indicates

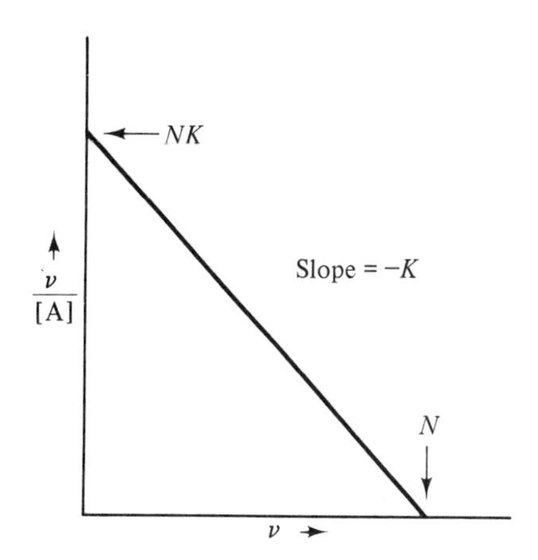

Fig. 5.3 Scatchard plot of the binding of substrate A to a macromolecule containing N binding sites, where ν is the average number of A molecules bound per macromolecule, and K is the equilibrium binding constant.

that the binding sites are not identical or not independent. One often finds a straight line for low ν but curvature for higher number of A's bound. The first few molecules bound may not change the properties of the macromolecule very much and not interact with each other; but as more molecules are bound, interactions between them become more likely. The consequences of these interactions between adjacent binding sites are discussed more fully in Chapter 11.

SURFACES, MEMBRANES, AND SURFACE TENSION

When we think of phases we think of the usual solid, liquid, or gaseous states. However, whenever there are two phases in contact, there is also a surface, or interface, between them. This surface has properties different from those in the two bulk phases and therefore will have different behavior. The surface has thermodynamic properties specified by its free energy, enthalpy, and so on, just as bulk phases have. However, there are differences. A surface is two-dimensional instead of three-dimensional. This means that concentration units for a surface are mol cm^{-2} instead of mol cm^{-3}. Furthermore, because the surface properties are determined relative to the bulk phase, the surface concentration is treated as an excess or a deficiency and can be either positive or negative. The surface pressure has dimensions of force per unit length instead of force per unit area; the surface free energy, surface enthalpy, and so on have units of cal cm^{-2}.

Usually, we ignore the surface contributions to the total thermodynamic properties of a system. However, we can focus attention on the properties of the surface if we choose to. Sometimes the surface area is so large that we cannot ignore its properties. This occurs for a system divided into very small pieces where a large fraction of the molecules are at the surface. One gram of finely divided carbon black can have a surface area of 500 m^2!

Biological cells are small volumes surrounded by membranes. (Membranes are present inside the cells, also.) The membranes are a different phase from the rest of the cell, and for some purposes they can be thought of as a surface phase. The membranes, of course, have a finite thickness and definite volumes, but it may be more useful to think about their concentrations in molecules cm^{-2} instead of cm^{-3}.

Surface Tension

Because of intermolecular attractions, the molecules at the surface of a liquid are attracted inward. This creates a force in the surface which tends to minimize the surface area. If the surface is stretched, the free energy of the system is increased. The free energy per unit surface area, or the force per unit length on the surface, is called the *surface tension*. Note that the units for energy per unit area and force per unit length are identical. Data in Table 5.1 illustrate the range of surface tensions that can occur. What happens when a substance is dissolved in the liquid, or added to the surface? The surface tension either decreases or does not change very much. It never increases greatly. There is a thermodynamic reason for this, but before stating it let us consider what would happen if we could greatly increase the surface tension of water. Water drops from your faucet could become the size of basketballs, because the size of the drop is directly proportional to the surface tension. You might have to cut water with a knife and chew it well before swallowing. Water would not wet anything, so processes depending on capillary action would not work.

Table 5.1 Surface tensions of pure liquids in air

Substance	Surface tension (dyn cm^{-1})	Temperature (°C)
Platinum	1819	2000
Mercury	487	15
Water	71.97	25
	58.85	100
Benzene	28.9	20
Acetone	23.7	20
n-Hexane	18.4	20
Neon	5.2	−247

Any substance that tends to raise the surface tension of a liquid raises the free energy of the surface. The substance will therefore not concentrate at the surface. Thus we are saved from these possible catastrophes because the surface tension, or the surface free energy, is just $(\partial G/\partial A)_{T,P}$, where A is the surface area. However, substances that lower the surface tension also lower the free energy of the surface; they will preferentially migrate to the surface. Thus substances that lower the surface tension will concentrate at the surface and can give large decreases in surface tension, but substances that raise the surface tension will avoid surfaces and can give only small increases in surface tension. The quantitative expression for this is called the *Gibbs adsorption isotherm*:

$$\Gamma = -\frac{1}{RT}\frac{d\gamma}{d \ln a} \cong -\frac{1}{RT}\frac{d\gamma}{d \ln c} \tag{5.13}$$

where Γ = adsorption (excess concentration) of solute at surface, mol cm^{-2}
γ = surface tension, dyn cm^{-1} = ergs cm^{-2}
R = gas constant = 8.314×10^7 erg deg^{-1} mol^{-1}
a = activity of solute in bulk solution
c = concentration of solute in bulk solution

The sign of the excess surface concentration, Γ, is opposite to the sign of the change of the surface tension with concentration (or activity) of solute in the solution. Figure 5.4 shows the change in surface tension of water when LiCl or ethanol is added. Ionizing salts are almost the only solutes that raise the surface tension of water.

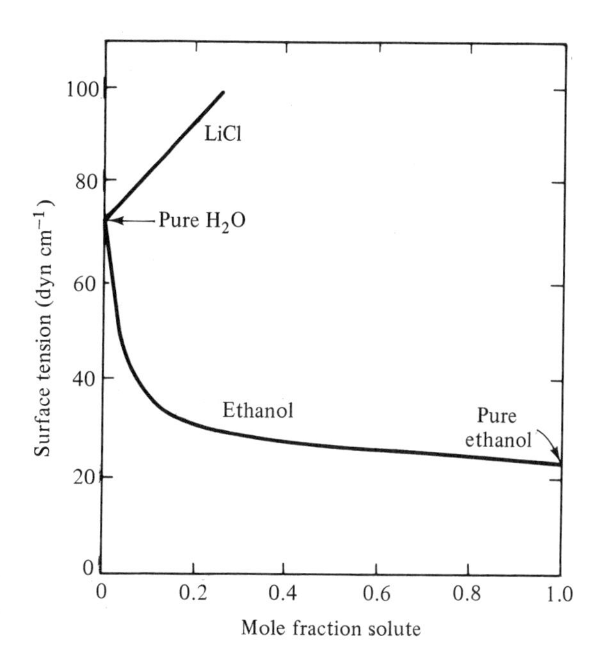

Fig. 5.4 Effect of solute concentration on surface tension in aqueous solution at 20°C.

Why do some molecules preferentially adsorb at a surface separating the two phases and thus reduce the surface tension? The molecule may consist of two parts or regions, one of which interacts primarily with one phase and the other with the second phase. As mentioned in Chapter 3, such molecules are said to be amphiphilic, and they tend to locate preferentially at the interface. This is what happens to detergent molecules that have a hydrocarbon tail and a polar head; they are called *surface-active molecules*, or *surfactants*. Sodium dodecylsulfate (see Fig. 5.5), a principal component of many detergents, orients at the surface of

$Na^+ \, ^-O_3SO(CH_2)_{11}CH_3$	Sodium dodecylsulfate
$Na^+ \, ^-O_2C(CH_2)_{14}CH_3$	Sodium palmitate
$(CH_3)_3\overset{+}{N}(CH_2)_2O\overset{-}{P}O_3CH_2$ $CHO_2C(CH_2)_{14}CH_3$ $CH_2O_2C(CH_2)_{14}CH_3$	Dipalmitoyl lecithin

Air

Sodium palmitate

Water phase

SURFACE TENISON LOWERING

Water phase

Sodium dodecylsulfate

Dirt phase

DETERGENT ACTION

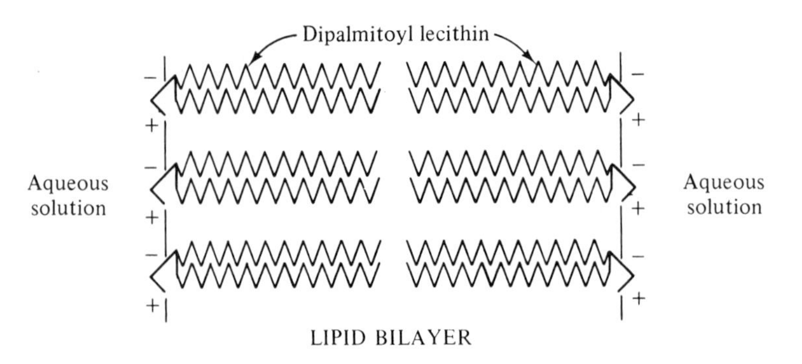

Fig. 5.5 Some common surface-active molecules and their orientations at interfaces.

dirt particles and makes them soluble in water. The hydrocarbon tail is attracted to the oily dirt "phase," and the polar sulfate head faces the water. The hydrocarbon part is hydrophobic; the polar group is hydrophilic, which is typical of the characteristics of amphiphilic molecules.

There are many naturally occurring detergents in plants and animals. Dipalmitoyl lecithin forms a layer (see Fig. 5.5) that lowers the surface tension at the surface of the alveoli* in the lung and allows you to breathe. A large surface area is necessary for the efficient exchange of gases in the lung. Dipalmitoyl lecithin lowers the surface tension of water to nearly zero and allows large aqueous surface areas to exist in the lung. Premature babies lack this vital surfactant and have great difficulty breathing effectively.

The surface-tension decrease caused by surface-active molecules is easily measured with a Langmuir film balance, as illustrated in Fig. 5.6. One adds some

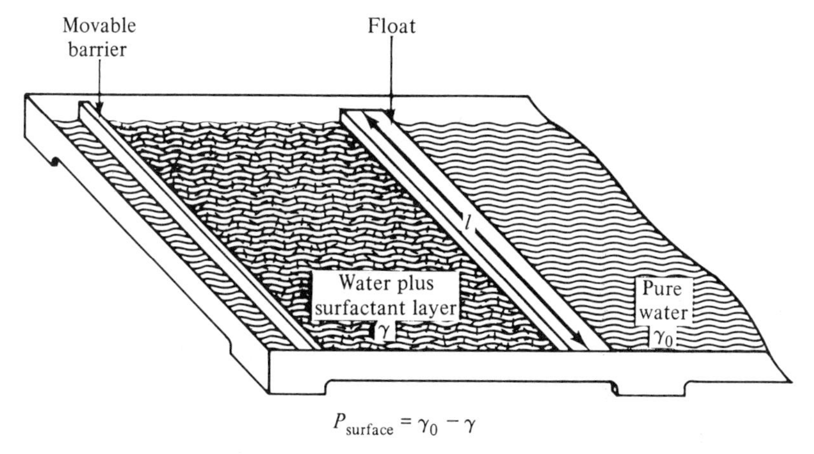

Fig. 5.6 Surface tension can be measured using a Langmuir film balance. The force per unit length, F/l, required to compress a surface containing a surfactant layer is known as the surface pressure, P_{surface}. The float is connected to a torsion wire (not shown) to measure the force.

molecules to the surface of water and uses the movable barrier to compress them. The surface pressure necessary to maintain the film in the compressed state is just the difference between the surface tension of the water and the surface tension of the coated water.

$$P_{\text{surface}}\,(\text{dyn cm}^{-1}) = \gamma_{H_2O} - \gamma_{\text{film}}$$

The surface concentration in mol cm^{-2} depends on the surface pressure. The experiment is the two-dimensional analog of a pressure versus volume experiment in three dimensions. It is traditional to plot surface pressure versus area per molecule in Å^2 molecule^{-1}. The area per molecule is just the reciprocal of the

*The alveoli are the smallest air compartments found in the lung.

surface concentration. Some representative data are given in Fig. 5.7. Note that the high surface pressures result from the low surface tensions produced by the surface-active molecules. The maximum possible surface pressure, equal to the surface tension of pure water, corresponds to zero surface tension of the film. This is implicit in the preceding equation.

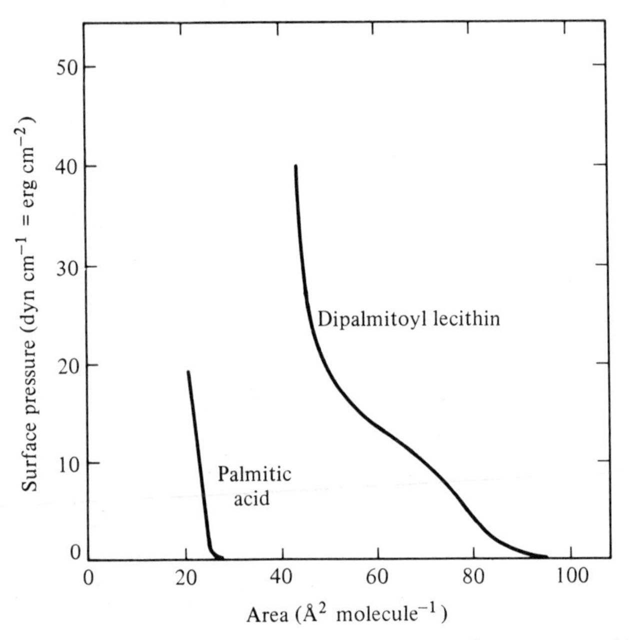

Fig. 5.7 Surface pressure versus area curves for two amphiphilic molecules (surfactants) at an air-water interface. The molecular areas correspond to average cross sections of the molecules projected on the surface. The steeply rising portions of the curves occur when the molecules in the surface layer have been compressed into a compact arrangement.

Biological Membranes

The presence of molecules at a surface markedly changes other surface properties besides the surface tension. The rate of transport of water, ions, or other solutes across the surface is strongly affected. The composition and orientation of the molecules in biological membranes determine their function. Along with the diversity in composition comes a richness of properties that correlates with the variety of membrane functions. Usually membranes contain an assortment of lipid molecules with diverse chemical structures, together with proteins and sometimes polysaccharides. The lipids are typically fatty acid esters that differ in the length of the fatty acid chain, the degree of unsaturation, the charge or polarity of the esterifying group, and the number of fatty acids esterified per molecule. The proteins may be intrinsic or integral, in the sense that they are incorporated directly into the membrane structure, or extrinsic, if they are attached to the membrane surface or interact strongly with it. Other components, such as cholesteryl esters, may also be present.

The variety of membrane composition results in a range of physical properties. For example, the mobility of the molecules in two dimensions in a membrane may correspond to that of a typical liquid or to that of a solid. In fact, it is relatively common to encounter phase transitions in membranes or artificial surface films, and these transitions are closely analogous to their three-dimensional counterparts. The transition temperature or melting point, for example, is sensitive to the lipid composition; and the thermodynamic analysis of such two-dimensional solutions can be carried out just as for three-dimensional systems.

A simple way of studying the interactions among surface active molecules is to measure surface pressure versus area curves for mixtures of the components. Since the surface free energy in ergs cm^{-2} is equal to the surface tension, the surface enthalpy can be obtained from the temperature dependence of the surface tension:

$$G_{\text{surface}} = \gamma \tag{5.14}$$

$$H_{\text{surface}} = \gamma - T \cdot \left(\frac{\partial \gamma}{\partial T}\right)_P \tag{5.15}$$

There will be enthalpy changes associated with transitions between surface phases, and these can be determined experimentally by careful calorimetric measurements.

The relation between the intrinsic proteins and the lipids is more complex and is not yet well understood. Many proteins are amphiphilic and interact both with the hydrophobic lipids and with the polar aqueous interface. These hydrophobic and hydrophilic sites are located in different regions of the protein molecule, resulting in a strong orientation with respect to the membrane surface. Apart from this orientation, the protein may behave much like a solute in the lipid, as solvent, in a kind of two-dimensional solution. There is even a two-dimensional analog of precipitation, in which the solution becomes saturated with respect to a protein constituent and the protein separates as a distinct phase. Electron microscopy of membrane surfaces provides a useful method for detecting and characterizing the phase changes.

ACTIVE AND PASSIVE TRANSPORT

Frogs, seaweed, and other organisms that live in contact with water have semipermeable skins. Water and some ions and small molecules pass through the skins; macromolecules generally do not. The frog or the seaweed can selectively concentrate certain molecules inside itself and selectively exclude or excrete other molecules. How do they do it? The thermodynamic question is: If a molecule can easily pass through the skin, how can the inside concentration be maintained at a value that is different from the outside concentration? The answer is to ensure that the free-energy change for the process is negative. For example, the presence of a protein inside seaweed which strongly binds iodide ion ensures that the iodide concentration in the seaweed is always higher than in the seawater. If the concentration of *free* iodide is the same inside and outside, the bound iodide would account for the concentrating effect of the seaweed. This effect, known as *passive transport*, does not depend on whether the seaweed is alive or dead; that is, metabolism is not involved in the concentrating effect. The presence of the specific binding protein lowers the chemical potential of the iodide, which therefore concentrates in the seaweed. There is another kind of transport, known as *active transport*, which is closely dependent on active cellular metabolism. Active transport is defined as the transport of a substance from a lower to a higher chemical potential. Because the total free energy of the process must be negative, active transport must be tied to a chemical reaction that has a negative free energy. In a biological system this means that metabolism must be occurring. Therefore, an experimental test of whether active transport is involved is to poison the metabolic activity and see if the transport also stops. A well-known example of active transport is the concentration of K^+ ions and the dilution of Na^+ ions that occurs inside the cells of animals. An illustration involving representative concentrations is shown in Fig. 5.8. Furthermore, a voltage difference of −60 mV is often present; the outside is positive relative to the inside. The net reaction for the active transport is

$$\left.\begin{matrix}\text{Na}^+(\text{inside}) \\ + \\ \text{K}^+(\text{outside})\end{matrix}\right\} + \text{ATP} \longrightarrow \text{ADP} + \text{phosphate} + \left\{\begin{matrix}\text{Na}^+(\text{outside}) \\ + \\ \text{K}^+(\text{inside})\end{matrix}\right.$$

The thermodynamics of the transport is

$$\Delta G = \Delta G(\text{transport of Na}^+) + \Delta G(\text{transport of K}^+) + \Delta G(\text{hydrolysis of ATP})$$

$$\Delta G\ (\text{transport of Na}^+) = RT \ln \frac{a_{\text{Na}^+}\ (\text{outside})}{a_{\text{Na}^+}\ (\text{inside})} + ZFV \qquad (5.6)$$

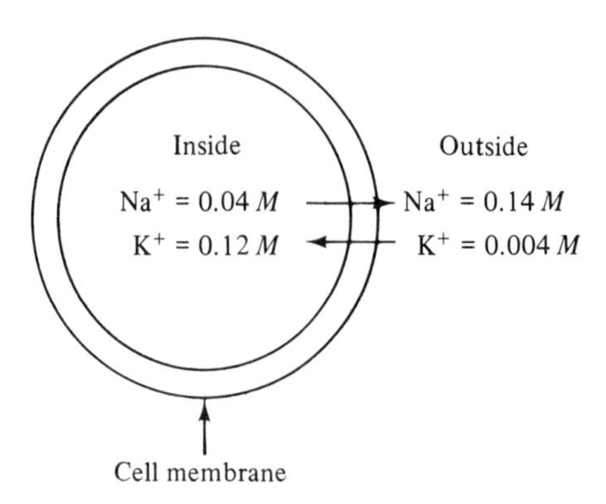

Fig. 5.8 Ion gradients occur typically across cell membranes in plants and animals. These are maintained by active transport.

At 25°C,

$$\Delta G\text{ (transport of Na}^+) = 1365\log\frac{0.14}{0.04} + (1)(23{,}060)(0.06)$$

$$= 2.1\text{ kcal}$$

$$\Delta G\text{(transport of K}^+) = 1365\log\frac{0.12}{0.004} + (1)(23{,}060)(-0.06)$$

$$= 0.6\text{ kcal}$$

Both transport steps have positive free energies. For the hydrolysis of the ATP, we have

$$\Delta G(\text{ATP}) = \Delta G^0(\text{ATP}) + RT\ln Q \tag{4.4}$$

ΔG^0(ATP) is large and negative ($-$ 7.4 kcal mol^{-1}); its sign will not be changed by the value of Q that corresponds to physiological conditions. The positive free-energy requirement for transporting the Na^+ and K^+ ions from low to high chemical potentials is compensated by the large negative free energy in hydrolyzing the ATP. A possible mechanism is shown in Fig. 5.9. The carrier, X, is

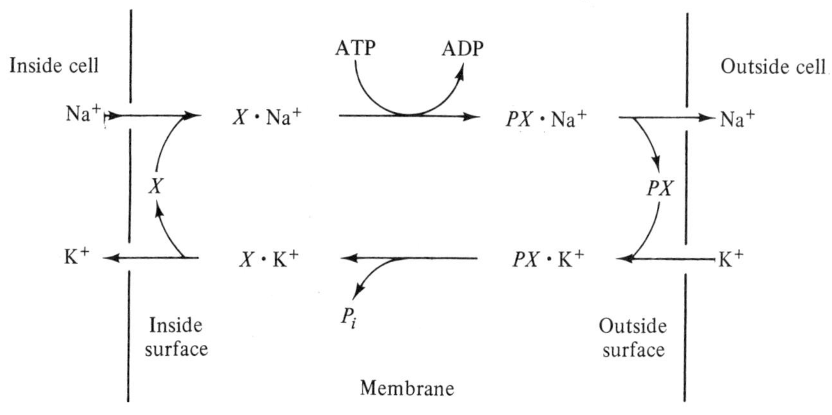

Binding equilibria:

$$Na^+ + X \xrightarrow{\Delta G\text{ negative}} Na^+ \cdot X$$
$$K^+ + X \xrightarrow{\Delta G\text{ positive}} K^+ \cdot X$$
at inside surface of membrane

$$Na^+ \cdot PX \xrightarrow{\Delta G\text{ negative}} Na^+ + PX$$
$$K^+ \cdot PX \xrightarrow{\Delta G\text{ positive}} K^+ + PX$$
at outside surface of membrane

Covalent reactions:

$$X \cdot Na^+ + \text{ATP} \xrightarrow{\Delta G\text{ negative}} PX \cdot Na^+ + \text{ADP}$$
$$PX \cdot K^+ \xrightarrow{\Delta G\text{ negative}} X \cdot K^+ + \text{phosphate}$$

Fig. 5.9 A plausible mechanism of active transport.

thought to be a membrane-bound enzyme called Na^+–K^+ ATPase. It has the property of preferentially binding K^+ when it is phosphorylated (PX), but preferring Na^+ when it is not phosphorylated (X). This is shown by the binding equilibria equations in the figure. Furthermore, in the presence of Na^+ the enzyme hydrolyzes ATP and becomes phosphorylated, whereas in the presence of K^+ the phosphorylated enzyme is hydrolyzed. The reactions are given in the figure. The final requirement is that the equilibrium between X and Na^+ or K^+ be confined to the surface of the membrane on the inside of the cell and the PX plus Na^+ or K^+ equilibrium be at the outside surface of the cell. This requirement is presumably met by a change in conformation of the enzyme that accompanies its phosphorylation.

This proposed mechanism for active transport of Na^+ and K^+ in certain cells is complicated, and it may not be correct. However, as usual, the thermodynamic conclusions are simple and correct. For transport to occur, the total free energy of the process must be negative. In *active* transport the transport is coupled to chemical reactions with negative free energies.

COLLIGATIVE PROPERTIES

The history of mountaineering and the thermodynamics of gas-liquid equilibria have always been closely linked. The reason is that the boiling point of water, or any liquid, depends on the pressure of its vapor in equilibrium with it. In the 1850s, intrepid mountaineers on first ascents carried 760-mm-long mercury manometers to record the pressure and thus the altitude at the peak. Many mountaineers failed or were hurt because of the bulky glass tube. The scientist mountaineer carried a thermometer to measure the boiling point of water at the peak. Present-day mountaineers who climb above 15,000 feet routinely carry a pressure cooker for meals and an aneroid barometer to measure altitudes.

Boiling Point and Freezing Point of a Pure Component

It is straightforward to derive the equation relating the boiling point of a pure liquid to the pressure of its gas phase in equilibrium. Consider the reaction

$$\text{A(pure liquid)} \longrightarrow \text{A(gas)}$$

The equilibrium constant is

$$K = \frac{a_A(\text{gas})}{a_A(\text{pure liquid})}$$

If we approximate the gas as an ideal gas, we replace the activity of gas by the pressure of the gas in atmospheres,

$$a_A(\text{gas}) = P(\text{atm})$$

We can ignore the effect of pressure on the activity of the liquid,

$$a_A(\text{pure liquid}) = 1$$

Therefore, the equilibrium constant is just the equilibrium gas pressure,

$$K = P$$

The temperature dependence of the equilibrium constant [Eq. (4.43)] gives

$$\ln \frac{P_2}{P_1} = \frac{-\Delta \bar{H}_{vap}}{R}\left(\frac{1}{T_2} - \frac{1}{T_1}\right) \quad (5.16)$$

where $\Delta \bar{H}_{vap} = \bar{H}_{gas} - \bar{H}_{liquid}$ = molar enthalpy (heat) of vaporization. This equation is known as the *Clausius-Clapeyron equation.* Let us consider its uses.

The pressures P_2 and P_1 are the equilibrium vapor pressures of a liquid at temperatures T_2 and T_1, respectively. The normal boiling point, T_b, is defined as the temperature where the vapor pressure is 1 atm. Therefore, from the measured boiling point of a pure liquid whose heat of vaporization and normal boiling point are known, one can obtain the vapor pressure.

Example 5.1 At 20,320 ft altitude at the summit of Mt. McKinley, pure water boils at only 75°C. We note in passing that "hot" tea will be tepid and weak at this altitude. What is the atmospheric pressure?

Solution The enthalpy of vaporization of water depends on temperature; for Eq. (5.16) we use an approximate value of $\Delta \bar{H}_{vap} = 10{,}000 \text{ cal mol}^{-1}$. $P_1 = 1$ atm, and $T_1 = 373$K (the normal boiling point of water).

$$\ln P = \frac{-10{,}000}{1.987}\left(\frac{1}{348} - \frac{1}{373}\right)$$

$$P = 0.38 \text{ atm}$$

This is the pressure of the water in equilibrium with the boiling liquid. Liquid boils when the vapor pressure of the liquid equals that of the surrounding atmosphere: therefore, the atmospheric pressure is 0.38 atm. The effect of this reduced pressure on the equilibrium binding of oxygen to hemoglobin in the lungs is another problem in phase equilibria that concerns mountaineers.

The Clausius-Clapeyron equation can be used to calculate a boiling point from two measured vapor pressures at different temperatures. First, the enthalpy of vaporization is calculated; then the temperature corresponding to a vapor pressure of 1 atm is calculated. An easy way to apply the Clausius-Clapeyron equation is to plot log P versus $1/T$. The slope is equal to $-\Delta \bar{H}_{vap}/2.303R$. The vapor pressure at any temperature, or the boiling point at any pressure, can be read from the graph.

In using Eq. (5.16) we should remember its limitations. The main one is the assumption that $\Delta\bar{H}_{vap}$ is independent of temperature. Over a wide-enough temperature range $\log P$ versus $1/T$ will be a curve; the slope at any point will give $\Delta\bar{H}_{vap}$ at that temperature. The assumption of gas ideality and the neglect of pressure dependence of the liquid activity are usually valid. We can increase the activity and therefore increase the vapor pressure of a liquid by applying an external pressure. However, this effect is small compared to the large effect of temperature on the vapor pressure.

For the effect of pressure on melting-point temperature, we need to use a somewhat different and even more general approach. Consider a solid and a liquid in equilibrium at temperature T and pressure P. Then

$$\bar{G}_s = \bar{G}_l$$

at T and P. Now increase the temperature by an amount dT and the pressure by an amount dP so as to maintain the solid and liquid in equilibrium. Under the new conditions,

$$\bar{G}_s + d\bar{G}_s = \bar{G}_l + d\bar{G}_l$$

at $T + dT$, $P + dP$. Subtracting the first equation from the second, we obtain

$$d\bar{G}_s = d\bar{G}_l$$

This states that if we change the pressure of a solid substance in equilibrium with its liquid, we must also change the temperature so as to keep the increment in free energy the same for the two phases. Otherwise, the one with higher free energy will be converted into the one with lower free energy.

Using

$$d\bar{G} = -\bar{S}\,dT + \bar{V}\,dP \tag{3.48}$$

we obtain

$$-\bar{S}_s\,dT + \bar{V}_s\,dP = -\bar{S}_l\,dT + \bar{V}_l\,dP$$

which can be rearranged to give

$$\frac{dT}{dP} = \frac{\Delta\bar{V}_{fus}}{\Delta\bar{S}_{fus}}$$

$$\Delta\bar{V}_{fus} = \bar{V}_l - \bar{V}_s \qquad \Delta\bar{S}_{fus} = \bar{S}_l - \bar{S}_s$$

Because the process is reversible,

$$\Delta\bar{S}_{fus} = \frac{\Delta\bar{H}_{fus}}{T}$$

and

$$\frac{dT}{dP} = \frac{T\Delta\bar{V}_{fus}}{\Delta\bar{H}_{fus}} \tag{5.17}$$

This is an example of a general expression that can be applied to equilibrium between any two phases. For example, Eq. (5.16) can be derived as a special case for liquid-vapor equilibrium.

The effect of external pressure on melting points is not large, so we can write Eq. (5.17) as

$$\frac{T_2 - T_1}{P_2 - P_1} = T\frac{\Delta \bar{V}_{fus}}{\Delta \bar{H}_{fus}} \tag{5.18}$$

where T = average of T_2 and T_1
$\Delta \bar{V}_{fus} = \bar{V}(\text{liquid}) - \bar{V}(\text{solid})$ = molar volume change of fusion
$\Delta \bar{H}_{fus} = \bar{H}(\text{liquid}) - \bar{H}(\text{solid})$ = molar enthalpy (heat) of fusion

We must be careful with units in this equation. If P is in atm, $\Delta \bar{V}_{fus}$ in $cm^3\ mol^{-1}$, and $\Delta \bar{H}_{fus}$ in $cal\ mol^{-1}$, the conversion factor of 41.29 $cm^3\ atm\ cal^{-1}$ must be used. The effect of pressure on a melting point depends on the signs and magnitudes of two thermodynamic variables: $\Delta \bar{H}_{fus}$ and $\Delta \bar{V}_{fus}$. The heat of fusion is always positive; heat is always absorbed on melting. But the volume of fusion can be positive or negative, although it is usually positive. Luckily, for ice skaters, H_2O is an exception. The volume of water decreases on melting. The pressure of the ice skate lowers the melting point of the ice and provides some liquid water as a lubricant.

Colligative Properties of Ideal Solutions

The vapor pressure, boiling point, freezing point, and osmotic pressure of a solution are known as *colligative properties*. For sufficiently dilute solutions, these properties are linearly dependent on the concentration of solutes present. Furthermore, they are all closely related to each other. If one is measured, the others can be calculated. For example, if the osmotic pressure of a solution is measured, the vapor pressure of the solvent is immediately known. For dilute or ideal solutions, from the measured boiling-point rise, the freezing-point lowering can be calculated. We find that in ideal solutions the colligative properties depend only on the number of solute molecules. So if we measure the molar concentration of the solute in solution, we can calculate any colligative properties. The effective link between the measured colligative properties is the activity of the solvent. For a real solution we can obtain the activity of the solvent in the solution from one measured colligative property. Then any other colligative property at the same temperature can be calculated.

In this section we shall consider only the empirical results found for ideal solutions. The general thermodynamic equations relating the activity of the solvent to colligative properties will be discussed in a later section.

Raoult's law

Adding a solute to a solvent decreases the vapor pressure of the solvent. Raoult found experimentally that the vapor pressure of the solvent in a solution is directly proportional to its mole fraction. *Raoult's law* is stated:

$$P_A = X_A P_A^0 \tag{5.19}$$

where P_A = vapor pressure of component A in solution
X_A = mole fraction of A
P_A^0 = vapor pressure of pure A

Raoult's law is like the ideal gas law, in that it applies approximately to many solutions and it applies exactly to all very dilute solutions. Therefore, for a dilute solution you can always calculate the vapor pressure of the solvent simply by knowing its mole fraction, X_A, and the vapor pressure of pure solvent at the same temperature. If Raoult's law applies, the solution is defined to be ideal.

Henry's law

The solutes in the solution may also be volatile. What about their vapor pressures? Henry found experimentally that the solute vapor pressure is proportional to the solute mole fraction, but the proportionality constant is not, in general, the vapor pressure of the pure solute. *Henry's law* is stated:

$$P_B = k_B X_B \tag{5.20}$$

where P_B = vapor pressure of component B
X_B = mole fraction of B
k_B = Henry's law constant for component B in solvent A

You can think of Henry's law as specifying the pressure dependence of the solubility of a gas in a solution. It says that the gas solubility should be directly proportional to pressure:

$$X_B = \frac{P_B}{k_B} \tag{5.21}$$

The solubility will also depend on temperature and the nature of the solvent. These must be specified for each value of the Henry's law constant k_B. Table 5.2

Table 5.2 Henry's law constants ($k_B = P_B/X_B$) for aqueous solutions at 25°C

Gas	k_B(atm)
He	131×10^3
N_2	86×10^3
CO	57×10^3
O_2	43×10^3
CH_4	41×10^3
Ar	40×10^3
CO_2	1.6×10^3
C_2H_2	1.3×10^3

gives Henry's law constants for various gases in water in order of decreasing vapor pressure of the gas and therefore in order of increasing solubility of the gas. Using Eq. (5.21), you will obtain the solubility as mole fraction, but because these are dilute aqueous solutions, it is easy to see that the molarity is just 55.6 (= 1000/18) times the mole fraction; for dilute aqueous solutions: $c = 55.6X$.

Example 5.2 Divers get the bends when bubbles of N_2 gas form in their bloodstream if they rise too rapidly from a deep dive. Calculate the solubility (mol ℓ^{-1}) of N_2 in water (this is roughly equal to the solubility in blood) at a depth of 300 ft below the surface of the ocean and at sea level.

Solution First, we need to find the pressure at a depth of 300 ft. We know that a column of mercury 760 mm (= 2.49 ft) high exerts a pressure of 1 atm. An equal column of seawater will exert a pressure proportional to the ratio of the density of seawater (1.01) to mercury (13.6); therefore, the pressure 300 ft deep in the ocean is

$$P = \frac{300\text{ ft}}{2.49\text{ ft atm}^{-1}}\left(\frac{1.01}{13.6}\right)$$

$$= 8.9\text{ atm}$$

The solubilities are obtained from Eq. (5.21) and $c = 55.6X$.

$$c_{N_2} = \frac{55.6P_{N_2}}{k_{N_2}} = \frac{55.6P_{N_2}}{86 \times 10^3}$$

$$= 0.65 \times 10^{-3}\text{ mol }\ell^{-1}\text{ at 1 atm}$$

$$= 5.7 \times 10^{-3}\text{ mol }\ell^{-1}\text{ at 8.9 atm}$$

The diver should rise slowly to allow plenty of time for the extra dissolved nitrogen in his blood to equilibriate with the decreasing external pressure.

Boiling-point rise and freezing-point lowering

When a small amount of solute is added to a solvent, the boiling point of the solution is raised and the freezing point of the solution is lowered relative to that of the solvent. The change in the boiling point and freezing point is found to be directly proportional to the concentration of the solute in dilute solution.

$$T_0 - T_f = K_f m \tag{5.22}$$

$$T_b - T_0 = K_b m \tag{5.23}$$

where T_0 = freezing point or boiling point of pure solvent (it depends on pressure)

T_f, T_b = freezing point and boiling point of solution

K_f, K_b = constants that depend only on the solvent ($K_f = 1.86$ and $K_b = 0.51$ for water)

m = molality of solution = number of moles of solute in 1000 g of solvent

Because K_f and K_b depend only on the solvent, Eqs. (5.22) and (5.23) can be used to measure total concentration of solutes in complicated mixtures of solutes in solution. For example, the freezing point of seawater immediately gives a measure of the total number of moles of ions and other solutes of all kinds present in the seawater. In using Eqs. (5.22) and (5.23), we must remember that the solutions should be dilute and that only the pure solvent must freeze or evaporate. The solutes should not be volatile at the boiling point or soluble in the solid solvent at the freezing point for the simple equations to apply.

We have been using three different concentration scales, so it is well to know the relations among them. From the definition of mole fraction, X, molality, m, and molarity, c, we can convert from one to the other. Here we shall give the relations that apply to dilute solutions (m or $c < 0.1$):

$$X_B = \frac{M_A m}{1000} \tag{5.24}$$

$$c = \rho_0 m \tag{5.25}$$

where X_B = mole fraction of solute
M_A = molecular weight of solvent
m = molality of solute
c = molarity of solute
ρ_0 = density of solvent

We note that for dilute aqueous solution,

$$X_B = \frac{m}{55.6} \tag{5.26}$$

$$c = m$$

Osmotic pressure

It has happened that a hospital patient died because pure water was accidently injected into the veins. The osmotic pressure of the solution inside the blood cells is high enough to cause them to break under these circumstances. Osmotic pressure is defined as the pressure that must be applied to a solution to keep its solvent in equilibrium with pure solvent at the same temperature. If a solution is separated from a solvent by a semipermeable membrane (such as the membrane of a red blood cell), solvent will flow into the solution side until equilibrium is reached. At constant temperature, equilibrium can be attained by increasing the pressure on the solution or by diluting the solution until it is essentially pure solvent. A red blood cell placed in pure water or in a dilute salt solution will burst before equilibrium is reached.

Osmotic pressure is a colligative property closely related to the vapor pressure, freezing point, and boiling point. The activity of the solvent is the parameter that is most easily related to the osmotic pressure.

Adding a solute to a solution lowers the activity of the solvent; by raising the pressure on the solution, the activity of the solvent is increased. Van't Hoff found empirically that for dilute solutions the osmotic pressure, π, is directly proportional to the concentration of the solute. Because of the importance of osmotic pressure in biology, we shall derive the relevant equations.

Figure 5.10 shows the condition at equilibrium for a solution in contact with pure solvent through a semipermeable membrane that allows only solvent to pass. The equilibrium occurs in an osmometer in which the extra pressure, π,

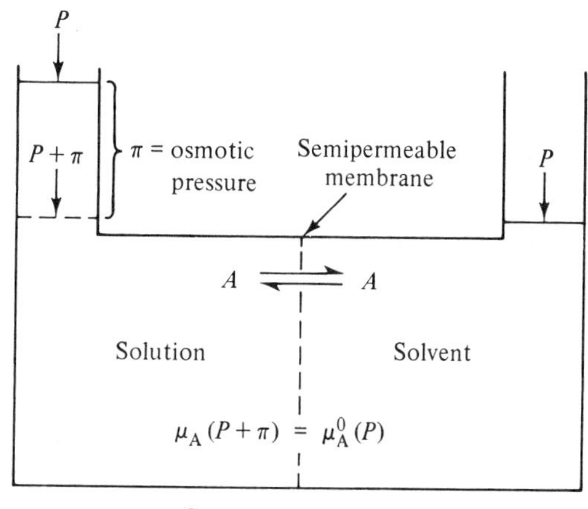

Fig. 5.10 Osmometer showing conditions at equilibrium. *A* is a solvent molecule.

necessary for equilibrium is produced by the hydrostatic pressure of a water column on the solution side. At equilibrium, there will be no net flow of solvent across the membrane, so the chemical potential of the pure solvent at pressure P must be equal to the chemical potential of the solvent in solution at pressure $P + \pi$:

$$\mu_A(\text{solution}, P + \pi) = \mu_A(\text{solvent}, P)$$

The temperature is constant throughout. If we let $P = 1$ atm, then at equilibrium,

$$\mu_A(\text{solution}, 1 + \pi) = \mu_A^0(\text{solvent}) \tag{5.27}$$

Because of the presence of the solute in solution, the chemical potential of A at 1 atm pressure is decreased by an amount that can be calculated using Eq. (4.14):

$$\Delta\mu_{\text{osm}} = \mu_A(\text{solution, 1 atm}) - \mu_A^0(\text{solvent}) = RT \ln a_A$$

To compensate for this change and restore equilibrium with the solvent, we need to increase the chemical potential by an equal amount by applying external pressure so that Eq. (5.27) will be satisfied. Remembering that

$$\bar{G}_2 - \bar{G}_1 = \int_{P_1}^{P_2} \bar{V}\,dP = \bar{V}\cdot(P_2 - P_1) = \Delta\mu \tag{3.33}$$

we obtain

$$\Delta\mu_{\text{osm}} = RT \ln a_A = \int_{1+\pi}^{1} \bar{V}_A \, dP = -\bar{V}_A \pi$$

and

$$\ln a_A = \frac{-\pi \bar{V}_A}{RT} \tag{5.28}$$

For dilute solutions, we can use the following approximation for $\ln a_A$:

$$\ln a_A = \ln X_A = \ln (1 - X_B) \cong -X_B = -\frac{n_B}{n_A + n_B} \cong -\frac{n_B}{n_A} \tag{5.29}$$

Then

$$\pi = \frac{RT}{n_A \bar{V}_A} n_B \tag{5.30}$$

Furthermore, in dilute solutions, $n_A \bar{V}_A$ is essentially equal to the volume of solution (the volume of solute is negligible). Therefore, for dilute solutions,

$$\pi V = n_B RT \tag{5.31a}$$

$$\pi = cRT \tag{5.31b}$$

where π = osmotic pressure, atm
V = volume of solution, ℓ
n_B = number of moles of solute in solution
R = gas constant = $0.08205 \ \ell \text{ atm deg}^{-1} \text{ mol}^{-1}$
c = concentration of solute, mol ℓ^{-1}

These equations state that the osmotic pressure of a solution is directly proportional to the concentration of solute in dilute solution. Equation (5.31a) is written to look like the ideal gas equation, to make it easy to remember. However, you must also remember that π is not the pressure of the solute, but rather the pressure that must be applied to the solution to keep solvent from flowing in through a semipermeable membrane.

To gain an appreciation for the magnitude of osmotic pressures, note that according to Eq. (5.31) a solution with $c = 0.5 \text{ mol } \ell^{-1}$ at 310 K (37°C) will have an osmotic pressure of 12.7 atm. This is typical of the osmotic concentration of the cytoplasm inside animal cells. When such cells come into contact with pure water, the osmotic pressure may cause destructive rupture of the cell membranes. This happens in the hemolysis of red blood cells when the blood serum is diluted with pure water.

MOLECULAR-WEIGHT DETERMINATION

The molecular weight is an important property of any molecule, but it is particularly important for proteins, nucleic acids, and other macromolecules. Just learning the rough size of the molecule may be sufficient for some purposes (M = 1000?, 10,000?, 1,000,000?), but usually we need more accurate molecular weights. The colligative properties that we have been discussing in the previous section are all used to measure molecular weights. Osmotic pressure is the most useful for macromolecules.

Vapor-pressure lowering, freezing-point lowering, boiling-point rise, and osmotic pressure are all related to the mole fraction of solvent in dilute solution. This means that they all depend on the number of solute molecules (or ions, in the case of electrolytes) added to the solvent. If we divide the number of solute molecules by the weight of the solute, we obtain the molecular weight.

Vapor-pressure Lowering

In dilute solution we can write Raoult's law,

$$X_A = \frac{P_A}{P_A^0} \tag{5.19}$$

as

$$X_B = 1 - X_A = \frac{P_A^0 - P_A}{P_A^0} \cong \frac{n_B}{n_A}$$

where X_B = mole fraction of solute
P_A^0 = vapor pressure of pure solvent
P_A = vapor pressure of solvent in solution
n_A, n_B = number of moles of solvent and solute, respectively

The number of moles of solute, n_B, is equal to wt_B/M_B; therefore,

$$M_B = \frac{\text{wt}_B}{n_A} \cdot \frac{P_A^0}{P_A^0 - P_A} \tag{5.32}$$

where M_B = molecular weight of solute
wt_B = weight of solute

For a solution of known weight concentration, vapor-pressure measurements provide the molecular weight of the solute. The solution should be dilute and ideal, so measurements are made as a function of concentration and extrapolated to zero concentration. Inaccuracies in measuring vapor pressures put an upper limit of a few thousand on the molecular weight that can be measured. For very high molecular weights, the vapor-pressure lowering is so small that slight temperature differences between solvent and solution can ruin the results.

A more significant problem for high-molecular-weight solutes is the effect of impurities. Because the vapor pressure depends on the mole fraction of solvent, each solute molecule contributes equally to decreasing the vapor pressure. In dilute solution a large molecule has the same effect on vapor pressure as a small molecule. On a weight basis, the small molecule produces a much larger effect. One milligram of solute with a molecular weight of 100 lowers the vapor pressure of solution as much as 1 g of solute with a 100,000 molecular weight. It is clear that vapor-pressure lowering cannot be used to measure the molecular weight of a protein dissolved in a buffer solution; the buffer ions would dominate the vapor-pressure lowering.

Boiling Points and Freezing Points

The change of the boiling point or freezing point of a dilute solution is inversely proportional to the molecular weight of the solute.

Equations (5.22) and (5.23) can be written as:

Boiling point:

$$M_B = \frac{\text{wt}_B}{\text{kg solvent}} \cdot \frac{K_b}{\Delta T_b} \tag{5.33}$$

Freezing point:

$$M_B = \frac{\text{wt}_B}{\text{kg solvent}} \cdot \frac{K_f}{\Delta T_f} \tag{5.34}$$

where M_B = molecular wt of solute

ΔT_b = increase in boiling-point temperature

ΔT_f = decrease in freezing-point temperature

$\frac{\text{wt}_B}{\text{kg solvent}}$ = weight of solute per 1000 g of solvent

These methods have the same limitations as do vapor-pressure measurements. Molecular weights should be no higher than about 10,000 and impurities must be absent. In addition, the measurements are limited to temperatures near the boiling point or freezing point of the solvent. These methods have limited applicability to purely biological systems, but in special cases they may be quite useful. A study of the molal concentration of total solute particles in different cells of the kidney was made by putting a kidney slice on the cold stage of a microscope. The freezing point determined for each separate cell immediately gave the total molality of its contents.

Osmotic Pressure

Osmotic-pressure measurement is the most useful colligative method for determining molecular weights. It is accurate for molecular weights approaching 1 million, and small molecule impurities, buffers, or salts do not interfere if suitable membranes are used. The molecular weight of only the solutes that do not pass the semipermeable membrane is obtained.

Equation (5.31b) can be rewritten as

$$\pi = \frac{w}{M} RT \tag{5.35}$$

where π = osmotic pressure, atm
w = concentration of solution, g ℓ^{-1}
R = gas constant = 0.08205 ℓ atm deg^{-1} mol^{-1}
M = molecular weight
T = absolute temperature

Usually, a series of concentrations is measured, and extrapolation is made to zero concentration. A plot of π/w versus w gives RT/M as the intercept; the slope characterizes deviations from ideality (see Fig. 5.11). A positive slope is usually

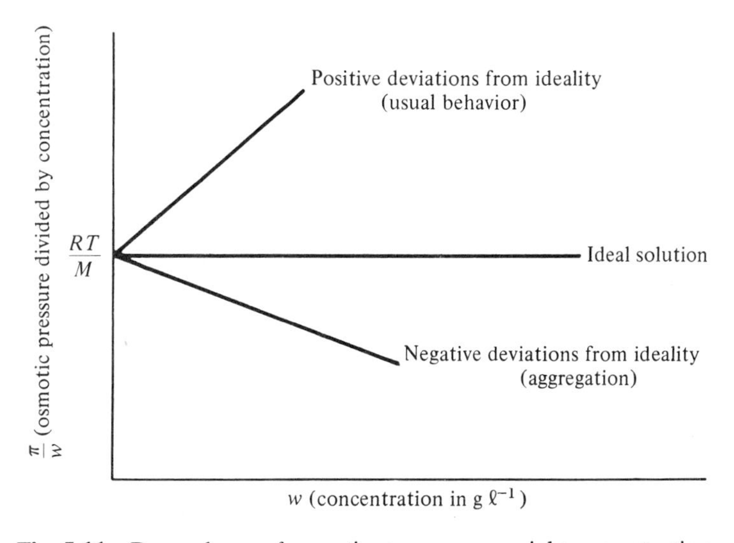

Fig. 5.11 Dependence of osmotic pressure on weight concentration.

found for these plots; it can be accounted for in terms of the activity coefficient of the solute. A negative slope indicates an increase in molecular weight with increasing concentration. This will occur if the solute undergoes aggregation. If we assume that the solution is ideal, then π/w can be equated to RT/M at any concentration; for real solutions this relation may begin to fail for concentrations greater than 0.01 M. Osmotic pressure, as well as the other colligative methods, can be used to study aggregation in solution at higher concentrations.

Number-average Molecular Weights

For aggregating molecules (or synthetic polymers or mixtures of molecules), the methods that we have been discussing will give average molecular weights. All colligative methods give a number-average molecular weight. This is defined as

$$M_n = \frac{\sum_i n_i M_i}{\sum_i n_i} \tag{5.36}$$

where M_n = number-average molecular weight
n_i = number of molecules with molecular weight M_i

The sums are over all solute species. For example, if we use vapor-pressure lowering or some other colligative method to measure the molecular weight of sodium chloride in water, we will obtain the arithmetic mean of the molecular weights of Na^+ and Cl^-:

$$M_n = \frac{M_{Na^+} + M_{Cl^-}}{2} = \frac{23 + 35.5}{2} = 29.25$$

Equation (5.36) can also be written in terms of the weight of each species present in solution:

$$M_n = \frac{\sum_i \text{wt}_i}{\sum_i \frac{\text{wt}_i}{M_i}} \tag{5.37}$$

where wt_i = weight of molecules with molecular weight M_i. This equation illustrates that for equal weights of various species the lower-molecular-weight species dominate the number average. This is why impurities can be so significant in molecular-weight determinations of macromolecules.

Example 5.3 How will 1.0 wt % impurity affect the determination of the number-average molecular weight of a macromolecule whose true molecular weight is 100,000? Assume that the impurity has a molecular weight of 100.

Solution

$$M_n = \frac{0.99 + 0.01}{\frac{0.99}{100{,}000} + \frac{0.010}{100}} = \frac{1.00}{0.000110}$$
$$= 9.1 \times 10^3$$

Although the sample is 99% pure, the molecular weight obtained is more than tenfold too small. Osmotic-pressure measurements that allow the small-molecule impurities to equilibrate and not contribute to the sums in Eqs. (5.36) and 5.37) are obviously preferable.

Weight-average Molecular Weights

Some molecular-weight methods such as equilibrium ultracentrifuge measurements give weight-average molecular weights

$$M_w = \frac{\sum_i \mathrm{wt}_i M_i}{\sum_i \mathrm{wt}_i} \tag{5.38}$$

where M_w = weight-average molecular weight
wt_i = weight of all molecules with molecular weight M_i

This can also be written as

$$M_w = \frac{\sum_i n_i M_i^2}{\sum_i n_i M_i} \tag{5.39}$$

This average depends more on the high-molecular-weight species than does the number-average weight.

Example 5.4 How will 1.0 wt % impurity affect the determination of the weight-average molecular weight of a macromolecule whose molecular weight is 100,000? Assume that the impurity has a molecular weight of 100.

Solution

$$M_w = \frac{(0.99)(100{,}000) + (0.010)(100)}{0.99 + 0.010} = \frac{99{,}001}{1.00}$$

$$= 0.99 \times 10^5$$

The weight average is almost unaffected by 1% impurity.

A good way of determining whether a sample is pure or a mixture is to measure both the number-average and the weight-average molecular weight. If the two averages agree, the sample is probably a single-molecular-weight species. If the averages disagree, either the sample is a mixture or one of the measurements is in error.

ACTIVITY OF THE SOLVENT AND COLLIGATIVE PROPERTIES

Thermodynamics can be used to relate the empirical Raoult's law and Henry's law to other properties of solutions, such as freezing-point lowering, boiling-point rise, or osmotic pressure. The activity of the solvent is the link that connects these seemingly unrelated colligative properties. The general definition of activity is given by

$$\mu = \mu^0 + RT \ln a \tag{4.14}$$

We use the solvent standard state for the activity of the solvent, so the chemical potential of the standard state, μ^0, is chosen to be that of the pure solvent.

Vapor Pressure

For equilibrium between any two phases, there is equality of the chemical potentials of each component in the two phases. For example,

$$\mu_A(\text{solution}) = \mu_A(\text{gas}, P_A)$$

$$\mu_A^0(\text{pure solvent}) = \mu_A^0(\text{gas}, P_A^0)$$

where P_A = vapor pressure of A for solution
P_A^0 = vapor pressure of pure solvent A

For an ideal gas, we can use Eq. (3.35) to obtain

$$\mu_A(\text{gas}, P_A) - \mu_A^0(\text{gas}, P_A^0) = RT \ln \frac{P_A}{P_A^0}$$

Any nonideality in the gas at pressures near 1 atm can be ignored, because it will be small compared to nonidealities in solution. Therefore,

$$\mu_A(\text{solution}) - \mu_A^0(\text{pure solvent}) = RT \ln \frac{P_A}{P_A^0}$$

We can now relate vapor pressures to the activity of the solvent. From Eq. (4.14),

$$\mu_A(\text{solution}) - \mu_A^0(\text{pure solvent}) = RT \ln a_A$$

Therefore, at equilibrium with gas, from Eq. (4.17),

$$RT \ln \left(\frac{P_A}{P_A^0}\right) = RT \ln a_A$$

$$a_A = \frac{P_A}{P_A^0} \tag{5.40}$$

where a_A = activity of solvent
P_A = vapor pressure of solvent in solution
P_A^0 = vapor pressure of pure liquid solvent

All of these must be measured at the same temperature. Vapor-pressure measurements thus provide a simple method for measuring the activity of the solvent.

Because the activity of the solvent can be written

$$a_A = \gamma_A X_A$$

it is clear that Eq. (5.40) reduces to Raoult's law for ideal solutions, where $\gamma_A \equiv 1$. It can be shown that when Raoult's law applies to the solvent in a solution, Henry's law must apply to the solute.

Boiling Point

Another method of determining the activity of the solvent in a solution is to measure the boiling-point rise. The normal boiling point of a solution is defined as the temperature where the vapor pressure of the solvent equals 1 atm. When a nonvolatile solute is added to a solvent, the vapor pressure of the solution decreases, and therefore the normal boiling point is raised. How much higher must the temperature of the solution be so that its vapor pressure is 1 atm, the same as that of the pure solvent at the normal boiling point, T_0? Because the vapor pressure is decreased in the solution, we need to raise the temperature from T_0 to T_b, which we choose so that $P_A = 1$ atm. The vapor pressure of the pure solvent at temperature T_b would be $P_A/a_A = (1\text{ atm})/a_A$. Now we can use Eq. (5.16), which gives the vapor pressure as a function of temperature. Let $P_1 =$ 1 atm at $T_1 = T_0$ and $P_2 = (1\text{ atm})/a_A$ at $T_2 = T_b$. After rearranging terms, we obtain

$$\ln a_A = \frac{\Delta \bar{H}_{vap}}{R} \cdot \left(\frac{1}{T_b} - \frac{1}{T_0}\right) \qquad (5.41)$$

where a_A = activity of solvent in solution; if the solution is ideal, $a_A = X_A$
$\Delta \bar{H}_{vap}$ = enthalpy of vaporization of solvent
T_b = boiling point of solution
T_0 = boiling point of pure solvent

Equation (5.41) looks different from the familiar equation (5.23) relating boiling point and concentration in dilute solution, but Eq. (5.41) reduces to the familiar result as follows. For dilute (two component) solutions we have the relations

$$\frac{1}{T_b} - \frac{1}{T_0} = \frac{T_0 - T_b}{T_b T_0} \cong \frac{T_0 - T_b}{T_0^2}$$

$$\ln a_A \cong -\frac{n_B}{n_A} \qquad (5.29)$$

$$\ln a_A \cong -\frac{mM_A}{1000} \qquad (5.42)$$

where a_A = activity of solvent
m = molality of solute (moles of solute per 1000 g of solvent)
M_A = molecular weight of solvent

Therefore, from Eqs. (5.41) and (5.42),

$$T_b - T_0 = \frac{RT_0^2 M_A}{1000\, \Delta \bar{H}_{vap}} m = K_b m \qquad (5.43)$$

The boiling-point rise is directly proportional to the molality, just as in Eq. (5.23). The proportionality constant depends only on the properties of the solvent. In fact, Eq. (5.43) allows us to calculate values of K_b from known thermodynamic properties of the solvent. For water, we can use $T_0 = 373$ K, $M_A = 18.0$, and $\Delta\bar{H}_{vap} = 9.70$ kcal mol^{-1} at 373 K to calculate a value of $K_b = 0.51$ deg m^{-1} (m = molal). This agrees well with the experimental value of 0.52 deg m^{-1}, obtained by extrapolating to infinite dilution.

Freezing Point

Adding a solute to a solution decreases the activity of the solvent and, thus, lowers the melting point. The temperature of the solution must be lowered to keep it in equilibrium with the pure solid solvent. The equation is

$$\ln a_A = \frac{\Delta\bar{H}_{fus}}{R} \cdot \left(\frac{1}{T_0} - \frac{1}{T_f}\right) \tag{5.44}$$

where a_A = activity of pure solvent; if solution is ideal, $a_A = X_A$
$\Delta\bar{H}_{fus}$ = enthalpy of fusion per mole
T_f = freezing point of solution
T_0 = freezing point of pure solvent

For a dilute solution,

$$T_0 - T_f = \frac{RT_0^2 M_A}{1000\,\Delta\bar{H}_{fus}} m \tag{5.45}$$

$$\Delta T_f = K_f m \tag{5.46}$$

Again, for water at $T_0 = 273$ K, $\Delta H_{fus} = 1.438$ kcal m^{-1}, and we can calculate $K_f = 1.857$ deg m^{-1}, compared with the experimental value of 1.852 deg m^{-1}.

When the solid in equilibrium with a solution is pure solute instead of pure solvent, we think of the equilibrium as a solubility. We can then derive an equation for the temperature dependence of solubility. For an ideal solution,

$$\ln \frac{X_B(T_2)}{X_B(T_1)} = \frac{-\Delta\bar{H}_{sat}}{R} \cdot \left(\frac{1}{T_2} - \frac{1}{T_1}\right) \tag{5.47}$$

where $X_B(T_2), X_B(T_1)$ = solubility of solute at temperatures T_2, T_1
$\Delta\bar{H}_{sat} = \bar{H}_B(\text{solution}) - \bar{H}_B(\text{solid})$ = molar enthalpy (heat) of solution of solute in saturated solution

This equation is most useful for slightly soluble solutes, because then the solutions will be dilute enough for the assumption of ideality to be valid. For dilute solutions the mole fraction is proportional to molarity and molality, so the ratio of solubilities in any convenient units can be used in Eq. (5.47). Usually, we expect solubility to increase with increasing temperature. However, we now see that the change of solubility with temperature actually depends on the sign of the heat of solution. Many ionic salts decrease in solubility with increasing temperature.

Osmotic Pressure

As derived earlier, the osmotic pressure of a solution is a measure of the activity of the solvent:

$$\ln a_A = \frac{-\pi \bar{V}_A}{RT} \tag{5.28}$$

Both vapor pressure and osmotic pressure can be measured at any temperature. They are therefore more useful methods than boiling-point or freezing-point measurements.

PHASE RULE

We have considered equilibria between various phases, but not the number of phases that may coexist or the constraints that apply to the thermodynamics of systems of more than one phase. Willard Gibbs deduced the thermodynamic implications of multiple phases in equilibrium and obtained the *phase rule*. This rule answers the questions in the first sentence above. The maximum number of phases that can coexist at equilibrium in the absence of external fields is equal to the number of components plus two. Pure water (one component) can have up to three phases in equilibrium. This can only occur at one temperature and pressure; it is called the *triple point*. For example, the triple point for the vapor, the liquid, and the solid of pure water occurs at $T = 0.0075°C$ and $P = 4.58$ torr. In addition to the common form of solid water, ice, five other solid phases of water with different crystal structures have been detected at high pressure. The phase rule states that no more than three of these can coexist in equilibrium. If liquid and gaseous water are also present, only one of the solid phases will be stable.

For a two-component system (NaCl in water) four phases can be in equilibrium. For example, at the freezing point of a saturated NaCl solution we could have solid NaCl, solid H_2O, the liquid solution, and H_2O vapor all in equilibrium, but again only at a unique value of T and P. A system that has the maximum number of coexistent phases is said to have no degrees of freedom; all the properties of the system are defined. The degrees of freedom of the systems that we normally consider are T and P and concentrations of the components. Every time a new phase appears in a system at equilibrium the number of degrees of freedom is reduced by one. We can adjust the T and P of pure gaseous water (one component, one phase) to any values that we choose over a wide range. But if we have liquid water in equilibrium with gaseous water (one component, two phases), the T and P are no longer independent. For each temperature we choose, only one pressure (the vapor pressure) will keep the liquid and gas in equilibrium. For equilibrium of solid, liquid, and gaseous water we cannot choose either T or P; the triple point is fixed by nature. The Gibbs phase rule that summarizes the discussion above is

$$F = C - P + 2 \tag{5.48}$$

where F = number of degrees of freedom of the system
C = number of components of the system
P = number of phases at equilibrium in the system

To find the maximum number of phases for a given number of components, just set F equal to zero. The value of 2 in Eq. (5.48) occurs because we consider only two variables (T and P) beside concentrations. For each additional variable other than T and P that we consider (electric fields, magnetic fields, etc.), the number 2 in Eq. (5.48) is increased by 1.

SUMMARY

Phase Equilibrium

Equilibrium conditions:

$$\begin{aligned} T(\text{phase 1}) &= T(\text{phase 2}) \\ P(\text{phase 1}) &= P(\text{phase 2}) \\ \mu_A(\text{phase 1}) &= \mu_A(\text{phase 2}) \end{aligned} \tag{5.1}$$

Free energy of transport of n moles of an uncharged molecule from phase 1 to phase 2:

$$\Delta G = nRT \ln \frac{a_A(\text{phase 2})}{a_A(\text{phase 1})} \cong nRT \ln \frac{c_A(\text{phase 2})}{c_A(\text{phase 1})} \qquad (5.4), (5.5)$$

Free energy of transport of n moles of an ion with charge Z from phase 1 to phase 2:

$$\Delta G = n\left[RT \ln \frac{a_A(\text{phase 2})}{a_A(\text{phase 1})} + ZFV\right] \tag{5.6}$$

F = Faraday = 23,060 cal $(\text{eV})^{-1}$
V = potential difference between phases, V

Gibbs' phase rule (for T and P as only external variables):

$$F = C - P + 2 \tag{5.48}$$

F = number of degrees of freedom
C = number of components
P = number of phases at equilibrium

Scatchard equation for binding to a macromolecule:

$$\frac{\nu}{[\mathrm{A}]} = K \cdot (N - \nu) \tag{5.12}$$

ν = average number of A's bound to macromolecule
N = number of binding sites on macromolecules

Solutions

Gibbs' adsorption isotherm for adsorption at a surface of a solution:

$$\Gamma = -\frac{1}{RT}\frac{\partial \gamma}{\partial \ln c} \tag{5.13}$$

Γ = adsorption (excess concentration) of solute at surface, mol cm^{-2}
γ = surface tension, ergs cm^{-2} $\equiv$ dyn cm^{-1}
R = gas constant = 8.314×10^7 erg deg^{-1} mol^{-1}
c = concentration of solute in bulk solution

Raoult's law for solvent:

$$P_\mathrm{A} = X_\mathrm{A} P_\mathrm{A}^0 \tag{5.19}$$

P_A = vapor pressure of solvent in solution with mole fraction X_A.

$$X_\mathrm{A} = \frac{n_\mathrm{A}}{n_\mathrm{A} + n_\mathrm{B}}$$

P_A^0 = vapor pressure of pure solvent

Henry's law for solute:

$$P_\mathrm{B} = k_\mathrm{B} X_\mathrm{B} \tag{5.20}$$

k_B = Henry's law constant (from tables)

Clausius-Clapeyron equation for the change of the boiling point with pressure and the change of the vapor pressure with temperature:

$$\ln\frac{P_2}{P_1} = \frac{-\Delta \bar{H}_\mathrm{vap}}{R} \cdot \left(\frac{1}{T_2} - \frac{1}{T_1}\right) \tag{5.16}$$

$\Delta \bar{H}_\mathrm{vap}$ = molar enthalpy of vaporization

Pressure dependence of the freezing point:

$$\frac{T_2 - T_1}{P_2 - P_1} = T\frac{\Delta \bar{V}_\mathrm{fus}}{41.29\,\Delta \bar{H}_\mathrm{fus}} \tag{5.18}$$

$\Delta \bar{V}_\mathrm{fus}$ = molar volume (cm^3 mol^{-1}) of fusion
$\Delta \bar{H}_\mathrm{fus}$ = molar heat (cal mol^{-1}) of fusion

P must be in atm to be consistent with the unit conversion factor of 41.29.

Concentration units:

mole fraction: $X_B = \frac{n_B}{n_A + n_B}$

molality: m = moles of solute per 1000 g of solvent

molarity: c = moles of solute per liter of solution

Dilute solutions:

$$X_B = \frac{M_A m}{1000} \tag{5.24}$$

$$c = \rho_0 m \tag{5.25}$$

M_A = molecular weight of solvent of density ρ_0

Boiling-point rise:

$$T_b - T_0 = K_b m = \frac{M_A R T_0^2}{1000\,\Delta \bar{H}_{vap}} m \qquad (5.23), (5.43)$$

For H_2O, $K_b = 0.51$.

Freezing-point lowering:

$$T_0 - T_f = K_f m = \frac{M_A R T_0^2}{1000\,\Delta \bar{H}_{fus}} m \qquad (5.22), (5.45)$$

For H_2O, $K_f = 1.86$.

Osmotic pressure (in atm):

$$\pi = \frac{n_B RT}{V} = cRT \qquad (5.31a), (5.31b)$$

R = 0.08205 ℓ atm deg^{-1} mol^{-1}
V = volume of solution containing n_B moles solute, ℓ
c = concentration in mol ℓ^{-1}

Number-average molecular weight:

$$M_n = \frac{\sum_i n_i M_i}{\sum_i n_i} \tag{5.36}$$

Weight-average molecular weight:

$$M_w = \frac{\sum_i \mathrm{wt}_i M_i}{\sum_i \mathrm{wt}_i} \tag{5.38}$$

n_i = number of molecules with molecular weight M_i
wt_i = weight of molecules with molecular weight M_i

Number-average molecular weights for solutes are obtained from colligative properties in dilute solution.

Osmotic-pressure molecular weight:

$$M = \frac{wRT}{\pi} \tag{5.35}$$

w = concentration of solution, g ℓ^{-1}

Vapor-pressure-lowering molecular weight:

$$M = \frac{\text{wt of solute}}{\text{moles of solvent}} \cdot \frac{P_A^0}{P_A^0 - P_A} \tag{5.32}$$

P_A = vapor pressure of solution
P_A^0 = vapor pressure of pure solvent

Boiling-point and freezing-point molecular weights:

$$M = \frac{\text{wt of solute}}{\text{kg solvent}} \cdot \frac{K_b}{\Delta T_b} \tag{5.33}$$

$$M = \frac{\text{wt of solute}}{\text{kg solvent}} \cdot \frac{K_f}{\Delta T_f} \tag{5.34}$$

$\Delta T_b, \Delta T_f$ = change in boiling point or freezing point
K_b, K_f = constants from tables

The activity of the solvent is obtained from colligative properties.

Osmotic pressure, π:

$$\ln a_A = \frac{-\pi \bar{V}_A}{RT} \tag{5.28}$$

$\bar{V}_A$ = molar volume of solvent

Vapor pressure:

$$a_A = \frac{P_A}{P_A^0} \tag{5.40}$$

Boiling point and freezing point:

$$\ln a_A = \frac{\Delta \bar{H}_{vap}}{R} \cdot \left(\frac{1}{T_b} - \frac{1}{T_0}\right) \tag{5.41}$$

$$\ln a_A = \frac{\Delta \bar{H}_{fus}}{R} \cdot \left(\frac{1}{T_0} - \frac{1}{T_f}\right) \tag{5.44}$$

$\Delta \bar{H}_{vap}, \Delta \bar{H}_{fus}$ = molar heats of vaporization, fusion

REFERENCE

TANFORD, C., 1973. *The Hydrophobic Effect: Formation of Micelles and Biological Membranes*, John Wiley, New York.

SUGGESTED READING

SINGER, S. J., and G. L. NICHOLSON, 1972. The Fluid Mosaic Model of the Structure of Cell Membranes, *Science 175*, 720–731.

PROBLEMS

1. An aqueous solution contains 50 g of solute per liter and has an osmotic pressure of 9 atm at 37°C. Assume that this is an ideal solution.
 (a) What is the molecular weight of the solute?
 (b) What would you predict for the freezing point of the solution?
 (c) If this solution were put in contact through a semipermeable membrane with a solution of osmotic pressure of 10 atm, which way would the solvent flow?
 (d) Which of the two solutions (osmotic pressure = 9 atm or osmotic pressure = 10 atm) will have the higher vapor pressure for the solvent?

2. The vapor pressure of pure, solid pyrene at 25°C is P_1 atm. The solubility at 25°C of pyrene in water is 10^{-4} mol ℓ^{-1}; the solubility of pyrene in ethanol at 25°C is greater than 10^{-4} mol ℓ^{-1}. Neither water nor ethanol dissolves significantly in pyrene.
 (a) What is the vapor pressure of pyrene above a saturated solution of pyrene in water at 25°C?
 (b) If ethanol is added to the saturated aqueous solution, more pyrene dissolves. When equilibrium with the solid pyrene is reached again, will the vapor pressure of the pyrene above the new solution be greater than, less than, or the same as in part (a)?
 (c) The solubility of pyrene at 25°C is 1.1×10^{-3} mol ℓ^{-1} in an aqueous solution of 0.1 mol ℓ^{-1} of cytosine. This increase in solubility is due to the formation of a complex between pyrene and cytosine. Calculate the values of the equilibrium constants for the following reactions at 25°C, where P = pyrene and C = cytosine. Use concentration in mol ℓ^{-1} instead of activities.
 (1) P(solution) + C(solution) = P · C(solution)
 (2) P(solid) + C(solution) = P · C(solution)

3. (a) The concentration of creatine in blood is $20\ \text{mg}\ \ell^{-1}$ and in urine is $750\ \text{mg}\ \ell^{-1}$. Calculate the free energy per mole required for its transfer from blood to urine at 37°C.

(b) To determine the molecular weight of creatine, a sample was purified and used to prepare a solution containing $0.1\ \text{g}\ \ell^{-1}$. This solution exhibited an osmotic pressure against pure water of 13.0 mm Hg at 25°C. What is the molecular weight of creatine?

(c) Using your answer to part (b), calculate the vapor-pressure lowering of water above a solution containing 0.1 g of creatine per liter at 25°C. The vapor pressure of pure water at 25°C is 23.8 mm Hg.

4. Myoglobin is a skeletal muscle protein that binds oxygen. The standard free energy for the reaction

$$\text{myoglobin} + O_2 \longrightarrow \text{oxymyoglobin}$$

is $\Delta G^{0\prime} = -7.18\ \text{kcal mol}^{-1}$ at 298° and pH 7.

What is the ratio (oxymyoglobin)/(myoglobin) in an aqueous solution at equilibrium with a partial pressure of oxygen $P(O_2) = 30$ mm Hg? Assume ideal behavior of O_2 gas, activity coefficient for the protein in solution = 1, and that Henry's law holds for O_2 dissolved in water. The standard state for O_2 is the dilute solution, molarity scale.

5. Certain bacteria that grow in salt flats have been discovered to have a membrane protein capable of utilizing light energy to create a concentration difference of hydrogen ions across the cell membrane. It has been proposed that the free energy provided by this concentration difference is directly utilized to synthesize ATP from ADP and inorganic phosphate.

$$\text{ATP} \longrightarrow \text{ADP} + \text{phosphate} \qquad \Delta G^{0\prime} = -7.4\ \text{kcal mol}^{-1}\ \text{at } 25°\text{C}$$

(a) What must be the ratio of the concentration of H^+ on one side to that on the other side of the membrane to provide sufficient free energy to synthesize ATP at 4 mM when the concentrations of ADP and phosphate are 10 mM and 1 mM? Assume that all activity coefficients = 1.

(b) If the salt flats where the bacterium lives have a constant pH = 7.0, what is the pH of the cell interior if H^+ are moved out of the cell in creating the concentration difference?

6. In living biological cells the concentration of sodium ions inside the cell is kept at a lower concentration than the concentration outside the cell, because sodium ions are actively transported from the cell.

(a) Consider the following process at 37°C and 1 atm.

$$1\ \text{mol NaCl}\ (0.04\ M\ \text{inside}) \longrightarrow 1\ \text{mol NaCl}\ (0.14\ M\ \text{outside})$$

Write an expression for the free-energy change for this process in terms of activities. Define all symbols used.

(b) Calculate $\Delta \bar{G}$ for the process. You may approximate the activities by concentrations in mol ℓ^{-1}. Will the process proceed spontaneously?

(c) Repeat the calculation in part (b), except move 3 mol of NaCl from inside to outside.

(d) Calculate $\Delta \bar{G}$ for the process if the activity of NaCl inside is equal to that outside.

(e) Calculate $\Delta \bar{G}$ for the process at equilibrium.

(f) The standard free energy for hydrolysis of ATP to ADP (ATP → ADP + phosphate) in solution is $\Delta G^{0\prime} = -7.3\ \text{kcal mol}^{-1}$ at 37°C, 1 atm. The free energy of this reaction can be used to power the sodium-ion pump. For a ratio of ATP to ADP of 10, what must be the concentration of phosphate to obtain $-10\ \text{kcal mol}^{-1}$ for the hydrolysis? Assume that activity coefficients are 1 for the calculation.

(g) If the ratio of ATP to ADP is 10, what is the concentration of phosphate at equilibrium? Assume that $\gamma = 1$ for the calculation.

7. Consider a dilute solution of NaCl in water at 27°C. The concentration can be given as 0.1 *m*, 0.1 *M*, or 0.0018 mole fraction in NaCl. The vapor pressure of pure water is 30 torr at 27°C. Use ideal solution behavior.

(a) Calculate the vapor pressure of the salt solution at 27°C.

(b) Calculate the boiling point of the salt solution.

(c) Calculate the freezing point of the salt solution.

(d) The salt solution is separated from pure water by a membrane that allows water to pass, but not salt. What is the osmotic pressure (in atm) of the salt solution?

(e) The osmotic pressure can support a column of water. How high is this column of water?

(f) What will be the measured osmotic pressure if the membrane is permeable to both salt and water?

8. (a) Using the Henry's law constants in Table 5.2, calculate the solubility (in $\text{mol}\ \ell^{-1}$) of each gas in water at 25°C if $P_{O_2} = 0.2$ atm, $P_{N_2} = 0.75$ atm, and $P_{CO_2} = 0.05$ atm.

(b) What will be the vapor pressure at 25°C of the water in this solution if Raoult's law holds? The vapor pressure of pure water at 25°C is 23.756 torr.

9. (a) Assume that Raoult's law holds and calculate the boiling point of a 2 *M* solution of urea in water.

(b) Urea actually forms complexes in solution in which two or more molecules hydrogen bond to form dimers and polymers. Will this effect tend to raise or lower the boiling point?

10. When cells of the skeletal vacuole of a frog were placed in a series of aqueous NaCl solutions of different concentrations at 25°C, it was observed microscopically that the cells remained unchanged in 0.7% (by weight) NaCl solution, shrank in more concentrated solutions, and swelled in more dilute solutions. Water freezes from the 0.7% salt solution at −0.406°C.

(a) What is the osmotic pressure of the cell cytoplasm at 25°C?

(b) Suppose that sucrose (mol wt 342) was used instead of NaCl (mol wt 58.5) to make the isoosmotic solution. Estimate the concentration (% by weight) of sucrose that would be sufficient to balance the osmotic pressure of the cytoplasm of the cell. Assume that sucrose solutions behave ideally.

11. On the imaginary planet Erehwon, ammonia plays a role similar to that of water on the earth. Ammonia has the following properties: Normal (1 atm) boiling point is −33.4°C, where its heat of vaporization is $327\ \text{cal g}^{-1}$. Normal freezing point is −77.7°C.

(a) Estimate the temperature at which the vapor pressure of NH_3(liquid) is 60 torr.

(b) Calculate the entropy of vaporization of NH_3 at its normal boiling point. Compare this value with that predicted by Trouton's rule ($\Delta S_{vap} \cong 21\ \text{cal deg}^{-1}$) for typical liquids. Give a likely reason for any discrepancy.

(c) Ammonia undergoes self-dissociation according to the reaction

$$2\,NH_3(l) \rightleftharpoons NH_4^+(am) + NH_2^-(am)$$

where (am) refers to the "ammoniated" ions in solution. At −50°C, the equilibrium constant for this reaction is

$$K = (NH_4^+)(NH_2^-) = 10^{-30}$$

Calculate ΔG^0_{223} for the self-dissociation. What is the standard state for which $a_{NH_3} = 1$?

(d) When 1 mol of ammonium chloride, NH_4Cl, is dissolved in 1 kg of liquid ammonia, the boiling point at 760 torr is observed to occur at −32.7°C. What conclusion can you reach about the nature of this solution? State the evidence for your answer.

12. Indicate whether each of the following statements is true or false as it is written. Alter each false statement so as to make it correct.

(a) If two aqueous solutions containing different nonvolatile solutes exhibit exactly the same vapor pressure at the same temperature, the activities of water are identical in the two solutions.

(b) The most important reason that foods cook more rapidly in a pressure cooker is that equilibrium is shifted in the direction of products at the higher pressure.

(c) If solutions of a single solute are prepared at equal molalities but in different solvents, the freezing-point lowering will be the same for the different solutions, assuming that each behaves ideally.

(d) If solutions of a single solute are prepared at equal mole fractions but in different solvents, the fractional vapor-pressure lowering will be the same for the different solutions, assuming that each behaves ideally.

(e) A saturated solution of sodium chloride in liquid water has one degree of freedom.

(f) If two liquids, such as benzene and water, are not completely miscible in one another, the mixture can never be at equilibrium.

13. A solution of the sugar mannitol (mol wt 182.2) is prepared by adding 54.66 g of mannitol to 1000 g of water. The vapor pressure of pure liquid water is 17.54 torr at 20°C. Mannitol is nonvolatile and does not ionize in aqueous solutions.

(a) Assuming that aqueous mannitol solutions behave ideally, calculate the vapor-pressure lowering (the difference between the vapor pressure of pure water and that of the solution) for the above solution at 20°C.

(b) The *observed* vapor-pressure lowering of the mannitol solution above is 0.0930 torr. Calculate the activity coefficient (based on mole fraction) of water in this solution.

(c) Calculate the osmotic pressure of the mannitol solution of part (b) when it is measured against pure water.

14. (a) Assuming that glucose and water form an ideal solution, what is the equilibrium vapor pressure at 20°C of a solution of 1.00 g of glucose (mol wt 180) in 100 g of water? The vapor pressure of pure water is 17.54 mm Hg at 20°C.

(b) What is the osmotic pressure, in mm Hg, of the solution in part (a) versus pure water?

(c) What is the activity of water as a solvent in such a solution?

(d) What would be the osmotic pressure, versus pure water, of a solution containing both 1.00 g of glucose and 1.00 g of sucrose (mol wt 342) in 100 g of water at 20°C?

15. Workers in underwater caissons or diving suits necessarily breathe air at greater than normal pressures. If they are returned to the surface too rapidly, N_2, dissolved in their blood at the previously higher pressure, comes out of solution and may cause emboli (gas bubbles in the bloodstream), bends, and decompression sickness. Blood plasma at 37°C and 1 atm pressure dissolves 1.3 mℓ of N_2 gas (measured for pure N_2 at 37°C and 1 atm) in 100 mℓ of plasma. Calculate the volume of N_2 likely to be liberated from the blood plasma of a caisson worker returned to 1 atm pressure after prolonged exposure to air pressure at 300 m of water (below the surface). The total blood volume of the average adult is 3.2 ℓ; density of mercury is 13.6 g mℓ^{-1}; air contains 78% N_2 by volume.

16. A biochemical researcher is investigating an antibiotic that is stable in boiling water. He decides to test his compound at 140°C by heating a very dilute water solution of it in a sealed tube, from which he has first pumped out all the air. What internal pressure will be the tube have to withstand? The normal boiling point of water is 100°C and its enthalpy of vaporization is 9.70 kcal mol^{-1}.

17. There are semipermeable membranes which are permeable to water but not sodium ions or chloride ions. They can be used to measure the osmotic pressure of sodium chloride solutions. Sodium chloride is completely ionized in water solutions.
(a) Calculate the osmotic pressure of 0.1 M NaCl solution at 25°C assuming an ideal dilute solution.
(b) The actual osmotic pressure is slightly less than the calculated answer to part (a). Explain why you think this happens.
(c) Assume that the actual osmotic pressure is π. Write an equation that would allow you to calculate the actual freezing point of the solution from this information.

18. The human red-blood-cell membrane is freely permeable to water but not at all to sucrose. The membrane forms a completely closed bag, and it is found by trial and error that adding the cells to a solution of a particular concentration of sucrose results in neither swelling nor shrinking of the cells. In a separate experiment, the freezing point of that particular sucrose solution is found to be −0.56°C.
(a) If the exterior of the cell contains only KCl solution, estimate the KCl concentration. Assume that the membrane is also impermeable to KCl.
(b) If these cells are suspended in distilled water at 0°C, what would be the internal hydrostatic pressure at equilibrium, assuming that the cell did not change volume?

19. (a) How many solute molecules are there in a small drop of aqueous solution whose volume is 10^{-15} ℓ and whose freezing point is -10^{-6} degrees C.
(b) What is the osmotic pressure in atm of a 0.1 M solution of urea in ethanol at 27°C? State what assumptions you made in the calculation.
(c) What is the osmotic pressure in atm of a 0.1 M solutions of NaCl in water at 27°C? State what assumptions you made in the calculation.
(d) An aqueous solution of osmotic pressure 5 atm is separated from an aqueous solution of osmotic pressure 2 atm. Which way will the solvent flow? How large a pressure must be applied to prevent this solvent flow?
(e) What is the free-energy change for transferring 1 mol of water from a solution of osmotic pressure 5 atm to a solution of osmotic pressure 2 atm?

20. A protein solution containing 0.6 g of protein per 100 mℓ of solution has an osmotic pressure of 22 mm of water at 25°C and at a pH where the protein has no net charge (the isoelectric point). What is the molecular weight of the protein?

21. A climber has carefully measured the boiling point of water at the top of a mountain and found it to be 95°C. Calculate the atmospheric pressure at the mountain top. The boiling point of water at 1 atm pressure is 100°C, and the heat absorbed in vaporizing 1 mol of water at constant pressure is 9.7 kcal mol^{-1} and independent of temperature.

22. A cell membrane at 37°C is found to be permeable to Ca^{2+}, and analysis shows the inside concentration to be 0.100 *M* and the outside concentration to be 0.001 *M*.

(a) What potential difference in volts must exist across the membrane for Ca^{2+} to be in equilibrium at the measured concentrations? Assume that activity coefficients are equal to 1. Give the sign of the potential with respect to the outside at zero potential.

(b) If the measured inside potential is +10 mV with respect to the outside, what is the minimum (reversible) work required to transfer 1 mol of Ca^{2+} from outside to inside under these conditions?

6

Transport Phenomena

With the rapid advances made in the biological sciences in the last few decades, many life processes have now been dissected to the molecular level. Attention has been focused on the chemical nature of the molecules that carry out the various biochemical processes: their sizes, shapes, structures, how they interact with each other, and with what rates.

In the preceding four chapters, we have discussed the laws of thermodynamics, which deal with the energetics of various processes and set the proper criteria for determining whether certain changes are spontaneous. Because of their general applicability, the thermodynamic laws are powerful and useful tools. Nevertheless, classical thermodynamics was developed for bulk matter, and its fundamental principles are not dependent on the details of molecular structure. Nor does classical thermodynamics deal with how fast a system changes from one state to another.

In this chapter, we shall discuss certain time-dependent processes from a molecular point of view. A simple kinetic-molecular model of gases will be described first. This model provides a relation between the pressure of a gas, a macroscopic property, and the properties of the molecules. The model also describes the collisions between molecules and how they are related to the diffusion of one gas in another. We shall then turn our attention to a number of transport properties of large molecules (macromolecules): diffusion coefficient, sedimentation coefficient, viscosity, and electrophoretic mobility. These properties are related to the motion of molecules, respectively, across a concentration gradient (diffusion), in a centrifugal field (sedimentation), in a velocity gradient in a fluid (viscosity), and in an electric field (electrophoresis). As we will see, these properties provide us with information about the size and shape of the kinds of macromolecules that play important roles in living cells.

KINETIC-MOLECULAR MODEL OF GASES

In a gas at ordinary pressures, the molecules occupy a volume much greater than their physical bulk. While the molecules in 1 g of liquid water or ice are crowded into a volume of approximately 1 cm^3, when vaporized at 100°C the same number of molecules occupy a volume of 1700 cm^3 at 1 atm. The molecules in a dilute gas are far apart from one another, and, consequently, they interact only weakly. Therefore, as an approximation we can assume that the molecules are moving about more or less randomly in the gaseous phase; the motion of one molecule is little affected by the others unless it encounters another molecule in a direct collision.

For the time being we shall not consider molecular collisions. We can assume that the molecules are vanishingly small, or that the gas is so dilute, that collisions between molecules are highly improbable. Molecules in such a gas are nevertheless colliding with the walls of the container, and thereby exerting pressure on the walls.

Velocities of Molecules

Consider a certain molecule in a cubic container of side a. One face of the cube perpendicular to the x axis is labeled S, as shown in Fig. 6.1(a). Let m and u be the mass and velocity of the molecule, respectively. The velocity u can be resolved into its three Cartesian components u_x, u_y, and u_z along the three axes, as indicated in Fig. 6.1(b). Let us first focus our attention on the collisions between this molecule and face S. If the x component of the velocity is u_x cm s^{-1}, then in 1 s the molecule travels a distance of u_x cm. If u_x is much greater than a, the dimension of the container, the molecule will traverse many times in 1 s between the two walls perpendicular to the x axis. Because each round trip spans a distance of $2a$, the number of times the molecule hits face S per second is $u_x/2a$. Now consider the momentum change per collision. Each time the molecule approaches wall S it has a perpendicular velocity component u_x and a corresponding momentum mu_x. As it departs in the opposite direction after a collision, the velocity component perpendicular to the wall changes to $-u_x$, with a corresponding momentum $-mu_x$. Thus the momentum change along the x direction is $2mu_x$ per collision.

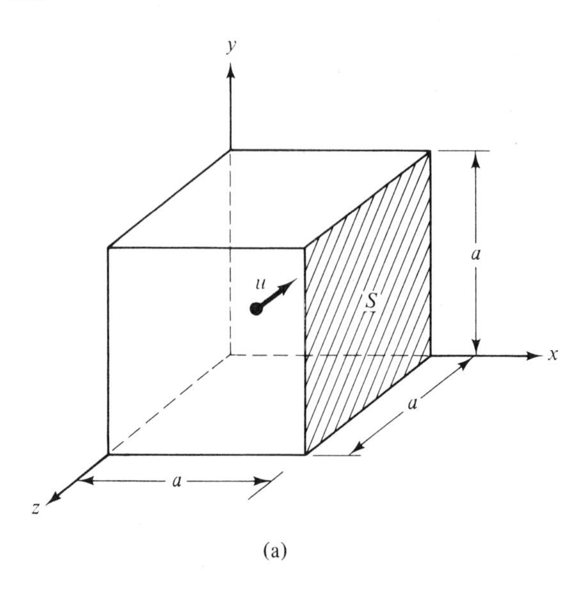

(a)

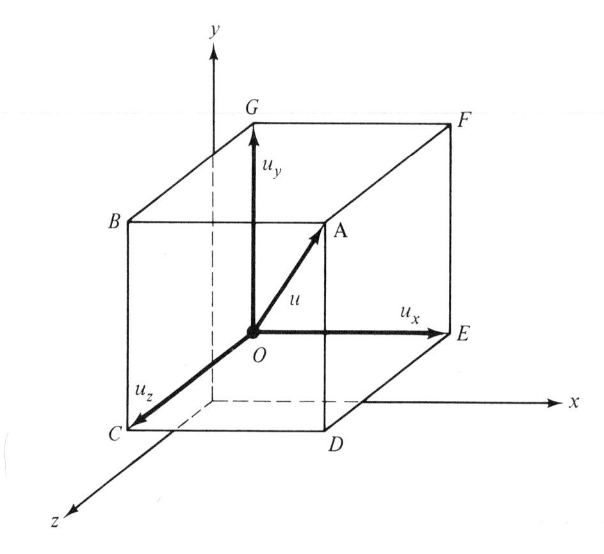

Fig. 6.1 (a) Molecule of mass m and velocity u in a cubic container of side a. We are interested in the number of collisions between this molecule and the crosshatched face labeled S. (b) Resolving the velocity u (OA) into its three Cartesian components, u_x (OE), u_y (OG), and u_z (OC). Note that the angle $\angle AGO = 90°$. Thus, $\overline{OA}^2 = \overline{GA}^2 + \overline{OG}^2$. Also, $\angle OED = 90°$. Thus, $\overline{OD}^2 = \overline{OE}^2 + \overline{OC}^2$. But $\overline{GA}^2 = \overline{OD}^2$. Therefore,

$$\overline{OA}^2 = \overline{OD}^2 + \overline{OG}^2$$
$$= \overline{OE}^2 + \overline{OC}^2 + \overline{OG}^2,$$

or $\quad u^2 = u_x^2 + u_y^2 + u_z^2.$

The total momentum change along the x axis per second, which according to Newton's second law is the force exerted by the molecule along the x axis, is then

$$f_x = (\text{momentum change per collision}) \times (\text{number of collisions per second})$$

$$= 2mu_x \frac{u_x}{2a}$$

$$= \frac{mu_x^2}{a} \tag{6.1}$$

If there are N molecules, each of mass m, in the container, each of them can be characterized by a velocity u_i. The total force in the direction perpendicular to wall S exerted by all the molecules is

$$F_x = \frac{mu_{x1}^2}{a} + \frac{mu_{x2}^2}{a} + \cdots + \frac{mu_{xN}^2}{a}$$

$$= \frac{m}{a} \sum_{i=1}^{N} u_{xi}^2 \tag{6.2}$$

The force per unit area, in the direction normal to wall S, is by definition the pressure P. Thus,

$$P = \frac{F_x}{a^2}$$

$$= \frac{m}{a^3} \sum_{i=1}^{N} u_{xi}^2$$

$$= \frac{m}{V} \sum_{i=1}^{N} u_{xi}^2 \tag{6.3}$$

where $V = a^3$ is the volume of the container.

We define the mean-square velocity of the molecules in the x direction as

$$\langle u_x^2 \rangle = \frac{1}{N} \sum_{i=1}^{N} u_{xi}^2 \tag{6.4}$$

The mean-square velocities in the other directions are similarly defined. Equation (6.3) becomes then

$$P = \frac{Nm}{V} \langle u_x^2 \rangle \tag{6.5}$$

Recall that, as explained in the legend to Fig. 6.1(b),

$$u_1^2 = u_{x1}^2 + u_{y1}^2 + u_{z1}^2$$

$$u_2^2 = u_{x2}^2 + u_{y2}^2 + u_{z2}^2$$

$$\cdots\cdots\cdots\cdots\cdots\cdots\cdots$$

$$u_N^2 = u_{xN}^2 + u_{yN}^2 + u_{zN}^2 \tag{6.6}$$

By adding all these equations and dividing each side by N, we obtain

$$\frac{1}{N} \sum_{i=1}^{N} u_i^2 = \frac{1}{N} \sum_{i=1}^{N} u_{xi}^2 + \frac{1}{N} \sum_{i=1}^{N} u_{yi}^2 + \frac{1}{N} \sum_{i=1}^{N} u_{zi}^2 \tag{6.7}$$

We define the mean-square velocity, $\langle u^2 \rangle$, as

$$\langle u^2 \rangle = \frac{1}{N} \sum_{i=1}^{N} u_i^2$$

Therefore, Eq. (6.7) becomes

$$\langle u^2 \rangle = \langle u_x^2 \rangle + \langle u_y^2 \rangle + \langle u_z^2 \rangle \tag{6.8}$$

Since the molecules are moving about in space randomly, we expect no bias for the x, y, or z directions. Therefore,

$$\left.\begin{aligned} \langle u_x^2 \rangle = \langle u_y^2 \rangle = \langle u_z^2 \rangle \\ \text{and} \qquad \\ \langle u_x^2 \rangle = \tfrac{1}{3}\langle u^2 \rangle \end{aligned}\right\} \tag{6.9}$$

Substituting Eq. (6.9) into (6.5), we obtain

$$P = \frac{1}{3}\frac{Nm}{V}\langle u^2 \rangle \tag{6.10}$$

or

$$PV = \tfrac{1}{3}Nm\langle u^2 \rangle \tag{6.11}$$

According to the ideal gas law,

$$PV = nRT = \frac{N}{N_0}RT \tag{6.12}$$

where $n = N/N_0$, N_0 is Avogadro's number, and n is the number of moles of gas. Combining Eqs. (6.11) and (6.12) gives

$$RT = \tfrac{1}{3}N_0 m\langle u^2 \rangle \tag{6.13}$$

$$= \tfrac{2}{3}(\tfrac{1}{2}N_0 m\langle u^2 \rangle) \tag{6.14}$$

Note that $\frac{1}{2}m\langle u^2 \rangle$ is the average translational kinetic energy per molecule, and that $\frac{1}{2}N_0 m\langle u^2 \rangle$ is the translational kinetic energy $\bar{U}_{tr}$ of 1 mol of gas. Therefore,

$$RT = \tfrac{2}{3}\bar{U}_{tr} \tag{6.15}$$

Note that $\bar{U}_{tr}$ is a function of T only. Also, since $N_0 m$ is the molecular weight, M, Eq. (6.13) gives

$$\langle u^2 \rangle = 3\frac{RT}{M} \tag{6.16}$$

Equation (6.16) is an important result. It states that for a gas consisting of molecules with molecular weight M, the mean-square velocity is a function of T only. In deriving Eq. (6.16), we have neglected molecular collisions. It can be shown that the same result is obtained without this omission.

The root-mean-square velocity

$$\langle u^2 \rangle^{1/2} = \left(\frac{3RT}{M}\right)^{1/2} \tag{6.17}$$

is one measure of the average speed of the molecules.

Example 6.1 Calculate the root-mean-square velocity of oxygen and geraniol vapor at 20°C. Geraniol, an alcohol responsible for the fragrance of roses, has a molecular weight of 154.2.

Solution The molecular weight of O_2 is 32.0. To obtain reasonable units ($cm\ s^{-1}$) for the velocity, we must express the gas constant R as 8.314×10^7 erg deg^{-1} mol^{-1}.

From Eq. (6.17),

$$\langle u^2 \rangle^{1/2} = \left(\frac{3RT}{M}\right)^{1/2}$$

$$= \left[3(8.314 \times 10^7 \text{ erg deg}^{-1} \text{ mol}^{-1}) \frac{(1 \text{ g cm}^2 \text{ s}^{-2} \text{ erg}^{-1})(293 \text{ K})}{32 \text{ g mol}^{-1}}\right]^{1/2}$$

$$= (2.284 \times 10^9 \text{ cm}^2 \text{ s}^{-2})^{1/2}$$

$$= 4.78 \times 10^4 \text{ cm s}^{-1}$$

Similarly, for geraniol,

$$\langle u^2 \rangle^{1/2} = 2.18 \times 10^4 \text{ cm s}^{-1}$$

This example shows that the velocities are rather high—several tenths of kilometers per second. The values that we have calculated for oxygen and geraniol are of the same order of magnitude as the speed of sound in the atmosphere. Thus one might think that if a bottle of perfume is opened in the center of a room in still air the fragrance would reach the four corners in a fraction of a second. Experience tells us that this does not occur and that a much longer time is required. To see why this is so, we must consider molecular collisions.

Molecular Collisions

We shall use a simple model for the calculation of the frequency of molecular collisions. We assume that the molecules in a gas are hard spheres of diameter σ and that each and every molecule is moving at a constant speed c. (We use the term "speed" to refer to the velocity without regard to direction. Thus the speed is the magnitude of the velocity.) There is actually a wide range of speeds; however, for this simple model we *assume* the same speed for all molecules.

Let us first consider two molecules A and B in a container of volume V. It is convenient to view molecule A as moving while molecule B is stationary. We can imagine that we are standing on molecule B. Then, as shown in Fig. 6.2, in our frame of reference molecule A is moving with a speed $2c \sin(\theta/2)$, where θ is the angle between the directions A and B. It can be shown that when averaged over all directions in space, the mean value of $\sin(\theta/2)$ is $\frac{2}{3}$. Thus the average speed of A *relative* to B is $4c/3$.

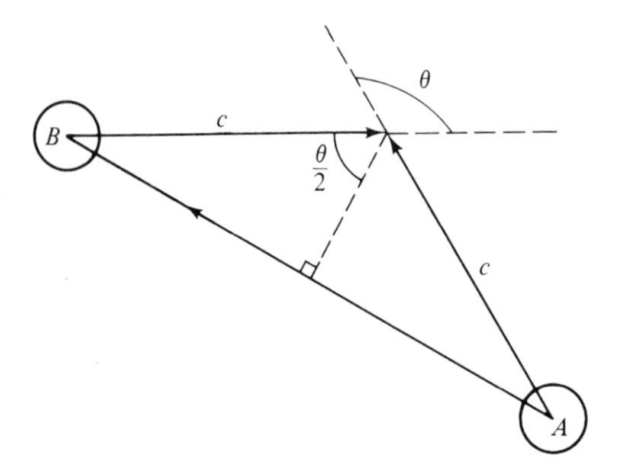

Fig. 6.2 Two molecules, A and B, are moving with the same speed c but in different directions. When viewed from B, A is approaching in the direction AB with a speed equal to the length $\overline{AB}$. If θ is the angle between the directions of motion of A and B, $\overline{AB} = 2c \sin(\theta/2)$. Note that when the two molecules are moving head on, $\theta = 180°$ and $2c \sin(\theta/2) = 2c$. When the two molecules are moving in the same direction, $\theta = 0°$, and $2c \sin(\theta/2) = 0$. The results obtained for these special cases are in agreement with our expectations.

Let us now define the circumstances under which a collision will occur between molecules A and B. As observed from molecule B, molecule A travels a distance $4c\ \Delta t/3$ during a small time interval Δt. As illustrated in Fig. 6.3, we draw a circle of diameter 2σ, centered on molecule A and in the plane perpendicular to its direction of motion. The area of this circle is $\pi\sigma^2$. As molecule A travels the distance $4c\ \Delta t/3$, the circle sweeps out a volume $\delta V = (\pi\sigma^2)4c\ \Delta t/3$ in space. If the center of molecule B happens to be within this volume, there will be a collision between it and molecule A. Because molecule B may be anywhere in the entire volume V, the chance that the center of B is within the small volume

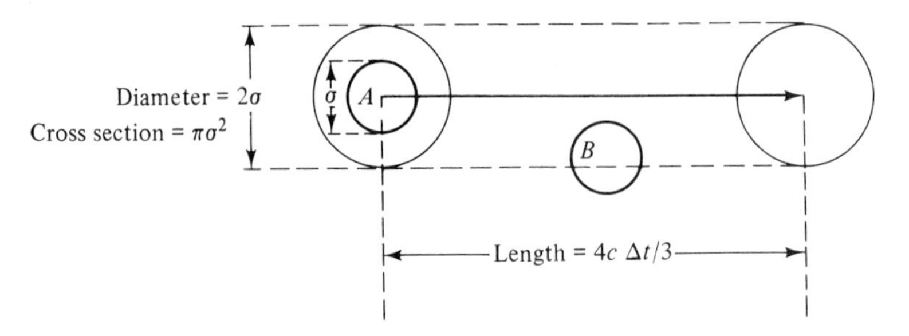

Fig. 6.3 Intermolecular collisions. In a small time interval Δt, molecule A will sweep out a volume $[(\pi\sigma^2)(4c\ \Delta t/3)]$, represented by the cylinder shown. Any other molecule, such as molecule B shown here, with its center inside the cylinder will collide with molecule A.

δV swept out by A during the time interval Δt is $\delta V/V$, and

$$\frac{\delta V}{V} = (\pi\sigma^2)\frac{4c\,\Delta t}{3V} \tag{6.18}$$

The number of collisions per unit time is

$$z = \frac{\delta V}{V}\cdot\frac{1}{\Delta t}$$

$$= (\pi\sigma^2)\frac{4c}{3V} \tag{6.19}$$

It is straightforward to generalize to the case of N molecules rather than two molecules. For $N \gg 1$,

$$z = N\pi\sigma^2\frac{4c}{3V} \tag{6.20}$$

We rearrange the equation to emphasize its dependence on volume:

$$z = 4\pi\sigma^2\frac{c}{3}\frac{N}{V} \tag{6.21}$$

In an actual gas there is, of course, a distribution of velocities. We can therefore equate the assumed constant speed, c, to different average velocities, such as $\langle u^2\rangle^{1/2}$ or $\langle u\rangle$. It is easy to see that unless all velocities are equal, $\langle u^2\rangle^{1/2}$ will never equal $\langle u\rangle$. If we equate c to $\langle u^2\rangle^{1/2}$ and use Eq. (6.17), we obtain

$$z = 7.26\frac{N}{V}\sigma^2\cdot\left(\frac{RT}{M}\right)^{1/2} \tag{6.22}$$

A more rigorous derivation, which takes proper consideration of the distribution in the velocities of the molecules rather than assuming that they all travel at the same speed, gives

$$z = \sqrt{2}\pi\frac{N}{V}\sigma^2\langle u\rangle \tag{6.23}$$

where $\langle u\rangle$ is the mean velocity of the molecules. It can be shown that $\langle u\rangle = (8RT/\pi M)^{1/2}$, which when substituted into the equation above gives

$$z = 7.09\frac{N}{V}\sigma^2\cdot\left(\frac{RT}{M}\right)^{1/2} \tag{6.24}$$

Note that Eq. (6.22), which results from the simpler derivation, is a good approximation to Eq. (6.24).

Since there are N/V molecules per unit volume, there are Nz/V molecules per unit volume undergoing collisions per unit time. The total number of collisions per unit volume per unit time is

$$Z = \frac{N}{V}\frac{z}{2} \tag{6.25}$$

The factor 2 results from the fact that each collision involves two molecules.

Mean Free Path

The mean free path, l, is defined as the average distance a molecule travels between two successive collisions with other molecules. Because $\langle u \rangle$ is the average distance a molecule travels per unit time and z is the number of intermolecular collisions it encounters during that time, the mean free path is given by

$$l = \frac{\langle u \rangle}{z} = \frac{1}{\sqrt{2}\pi(N/V)\sigma^2} \tag{6.26}$$

using Eq. (6.23).

Example 6.2 For N_2 gas at 1 atm and 25°C, calculate (a) the number of collisions each N_2 molecule encounters in 1 s; (b) the total number of collisions in a volume of 1 cm^3 in 1 s; (c) the mean free path of a N_2 molecule. The diameter of a N_2 molecule can be taken as 3.74 Å (3.74×10^{-8} cm).

Solution

(a) For N_2 gas at 1 atm and 25°C, the number of moles per unit volume is P/RT, and the number of molecules per unit volume is

$$\frac{N}{V} = \frac{N_0 P}{RT}$$

$$= \frac{(6.023 \times 10^{23} \text{ molecules mol}^{-1})(1 \text{ atm})}{(0.08205\ \ell \text{ atm deg}^{-1} \text{ mol}^{-1})(298 \text{ deg})}$$

$$= 2.46 \times 10^{22} \text{ molecules } \ell^{-1}$$

$$= 2.46 \times 10^{19} \text{ molecules cm}^{-3}$$

From Eq. (6.24), the number of collisions each molecule encounters per unit time is

$$z = 7.09(2.46 \times 10^{19} \text{ cm}^{-3})(3.74 \times 10^{-8} \text{ cm})^2 \times \left[(8.31 \times 10^7 \text{ erg deg}^{-1} \text{ mol}^{-1})\left(\frac{298 \text{ deg}}{28 \text{ g mol}^{-1}}\right)\right]^{1/2}$$

$$= 7.26 \times 10^9 (\text{erg cm}^{-2} \text{ g}^{-1})^{1/2}$$

$$= 7.26 \times 10^9 \text{ cm}^{-1} (\text{erg g}^{-1})^{1/2} \left(\frac{\text{g cm}^2}{\text{s}^2 \text{ erg}}\right)^{1/2}$$

$$= 7.26 \times 10^9 \text{ s}^{-1}$$

If Eq. (6.22) is used instead of Eq. (6.24), we obtain

$$z = 7.43 \times 10^{9}\ \text{s}^{-1}$$

(b) From Eq. (6.25), the total number of collisions is

$$Z = \frac{N}{V}\frac{z}{2}$$

$$= \frac{(2.46 \times 10^{19}\ \text{cm}^{-3})(7.26 \times 10^{9}\ \text{s}^{-1})}{2}$$

$$= 8.93 \times 10^{28}\ \text{collisions cm}^{-3}\ \text{s}^{-1}$$

(c) From Eq. (6.26), the mean free path is

$$l = \frac{1}{\sqrt{2}\pi(2.46 \times 10^{19}\ \text{cm}^{-3})(3.74 \times 10^{-8}\ \text{cm})^2}$$

$$= 6.54 \times 10^{-6}\ \text{cm}$$

Thus at 1 atm and 25°C a typical N_2 molecule goes a distance almost 200 times its diameter between collisions.

DIFFUSION

Diffusion in a Gas

If a certain molecule in a gas at equilibrium is at the origin O at time zero, on the average how far from the origin will it be after 1 s? In 1 s the molecule undergoes z collisions with other molecules. The distance it travels between two successive collisions is, on the average, l. Since molecules in a gas at equilibrium are moving randomly, the molecules colliding with our selected molecule come from random directions. Therefore, after each collision, we can expect that our selected molecule will change its direction in a random manner. This is illustrated in Fig. 6.4.

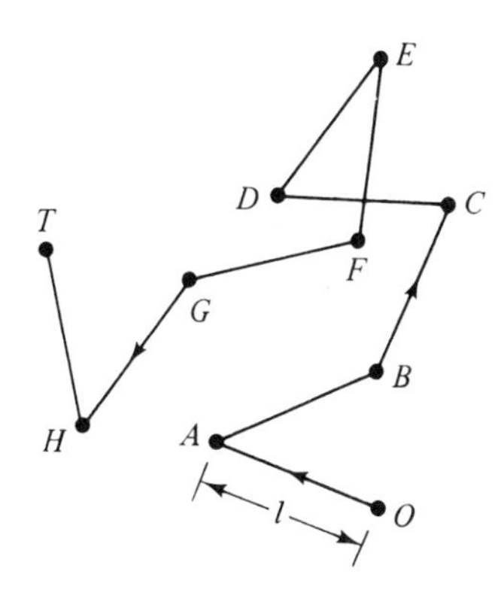

Fig. 6.4 Random path of a molecule. At time zero, its position is at O, and collisions are encountered at points A, $B, \ldots, H$. At time t, its position is at T. The average distance it travels between two successive collisions is the mean free path, l. In our illustration, the distances $\overline{OA}$, $\overline{AB}$, $\overline{BC}$, etc., are depicted as equal to l. The displacement, d, that occurs during time t is the distance $\overline{OT}$.

If d is the distance from its origin after 1 s or z collisions, it can be shown [Eq. (11.45)] that the average of d^2 ($cm^2\ s^{-1}$) is

$$\langle d^2 \rangle = zl^2 \tag{6.27a}$$

$$= \frac{\langle u \rangle^2}{z} \tag{6.27b}$$

The square root of $\langle d^2 \rangle$ is the root-mean-square displacement,

$$\langle d^2 \rangle^{1/2} = \sqrt{z}l \tag{6.28a}$$

$$= \frac{\langle u \rangle}{\sqrt{z}} \tag{6.28b}$$

which is a measure of the average displacement per unit time.

We mentioned earlier that if a bottle of perfume is opened in the middle of a windless room, the scent takes much longer to reach the other parts of the room than one might naively anticipate from the high value of $\langle u \rangle$. The explanation lies in Eq. (6.28b). While $\langle u \rangle$ is large, any molecule encounters many collisions per unit time, and each collision changes its direction. Thus, the root-mean-square displacement per unit time is $\langle u \rangle/\sqrt{z}$. At atmospheric pressure, $\sqrt{z}$ is a large number, as we found in Example 6.2.

Example 6.3 Consider a molecule of N_2 gas at 1 atm and 25°C.

(a) Estimate how far away it is from its position 1 s ago.
(b) What is the total distance traveled along the zigzag path in 1 s?

Solution

(a) The root-mean-square displacement $\langle d^2 \rangle^{1/2}$ is a measure of its displacement. From Eq. (6.28a) and using the results of Example 6.2,

$$\begin{aligned}\langle d^2 \rangle^{1/2} &= \sqrt{z}l \\ &= (7.26 \times 10^9)^{1/2}(6.54 \times 10^{-6}\ \text{cm}) \\ &= 0.557\ \text{cm}\end{aligned}$$

(b) The total distance traveled per second is

$$\begin{aligned} zl &= (7.26 \times 10^9)(6.54 \times 10^{-6}\ \text{cm}) \\ &= 4.75 \times 10^4\ \text{cm}\end{aligned}$$

Note that these distances differ by a factor of almost 10^5 for this example.

Diffusion Coefficient and Fick's First Law

We shall now investigate the net transport of material by diffusion. Consider a 1-cm^2 cross-sectional area in the yz plane (Fig. 6.5). We define the flux J_x as the net amount of substance that diffuses through this unit area, per unit time, in the x direction. Units of flux, for example, would be mol cm^{-2} s^{-1}. If there is no concentration gradient in the x direction, one expects equal numbers of molecules crossing this area from the left and from the right. Therefore, the net flux, J_x, is zero. Suppose that there is a concentration gradient in the x direction, say $dc/dx > 0$. The concentration increases with x; therefore, there are more molecules per unit volume to the right of the area than to the left. One expects,

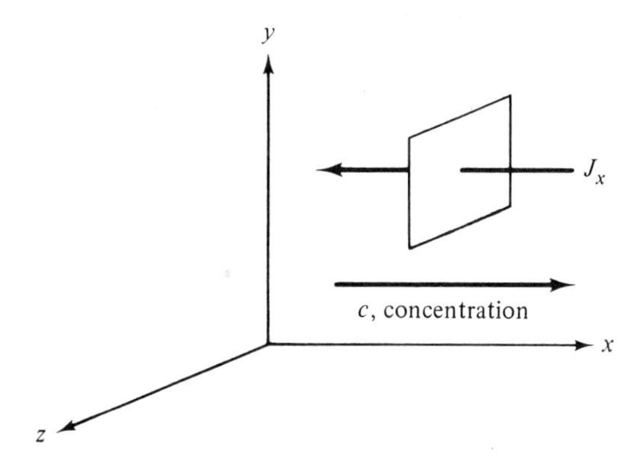

Fig. 6.5 Flux J_x is the net diffusional transport of material per unit time, in the x direction, across a unit cross-sectional area perpendicular to x. If the concentration c increases with x, the flux J_x is in the direction opposite to the concentration gradient.

then, that more molecules per unit time will diffuse across the area from the right than from the left. In other words, there will be a net transport of material in the direction opposite to the concentration gradient, as shown in Fig. 6.5. The steeper the concentration gradient, the larger the net flux. These considerations lead to the equation for *Fick's first law*:

$$J_x = -D\frac{dc}{dx} \tag{6.29}$$

where D, the proportionality constant, is called the *diffusion coefficient.* The negative sign indicates that the net transport by diffusion is in a direction opposite to the concentration gradient. The units of D are cm^2 s^{-1}; the concentration gradient is in mol cm^{-4}.

We note further that, because there is net transport of material, dc/dx is itself a function of time. As time goes on, the system gradually approaches homogeneity, and dc/dx approaches zero. Thus Eq. (6.29) is given in terms of the instantaneous flux at any time t. Equation (6.29), Fick's first law of diffusion, has been shown experimentally to be correct.

Fick's Second Law

It is also possible to describe how the concentration in a gradient changes with time. In a uniform concentration gradient, where dc/dx = constant for all values of x, Eq. (6.29) tells us that the flux J_x is the same at all positions and, therefore, c will not change with time. (The flux into every volume element from one side is exactly equal to that going out the other side.) This describes a steady-state flow of material by diffusion.

If the concentration gradient is not the same everywhere, it is possible to show by a straightforward derivation that

$$\frac{\delta c}{\delta t} = -\left(\frac{\partial J}{\partial x}\right)_t$$

where $\delta c/\delta t$ is the rate of change of concentration with respect to time, in a small volume between planes at x and $x + \delta x$. In the notation of partial differentials, we can write

$$\left(\frac{\partial c}{\partial t}\right)_x = -\left(\frac{\partial J}{\partial x}\right)_t \qquad (6.30)$$

This equation states that the change of concentration with time at position x is proportional to the gradient in the flux at that point. Intuitively, if more material is diffusing into the volume element from the left than is diffusing out to the right, then $(\partial J/\partial x) < 0$, and the concentration inside should increase with time, $(\partial c/\partial t) > 0$. This is in accord with Eq. (6.30).

Using Fick's first law,

$$J_x = -D \cdot \left(\frac{\partial c}{\partial x}\right)_t \qquad (6.29)$$

combined with Eq. (6.30), we obtain

$$\left(\frac{\partial c}{\partial t}\right)_x = \left[\frac{\partial}{\partial x} D \frac{\partial c}{\partial x}\right]_t$$

If D is a constant independent of concentration, then

$$\left(\frac{\partial c}{\partial t}\right)_x = D \cdot \left(\frac{\partial^2 c}{\partial x^2}\right)_t \qquad (6.31)$$

This is *Fick's second law*. An illustration of the application of Fick's second law to a step gradient is shown in Fig. 6.6.

Determination of the Diffusion Coefficient

In principle, D can be determined by the use of either Fick's first law [Eq. (6.29)] or Fick's second law [Eq. (6.31)]. Direct measurement of J is difficult, however,

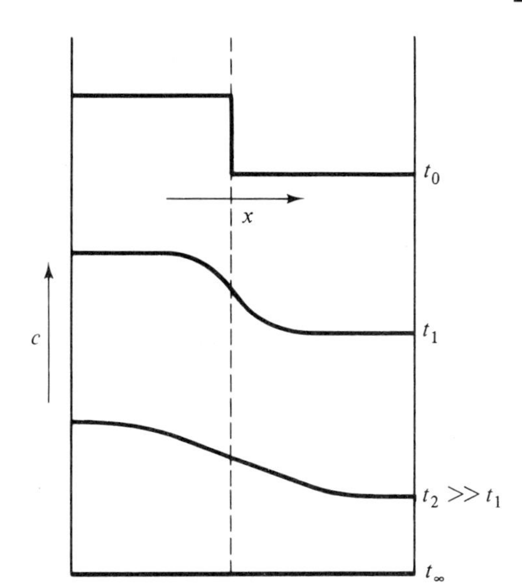

Fig. 6.6 Diffusion of a material with a concentration gradient that is initially a step. Profiles are shown at time zero (t_0) and at successive times, t_1, t_2, and t_∞. The exact behavior is described by Fick's second law [Eq. (6.31)].

and D is usually evaluated by the use of Fick's second law. The concentration c or the concentration gradient dc/dx is experimentally determined as a function of x and t. When compared with the solution of Eq. (6.31), this allows the determination of the diffusion coefficient D. Recall that Eq. (6.31) is obtained if D is independent of concentration. *This is true only at very low concentrations.* Thus, experimentally, D is obtained either from a measurement at a very low concentration, or from measurements at different concentrations followed by extrapolation to zero concentration.

The design of the experimental procedure usually involves setting up well-defined initial conditions to facilitate the solution of Eq. (6.31). A common procedure involves initially setting up a sharp boundary between a solution of uniform concentration c_0 and the solvent. A discussion of how to solve a second-order differential equation [Eq. (6.31)] is beyond the level of this book. The detailed procedures can be found in several of the references listed at the end of the chapter.

Relation Between the Diffusion Coefficient and the Mean-Square Displacement

In our discussion on diffusion in a gas, we have used the quantity $\langle d^2 \rangle$, which is the average of the square of the displacement of a molecule from its position 1 s ago. Since diffusion is due to the random motion of molecules, one would expect that the larger $\langle d^2 \rangle$ is, the larger the diffusion coefficient D. That is, the faster the molecules can move from one point to another, the greater the flux, J, for a given concentration gradient.

We shall consider a very thin layer of a solution containing g grams of solute per unit area sandwiched between layers of solvent at time $t = 0$, as illustrated in Fig. 6.7(a).

For this case, Eq. (6.31) can be solved to give c in g cm^{-3} at time t,

$$c = \frac{g}{(4\pi Dt)^{1/2}} e^{-x^2/4Dt} \tag{6.32}$$

This is plotted in Fig. 6.7(b). We can consider the concentration profile as a distribution in x. That is, during time t some solute molecules have moved far away from their original positions ($x = 0$), but others have hardly moved away at all. The mean-square displacement $\langle x^2 \rangle$ can be evaluated directly from this distribution. We shall omit the mathematics but give the result:

$$\langle x^2 \rangle = 2Dt$$

$$D = \frac{\langle x^2 \rangle}{2t} \tag{6.33}$$

The diffusion coefficient is one half of the mean-square displacement per unit time. This relation was first derived by Einstein in 1905.

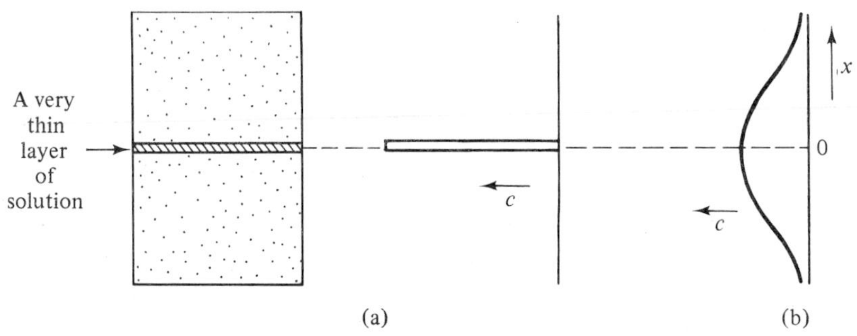

Fig. 6.7 Diffusion from a very thin layer of solution: (a) initial state; (b) concentration profile after time t.

Example 6.4 The diffusion coefficient of the oxygen storage protein myoglobin is $D_{20,w} = 11.3 \times 10^{-7}$ cm^2 s^{-1}, at 20°C in water. Estimate a mean value for the time required for a myoglobin molecule to diffuse a distance of 10 μm, which is the order of the size of a cell.

Solution The mean-square displacement, $\langle x^2 \rangle$, is a measure of the displacement of an average molecule. From Eq. (6.33),

$$t = \frac{\langle x^2 \rangle}{2D} = \frac{(10 \times 10^{-4}\ \text{cm})^2}{2(11.3 \times 10^{-7}\ \text{cm}^2\ \text{s}^{-1})} = 0.44\ \text{s}$$

Although the viscosity of cytoplasm is greater than that of water, nevertheless the diffusion of even relatively large molecules such as myoglobin (mol wt 17,000) occurs rapidly across dimensions comparable to the size of a cell.

Determination of the Diffusion Coefficient by Laser Light Scattering

The method described above starts with a nonequilibrium system, and the diffusion coefficient is evaluated from the way the system moves towards equilibrium. But since the diffusion coefficient D is related to the random motion of the molecules, which occurs in equilibrium as well as nonequilibrium systems, one should be able to measure D for a system at equilibrium by monitoring the random motion of molecules directly. One such method is laser light scattering, sometimes called quasi-elastic light scattering.

There is random motion of molecules in a homogeneous solution. If a beam of monochromatic light of frequency ν_0 passes through the solution, the scattered light is no longer monochromatic. Some molecules will be moving toward the light and some away; this causes a Doppler broadening of the scattered light. The intensity of the scattered light, when plotted as a function of frequency, has a maximum at ν_0, and its width at half-height is directly proportional to D (for molecules whose characteristic dimensions are small compared with the wavelength of the light). For a typical protein molecule with a D of 10^{-6} cm^2 s^{-1}, the width of the spectrum of the scattered light is of the order of 10^4 Hz. This is still an extremely sharp line, considering the fact that ν_0 is of the order of 10^{15} Hz. The development of highly monochromatic light from a laser source has made it possible to measure diffusion coefficients of macromolecules by this method.

Some representative results are given in Table 6.1. For small rigid proteins the precision of the measurements is about $\pm$ 3%, and the agreement with literature values is quite good. The values for tobacco mosaic virus and for DNA are less reliable because the molecules are not small compared with the wavelength of light. Also, the DNA molecule is not rigid in solution, and the internal motions contribute to the larger relative range of uncertainty. Nevertheless, for suitable systems, the light-scattering method is relatively rapid and convenient.

Table 6.1 Diffusion coefficients for macromolecules measured using laser light-scattering (LS) or conventional (Lit) methods

			$D \times 10^7$ (cm^2 s^{-1})	
	pH	Salt	LS	Lit
Bovine serum albumin	6.80	0.5 *M* KCl	6.7 ± 0.1	6.7
Ovalbumin	6.80	0.5 *M* KCl	7.1 ± 0.2	8.3
Lysozyme	5.60	—	11.5 ± 0.3	11.6
Tobacco mosaic virus	7.20	0.01 *M* Na phosphate	0.40 ± 0.02	0.3
DNA, calf thymus	7.00	0.15 *M* NaCl	0.2 ± 0.1	0.13

SOURCE: S. B. Dubin, J. H. Lunacek, and G. B. Benedek, *Proc. Nat. Acad. Sci., U.S. 57*, 1164–1171 (1967).

Because of the importance of large molecules in biological systems, an important aspect of transport measurements is to provide information on the size and shape of such macromolecules. Before discussing molecules, let us first consider a simpler case: the free fall of a ball of mass m and specific volume $\bar{v}$ through a viscous medium of density ρ. This is illustrated in Fig. 6.8. The driving force that causes the ball to sink is the difference between the gravitational force and the buoyant force:

$$\text{driving force} = mg - m\bar{v}\rho g = m \cdot (1 - \bar{v}\rho)g$$

where g is the gravitational acceleration.

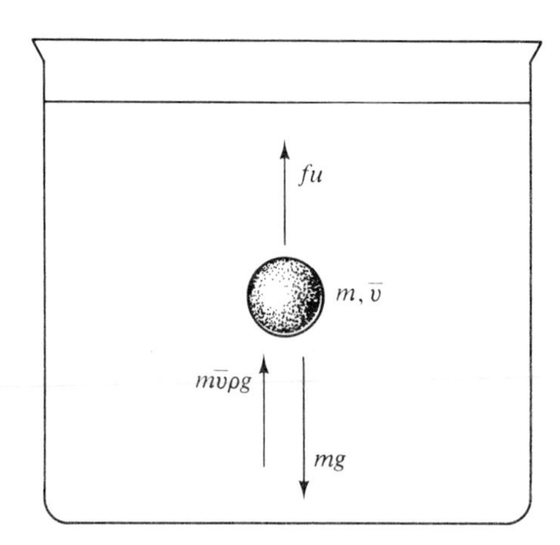

Fig. 6.8 Free fall of a ball, of mass m and specific volume $\bar{v}$, through a viscous medium of density ρ. The forces acting on the ball are, in one direction, the gravitational force mg, and in the other direction, the buoyant force $m\bar{v}\rho g$ and the frictional force fu, where u is the velocity of the ball and f is the frictional coefficient.

At time zero, when the ball is first released, its velocity is zero. As it falls, it picks up speed. Let u be its velocity at t. It can be shown that, as the ball moves through the medium with a velocity u, it experiences a frictional force proportional to u:

$$\text{frictional force} = fu$$

where f is called the *frictional coefficient.* From Newton's law we obtain

$$m \cdot (1 - \bar{v}\rho)g - fu = m\frac{du}{dt}$$

where du/dt is the acceleration.

Since the frictional force increases with u, a velocity is soon reached at which the frictional force balances the driving force, and the acceleration becomes zero. This velocity is called the terminal velocity, u_t:

$$m \cdot (1 - \bar{v}\rho)g - fu_t = 0$$

or

$$u_t = \frac{m \cdot (1 - \bar{v}\rho)g}{f} \tag{6.34}$$

In a viscous medium the terminal velocity is quickly reached. It was shown by Stokes that if the radius of the ball is r, then f is related to r by the equation

$$f = 6\pi\eta r \tag{6.35}$$

where η is the viscosity coefficient, a measure of how viscous the medium is. We shall discuss viscosity in more detail in a later section.

Now let us come back to the diffusion of molecules. The basis for relating the diffusion coefficient to the molecular parameters is the equation

$$D = \frac{kT}{f} \tag{6.36}$$

where k, the Boltzmann constant, is equal to the gas constant R divided by Avogadro's number N_0. This equation was first derived by Einstein. We shall discuss this equation later.

If the macromolcule is spherical and unsolvated, we can readily calculate its radius from its molecular weight, M, and partial specific volume, $\bar{v}_2$;

$$\text{radius} = \left(\frac{3M\bar{v}_2}{4\pi N_0}\right)^{1/3}$$

The corresponding frictional coefficient for this sphere can then be calculated from the Stokes equation (6.35). The frictional coefficient so calculated is usually found to be lower than f obtained from the diffusion coefficient [Eq. (6.36)]. We denote the frictional coefficient calculated for the unsolvated sphere as f_{min}, the minimum frictional coefficient. The frictional ratio f/f_{min} differs from unity for two reasons: solvation and any nonspherical shape of the macromolecule. These are discussed in the following sections.

Solvation

A certain amount of solvent is usually associated with a macromolecule in solution. This solvation increases the effective or hydrodynamic volume of the macromolecule and, therefore, its frictional coefficient. We must now distinguish between the postulated specific volume of the unsolvated macromolecule and the measured partial specific volume of the macromolecule in a particular solution. The effect of solvation can be calculated as follows. Let

v_2 = specific volume of an unsolvated macromolecule of molecular weight M
v_1 = specific volume of the solvent which is associated hydrodynamically with the macromolecule
v_1^0 = specific volume of the bulk solvent = density^{-1}
δ_1 = grams of solvent associated with each gram of the macromolecule

The volume of a solvated macromolecule is then

$$v_s = \frac{M}{N_0} \cdot (v_2 + \delta_1 v_1) \tag{6.37}$$

For a solution containing g_2 grams (dry weight) of macromolecules and g_1 grams of solvent, the total volume is

$$V = g_2 v_2 + g_2 \delta_1 v_1 + (g_1 - g_2 \delta_1) v_1^0$$

The partial specific volume of the macromolecule is, by definition,

$$\bar{v}_2 = \left(\frac{\partial V}{\partial g_2}\right)_{T, P, g_1} = v_2 + \delta_1 v_1 - \delta_1 v_1^0$$

$\bar{v}_2$ is the change in volume of the solution of a given concentration when a macromolecule is added. Substituting into Eq. (6.37) gives

$$v_s = \frac{M}{N_0} \cdot (\bar{v}_2 + \delta_1 v_1^0) \tag{6.38}$$

Thus the effect of solvation is to increase the volume of the macromolecule and therefore the frictional coefficient. If the macromolecule is spherical, it is straightforward to show that $f_{\min}$ is increased by a factor of $[(\bar{v}_2 + \delta_1 v_1^0)/\bar{v}_2]^{1/3}$.

Shape Factor

It is convenient to partition the frictional ratio $f/f_{\min}$ into two parts:

$$\frac{f}{f_{\min}} = \left(\frac{f}{f_0}\right)\left(\frac{f_0}{f_{\min}}\right) \tag{6.39}$$

where f_0 *is the frictional coefficient of a sphere with a volume equal to that of the solvated macromolecule.* On the basis of the previous discussion, we assign

$$\frac{f_0}{f_{\min}} = \left(\frac{\bar{v}_2 + \delta_1 v_1^0}{\bar{v}_2}\right)^{1/3} \tag{6.40}$$

as the factor due to solvation. The other factor (f/f_0) is due to deviation from spherical shape. So far, (f/f_0) has been calculated only for simple geometric forms, notably for ellipsoids of revolution. Some calculated results are given in Table 6.2.

Example 6.5 The diffusion coefficient D and the partial specific volume of an enzyme (ribonuclease) from bovine pancreas, which digests RNA, have been measured in a dilute buffer at 20°C. The values are:

$$D = 13.1 \times 10^{-7} \text{ cm}^2 \text{ s}^{-1}$$

$$\bar{v}_2 = 0.707 \text{ cm}^3 \text{ g}^{-1}$$

The molecular weight of the protein is 12,640.

(a) Calculate the frictional coefficient f.

Table 6.2 Ratio f/f_0 as a function of the axial ratio for ellipsoids of revolution

Major axis / Minor axis	f/f_0 (a) *Prolate ellipsoids* Ellipsoid of revolution generated by an ellipse of major axis a and minor axis b revolving about its major axis	f/f_0 (b) *Oblate ellipsoids* Ellipsoid of revolution generated by an ellipse of major axis b and minor axis a revolving about its minor axis
1	1.000	1.000
2	1.044	1.042
3	1.113	1.105
4	1.182	1.166
5	1.250	1.223
6	1.314	1.277
8	1.433	1.373
10	1.543	1.458
12	1.644	1.534
14	1.739	1.603
16	1.829	1.667
18	1.914	1.727
20	1.995	1.783
25	2.184	1.909
30	2.357	2.020
35	2.518	2.120
40	2.668	2.212
45	2.811	2.293
50	2.947	2.375
60	3.201	2.519
70	3.437	2.648
80	3.658	2.765
90	3.867	2.873
100	4.066	2.974

SOURCE: Data calculated using the equations of F. Perrin, *J. Phys. Radium* 7, 1 (1936).

(b) Assuming that the protein molecule is an unhydrated sphere, calculate its volume, radius r, and the frictional coefficient f_{min} corresponding to this unhydrated sphere.

(c) From the ratio f/f_{min}, estimate the extent of hydration if the difference from unity is due entirely to hydration. That is, we assume the molecule is spherical.

Solution

(a) From Eq. (6.36),

$$f = \frac{kT}{D}$$

$$k = \frac{R}{N_0} = \frac{8.314 \times 10^7 \text{ erg deg}^{-1} \text{ mol}^{-1}}{6.023 \times 10^{23} \text{ molecules mol}^{-1}}$$

$$= 1.380 \times 10^{-16} \text{ erg deg}^{-1} \text{ molecule}^{-1}$$

$$= 1.380 \times 10^{-16} \text{ g cm}^2 \text{ s}^{-2} \text{ deg}^{-1} \text{ molecule}^{-1}$$

Therefore,

$$f = \frac{(1.380 \times 10^{-16} \text{ g cm}^2 \text{ s}^{-2} \text{ deg}^{-1})(293 \text{ deg})}{13.1 \times 10^{-7} \text{ cm}^2 \text{ s}^{-1}}$$

$$= 3.09 \times 10^{-8} \text{ g s}^{-1}$$

(b) The mass of a ribonuclease molecule is ($12{,}640 \text{ g}/6.023 \times 10^{23}$), or 2.10×10^{-20} g. The volume is the mass times the partial specific volume $\bar{v}_2$:

$$V = (2.10 \times 10^{-20} \text{ g})(0.707 \text{ cm}^3 \text{ g}^{-1})$$

$$= 1.48 \times 10^{-20} \text{ cm}^3$$

Since

$$\tfrac{4}{3}\pi r^3 = 1.48 \times 10^{-20} \text{ cm}^3$$

$$r = 1.52 \times 10^{-7} \text{ cm}$$

$$= 15.2 \text{ Å}$$

$$f_{\min} = 6\pi\eta r$$

The viscosity coefficient for a dilute aqueous buffer is approximately equal to that of water, which is, at 20°C, 1.00 centipoise or $1.00 \times 10^{-2} \text{ g s}^{-1} \text{ cm}^{-1}$. Thus,

$$f_{\min} = 6\pi(1.00 \times 10^{-2} \text{ g s}^{-1} \text{ cm}^{-1})(1.52 \times 10^{-7} \text{ cm})$$

$$= 2.87 \times 10^{-8} \text{ g s}^{-1}$$

(c) $f/f_{\min} = (3.09 \times 10^{-8})/(2.87 \times 10^{-8}) = 1.08$. As we are assuming a sphere, $f = f_0$ and from Eq. (6.40) we have

$$\frac{\bar{v}_2 + \delta_1 v_1^0}{\bar{v}_2} = \left(\frac{f}{f_{\min}}\right)^3$$

$$= 1.26$$

Thus, $\delta_1 v_1^0 = 1.26\bar{v}_2 - \bar{v}_2 = 0.26\,\bar{v}_2 = 0.184 \text{ cm}^3 \text{ g}^{-1}$. The specific volume of bulk water, v_1^0, is 1.00 cm^3 g^{-1}. Thus,

$$\delta_1 = 0.184$$

That is, the amount of hydration is 0.184 g of water per g of protein.

Example 6.6 Protein L7, a protein of the 50 S subunit of the *Escherichia coli* ribosome,* exists as a dimer of molecular weight 24,000 in a dilute aqueous buffer. The following data have been obtained at 20°C (K. P. Wong and H. H. Paradies, *Biochem. Biophys. Res. Commun. 61*, 178, 1974):

$$D = 5.59 \times 10^{-7} \text{ cm}^2 \text{ s}^{-1}$$

$$\bar{v}_2 = 0.735 \text{ cm}^3 \text{ g}^{-1}$$

(a) Calculate f/f_{min}.
(b) Calculate the degree of hydration if the molecule is spherical in shape.
(c) Calculate the axial ratio of the protein molecule if it is an unhydrated prolate ellipsoid.
(d) Calculate the dimensions of the prolate ellipsoid.

Solution

(a)

$$f = \frac{kT}{D} = 7.23 \times 10^{-8} \text{ g s}^{-1}$$

$$\frac{4}{3}\pi r^3 = \left(\frac{24{,}000}{6.023 \times 10^{23}}\right) 0.735 \text{ cm}^3$$

$$= 2.93 \times 10^{-20} \text{ cm}^3$$

$$r = 1.91 \times 10^{-7} \text{ cm}$$

$$f_{\text{min}} = 3.60 \times 10^{-8} \text{ g s}^{-1}$$

Thus,

$$\frac{f}{f_{\text{min}}} = 2.00$$

(b)

$$\frac{\bar{v}_2 + \delta_1 v_1^0}{\bar{v}_2} = (2.00)^3 = 8.00$$

$$\delta_1 v_1^0 = 5.14 \text{ cm}^3 \text{ g}^{-1}$$

$$\delta_1 = 5.14 \text{ g of water/g of protein}$$

* Ribosomes are particles of protein and nucleic acid on which protein synthesis occurs. They are present in all cells including the bacterium *E. coli*. The *E. coli* ribosome has two subunits; the larger one has a sedimentation coefficient of 50 svedbergs and is thus called the 50 S subunit.

(c) If the protein is unhydrated,

$$\frac{f_0}{f_{\min}} = 1$$

and

$$\frac{f}{f_0} = \frac{f}{f_{\min}} = 2.00$$

For a prolate ellipsoid, Table 6.2 gives the axial ratio $a/b = 20.1$ by interpolation.

(d) The volume of a prolate ellipsoid is $\frac{4}{3}\pi ab^2$. From part (a), the volume of the molecule is $2.93 \times 10^{-20}\ \text{cm}^3$. From part (c), $a = 20.1b$. Thus,

$$\tfrac{4}{3}\pi(20.1)b^3 = 2.93 \times 10^{-20}\ \text{cm}^3$$

or

$$\begin{aligned} b &= \left(2.93 \times 10^{-20} \times \frac{3}{20.1 \times 4\pi}\right)^{1/3} \\ &= 7.03 \times 10^{-8}\ \text{cm} \\ &= 7.03\ \text{Å} \\ a &= 20.1b = 141\ \text{Å} \end{aligned}$$

The length of the molecule along the long axis is $2a$ or 282 Å; the length along the short axis is $2b$ or 14.1 Å.

From this example we see that, because $f/f_{\min}$ is affected by both shape and solvation, without additional information the ratio cannot be interpreted unambiguously. For the L7 protein discussed, measurements by low-angle x-ray scattering, a technique that we shall mention in Chapter 12, reveal that the degree of hydration is about 0.26 g of water per g of protein. With this additional piece of information, an improved estimation of the dimensions of the protein can be made. This is left to the reader (Problem 4).

Diffusion Coefficients of Random Coils

The concepts underlying our discussions in the preceding section are useful for macromolecules which are reasonably rigid. The situation is different when very flexible macromolecules are involved. Consider, for example, a large DNA molecule. Such a molecule, with its length much greater than its diameter, resembles somewhat a loose ball of thread in solution. As the macromolecule moves through solution, two limiting possibilities arise with regard to the solvent molecules within the domain of the coiled macromolecules. They can move freely, independent of the motion of the macromolecule, or they can be "trapped" by the macromolecule and move with it. These two limiting possibilities are called "free draining" and "nondraining," respectively. For DNA molecules in aqueous media, the "nondraining" model describes their hydrodynamic properties adequately. The amount of solvent hydrodynamically associated with a DNA molecule is large. Therefore, as an approximation the DNA molecule can be considered as a highly hydrated sphere. The radius of this sphere is expected to be directly proportional to the average three-dimensional size of the molecule.

In Chapter 11 we describe a quantity called the mean-square radius, which is a measure of the average dimension of the molecule. The mean-square radius is proportional to the square root of the molecular weight of the macromolecule, if the macromolecule is a very flexible coil.

For such flexible coils the frictional coefficients are expected to be proportional to $M^{1/2}$, and therefore the diffusion coefficients are expected to be proportional to $M^{-1/2}$. This is found to be approximately true.

More exact theories for the hydrodynamic properties of flexible coils are available, but their discussion is beyond the scope of this book.

SEDIMENTATION

We have described the terminal velocity u_t of a sphere of mass m and specific volume $\bar{v}$ falling through a medium of density ρ in the gravitational field of the earth:

$$u_t = \frac{m \cdot (1 - \bar{v}\rho)g}{f} \tag{6.34}$$

In a centrifugal field, the centrifugal acceleration is $\omega^2 x$, where ω is the angular velocity of the centrifuge in units of rad s^{-1}, and x is the distance from the center of rotation. By analogy to the free fall of a sphere in a gravitational field, a molecule (of mass m and partial specific volume $\bar{v}$) sedimenting through a viscous medium in a centrifugal field also reaches a terminal velocity u_t:

$$u_t = \frac{m \cdot (1 - \bar{v}\rho)}{f}\omega^2 x \tag{6.41}$$

The centrifugal acceleration $\omega^2 x$ now replaces the gravitational acceleration g in Eq. (6.34).

The quantity $u_t/\omega^2 x$, which is the velocity per unit acceleration, is called the *sedimentation coefficient.* The symbol s is usually used for the sedimentation coefficient. The dimension of s is $(\text{cm s}^{-1})/(\text{cm s}^{-2})$ or s. A convenient unit for s is the *Svedberg*, named in honor of T. Svedberg, who pioneered much research on sedimentation in an ultracentrifuge. One Svedberg, or 1 S, is defined as 10^{-13} s. From the definition of s and Eq. (6.41), we have

$$s = \frac{m \cdot (1 - \bar{v}_2\rho)}{f} \tag{6.42}$$

Determination of the Sedimentation Coefficient

Two methods can be used to measure the sedimentation coefficients of macromolecules. In one, a homogeneous solution is spun in an ultracentrifuge. As the macromolecules move down the centrifugal field, a solution-solvent boundary is generated. By following the movement of the boundary with time, the sedimentation coefficient can be calculated. This method is called *boundary sedimentation* and is illustrated in Fig. 6.9.

The generation of a boundary means that a concentration gradient is also generated, which according to Eq. (6.29) results in diffusional transport of solute molecules. Thus, transport by sedimentation is coupled with transport by diffusion. If the macromolecules are very large or the centrifugal acceleration is very high, transport by sedimentation is much greater than transport by diffusion. Under such conditions, the boundary is very sharp. If transport by diffusion is significant, the boundary broadens as it moves downfield. As an approximation, we can assume that the diffusion process does not affect the rate of movement of the boundary. It is usually sufficiently accurate to calculate s from the position of the midpoint of the boundary, $x_{1/2}$:

$$s = \frac{u_t}{\omega^2 x} = \frac{dx_{1/2}/dt}{\omega^2 x_{1/2}} = \frac{1}{\omega^2}\frac{d\ln x_{1/2}}{dt} = \frac{2.303}{\omega^2}\frac{d\log x_{1/2}}{dt} \qquad (6.43)$$

For a more rigorous discussion of the calculation of s, a more advanced text should be consulted. Several references are listed at the end of this chapter.

Example 6.7 *Escherichia coli* DNA ligase, an enzyme that catalyzes the formation of a phosphodiester bond from a pair of 3′-hydroxyl and 5′-phosphoryl groups in a double-stranded DNA, has a molecular weight of 74,000 and a $\bar{v}_2$ of 0.737 $cm^3\,g^{-1}$ at 20°C. Boundary sedimentation in a dilute aqueous buffer (0.02 M potassium phosphate, 0.01 M NH_4Cl, and 0.2 M KCl, pH 6.5) at 20.6°C and a speed of rotation of 56,050 rpm gave the following results (data courtesy of P. Modrich):

Time (min)	$x_{1/2}$ (cm)	$\log x_{1/2}$
0	5.9110	0.7717
20	6.0217	0.7797
40	6.1141	0.7863
60	6.2068	0.7929
80	6.3040	0.7996
100	6.4047	0.8065
120	6.5133	0.8138
140	6.6141	0.8205

(a) Calculate s.
(b) Calculate the frictional factor f in the dilute aqueous buffer; the density of the buffer at 20.6°C is 1.010 $g\,cm^{-3}$.

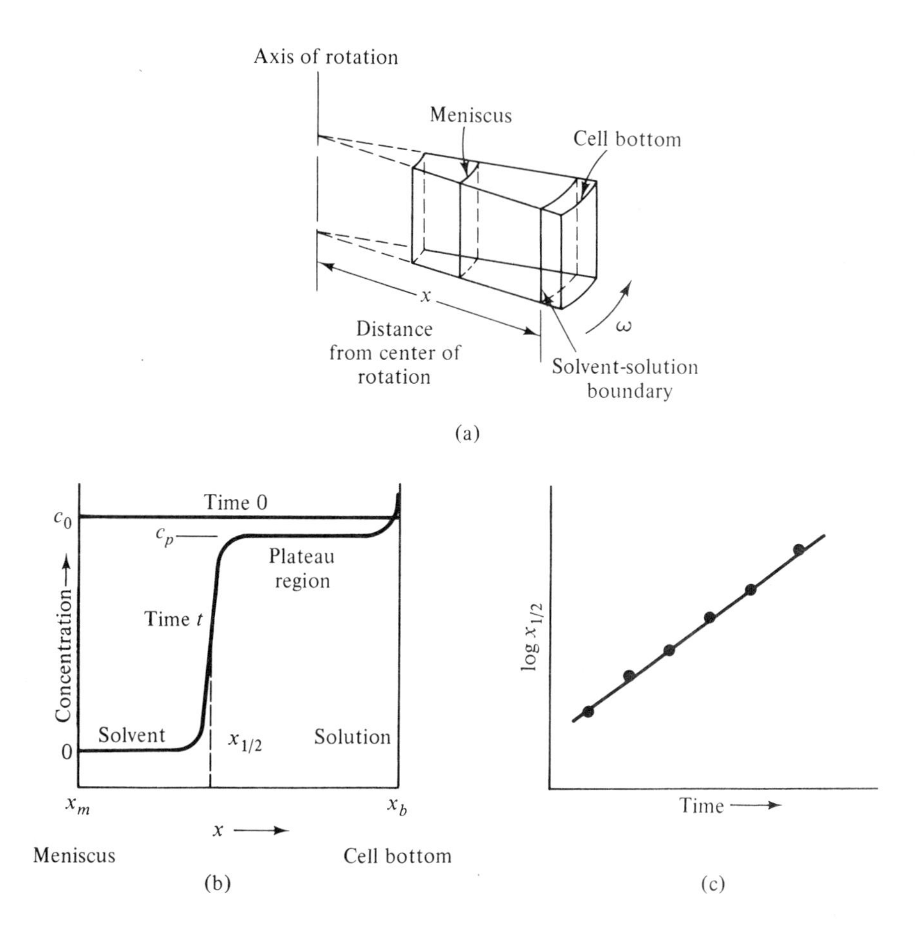

Fig. 6.9 Boundary sedimentation. (a) The compartment of the centrifuge cell containing the solution is sector-shaped, with the center of the sector at the axis of rotation. The dimensions of the cell are exaggerated for clarity. In recent years, double-sector cells have been frequently used. Each cell contains two adjacent sectors; solution is placed in one sector and solvent in the other. The detecting system is designed to measure the difference between the two sectors. (b) Concentration as a function of the distance from the center of rotation. At time 0, the concentration (c_0) is uniform across the cell. At time t, owing to sedimentation of the macromolecular solute molecules, a sharp boundary has been generated, with solvent to its left and solution to its right. The concentration in the plateau region, c_p, is independent of position; c_p is lower than c_0 because of the sector shape of the cell compartment and because the centrifugal field increases with increasing x. It can be shown that $c_p/c_0 = (x_m/x_{1/2})^2$, where x_m is the position of the meniscus and $x_{1/2}$ is the position of the boundary. (c) Plot of $\log x_{1/2}$ versus t gives a straight line. The sedimentation coefficient can be obtained from the slope of this line [Eq. (6.43)].

Solution

(a) A plot of $\log x_{1/2}$ versus t gives a straight line with a slope of $3.42 \times 10^{-4}\ \text{min}^{-1}$. The angular velocity ω is

$$\omega = (56{,}050 \text{ revolutions min}^{-1})\,(2\pi \text{ rad revolution}^{-1})\,(1\,\text{min}/60\,\text{s})$$
$$= 5.870 \times 10^{3} \text{ rad s}^{-1}$$

(note that the unit radian is defined as an arc length divided by a radius and is therefore dimensionless)

From Eq. (6.43),

$$s = \frac{2.303}{\omega^2}\,(3.42 \times 10^{-4}\,\text{min}^{-1})$$

$$= \frac{2.303(3.42 \times 10^{-4}\,\text{min}^{-1})}{(5.870 \times 10^{3}\,\text{s}^{-1})^2(60\,\text{s min}^{-1})}$$

$$= 3.81 \times 10^{-13}\,\text{s}$$

$$= 3.81\ \text{S}$$

(b) From Eq. (6.42),

$$f = \frac{m \cdot (1 - \bar{v}_2\rho)}{s}$$

$$m = \frac{74{,}000}{6.023 \times 10^{23}} = 1.23 \times 10^{-19}\,\text{g}$$

$$f = \frac{(1.23 \times 10^{-19}\,\text{g})(1 - 0.737 \times 1.010)}{3.81 \times 10^{-13}\,\text{s}}$$

$$= 8.24 \times 10^{-8}\,\text{g s}^{-1}$$

A second method of sedimentation is called *zone sedimentation*. A thin layer of a solution of macromolecules is placed on top of a solvent at the beginning of centrifugation. As centrifugation continues, the macromolecules sediment through the solvent as a zone or band. However, as the density of the macromolecular solution will be higher than that of the solvent, an instability will occur. The sedimenting zone will be denser than the solvent below it and will tend to mix. To avoid such instability, a solvent gradient is usually imposed so that the net density gradient in the direction of the centrifugal field is always positive. For example, a pre-formed sucrose gradient can be used, as illustrated in Fig. 6.10. Alternatively, if a concentrated salt solution is used as the medium for sedimentation measurements, the density gradient generated by the sedimentation of the salt is sufficient to provide the stabilization.

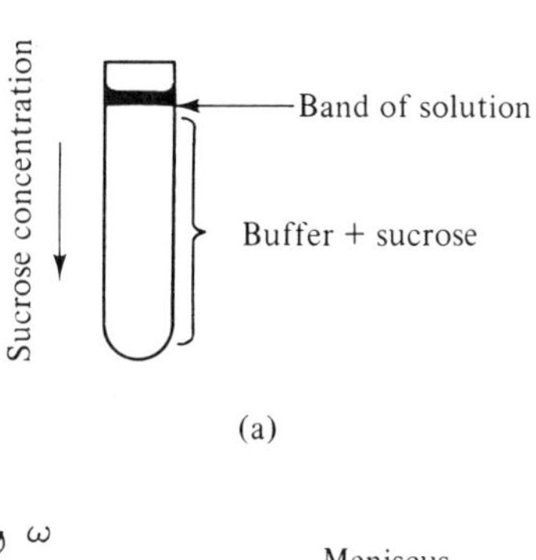

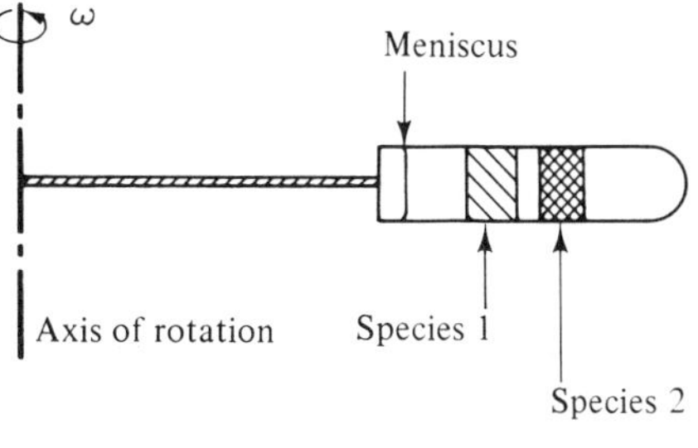

(b)

Species 1
Species 2
Concentration
Distance from center of rotation
(c)

Fig. 6.10 Zone sedimentation through a pre-formed density gradient. (a) At time 0, a layer of solution is placed on a pre-formed gradient. In this illustration, a sucrose gradient (for example, a linear gradient from 5% to 20% sucrose) is employed. The buffer concentration is constant along the tube. (b) After spinning the tube in a centrifuge at an angular speed ω for a certain time, t, the macromolecular species have sedimented. In this example, the original solution contained two macromolecular species with different sedimentation coefficients. (c) The concentration profile after time t. A typical measurement technique involves the use of radioactively labeled macromolecules. The centrifuge is stopped after time t, a hole is punched through the bottom of the tube, and fractions are collected. The radioactivity of each fraction is then determined.

Standard Sedimentation Coefficient

From Eq. (6.42),

$$s = \frac{m \cdot (1 - \bar{v}_2\rho)}{f}$$

Since f is proportional to the solvent viscosity coefficient η, we write

$$s = \frac{m}{f'} \cdot \frac{1 - \bar{v}_2\rho}{\eta}$$

where $f' = \dfrac{f}{\eta}$

To facilitate comparison of values of s measured in solvents with different values of ρ and η, it is a common practice to standardize the measured s. If $s_{20,w}$ is the value at 20°C in water, we can write

$$s_{20,w} = \frac{m}{f'} \cdot \frac{(1 - \bar{v}_2\rho)_{20,w}}{\eta_{20,w}}.$$

if f' remains constant. The subscripts 20, w refer to the quantities in water at 20°C. The relation between $s_{20,w}$ and s is therefore

$$s_{20,w} = s\frac{\eta}{\eta_{20,w}} \frac{(1 - \bar{v}_2\rho)_{20,w}}{1 - \bar{v}_2\rho} \tag{6.44}$$

It is a common practice to convert the measured s to $s_{20,w}$ by Eq. (6.44). We note, however, that f' may not remain constant in different media and Eq. (6.42) is itself formulated for a two-component system (the macromolecular species in a pure solvent). Since biological macromolecules are usually studied in a multicomponent solution containing buffer molecules and other electrolytes in addition to the macromolecules and water, Eq. (6.42) is an approximation. These limitations should be kept in mind in the interpretation of $s_{20,w}$ values obtained from s values measured in very different media under different conditions.

Example 6.8 Convert the value of s obtained in Example 6.7(a) to $s_{20,w}$. At 20°C the ratio of the viscosity of the buffer, η_b, to that of water, η_w, has been measured to be $\eta_b/\eta_w = 1.003$.

Solution For a dilute buffer, it is sufficiently accurate to assume that its temperature dependence of viscosity is the same as that of water. Thus,

$$\frac{\eta_b}{\eta_w} = 1.003$$

From a handbook we obtain $\eta_{20.6,w} = 0.9906$ centipoise (cP) and $\eta_{20,w} =$ 1.0050 cP. Therefore,

$$\frac{\eta_{b,20.6}}{\eta_{w,20}} = \frac{1.003 \times 0.9906}{1.0050} = 0.989$$

We assume for the protein that $\bar{v}_2$ at 20.6°C is the same as $\bar{v}_2$ at 20°C. The density of water at 20°C is 0.9982 g cm^{-3} from Table 2.2. Thus,

$$\frac{(1 - \bar{v}_2\rho)_{20,w}}{1 - \bar{v}_2\rho} = \frac{1 - 0.737 \times 0.9982}{1 - 0.737 \times 1.010} = 1.034$$

Substituting these values into Eq. (6.44), we obtain

$$s_{20,w} = 3.81\ \text{S} \times 0.989 \times 1.034 = 3.90\ \text{S}$$

Dependence of s on Molecular Size and Shape

Since s, m, and ρ can all be determined experimentally, we can calculate the frictional coefficient f from Eq. (6.42). We have already discussed the relation between f and molecular size and shape in connection with our discussions on diffusion.

In a given medium for a family of macromolecules of the same $\bar{v}_2$, we expect s to be proportional to M/f. We discussed previously that, for random coils in the nondraining limit, f is proportional to $M/M^{1/2}$ or $M^{1/2}$. Experimentally, for large DNA molecules s is found to be proportional to $M^{0.44}$, close to our expectation. For spherical, unsolvated molecules of the same partial specific volume, f is proportional to r and therefore to $M^{1/3}$, and s is expected to be proportional to $M^{2/3}$.

Sedimentation-diffusion Equilibrium

Suppose that a homogeneous two-component solution (a solute plus a solvent) is spun in an ultracentrifuge. Because of sedimentation, a concentration gradient is generated. Diffusion then sets in. Since transport by sedimentation and by diffusion go in opposite directions, an equilibrium concentration gradient can be generated by centrifugation, in which transport by sedimentation exactly balances transport by diffusion. Similar concentration gradients would develop in the earth's atmosphere and in the oceans, if there were no convective currents to disrupt the balance between gravitational and diffusive transport.

This equilibrium concentration gradient can be derived from rigorous thermodynamic considerations that are independent of how the equilibrium is established (see, for example, the discussions in Tanford, 1961). For an ideal solution or for real solutions at low concentrations, the result is

$$\frac{RT}{c}\frac{dc}{dx} = M \cdot (1 - \bar{v}_2\rho)\omega^2 x \tag{6.45}$$

where c, M, and $\bar{v}_2$ are the concentration, molecular weight, and partial specific volume of the solute, and ρ is the density of the solution.

Note that $dc/c = d\ln c = 2.303\, d\log c$ and $x\,dx = \frac{1}{2}d(x^2)$. We can write Eq. (6.45) in the form

$$M = \frac{4.606RT}{\omega^2 \cdot (1 - \bar{v}_2\rho)}\frac{d\log c}{d(x^2)} \tag{6.46}$$

For dilute solutions, ρ is essentially the density of the solvent, a constant. Thus at constant temperature and constant ω, when equilibrium is reached a plot of $\log c$ versus x^2 should give a straight line. The slope, when multiplied by $4.606RT/\omega^2 \cdot (1 - \bar{v}_2\rho)$, gives the molecular weight M. Thus, equilibrium centrifugation can be used to measure M.

Let us now come back to Eq. (6.45), which gives the concentration gradient when equilibrium is reached in a centrifugal field. We take an arbitrary cross-sectional area A perpendicular to the centrifugal field and at a position x from the center of rotation. This is illustrated in Fig. 6.11. We ask how many solute molecules will be transported across this area by sedimentation in a small time interval Δt. The centrifugal acceleration at position x is $\omega^2 x$ and, therefore, the sedimentation velocity of a molecule at this position is $s\omega^2 x$. In a time Δt, the distance a molecule sediments is $s\omega^2 x\ \Delta t$. As depicted in the figure, all molecules within the volume $s\omega^2 x\ \Delta t A$ will sediment through A. If the concentration at x is c, the amount of solute in this volume is $cs\omega^2 x\ \Delta t A$, and it will all be transported across A in time Δt.

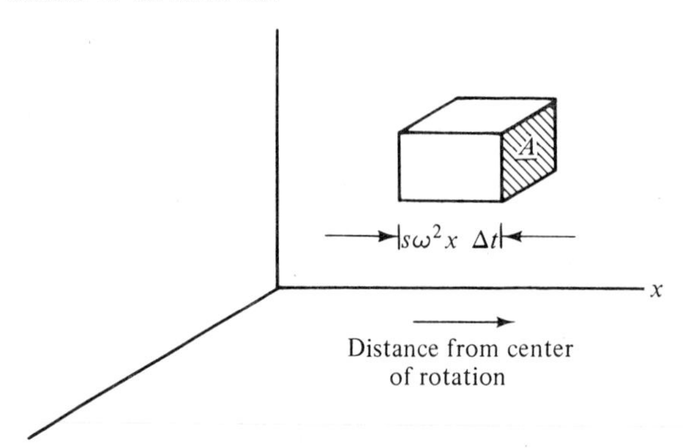

Fig. 6.11 All macromolecules of sedimentation coefficient s in the volume element shown will sediment through the area A in a time interval Δt.

How much solute will be transported across A, in the reverse direction, by diffusion? Because the concentration gradient is dc/dx, the flux J by diffusion in the opposite direction is $D \cdot (dc/dx)$ from Fick's first law Eq. (6.29). Therefore, the amount of solute transported in time Δt is $D \cdot (dc/dx)A\ \Delta t$.

Equilibrium is reached when transport by sedimentation is balanced by transport by diffusion in the opposite direction,

$$cs\omega^2 x\ \Delta t A = D\frac{dc}{dx}A\ \Delta t$$

Upon canceling $A\ \Delta t$ and dividing by c, we obtain

$$D\frac{1}{c}\frac{dc}{dx} = s\omega^2 x$$

But

$$s = \frac{m \cdot (1 - \bar{v}_2\rho)}{f} \tag{6.42}$$

$$= \frac{M \cdot (1 - \bar{v}_2\rho)}{N_0 f}$$

Substituting into the equation above and multiplying both sides by $N_0 f$,

$$N_0 f D\frac{1}{c}\frac{dc}{dx} = M \cdot (1 - \bar{v}_2\rho)\omega^2 x \tag{6.47}$$

Comparing Eq. (6.47) with Eq. (6.45) gives

$$N_0 f D = RT$$

or

$$D = \frac{RT}{N_0 f} = \frac{kT}{f}$$

which is Eq. (6.36) given previously without derivation.

Note also, if we cancel f from Eqs. (6.36) and (6.42), we obtain

$$\frac{s}{D} = \frac{M \cdot (1 - \bar{v}_2 \rho)}{N_0 kT} = \frac{M \cdot (1 - \bar{v}_2 \rho)}{RT}$$

or

$$M \doteq \frac{RTs}{D \cdot (1 - \bar{v}_2 \rho)} \tag{6.48}$$

Equation (6.48) is an important result, which allows the determination of the molecular weight from the experimentally measured values of s, D, and $\bar{v}_2$.

Example 6.9 The following data have been obtained for ribosomes from a paramecium: $s_{20,w} = 82.6\,\text{S}$, $D_{20,w} = 1.52 \times 10^{-7}\,\text{cm}^2\,\text{s}^{-1}$, $\bar{v}_{20} = 0.61\,\text{cm}^3\,\text{g}^{-1}$ (from A. H. Reisner, J. Rowe, and H. M. Macindoe, *J. Mol. Biol. 32*, 587, 1968). Calculate the molecular weight of the ribosomes.

Solution From Eq. (6.48),

$$\begin{aligned} M &= \frac{(8.314 \times 10^7\,\text{erg deg}^{-1}\,\text{mol}^{-1})(293\,\text{deg})(82.6 \times 10^{-13}\,\text{s})}{(1.52 \times 10^{-7}\,\text{cm}^2\,\text{s}^{-1})(1 - 0.61 \times 1.00)} \\ &= 3.4 \times 10^6\,\text{erg mol}^{-1}\,\text{cm}^{-2}\,\text{s}^2 \\ &= 3.4 \times 10^6\,\text{erg}\,(1\,\text{g cm}^2\,\text{s}^{-2}/\text{erg})\,\text{mol}^{-1}\,\text{cm}^{-2}\,\text{s}^2 \\ &= 3.4 \times 10^6\,\text{g mol}^{-1} \end{aligned}$$

A second application of equilibrium centrifugation involves the spinning of a concentrated salt solution, such as CsCl, at high speed. A concentration gradient is generated, which results in a density gradient, because the density of a CsCl solution increases with increasing concentration. If a macromolecular species is also present in the solution, it will form a band at a point in the salt gradient where the macromolecules are "buoyant." This is illustrated in Fig. 6.12. The higher the molecular weight of the macromolecular species, the sharper will be the band that forms in the density gradient. Many biological macromolecules have "buoyant densities" sufficiently different that they can be resolved by the density-gradient centrifugation method.

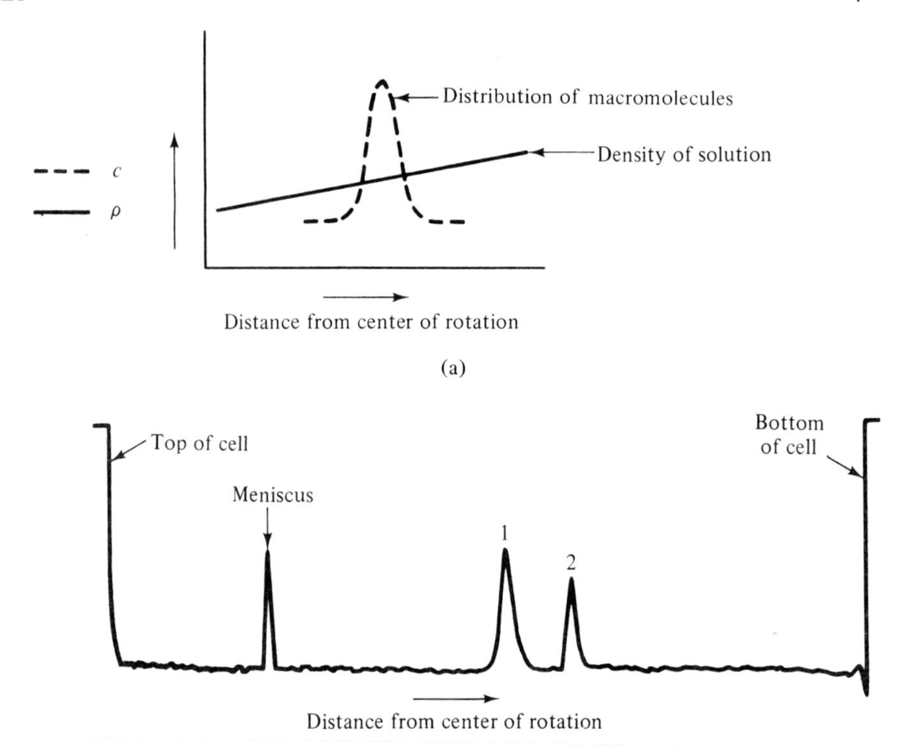

Fig. 6.12 Density-gradient centrifugation. (a) A macromolecular species in a concentrated salt solution of an appropriate density is spun in an ultracentrifuge. The solution was initially homogeneous. After a certain time, equilibrium is reached. The concentration of the salt, and consequently, the density of the solution, increases with increasing distance from the center of rotation. The macromolecular species forms a band at a position at which the solvated molecules are buoyant. (b) An actual tracing of two DNA species in a CsCl solution. The initial homogeneous solution has a density of 1.739 g cm^{-3}. After 17 h at 44,770 revolutions min^{-1} and 25°C, the DNA species form two sharp bands. Species 1 is a bacterial virus DNA with a molecular weight 20×10^6. Species 2 is the same DNA except that it contains a heavier isotope of nitrogen (^{15}N rather than the usual ^{14}N). The substitution of ^{14}N by ^{15}N increases the buoyant density of this DNA by 0.012 g cm^{-3}.

VISCOSITY

When an external force F is applied to a solid particle to make it move through a liquid with a velocity u_t, the molecules of the liquid at the interface with the particle move at the same velocity u_t because of the attractive forces between the two. Far away from the particle, the liquid remains stationary. Thus the movement of the particle through the liquid generates a velocity gradient in the medium.

Whenever a velocity gradient is maintained in a liquid, momentum is constantly transferred in a direction opposite to the direction of the velocity gradient. This is illustrated in Fig. 6.13. Here the liquid is moving in the direction x. The x components of the average velocities of molecules in different layers are represented by arrows of different lengths. The velocity gradient is in the y direction; that is, the x component of the velocity, u_x, increases with increasing y.

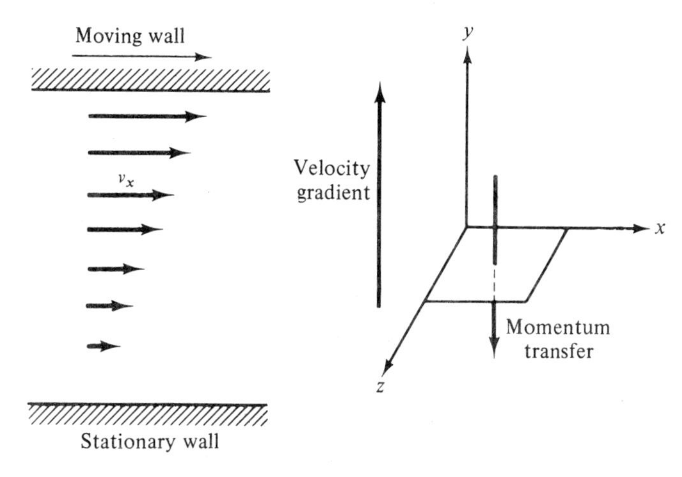

Fig. 6.13 Uniform velocity gradient produced in a fluid placed between a moving wall and a parallel stationary wall. The velocity gradient is a vector in the direction of increasing velocity, y, and is perpendicular to the direction of flow, x. Momentum transfer occurs in the direction $-y$, opposite to the velocity gradient.

The molecules are also moving in the y and z directions. However, because there is no net flow in the y and z directions, the movement of the molecules in these directions is random. Now consider a unit cross-sectional area in the xz plane. In 1 s, a certain number of molecules will move across this area from below, and an equal number of molecules will move across this area from above. But since molecules from below have lower u_x and consequently lower momentum in the x direction, there is a net transfer of momentum in the x direction from above to below, that is, in a direction opposite to the velocity gradient. The steeper the velocity gradient, the larger the net momentum transfer. Mathematically, we write

$$J_{mu} \propto -\frac{du_x}{dy}$$

or

$$J_{mu} = -\eta \frac{du_x}{dy} \tag{6.49}$$

where J_{mu} is the rate of momentum transfer per unit time per unit cross-sectional area. Because the rate of momentum transfer per unit time is the force, according to Newton's law J_{mu} is also the force per unit area in the x direction.

The quantity η in Eq. (6.49) is called the *viscosity coefficient* or *viscosity*. If η is a constant independent of du_x/dy, the fluid is called a *Newtonian fluid*. If η is itself dependent on du_x/dy, the fluid is *non-Newtonian*.

Thus, when a particle moves through a stationary fluid under an external force F, a velocity gradient is generated and the velocity gradient, in turn, imposes a viscous drag F' on the particle. The final velocity gradient is such that F and F' balance each other, and the particle moves at the terminal velocity u_t.

Measurement of Viscosity

In our discussion of the free fall of a ball through a viscous medium, we obtained

$$u_t = \frac{m \cdot (1 - \bar{v}\rho)g}{f} \tag{6.34}$$

Substituting the Stokes equation for a sphere, $f = 6\pi\eta r$, we obtain, upon rearranging,

$$\eta = \frac{m \cdot (1 - \bar{v}\rho)g}{6\pi r u_t} \tag{6.50}$$

Thus η for a fluid can be determined by measuring u_t for a sphere falling through it. Note also that from measurements using the same ball, the relative viscosities of two liquids η_2 and η_1 can be calculated from

$$\frac{\eta_2}{\eta_1} = \frac{1 - \bar{v}\rho_2}{1 - \bar{v}\rho_1} \frac{u_{t1}}{u_{t2}} \tag{6.51}$$

where the subscripts 1 and 2 refer to the quantities for liquids 1 and 2, respectively.

A more convenient method is to measure the volume rate of flow through a capillary. A simple viscometer, called an Ostwald viscometer, is shown in Fig. 6.14. Here the time t required for the liquid level to drop from mark 1 to mark 2 is measured. At a flow rate sufficiently slow to avoid turbulence, Eq. (6.49) can be applied and this leads after some mathematical manipulation to the result that

$$\eta = \frac{\pi r^4 \rho g t \, \Delta h}{8lV} \tag{6.52}$$

where r is the radius of the capillary, ρ is the density of the liquid, g is the gravitational acceleration, t is the measured time, l is the length of the capillary, and V is the volume of the fluid between the two marks. The quantity Δh is the (time) average difference in height between the liquid levels. If Δh_1 and Δh_2 are

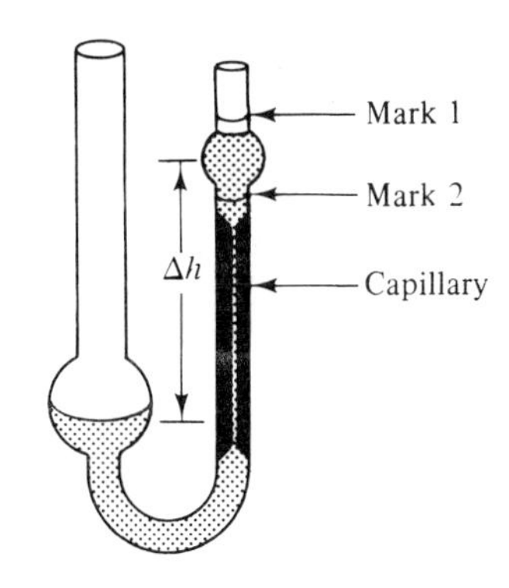

Fig. 6.14 Simple Ostwald viscometer. Liquid is drawn up initially into the arm on the right to above mark 1. Upon release, it begins to flow back under the influence of gravity but restricted by viscosity primarily in the capillary region. The elapsed time, t, is measured from when the meniscus passes mark 1 until it passes mark 2.

the heights when the liquid level in the bulb is at mark 1 and mark 2, respectively, it can be shown that

$$\Delta h = \frac{\Delta h_1 - \Delta h_2}{\ln(\Delta h_1/\Delta h_2)}$$

Again, if the same viscometer is used to measure the relative viscosities of two liquids,

$$\frac{\eta_2}{\eta_1} = \frac{\rho_2 t_2}{\rho_1 t_1} \tag{6.53}$$

because all other quantities are either constants or instrument parameters.

A third type of viscometer is shown in Fig. 6.15. It consists of two concentric cylinders. The outer cylinder contains the solution for which the viscosity is to be measured. The inner cylinder contains a piece of metal at its bottom. In one design, the weight of the inner cylinder is adjusted to make it barely float, as illustrated in Fig. 6.15(a). In a second design, the inner cylinder is submerged completely. Its vertical position is controlled either by making it a "Cartesian diver" [Fig. 6.15(b)] or by suspending it with a magnetic field [M. G. Hodgin and J. W. Beams, *Rev. Sci. Instr. 42,* 1455 (1971)]. A rotating magnetic field, which can be a rotating magnet as shown in Fig. 6.15(a), is used to make the inner cylinder rotate. The rotating magnetic field generates eddy currents inside the metal piece, which in turn induce a magnetic field. The interaction between the rotating magnetic field and the induced field causes the inner cylinder to rotate. How fast the cylinder rotates, for a given apparatus, is inversely proportional to the viscosity of the solution.

Solutions of macromolecules are usually non-Newtonian, because the velocity gradient tends to orient or distort the macromolecules and therefore changes their interactions with the flow. Viscosity measurements for such solutions should be made at several velocity gradients and extrapolated to zero velocity gradient. For solutions of large random coils such as DNA, the non-Newtonian behavior is severe, and measurements to very low du_x/dy values are necessary. The types of viscometer shown in Fig. 6.15 can be operated at very low velocity gradients and are especially useful.

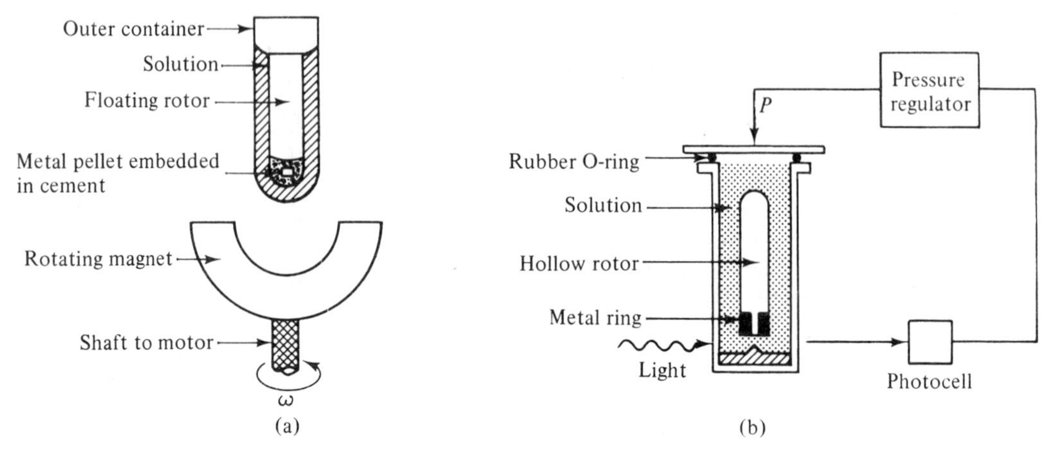

Fig. 6.15 (a) Concentric cylinder viscometer designed by B. H. Zimm and D. M. Crothers [*Proc. Natl. Acad. Sci. U.S. 48*, 905 (1962)]. The outer container is usually jacketed so that the temperature of the solution can be regulated. The motor speed can be changed so that measurements can be made at different velocity gradients to allow extrapolation to zero velocity gradient. (b) Cartesian diver viscometer [S. J. Gill and D. S. Thompson, *Proc. Natl. Acad. Sci. U.S. 57*, 562 (1967)]. The hollow rotor with air inside is completely submerged in the solution. A light beam traverses the solution at the bottom of the rotor. The amount of light reaching the photocell is dependent on the vertical position of the rotor. The signal from the photocell controls the hydrostatic pressure in the solution, which in turn determines the position of the rotor. The rotor is driven by a rotating magnet, as shown in Fig. 6.15(a).

Viscosity of Solutions of Rigid Macromolecules

If the viscosity of a solution of macromolecules is η' and that of the solvent is η, the quantity

$$\frac{\eta' - \eta}{\eta} \equiv \eta_{\text{sp}} \tag{6.54}$$

is termed the *specific viscosity*. For spherical particles, Einstein derived a relation for very low concentrations,

$$\eta_{\text{sp}} = 2.5\phi \tag{6.55}$$

where ϕ is the fractional volume occupied by the particles.

As we have already discussed, for a rigid macromolecule the volume occupied by a solvated molecule is

$$v_s = m \cdot (\bar{v}_2 + \delta_1 v_1^0) \tag{6.38}$$

The hydrodynamic volume per gram dry weight of the macromolecule is

$$\frac{v_s}{m} = \bar{v}_2 + \delta_1 v_1^0$$

If a solution contains c grams of macromolecules (dry weight) per cm^3, the fractional volume occupied by the solvated macromolecules is

$$\phi = c \cdot (\bar{v}_2 + \delta_1 v_1^0)$$

Substituting into Eq. (6.55) gives, for solvated macromolecules which are spherical in shape,

$$\eta_{sp} = 2.5c(\bar{v}_2 + \delta_1 v_1^0) \tag{6.56a}$$

or

$$\frac{\eta_{sp}}{c} = 2.5(\bar{v}_2 + \delta_1 v_1^0) \tag{6.56b}$$

Equation (6.55) is true, however, only for very dilute solutions. Therefore, η_{sp}/c is commonly determined for several concentrations, and the extrapolated value at $c = 0$ is used in Eq. (6.56b). This extrapolated value at infinite dilution is mathematically expressed as

$$[\eta] \equiv \lim_{c \to 0} \frac{\eta_{sp}}{c} \tag{6.57}$$

The term $[\eta]$ is called the *intrinsic viscosity* or the *limiting viscosity number.*

The dimension of $[\eta]$ is the reciprocal of c, the concentration. In this chapter, c will be given in g cm^{-3} and the unit of $[\eta]$ is $cm^3\,g^{-1}$. In some other sources c is expressed in g per 100 ml (g $deciliter^{-1}$) and the corresponding unit of $[\eta]$ is $10^2\,cm^3\,g^{-1}$.

Thus, for spherical solvated macromolecules,

$$[\eta] = 2.5(\bar{v}_2 + \delta_1 v_1^0) \tag{6.58}$$

This equation has been extended by Simha to particles whch are ellipsoids of revolution. The result is

$$[\eta] = \nu(\bar{v}_2 + \delta_1 v_1^0) \tag{6.59}$$

where ν is a shape factor. For spheres, $\nu = 2.5$. Values for prolate and oblate ellipsoids are plotted in Fig. 6.16. Note that the molecular weight does not enter Eq. (6.59). Therefore, for rigid macromolecules of the same shape, the intrinsic viscosity is insensitive to the molecular weight.

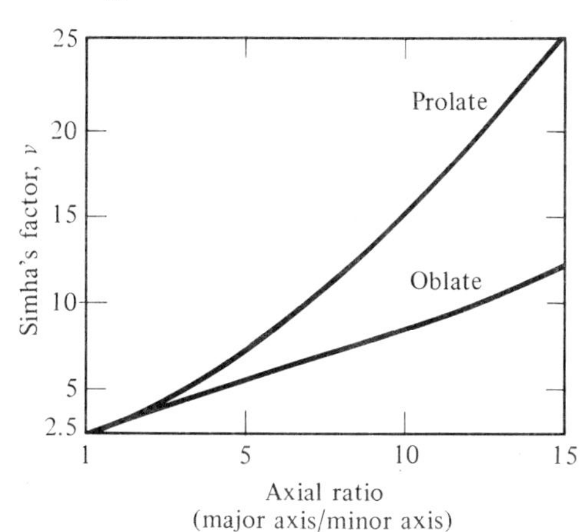

Fig. 6.16 Simha's factor for the viscosity increment of ellipsoids, for relatively low values of the axial ratio (major axis/minor axis).

Viscosity of Random Coils

For a random coil, we can again view the molecule as a highly solvated, nondraining sphere of a dimension proportional to the mean-square radius of the flexible coil

$$[\eta] = 2.5\frac{v_s}{m}$$

$$v_s = \tfrac{4}{3}\pi R_e^3$$

$$= \tfrac{4}{3}\pi \cdot (\xi R_g)^3$$

where R_e is the radius of the equivalent sphere and is proportional to the root mean square radius R_g. The parameter ξ (Greek letter xi) is the proportionality constant. For a large random coil, theoretical analyses yield $\xi \sim 0.88$.

Combining the preceding two equations gives

$$[\eta] = 10\pi\xi^3 \frac{R_g^3}{3m}$$

$$= 10\pi N_0 \xi^3 \frac{R_g^3}{3M} \tag{6.60}$$

According to Eq. (11.49), the mean-square radius is proportional to $M^{1/2}$. Thus $[\eta]$ for random coils is expected to be proportional to $M^{3/2}/M$ or $M^{1/2}$, from Eq. (6.60). This is in contrast with rigid molecules, in which there is no molecular-weight dependence.

COMBINATION OF HYDRODYNAMIC MEASUREMENTS

We have already seen that data on s and D can be combined to give the molecular weight of a macromolecule (Eq. 6.48). Similar combinations can be formulated for D and $[\eta]$ and for s and $[\eta]$.

From Eq. (6.59)

$$\bar{v}_2 + \delta_1 v_1^0 = \frac{[\eta]}{\nu}$$

Substituting $[\eta]/\nu$ for $(\bar{v}_2 + \delta_1 v_1^0)$ into the appropriate equations for D and s, and after some straightforward but tedious algebra, we obtain

$$M = \frac{4\pi N_0}{3}\left(\frac{kT}{6\pi\eta}\right)^3 \frac{\nu}{(f/f_0)^3[\eta]D^3} \tag{6.61a}$$

and

$$M = \frac{9\sqrt{2}\pi N_0 \eta^{3/2}}{[(f/f_0)^{-3}\nu]^{1/2}} \frac{s^{3/2}[\eta]^{1/2}}{(1 - \bar{v}_2\rho)^{3/2}} \tag{6.62a}$$

If measurements are done at 20°C in a dilute aqueous buffer, $T = 293$ K and $\eta = 1.005 \times 10^{-2}$ P. Substituting these values into the equations above gives

$$M = 2.45 \times 10^{-14} \frac{[(f_0/f)\nu^{1/3}]^3}{[\eta]D_{20,w}^3} \tag{6.61b}$$

and

$$M = 2.43 \times 10^{22} \frac{s_{20,w}^{3/2}[\eta]^{1/2}}{[(f_0/f)\nu^{1/3}]^{3/2}(1 - \bar{v}_2\rho)^{3/2}} \tag{6.62b}$$

The units of $[\eta]$, $D_{20,w}$ and $s_{20,w}$ for these two equations are $cm^3\ g^{-1}$, $cm^2\ s^{-1}$, and s, respectively. The factor $(f_0/f)\nu^{1/3}$ appears in both equations above. Its values for ellipsoids of revolution can be calculated and are tabulated in Table 6.3.

Table 6.3 Factor $(f_0/f)\nu^{1/3}$ as a function of the axial ratio for prolates and oblates

	$(f_0/f)\nu^{1/3}$	
Axial ratio	Prolate	Oblate
1	1.36	1.36
2	1.37	1.36
3	1.39	1.37
4	1.41	1.37
5	1.43	1.37
6	1.46	1.38
8	1.51	1.38
10	1.55	1.38
12	1.59	1.38
15	1.63	1.38
20	1.70	1.38
25	1.75	1.38
30	1.79	1.38
40	1.86	1.38
50	1.91	1.38
60	1.95	1.38
80	2.02	1.38
100	2.07	1.38
200	2.24	1.38
300	2.31	1.38

SOURCE: Calculated from the tabulation of H. A. Scheraga and L. Mandelkern, *J. Amer. Chem. Soc. 75*, 179 (1953).

If M is known, $(f_0/f)\nu^{1/3}$ can be calculated from $[\eta]$ and $D_{20,w}$ using Eq. (6.61b), or from $s_{20,w}$ and $[\eta]$ using Eq. (6.62b). In principle, the axial ratio can be calculated using Table 6.3 without any knowledge of solvation. In practice, this is seldom done, because $(f_0/f)\nu^{1/3}$ is not very sensitive to the axial ratio, and a large uncertainty results when Table 6.3 is used to calculate the axial ratio. Instead, we note that for an oblate ellipsoid with any axial ratio or for a prolate ellipsoid with an axial ratio less than 5, $(f_0/f)\nu^{1/3}$ is approximately equal to 1.38. Thus, with the exception of very asymmetric prolates, we can calculate M from these equations by taking $(f_0/f)\nu^{1/3} = 1.38$ as an approximation.

Example 6.10 Estimate the intrinsic viscosity, $[\eta]$, of the ribosomes described in Example 6.9. The axial ratio of ribosomes is sufficiently close to unity that we may take $(f_0/f)\nu^{1/3} = 1.38$.

Solution From Eq. (6.61b) we obtain

$$[\eta] = \frac{(2.45 \times 10^{-14})(1.38)^3}{MD^3_{20,w}}$$

Using the data in Example 6.9, we have

$$[\eta] = \frac{(2.45 \times 10^{-14})(1.38)^3}{(3.39 \times 10^6)(1.52 \times 10^{-7})^3}$$
$$= 5.4 \text{ cm}^3 \text{ g}^{-1}$$

Show that the same answer is obtained using Eq. (6.62b).

ELECTROPHORESIS

Macromolecules involved in biological processes are usually charged. Therefore, they migrate in an electric field. For a rigid particle in a nonconducting solvent, the velocity u of migration is proportional to the charge of the particle and to the magnitude of the electrical field, and is inversely proportional to the frictional coefficient of the particle:

$$u = \frac{ZeE}{f}$$

where Z is the number of charges per particle, $e = 4.8 \times 10^{-10}$ is the charge per electron in electrostatic units (esu), E is the field in electrostatic units per centimeter, and f is the frictional coefficient defined as before.

The velocity per unit field, u/E, is called the *electrophoretic mobility* μ:

$$\mu \equiv \frac{u}{E} \tag{6.63}$$

Substitution into the equation above gives

$$\mu = \frac{Ze}{f} \tag{6.64}$$

Thus, if we know the number of charges that a macromolecule carries, a measurement of its electrophoretic mobility should in principle allow us to calculate the frictional coefficient and to gain information on its size and shape. Conversely, if we know its frictional coefficient, we can calculate the number of charges.

We are rarely, however, able to obtain biological macromolecules in nonconducting solvent. Small ions are always present as counterions, buffers, and so on. Therefore, the charged macromolecule is surrounded by an atmosphere of small ions. This ion atmosphere greatly complicates the interpretation of electrophoretic mobility. The shielding effect of the ion atmosphere reduces the electric field experienced by the macromolecule. Furthermore, when the macromolecule moves, it drags its ion atmosphere with it. Therefore, the frictional coefficient is changed. The complications make it very difficult to interpret electrophoretic mobility quantitatively as we have done for the other hydrodynamic measurements. Electrophoresis is a powerful analytical tool, however, and macromolecules with very small differences in their properties can be resolved. One example is the separation of normal hemoglobin and hemoglobin from patients who suffer from sickle-cell anemia. Sickle-cell hemoglobin S differs from normal hemoglobin A by a single interchange of valine for glutamic acid in each of the two β-chain peptides. As a consequence, the proteins differ by two charges per molecule, which is sufficient to effect a clean separation by electrophoresis.

In the sections below, we shall treat electrophoresis in a qualitative and empirical way.

Isoelectric Point

The number of charges on a macromolecule can be changed by the binding of small ions. For example, the number of protonated groups of a protein molecule is dependent on the pH of the solution. If at a certain pH, the net charge on a macromolecule is zero, its electrophoretic mobility will be zero, and the macromolecule is said to be at its *isoelectric point.* Different proteins usually have different isoelectric points. A separation technique based on such differences is called *isoelectric focusing.* A pH gradient is set up between two electrodes. When a mixture of proteins is introduced, they migrate to their respective isoelectric points and come to rest there. The separated proteins can then be recovered easily from the gradient.

Gel Electrophoresis

In our discussion on zone sedimentation, we pointed out the necessity of stabilizing the moving zone. Similarly, if a layer of macromolecules is electrophoresed through a solution, a density gradient is necessary to stabilize the zone against convection.

An alternative is to perform the electrophoresis through a gel. A gel is a three-dimensional polymer network dispersed in a solvent. A variety of gels have been used. For example, an agarose gel consists of an aqueous medium and a polysaccharide obtained from agar. An acrylamide gel consists of an aqueous medium and a copolymer of acrylamide ($CH_2{=}CH{-}CO{-}NH_2$) and *N,N'*-methylenebisacrylamide ($CH_2{=}CH{-}CO{-}NH{-}CH_2{-}NH{-}CO{-}CH{=}CH_2$). In the latter gel, the water-soluble acrylamide, with its reactive double bond, can polymerize into a linear chain

$$-\!(CH-\underset{\underset{NH_2}{|}}{\underset{CO}{|}}{CH}-CH-\underset{\underset{NH_2}{|}}{\underset{CO}{|}}{CH})\!-$$

Adding the bisacrylamide, which has two double bonds, results in the formation of cross-links between different chains. The degree of cross-linking can be controlled by the ratio of the bis compound to acrylamide. In a typical gel, over 90% of the space is occupied by the aqueous medium, but the presence of the three-dimensional polymer network prevents convectional flow.

The gel contributes additional factors which affect the electrophoretic mobility: (1) the actual path length traveled by a macromolecule through the porous gel is likely to be considerably longer than the gel tube; (2) the gel imposes additional frictional resistance to the macromolecules as they move through pores of comparable dimensions; (3) the macromolecules may interact with the gel network (for example, through electrostatic interactions, if the gel network contains charged groups); and (4) the parts of the gel with pores smaller than a particular macromolecular species are inaccessible to that species. The combination of these factors further improves the resolution of electrophoresis.

Gel Electrophoresis of Proteins

In a given buffer, proteins differ both in the numbers of charges and in their frictional coefficients, so their mobilities are not easily interpretable. A popular procedure to avoid this problem involves treatment with an anionic detergent, sodium dodecylsulfate (SDS), and 2-mercaptoethanol; the latter disrupts sulfur-sulfur linkages in proteins. The combined action of these reagents causes the proteins to unfold. Furthermore, for most proteins at an SDS concentration greater than 10^{-3} M, a nearly constant amount of SDS is bound per unit weight of protein (approximately 0.5 detergent molecule is bound per amino acid residue). Thus, the charge of the protein-dodecylsulfate complex is due primarily to the charges of the bound dodecylsulfate groups, making the charge per unit weight approximately the same for most proteins.

Under these conditions the mobilities of the SDS-treated proteins are determined by their molecular weights. If M is the molecular weight of a protein and x is the distance migrated in the gel, the relation

$$\log M = a - bx \tag{6.65}$$

is usually observed, where a and b are constants for a given gel at a given electric field. Usually, a set of proteins of known molecular weight is used for calibration in the determination of a protein of unknown molecular weight. An example is shown in Fig. 6.17.

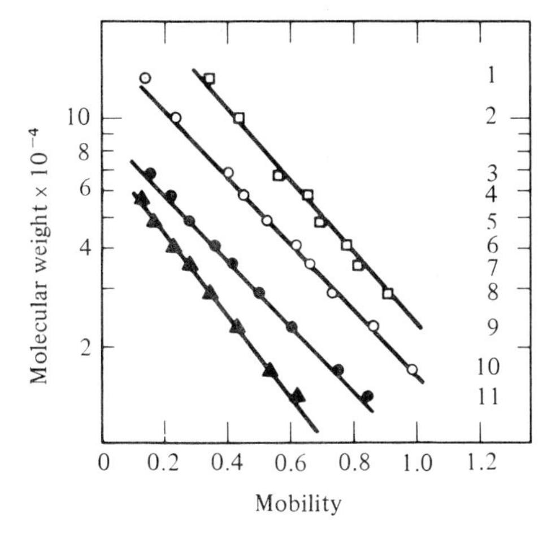

Fig. 6.17 Electrophoretic mobilities of proteins as a function of the acrylamide concentration in the gel. The acrylamide concentrations illustrated are 15% (▲), 10% (●), 7.5% (○), and 5% (□). The weight ratio of acrylamide to methylenebisacrylamide is 37:1. The numbers 1 to 11 refer to the following proteins: β-galactosidase, phosphorylase a, serum albumin, catalase, fumarase, aldolase, glyceraldehyde-phosphate dehydrogenase, carbonic anhydrase, trypsin, myoglobin, and lysozyme. [From K. Weber, J. R. Pringle, and M. Osborn, *Methods Enzymol. 26*, 3 (1972).]

We emphasize that Eq. (6.65) is dependent on two factors: a constant amount of bound detergent per unit weight of protein, and charges due to bound detergent dominate those carried by the protein itself. Deviations from this type of relation occur when a protein binds an abnormal amount of dodecylsulfate (such as glycoproteins)* or carries a large number of charges (such as the histones).*

Gel Electrophoresis of Double-stranded DNA

At neutral pH's, large DNA's behave like random coils with charge densities that are uniform along the length of the molecules. Electrophoresis measurements in solution for molecules in the molecular-weight range 2.5×10^5 to 1.3×10^8 showed that the mobility is independent of the molecular weight. Rather different results are obtained for gel electrophoresis. Although above a molecular weight of $\sim 10^7$ the electrophoretic mobility is insensitive to M, very good resolution can

* A glycoprotein is a protein that has oligosaccharide groups attached. Histones are highly positively-charged proteins rich in lysine and arginine residues.

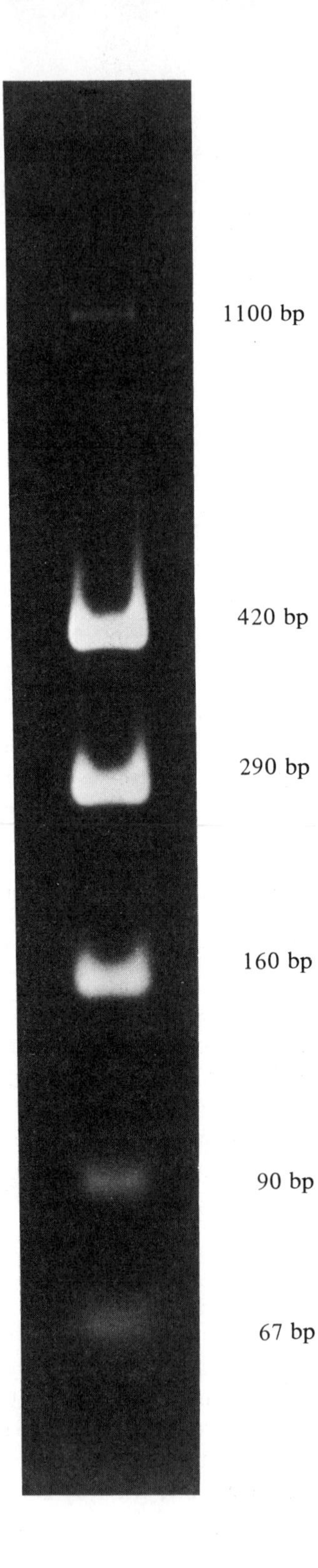

Fig. 6.18 Electrophoresis of double-stranded DNA fragments in 3.5% acrylamide gel (acrylamide/bisacrylamide ≈ 30 wt/wt). The mixture originally started as a thin band at the top of the picture and was electrophoresed toward the positive electrode. After electrophoresis, the gel was immersed in a staining solution containing a fluorescent dye, ethidium, and then photographed under uv light. The sizes of the fragments are given in terms of base pairs (bp). One base pair corresponds to a molecular weight of ~660. For a calibration of mobility versus fragment size, see the reference given in the legend to Fig. 6.19. (Unpublished data of T-S. Hsieh and J. C. Wang.)

be obtained below this value. This is illustrated in Fig. 6.18. For the smaller fragments, gel electrophoresis provides a versatile and accurate method of molecular-weight measurement.

DNA's of different topological forms also have different electrophoretic mobilities. Rings and twisted rings (superhelical DNA's) have different electrophoretic mobilities from those of linear molecules of the same molecular weight.

Gel Electrophoresis of Single-stranded Nucleic Acids

Single-stranded nucleic acids usually contain regions of complementary base sequences. As a consequence, intramolecular base pairing can occur. Since such pairing affects the frictional coefficients, the electrophoretic mobility of single-stranded nucleic acids is not a simple function of molecular weight.

If electrophoresis is carried out in a gel containing a high concentration of a neutral reagent (such as urea) which disrupts base pairing, the mobility of a nucleic acid is determined primarily by its size. For example, in a 12% acrylamide gel (with an acrylamide/bis acrylamide weight ratio of ~30) containing 7 *M* urea, the resolution is such that polynucleotide chains from approximately 100 residues down can all be resolved, as illustrated in Fig. 6.19. Chains of higher molecular weights can be resolved in lower-percentage acrylamide or agarose gels.

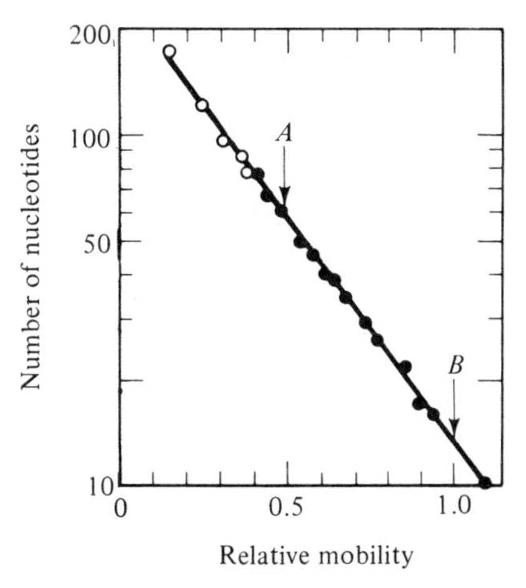

Fig. 6.19 Relative mobility of single-stranded nucleic acids in 12% acrylamide gel containing 7 *M* urea. Note that the number of nucleotides is plotted on a logarithmic scale. The relative mobility of a species is obtained from the ratio of the distances migrated by this species and by a dye marker. Two dye markers, bromophenol blue and xylene cyanol FF, were used in obtaining these data. The relative mobilities of these two markers are indicated by the arrows *B* and *A*, respectively. [Reprinted with permission from T. Maniatis, A. Jeffrey, and H. van deSande, *Biochemistry 14*, 3787 (1975). Copyright by the American Chemical Society.]

SIZE AND SHAPE OF MACROMOLECULES

We have seen that measurements of the transport properties of a macromolecule allow the evaluation of its molecular weight M. Relevant properties for some representative macromolecules and particles are given in Table 6.4. From the diffusion coefficient or the sedimentation coefficient, the frictional coefficient f can be calculated. If the macromolecule is reasonably rigid, we can calculate for comparison the frictional coefficient f_{min} for an unsolvated spherical molecule with molecular weight M and partial specific volume $\bar{v}_2$ that are the same as those of the macromolecule of interest. If f/f_{min} is close to 1, we conclude that the macromolecule is approximately spherical and only lightly solvated. If f/f_{min} is much greater than 1, the molecule is either highly solvated or asymmetric or both. Rough estimates of the extent of solvation can usually be made. For most proteins in dilute aqueous buffers, the amount of hydrodynamically associated water is in the range 0.2–0.8 g per gram of protein. Measurements, such as that of low-angle x-ray scattering, also provide determinations of the extent of solvation. Once a reasonable estimate or measurement is available for the solvation, the value of f/f_{min} permits determination of the asymmetry of the macromolecule. We

Table 6.4 Transport and related properties of some proteins

	$s_{20,w} \times 10^{13}$ (s)	$D_{20,w} \times 10^{7}$ ($\text{cm}^2\ \text{s}^{-1}$)	$\bar{v}_{20,w}$ ($\text{cm}^3\ \text{g}^{-1}$)	M	f/f_{min}
Ferricytochrome c (bovine heart)	1.91	13.20	0.707	11,980	1.077
Ribonuclease (bovine pancreas)	2.00	13.10	0.707	12,640	1.066
Insulin	1.95	7.30	0.735	24,430	1.516
Myoglobin (horse heart)	2.04	11.30	0.741	16,890	1.105
Lysozyme (chicken egg white)	1.91	11.20	0.703	13,930	1.210
Chymotrypsinogen (bovine pancreas)	2.54	9.50	0.721	22,240	1.193
γ-Globulin (human)	7.12	4.00	0.718	153,100	1.513
Hemocyanin (squid)	19.50	2.80	0.724	611,800	1.358
Turnip yellow mosaic virus protein	48.80	1.51	0.740	3,013,000	1.470

SOURCE: Selected from extensive tables in the *Handbook of Biochemistry and Molecular Biology*, 2nd ed. (Chemical Rubber Co., Cleveland, Ohio, 1970).

can calculate the dimensions of a prolate or oblate ellipsoid that fit the values the best. Additional information is provided by the intrinsic viscosity. The shape factor ν contains information which complements that obtained from the ratio $f/f_{\min}$.

While the transport properties provide important information on the size and shape of a macromolecule, more detailed knowledge on the structural features and functional aspects of the macromolecule can only be obtained by other physical-chemical methods, some of which we will discuss in later chapters. The understanding of how molecules function in biological processes is usually achieved by a combination of many methods.

Lastly, since the transport properties are sensitive to the sizes and shapes of molecules, they can be used to study interactions between molecules, and conformational changes of molecules. Some examples are interactions between subunits of a macromolecular structure (such as a multi-subunit protein or a ribosome), conformational change of a protein due to the binding of a substrate, circularization of a linear DNA (Chapter 11), unfolding of a protein chain, and antibody-antigen interactions.

SUMMARY

Kinetic Model of Gases

Mean-square velocity:

$$\langle u^2 \rangle = \frac{3RT}{M} \tag{6.16}$$

$\langle u^2 \rangle \equiv$ average of the square of the velocities of molecules in a gas of molecular weight M at T K

R = gas constant = 8.314×10^7 erg deg^{-1} mol^{-1}

1 erg = 1 g cm^2 s^{-2}

Collision frequency:

If every molecule moves at a speed equal to $\langle u^2 \rangle^{1/2}$,

$$z = 7.26 \frac{N}{V} \sigma^2 \cdot \left(\frac{RT}{M}\right)^{1/2} \tag{6.22}$$

$z \equiv$ number of collisions a molecule encounters per second

$\dfrac{N}{V} \equiv$ number of molecules per cm^3

$\sigma \equiv$ diameter of a molecule (each molecule is considered a hard sphere)

When the distribution in molecular speed is taken into consideration,

$$z = 7.09 \frac{N}{V}\sigma^2 \cdot \left(\frac{RT}{M}\right)^{1/2} \tag{6.24}$$

Total number of collisions Z per $cm^3\, s^{-1}$:

$$Z = \frac{N}{V}\frac{z}{2} \tag{6.25}$$

Mean free path:

$$l = \frac{1}{\sqrt{2}\,\pi(N/V)\sigma^2} \tag{6.26}$$

$l \equiv$ average distance a molecule travels between two successive collisions

Root-mean-square displacement per second:

$$\langle d^2\rangle^{1/2} = \sqrt{z}\, l \tag{6.28a}$$

$$= \frac{\langle u\rangle}{\sqrt{z}} \tag{6.28b}$$

$\langle d^2\rangle \equiv$ average of the square of displacement of a molecule per second
$\langle u\rangle \equiv$ average speed of the molecule

Diffusion

Fick's first law:

$$J_x = -D \cdot \left(\frac{\partial c}{\partial x}\right)_t \tag{6.29}$$

$J_x \equiv$ net amount of material that diffuses in the x direction per second, across an area 1 cm^2 perpendicular to x

$D \equiv$ diffusion coefficient, which is directly related to the mean-square displacement per unit time; the units are $cm^2\, s^{-1}$

$\left(\frac{\partial c}{\partial x}\right)_t$ = concentration gradient; it is the change of concentration with respect to x at a specified time t

Fick's second law:

$$\left(\frac{\partial c}{\partial t}\right)_x = D \cdot \left(\frac{\partial^2 c}{\partial x^2}\right)_t \tag{6.31}$$

$\left(\frac{\partial c}{\partial t}\right)_x$ = change of concentration versus time at a given position x

$\left(\frac{\partial^2 c}{\partial x^2}\right)_t$ = second derivative of the concentration c with respect to the position x, at a given time t

Diffusion coefficient D and frictional coefficient f:

$$D = \frac{kT}{f} \tag{6.36}$$

$k \equiv$ Boltzmann constant $\equiv R/N_0$
$= 1.380 \times 10^{-16}\ \text{g cm}^2\ \text{s}^{-2}\ \text{deg}^{-1}\ \text{molecule}^{-1}$

Sedimentation

$$s = \frac{m \cdot (1 - \bar{v}_2\rho)}{f} \tag{6.42}$$

$s \equiv \frac{u_t}{\omega^2 x}$ = velocity of sedimentation per unit centrifugal acceleration

The dimension of s is seconds. A convenient unit for s, the Svedberg (S), is equal to 10^{-13} s (1 S = 10^{-13} s).

m = mass of the unsolvated macromolecule
$\bar{v}_2$ = partial specific volume of the macromolecule
ρ = density of the solution
f = frictional coefficient

$$s_{20,w} = s \frac{\eta}{\eta_{20,w}} \frac{(1 - \bar{v}_2\rho)_{20,w}}{1 - \bar{v}_2\rho} \tag{6.44}$$

$s_{20,w} \equiv$ sedimentation coefficient corrected to give the expected value at 20°C in water

The subscript $_{20,w}$ refers to quantities in water at 20°C.

Relation Between *f* and Molecular Parameters

f can be obtained from either s or D. For a sphere of radius r,

$$f = 6\pi\eta r \tag{6.35}$$

$\eta \equiv$ viscosity of medium

For an unsolvated, spherical molecule of mass m and partial specific volume $\bar{v}_2$, its radius r is

$$r = \left(\text{volume} \cdot \frac{3}{4\pi}\right)^{1/3}$$

$$= \left(\frac{3m\bar{v}_2}{4\pi}\right)^{1/3}$$

The frictional coefficient $f_{\min}$ for this unsolvated sphere is

$$f_{\min} = 6\pi\eta \cdot \left(\frac{3m\bar{v}_2}{4\pi}\right)^{1/3}$$

The deviation of $f/f_{\min}$ from unity for a macromolecule can be due to either solvation or asymmetric shape:

$$\frac{f}{f_{\min}} = \frac{f}{f_0}\frac{f_0}{f_{\min}} \tag{6.39}$$

$f_0 \equiv$ frictional coefficient of a sphere of the same $\bar{v}_2$ and degree of solvation as the macromolecule

$$\frac{f_0}{f_{\min}} = \left(\frac{(\bar{v}_2 + \delta_1 v_1^0)}{\bar{v}_2}\right)^{1/3} \tag{6.40}$$

$\delta_1 \equiv$ grams of solvent hydrodynamically associated with each gram of macromolecule
$v_1^0 \equiv$ specific volume of the solvent; for dilute aqueous buffer, $v_1^0 = 1$

f/f_0 is determined by the shape of the molecule; it is 1 for a sphere, and values are tabulated for ellipsoids of revolution; for very flexible coils (random coils), f is expected to be proportional to $M^{1/2}$.

Combination of Diffusion and Sedimentation

$$M = \frac{RTs}{D \cdot (1 - \bar{v}_2\rho)} \tag{6.48}$$

Molecular weight M can be calculated from s, D, and the partial specific volume $\bar{v}_2$. ρ is the density of the solution.

Equilibrium centrifugation:

$$M = \frac{4.606RT}{\omega^2 \cdot (1 - \bar{v}_2\rho)} \frac{d \log c}{d(x^2)} \tag{6.46}$$

$\omega \equiv 2\pi\nu$ = angular speed, rad s^{-1} (ν = revolutions s^{-1})
$c \equiv$ concentration
$x \equiv$ distance from the center of rotation

When centrifuged to equilibrium, the molecular weight of a macromolecule can be calculated from the slope of a plot of $\log c$ versus x^2.

Viscosity

$$J_{mu} = -\eta \frac{du_x}{dy} \tag{6.49}$$

$J_{mu} \equiv$ rate of transfer of momentum in the x direction per second across a cross-sectional area 1 cm^2 in the xz plane

$\frac{du_x}{dy} \equiv$ velocity gradient, or the rate of change of the x component of the velocity with respect to y

$\eta \equiv$ viscosity coefficient

Newtonian fluid: η is independent of du_x/dy.
Non-Newtonian fluid: η is dependent on du_x/dy.
Solutions of macromolecules are often non-Newtonian and viscosity measurements are usually done at several values of du_x/dy and extrapolated to $du_x/dy = 0$.

Specific viscosity:

$$\eta_{sp} \equiv \frac{\eta' - \eta}{\eta} \tag{6.54}$$

$\eta' \equiv$ viscosity coefficient of a solution of a macromolecular species
$\eta \equiv$ viscosity coefficient of solvent

Intrinsic viscosity is the limiting vaue of η_{sp}/c as c approaches 0, where c is the concentration of the macromolecules in g cm^{-3}.

$$[\eta] \equiv \lim_{c \to 0} \frac{\eta_{sp}}{c} \tag{6.57}$$

Relation between $[\eta]$ *and molecular parameters:*

Rigid molecules:

$$[\eta] = \nu \cdot (\bar{v}_2 + \delta_1 v_1^0) \tag{6.59}$$

$\nu \equiv$ shape factor; $\nu = 2.5$ for a sphere; values of ν for ellipsoids of revolution are given in Fig. 6.16

$\bar{v}_2 \equiv$ partial specific volume of the macromolecule
$\delta_1 \equiv$ grams of solvent molecules hydrodynamically associated with each gram of macromolecule
$v_1^0 \equiv$ specific volume of the solvent; for a dilute aqueous solution, $v_1^0 = 1$

Very flexible coils (random coils):

$$[\eta] \text{ approximately} \propto M^{1/2}$$

Combination of $[\eta]$ and D or $[\eta]$ and s Measurements

$$M = 2.45 \times 10^{-14} \frac{[(f_0/f)\nu^{1/3}]^3}{[\eta]D_{20,w}^3} \quad (6.61b)$$

with $[\eta]$ in $cm^3\ g^{-1}$

$$M = 2.43 \times 10^{22} \frac{s_{20,w}^{3/2}[\eta]^{1/2}}{[(f_0/f)\nu^{1/3}]^{3/2}(1 - \bar{v}_2\rho)^{3/2}} \quad (6.62b)$$

with s in seconds, $[\eta]$ in $cm^3\ g^{-1}$

$[(f_0/f)\nu^{1/3}] \approx 1.38$ except for very asymmetric prolate ellipsoids.

Electrophoresis

Electrophoretic mobility:

$$\mu = \frac{u}{E} \quad (6.63)$$

u = velocity of charged particle, $cm\ s^{-1}$
E = electric field strength, $V\ cm^{-1}$

$$\mu = \frac{Ze}{f} \quad (6.64)$$

Z = number of charges on particle
f = frictional coefficient
e = electronic charge

Electrophoretic mobility $\equiv \mu/E$, or velocity per unit electric field.

REFERENCES

The following book is an introductory text similar in level to this chapter.

VAN HOLDE, K. E., 1971. *Physical Biochemistry*, Prentice-Hall, Englewood Cliffs, N.J.

The material in this chapter is covered at a more advanced level in:

CANN, J. R., 1970. *Interacting Macromolecules*, Academic Press, New York.
KAUZMAN, W., 1966. *Kinetic Theory of Gases*, W. A. Benjamin, Menlo Park, Calif.
PRESENT, R. D., 1958. *Kinetic Theory of Gases*, McGraw-Hill, New York.
SCHACHMAN, H. K., 1959. *Ultracentrifugation in Biochemistry*, Academic Press, New York.
TANFORD, C., 1961. *Physical Chemistry of Macromolecules*, John Wiley, New York.

PROBLEMS

1. The collisional diameter σ of a H_2 molecule is about 2.5 Å or 2.5×10^{-8} cm. For H_2 gas at 0°C and 1 atm, calculate
(a) The root-mean-square velocity.
(b) The translational kinetic energy of 1 mol of H_2 molecules.
(c) The number of H_2 molecules in 1 cm^3 of the gas.
(d) The mean free path.
(e) The number of collisions each H_2 molecule encounters in 1 s.
(f) The total number of intermolecular collisions in 1 s in 1 cm^3 of the gas.
(g) The average displacement of a H_2 molecule from its position 1 s ago.

2. A gaseous mixture at a temperature of 50°C contains H_2 at a partial pressure of 100 torr and I_2 at a partial pressure of 1 torr. The collision diameters σ_{H_2} and σ_{I_2} of H_2 and I_2 are approximately 2.5 Å and 4.6 Å, respectively. The molecular weight of I_2 is 252.8.
(a) Calculate the ratio of the root-mean-square velocity of H_2 to that of I_2.
(b) The calculation made in part (a) should indicate that H_2 moves much faster than I_2 in the mixture. Therefore, when we consider collisions between H_2 and I_2, we can assume that the I_2 molecules are essentially not moving. Derive an expression for the number of times each H_2 molecule collides with I_2 molecules in 1 s. Denote the number of I_2 molecules per cm^3 as n_{I_2}.
(c) For the mixture given, calculate the total number of collisions between H_2 and I_2 in 1 $cm^3\ s^{-1}$.

3. The following data were reported for human immunoglobulin G (IgG) at 20°C in a dilute aqueous buffer,

$$M = 156{,}000$$

$$D_{20,w} = 4.0 \times 10^{-7}\ cm^2\ s^{-1}$$

$$\bar{v}_2 = 0.739\ cm^3\ g^{-1}$$

(a) Calculate f, the frictional coefficient.
(b) Calculate f_{min}, the frictional coefficient of an unhydrated sphere of the same M and $\bar{v}_2$ as IgG.
(c) From the ratio f/f_{min}, estimate the extent of hydration of IgG if it is a spherical molecule.
(d) If IgG is not significantly hydrated, find the dimensions of a prolate ellipsoid that best fit the data.

4. For the dimer of protein L7 from the 50 S subunit of *E. coli* ribosomes, the hydration parameter δ_1 is estimated to be 0.26 g of H_2O per gram of protein from small-angle x-ray scattering measurements. The molecular weight, diffusion coefficient, and partial specific volume for this protein have been given in Example 6.6. Calculate the dimensions of a prolate ellipsoid that best fit the data.

5. The sedimentation coefficient of a certain DNA in 1 *M* NaCl at 20°C was measured by boundary sedimentation at 24,630 rpm. The following data were recorded:

Time, t (min)	Distance of boundary from center of rotation, x (cm)
16	6.2687
32	6.3507
48	6.4380
64	6.5174
80	6.6047
96	6.6814

(a) Plot $\log x$ versus t. Calculate the sedimentation coefficient s.

(b) The partial specific volume of the sodium salt of DNA is 0.556 $cm^3\,g^{-1}$. The viscosity and density of 1 *M* NaCl and the viscosity of water can be found in the International Critical Tables; the values are 1.104 cP, 1.04 $g\,cm^{-3}$, and 1.005 cP, respectively. Calculate $s_{20,w}$ of the DNA.

6. For a bacteriophage T7, the following data at zero concentration have been obtained [Dubin et al., *J. Mol. Bio. 54,* 547 (1970)]:

$$s^0_{20,w} = 453\ \text{S}$$

$$D^0_{20,w} = 6.03 \times 10^{-8}\ \text{cm}^2\,\text{s}^{-1}$$

$$\bar{v}_2 = 0.639\ \text{cm}^3\,\text{g}^{-1}$$

(a) Calculate the molecular weight of the phage.

(b) It can be calculated from phosphorus and nitrogen analyses of the phage that 51.2% (by weight) of the phage is DNA. Calculate the molecular weight of T7 DNA. Each phage contains one DNA molecule.

7. Two *spherical* viruses of molecular weights M_1 and M_2, respectively, happen to have the same partial specific volume $\bar{v}$. Neglecting hydration, what are the expected values for the ratios s_2/s_1, D_2/D_1, and $[\eta]_2/[\eta]_1$, where s_1, D_1, and $[\eta]_1$ are the sedimentation coefficient, diffusion coefficient, and the intrinsic viscosity, respectively, of the virus of molecular weight M_1, and s_2, D_2, and $[\eta]_2$ are the corresponding parameters of the virus of molecular weight M_2.

8. For a rodlike particle with length L and diameter d, its hydrodynamic properties are similar to those of a prolate ellipsoid of the same length and volume. Show that

$$\frac{L}{d} = \left(\frac{3}{2}\right)^{1/2} \frac{a}{b}$$

where a and b are the long and short semiaxes of the ellipsoid, respectively.

9. For prolates of large axial ratios, say with $a/b > 20$, it can be shown that the shape factor $\nu \approx 0.207(a/b)^{1.732}$. A certain rodlike macromolecule is believed to form an end-to-end dimer. Show that the intrinsic viscosity of the dimer is expected to be 3.32 times that of the monomer.

10. The DNA from an animal virus, polyoma, has a sedimentation coefficient $s_{20,w}$ of 20 S. Digesting the DNA very briefly with pancreatic DNase I, an enzyme that introduces single-chain breaks into a double-stranded DNA, converts it to a species sedimenting at 16 S. This reduction in s could be due to either a reduction in molecular weight or a conformational change of the DNA (so that its frictional coefficient is increased). How would you design an experiment to decide between these possibilities?

11. The following data have been obtained for human serum albumin:

$$s_{20,w} = 4.6 \text{ S}$$

$$D_{20,w} = 6.1 \times 10^{-7} \text{ cm}^2 \text{ s}^{-1}$$

$$[\eta] = 4.2 \text{ cm}^3 \text{ g}^{-1}$$

$$\bar{v}_2 = 0.733 \text{ cm}^3 \text{ g}^{-1}$$

Calculate the molecular weight of this protein by using (a) Eq. (6.48), (b) Eq. (6.61b), and (c) Eq. (6.62b).

12. The DNA from a simian virus SV 40 (simian viruses infect monkeys) is a twisted, double-stranded ring. A restriction enzyme Eco RI cleaves this ring structure at a unique location, and converts the DNA to a linear form. The ratio of the sedimentation coefficients of the two forms is measured to be 1.45. Estimate the ratio of the intrinsic viscosities of the two forms.

13. In 6 M guanidine hydrochloride and in the presence of 2-mercaptoethanol, it is generally believed that complete unfolding of proteins occurs. Could you test whether this is true by viscosity measurements? Give a brief and concise discussion. Some experimental data are listed below [taken from C. Tanford, *Adv. Protein Chem.* 23, 121 (1968)]:

		$[\eta]$ (cm^3 g^{-1})	
Protein	M* Molecular weight	In dilute aqueous buffer	In 2 M guanidine · HCl and in the presence of 2-mercaptoethanol
Ribonuclease	13,700	3.3	16.6
Myoglobin	17,800	3.1	20.9
Chymotrypsinogen	25,700	2.5	26.8
Serum albumin	69,000	3.7	52.2

* These proteins have only one polypeptide chain per molecule; therefore, the molecular weight does not change upon unfolding of the protein.

14. The O_2-carrying protein, hemoglobin, contains a total of four polypeptide chains: two α chains and two β chains per molecule. The hemoglobin of a certain person from Boston, designated hemoglobin M Boston, differs from normal hemoglobin (hemoglobin A) in that a histidine residue in each of the α chains of hemoglobin A is substituted by a tyrosine residue. From the ionization constants given in Table 4.1, do you expect the electrophoretic mobilities at pH 7.0 of the two proteins hemoglobin M Boston and hemoglobin A to differ? Give your reasons. What would be a reasonable pH to use to separate the two by electrophoresis? Which is more negatively charged at this pH?

7

Kinetics

Chemical kinetics is the study of the rates of reactions. Some reactions, such as that between hydrogen gas and oxygen gas in an undisturbed clean flask, occur so slowly as to be unmeasurable. Radioisotopes of some nuclei have very long lifetimes; the carbon isotope $^{14}C_6$ decays so slowly that half of the initial amount is still present after 5770 years. Other radionuclei have half-lives that are orders of magnitude longer than this. Processes such as the growth of bacterial cells are slow but easily measurable. The rate of reaction between H^+ and OH^- to form water is so fast that it has been measured only recently and by special techniques. The frontier is expanding to include still faster processes, and reaction times of fractions of a picosecond (1 ps = 10^{-12} s) are currently being reported using apparatus based on the mode-locked laser. Nuclear reactions involving species with lifetimes shorter than 10^{-20} s are known. Clearly, the methods of observation are very different—to include processes over such an enormous range of time.

The first chemical reaction whose rate was studied quantitatively involved a compound of biological origin. In 1850 L. Wilhelmy reported that the hydrolysis of a solution of sucrose to glucose and fructose occurred at a rate that decreased steadily with time but always remained proportional to the concentration of sucrose remaining in the solution. He was able to follow the reaction indirectly by measuring the change with time of the optical rotation—the rotation of the direction of plane-polarized light passing through the solution. The phenomenon of optical rotation results from the molecular asymmetry of sugars such as sucrose, glucose, and fructose; asymmetry refers to the fact that these molecules are structurally (and chemically) distinct from their mirror-image molecules (see Chapter 10).

Subsequent pioneering work by Guldberg and Waage (1863) and by van't Hoff demonstrated that systems at equilibrium are not static but are undergoing transformations between reactants and products in both directions and at equal rates.

The role of catalysts, substances that increase (or decrease) the rates of reactions without themselves being consumed, was recognized early from the influence of hydrogen ion on the rate of sucrose inversion. The early "ferments" used to convert sugars from grain or grapes into stimulating beverages are now known to contain enzyme catalysts that greatly speed up the rates of these processes. Biological organisms contain thousands of different enzymes—protein molecules that not only can provide increases of many millionfold in rates of reaction but also can be highly selective in their choice of reactant molecules from the large assortment present in live cells. The dramatic effect of some catalysts is easily seen in such cases as the hydrogen-oxygen gas mixture, where the introduction of a trace of finely divided metal such as platinum leads to a violent explosion. Ostwald pointed out that catalysts that affect the rates of reactions nevertheless have no effect on the position of equilibrium. We can conclude that the hydrogen-oxygen mixture in the absence of a catalyst is not at equilibrium.

The ability of increases in temperature to speed up most chemical reactions was put on a quantitative basis by Arrhenius (1889). Subsequent studies have made important contributions to our understanding of the mechanisms of reactions, which is a primary goal of kinetics. The mixture of hydrogen and oxygen, which is stable indefinitely at room temperature, can be made to explode by heating it to temperatures in excess of 400°C. In other cases, such as the processes that occur in biological cells, an increase in temperature causes the enzyme-catalyzed reactions to cease altogether.

Light or electromagnetic radiation serves as a "reagent" in some biological processes that are essential to our survival. Photosynthesis and vision are just two of the most obvious examples. Other processes that are not chemical in nature, such as the nuclear reactions that provide the energy source of the sun and have been the major source of heat within the earth, can nevertheless be profitably described using the methods of chemical kinetics. Population dynamics, ecological changes and balance, atmospheric pollution, and biological waste disposal are just a few of the relatively new applications of this powerful approach, which has its origins in physical chemistry.

Dynamic processes leading to growth, change, and evolution permeate all of biology. Some of these, such as our visual response or those of the involuntary muscles, are exceedingly fast even though they represent the consequences of an extensive chain of events. On a more leisurely time scale are the processes of cell division, growth, and death of organisms and the associated steps of metabolism. Extremely slow changes are of significance in genetic evolution and in adaptation to the modifications of our environment. The trend from reducing conditions to oxidizing conditions during the time since life first appeared on the earth can be traced in the response of photosynthetic organisms. Initially, there were much better reducing agents than water available in the environment, and there was no need to build up protection against the potentially toxic effects of oxygen. When the reductants became depleted and oxygen became abundant in the atmosphere, the modified characteristics of higher plants developed and new forms of photosynthesis continued to support life. In more recent times, such phenomena as the fluctuations in climatic temperature associated with the ice ages have led to important biological responses that can be investigated from fossil remains.

The methods of chemical kinetics or reaction-rate analysis were developed for the resolution and understanding of the relatively simple systems encountered by chemists. It may be surprising to some people to find that these approaches are also valuable in analyzing the much more complex processes of biology. The reason that the methods do work is often because one or a few steps control the rate of an extensive chain of reactions. All the steps involved in metabolism, cell division and replication, muscular contraction, and so on are subject to the same basic principles as are the elementary reactions of the chemist.

The rate or velocity, v, of a reaction or process describes how fast it is occurring. Usually the velocity is expressed as a change in concentration per unit time,

$$v = \frac{dc}{dt} = \text{rate of reaction} \tag{7.1}$$

but it may alternatively express the change of a population of cells with time, the increase or decrease in the pressure of a gas with time, or a change in the absorption of light by a colored solution with time. In general, the rate of a process depends in some way on the concentrations or amounts involved; the rate is a function of the concentrations. This relation is known as the *rate law*:

$$v = f\,(\text{concentrations}) \tag{7.2}$$

It may be simple (v = constant) or complex, but it gives important information about the mechanism of the process. One of the main objectives of research in kinetics is the determination of the rate law for the phenomena under investigation.

RATE LAW

Substances that influence the velocity of a reaction can be grouped into two categories:

1. Those whose concentration changes with time during the course of reaction:
   A. Reactants—decrease with time.
   B. Products—increase with time.

C. Intermediates—increase and then decrease during the course of the reaction.

$$A \longrightarrow C \longrightarrow B$$

2. Those whose concentrations do not change with time:
   A. Catalysts (both promoters and inhibitors), including enzymes and active surfaces.
   B. Intermediates in a steady-state process, including reactions under flowing conditions.
   C. Components that are buffered by means of equilibrium with large reservoirs.
   D. Solvents and the environment in general.

   These concentrations or influences do not change during a single run, but they can be changed from one experiment to the next.

Order of a Reaction

It is important in kinetics to learn immediately the vocabulary that kineticists use. We need to distinguish the stoichiometric reaction, the order of the reaction, and the mechanism of the reaction. It is essential to understand these terms. The *stoichiometry* of the reaction tells you how many moles of each reactant are needed to form each mole of products. Only ratios of moles are significant. For example,

$$H_2 + \tfrac{1}{2}O_2 = H_2O$$

and

$$2\,H_2 + O_2 = 2\,H_2O$$

are both correct stoichiometric reactions. The *mechanism* of a reaction tells you how the molecules react to form products. The mechanism is, in general, a set of *elementary reactions* consistent with the stoichiometric reaction. For the reaction of H_2 and O_2 in the gas phase, the reaction is thought to be a chain involving H, O, and OH radicals:

$$\begin{aligned} H_2 &\longrightarrow 2\,H \\ H + O_2 &\longrightarrow OH + O \\ OH + H_2 &\longrightarrow H_2O + H \\ O + H_2 &\longrightarrow OH + H \end{aligned}$$

Each reaction is an elementary reaction; the four reactions describe the proposed mechanism.

The kinetic *order* of a reaction describes the way in which the velocity of the reaction depends on the concentration. Consider a reaction that can be written

$$A + B \longrightarrow P$$

For many such reactions the rate law is of the form

$$v = kc_A^m c_B^n c_P^q \tag{7.3}$$

where the concentrations c_A, c_B, etc., are raised to powers m, n, etc., that are usually integers or zero (c_A^0 = constant), but may be nonintegral as well. The *order* of the reaction with respect to a particular component A, B, P, etc., is just the exponent of the concentration term. Because the velocity may depend on the concentrations of several species, we need to distinguish between the order with respect to a particular component and the overall order, which is the sum of the exponents of all components. Some representative examples are listed in Table 7.1.

Table 7.1 Rate laws and kinetic order for some reactions

Stoichiometric reaction	Rate law	Kinetic order
Sucrose + $H_2O \xrightarrow{H^+}$ fructose + glucose	$v = k[\text{sucrose}]$	1
L-Isoleucine → D-isoleucine	$v = k[\text{L-isoleucine}]$	1
$^{14}C_6 \rightarrow {}^{14}N_7 + \beta^-$	$v = k[^{14}C_6]$	1
$2\,N_2O_5 \rightarrow 4\,NO_2 + O_2$	$v = k[N_2O_5]$	1
$2\,NO_2 \rightarrow 2\,NO + O_2$	$v = k[NO_2]^2$	2
Hemoglobin · 3 O_2 + O_2 → Hb · 4 O_2	$v = k[\text{Hb} \cdot 3\,O_2][O_2]$	2 (overall)
$H_2 + I_2 \rightarrow 2\,HI$	$v = k[H_2][I_2]$	2 (overall)
$H_2 + Br_2 \rightarrow 2\,HBr$	$v = \dfrac{k[H_2][Br_2]^{1/2}}{k' + [HBr]/[Br_2]}$	Complex
$CH_3CHO \rightarrow CH_4 + CO$	$v \cong k[CH_3CHO]^{3/2}$	$\frac{3}{2}$ (approx.)
$C_2H_5OH \xrightarrow[\text{enzymes}]{\text{liver}} CH_3CHO$	v = constant	0

Several points are worth noting. It is never possible to *deduce* the order of the reaction by inspection of the stoichiometric equation. Kinetic experiments must be done to measure the order of the reaction. The reaction of H_2 and I_2 is first order with respect to each reactant over a wide range of conditions, whereas the superficially similar reaction of $H_2 + Br_2$ exhibits a complex rate law that cannot be described by a single "order" under all conditions. Note that in this case the velocity depends on the concentration of a product as well as on reactants. This comes about because of a reverse step in the mechanism that becomes important as the product concentration builds up. Reaction orders may be nonintegral, as in the case of the thermal decomposition of acetaldehyde, and they may be significantly different during the initial stages, when the reaction is getting under way, or at the end, when other complications set in.

The significance of a zero-order reaction is that the velocity is constant and independent of the concentration of the reactants. This is characteristic of reactions catalyzed by enzymes, such as liver alcohol dehydrogenase, under the special conditions where the enzyme is saturated with reactants (called substrates in enzyme reactions). The role of enzymes and other catalysts, such as H^+ in the sucrose inversion reaction, can be detected by observing that changes in the catalyst concentration are reflected by changes in the experimental rate constants for the reaction, even though the catalyst concentration does not vary with time during the course of any single experiment. Even these preliminary comments indicate that the rate law contains important information about the mechanism of the reaction.

Experimental Rate Data

The rate law for a reaction must be determined from experimental data. We may simply want to know how the rate depends on concentrations for practical reasons, or we may want to understand the mechanism of the reaction. For example, if we are inactivating a virus by reaction with formaldehyde, it is vital for us to know how the rate depends on the concentrations of virus and formaldehyde. If the rate is first order in virus concentration, the rate will be one tenth as fast for 10^5 viruses per milliliter as for 10^6 viruses per milliliter. If the rate is second order in virus concentration (unlikely), the rate will be only one hundredth as fast. Actually, the rate may be effectively zero order in virus concentration. However, knowledge of the rate law is necessary to determine the time of treatment with formaldehyde needed to ensure that all the viruses are inactivated before they are used to immunize a population. We also must know how temperature, pH, and solvent affect the rate.

There are many convenient ways to obtain the rate data. A usual method is to obtain concentrations of reactants and products at different times during the reaction. For the virus inactivation we would presumably measure live virus versus time by an infectivity assay. In general, any analytical method that determines concentration can be used. For first-order reactants we will see that we do not even have to measure absolute concentrations. Measurement of any property proportional to concentration is sufficient to determine the rate law and the rate constant for a first-order reaction. Changes in light absorption, electrical conductivity, radioactivity, and so on, can be used to measure rates. The time required for the reaction to reach 50%, or some other fraction, of completion is sometimes convenient to determine. The rate of action of an anesthetic can be determined by having the patient count backward from 100. The kinetics of the conversion of fibrinogen to fibrin (the blood-clotting protein) is studied by measuring a gelation time. The most direct way to obtain the order of reaction is to measure the initial rate for different initial concentrations. This directly gives the exponents in Eq. (7.3).

Zero-order Reactions

A *zero-order reaction* corresponds to the rate law:

$$\frac{dc}{dt} = k_0 \tag{7.4}$$

The units of k_0 for a zero-order reaction are obviously concentration per time, such as $M\,s^{-1}$. This expression can readily be integrated by writing it in the differential form with the variables c and t separated on each side of the equality:

$$dc = k_0\,dt \tag{7.5}$$

Integrating both sides, we obtain

$$\int dc = k_0 \int dt + \text{constant} \tag{7.6}$$

where the rate constant is placed outside the integral sign because it does not depend on time. Once the initial and final conditions are specified, Eq. (7.6) can be written as a definite integral; that is, the constant can be evaluated. If the concentration is c_1 at time t_1 and c_2 at time t_2, then

$$\int_{c_1}^{c_2} dc = k_0 \int_{t_1}^{t_2} dt$$

$$c_2 - c_1 = k_0 \cdot (t_2 - t_1) \tag{7.7}$$

A different form is commonly used, where c_0 refers to the initial concentration at zero time, and the concentration is c at any later time t. Then

$$c - c_0 = k_0 t \tag{7.8}$$

This is the equation for a straight line giving the dependence of c on t; the slope is the rate constant, k_0.

This behavior is illustrated by the conversion of ethanol to acetaldehyde by the enzyme liver alcohol dehydrogenase (LADH). The oxidizing agent is nicotinamide adenine dinucleotide (NAD^+) and the reaction can be written

$$CH_3CH_2OH + NAD^+ \xrightarrow{\text{LADH}} CH_3CHO + NADH + H^+$$

In the presence of an excess of alcohol over the enzyme (a beer or two will suffice) and with the NAD^+ buffered via metabolic reactions that rapidly restore it, the rate of this reaction in the liver is zero-order over most of its course,

$$v = -\frac{d[CH_3CH_2OH]}{dt} = \frac{d[CH_3CHO]}{dt} = k_0 \tag{7.9}$$

The negative sign is used with the reactant, ethanol, because its concentration decreases with time; the concentration of the product, acetaldehyde, increases with time. This behavior is illustrated in Fig. 7.1.

The reaction cannot be of zero order for all times; because obviously the reactant concentration cannot become less than zero, and the product concentration must also reach a limit. For the oxidation of alcohol by LADH the reaction is zero order only while alcohol is in excess.

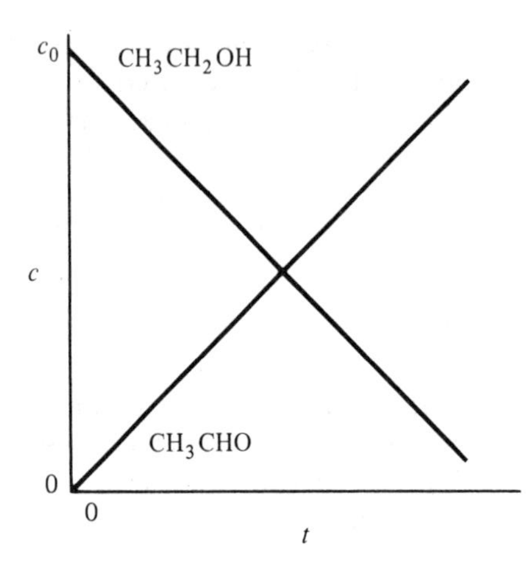

Fig. 7.1 Plot of concentration versus time for a zero-order reaction. The magnitude of the slope of each straight line is equal to the rate constant, k_0. The order must eventually change from being zero-order as the concentration of CH_3CH_2OH approaches zero.

First-order Reactions

A *first-order reaction* corresponds to the rate law:

$$\frac{dc}{dt} = k_1 c \tag{7.10}$$

The units of k_1 are time^{-1}, such as s^{-1}. There are no concentration units in k_1, so it is clear that we do not need to know absolute concentrations; only relative concentrations are needed. An elementary step in a reaction of the form

$$A \longrightarrow B$$

has a rate law of the form

$$v = -\frac{d[A]}{dt} = \frac{d[B]}{dt} = k_1[A] \tag{7.11}$$

where k_1 is the rate constant for the particular reaction. The velocity of the reaction can be expressed in terms of either the rate of disappearance of reactant, $-d[A]/dt$, or the rate of formation of product, $d[B]/dt$. The stoichiometric equation assures us that these two quantities will always be equal to one another. To solve the rate-law expression we choose the form involving the smallest number of variables.

$$-\frac{d[A]}{dt} = k_1[A] \tag{7.12}$$

Here time is one variable and the concentration of A is the other. This is a derivative equation. By rewriting it in the differential form and dividing both sides by [A], we obtain

$$\frac{d[\mathrm{A}]}{[\mathrm{A}]} = -k_1\, dt \tag{7.13}$$

In this form the variables are *separated* in the sense that the left side depends only on [A] and the right side only on t. Once the variables are separated, the equation can be integrated, separately, on each side.

$$\int \frac{d[\mathrm{A}]}{[\mathrm{A}]} = \int -k_1\, dt = -k_1 \int dt \tag{7.14}$$

$$\ln[\mathrm{A}] = -k_1 t + C \tag{7.15}$$

where C is a constant of integration. This states that for a first-order reaction, the *logarithm* of the concentration will be a linear function of time, as shown in Fig. 7.2. To evaluate the constant C we need to know one specific concentration at a

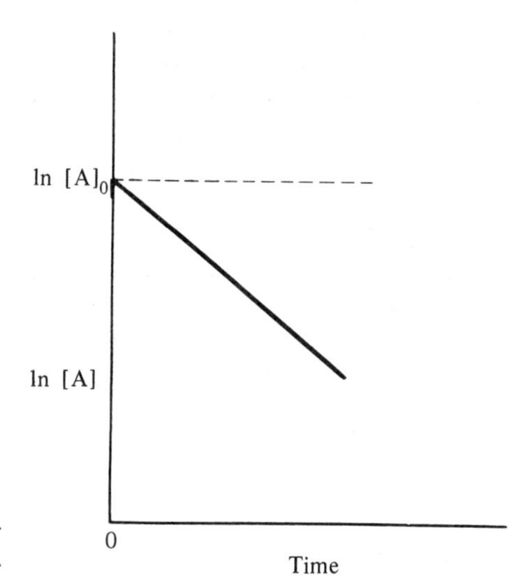

Fig. 7.2 Semilogarithmic straight-line plot for a first-order reaction.

particular time. For example, if $[\mathrm{A}]_0$ is the value of the concentration initially when $t = 0$, it, too, must satisfy Eq. (7.15). Substitution gives

$$\ln[\mathrm{A}]_0 = C \tag{7.16}$$

which provides an alternative form of Eq. (7.15),

$$\ln\frac{[\mathrm{A}]}{[\mathrm{A}]_0} = 2.303 \log\frac{[\mathrm{A}]}{[\mathrm{A}]_0} = -k_1 t \tag{7.17}$$

A more general form involving any two points during the course of a first-order reaction is

$$\ln \frac{[A]_2}{[A]_1} = -k_1(t_2 - t_1) \tag{7.18}$$

An alternative form of Eq. (7.17) is

$$[A] = [A]_0\, e^{-k_1 t} = [A]_0 10^{-k_1 t/2.303} \tag{7.19}$$

This says that the concentration of A decreases exponentially with time for a first-order reaction. It starts at an initial value $[A]_0$, since $e^0 = 1$, and reaches zero only after infinite time! Strictly speaking, the reaction is never "finished." Before worrying about the philosophical implications of this, be assured that the inability to detect any remaining [A] will occur in finite time even using the most sensitive analytical methods.

Let us consider in detail a first-order reaction; the hydrolysis of sucrose:

$$\text{sucrose} + H_2O \longrightarrow \text{glucose} + \text{fructose}$$

Experimental data for the reaction are given in Table 7.2 and plotted in Fig. 7.3. The concentration of sucrose could be determined at each time by stopping the reaction by neutralizing the acid catalyst and analytically determining the sucrose. An easier method would be to measure the rotation by the solution of plane polarized light (see Chapter 10) in a polarimeter. As the rotation is proportional to the concentration of sucrose, the rotation values can be used to determine the rate.

Table 7.2 Concentration of sucrose at successive times (temperature 23°C; 0.5 M HCl)

Time (min)	Sucrose concentration, [S] (mol ℓ^{-1})	$\bar{v} = \frac{\Delta[S]}{\Delta t}$ (mol ℓ^{-1} min^{-1})	$[\bar{S}] = \frac{[S]_2 + [S]_1}{2}$ (mol ℓ^{-1})	$\frac{\bar{v}}{[\bar{S}]}$ (min^{-1})
0	0.316			
		1.14×10^{-3}	0.308	3.70×10^{-3}
14	0.300			
		1.04×10^{-3}	0.287	3.62×10^{-3}
39	0.274			
		0.86×10^{-3}	0.265	3.25×10^{-3}
60	0.256			
		0.90×10^{-3}	0.247	3.64×10^{-3}
80	0.238			
		0.90×10^{-3}	0.225	4.00×10^{-3}
110	0.211			
		0.70×10^{-3}	0.201	3.48×10^{-3}
140	0.190			
		0.67×10^{-3}	0.180	3.72×10^{-3}
170	0.170			
		0.60×10^{-3}	0.158	3.80×10^{-3}
210	0.146			
			Average $\bar{k} = \frac{\bar{v}}{[\bar{S}]}$	$= 3.65 \times 10^{-3}$ min^{-1}

The data given in Table 7.2 for the hydrolysis of sucrose can be plotted according to Eq. (7.17) to test whether the reaction exhibits first-order kinetics. This is done in Fig. 7.4, where it can be seen that the data do indeed fall on a straight line. The use of the integrated rate expression to test the order of a

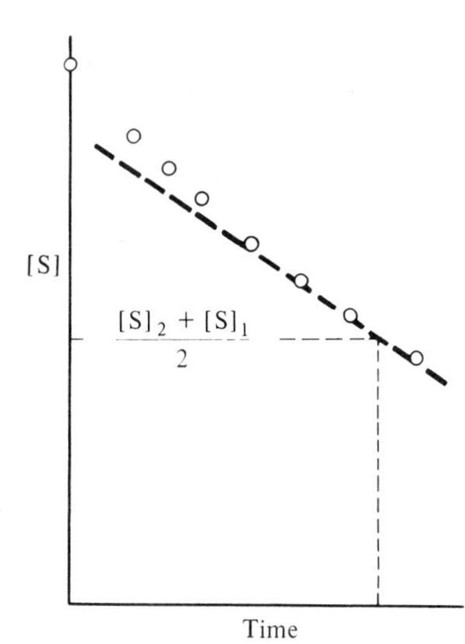

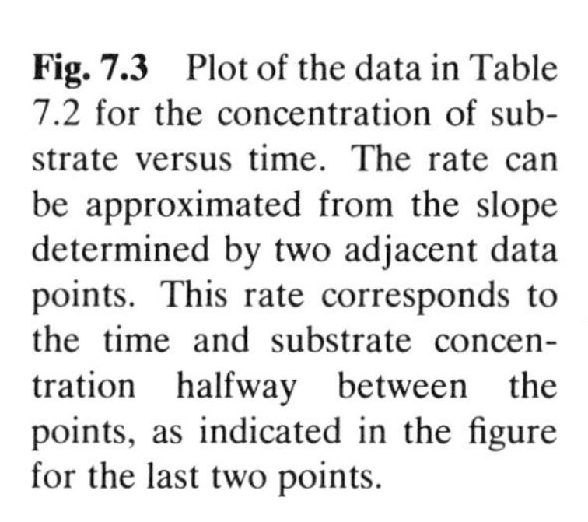

Fig. 7.3 Plot of the data in Table 7.2 for the concentration of substrate versus time. The rate can be approximated from the slope determined by two adjacent data points. This rate corresponds to the time and substrate concentration halfway between the points, as indicated in the figure for the last two points.

reaction is usually the best procedure and gives the most accurate value for the rate constant.

Other methods can be used to test whether the kinetics is first order and to obtain an approximate rate constant. The velocity of the reaction can be estimated using the approximation

$$v = \frac{d[S]}{dt} \cong \frac{\Delta[S]}{\Delta t} = \frac{[S]_2 - [S]_1}{t_2 - t_1} = \bar{v} \qquad (7.20)$$

This approximation works best if the points are not too far apart along the curve and if the data are reasonably precise. Values for $\Delta[S]/\Delta t$ are given in the third

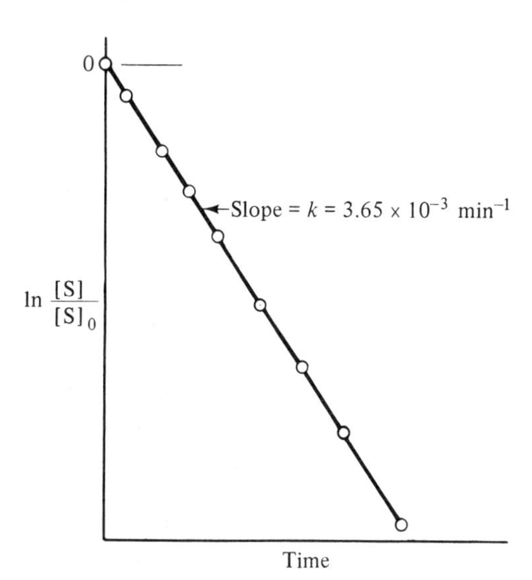

Fig. 7.4 First-order plot of the sucrose concentration data given in Table 7.2. The rate constant k found from the slope is 3.65×10^{-3} min^{-1}.

column of Table 7.2. These correspond most closely to the velocity (slope) at the midpoint of the concentration interval. Thus the mean concentrations for each interval are tabulated in the fourth column of Table 7.2. We can now proceed to plot $\Delta[S]/\Delta t$ versus $[\bar{S}]$, as seen in Fig. 7.5. The straight-line relation passing through the origin indicates first-order kinetics.

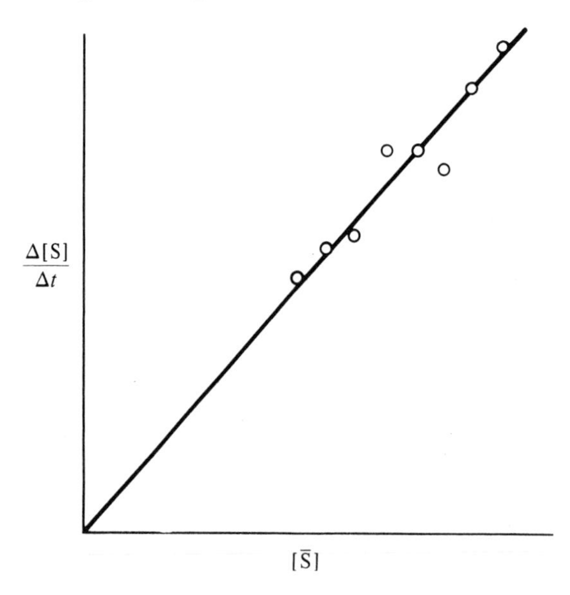

Fig. 7.5 Plot of data for sucrose hydrolysis from Table 7.2.

Alternatively, we can examine the values of the ratio $\bar{v}/[\bar{S}]$ in the fifth column of Table 7.2 and see that this value is essentially independent of concentration, as required for a first-order reaction. The ratio $\bar{v}/[\bar{S}]$ is just the approximate rate constant, $\bar{k}$, for a first-order reaction. Because of the approximations involved in this method, we should recognize that the value of a rate constant obtained in this way may incorporate certain systematic errors. In our example, the value from Table 7.2 agrees with that from Fig. 7.4.

The measured value of the rate constant determined from Fig. 7.4 can be compared with the published value of $k = 3.74 \times 10^{-3}\ \text{min}^{-1}$, for the same temperature and acid concentration. The agreement is good, suggesting (but not proving) that the measurements were carefully performed. The differences arise from random uncertainties of the individual measurements (random errors) as well as persistent flaws or repeatable biases incorporated into the measuring procedure (systematic errors). Many of these errors tend to have similar absolute values (with respect to the concentration of S or with respect to the time). As a consequence, the *relative* errors tend to be greater for small concentrations than for large ones. Thus, the points at the lower right in Fig. 7.4 have larger errors than those at the upper left. In drawing the best straight line through the points, we give greater weight to the points with the smaller errors. This procedure also compensates somewhat for side reactions or back reactions that tend to become more important as the reaction nears completion.

It is interesting to note the significance of the rate constant in terms of the direct plot shown in Fig. 7.3. A point of reference occurs at the time $t = 1/k_1 = \tau$. The exponent in Eq. (7.19) is then -1, and produces $[A]_\tau = [A]_0/e = 0.368[A]_0$. The quantity τ, known as the *relaxation time* or *time constant* for the reaction, is the time required for the concentration to fall to $1/e$ (numerically, 0.368) of its initial value. Since τ is just the reciprocal of the rate constant, it follows that the fast reactions have small values of τ.

Another point worth noting is that Eq. (7.17) depends only on the ratio $[A]/[A]_0$, that is, on the fraction of A remaining at time t. We do not need to know the concentrations themselves to determine the rate constant by means of Eq. (7.17) or the plot shown in Fig. 7.4. (This is true only for first-order reactions, not for those of higher or lower orders.) Frequently, we measure some property, such as the total pressure, the optical absorbance, or the refractive index, that we know to be proportional to the concentration. The concentration data listed in Table 7.2 were actually calculated from measurements of the optical rotation of a reacting sucrose solution. It is sufficient to plot the appropriate logarithmic function of the measured property itself in order to test for first-order kinetics and to determine the first-order rate constant. If the sucrose is completely hydrolyzed to fructose and glucose, one can derive the fact that

$$\frac{\theta(t) - \theta(\infty)}{\theta(0) - \theta(\infty)} = \frac{[A]}{[A]_0} \tag{7.21}$$

where $\theta(t)$ is the measured rotation angle at time t, $\theta(\infty)$ is the measured rotation at a time long enough for the reaction to go to completion, and $\theta(0)$ is the rotation at zero time.

A common parameter used to describe the kinetics of rate processes is the half-life $t_{1/2}$. This is simply the time required for half the initial concentration to react, that is, when $[A] = \frac{1}{2}[A]_0$. Put into Eq. (7.19), this results in

$$\tfrac{1}{2}[A]_0 = [A]_0\, e^{-k_1 t_{1/2}}$$

$$\tfrac{1}{2} = e^{-k_1 t_{1/2}}$$

Taking logarithms of both sides,

$$\ln \tfrac{1}{2} = -k_1 t_{1/2}$$

Since $\ln \frac{1}{2} = -0.693$, we see that

$$t_{1/2} = \frac{0.693}{k_1} \tag{7.22}$$

Exercise Show that Eqs. (7.19) and (7.22) taken together lead to the result

$$\frac{[A]}{[A]_0} = 2^{-t/t_{1/2}} \tag{7.23}$$

We notice that Eqs. (7.19), (7.22), and (7.23) apply only to first-order processes. They were derived specifically using the integrated form of the first-order rate law. Other reaction orders give different forms for the half-life and for the time constant. A unique and useful property of the half-life (or time constant) *of a first-order reaction* is that it is independent of the initial concentration. (This is not the case for other kinetic orders.) A familiar example is radioactive decay. It does not matter whether we have a large or a small amount (concentration) of a particular radioisotope: The half-life for any size sample is always the same.

Exercise Show that $t_{1/2}$ is independent of the choice of starting point for a first-order decay process such as that shown in Fig. 7.3. For example, show using Eq. (7.19) that the time required to go from $\frac{1}{2}[A]_0$ to half that value, $\frac{1}{4}[A]_0$, is also $t_{1/2}$.

Example 7.1 Carbon dioxide in the atmosphere contains a small but readily detectable amount of the radioactive isotope $^{14}C_6$. This isotope is produced by high-energy neutrons (in cosmic rays) that transform nuclei of nitrogen atoms by way of the process

$$^{14}N_7 + {}^1n_0 \longrightarrow {}^{14}C_6 + {}^1H_1$$

The $^{14}C_6$ nucleus is unstable and decays by the first-order process

$$^{14}C_6 \longrightarrow {}^{14}N_7 + {}^0\beta_{-1}$$

with a half-life of 5770 years. The production and decay of $^{14}C_6$ leads to a constant steady-state concentration. The amount of $^{14}C_6$ present in a carbon-containing sample can be determined by measuring its radioactivity (the rate of production of high-energy electrons, $^0\beta_{-1}$ particles) using a suitable detector.

In the atmosphere CO_2 contains a constant amount of $^{14}C_6$. Once CO_2 is "fixed" by photosynthesis, it is taken out of the atmosphere and new $^{14}C_6$ is no longer added to it. The level of radioactivity then decreases by a first-order process with a 5770-year half-life.

A sample of wood from the core of an ancient bristlecone pine in the White Mountains of California shows a ^{14}C content that is 54.9% as great as that of atmospheric CO_2. What is the approximate age of the tree?

Solution Since the process is first order, we need know only the *ratio* of the present radioactivity to the original value. Assume that the level of ^{14}C in the atmosphere has not changed over the life of the tree. (This is not strictly true. Corrections can be made, but they are small.) Thus,

$$\frac{[^{14}C]}{[^{14}C]_0} = 0.549$$

Using Eqs. (7.17) and (7.22), we obtain

$$\ln \frac{[^{14}\mathrm{C}]}{[^{14}\mathrm{C}]_0} = -\frac{0.693}{t_{1/2}} t$$

where t is the age of the wood. Therefore,

$$\ln (0.549) = -\frac{0.693}{5770 \text{ yr}} t$$

and

$$t = \frac{(0.600)(5770 \text{ yr})}{0.693} = 4990 \text{ yr}$$

Second-order Reactions

A *second-order reaction* corresponds to the rate law:

$$v = k_2 c^2 \quad \text{or} \quad v = k_2 c_A c_B$$

The units of k_2 are concentration^{-1} time^{-1} such as $M^{-1}\,\mathrm{s}^{-1}$.

It is useful to separate the treatment of second-order reactions into two major classifications depending on whether the rate law depends on (I) the second power of a single reactant species, or (II) the product of the concentrations of two different reagents.

Class I (A + A ⟶ P)

$$v = k_2[\mathrm{A}]^2$$

Although the stoichiometric equation may involve either one or several components, the rate law for many reactions depends only on the second power of a single component. Some examples are:

$$2\ \text{proflavin} \longrightarrow [\text{proflavin}]_2 \ : \ v = k_2\,[\text{proflavin}]^2$$

$$\underset{\text{Ammonium cyanate}}{NH_4OCN} \rightleftharpoons NH_4^+ + OCN^- \longrightarrow \underset{\text{Urea}}{NH_2CONH_2} : v = k_2[NH_4OCN]^2$$

$$\underset{\text{Hexanucleotide*}}{2\ \text{A–A–G–C–U–U}} \longrightarrow \underset{\text{Hexanucleotide dimer}}{\begin{matrix}\text{A–A–G–C–U–U}\\ \vdots\ \vdots\ \vdots\ \vdots\ \vdots\ \vdots\\ \text{U–U–C–G–A–A}\end{matrix}} : v = k_2[A_2GCU_2]^2$$

In each case the rate law is of the form

$$v = -\frac{d[\mathrm{A}]}{dt} = k_2[\mathrm{A}]^2 \tag{7.24}$$

* See Appendix for the structure of the nucleotide.

It is useful to rewrite this in terms of the reaction variable x, where x is the amount of A reacted. If

$$[A]_0 = a$$

then

$$[A] = a - x$$

and

$$d[A] = -dx$$

Therefore,

$$v = \frac{dx}{dt} = k_2 \cdot (a - x)^2 \tag{7.25}$$

The variables can be separated,

$$\frac{dx}{(a - x)^2} = k_2\, dt \tag{7.26}$$

and each side integrated, to give

$$\frac{1}{a - x} = k_2 t + C \tag{7.27}$$

Since $x = 0$ when $t = 0$, the constant of integration is $1/a$ and we obtain an integrated form of the second-order rate law:

$$\frac{1}{a - x} - \frac{1}{a} = k_2 t \tag{7.28}$$

An alternative form in terms of reactant concentrations is

$$\frac{1}{[A]} - \frac{1}{[A]_0} = k_2 t \tag{7.29}$$

Note that for this kinetic behavior one expects a linear relation between the reciprocal of the reactant concentration and time. This is in contrast with first-order kinetics, where a logarithmic relation obtains.

The half-life for a second-order (class I) reaction is the time required for half of $[A_0]$ to react. In terms of Eq. (7.28), this means that $a - x = \frac{1}{2}a$ at $t = t_{1/2}$ and

$$\frac{1}{\frac{1}{2}a} - \frac{1}{a} = k_2 t_{1/2}$$

$$\frac{1}{a} = k_2 t_{1/2}$$

$$t_{1/2} = \frac{1}{k_2 a} = \frac{1}{k_2 [A]_0} \tag{7.30}$$

For second-order reactions, the half-life depends on the initial concentration as well as on the rate constant. It is only first-order reactions where the half-life (or any other fractional life) is independent of initial concentration.

A graphical comparison of first- and second-order kinetics is instructive. Figure 7.6 shows concentration versus time plots to illustrate the differences. The time scales have been adjusted so that the two half-lives are the same. Note that the second-order reaction proceeds more rapidly during the early stages of the reaction, but slows down and persists to longer times than does the first-order reaction. The difference in the two curves is subtle, however, and it is best seen using this superposition approach.

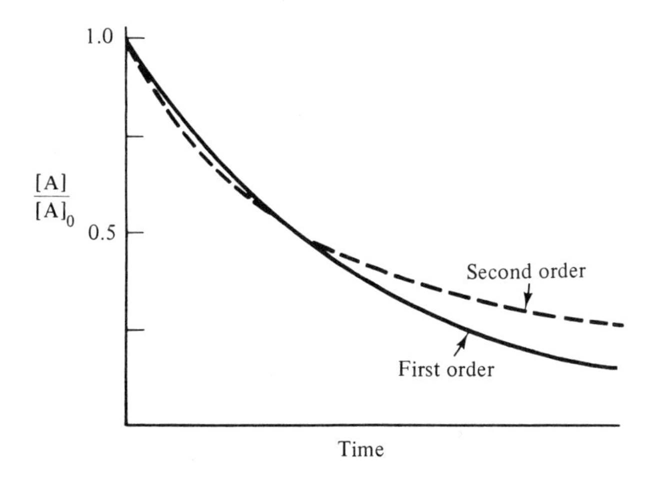

Fig. 7.6 Comparison of normalized first- and second-order kinetics with time scales adjusted to give the same $t_{1/2}$.

Class II (A + B ⟶ P)

$$v = k_2[\mathrm{A}][\mathrm{B}]$$

A reaction that is second order overall may be first order with respect to *each* of the two reactants. Some examples are

$$CH_3COOC_2H_5 + OH^- \longrightarrow CH_3COO^- + C_2H_5OH: \quad v = k_2[CH_3COOC_2H_5][OH^-]$$

$$NO(g) + O_3(g) \longrightarrow NO_2(g) + O_2(g): \quad v = k_2[NO][O_3]$$

$$H_2O_2 + 2\,Fe^{2+} + 2\,H^+(\text{excess}) \longrightarrow 2\,H_2O + 2\,Fe^{3+}: \quad v = k_2[H_2O_2][Fe^{2+}]$$

Again we see that the overall stoichiometric equation is not a valid indicator of the rate law. We may expect that there are significant underlying differences in the mechanisms of these reactions.

$[a \neq b]$ In general, for reactions of the type

$$\mathrm{A} + \mathrm{B} \longrightarrow \text{products}$$

exhibiting class II kinetics, the initial concentrations of the reactants need not be the stoichiometric ratio. For this reason, we rewrite the rate law

$$v = k_2[\mathrm{A}][\mathrm{B}] \tag{7.31}$$

in terms of the reaction variable, where

$$[\mathrm{A}] = a - x; \qquad [\mathrm{A}]_0 = a$$

$$[\mathrm{B}] = b - x; \qquad [\mathrm{B}]_0 = b$$

Following substitution into Eq. (7.31) and separation of variables, we obtain

$$\frac{dx}{(a - x)(b - x)} = k_2\, dt \tag{7.32}$$

This can be integrated using the method of partial fractions to give the result

$$\frac{1}{a - b} \ln \frac{b(a - x)}{a(b - x)} = k_2 t: \qquad [a \neq b] \tag{7.33}$$

or, alternatively,

$$\frac{1}{[\mathrm{A}]_0 - [\mathrm{B}]_0} \ln \frac{[\mathrm{B}]_0[\mathrm{A}]}{[\mathrm{A}]_0[\mathrm{B}]} = k_2 t \tag{7.34}$$

After separating the constant terms involving only initial concentrations, we see that class II second-order reactions exhibit a linear relation between $\ln [\mathrm{A}]/[\mathrm{B}]$ and time.

$[a = b]$ Note that when $a = b$ (when the initial concentrations are stoichiometric), then Eqs. (7.33) and (7.34) do not apply. In this case, however, the method of class I is appropriate, since the values of [A] and [B] will be in a constant ratio throughout the entire course of the reaction. That is, if $a = b$, then $[\mathrm{A}] = [\mathrm{B}]$ at all times and

$$v = k_2[\mathrm{A}][\mathrm{B}] = k_2[\mathrm{A}]^2$$

which can be integrated exactly as in class I to give Eq. (7.28).

Example 7.2 Hydrogen peroxide reacts with ferrous ion in acidic aqueous solution according to the reaction

$$H_2O_2 + 2\,Fe^{2+} + 2\,H^+ \longrightarrow 2\,H_2O + 2\,Fe^{3+}$$

From the following data, obtained at 25°C, determine the order of the reaction and the rate constant.

$$[H_2O_2]_0 = 1 \times 10^{-5}\, M$$

$$[Fe^{2+}]_0 = 1 \times 10^{-5}\, M$$

$$[H^+]_0 = 1\, M$$

Time (min)	0	5.3	8.7	11.3	16.2	18.5	24.6	34.1
$10^5 \times [Fe^{3+}]$ M	0	0.309	0.417	0.507	0.588	0.632	0.741	0.814

Solution Since the H^+ concentration is in huge excess, it will not change significantly during the course of the reaction. As a consequence, we can write the velocity expression as

$$v = k[H_2O_2]^i[Fe^{2+}]^j$$

recognizing that this is the simplest general form. Note also that the initial concentration of H_2O_2 is twice the amount needed to react with all of the Fe^{2+} present.

1. Test for first order in $[Fe^{2+}]$, where $i = 0$ and $j = 1$. In this case, we expect from Eq. (7.17) that

$$\ln\frac{[Fe^{2+}]_0}{[Fe^{2+}]} = kt \quad \text{or} \quad \frac{1}{t}\ln\frac{[Fe^{2+}]_0}{[Fe^{2+}]} = k$$

2. Test for first order in $[H_2O_2]$, where $i = 1$ and $j = 0$. Then

$$\frac{1}{t}\ln\frac{[H_2O_2]_0}{[H_2O_2]} = k$$

The data for testing proposals 1 and 2 are as follows:

Time (min)	0	5.3	8.7	11.3	16.2	18.5	24.6	34.1
$10^5 \times [Fe^{2+}](M)$	1.00	0.691	0.583	0.493	0.412	0.368	0.259	0.186
$\ln\frac{[Fe^{2+}]_0}{[Fe^{2+}]}$	0	0.370	0.540	0.707	0.887	1.000	1.351	1.682
(1) $\frac{10^3}{t}\ln\frac{[Fe^{2+}]_0}{[Fe^{2+}]}$	—	69.8	62.0	62.6	54.7	54.0	54.9	49.3
$10^5 \times [H_2O_2](M)$	1.00	0.845	0.791	0.741	0.706	0.684	0.630	0.593
(2) $\frac{10^3}{t}\ln\frac{[H_2O_2]_0}{[H_2O_2]}$	—	31.8	26.9	26.5	21.5	20.5	18.8	15.3

The fact that the rows listed as (1) and (2) are not constant, but show a trend, indicates that neither of these predictions is followed by the reaction.

3. Test for second order overall, where $i = 1$ and $j = 1$. Since the stoichiometric coefficients are not unity, we need to modify the development

of Eq. (7.33). For a reaction with stoichiometry A + 2B → product, we define

$$v = -\frac{d[A]}{dt} = k_2[A][B] = -\frac{1}{2}\frac{d[B]}{dt}$$

where $[A] = a - x \qquad [A]_0 = a$

and $[B] = b - 2x \qquad [B]_0 = b$

The equation analogous to (7.32) is now

$$\frac{dx}{(a - x)(b - 2x)} = k_2\,dt$$

or

$$\frac{dx}{(a - x)(b/2 - x)} = 2k_2\,dt$$

This is readily solved by noting that with respect to Eq. (7.32) we have just replaced b by $b/2$ and k_2 by $2k_2$. We simply make the same replacements in the solution, Eq. (7.33), to obtain the result

$$\frac{1}{a - \frac{b}{2}}\ln\frac{\frac{b}{2}(a - x)}{a\left(\frac{b}{2} - x\right)} = 2k_2t$$

or

$$\frac{1}{2a - b}\ln\frac{b(a - x)}{a(b - 2x)} = k_2t$$

or

$$\frac{1}{2[H_2O_2]_0 - [Fe^{2+}]_0}\ln\frac{[Fe^{2+}]_0[H_2O_2]}{[H_2O_2]_0[Fe^{2+}]} = k_2t$$

Since $[Fe^{2+}]_0/[H_2O_2]_0 = 1.0$ from the initial conditions, we expect the relation

$$\frac{1}{t}\ln\frac{[H_2O_2]}{[Fe^{2+}]} = \text{constant}$$

to hold if the rate law is first order with respect to each reactant.

Proposal 3 is tested by the following tabulation:

Time (min)	0	5.3	8.7	11.3	16.2	18.5	24.6	34.1
$\frac{[H_2O_2]}{[Fe^{2+}]}$	1.00	1.223	1.357	1.503	1.714	1.859	2.432	3.188
(3) $\frac{1}{t}\ln\frac{[H_2O_2]}{[Fe^{2+}]}$ (min^{-1})	—	0.0380	0.0351	0.0361	0.0332	0.0335	0.0361	0.0340

The entries in the bottom row are approximately constant and show that the results do correspond to this rate law. Therefore,

$$v = k_2[H_2O_2][Fe^{2+}]$$

The value of k can be determined from

$$k_2 = \frac{1}{2[H_2O_2]_0 - [Fe^{2+}]_0}\left[\frac{1}{t}\ln\frac{[H_2O_2]}{[Fe^{2+}]}\right]$$

The average value of $(1/t)\ln[H_2O_2]/[Fe^{2+}]$ from the table is 0.0351 min^{-1}. Therefore,

$$k_2 = \frac{0.0351\ \text{min}^{-1}}{(2 \times 10^{-5} - 1 \times 10^{-5})\,M} = 3.51 \times 10^3\,M^{-1}\,\text{min}^{-1}$$

We could, of course, have solved this problem by graphical rather than numerical methods.

Reactions of Other Orders

It is easy to generalize the treatment of class I reactions of order n, where n is any positive or negative number except +1. In this case the rate law is

$$v = k[A]^n \tag{7.35}$$

which can be integrated to give

$$\frac{1}{n-1}\left[\frac{1}{[A]^{n-1}} - \frac{1}{[A]_0^{n-1}}\right] = kt: \quad n \neq 1 \tag{7.36}$$

The half-life will be

$$t_{1/2} = \frac{2^{n-1} - 1}{(n-1)k[A]_0^{n-1}} \tag{7.37}$$

For example, the thermal decomposition of acetaldehyde is $\frac{3}{2}$ order over most of the course of the reaction. A linear plot should be obtained, according to Eq. (7.36), if $1/\sqrt{[A]}$ is plotted versus time. The half-life will be given by

$$t_{1/2} = \frac{\sqrt{2} - 1}{\frac{1}{2}k[A]_0^{1/2}}$$

The examples of integrated rate expressions derived here are among the most important, but many others occur as well. Often rate laws that can be written are impossible to integrate as we have done here. In practice, the kineticist usually faces a more complex problem than is indicated by these derivations. Interest in learning the *mechanism* of the reaction is what prompts the search to find the rate law that best fits the data.

Determining the Order and Rate Constant of a Reaction

We have discussed various simple orders which can represent reaction kinetics and we have given specific illustrations. Here we shall mention some general methods that can be used to determine the order and rate constant for a reaction. We need to know the stoichiometry of the reaction and we need to be able to measure the concentration of a product or reactant. To determine the order of the reaction with respect to each reactant, we actually only need to measure a property that is directly proportional to concentration. Except for first-order reactions, we do need to know concentrations to obtain rate constants.

Direct data plots

A plot of concentration versus time gives a clue to the order. If the plot is linear, a zero-order reaction is indicated. If a curved plot results, other plots, such as log c versus time or $1/c$ versus time, can be tried. One looks for a linear plot that indicates the order and whose slope is proportional to the rate constant.

Velocity versus concentration plots (differentiation of the data)

From measurements of the slopes of concentration versus time plots, one can determine the velocity of the reaction at different times. These velocities can then be plotted versus the corresponding reactant concentration in an attempt to deduce the form of the rate law. This approach requires many experimental points or a continuous curve.

Method of initial rates

The velocity is measured during the earliest stages of the reaction. The concentrations do not change much with time, and they can be replaced in the rate law by the initial concentrations. This method is particularly useful when the rate law is not of the simplest form. For example, in many enzyme-catalyzed reactions the method of initial rates was used to find that

$$v = \frac{k[\mathrm{S}]}{K + [\mathrm{S}]}$$

where k and K are constants and [S] is the substrate concentration. We shall discuss this Michaelis-Menten type of kinetics in Chapter 8.

Changes in initial concentrations

The influence of different initial concentrations on the initial rate, the half-life, or some other parameter can be determined. The method measures the order of the reaction with respect to a particular component directly. To find the order with respect to component A in the expression

$$v_1 = k[A]^a[B]^b[C]^c$$

we measure the initial rate v_1 with initial concentrations A, B, and C.

$$v_1 = k[A]^a[B]^b[C]^c$$

We then double the concentration of A, keeping the others constant, and measure the new initial rate v_2,

$$v_2 = k(2[A])^a[B]^b[C]^c$$

The ratio of these two rates leads to the value of a,

$$\frac{v_2}{v_1} = 2^a$$

because all other factors cancel.

Method of reagents in excess

We can decrease the kinetic order of a reaction by choosing one or more of the reactant concentrations to be in large excess. Normally, the concentration of a reactant decreases steadily during the course of a reaction. This complicates the kinetic analysis, especially if the rate law is not a simple one. By using a large excess of one component, its concentration is maintained nearly constant and the corresponding term in the rate law now hardly changes during the course of the reaction. In the example above we could choose B and C to be 1 M and A to be 0.01 M. Concentrations of B and C would change only slightly (depending on the stoichiometry) during the complete reaction for A. We could essentially study the kinetics with respect to A independently of the B and C kinetics.

Once the order of the reaction is known with respect to each component, the value of the rate constant can be calculated. It is important to specify what rate and what species are being described. Uncertainty can occur when the stoichiometric coefficients for the reaction are not all the same. For example, the reaction

$$N_2O_5 \longrightarrow 2\,NO_2 + \tfrac{1}{2}O_2$$

exhibits first-order kinetics. If we express concentrations of these substances in mol ℓ^{-1}, the rate of disappearance of reactant N_2O_5 is half as great as the rate of formation of the product NO_2 and twice as great as the rate of formation of O_2. Expressed mathematically, this is

$$-\frac{d[N_2O_5]}{dt} = \frac{1}{2}\frac{d[NO_2]}{dt} = 2\frac{d[O_2]}{dt}$$

The coefficients in this velocity equation are simply the reciprocals (apart from the signs) of the stoichiometric coefficients. There is no universal rule among kineticists as to which of these to use to express the velocity of the reaction quantitatively. Usually, the specific substance measured is used. In this book we shall adopt the convention that, for a reaction whose stoichiometric equation is written as

$$mM + nN \longrightarrow pP + qQ$$

the velocity of the reaction will be taken as

$$v = -\frac{1}{m}\frac{d[M]}{dt} = -\frac{1}{n}\frac{d[N]}{dt} = \frac{1}{p}\frac{d[P]}{dt} = \frac{1}{q}\frac{d[Q]}{dt}$$

and the rate-law expression

$$v = k[M]^i[N]^j \cdots$$

will serve to define the relation of the rate constant to the velocity of the reaction. (Remember that the exponents i and j in the rate law are not related to the stoichiometric coefficients.)

REACTION MECHANISMS AND RATE LAWS

The form of the rate law determined from the kinetics of a reaction does not tell us the actual mechanism, although it gives important clues. A very complex process, such as the growth and multiplication of bacterial cells, can exhibit simple first-order kinetics. Growth, as measured by the increase in the number of cells, occurs in proportion to the number of cells present at any given time and with a rate constant determined by the turnaround time for the complex events in the cell cycle. We cannot hope for this single parameter, the rate constant, to give us detailed information about the individual steps of DNA replication, transcription into RNA, synthesis of protein, mitotic division, and so on, that are involved in the cell cycle. What the rate law does do is to make a statement in a quantitative form about how fast the cells grow.

Typically, a complex reaction sequence or mechanism is made up of a set of elementary reactions. Consider, for example, the process by which the ozone layer of the upper atmosphere is maintained. This ozone layer acts as an important shield, absorbing potentially harmful ultraviolet (uv) radiation from the sun and protecting organisms on the surface of the earth from dangerous biological consequences. Ozone is formed from atmospheric oxygen by a process that can be described by the stoichiometric equation

$$3\,O_2 \longrightarrow 2\,O_3$$

This reaction does not occur as such in the atmosphere. Instead, ozone is formed as a consequence of an elementary photochemical step resulting from the absorption of short uv radiation followed by a second elementary reaction in which the oxygen atoms produced in the first step go on to react with O_2 to form ozone.

$$O_2 \xrightarrow{h\nu \text{ (below 242 nm)}} O + O$$

$$O + O_2 + M \longrightarrow O_3 + M$$

where M is a third body (any other gas molecule) whose function is to remove excess energy from the activated ozone initially formed. The protective role of ozone results from its absorption of somewhat longer-wavelength ultraviolet light not absorbed by O_2, and this radiation leads to the decomposition of ozone by the elementary reaction

$$O_3 \xrightarrow{h\nu \text{ (190 to 300 nm)}} O_2 + O$$

The balance in the ozone concentration is achieved by these three elementary steps operating together, supplemented by lesser contributions of additional reactions involving other elements.

Johnston (1971) has pointed out that the proposed flights of supersonic transports (SST's) in the upper atmosphere pose a threat to the stability of the protective ozone layer. This comes about because of the role of oxides of nitrogen (NO_x) in the exhaust gases of the SST. The oxides catalyze the decomposition of ozone:

$$O + O_3 \xrightarrow{NO_x} 2\,O_2$$

This process can be understood in terms of the elementary reactions involved, which are

$$NO + O_3 \longrightarrow NO_2 + O_2$$

$$NO_2 + O \longrightarrow NO + O_2$$

Not only does this show in molecular detail how the oxides of nitrogen produce the ozone decomposition, but it also shows that the NO and NO_2 are not used up in the process. Together they serve a true catalytic function.

It often happens that more than one mechanism can be written consistent with a given rate law. The correct one can be chosen only if we learn more about the intermediates, the energetics of the process, and the rates of elementary steps determined from other sources (different reaction systems). The concept of elementary reactions is important to chemical kinetics for the same reason that the study of chemical bonds or of functional groups is important in simplifying and organizing the chemistry of the hundreds of thousands of different organic compounds. The kinetic parameters of an elementary step are the same regardless of what overall reaction it is participating in.

Elementary reactions are described in terms of their *molecularity*: the number of reactant particles involved in the elementary reaction. Reactions between two partices are called *bimolecular*. They may be thought of in terms of a collision between the two reactants to form an *activated complex*. For example,

$$O_3 + NO \xrightarrow{\text{collision}} \left[O{=}\overset{+}{O}{-}O^{-}\cdots\overset{+}{N}{-}O^{-} \right] \longrightarrow O_2 + NO_2$$

Activated complex

the reaction of ozone with nitric oxide in the gas phase occurs as a simple bimolecular process. Since the process is bimolecular, it follows that it must exhibit second-order kinetics. The converse is not true, however; many kinetically second-order reactions do not have simple bimolecular mechanisms but are more complex. *Hence one cannot use the form of the rate law alone to predict the mechanism; whereas knowing the mechanism does permit the rate law to be deduced.* This is an important distinction to learn.

From the example given above we can define molecularity as the number of particles that come together to form the activated complex in an elementary reaction. Note that the term "molecularity" applies only to elementary reactions and not to the overall mechanism. *Unimolecular* reactions are those that involve only a single reactant particle, for example radioactive decay. True unimolecular processes are rare, because most reactions require some *activation energy* in order to proceed. This energy is commonly picked up by way of collisions with the surroundings (other molecules, walls of the vessel). *Termolecular* reactions involve collisions of three molecules and are more likely to occur at high pressures or in liquid solution.

The concept of molecularity was developed largely for gas-phase reactions. It is a difficult concept to carry over to liquid solutions, where the role of the solvent and other closely associated molecules in the medium makes it difficult to define or determine the reacting "particle." We will, nevertheless, make considerable use of the idea despite this ambiguity.

The simplest examples of reaction mechanisms are those that involve only a single elementary reaction. In this case the rate law can be written by inspection. The velocity is proportional to the concentration of each of the species involved in forming the activated complex, and the exponents are determined by the participating number of each kind of particle. For example, if the reaction

$$2A + B \longrightarrow \text{products}$$

is an *elementary* reaction, its velocity will be given by

$$v = k[A]^2[B]$$

The real problem occurs when the overall reaction is complex and consists of several connected elementary reactions. The overwhelming majority of reactions and processes in nature are complex, and it is a relatively rare one that turns out to be elementary.

To account for the kinetic observations, it is useful to formulate ("hypothesize") the mechanisms in terms of a set of elementary reactions. These often involve one or more intermediates, which may have been observed previously or may be postulated uniquely for this reaction. This proposed mechanism can then be used to predict a rate law that can be tested against experimental findings. Furthermore, one can look for any postulated intermediates by standard physical or chemical methods.

A few examples will illustrate some important methods of dealing with complex reactions. One general method is to see if one elementary reaction controls the overall kinetics, that is, to look for the *rate-determining step.* For certain mechanisms one elementary reaction is found to dominate the kinetics of the complex reaction. When this is so, the analysis of the mechanism is greatly simplified. We can write the rate for the rate-determining step and equate it to the overall rate. In the following examples we shall discuss the general case first and then indicate the simplification that occurs if there is a rate-determining step.

Parallel Reactions

Parallel reactions are of the following type:

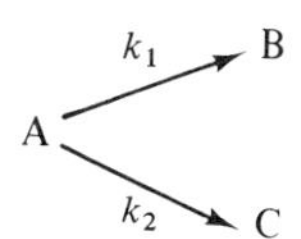

The substance A can decompose by either of two paths, giving rise to different products, B and C. The rate expressions are

$$-\frac{d[\mathrm{A}]}{dt} = k_1[\mathrm{A}] + k_2[\mathrm{A}] = (k_1 + k_2)[\mathrm{A}] \tag{7.38}$$

$$\frac{d[\mathrm{B}]}{dt} = k_1[\mathrm{A}] \tag{7.39}$$

$$\frac{d[\mathrm{C}]}{dt} = k_2[\mathrm{A}] \tag{7.40}$$

The solution to the first equation, which has the *form* of the first-order rate law, Eq. (7.17), is

$$\ln\frac{[\mathrm{A}]}{[\mathrm{A}]_0} = -(k_1 + k_2)t \tag{7.41}$$

$$[\mathrm{A}] = [\mathrm{A}]_0 \exp[-(k_1 + k_2)t] \tag{7.42}$$

Thus [A] decays exponentially as illustrated in Fig. 7.7. To find out how [B] and [C] increase with time, we need to solve Eqs. (7.39) and (7.40). This appears to present a problem, because each of these equations contains three variables; it cannot be solved as it stands. However, we can substitute for A using Eq. (7.42). Thus,

$$\frac{d[\mathrm{B}]}{dt} = k_1[\mathrm{A}] = k_1[\mathrm{A}]_0 \exp[-(k_1 + k_2)t] \tag{7.43}$$

and

$$\frac{d[\mathrm{C}]}{dt} = k_2[\mathrm{A}] = k_2[\mathrm{A}]_0 \exp[-(k_1 + k_2)t] \tag{7.44}$$

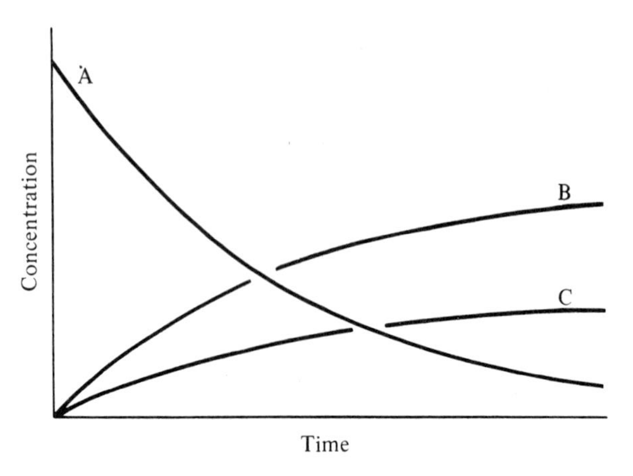

Fig. 7.7 Kinetic curves for reactant and products for two first-order reactions in parallel.

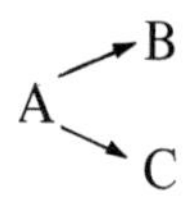

We now separate variables and integrate. If $[B]_0 = [C]_0 = 0$, the solutions are

$$[B] = \frac{k_1[A]_0}{k_1 + k_2}\{1 - \exp[-(k_1 + k_2)t]\} \tag{7.45}$$

$$[C] = \frac{k_2[A]_0}{k_1 + k_2}\{1 - \exp[-(k_1 + k_2)t]\} \tag{7.46}$$

Note that B and C are formed in a constant ratio, $[B]/[C] = k_1/k_2$ throughout the course of the reaction. The extension of this method to three or more parallel reactions is straightforward.

Example 7.3 Radioactive decay frequently occurs simultaneously by two separate first-order pathways. For example, $^{64}Cu_{29}$ is unstable against decay, giving either an electron or a positron. The scheme can be written as follows:

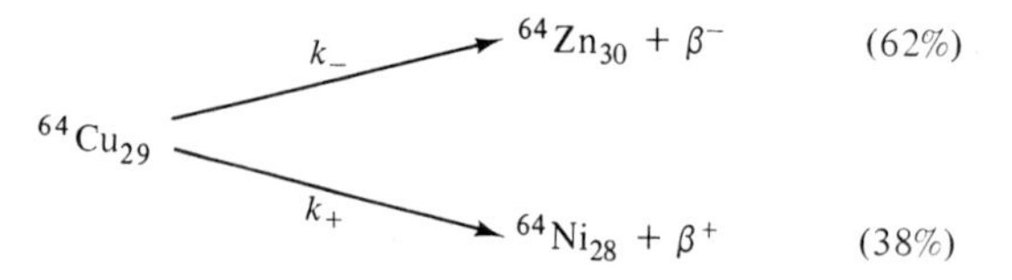

The decay of $^{64}Cu_{29}$ occurs exponentially with a single half-life of 12.80 h. Knowing the fraction of decays by each pathway then permits one to calculate k_+ and k_-. Do this for the example.

In parallel reactions, if one step is much faster than the others, this fast step dominates the reaction. In the example above, if 99% of the reaction goes through the β^- path, we can ignore the β^+ path in calculating the decay of ^{64}Cu. We can see this quantitatively in Eq. (7.42). When $k_1 \gg k_2$, the decay of [A] with time depends only on k_1. $[A] = [A]_0 \exp(-k_1 t)$.

Series Reactions (First Order)

Series reactions are of the type

$$\mathrm{A} \xrightarrow{k_1} \mathrm{B} \xrightarrow{k_2} \mathrm{C}$$

Compound A reacts to form B, which then goes on to form C. The rate expressions are

$$v_1 = -\frac{d[\mathrm{A}]}{dt} = k_1[\mathrm{A}]$$

$$[\mathrm{A}] = [\mathrm{A}]_0 \exp(-k_1 t) \tag{7.47}$$

$$\frac{d[\mathrm{B}]}{dt} = k_1[\mathrm{A}] - k_2[\mathrm{B}] = k_1[\mathrm{A}]_0 \exp(-k_1 t) - k_2[\mathrm{B}] \tag{7.48}$$

This differential equation is slightly more difficult to solve. We shall give the result. If $[\mathrm{B}]_0 = 0$,

$$[\mathrm{B}] = \frac{k_1[\mathrm{A}]_0}{k_2 - k_1}[\exp(-k_1 t) - \exp(-k_2 t)] \tag{7.49}$$

Similarly,

$$v_2 = \frac{d[\mathrm{C}]}{dt} = k_2[\mathrm{B}]$$

$$= \frac{k_1 k_2[\mathrm{A}]_0}{k_2 - k_1}[\exp(-k_1 t) - \exp(-k_2 t)] \tag{7.50}$$

Variables [C] and t can be separated here to give, for $[\mathrm{C}]_0 = 0$,

$$[\mathrm{C}] = [\mathrm{A}]_0\left\{1 - \frac{1}{k_2 - k_1}[k_2 \exp(-k_1 t) - k_1 \exp(-k_2 t)]\right\} \tag{7.51}$$

Where such complex formulas are involved, it is important to get some intuitive feeling for how these reactions behave. The graph in Fig. 7.8 illustrates the pattern for series first-order reactions. As the concentration of A decreases, it does so in a simple exponential fashion. If there is no B present initially, the rate of the second reaction will be negligible at the beginning, $[v_2]_0 = 0$. As B is formed by the first step, its concentration rises, reaches a maximum, and then declines to zero as the second step removes it from the reaction mixture. In accordance with the rate law, the curve for [C] starts out with zero slope, where $[v_2]_0 = 0$. The rising portion of curve [C] then undergoes an inflection (the point where $d^2[\mathrm{C}]/dt^2 = 0$) when [B] has its maximum value. Finally, [C] approaches a limiting value as the reaction nears completion. These properties are useful diagnostics for series reactions.

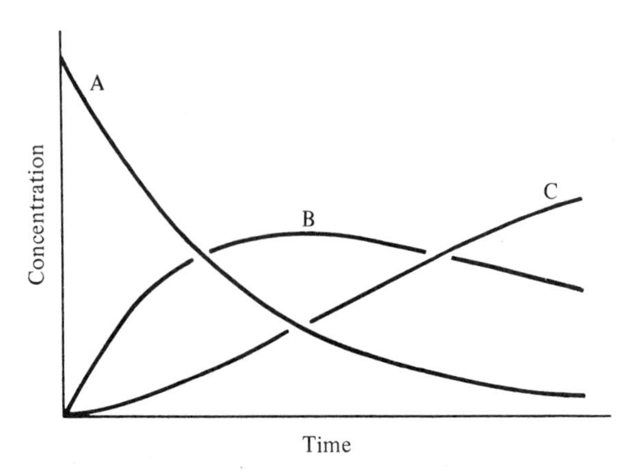

Fig. 7.8 Kinetic curves for reactant, A, intermediate, B, and product, C, for two first-order reactions in series. $A \rightarrow B \rightarrow C$.

When two or more reactions occur in series, the slow reaction is the *rate-determining step.* It dominates in the kinetic control of the overall process. The analogy with a city water distribution system is entirely appropriate. It does not matter how large the reservoir, water mains, and distribution pipes are. If the smallest cross section of the whole system is the $\frac{1}{2}$-inch pipe coming into your house, that will limit the maximum possible flow, and the only practical way to increase the flow is to replace the small pipe with one of larger diameter.

For the series reactions

$$A \xrightarrow{k_1} B \xrightarrow{k_2} C$$

we can consider the two cases $k_1 \gg k_2$ and $k_1 \ll k_2$. The first situation means that the first reaction is much faster than the second, and the second is the rate-determining step. Once the reaction is started, A will be converted to B rapidly in comparison with the subsequent reaction of B to C. During most of the course of the reaction, B undergoes a first-order conversion to C which is controlled by the rate constant k_2. If the formation of C is our measure of the velocity of the reaction, its appearance is nearly identical to that of a single-step (elementary) first-order reaction. This is illustrated schematically in Fig. 7.9. We arrive at the same conclusion from examination of Eq. (7.51) in the limit where $k_1 \gg k_2$. Under those conditions, Eq. (7.51) reduces to

$$[C] = [A]_0[1 - \exp(-k_2 t)]$$

for $k_1 \gg k_2$, which has the same form as the simple first-order formation kinetics.

The second situation, where $k_1 \ll k_2$, starts out with a slow conversion of A to B and is followed by a very rapid reaction for B going on to C. In this case the concentration of B will remain low throughout the course of the reaction, and C will appear essentially as A disappears. This is illustrated in Fig. 7.10. It should be obvious that the velocity of the reaction can be measured either by the

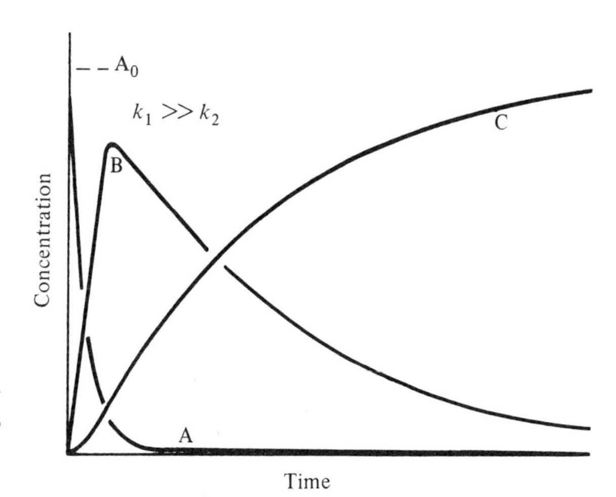

Fig. 7.9 Two first-order reactions in series, with the *second* as the rate-determining step. $A \rightarrow B \rightarrow C$.

formation of C or by the disappearance of A, in this case. Thus Eq. (7.47) applies to the overall reaction.

The chief reason for emphasizing this point is because complex processes in biology frequently have rate-limiting steps. For our analysis, it does not matter how many steps are involved in the series mechanism; all that is required is that one step should be appreciably slower than any of the others. If the rate-limiting step is not first order, as in the example treated here, it must be treated using the kinetic equations for the appropriate order.

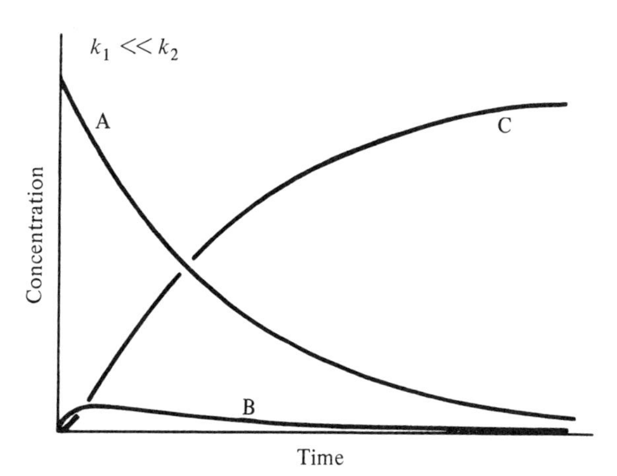

Fig. 7.10 Two first-order reactions in series, with the *first* as the rate-determining step. $A \rightarrow B \rightarrow C$.

Example 7.4 A complex but illustrative example of the kinetic behavior of series reactions occurs in the clotting of blood. The final step in the process is the conversion of fibrinogen to fibrin (the clotting material) catalyzed by the proteolytic (protein cutting) enzyme thrombin. Thrombin is not normally present in the blood, but is converted from a precursor, prothrombin, by the action of another activated proteolytic enzyme and calcium ion. The overall process involves a cascade of at least eight such steps, as shown.

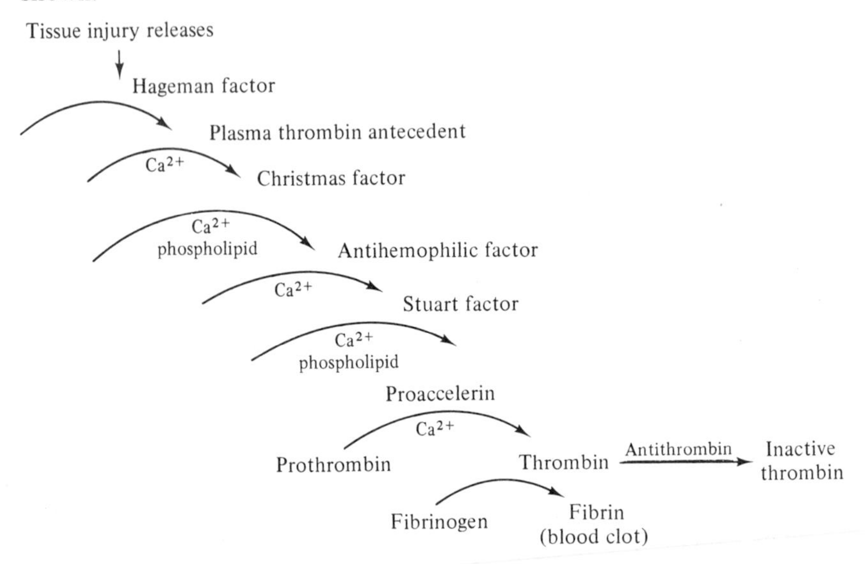

Blood clotting mechanism: sequence of stages in cascade process

A consequence of this cascade of steps is the fact that the thrombin concentration, as measured by the decrease in clotting time for blood samples, increases dramatically following wounding and then decreases again once the clot is formed. The behavior, shown in Fig. 7.11, is very much like that of an intermediate in a series of reactions. In the mechanism outlined above, the injury sets off the cascade of reactions involving a series of proteolytic enzymes, each of which activates a target protein. This cascade phenomenon serves to accelerate the mobilization of the clotting mechanism so that it can be switched from fully off to fully on in less than 1 min. The conversion of prothrombin to thrombin,

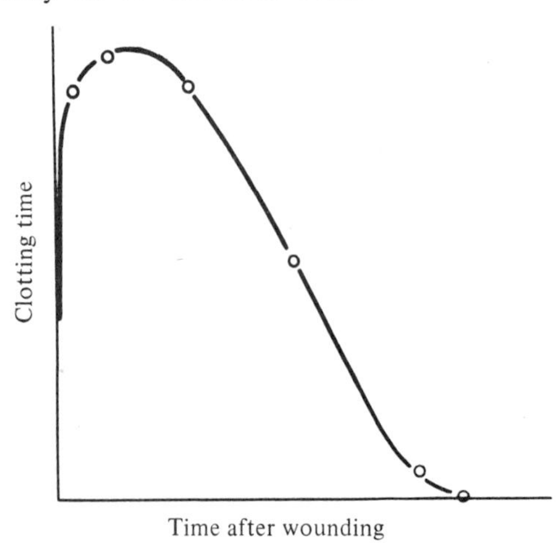

Fig. 7.11 Concentration of active thrombin after wounding. The clotting time for the conversion of fibrinogen to fibrin is a direct measure of the thrombin concentration. [From R. Biggs, *Analyst 78*, 84 (1953).]

catalyzed by active proaccelerin and Ca^{2+}, is the immediate cause of clotting. The level of thrombin in the blood must not remain high, however, or it will cause fibrin to form in the circulatory system and block the normal flow of blood. Antithrombin is an additional factor that inactivates thrombin, usually within a few minutes after it has reached its maximum level. In this way the action of thrombin is restricted to the period when it is critically needed.

While this overall process is much more elaborate than the simple example of series reactions considered above, a similarity is seen when we inspect the formation and disappearance of thrombin as rate-limiting steps. For this purpose, prothrombin plays the role of A, thrombin is B, and inactive thrombin is C. The appearance of active proaccelerin turns on the first step ($A \rightarrow B$), and the subsquent formation of antithrombin turns on the second step ($B \rightarrow C$). The rate of clotting of fibrinogen serves (kinetically speaking) as a simple indicator of the level of thrombin present at any time in the course of this process.

Equilibrium and Kinetics

All reactions approach equilibrium. Therefore, in principle, for every forward reaction step, there is a backward step. In practice, we can sometimes ignore the backward step, because either the equilibrium constant is very large or the concentrations of products are kept very small. Nevertheless, it is important to know the general relation between kinetic rate constants (k) and thermodynamic equilibrium constants (K).

Let us consider an elementary first-order reversible reaction.

$$\mathrm{A} \underset{k_{-1}}{\overset{k_1}{\rightleftharpoons}} \mathrm{B}$$

The rate of disappearance of A is

$$-\frac{d[\mathrm{A}]}{dt} = k_1[\mathrm{A}] - k_{-1}[\mathrm{B}]$$

At equilibrium, $-d[\mathrm{A}]/dt = 0$; therefore,

$$\frac{[\mathrm{B}]^{\mathrm{eq}}}{[\mathrm{A}]^{\mathrm{eq}}} = \frac{k_1}{k_{-1}} = K$$

Exercise Show that for the elementary first-order reversible reaction

$$\mathrm{A} \underset{k_{-1}}{\overset{k_1}{\rightleftharpoons}} \mathrm{B}$$

$$\ln\frac{[\mathrm{A}]-[\mathrm{A}]^{\mathrm{eq}}}{[\mathrm{A}]_0-[\mathrm{A}]^{\mathrm{eq}}} = \ln\frac{[\mathrm{B}]-[\mathrm{B}]^{\mathrm{eq}}}{[\mathrm{B}]_0-[\mathrm{B}]^{\mathrm{eq}}} = -(k_1 + k_{-1})t$$

where $[\mathrm{A}]_0$ and $[\mathrm{B}]_0$ are the concentrations of A and B respectively at $t = 0$.

Often there is more than one path for the reaction of A to form B. To be consistent with the principles of equilibrium thermodynamics, we must apply the principle of *microscopic reversibility*. This states that if A can react to form B by two or more different paths, we cannot have a mechanism by which A → B occurs only by one path and B → A occurs only by another. That is, the following mechanism is *not* possible:

Each elementary step in the reaction must be reversible to be consistent with the principle of microscopic reversibility. Thus we must formulate the mechanisms as follows:

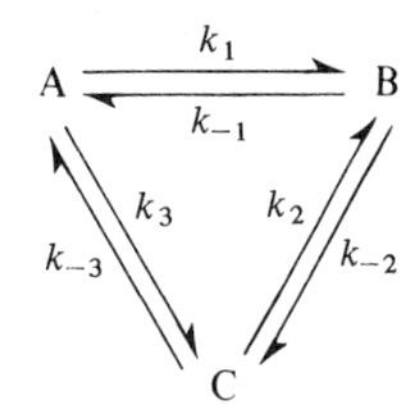

The relation to thermodynamics requires further that

$$K = \frac{[\mathrm{B}]^{\mathrm{eq}}}{[\mathrm{A}]^{\mathrm{eq}}} = \frac{k_1}{k_{-1}} = \frac{k_2 k_3}{k_{-2} k_{-3}} \tag{7.52}$$

The six rate constants are therefore not independent.

Complex Reactions

For many reactions, particularly those involving enzymes as catalysts, a series of reversible steps is involved. An example is the following set of coupled elementary reactions:

$$\mathrm{A} + \mathrm{B} \underset{k_{-1}}{\overset{k_1}{\rightleftharpoons}} \mathrm{X}$$

$$\mathrm{X} \underset{k_{-2}}{\overset{k_2}{\rightleftharpoons}} \mathrm{P} + \mathrm{Q}$$

In this case the exact solution to the rate equations is complex, because two of the elementary reactions are bimolecular and a total of five molecular species are involved. It is useful to learn some approximations that can be applied in cases like this.

Initial-velocity approximation

At the beginning of a reaction the product concentrations are usually very small or zero. For the set of opposed reactions above, the step designated −2 can be neglected in an analysis of *initial* velocities. This simplifies the kinetic expressions considerably. As the extent of reaction increases, the experimental results will begin to differ from the prediction of the approximate theory.

Prior equilibrium approximation

This approach is valuable when steps 1 and −1 are rapid in comparison with step 2.

$$\mathrm{A} + \mathrm{B} \underset{k_{-1}}{\overset{k_1}{\rightleftharpoons}} \mathrm{X} \qquad \text{(fast, equilibrium)}$$

$$\mathrm{X} \xrightarrow{k_2} \mathrm{P} + \mathrm{Q} \qquad \text{(slow)}$$

A, B, and X rapidly attain a state of quasi-equilibrium, such that

$$v_1 = v_{-1}$$

and

$$k_1[\mathrm{A}][\mathrm{B}] = k_{-1}[\mathrm{X}]$$

We can write this as an equilibrium expression:

$$K = \frac{k_1}{k_{-1}} = \frac{[\mathrm{X}]}{[\mathrm{A}][\mathrm{B}]}$$

Step 2 is the rate-limiting step, and the rate of formation of product is given by

$$v = \frac{d[\mathrm{P}]}{dt} = \frac{d[\mathrm{Q}]}{dt} = k_2\mathrm{X} \tag{7.53}$$

Substituting from above, we see that the velocity

$$v = \frac{k_2 k_1}{k_{-1}}[\mathrm{A}][\mathrm{B}] \tag{7.54}$$

can be expressed in terms of reactant concentrations only. This is particularly valuable if the relative concentration of X remains very small ($k_{-1} \gg k_1$), but it is just as valid when X is fairly large ($k_{-1} \sim k_1$). The important criterion for the prior equilibrium approximation to apply can be written

$$v \cong v_2 \ll v_1 \cong v_{-1} \tag{7.55}$$

This should be read: "The overall velocity of the reaction is limited by the slow step 2, and the velocity is much slower than the forward and reverse reactions of step 1, which are essentially in equilibrium."

Steady-state approximation

It frequently happens that an intermediate is formed that is very reactive. As a consequence, it never builds up to any significant concentration during the course of the reaction.

$$A + B \xrightarrow{k_1} X \qquad \text{(slow)}$$

$$X + D \xrightarrow{k_2} P \qquad \text{(fast)}$$

To a first approximation, X reacts as rapidly as it is formed:

$$v_1 = v_2$$

$$k_1[A][B] = k_2[X][D] \tag{7.56}$$

Hence,

$$v = \frac{d[P]}{dt} = k_1[A][B] \tag{7.57}$$

in terms of reactant concentrations only. A more general way of formulating the steady-state approximation is to consider all of the steps involving formation and disappearance of the reactive intermediate and to set the sum of their rates equal to zero. In this example X is formed in step 1 and it disappears in step 2. Thus

$$\frac{d[X]}{dt} = k_1[A][B] - k_2[X][D] \cong 0$$

This gives the same result [Eq. (7.56)]. However, this second approach is more general. The significance of this equation is not that (X) is constant during the course of the reaction. Such is never the case. It is true, however, that the *slope*, $d[X]/dt$, of the curve of [X] versus time is much smaller than those typical of reactants or products. For example, this can be seen for component B for the scheme shown in Fig. 7.10. It is this nearly zero slope throughout the course of the reaction that is the characteristic of the steady-state condition.

The steady-state approximation can be applied in many chemical and biochemical reactions. Care must be taken that the intermediate satisfies the criterion that $d[X]/dt \cong 0$. Note that this is not necessarily the case where prior equilibrium is involved. It is also not true for the intermediate B illustrated in Fig. 7.8.

Deducing a Mechanism from Kinetic Data

There is no straightforward way to obtain a mechanism from kinetic data. There is always more than one mechanism which will be consistent with the kinetic data. Therefore, why propose a mechanism at all? The answer is that we like to think that we understand a reaction and that we can guess what the molecules are doing. A mechanism gives us some basis for predicting what should happen for other reactions.

The best way to illustrate how one hypothesizes mechanisms is to discuss specific examples. The only general advice that can be given is to think of a simple and plausible mechanism and then calculate the kinetics and see if they are consistent with the data. A unique rate equation can always be obtained from a proposed mechanism. If the rate law is simple first or second order, assume a unimolecular or bimolecular rate-determining step. Then, if necessary, add other steps to agree with the stoichiometry. For example, suppose that for the stoichiometric reaction

$$A + B \longrightarrow C$$

the rate law is found to be

$$-\frac{d[A]}{dt} = k[A][OH^-]$$

We can assume for a mechanism

$$A + OH^- \xrightarrow{k_1} M^-$$

$$M^- + B \xrightarrow{k_2} C + OH^-$$

The first elementary reaction postulates that $A + OH^-$ react to form an intermediate, M^-; this gives the correct rate law. The second elementary reaction must be added to account for the stoichiometry. From the proposed mechanism we can now predict rate laws for [B] and [C] and experimentally check the predictions. These rate laws are complicated in the general case, so we shall not present them. Another mechanism consistent with the data is

$$A + OH^- \underset{}{\overset{K}{\rightleftharpoons}} N^- + H_2O \qquad \text{(fast to equilibrium)}$$

$$N^- \xrightarrow{k_1} P^- \qquad \text{(slow, rate determining)}$$

$$P^- + B \xrightarrow{k_2} C + OH^- \qquad \text{(fast)}$$

The rate-determining step is

$$-\frac{d[N^-]}{dt} = k_1[N^-]$$

but

$$\frac{[N^-]}{[A][OH^-]} = K$$

Therefore,

$$-\frac{d[N^-]}{dt} = k_1K[A][OH^-]$$

Because all the other elementary reactions are fast, each time N^- reacts to form P^-, A and B react and C is formed. The rate laws are then

$$-\frac{d[\mathrm{B}]}{dt} = \frac{d[\mathrm{C}]}{dt} = k_1 K[\mathrm{A}][\mathrm{OH}^-]$$

To decide between the two mechanisms, other kinetic experiments can be done and attempts to detect the postulated intermediates can be made. Many other mechanisms could also be written. It should be clear that proposing reasonable mechanisms requires practice and experience.

TEMPERATURE DEPENDENCE

It seems almost intuitive that if you want a reaction to proceed faster, you should heat it. It came as a surprise when some chemical reactions were found to have negative temperature coefficients; that is, the reaction went *slower* at higher temperature. In biology this is a common occurrence. Few organisms or biochemical processes can withstand temperatures above 50°C for very long in the first place, but this results from irreversible structural changes that alter the mechanisms or destroy the organism. Processes like the coagulation of the protein egg albumin (egg white) occur during cooking, and they occur more rapidly at high temperatures. This seems normal. If, however, a protein is first carefully denatured, it can be renatured more rapidly at a lower temperature than at a higher one! There is a kinetic effect involved in addition to the shift in equilibrium constant for this process.

Observations of Arrhenius and others showed that for most chemical reactions there is a logarithmic relation between the rate constant and the reciprocal of the temperature in degrees Kelvin:

$$\ln k = -\left(\frac{E_a}{RT}\right) + \text{constant} \tag{7.58}$$

The proportionality constant E_a is known as the *activation energy* for the reaction; alternatively, E_a/R can be thought of as the "temperature" at which the reaction occurs at 37% ($1/e$) of its maximum possible rate. To interpret the Arrhenius expression further, let us write it in the form

$$k = A \exp\left(-\frac{E_a}{RT}\right) \tag{7.59}$$

where A is called the preexponential factor. This is an empirical equation which represents many chemical systems and even some rather complex biological processes (see Fig. 7.12). The usual way of obtaining the activation energy is to plot $\log k$ versus $1/T$. The slope of the line is equal to $-E_a/2.303R$, with the units of E_a dependent on the value of R. Once E_a is known, Eq. (7.59) can be solved for A. An alternative method of obtaining E_a and A is illustrated below.

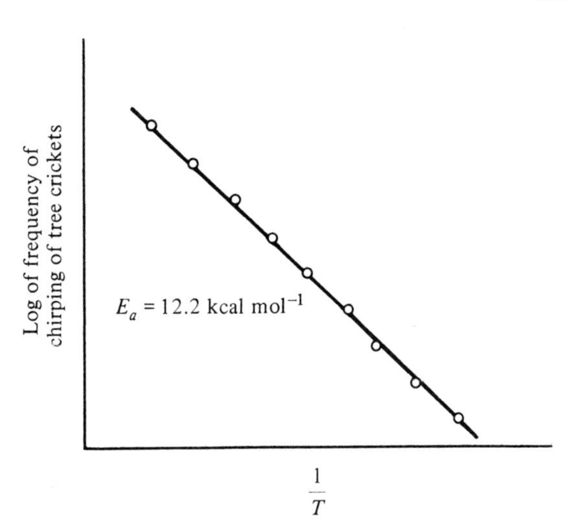

Fig. 7.12 Arrhenius plot of the chirping rate of tree crickets at different temperatures. [From K. J. Laidler, *J. Chem. Educ.* *49*, 343 (1972).]

Example 7.5 The decomposition of urea in 0.1 M HCl occurs according to the reaction

$$NH_2CONH_2 + 2\,H_2O \longrightarrow 2\,NH_4^+ + CO_3^{2-}$$

The first-order rate constant for this reaction was measured as a function of temperature, with the following results:

Expt.	Temperature (°C)	k (min^{-1})
1	61.05	0.713×10^{-5}
2	71.25	2.77×10^{-5}

Calculate the activation energy, E_a, and the A factor in Eq. (7.59) for this reaction.

Solution This problem can be solved either numerically or graphically. The numerical solution is given here. Since $\log k = -E_a/2.30RT + \log A$, we will first calculate $\log k$ and $1/T$ for each experiment.

Expt.	$\log k$	$\frac{1}{T}$ (K)$^{-1}$
1	−5.147	2.992×10^{-3}
2	−4.558	2.904×10^{-3}

We can write

$$\log k_2 - \log k_1 = -\frac{E_a}{2.30R} \cdot \left(\frac{1}{T_2} - \frac{1}{T_1}\right)$$

Combining experiments 1 and 2 in this way, we obtain

$$E_a = -\frac{2.30R \cdot (\log k_2 - \log k_1)}{1/T_2 - 1/T_1}$$

$$= -\frac{2.30(1.99 \text{ cal deg}^{-1} \text{ mol}^{-1})(-4.558 + 5.147)}{(2.904 - 2.992) \times 10^{-3} \text{ deg}}$$

$$= 30.6 \text{ kcal mol}^{-1}$$

Using the data from either experiment, we can calculate a value for A from

$$\log A = \log k + \frac{E_a}{2.30RT}$$

Using the data of experiment 2,

$$\log A = -4.558 + \frac{30{,}600}{(2.30)(1.99)(344.4)}$$

$$= -4.558 + 19.412 = 14.854$$

Therefore,

$$A = 7.1 \times 10^{14} \text{ min}^{-1}$$

$$= 1.2 \times 10^{13} \text{ s}^{-1}$$

To understand the meaning of the activation energy, consider the elementary reaction M + N → P with an energy barrier between reactants and products as illustrated in Fig. 7.13. In this figure we plot the energy versus the *reaction*

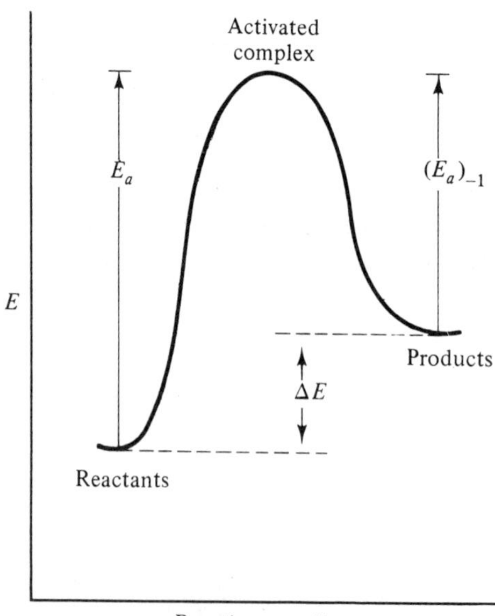

Fig. 7.13 Energy changes that occur throughout the course of a hypothetical reaction. The reaction coordinate is a fictitious variable that measures the progress from reactants to activated complex to product. E_a is the activation energy for the forward reaction and $(E_a)_{-1}$ is the activation energy for the reverse reaction.

coordinate. The reaction coordinate is a convenient way to represent the change of the reactants into products as a reaction takes place. It is not a simple coordinate, such as the x coordinate for a point in space; instead, it can represent the positions of all the significant atoms in the reacting molecules. For example, in the reaction

$$OH^- + CH_3Br \longrightarrow CH_3OH + Br^-$$

the reaction coordinate represents the concerted motions of the OH^- approaching the central carbon, while the Br^- is leaving and the three hydrogen atoms are also moving. We can plot any property versus the reaction coordinate, but energy or free energy is particularly interesting.

The molecules M and N must together have sufficient energy to react. Because the process must be reversible at the microscopic level, the same barrier will be present for the reverse reaction. The energy of the products of a reaction is usually different from that of the reactants

$$\Delta E = E_P - (E_M + E_N)$$

Therefore the activation energy for the reverse reaction is usually not the same as for the forward reation. The relation

$$(E_a)_{\text{forward}} - (E_a)_{\text{reverse}} = \Delta E \tag{7.60}$$

does hold, however, at least for elementary reactions.

As the temperature is increased, the *fraction* of molecules having energy greater than E_a increases, and the rate of reaction correspondingly increases. In some cases, correlations can be made between the activation energy and the energy of a bond that must be broken in order for a reaction to proceed.

Very detailed theories of chemical kinetics exist, or are being developed, which can predict rate constants for simple, gas-phase reactions. We shall discuss more general theories which are useful mainly in helping us understand the factors that affect the kinetics.

In the *collision theory* of chemical kinetics we assume that, to react, molecules must collide (including collisions at the walls of the vessel in many cases). The rate of a reaction whose mechanism is

$$M + N \longrightarrow P$$

is given by

$$v = k[M][N] = A[M][N] \exp\left(-\frac{E_a}{RT}\right) \tag{7.61}$$

At high temperatures the exponential approaches unity, and $k \cong A$:

$$\lim_{T\to\infty} A \exp\left(-\frac{E_a}{RT}\right) = A$$

The maximum rate of this reaction at high temperature would be $A[M][N]$, which is therefore interpretable as a collision frequency for the molecules M and N. The exponential factor resembles a Boltzmann probability distribution (see Chapter 11). It can be considered to represent the fraction of molecules that have sufficient energy to react.

Examined in detail, the *collision theory* has serious shortcomings. Collision frequencies of molecules in the gas phase can be calculated quite well using the kinetic theory of gases (see Chapter 6). From Eq. (7.61) theoretical values for the rate constant of a bimolecular gas-phase reaction can be calculated. These are on the order of 10^{11} $\text{mol}^{-1}\,\ell\,\text{s}^{-1}$ for small molecules, somewhat dependent on the molecular radii. Experimentally, values of A range widely, from 10^2 to 10^{13} $\text{mol}^{-1}\,\ell\,\text{s}^{-1}$ for reactions that are thought to be relatively simple. While there are few reactions that proceed much faster than at collision rates, there are many that are orders of magnitude slower. Attempts to correct this by introducing "steric factors" are simply assigning an adjustable parameter to our ignorance.

The situation is worse for liquid-phase reactions, where the nature of collisions is harder to describe or treat theoretically. Furthermore, it is impossible to interpret an inverse temperature dependence (negative activation energy) in a meaningful way by collision theory. The difficulty here arises from the fact that collision theory assumes that essentially only the molecular kinetic energy (translation, vibration, rotation) is involved in the activation process. There is no rational way of handling configurational or entropic contributions in the collision theory framework.

Transition-state Theory

An alternative model was introduced in 1935 by Eyring and others. Its significant new feature was that it considered the reactive intermediate or *transition state* between reactants and products to be a real chemical molecule, even if only of fleeting lifetime. In the transition-state theory, every elementary reaction

$$\text{M} + \text{N} \underset{k_-}{\overset{k}{\rightleftharpoons}} \text{P} + \text{Q}$$

is rewritten as

$$\text{M} + \text{N} \underset{k_{-1}^{\ddagger}}{\overset{k_1^{\ddagger}}{\rightleftharpoons}} [\text{MN}]^{\ddagger} \underset{k_{-2}^{\ddagger}}{\overset{k_2^{\ddagger}}{\rightleftharpoons}} \text{P} + \text{Q} \tag{7.62}$$

It is assumed that reactants are in rapid equilibrium with $[\text{MN}]^{\ddagger}$, thus, the initial rate of the reaction to form products is

$$v = k_2^{\ddagger}[\text{MN}]^{\ddagger} \tag{7.63}$$

$[\text{MN}]^{\ddagger}$ is the concentration of the transient intermediate that occurs at the top of the barrier; but, as we shall see, the barrier here is one in ΔG rather then E. It should be clear that $k_1^{\ddagger}$, $k_2^{\ddagger}$, and so on, are not directly measurable; $[\text{MN}]^{\ddagger}$ is not a detectable intermediate.

At the pass, the summit of the barrier, the transition-state species has about equal probability of going forward to form products or backward again to reactants. The transition-state species is considered to be in equilibrium with the reactants. This is expressed in terms of an equilibrium constant of activation

$$K^{\ddagger} = \frac{[\mathrm{MN}]^{\ddagger}}{[\mathrm{M}][\mathrm{N}]} \tag{7.64}$$

and an associated free energy of activation

$$\Delta G^{\ddagger} = -RT \ln K^{\ddagger} \tag{7.65}$$

A similar argument can be applied to the overall reverse process, and these are illustrated together in Fig. 7.14. Before considering the consequences of the fact

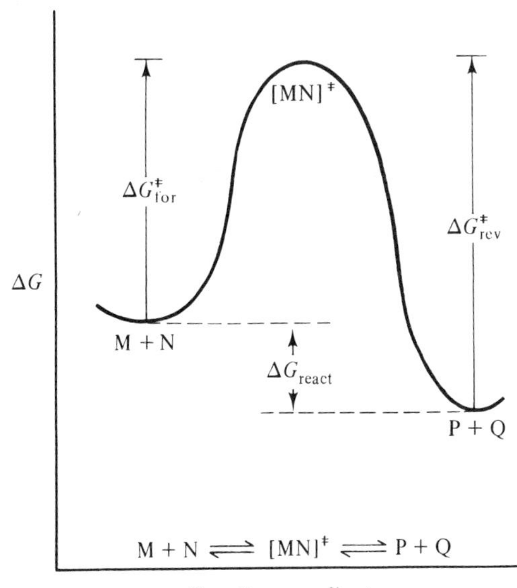

Fig. 7.14 Progress of a reaction according to the transition-state theory.

that ΔG is now involved instead of E, we need first to estimate values of $k_2^{\ddagger}$ (or $k_{-1}^{\ddagger}$ for the reverse reaction). This is simply the rate of dissociation of the transition-state species once it is formed—roughly the inverse of the lifetime of $[\mathrm{MN}]^{\ddagger}$. This species is at the top of the barrier (therefore not a stable species); hence, it is thought to be held together only by a very weak link or bond, $[\mathrm{M}\cdots\mathrm{N}]^{\ddagger}$, which we can associate with the reaction coordinate. In fact, the bond is so weak that it flies apart essentially during the first vibration. We assume that the energy of the vibration, $h\nu$, is sufficiently low that we can equate it to its classical value $h\nu = kT$, where h is Planck's constant, k is the Boltzmann constant,* and the classical energy of a vibrational degree of freedom is kT. Then

* In this section only, we shall use k for the Boltzmann constant to distinguish it from rate constants.

we can solve for the "frequency" ν of the vibration, and hence the dissociation rate constant. Thus,

$$\nu = \frac{kT}{h} = k_2^{\ddagger} = k_{-1}^{\ddagger} \tag{7.66}$$

This is not a rigorous derivation, but the value of ν at a temperature of 298 K of $6 \times 10^{12}\ \mathrm{s}^{-1}$ is reasonable for the vibrating analog of $[\mathrm{M}\cdots\mathrm{N}]^{+}$ dissociation that we are considering here. The rate of the reaction can now be expressed as

$$v = \frac{d[\mathrm{P}]}{dt} = \frac{kT}{h}[\mathrm{MN}]^{+} = \frac{kT}{h}K^{+}[\mathrm{M}][\mathrm{N}] = k[\mathrm{M}][\mathrm{N}] \tag{7.67}$$

Therefore, the rate constant for the overall reaction is

$$k = \frac{kT}{h}K^{+} = \frac{kT}{h}\exp\left(-\frac{\Delta \mathrm{G}^{+}}{\mathrm{RT}}\right) \tag{7.68}$$

This equation permits a calculation of free energy of activation, ΔG^{+}, from a measured rate constant. The rate constant must have time units of seconds; the concentration units (usually mol ℓ^{-1}) specify the standard state. Using $k = 1.380 \times 10^{-16}$ erg deg^{-1} and $h = 6.626 \times 10^{-27}$ erg s^{-1} gives $kh^{-1} = 2.083 \times 10^{10}$ deg s^{-1}. Additional insight comes if we consider the distinction between energy (enthalpy) and entropy contributions. At constant temperature,

$$\Delta G^{+} = \Delta H^{+} - T\,\Delta S^{+} \tag{7.69}$$

where ΔH^{+} and ΔS^{+} are the activation enthalpy and entropy, respectively. Substituting into Eq. (7.68),

$$k = \frac{kT}{h}\exp\left(\frac{\Delta S^{+}}{R}\right)\exp\left(-\frac{\Delta H^{+}}{RT}\right) \tag{7.70}$$

Transition-state theory predicts a slightly different temperature dependence from that of Arrhenius. However, the main temperature dependence is the exponential term. One can show that

$$\Delta H^{+} = E_a - RT \tag{7.71}$$

Often E_a is large compared to RT and we can use ΔH^{+} and E_a interchangeably. To obtain E_a and ΔH^{+} we usually plot ln k versus $1/T$ to obtain E_a [Eq. (7.59)], then Eq. (7.71) gives ΔH^{+}. Substituting Eq. (7.71) into Eq. (7.70), we obtain ΔS^{+}, from the Arrhenius preexponential term A.

$$\Delta S^{+} = R \ln \frac{Ah}{kTe} \tag{7.72}$$

$$= R \ln (1.76 \times 10^{-11}\ \mathrm{deg\ s^{-1}} \cdot A/T)$$

For both Eqs. (7.71) and (7.72) we use an average T in the temperature range of interest.

The chief advantage of the transition-state theory is that it relates kinetic rates to thermodynamic properties of reactants and a transition intermediate. This can help us to understand some of the factors that influence the rates. The entropy of activation characterizes the changes in configurations of the reactants along the reaction path. Entropies of activation are typically negative for simple gas-phase reactions. They reflect the loss in randomness in the transition state relative to the reactants.

Consider the bimolecular dimerization of butene-1 in the gas phase

$$2\,C_4H_8 \xrightarrow[\text{cal mol}^{-1}\text{ deg}^{-1}]{-\Delta S^{\ddagger} = +16.3} [C_4H_8\cdots C_4H_8]^{\ddagger} \longrightarrow C_8H_{16}$$

The fairly large negative entropy of activation corresponds to the loss of three translational degrees of freedom in going from two independent reactant particles to the single one in the associated transition state. Theoretical calculations of $\Delta S^{\ddagger}$ have been attempted with only moderate success, even for fairly simple reactions. Theoretical estimates of $\Delta H^{\ddagger}$ values are even less reliable. Nevertheless, the transition state and its associated "thermodynamic" variables have great intuitive value.

Example 7.6 Calculate the entropy of activation at 71.25°C for the previous example involving the decomposition of urea.

Solution Using the approximation that $\Delta H^{\ddagger} = E_a$, we have

$$A = \frac{kT}{h}\exp\left(\frac{\Delta S^{\ddagger}}{R}\right)$$

or

$$\log A = \log\frac{kT}{h} + \frac{\Delta S^{\ddagger}}{2.30R}$$

Therefore,

$$\Delta S^{\ddagger} = 2.30R\cdot\left(\log A - \log\frac{kT}{h}\right)$$

$$\Delta S^{\ddagger} = (2.30)(1.99\text{ cal mol}^{-1}\text{ deg}^{-1})\left[13.07 - \log\frac{(1.38\times 10^{-16})(344.4)}{6.62\times 10^{-27}}\right]$$

$$= (2.30)(1.99\text{ cal mol}^{-1}\text{ deg}^{-1})(13.07 - 12.86)$$

$$= 0.96\text{ cal mol}^{-1}\text{ deg}^{-1}$$

The entropy of activation for this reaction is small. For reactions in which the solvent water is able to interact less favorably with the transition state than with the reactant, positive entropy of activation is often observed.

IONIC REACTIONS AND SALT EFFECTS

The effect of the nonideal behavior of reactants in solution can be treated by extending the transition-state theory to include activities and activity coefficients explicitly. Equation (7.63) is written as

$$v = k_2^{\ddagger} c^{\ddagger} \tag{7.73}$$

and Eq. (7.64) can be rewritten as

$$K^{\ddagger} = \frac{[\text{MN}]^{\ddagger}}{[\text{M}][\text{N}]} = \frac{c^{\ddagger}\gamma^{\ddagger}}{c_{\text{M}}\gamma_{\text{M}}c_{\text{N}}\gamma_{\text{N}}} \tag{7.74}$$

where the c's are concentrations and the γ's are activity coefficients. Combining Eqs. (7.73) and (7.74) and substituting into Eq. (7.67), we obtain the result

$$v = \frac{kT}{h} K^{\ddagger} \frac{\gamma_{\text{M}}\gamma_{\text{N}}}{\gamma^{\ddagger}} c_{\text{M}}c_{\text{N}} \tag{7.75}$$

In this case the rate constant can be written as

$$k = \frac{kT}{h} K^{\ddagger} \frac{\gamma_{\text{M}}\gamma_{\text{N}}}{\gamma^{\ddagger}} = k_0 \frac{\gamma_{\text{M}}\gamma_{\text{N}}}{\gamma^{\ddagger}} \tag{7.76}$$

where k_0 is the rate constant extrapolated to infinite dilution of all species in solution.

For reactions involving ionic species, we can use the Debye-Hückel limiting law (Chapter 4) to express the dependence of the rate constant, k, on ionic strength. For each ionic species

$$\log \gamma_i = -0.51 Z_i^2 \sqrt{I} \tag{7.77}$$

$$I = \tfrac{1}{2} \sum_{i}^{\text{all species}} c_i Z_i^2$$

where the constant 0.51 applies to water solutions at 25°C, Z_i is the charge on species i and I is the ionic strength. The charge relation for the reaction $\text{M} + \text{N} \rightarrow [\text{MN}]^{\ddagger}$ is clearly $Z_{\text{M}} + Z_{\text{N}} = Z^{\ddagger}$, and we can write Eq. (7.76) as

$$\begin{aligned} \log k &= \log k_0 - 0.51[Z_{\text{M}}^2 + Z_{\text{N}}^2 - (Z_{\text{M}} + Z_{\text{N}})^2]\sqrt{I} \\ &= \log k_0 + 2(0.51) Z_{\text{M}} Z_{\text{N}} \sqrt{I} \end{aligned} \tag{7.78}$$

Equation (7.78) predicts that a plot of $\log k$ versus $\sqrt{I}$ for dilute ionic solutions should give a straight line with a slope $2(0.51)Z_{\text{M}}Z_{\text{N}} \cong Z_{\text{M}}Z_{\text{N}}$. The slope then gives direct information about the charges of the reacting species. This has been verified for a variety of ionic reactions in dilute aqueous solutions. For example, when the rate of decomposition of H_2O_2 in the presence of HBr was studied as a function of ionic strength, it was found that the data fit Eq. (7.78) with a slope of −1. The stoichiometric reaction

$$H_2O_2 \longrightarrow H_2O + \tfrac{1}{2}O_2$$

does not involve ionic species, but the kinetics imply that H^+ and Br^- are involved in the formation of the transition-state species.

Example 7.7 The acid denaturation of CO-hemoglobin has been studied at pH 3.42 in a formic acid–sodium formate buffer as a function of sodium formate ($Na^+ + OOCH^-$) concentration (Zaiser and Steinhardt, 1951). The half-times for the first-order denaturation are as follows:

NaOOCH (*M*)	0.007	0.010	0.015	0.020
$t_{1/2}$ (min)	20.2	13.6	8.1	5.9

Determine whether these results follow the Debye-Hückel limiting law in the dependence of the rate constant on ionic strength and, if so, what is the apparent charge on the CO-hemoglobin?

Solution Since NaOOCH is a uni-univalent electrolyte, the ionic strength, I, is equal to the molar concentration, *c*, of sodium formate (we ignore the small H^+ concentration). For a first-order reaction,

$$k = \frac{\ln 2}{t_{1/2}} = \frac{0.693}{t_{1/2}}$$

Therefore:

$\sqrt{I}$	0.084	0.100	0.122	0.141
k (min^{-1})	0.0343	0.0510	0.0856	0.117
$\log k$	−1.465	−1.293	−1.068	−0.930

A plot of these data is shown in Figure 7.15. Although the data show some curvature, there is a reasonably good fit to a straight line with a slope of +9. Since the denaturation occurs by reaction with monovalent H^+, this suggests a charge of approximately +9 for the CO-hemoglobin at pH 3.42.

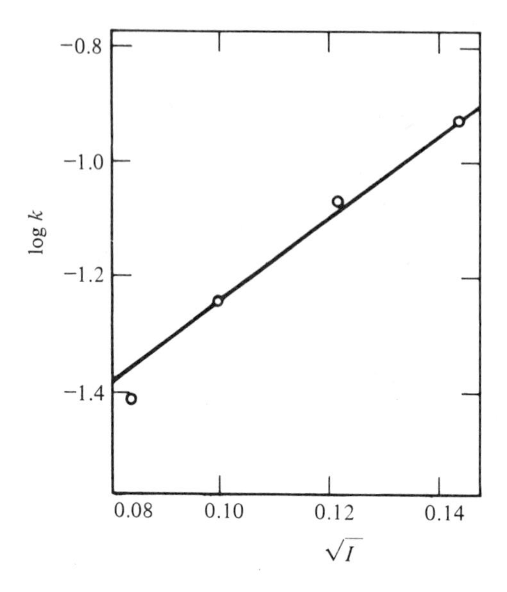

Fig. 7.15 Data for the rate of acid denaturation of CO-hemoglobin. [From E. M. Zaiser and J. S. Steinhardt, *J. Am. Chem. Soc.* *73*, 5568 (1951).]

DIFFUSION-CONTROLLED REACTIONS

The concept of a diffusion-controlled reaction is useful for interpreting the kinetics of fast reactions. Imagine a simplified system where reactant species M and N initially move about in the system independently of one another and far enough apart so that they do not interact. In the gas phase the intervening space is mostly empty, and the molecules would come together, react, and the product(s) depart without interference from other collisions.

The situation in the liquid (or solid phase) is quite different. The reactant species are always in contact with the solvent or other solutes, and they are constantly being bumped and jostled by their neighbors (Fig. 7.16). As a

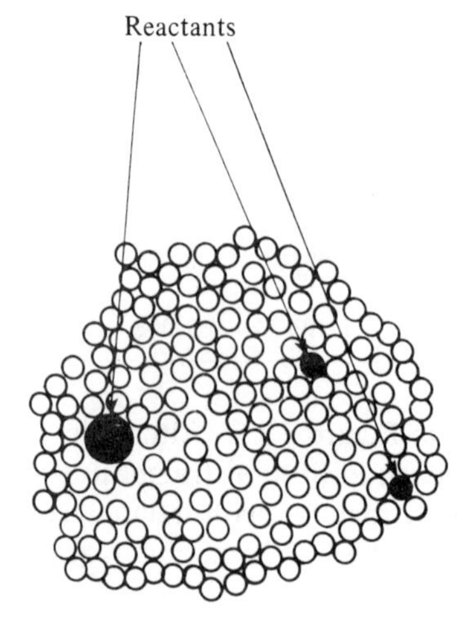

Fig. 7.16 Schematic diagram of the environment of two reactant species in tight "cages" in liquid solution.

consequence, the motion of any one reactant molecule is quite haphazard. This is known formally as a random-walk process. Because the spaces between molecules in the liquid phase are small (typically 10% or so of the molecular diameters), there results a kind of cage effect in which the molecule makes many collisions with its neighbors before the picture changes sufficiently to say that the molecule has moved away to a new site, where it encounters new neighbors. The motion is similar to that of people on a crowded dance floor. The time required to move across the dance floor by relying only on random changes in direction is very great when the floor is crowded, whereas a nearly empty dance floor is easily traversed.

The consequences of random motions of molecules are described in terms of their diffusion constants (see Chapter 6). Since the diffusion constant, D, is defined as the proportionality between the flux or velocity of material flow, J_x in units of (concentration) $\times$ (cm s^{-1}), across a unit area in response to a concentration gradient dc/dx, we can write the equation for Fick's first law:

$$J_x = -D\frac{dc}{dx} \tag{6.29}$$

The diffusion constant is a measure of the ability of reactive species to move toward one another to react. Diffusion constants for small molecules are typically on the order of 10^{-5} cm^2 s^{-1}. Large molecules have smaller diffusion constants compared with those of small species; but the numerical values depend on the nature of both the diffusing species and the medium (solvent) through which they are moving. Hydrogen ions in water diffuse with unusual ease.

This picture can be developed mathematically so as to calculate an encounter frequency for two solute species, M and N. Put into the units of the rate expression, we obtain a value for the A factor in the Arrhenius expression

$$A_{\text{diffusion}} = \frac{4\pi(r_M + r_N)(D_M + D_N)N_0}{1000} \tag{7.79}$$

for reactions in liquid solution. Here r_M and r_N are the effective radii of the reactant molecules, including participating solvent, and N_0 is Avogadro's number. For $D_M = D_N = 1.5 \times 10^{-5}$ cm^2 s^{-1} and $r_M + r_N = 4$ Å $= 4 \times 10^{-8}$ cm, the value of $A_{\text{diffusion}}$ is about 10^{10} ℓ mol^{-1} s^{-1}. For a reaction with no activation energy, this would be the value of the rate constant if every encounter led to reaction. It constitutes a kind of upper bound for reactions that are limited only by the frequency of encounters. These are known as *diffusion-controlled reactions.*

Rate constants for a variety of representative second-order reactions are listed in Table 7.3. Note that these are all rate constants at room temperature. Where activation energies are not zero, the Arrhenius A factors will be somewhat larger than the rate constants given in Table 7.3. Despite this limitation it is clear that many reactions involving protonation or reaction with hydroxide operate near the diffusion limit. Reactions involving oppositely charged ions or of ions with polar molecules are somewhat faster than that calculated from simple diffusion theory, because of electrostatic attractions. While a few enzyme-substrate complex formation reactions appear to be diffusion-limited, many others

Table 7.3 Rate constants for some second-order reactions in liquid solution

Reaction		k ($M^{-1}\,s^{-1}$)
$H^+ + OH^-$		1.3×10^{11}
$H^+ + NH_3$		4.3×10^{10}
H^+ + imidazole		1.5×10^{10}
$OH^- + NH_4^+$		3.4×10^{10}
OH^- + imidazole$^+$		2.3×10^{10}
Enzyme-substrate complex formation	ribonuclease + cytidine 3′-phosphate	6×10^{7}
	lactate dehydrogenase + NADH	1×10^{9}
	creatine kinase + ADP	2.2×10^{7}
	aspartate aminotransferase	
	+ α-methylaspartate	1.2×10^{4}
	+ β-erythrohydroxyaspartate	3.1×10^{6}
	+ aspartate	$>1 \times 10^{8}$
	catalase + H_2O_2	5×10^{6}
O_2 + hemoglobin		4×10^{7}

are slower by several orders of magnitude. It should be obvious that this difference is at least partly associated with the appreciable negative entropy of activation [Eq. (7.72)] that is typically found for such reactions.

Example 7.8 Glucose binds to the enzyme hexokinase from yeast with a rate constant $k = 3.7 \times 10^6\ M^{-1}\,s^{-1}$. Obtain the necessary data to estimate the rate constant if the reaction were diffusion-limited, and compare this calculation with the experimental value.

Solution To use Eq. (7.79) we need to know the diffusion coefficients and molecular radii of glucose and hexokinase.

$$D(\text{glucose}) = 0.673 \times 10^{-5}\ \text{cm}^2\,\text{s}^{-1}$$

$$D(\text{hexokinase}) = 2.9 \times 10^{-7}\ \text{cm}^2\,\text{s}^{-1}$$

The molecular radii can be estimated by assuming that the molecules are unhydrated spheres. The value for glucose can be determined from its partial molar volume, $\bar{v}_{\text{glucose}} = 112\ \text{cm}^3\,\text{mol}^{-1}$, calculated from the densities of aqueous solutions given in the *Handbook of Chemistry and Physics* (Chemical Rubber Co., Cleveland, Ohio). The value for hexokinase is obtained from the partial specific volume, $\bar{v} = 0.740\ \text{cm}^3\ \text{g}^{-1}$, and the molecular weight, 96,600:

$$r = \sqrt[3]{\frac{3\bar{v}}{4\pi N_0}}$$

$$r_{\text{glucose}} = \sqrt[3]{\frac{3(112)}{4\pi(6.0 \times 10^{23})}} = 3.5 \times 10^{-8}\ \text{cm} = 3.5\ \text{Å}$$

$$r_{\text{hexokinase}} = \sqrt[3]{\frac{3(0.740)(96{,}600)}{4\pi(6.0 \times 10^{23})}} = 30 \times 10^{-8}\ \text{cm} = 30\ \text{Å}$$

[Note that the diffusion constant of the small glucose molecule and the radius of the large hexokinase molecule are the dominant terms contributing to Eq. (7.79).] Thus, from (Eq. 7.79),

$$A_{\text{diff}} = \frac{4\pi(30 + 3.5)(10^{-8}\ \text{cm})(0.673 + 0.029)(10^{-5}\ \text{cm}^2\ \text{s}^{-1})(6.0 \times 10^{23})}{1000}$$

$$= 1.78 \times 10^{10} M^{-1}\ \text{s}^{-1}$$

which is nearly 5000 times faster than the observed rate constant. Thus the reaction is not diffusion-limited under these conditions.

PHOTOCHEMISTRY AND PHOTOBIOLOGY

Light is a factor in our environment having both highly beneficial and potentially hazardous aspects. We survive on planet earth because of the process of photosynthesis, which uses solar energy to form the chemical substances of the plants that are suitable as food. Our eyes employ rhodopsin to convert light stimuli into neural signals to the brain, enabling us to see. Solar irradiation of the upper atmosphere induces photochemistry that results in the formation of a complex distribution of oxygen, nitrogen, and hydrogen compounds. One consequence is the presence of the ozone shield, which screens out harmful ultraviolet radiation. The reason for concern about uv light is that its photons are of sufficiently high energy to induce mutations in both microorganisms and higher organisms. Such mutations are more likely to be harmful than beneficial. Furthermore, prolonged exposure to sunlight rich in uv components is known to lead to an increase in the incidence of skin cancer. Ultraviolet germicidal lamps are potent lethal weapons for microorganisms and are used for sterilizing food, drugs, serum solutions, and operating rooms. Ordinary fluorescent light has been used therapeutically for babies suffering from hyperbilirubinemia, a malfunctioning of the process of disposal of products of heme metabolism. Even this short list gives an indication of the wide variety of roles played by light in biology.

It is possible to look at the role of light in photoreactions from two rather different points of view. On the one hand, we can consider light to be simply another reagent that can be added to the system and that may influence the kinetics of reactions accordingly. This view is consistent with the *law of Grotthus and Draper*, which says that only radiation *absorbed* by the system is effective in producing photochemical change.

The quantitative measure of light is its intensity, I, usually given in photons $cm^{-2}\,s^{-1}$, einsteins $cm^{-2}\,s^{-1}$, or ergs $cm^{-2}\,s^{-1}$. The relation among these quantities can be derived from the *Planck equation*,

$$\varepsilon = h\nu = \frac{hc}{\lambda} \tag{7.80}$$

which states that the energy of a single photon, ε, is proportional to its frequency ν (in s^{-1}) or inversely proportional to its wavelength λ (in cm), h is Planck's constant (6.62×10^{-27} erg s), and c is the velocity of light ($3.0 \times 10^{10}\,cm\,s^{-1}$ in vacuum). A mole of photons (6.023×10^{23} photons) is known as an *einstein*. Thus, for an einstein of radiation, the energy is

$$\bar{E} = N_0 h\nu = \frac{N_0 hc}{\lambda} \tag{7.81}$$

Intensities of light are measured as the number of photons (or of einsteins, or of energy in ergs) crossing a 1-cm^2 cross section in each second of time. Alternative measures of intensity are often used in different applications.

Light that is transmitted through the sample or is scattered by it does not become absorbed and, therefore, will not produce any photochemistry. For a photochemical reaction

$$\mathrm{B} \xrightarrow{h\nu} \mathrm{P}$$

the velocity or photochemical rate is given by

$$v = -\frac{d[\mathrm{B}]}{dt} = \phi I_{\text{abs}} \tag{7.82}$$

where I_{abs} is the light absorbed by the sample in einsteins $\ell^{-1}\,s^{-1}$ and ϕ is the quantum yield. The quantum yield is defined as the number of molecules reacted per photon absorbed; it is obviously unitless.

$$\phi = \frac{\text{number of molecules reacted}}{\text{number of photons absorbed}} \tag{7.83}$$

Some representative values are given in Table 7.4.

According to the *Einstein law of photochemistry*, in a primary photochemical process each molecule is activated by the absorption of one photon. Secondary processes often obscure the Einstein law, and ϕ is an important quantity that needs to be determined for a photochemical reaction. Recently, exceptions to the Einstein law have been demonstrated at very high intensities using pulsed lasers. At super-high intensities two or three photons can be absorbed simultaneously by the same molecule.

Table 7.4 Some representative quantum yields

Reaction	Quantum yield, ϕ*
$NH_3(g) \rightarrow \frac{1}{2}N_2(g) + \frac{3}{2}H_2(g)$	0.2
$S_2O_8^{2-}(aq) + H_2O \rightarrow 2\,SO_4^{2-}(aq) + 2\,H^+(aq) + \frac{1}{2}O_2(g)$	1
$H_2(g) + Cl_2(g) \rightarrow 2\,HCl(g)$	10^5
$H_2O + CO_2 \xrightarrow{\text{chloroplasts}} \frac{1}{x}[CH_2O]_x + O_2$ (in photosynthesis)	10^{-1}
Rhodopsin → retinal + opsin (in mammalian eyes)	1
2 Thymine → thymine dimer (in DNA)	10^{-2}
Hemoglobin · CO → hemoglobin + CO(g)	1
Hemoglobin · O_2 → hemoglobin + $O_2(g)$	10^{-2}

* Values depend on wavelength and other experimental conditions.

The relation between the rate of a photochemical reaction and the absorbed light [Eq. (7.82)] can be written in a more useful form by introducing the Beer-Lambert law (see Chapter 10) for light absorption:

$$I = I_0 10^{-\varepsilon lc} = I_0 10^{-A} \qquad (7.84)$$

where I_0 = intensity of incident light
I = intensity of transmitted light
l = path length of light in sample
c = concentration of absorbing molecules, mol ℓ^{-1}
ε = molar absorptivity, or molar extinction coefficient, ℓ mol^{-1} cm^{-1}
A = absorbance; it is defined by Eq. (7.84)

The intensity of light absorbed is $(I_0 - I)$; therefore,

$$-\frac{d[\mathrm{B}]}{dt} = \phi(I_0 - I) = \phi I_0(1 - 10^{-A}) \qquad (7.85)$$

In this and the following kinetic equations, I_0 must have units einstein vol^{-1} s^{-1}. The exponential term can be rewritten using its series expansion, as in:

$$10^{-A} = 1 - 2.303A + \frac{(2.303A)^2}{2!}\cdots$$

For *dilute solutions*, where $A \ll 1$, the terms involving higher powers are negligible in comparison with the first-order term, and we can substitute:

$$1 - 10^{-A} \cong 2.303A$$

into Eq. (7.85) to obtain, for dilute solutions,

$$-\frac{d[\mathrm{B}]}{dt} = 2.303\phi I_0 A = 2.303\phi I_0 \varepsilon l[\mathrm{B}] \tag{7.86}$$

For constant incident light intensity, this has the form of the usual first-order rate expression. The concentration and time variables can be separated, and the equation can be integrated in the usual manner to give

$$\ln\frac{[\mathrm{B}]}{[\mathrm{B}]_0} = -2.303\phi I_0 \varepsilon l t = -kt \tag{7.87}$$

Most, but not all, photochemical reactions are first order in reactant concentration at sufficiently low concentration. A common test that this is so consists of illuminating identical dilute samples at a set of different intensities and for different times such that $I_0 \cdot t$ is a single constant for all the experiments. If the reaction is of first order in the absorbing species, the extent of photochemical conversion will be the same for each of the $I_0 \cdot t$ experiments.

Example 7.9 In the presence of a reducible dye D, chloroplasts from spinach are able to photocatalyze the oxidation of water to O_2 by the reaction

$$2\,\mathrm{H_2O} + 2\,\mathrm{D} \xrightarrow[\text{chloroplasts}]{h\nu} \mathrm{O_2} + 2\,\mathrm{DH_2}$$

The rate of this process at an incident intensity of 40×10^{-12} einstein $\mathrm{cm^{-2}\,s^{-1}}$ of 650-nm light was 6.5×10^{-12} equivalent $\mathrm{cm^{-3}\,s^{-1}}$ in a solution of 1-cm path length. The chloroplasts exhibited an absorbance of $A_{650}^{1\mathrm{cm}} = 0.140$. Calculate the quantum efficiency for this reaction.

Solution Because the solution is optically dilute ($A_{650} \ll 1$), we can use Eq. (7.86). In a photocatalytic reaction the concentration of the absorbing chlorophyll in the chloroplasts does not change with time; that is, A is a constant during the process. Therefore,

$$v = 2.303\phi I_0 A$$

and

$$\phi = \frac{v}{2.303 I_0 A} = \frac{6.5 \times 10^{-12}\ \text{equivalent cm}^{-3}\,\text{s}^{-1}}{2.303(40 \times 10^{-12}\ \text{einstein cm}^{-2}\,\text{s}^{-1})(0.140\ \text{cm}^{-1})}$$

$$= 0.50\,\frac{\text{equivalent}}{\text{einstein}} = 0.50\,\frac{\text{electron}}{\text{photon}}$$

(Note that for this calculation we used a value of $A = 0.140\ \mathrm{cm^{-1}}$, to make the equation dimensionally correct. If the path length were 2.0 cm, the absorbance would have been twice as great but the velocity measured in equivalent $cm^{-3} s^{-1}$ would have been the same and, of course, the quantum yield should be independent of path length. Expressing the absorbance per centimeter of path gets around this apparent inconsistency. This is equivalent to using I_0 in einstein $cm^{-3} s^{-1}$.)

We shall complete our discussion of photochemcal kinetics by describing some biologically important processes.

Vision

Visible and near-ultraviolet light absorbed by the eye pigment rhodopsin produces a sequence of photoreactions that is still under active investigation. A current proposal is shown in Fig. 7.17. The normal form of rhodopsin in dark-adapted mammalian rod outer segments has retinal in the 11-cis form covalently attached as a chromophore ("color bearer") to the protein opsin. Following illumination, the characteristic absorption spectrum of rhodopsin undergoes a series of changes associated with a set of intermediates, as shown in Fig. 7.17. Some of these steps are very fast at room temperature and have been studied using either flash relaxation methods or by freezing them at low

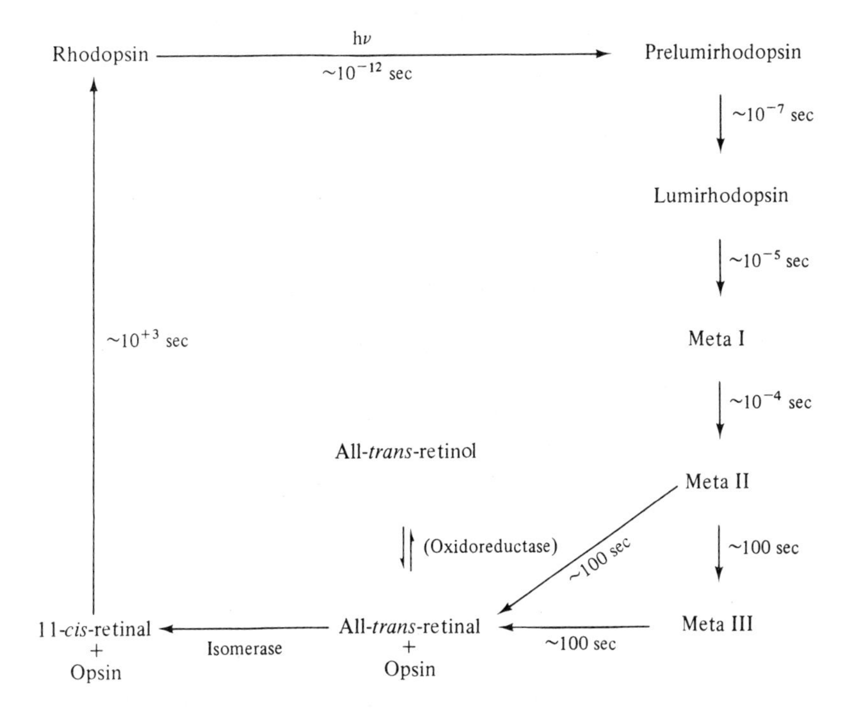

Fig. 7.17 Photobleaching and regeneration of rhodopsin. [From T. G. Ebrey and B. Honig, *Quarterly Reviews of Biophysics 8*, 129 (1975).]

adenosine triphosphate (ATP):

$$2\,H_2O + 2\,NADP^+ \xrightarrow[\text{chloroplast membranes}]{h\nu} O_2 + 2\,NADPH + 2\,H^+$$

$$2(ADP + P_i) \longrightarrow 2\,ATP$$

The NADPH and ATP are then used in a series of dark enzymatic reactions to fix CO_2.

$$CO_2 + 2\,NADPH + 2\,ATP + 2\,H^+ \xrightarrow[\text{enzymes}]{\text{dark}} (CH_2O) + 2\,NADP^+ + 2\,ADP + 2\,P_i + H_2O$$

The sum of these reactions is the overall process presented above.

Detailed and extensive studies of the kinetics of photosynthetic light reactions have been made, and the following list summarizes the current state of our knowledge of this subject:

1. The light reactions occur in lipoprotein membranes within chloroplasts in higher plants or within the cells of blue-green algae or photosynthetic bacteria.
2. Chlorophyll or bacteriochlorophyll are the essential pigments that absorb the light, although carotenoids and other accessory pigments can transfer absorbed photon excitation to the chlorophylls.
3. All wavelengths from the near-uv to the near-infrared can be effective. In particular, wavelengths as great as 700 nm (41 kcal einstein^{-1}) are effective in higher plants and as great as 1000 nm (29 kcal einstein^{-1}) in certain photosynthetic bacteria.
4. The quantum yield varies with growth conditions. Under optimal levels of CO_2 pressure, relative humidity, and soil nutrients, the yield at low light intensities corresponds to 1 mol of CO_2 fixed or O_2 evolved for each 8 to 9 einsteins absorbed. This represents a higher efficiency than it would seem, because each O_2 molecule is produced by the removal of 4 electrons from 2 water molecules and it is known that there are 2 photon acts or light reactions that operate in series for the transfer of each electron. A current view of the mechanism of photosynthetic electron transport is shown in Fig. 7.18.
5. The rates of photosynthesis reach a maximum at higher light intensities. In simple chloroplast photoreactions not connected to CO_2 fixation, this light saturation is hyperbolic in intensity and resembles the dependence of rates of enzymatic reactions on substrate concentration. When CO_2 fixation is included, the rates follow the same hyperbolic curve at low light intensities but then change abruptly to a constant, light-independent rate at higher intensities. This occurs when one of the steps in the carbon-fixation biochemistry becomes rate-limiting. The contrast between these two types of behavior is illustrated in Fig. 7.19.

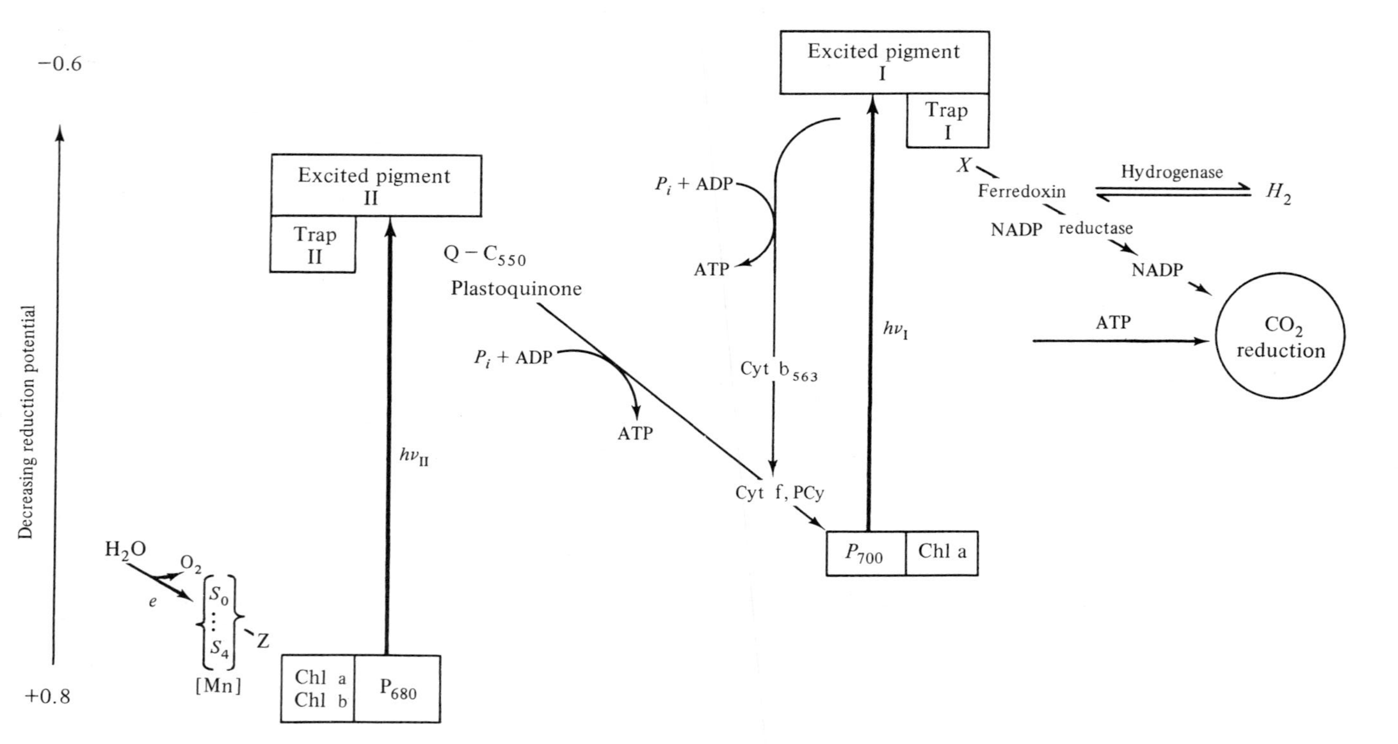

Fig. 7.18 Schematic diagram of the mechanism of photosynthetic electron transport involving two light reactions operating in series. P700 and P680 are the reaction-center chlorophylls of photosystems I and II, respectively. The electron transport cofactors include cytochromes (Cyt f, Cyt b_{563}), plastocyanin (PCy), etc., and a variety of components ($S_0 \cdots S_4$, Z, Q, C-550, X) whose chemical identity is unknown.

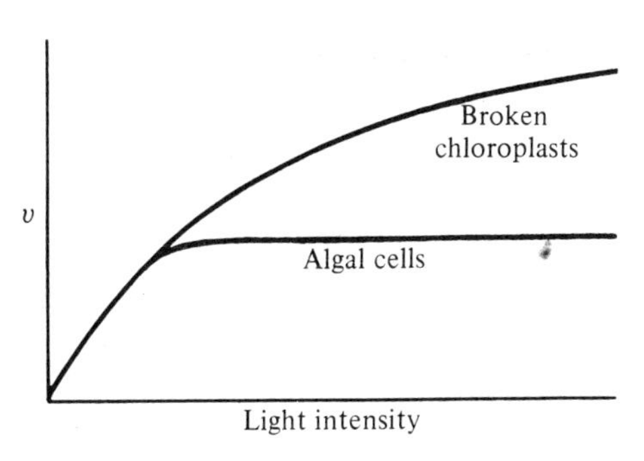

Fig. 7.19 Velocity of oxygen evolution by broken chloroplasts using an added oxidant (ferricyanide) is compared with that for whole cells of algae, where the CO_2-fixing reactions are coupled.

SUMMARY

Zero-order Reactions

Rate is constant for zero-order reactions:

$$\frac{dc}{dt} = k \tag{7.4}$$

$k \equiv$ rate constant; units are concentration time^{-1}, for example, mol ℓ^{-1} s^{-1}

Concentration is linear in time for zero-order reactions:

$$c_2 - c_1 = k(t_2 - t_1) \tag{7.7}$$

Plot of concentration versus time for zero-order reactions:

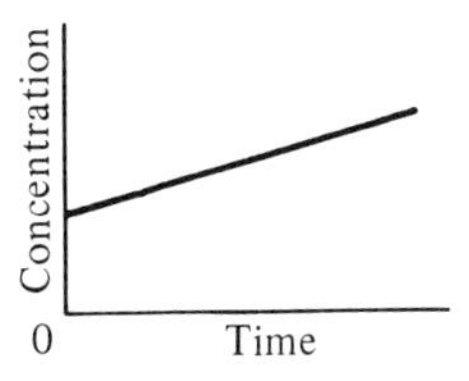

k = slope; concentration at zero time = y intercept

Possible mechanism for zero-order reaction:

$$\mathrm{A} \longrightarrow \mathrm{B} + \mathrm{C}$$

but concentration of A is held constant:

$$\frac{dc_\mathrm{B}}{dt} = k$$

$$c_\mathrm{B} = kt$$

First-order Reactions

Rate is proportional to concentration for first-order reactions:

$$-\frac{dc}{dt} = kc \tag{7.10}$$

$k \equiv$ rate constant; units are time^{-1}, for example, s^{-1}.

Logarithm of concentration linear in time for first-order reactions:

$$\ln c = -kt + \text{constant}$$

$$\ln c \equiv \text{natural logarithm of concentration}$$

Plot logarithm of concentration versus time for first-order reactions:

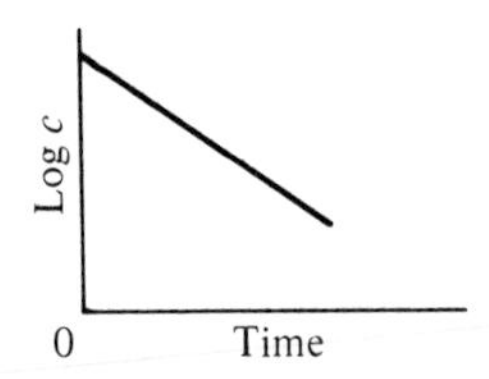

$k = -2.303 \cdot$ slope; logarithm of concentration at zero time is y intercept. Any concentration units can be used for the plot.

Concentration is exponential in time:

$$\frac{c}{c_0} = e^{-kt} \quad \text{or} \quad \frac{c}{c_0} = 10^{-0.4343kt} \tag{7.19}$$

c = concentration at time t

c_0 = concentration at zero time (note that only ratio of concentrations is important)

Half-life for first-order reactions:

$$\frac{c}{c_0} = 2^{-t/t_{1/2}} \tag{7.23}$$

$t_{1/2} \equiv$ half-life = time necessary for concentration to become half of its original value, c_0

$$= 0.693/k \tag{7.22}$$

Relaxation time for first-order reactions:

$$\frac{c}{c_0} = e^{-t/\tau}$$

$\tau \equiv$ relaxation time = time necessary for concentration to become $1/e$ of its original value; $1/e = 0.368$

Possible unimolecular mechanism for first-order reactions:

$$A \longrightarrow B$$

$$-\frac{dc_A}{dt} = kc_A \quad \text{and} \quad \frac{dc_B}{dt} = kc_A$$

$$\frac{c_A}{[c_A]_0} = e^{-kt} \qquad \frac{c_B}{[c_A]_0} = 1 - e^{-kt}$$

c_A, c_B = concentrations of A, B at any time t
$[c_A]_0$ = concentration of A at zero time

Second-order Reactions

Rate is proportional to concentration squared for second-order reactions:

$$\frac{-dc}{dt} = kc^2 \text{ (class I)} \quad \text{or} \quad \frac{-dc}{dt} = kc_Ac_B \text{ (class II)}$$

$k \equiv$ rate constant; units are concentration^{-1} time^{-1}, for example, ℓ mol^{-1} s^{-1}

Reciprocal of concentration linear in time for second-order reactions (class I):

$$\frac{1}{c} = kt + \text{constant}$$

Plot of reciprocal of concentration versus time for second-order reactions (class I):

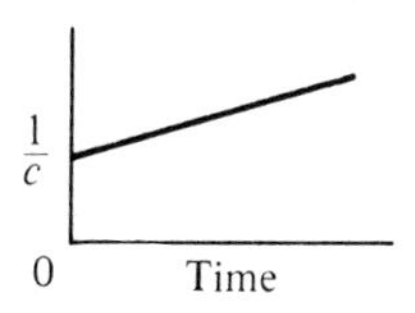

k = slope; reciprocal of concentration at zero time = y intercept

Possible bimolecular mechanism for second-order reactions (class I):

$$A + A \longrightarrow B$$

$$-\frac{dc_A}{dt} = kc_A^2 \quad \text{and} \quad \frac{dc_B}{dt} = \tfrac{1}{2}kc_A^2$$

$$\frac{1}{c_A} - \frac{1}{[c_A]_0} = kt \qquad c_B = \frac{[c_A]_0}{2} - \frac{c_A}{2}$$

c_A, c_B = concentrations of A, B at any time t
$[c_A]_0$ = concentration of A at zero time

Possible bimolecular mechanism for second-order reactions (class II):

$$A + B \longrightarrow C$$

$$-\frac{dc_A}{dt} = -\frac{dc_B}{dt} = kc_Ac_B$$

$$\ln\frac{[c_A][c_B]_0}{[c_A]_0[c_B]} = ([c_A]_0 - [c_B]_0)kt \tag{7.34}$$

c_A, c_B = concentrations of A, B at any time t
$[c_A]_0$ = concentration of A at zero time
$[c_B]_0$ = concentration of B at zero time

If $[c_A]_0 = [c_B]_0$, then $c_A = c_B$ at all times, and

$$\frac{1}{c_A} - \frac{1}{[c_A]_0} = \frac{1}{c_B} - \frac{1}{[c_B]_0} = kt \qquad \text{as in Class I.}$$

Temperature Dependence

Arrhenius equation:

$$k = Ae^{-E_a/RT} \tag{7.59}$$

k ≡ rate constant
A ≡ preexponential factor; units are same as rate constant
E_a = activation energy, units are cal mol^{-1} or J mol^{-1}
R = gas constant = 1.987 cal deg^{-1} mol^{-1} = 8.314 J deg^{-1} mol^{-1}
T = absolute temperature

Plot of logarithm of k versus reciprocal of absolute temperature:

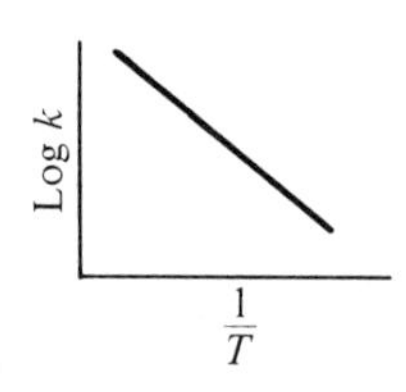

$$E_a = -2.303R \text{ slope}$$

$\log A$ = y intercept, but extrapolation is often impractical; therefore,
$\log A$ = $\log k + (1/T)(\text{slope})$ where $\log k$ and $(1/T)$ are the coordinates of any point on the line
$2.303R$ = 4.576 cal deg^{-1} mol^{-1} = 19.14 J deg^{-1} mol^{-1}

Activation energy:

$$E_a = \frac{2.303RT_2T_1}{T_2 - T_1}\log\frac{k(T_2)}{k(T_1)}$$

E_a = activation energy, which is assumed independent of temperature
$k(T_2), k(T_1)$ = rate constants at T_2, T_1
R = gas constant = 1.987 cal $\text{deg}^{-1}\ \text{mol}^{-1}$ = 8.314 J $\text{deg}^{-1}\ \text{mol}^{-1}$

Eyring equation:

$$k = \frac{\ell T}{h} e^{\Delta S^{\ddagger}/R} e^{-\Delta H^{\ddagger}/RT} \tag{7.70}$$

k = rate constant; its units of time must be seconds; its units of concentration specify the standard state for $\Delta H^{\ddagger}$, $\Delta S^{\ddagger}$
$\Delta H^{\ddagger}$ = enthalpy of activation = enthalpy difference per mole between activated complex and reactants, each in their standard states
$\Delta S^{\ddagger}$ = entropy of activation
T = absolute temperature
ℓ = Boltzmann constant = R/N_0; its units are consistent with Planck's constant, h.

$$\frac{\ell}{h} = \frac{R}{N_0 h} = \text{gas constant(Avogadro's number)}^{-1}\text{(Planck's constant)}^{-1}$$

$$= \frac{8.3143 \times 10^{7}\ \text{erg deg}^{-1}\ \text{mol}^{-1}}{(6.0225 \times 10^{23}\ \text{molecules mol}^{-1})(6.6256 \times 10^{-27}\ \text{erg s})}$$

$$= 2.0836 \times 10^{10}\ (\text{s deg})^{-1}$$

Relation of $\Delta H^{\ddagger}, \Delta S^{\ddagger}$ to E_a, A:

$$\Delta H^{\ddagger} = E_a - RT \tag{7.71}$$

$$\Delta S^{\ddagger} = R\left(\ln\frac{A}{T} - 1 - \ln\frac{R}{N_0 h}\right) \tag{7.72}$$

$$= 4.5756\left(\log\frac{A}{T} - 10.753\right) \text{ in cal mol}^{-1}$$

A = Arrhenius preexponential factor; its units of time must be s^{-1}; its concentration units specify the standard states of $\Delta H^{\ddagger}, \Delta S^{\ddagger}$

$$\log\frac{R}{N_0 h} = 10.32$$

Free energy of activation:

$$\Delta G^{\ddagger} = \Delta H^{\ddagger} - T\,\Delta S^{\ddagger} \tag{7.69}$$

$$k = \frac{RT}{N_0 h} e^{-\Delta G^{\ddagger}/RT} \tag{7.68}$$

$$\Delta G^{\ddagger} = -2.303RT\left(\log\frac{k}{T} - \log\frac{R}{N_0 h}\right)$$

$$= -RT\ln K^{\ddagger} \tag{7.65}$$

$\Delta G^{\ddagger}$ = free energy of activation = free energy difference between activated complex and reactants each in their standard states; the standard states are specified by the concentration units used in the rate constant k; if mol ℓ^{-1} is used, the standard state is the infinitely dilute solution state on the molarity scale

$K^{\ddagger}$ = equilibrium constant for the formation of the activated complex

Diffusion-controlled Reactions

Diffusion-controlled reaction M + N → *products:*

$$A_{\text{diffusion}} = \frac{4\pi(r_{\text{M}} + r_{\text{N}})(D_{\text{M}} + D_{\text{N}})N_0}{1000} \tag{7.79}$$

$A_{\text{diffusion}}$ = preexponential factor, in ℓ mol^{-1} s^{-1}

r = molecular radius, cm

D = diffusion constant, cm^2 s^{-1}

N_0 = Avogadro's number = 6.02×10^{23} mol^{-1}

Absorption of Light

Beer-Lambert law:

$$A = \log\frac{I_0}{I} = \varepsilon cl \tag{7.84}$$

A = absorbance (or optical density)

I_0 = incident intensity, I = transmitted intensity — same units for both

ε = molar absorptivity, ℓ mol^{-1} cm^{-1}

c = concentration, mol ℓ^{-1}

l = path length, cm

Photochemistry

Planck equation:

$$\varepsilon = h\nu = \frac{hc}{\lambda} \tag{7.80}$$

ε = photon energy, erg photon^{-1}
h = Planck's constant = 6.6256×10^{-27} erg s
ν = frequency of radiation, s^{-1}
c = velocity of light = 2.9979×10^{10} cm s^{-1}
λ = wavelength of radiation, cm

$$E = N_0 h\nu = \frac{N_0 hc}{\lambda} \tag{7.81}$$

E = energy per einstein
N_0 = Avogadro's number = 6.023×10^{23} mol^{-1}
= 6.023×10^{23} photons einstein^{-1}

Velocity of a photochemical reaction:

$$v = \phi I_{\text{abs}} \tag{7.82}$$

v = velocity of reaction, mol cm^{-3} s^{-1}
ϕ = quantum yield, moles reacting · (einstein absorbed)$^{-1}$
I_{abs} = flux of light absorbed, einstein cm^{-3} s^{-1}

Dilute solution photochemistry:

$$v = 2.30\phi I_0 A \qquad (\text{where } A \ll 1)$$

v = velocity of reaction, mol cm^{-3} s^{-1}
ϕ = quantum yield, moles reacting · (einstein absorbed)$^{-1}$
I_0 = incident intensity, einstein cm^{-2} s^{-1}
A = absorbance per cm

MATHEMATICS NEEDED FOR CHAPTER 7

A differential equation is an equation containing derivatives. In this chapter, we apply only differential equations that can be solved by separating the variables and integrating each side of the equation separately. Consider a differential equation of the form

$$\frac{dx}{dt} = f(x) \cdot f(t)$$

where $f(x)$ is a function of x only and $f(t)$ is a function of t only. We separate the variables x and t,

$$\frac{dx}{f(x)} = f(t)\,dt$$

and integrate each side of the equation

$$\int \frac{dx}{f(x)} = \int f(t)\,dt$$

The integrals we use are

$$\int x^n\,dx = \frac{x^{n+1}}{n+1} + C \qquad \int_{x_1}^{x_2} x^n\,dx = \frac{x_2^{n+1} - x_1^{n+1}}{n+1} \qquad (n \neq -1)$$

$$\int \frac{dx}{x} = \ln x + C \qquad \int_{x_1}^{x_2} \frac{dx}{x} = \ln \frac{x_2}{x_1}$$

$$\int e^{ax}\,dx = \frac{e^{ax}}{a} + C \qquad \int_{x_1}^{x_2} e^{ax}\,dx = \frac{e^{ax_2} - e^{ax_1}}{a}$$

In addition,

$$e^a = 10^{0.4343a}$$

$$10^a = e^{2.303a}$$

$$\ln a = 2.303 \log a$$

We use the expansion

$$e^x = 1 + x + \frac{x^2}{2!} + \frac{x^3}{3!} + \cdots$$

REFERENCES

Kinetics Textbooks

AMDUR, I., and G. G. HAMMES, 1966. *Chemical Kinetics: Principles and Selected Topics*, McGraw-Hill, New York.

BENSON, S. W., 1960. *The Foundations of Chemical Kinetics*, McGraw-Hill, New York.

BRAY, H. G., and K. WHITE, 1966. *Kinetics and Thermodynamics in Biochemistry*, Academic Press, New York.

CALVERT, J. G., and J. N. PITTS, JR., 1966. *Photochemistry*, John Wiley, New York.

FROST, A. A., and R. G. PEARSON, 1961. *Kinetics and Mechanism*, 2nd ed., John Wiley, New York.

GARDINER, W. C., JR., 1969. *Rates and Mechanisms of Chemical Reactions*, W. A. Benjamin, Menlo Park, Calif.

SUGGESTED READINGS

CLAYTON, R. K., 1970, 1971. *Light and Living Matter*, Vols. 1 and 2, McGraw-Hill, New York.

JOHNSTON, H. S., 1971. Reduction of Stratospheric Ozone by Nitrogen Oxide Catalysts from Supersonic Transport Exhaust, *Science 173*, 517–522.

JOHNSTON, H. S., 1975. Ground-Level Effects of Supersonic Transports in the Stratosphere, *Accounts Chem. Res. 8*, 289.

ROSS, R. T., and M. CALVIN, 1967. Thermodynamics of Light Emission and Free Energy Storage in Photosynthesis, *Biophys. J. 7*, 595–614.

PROBLEMS

1. Iodine reacts with a ketone in aqueous solution to give an iodoketone. The stoichiometric equation is

$$I_2 + \text{ketone} \longrightarrow \text{iodoketone} + H^+ + I^-$$

The rate of the reaction can be measured by measuring the disappearance of I_2 with time. Some data for initial rates and initial concentrations follow:

$-d(I_2)/dt$ (mol ℓ^{-1} s^{-1})	(I_2) (mol ℓ^{-1})	(ketone) (mol ℓ^{-1})	(H^+) (mol ℓ^{-1})
7×10^{-5}	5×10^{-4}	0.2	10^{-2}
7×10^{-5}	3×10^{-4}	0.2	10^{-2}
1.7×10^{-4}	5×10^{-4}	0.5	10^{-2}
5.4×10^{-4}	5×10^{-4}	0.5	3.2×10^{-2}

(a) Find the order of the reaction with respect to I_2, ketone, and H^+.

(b) Write a differential equation expressing your findings in part (a) and calculate the average rate constant.

(c) How long will it take to synthesize 10^{-4} mol ℓ^{-1} of the iodoketone starting with 0.5 mol ℓ^{-1} of ketone and 10^{-3} mol ℓ^{-1} of I_2, if the H^+ concentration is held constant at 10^{-1} mol ℓ^{-1}? Will the reaction go faster if we double the concentration of ketone? of iodine? of H^+? How long will it take to synthesize 10^{-1} mol ℓ^{-1} of the iodoketone if all conditions are the same as above?

2. $A \xrightarrow{k_1} B \xrightarrow{k_2} C$ is the mechanism for a set of reactions.

(a) Write a differential equation for the disappearance of A.

(b) Write a differential equation for $d[B]/dt$

(c) Write a differential equation for the appearance of C.

(d) If $[A]_0$ is the concentration of A at zero time, write an equation which gives [A] at any later time.

3. The stoichiometric equation for a reaction is

$$A + B \longrightarrow C + D$$

The initial rate of formation of C is measured with the following results:

Initial concentration of A (M)	Initial concentration of B (M)	Initial rate ($M\,s^{-1}$)
1	1	10^{-3}
2	1	4×10^{-3}
1	2	10^{-3}

(a) What is the order of the reaction with respect of A?
(b) What is the order of the reaction with respect to B?
(c) Use your conclusions in parts (a) and (b) to write a differential equation for the appearance of C.
(d) What is the rate constant k for the reaction? Do not omit the units of k.
(e) Give a possible mechanism for the reaction and discuss in words, or give equations to show, how the mechanism is consistent with the experiment.

4. Write an expression for the rate of appearance of D which follows from the following mechanisms:

(a)
$$A \underset{k_2}{\overset{k_1}{\rightleftharpoons}} B$$
$$B + C \xrightarrow{k_3} D \quad \text{(assuming steady state)}$$

(b)
$$A + B \overset{K}{\rightleftharpoons} AB \text{ (fast to equilibrium)}$$
$$AB + C \xrightarrow{k} D$$

5. Equal volumes of two equimolar solutions of reactants A and B are mixed, and the reaction $A + B \rightarrow C$ occurs. At the end of 1 h, A is 75% reacted. How much of A will be left *unreacted* at the end of 2 h if the reaction is:
(a) First order in A and zero order in B?
(b) First order in A and first order in B?
(c) Zero order in both A and B?
(d) First order in A and one-half order in B?

6. In a second experiment the same reaction mixture of problem 5 is diluted by a factor of two at the time of mixing. How much of A will be left unreacted after 1 h for each of the assumptions (a) – (d) of Problem 5?

7. The age of water or wine may be determined by measuring its radioactive tritium ($^3H_1$) content. Tritium is present in a steady state in nature. It is formed primarily by cosmic irradiation of water vapor in the upper atmosphere, and it decays spontaneously by a first-order process with a half-life of 12.5 yr. The formation reaction does not occur significantly inside a glass bottle at the surface of the earth. Calculate the age of a suspected vintage wine that is 20% as radioactive as a freshly bottled specimen. Would you recommend to a friend that he consider paying a premium price for the "vintage" wine?

8. A radioactive sample produced 1×10^5 disintegrations min^{-1}; 28 days later it produced only 0.25×10^5 disintegrations min^{-1}. (1 day = 1440 min.)

(a) What is the half-life of the sample?

(b) How many radioactive atoms were there in the sample that had 10^5 disintegrations min^{-1}?

9. Compounds A and B react to form C. When 1 *M* A and 1 *M* B were mixed, the following data were obtained

Time (s)	Concentration of A (*M*)
0	1
100	0.9
200	0.8
300	0.7
400	0.6

(a) What is the order of the reaction with respect to A?

(b) What is the rate constant for the reaction of A? Specify units.

(c) Write a differential equation for the rate of disappearance of A which is consistent with the data.

(d) Propose a mechanism that is consistent with the data. Use equations or a few sentences or both.

10. If you put 100 bacteria into a 1-ℓ flask containing appropriate growth medium at 40°C, you might find the following ideal results:

Time (min)	Bacteria (number)
0	100
30	200
60	400
90	800
120	1600

(a) What would you predict for the number of bacteria after 150 min?

(b) What is the order of the kinetics of this process?

(c) What is the doubling time of the reaction?

(d) How long would it take to get 10^6 bacteria?

(e) What is the rate constant for the reaction?

11. Calculate the activation energy which leads to a doubling of the rate of a reaction with an increase in temperature from 25°C to 35°C.

12. The mechanism for a reaction is assumed to be:

$$A + B \xrightarrow{k} P$$

$$k = 10^5\ \ell\ \text{mol}^{-1}\ \text{s}^{-1} \text{ at } 27°\text{C}$$

(a) Calculate the initial rate of formation of P if 0.1 *M* A is mixed with 0.1 *M* B at 27°C. State the units.

(b) Calculate the initial rate of formation of P if 10^{-4} M A is mixed with 10^{-6} M B at 27°C. State the units.
(c) How long would it take (in seconds) to form 0.05 M product from 0.1 M A and 0.1 M B at 27°C?
(d) At 127°C the rate constant of the reaction increases by a factor of 10^3. Calculate E_a and $\Delta H^{\ddagger}$ at 27°C.

13. If A and B are mixed together in solution, it is found that the concentration of A decreases with time, but B remains constant. The stoichiometric equation is A → P.
(a) If the initial concentration of A is less than 0.01 M, it is found that the initial rate is $-d[A]/dt = k_1[A][B]$. Write the integrated form of this equation. What is the order of the reaction with respect to A? What is the order of the reaction with respect to B?
(b) If the initial concentration of A is greater than 1 M, it is found that the initial rate is nearly independent of A:

$$-\frac{d[A]}{dt} = k_2[B]$$

Write the integrated form of this equation. What is the order of the reaction with respect to A? What is the order of the reaction with respect to B?
(c) Sketch a plot of the initial rate $-d[A]/dt$ versus initial concentration [A]. Write one rate equation which is consistent with experiment at both high and low concentrations of A.

14. A reacts to form P. A plot of the reciprocal of the concentration of A versus time is a straight line. When the initial concentration of A is 1.0×10^{-2} M, its half-life is found to be 20 min.
(a) What is the order of the reaction?
(b) Write a one-line mechanism that is consistent with the kinetics.
(c) What is the value of the rate constant for your mechanism of part (b)?
(d) When the initial concentration of A is 3×10^{-3} M, what will be the half-life?

15. A and B react stoichiometrically to form P. If 0.01 M A and 10 M B are mixed, it is found that the log of the concentration of A versus time is a straight line.
(a) What is the order of the reaction with respect to A?
(b) Write a one-line mechanism that is consistent with the kinetics and stoichiometry.
(c) According to your mechanism, what is the order of the reaction with respect to B?
(d) The half-life for the disappearance of A is 100 s. What would be the predicted half-life if the concentration of B is changed from 10 M to 20 M?
(e) From the half-life given in part (d), calculate the rate constant for the reaction of A with B. Specify the units of the rate constant.

16. In a paper by Bada, Protsch, and Schroeder [*Nature 241,* 394 (1973)], the rate of isomerization of isoleucine in fossilized bone is used as an indication of the average temperature of the sample since it was deposited. The reaction

$$\underset{\text{iso}}{\text{L-isoleucine}} \rightleftharpoons \underset{\text{allo}}{\text{D-alloisoleucine}}$$

produces a nonbiological amino acid, D-alloisoleucine, that can be measured using an automatic amino acid analyzer. At 20°C, this first-order reaction has a half-life of

125,000 years and its activation energy is 33.4 kcal mol^{-1}. After a very long time, the ratio allo/iso reaches an equilibrium value of 1.38. You may assume that this equilibrium constant is temperature independent.

(a) For a hippopotamus mandible found near a warm spring in South Africa, the allo/iso ratio was found to be 0.42. Assuming that no allo was present initially, calculate the ratio of the concentration of allo now present to the concentration of allo after a very long time. (*Note*: The correct answer is between 0.40 and 0.60.)

(b) Radiocarbon dating, which is temperature independent, indicated an age of 38,600 years for the hippo tooth. Using the result of part (a), estimate the half-life for the process.

(c) Calculate the average temperature of this specimen during its residence in the ground. (The present mean temperature of the spring is 28°C.)

17. Gaseous ozone, O_3, undergoes decomposition according to the stoichiometric equation

$$2\,O_3(g) \longrightarrow 3\,O_2(g)$$

Two alternative mechanisms have been proposed to account for this reaction:

(I) $$2\,O_3 \xrightarrow{k} 3\,O_2 \qquad \text{(bimolecular)}$$

(II) $$O_3 \underset{k_{-1}}{\overset{k_1}{\rightleftharpoons}} O_2 + O \qquad \text{(fast, equilibrium)}$$

$$O + O_3 \xrightarrow{k_2} 2\,O_2 \qquad \text{(slow)}$$

(a) Derive rate laws for the formation of O_2 for each of these mechanisms.

(b) Thermodynamic measurements give standard enthalpies of formation for each of the following species at 298 K:

Substance	ΔH^0_{298}(kcal mol^{-1})
$O_2(g)$	0.0
$O_3(g)$	34.0
$O\ (g)$	59.6

The observed activation enthalpy, ΔH^{+}, for the overall reaction $2\,O_3 \rightarrow 3\,O_2$ is 30 kcal mol^{-1} of O_3. Sketch a curve of enthalpy (per mole of O_3) versus reaction coordinate for each of the two proposed mechanisms. Label the curves with numerical values for the $\Delta\bar{H}$ between reactants, products, intermediates, and transition states.

(c) Can you exclude either of these mechanisms on the basis of the thermodynamic and activation enthalpy values of part (b)? Explain your answer.

(d) Devise a kinetic procedure for distinguishing between the two mechanisms. State clearly the nature of the experiments you would perform and what results you would look for to make the distinction.

18. Imidazole (Im) can react with H^+ or H_2O to form positively charged imidazole (ImH^+). The reaction mechanisms are

$$Im + H^+ \underset{k_{-1}}{\overset{k_1}{\rightleftharpoons}} ImH^+$$

$$Im + H_2O \underset{k_{-2}}{\overset{k_2}{\rightleftharpoons}} ImH^+ + OH^-$$

The rate constants are: $k_1 = 1.5 \times 10^{10}\ M^{-1}\ s^{-1}$; $k_{-1} = 1.5 \times 10^3\ s^{-1}$; $k_2 = 2.5 \times 10^3\ M^{-1}\ s^{-1}$, $k_{-2} = 2.5 \times 10^{10}\ M^{-1}\ s^{-1}$.

(a) What is the value of the equilibrium constant for the ionization of imidazole ($ImH^+ \rightleftharpoons Im + H^+$)?
(b) Write the differential equation for the rate of formation of ImH^+.
(c) If the pH is suddenly changed for a solution of 0.1 M imidazole in water from pH 7 to pH 4, what is the rate-determining step for the formation of ImH^+ at pH 4?
(d) What is the value of the initial rate of formation of ImH^+ at pH 4?
(e) The rate constants k_1 and k_{-1} both depend on temperature. Would you expect them to decrease or increase with increasing temperature? Which would you expect to change most with temperature and why?
(f) Predict the sign of the heat ionization for imidazole based on your answer to part (e). The experimental heat of ionization is positive.

19. The population of the earth in 1977 was 4.0 billion, and it is doubling every 35 years. Assuming that this doubling time remains constant, answer the following.
(a) By what year will the population of the earth reach 100 billion?
(b) How long will it take before the density of population gets so great that each person will have an average of only 1 m^2 of the earth's land surface to occupy? (Approximately 70% of the earth is covered by oceans.)
(c) Using the answer to part (b), calculate the population of the earth the same number of years *before* the present. (This answer is too low, because the doubling time has *decreased* dramatically during that interval.)

20. In view of the current shortage of fossil fuels (chiefly petroleum and natural gas), alternative sources of energy are being sought. A recent imaginative proposal is to use rubber (latex), which is a plant hydrocarbon, polyisoprene. Latex represents about 50% of the carbon fixed in photosynthesis by rubber trees! Presumably this material could be cracked and processed in refineries in a fashion quite analogous to that used for petroleum. Under good conditions a rubber tree in Brazil can produce 10 kg of usable latex per year, growing at a density of about 100 trees per acre. The heat of combustion (fuel value) of hydrocarbons is about 154 kcal mol^{-1} of CH_2 or about 11 kcal g^{-1}.
(a) What fraction of a year's solar energy is stored in rubber crop if the incident solar energy is 1 kcal min^{-1} ft^{-2} and the sun is shining 500 min day^{-1}, on the average? 1 acre = 43,560 ft^2. [Compare your answer with the value (energy stored)/(solar energy) = 0.3 that can be obtained from photosynthesis in the laboratory under optimal conditions.]
(b) In 1974 the total fossil fuel consumption in the United States was 9×10^{15} kcal yr^{-1}. What fraction of the area of the United States (3,000,000 mi^2) would have to be devoted to growing rubber trees (if they would grow in the climate of the United States, which they will not) to replace our present sources of fossil fuel by latex? (1 mi^2 = 640 acres.)
(c) In laboratory experiments it is found that photosynthesis occurs optimally with about 3% CO_2 in the (artificial) atmosphere. CO_2 occurs at only 0.03% in the natural atmosphere. Propose a scheme for increasing the energy yield of agricultural crops (not necessarily rubber trees) based on this observation. List some possible additional advantages and disadvantages of your scheme based on your background in kinetics and thermodynamics.

21. Consider the following reactions involved in the formation and disappearance of ozone in the upper atmosphere.

$$O_2 \xrightarrow{h\nu \text{ (below 242 nm)}} O + O \quad \text{(A)}$$

$$O + O_2 + M \longrightarrow O_3 + M \quad \text{(B)}$$

$$O_3 \xrightarrow{h\nu \text{ (190–300 nm)}} O_2 + O \quad \text{(C)}$$

where the first and third reactions are photochemical and driven by sunlight, and the second reaction is termolecular. (The third body, M, may be N_2, O_2, or any other gas molecule in the atmosphere.) The potential danger of reducing the "ozone shield" by supersonic transports comes from the presence of nitrogen oxides (NO, NO_2, etc.) in the engine exhaust. Although the actual mechanism is more complex, two important reactions in the proposed ozone reaction are

$$NO + O_3 \xrightarrow{k_1} NO_2 + O_2 \quad \text{(fast)} \quad (1)$$

$$O + NO_2 \xrightarrow{k_2} NO + O_2 \quad \text{(fast)} \quad (2)$$

Note that NO is not used up in this pair of reactions, but it continues to decompose ozone in a pseudo-catalytic fashion (until it eventually diffuses out of the ozone layer, which may be a very slow process taking many months).

(a) At a temperature of 217 K, characteristic of the 20-km altitude at which an SST might fly, the value of k_1 is 4.0×10^{-15} cm^3 molecule^{-1} s^{-1}. From data on exhaust composition, it is estimated that the steady-state concentration of NO would be about 10^{10} molecules cm^{-3}. If the NO were suddenly introduced at this level, what would be the half-time (in hours) for the reduction of ozone in the region of the SST flight paths? (*Note*: You do not need to know the ozone concentration to solve this problem.)

(b) The molecule NO_2 is unstable in the presence of sunlight, and decomposes photochemically according to the following reaction:

$$NO_2 \xrightarrow{h\nu \text{ (260–400 nm)}} NO + O \quad (3)$$

The oxygen atoms are highly reactive and disappear by reactions (B) and (2) above. Consider carefully the contribution of reaction (3) to the overall rate of ozone disappearance and describe qualitatively the effect that you would expect if this reaction were added to the situation approximated in part (a).

22. The photosynthesis efficiency of a strain of algae was measured by irradiating it for 100 s with an absorbed intensity of 10 W and an average wavelength of 550 nm. The yield of O_2 was 5.75×10^{-4} mol. Calculate the quantum yield of O_2 formation. (The quantum yield per "equivalent" of photochemistry is four times this value, since four electrons must be removed from two water molecules to produce each O_2.)

23. The human eye, when completely adapted to darkness, is able to perceive a point source of light against a dark background when the rate of incidence of radiation on the retina is greater than 2×10^{-16} W. (1W = 10^7 erg s^{-1}.)

(a) Find the minimum rate of incidence of quanta of radiation on the retina necessary to produce vision, assuming a wavelength of 550 nm.

(b) Assuming that all of this energy is converted ultimately to heat in visual cells having a total volume of 10^{-2} cm^3 and assuming that there are no losses of heat, by how much would the temperature of these cells rise in 1 s? Assume a heat capacity and density equal to that of water.

(c) Estimate the safe range of visual intensities (those which you would not expect to cause permanent eye damage).

24. An electronically excited molecule A* can either emit fluorescence to return to its ground state, or it can lose its excitation by collision.

$$A^* \xrightarrow{k_r} A + \text{photon}$$

$$A^* \xrightarrow{k_T} A + \text{heat}$$

(k_r and k_T are rate constants with units of s^{-1})

(a) The fluorescence intensity of A* can be measured as the number of photons emitted per second. Write an equation relating the fluorescence intensity to the concentration of A*.

(b) Derive an equation for the concentration of A* as a function of time, k_r, k_T, and $[A^*]_0$ (the concentration of A* at zero time).

25. Sunlight between 290 and 313 nm can produce sunburn (erythema) in 30 min. The intensity of radiation between these wavelengths in summer, at 45° latitude and at sea level, is about 50 μW cm^{-2}. Assuming that each incident photon is absorbed and produces a chemical change in 1 molecule, how many molecules per square centimeter of human skin must be photochemically affected to produce evidence of sunburn?

8

Enzyme Kinetics

Enzymes are proteins that increase the rates of reactions, often by many orders of magnitude, without themselves being consumed by the reaction. At the same time enzymes can be highly selective. Only certain reactions are affected by a particular enzyme; reactions involving other substrates that are similar may be totally unaffected. Enzymes are simply biological catalysts.

The role of enzymes as unusual catalysts is well illustrated by the decomposition of hydrogen peroxide:

$$2\,H_2O_2 \longrightarrow 2\,H_2O + O_2$$

This reaction occurs only very slowly in pure aqueous solution, but its rate is greatly increased by a large variety of catalysts. The increase in rate is typically first order in concentration of catalyst. It may be approximately first order in concentration of H_2O_2 as well. However, with catalysis by the enzyme catalase, the rate becomes zero order in H_2O_2 at higher initial concentrations of H_2O_2. This is common for enzyme-catalyzed reactions.

Table 8.1 gives approximate values for the kinetics of the rate expression

$$-\frac{d[H_2O_2]}{dt} = k[H_2O_2][\text{catalyst}]$$

In the table the velocities for the extrapolated conditions of $1\,\text{mol}\,\ell^{-1}$ of H_2O_2 and $1\,\text{mol}\,\ell^{-1}$ of catalyst are given for comparison. For catalase, the

Table 8.1 Catalase and H_2O_2 decomposition: Comparison of velocities and activation energies at 25°C*

Catalyst	Velocity, $-d[H_2O_2]/dt$ (mol ℓ^{-1} s^{-1})	E_a (kcal mol^{-1})
None	10^{-8}	17
HBr	10^{-4}	12
Fe^{2+}/Fe^{3+}	10^{-3}	10
Hematin or hemoglobin	10^{-1}	—
$Fe(OH)_2TETA^+$	10^3	7
Catalase	10^7	2

* The velocity is calculated for 1 *M* of H_2O_2 and 1 *M* catalyst (except for catalase). For catalase, the maximum velocity is given for 1 *M* concentration of active sites. This velocity is numerically equal to the turnover number for catalase. TETA is triethylenetetramine. For a discussion of the mechanism of the reaction and reference to the earlier literature, see J. H. Wang, *J. Am. Chem. Soc. 77,* 4715 (1955).

maximum velocity per active site is given. The rate for the uncatalyzed reaction, as shown in the table, corresponds to a decomposition of only 1% after 3 days at 25°C. Even then the reaction is probably "catalyzed" by dust particles that are impossible to remove completely from the solutions. Inorganic catalysts, such as iron salts or hydrogen halides, increase the velocity of H_2O_2 decomposition by four to five orders of magnitude per mole of catalyst. The enzyme catalase, which occurs in blood and a variety of tissues (liver, kidney, spleen, etc.), increases the velocity by more than 15 powers of 10 over the uncatalyzed rate! To appreciate the speed of this enzymatic reaction, consider that for the maximum velocity each molecule of catalase is able to decompose more than 10 million molecules of H_2O_2 per second!

We note that catalase is a hemoprotein with ferriprotoporphyrin (hematin) as a prosthetic group (the hematin group provides the catalytic function) at the active site. The hematin can be separated from the protein. Alone in solution, hematin exhibits catalytic activity toward H_2O_2 decomposition that is two orders of magnitude higher than that of the inorganic catalysts, but still more than a millionfold smaller than that of catalase. Not just any protein will enhance the catalytic activity of the iron protoporphyrin either. Heme (like hematin, but with the iron in the ferrous state) is the prosthetic group present in hemoglobin, but there is no catalytic enhancement of H_2O_2 decomposition in comparison with free hematin.

Chemists have risen to the challenge of trying to simulate the catalytic activity of enzymes, short of reproducing the entire biological molecule. One such attempt by Wang (1955) produced an iron complex $Fe(OH)_2TETA^+$ (TETA = triethylenetetramine), which exhibited a velocity of 1.2×10^3 mol s^{-1} ℓ^{-1} for the decomposition of $1\,M$ H_2O_2. That is pretty good compared with the other nonenzymatic catalysts listed in Table 8.1, but still less than 10^{-3} of the value for catalase.

The thermodynamics of the decomposition of hydrogen peroxide is also relevant to the function of catalysts. The reaction

$$H_2O_2(aq) \longrightarrow H_2O(l) + \tfrac{1}{2}O_2(g)$$

is exergonic, with $\Delta G^0_{298} = -24.64\ \text{kcal mol}^{-1}$. The major contribution comes from the enthalpy change, which is $\Delta H^0_{298} = -22.63\ \text{kcal mol}^{-1}$. Thus the reaction should go strongly to the right, if there is a suitable reaction path. The slowness of the uncatalyzed reaction must be associated with a large activation barrier. The experimental activation energy is $17\ \text{kcal mol}^{-1}$ (Table 8.1). Using the observed rate of the uncatalyzed decomposition, $v < 4 \times 10^{-8}\ M\,\text{s}^{-1}$, we can calculate an upper limit for the Arrhenius preexponential factor of $A < 1 \times 10^5\ M\,\text{s}^{-1}$. While the A factor is significantly less than the diffusion-limited rate, it is clear that the activation energy provides the larger contribution to the reaction barrier. This is illustrated in Figure 8.1.

By contrast with the uncatalyzed reaction, the activation energy for the H_2O_2 reaction catalyzed by Fe^{2+}/Fe^{3+} or HBr is only two thirds as large. In the presence of catalase, the activation energy is only $2\ \text{kcal mol}^{-1}$, and the Arrhenius preexponential factor is $1.6 \times 10^8\ M\,\text{s}^{-1}$. It appears that the enzyme (and to a lesser extent, the other catalysts) succeeds in speeding up the reaction by providing a path with a substantially smaller energy of activation. This is also shown in Figure 8.1. The entropy barrier is also reduced (hence the larger A factor), but its effect on the reaction velocity is less significant at room temperature. Note that the decrease in barrier height for the forward reaction implies a decrease in activation energy for the reverse reaction as well. For this example, the reverse reaction is strongly endergonic and it is very slow for that reason.

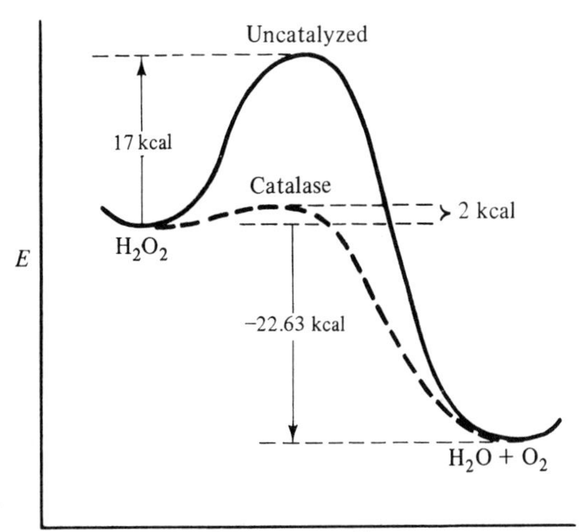

Fig. 8.1 Role of catalase in lowering the activation energy for the decomposition of H_2O_2.

However, for most other biochemical reactions that we will be considering, the overall reaction has a standard free energy change closer to zero. In those cases the enzymic pathway appreciably speeds both the forward and reverse reactions. This illustration provides a picture that helps us to understand the role that catalysts play.

A typical behavior for enzyme-catalyzed reactions is that the reaction is first order in substrate (reactant) at low substrate concentrations and then becomes zero order in substrate at high substrate concentration. This means the reaction reaches a maximum velocity for a constant enzyme concentration. The *turnover number* is defined as the maximum velocity (mol ℓ^{-1} s^{-1}) divided by the concentration of enzyme active sites (mol ℓ^{-1}). Its units are obviously s^{-1}. Concentration of active sites is used rather than concentration of enzyme so that a fair comparison can be made with enzymes that have more than one active site. Catalase has four active sites per molecule.

It would be misleading to give the impression that catalase activity is entirely typical of enzyme-catalyzed reactions (Table 8.2). Catalase is the fastest by several orders of magnitude. A more typical value for the turnover number of enzymes is 10^3 s^{-1}, and most values fall within a factor of 10 of that value. This is not to say that these other enzymes are only middling catalysts. The enzyme

Table 8.2 Maximum turnover numbers for several enzymes

Enzyme	Substrate	Turnover number (s^{-1})	Reference*
Catalase	H_2O_2	9×10^6	a
Acetylcholinesterase	Acetylcholine	1.2×10^4	b
Lactate dehydrogenase (chicken)	Pyruvate	6×10^3	c
Chymotrypsin	Acetyl-L-tyrosine ethyl ester	4.3×10^2	d
Myosin	ATP	3	e
Fumarase	L-Malate	1.1×10^3	f
	Fumarate	2.5×10^3	
Carbonic anhydrase (bovine)	CO_2	8×10^4	g
	HCO_3^-	3×10^4	

* (a) Bonnichsen, R. K., B. Chance, and H. Theorell, *Acta Chem. Scand. 1*, 685, (1947); (b) Froede, H. C., and I. B. Wilson, *The Enzymes 5*, 87 (1971); (c) Everse, J., and N. O. Kaplan, *Advan. Enzymol. 37*, 61 (1973); (d) Kaufman, S., H. Neurath, and G. W. Schwert, *J. Biol Chem. 177*, 793 (1949); (e) Brahms, J., and C. M. Kay, *J. Biol. Chem. 238*, 198, (1963); (f) Brant, D. A., L. B. Barnett, and R. A. Alberty, *J. Am. Chem. Soc. 85*, 2204 (1963); (g) DeVoe, H., and G. B. Kistiakowsky, *J. Amer. Chem. Soc. 83*, 274 (1961).

fumarase catalyzes the hydration of fumarate to L-malate with a turnover number of $2.5 \times 10^3\ s^{-1}$ at 25°C:

$$\underset{\text{Fumarate}}{{}^{-}OOC{-}CH{=}CH{-}COO^{-}} + H_2O \xrightleftharpoons{\text{fumarase}} \underset{\text{L-Malate}}{{}^{-}OOC{-}CH(OH){-}CH_2{-}COO^{-}}$$

The rate constants for 1 *M* fumarate hydrolysis catalyzed by acid (1 *M*) or base (1 *M*) are about $10^{-8}\ s^{-1}$. This "ordinary" enzyme still wins by eleven powers of 10.

KINETICS OF ENZYMATIC REACTIONS

Because enzymes are such enormously effective catalysts, they are able to exert their influence at exceedingly low concentrations, typically 10^{-10} to 10^{-8} *M*. At this level, especially if the enzyme is present in a complex cellular soup, it is difficult to make direct measurements of what the enzyme itself is doing. Enzymology began, therefore, with a study of the kinetics of the disappearance of substrate and formation of products, because their concentrations are typically 10^{-6} to 10^{-3} *M*. Early observations resulted in the following generalizations about enzyme reactions:

1. The rate of substrate conversion increases linearly with increasing enzyme concentration (Fig. 8.2).

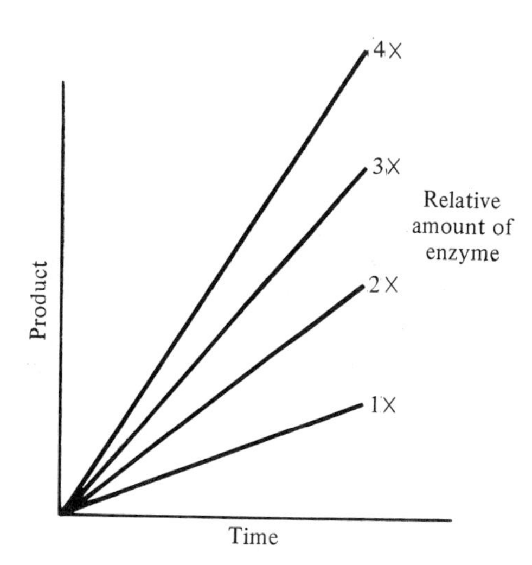

Fig. 8.2 Increase of product concentration with time for four different amounts of enzyme.

2. For a fixed enzyme concentration, the velocity is linear in substrate concentration [S] at low values of [S] (Fig. 8.3).
3. The rates of enzyme reactions approach a maximum or saturation velocity at high substrate concentration (Fig. 8.3).

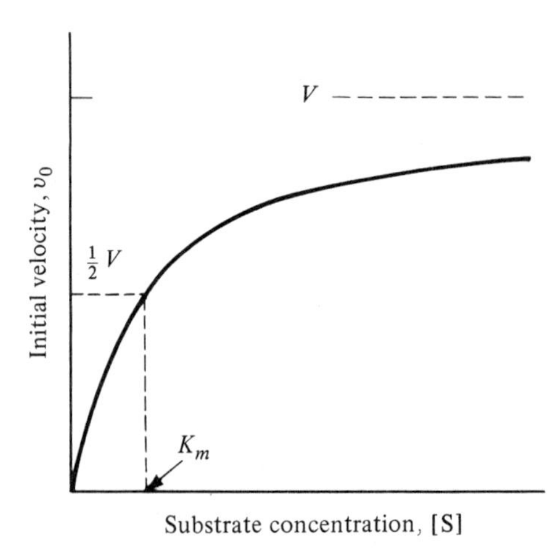

Fig. 8.3 Initial velocity, v_0, of an enzyme-catalyzed reaction as a function of substrate concentration [S] for a fixed amount of enzyme.

At low substrate concentrations,

$$v_0 = k[\mathrm{S}] \tag{8.1}$$

where v_0 is the initial velocity. At high substrate concentrations,

$$v_0 = V \qquad \text{(maximum velocity)} \tag{8.2}$$

This behavior is expected if the enzyme forms a complex with the substrate. At high substrate concentrations essentially all of the enzyme is tied up in the enzyme-substrate complex. Under these conditions the enzyme is working at full capacity, as measured by its

$$\text{turnover number} = \frac{V}{[\mathrm{E}]_0} \tag{8.3}$$

where $[\mathrm{E}]_0$ is the total enzyme site concentration. At low substrate concentration the enzyme is not saturated, and turnover is limited partly by the availability of substrate.

A kinetic formulation of these ideas was presented by Michaelis and Menten, with significant improvements by Briggs and Haldane (1925). In the simplest picture, the enzyme and substrate reversibly form a complex, followed by dissociation of the complex to form the product and regenerate the free enzyme.

$$\mathrm{E} + \mathrm{S} \underset{k_{-1}}{\overset{k_1}{\rightleftharpoons}} \mathrm{ES} \tag{8.4}$$

$$\mathrm{ES} \xrightarrow{k_2} \mathrm{E} + \mathrm{P} \tag{8.5}$$

Because the second step must also be reversible, this mechanism applies strictly only to the initial stages of the reaction before the product concentration has become significant. Under these conditions

$$v_0 = \left(\frac{d[\mathrm{P}]}{dt}\right)_0 = k_2[\mathrm{ES}] \tag{8.6}$$

We apply the steady-state approximation:

$$\frac{d[\mathrm{ES}]}{dt} = k_1[\mathrm{E}][\mathrm{S}] - k_{-1}[\mathrm{ES}] - k_2[\mathrm{ES}] \cong 0$$

Therefore,

$$[\mathrm{ES}] = \frac{k_1[\mathrm{E}][\mathrm{S}]}{k_{-1} + k_2} \tag{8.7}$$

[E] and [S] refer to the *free* concentrations of enzyme and substrate; however, these are difficult to measure, so we write the equation in terms of the measurable *total* enzyme, $[\mathrm{E}]_0$, and substrate, $[\mathrm{S}]_0$, concentrations:

$$\begin{aligned} [\mathrm{E}]_0 &= [\mathrm{E}] + [\mathrm{ES}] \\ [\mathrm{S}]_0 &\cong [\mathrm{S}] \end{aligned} \tag{8.8}$$

We can equate the total substrate concentration to the free substrate concentration, because the enzyme-substrate concentration is typically small compared to the substrate concentration. By substitution into Eq. (8.7) and rearranging terms, we obtain

$$[\mathrm{ES}] = \frac{[\mathrm{E}]_0}{1 + \dfrac{k_{-1} + k_2}{k_1[\mathrm{S}]}} \tag{8.9}$$

Substitution into Eq. (8.6) gives

$$v_0 = \frac{k_2[\mathrm{E}]_0}{1 + \dfrac{k_{-1} + k_2}{k_1[\mathrm{S}]}} = \frac{V}{1 + \dfrac{K_m}{[\mathrm{S}]}} \tag{8.10}$$

where $K_m = (k_{-1} + k_2)/k_1$ is the *Michaelis constant* for the enzyme and substrate combination, and the substitution $k_2[\mathrm{E}]_0 = V$ is the form of Eq. (8.6) appropriate to high substrate concentrations. The Michaelis constant is the ratio of rate constants for reactions involving dissociation of ES to those involving formation of ES. One interpretation is that $1/K_m$ is a measure of the affinity of an enzyme for its substrate. Another interpretation can be seen from the properties of Eq. (8.10). K_m has the dimensions of concentration, and when $[\mathrm{S}] = K_m$, then $v_0 = \frac{1}{2}V$. In other words, the Michaelis constant is equal to the substrate concentration that is sufficient to give half the maximum velocity for the enzyme. This is shown graphically in Fig. 8.3. One should get an intuitive feeling for the magnitude of K_m. A small value of K_m means that the enzyme binds the substrate tightly and small concentrations of substrate are sufficient to saturate the enzyme and to reach the maximum catalytic efficiency of the enzyme.

During the past two decades it has become increasingly clear that very few enzymes act by a simple mechanism with a single enzyme-substrate complex. Complications occur when more than two reactants are involved or when the enzyme has two or more active sites that are not independent. Many of these complexities are more easily understood if they are considered as variations on the theme of Michaelis-Menten. Therefore, it is useful to explore this basic pattern further.

Kinetic Data Analysis

For quantitative purposes it is useful to rewrite Eq. (8.10) in a form that suggests a straight-line plot of the data. Several such approaches are used; the most popular was proposed by Lineweaver and Burk (1934).

$1/v_0$ versus $1/[S]$: Lineweaver-Burk

Taking the reciprocal of both sides of Eq. (8.10) and rearranging, we obtain

$$\frac{1}{v_0} = \frac{1}{V} + \frac{K_m}{V} \cdot \frac{1}{[S]} \tag{8.11}$$

Thus a plot of the reciprocal of initial velocity versus the reciprocal of initial substrate concentration for experiments at fixed enzyme concentration should give a straight line. Furthermore, the intercept with the ordinate gives $1/V$, and the slope is K_m/V, from which these two constants can be determined. Alternatively, K_m can be determined by extrapolation to the abscissa intercept, to give $-1/K_m$, as shown in Fig. 8.4.

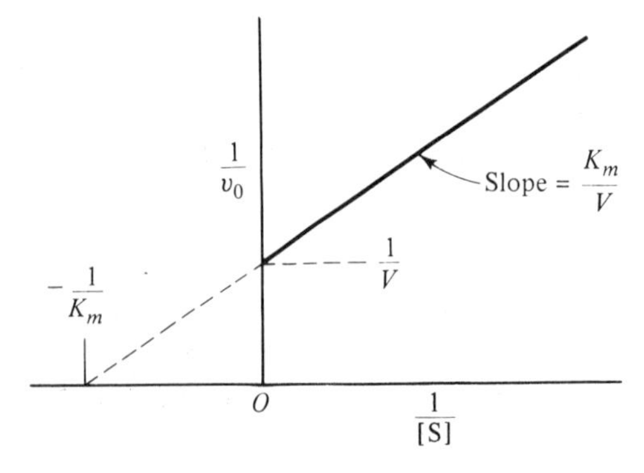

Fig. 8.4 Lineweaver-Burk plot of the reciprocal of initial reaction velocity versus reciprocal of initial substrate concentration for a series of experiments at fixed enzyme concentration.

Exercise Show on a graph such as Fig. 8.4 that $1/v_0 = 2/V$ when $[S] = K_m$, and explain how this could be used to determine V and K_m by a simple procedure.

Other procedures for manipulating the data are also used.

$[S]/v_0$ versus [S]: Dixon

Starting with Eq. (8.11), multiply both sides by [S] to obtain

$$\frac{[S]}{v_0} = \frac{[S]}{V} + \frac{K_m}{V} \tag{8.12}$$

v_0 versus $v_0/[S]$: Eadie-Hofstee

Multiply both sides of Eq. (8.11) by $V \cdot v_0$ and rearrange to

$$v_0 = -K_m \cdot \frac{v_0}{[S]} + V \tag{8.13}$$

Ideally, with no experimental error each of these equations gives exactly the same desired information from a straight-line plot. In practice, one or the other may be preferable because of the nature of the data involved. Qualitatively, the Eadie-Hofstee plot spreads the values at high substrate concentration (where $v_0 \to V$), whereas the Lineweaver-Burk plot compresses the points in this region. Quantitative analysis of the different plots may provide somewhat different values for K_m and V. An illustration of this is shown in the following example.

Example 8.1 The hydrolysis of carbobenzoxyglycyl-L-tryptophan catalyzed by pancreatic carboxypeptidase occurs according to the reaction

Carbobenzoxyglycyl-L-tryptophan + H_2O $\longrightarrow$
carbobenzoxyglycine + L-tryptophan

The following data on the rate of formation of L-tryptophan at 25°C, pH 7.5, were obtained by R. Lumry, E. L. Smith, and R. R. Glantz, *J. Amer. Chem. Soc. 73*, 4330 (1951):

Substrate concentration (mM)	2.5	5.0	10.0	15.0	20.0
Velocity (mM s^{-1})	0.024	0.036	0.053	0.060	0.064

(a) Plot these data according to the Lineweaver-Burk method and determine values for K_m and V. (b) Repeat the analysis using the Eadie-Hofstee method.

Solution For the purpose of making these plots, the data are reformulated in the following table:

[S] (mM)	2.5	5.0	10.0	15.0	20.0
$\frac{1}{v_0}$ (mM^{-1} s)	41.7	27.8	18.9	16.7	15.6
$\frac{1}{[S]}$ (M^{-1})	400	200	100	66.7	50
$10^3 \times \frac{v_0}{[S]}$ (s^{-1})	9.6	7.2	5.3	4.0	3.2

(a) The Lineweaver-Burk plot is as shown.

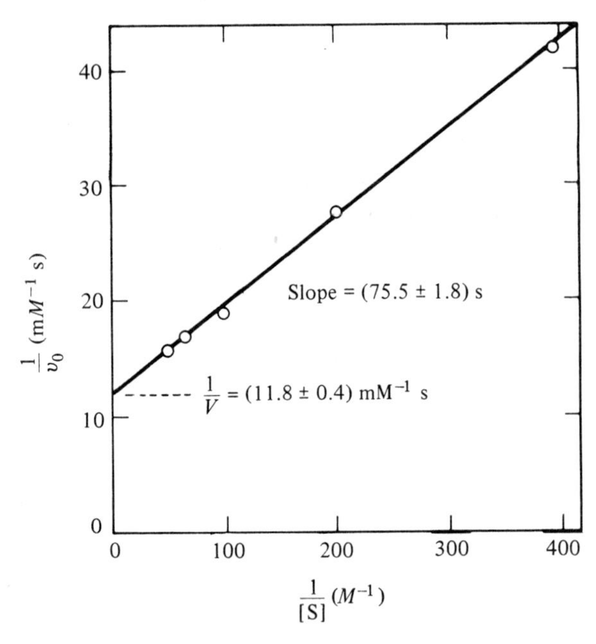

Lineweaver-Burk plot

From the slope and intercept, we obtain

$$V = (0.0847 \pm 0.0027)\,\text{m}M\,\text{s}^{-1}$$

$$\frac{K_m}{V} = (75.5 \pm 1.8)\,\text{s} \quad \text{and} \quad K_m = (6.39 \pm 0.36)\,\text{m}M$$

(b) The Eadie–Hofstee plot gives

$$V = (0.0856 \pm 0.0022)\,\mathrm{m}M\,\mathrm{s}^{-1}$$

$$K_m = (6.52 \pm 0.35)\,\mathrm{m}M$$

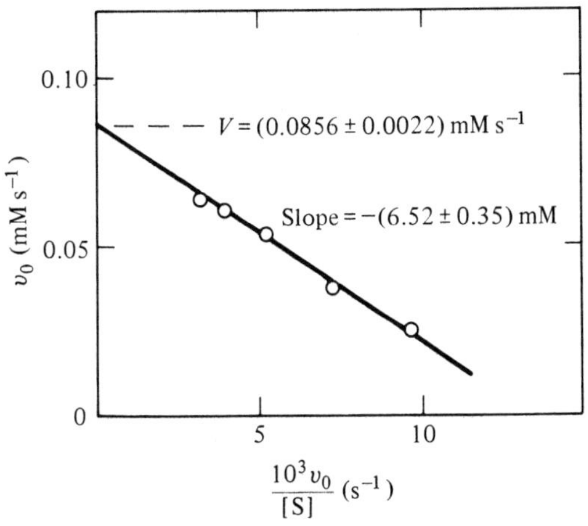

Eadie-Hofstee plot

Each of these values is given with its standard deviation, calculated for a least-squares fit of a straight line to the data. Although the values of V and K_m differ for the two methods, each best value lies within 1 standard deviation of the other.

Two intermediate complexes

Because most enzymatic reactions are readily reversible, it should not be surprising that there is also evidence for an enzyme-product complex. This is often kinetically distinct from the enzyme-substrate complex. The action of the enzyme fumarase illustrates this nicely.

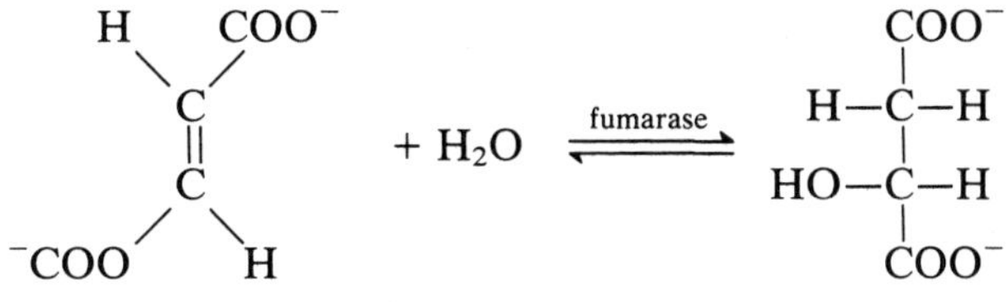

$$\text{Fumarate} + H_2O \xrightleftharpoons{\text{fumarase}} \text{L-Malate}$$

Fumarate L-Malate

The enzyme fumarase catalyzes the hydration of fumarate to form malate, and of course it must also catalyze the reverse reation. At equilibrium, the mixture consists of about 20% fumarate and 80% malate at room temperature.

Alberty and collaborators (see Alberty and Peirce, 1957) considered the following mechanism:

$$\mathrm{E + F} \underset{k_{-1}}{\overset{k_1}{\rightleftharpoons}} \mathrm{EF} \underset{k_{-2}}{\overset{k_2}{\rightleftharpoons}} \mathrm{EM} \underset{k_{-3}}{\overset{k_3}{\rightleftharpoons}} \mathrm{E + M} \tag{8.14}$$

There are six kinetic constants in this formalism and, as in the case of the simpler mechanism [Eqs. (8.4) and (8.5)], we cannot determine all of them by steady-state experiments. By fairly straightforward treatment similar to that of the earlier section, we can develop the following relations:

Forward reaction, $[\mathrm{M}]_0 = 0$

$$v_\mathrm{F} = \left(\frac{d[\mathrm{M}]}{dt}\right)_0 = k_3[\mathrm{EM}]$$

$$= \frac{k_2 k_3 [\mathrm{E}]_0 [\mathrm{F}]}{(k_2 + k_{-2} + k_3)\left\{\dfrac{k_{-1}k_{-2} + k_{-1}k_3 + k_2 k_3}{k_1(k_2 + k_{-2} + k_3)} + [\mathrm{F}]\right\}} \tag{8.15}$$

which can be written

$$v_\mathrm{F} = \frac{V_\mathrm{F}[\mathrm{F}]}{K_m^\mathrm{F} + [\mathrm{F}]} \tag{8.16}$$

where

$$V_\mathrm{F} = \frac{k_2 k_3}{k_2 + k_{-2} + k_3}[\mathrm{E}]_0 \tag{8.17}$$

and

$$K_m^\mathrm{F} = \frac{k_{-1}k_{-2} + k_{-1}k_3 + k_2 k_3}{k_1(k_2 + k_{-2} + k_3)} \tag{8.18}$$

Despite the seeming differences, Eq. (8.16) has precisely the same form as Eq. (8.10), and there is no way to detect the presence of the additional intermediate by simple velocity measurements *in steady-state experiments.* Methods like those of relaxation kinetics are needed to resolve this question.

Reverse reaction, $[\mathrm{F}]_0 = 0$

$$v_\mathrm{M} = \left(\frac{d[\mathrm{F}]}{dt}\right)_0 = k_{-1}[\mathrm{EF}] \tag{8.19}$$

which can be written

$$v_\mathrm{M} = \frac{V_\mathrm{M}[\mathrm{M}]}{K_m^\mathrm{M} + [\mathrm{M}]} \tag{8.20}$$

where

$$V_M = \frac{k_{-1}k_{-2}}{k_{-1} + k_2 + k_{-2}}[E]_0 \tag{8.21}$$

and

$$K_m^M = \frac{k_{-1}k_{-2} + k_{-1}k_3 + k_2k_3}{k_{-3}(k_{-1} + k_2 + k_{-2})} \tag{8.22}$$

It is also possible to formulate an expression for the net velocity as

$$v = \frac{V_F K_m^M[F] - V_M K_m^F[M]}{K_m^F K_m^M + K_m^M[F] + K_m^F[M]} \tag{8.23}$$

This reduces to the previous equations in the appropriate limits. Note that when equilibrium is reached, $v = 0$, and therefore

$$V_F K_m^M[F]^{eq} - V_M K_m^F[M]^{eq} = 0$$

The equilibrium constant K for the reaction is

$$K = \frac{[M]^{eq}}{[F]^{eq}} = \frac{V_F K_m^M}{V_M K_m^F} \tag{8.24}$$

In a process involving a complex sequence of steps, such as occurs in enzymatic reactions, it is difficult to determine the detailed mechanism from steady-state kinetic studies. The reason is that only one, or a very few, of the steps in the sequence is rate determining. Changes in concentration of substrate or enzyme will produce consequences only in terms of the effect on this rate-determining step.

It is difficult, but sometimes possible, to design steady-state studies so as to extract further information about other steps. For example, some of the individual rate constants in the mechanism of the fumarase reaction were estimated from a study of the temperature and pH dependence of the steady-state velocities (Brant et al., 1963). In other cases, direct measurement of an enzyme complex is possible, and this provides additional information which can be used to establish relations among the rate constants.

RELAXATION METHODS

Although steady-state kinetic studies cannot separate the many rate constants involved in a complicated mechanism, methods for studying very fast reactions can help to distinguish some of the steps. The relaxation methods we shall discuss here can be applied to any type of reaction, but applications to enzymes have been very fruitful. Hammes's (1974) analysis of ribonuclease is a good example. We shall discuss relaxation methods with special emphasis on temperature-jump kinetics.

Steady State

Consider a reversible system

$$A \underset{k_{-1}}{\overset{k_1}{\rightleftharpoons}} B$$

At equilibrium, the rates of the forward and back reactions are the same. The position of equilibrium is represented in terms of an equilibrium constant,

$$K = \frac{k_1}{k_{-1}} = \frac{[B]^{eq}}{[A]^{eq}}$$

which is the ratio of the rate constants k_1 and k_{-1}. The usual measurements of the "average" composition of the system at equilibrium can only provide a value for the ratio, not for the individual rate constants. Clearly, k_1 and k_{-1} can range from very large to very small and still give the same ratio (the equilibrium constant K). If we change the temperature or pressure of the system so as to attain a new equilibrium state, the new state will in general have a different value for the equilibrium constant. We are still completely ignorant of the individual rate constants, however.

Several approaches are possible to extract the desired rate constants and mechanistic information. One category involves the use of *relaxation* methods, in which the system at equilibrium is suddenly perturbed in such a way that it finds itself out of equilibrium under the new conditions. It then relaxes to a state of equilibrium under the new conditions, and the rate of relaxation is governed by the rate constants and mechanism of the process. Examples of relaxation processes include:

1. Temperature jump, which can be achieved by the discharge of an electric capacitor, the use of a rapid laser pulse, or by addition of hot water to produce a sudden increase in temperature of a system.
2. Pressure jump, where a restraining diaphragm is suddenly ruptured to release pressurization of the system.
3. Flash or laser pulse photolysis, where a pulse of light produces a sudden change in the population of electronically excited states of absorbing molecules.

In each of these cases the system suddenly finds itself out of equilibrium at the instant of the pertubation, and the experimenter uses a suitable experimental method to monitor the approach to a new equilibrium state under the altered conditions.

To convince yourself that there really is new information available from relaxation methods, consider the analogous situation of musicians with respect to their instruments. When a guitar player takes the instrument from its case, there is no way of knowing whether its strings are in tune without plucking them and listening to them as they relax. No amount of measurement of static properties, such as string tension, cross section, ambient temperature, and so on, will ever allow a musician to omit tuning an instrument *dynamically* before beginning to play.

A second major category of methods for dealing with systems at equilibrium involves the study of fluctuations about the equilibrium state. As a consequence of both the dynamic and the statistical nature of systems at equilibrium, the system and many of its properties undergo constant and rapid fluctuations about some average or median state (see Chapter 3). Any normal (long-term) measurements always give the same result if the system is truly at equilibrium, and these results serve to characterize the average equilibrium state. When we are able to study the system on a short-enough time scale, however, we observe that its properties exhibit a kind of Brownian motion, in the sense that they are constantly moving one way or the other with respect to their average equilibrium value. This can be true for such properties as

1. Refractive index, where fluctuations result in variations in scattering of light from laser beams.
2. Concentration, which can be detected as fluctuations in the absorption or fluorescence of particular species.
3. Position, involved in scattering of fine laser beams off single particles.
4. Pressure, where fluctuations can produce "noise" in a suitable acoustic detector.

In each of these cases there is no need to perturb the system artificially; one simply takes advantage of the fluctuations that occur naturally because of the particulate and statistical nature of matter.

Relaxation Kinetics

As an example of relaxation kinetics, consider an equilibrium of the type

$$A + B \underset{k_{-1}}{\overset{k_1}{\rightleftharpoons}} P$$

The rate expression is

$$\frac{d[P]}{dt} = k_1[A][B] - k_{-1}[P] \tag{8.25}$$

At equilibrium, $d[P]/dt$ is zero, and

$$K = \frac{k_1}{k_{-1}} = \frac{[\bar{P}]}{[\bar{A}][\bar{B}]} \tag{8.26}$$

where the bar over the concentration terms emphasizes the equilibrium values. Now generate a nearly discontinuous change (typically within 10^{-6} s, but it can be as short as 10^{-8} s) of temperature or pressure. Changing temperature or pressure will in general cause the equilibrium concentrations to change; we can use thermodynamics to tell us how large a change to expect. From Chapter 3 we remember that

$$\left(\frac{\partial \Delta G}{\partial T}\right)_P = -\Delta S \quad \text{and} \quad \left(\frac{\partial \Delta G}{\partial P}\right)_T = \Delta V \tag{8.27}$$

Combining these equations with $\Delta G^0 = -RT\ln K$, we obtain

$$\left(\frac{\partial \ln K}{\partial T}\right)_P = \frac{\Delta H^0}{RT^2} \quad \text{and} \quad \left(\frac{\partial \ln K}{\partial P}\right)_T = -\frac{\Delta V^0}{RT} \tag{8.28}$$

Thermodynamic standard states for solutions include the requirement that the pressure is 1 atm. Therefore, we note that the pressure dependence of K refers to change in equilibrium concentrations, not activities. So Eq. (8.28) tells us that the equilibrium constant, K, will be different at the new temperature or pressure immediately after the pulse. How much different will depend on the enthalpy change of the reaction for a T jump, or the volume change of the reaction for a P jump. The system finds itself suddenly out of equilibrium under the new conditions, and we can follow the concentration changes during the subsequent relaxation to the new equilibrium state.

Conditions are usually chosen so that the T jump or P jump produces relatively small displacements from equilibrium, that is, ΔT of 5 to 10 degrees or ΔP of 100 to 1000 atm. This simplifies not only the mathematics, but also the thermodynamic analysis of the results. Suppose that, following the sudden perturbation, the system relaxes so as to change the concentration of A by a small amount $\Delta[\mathrm{A}]$, B by $\Delta[\mathrm{B}]$, and P by $\Delta[\mathrm{P}]$. The perturbation is small enough so that $(\Delta[\mathrm{A}])^2$ is negligible compared to $\Delta[\mathrm{A}]$, and so on. It is then convenient to rewrite the instantaneous, time-dependent concentrations as

$$[\mathrm{P}] = [\bar{\mathrm{P}}] + \Delta[\mathrm{P}]$$
$$[\mathrm{A}] = [\bar{\mathrm{A}}] + \Delta[\mathrm{A}] = [\bar{\mathrm{A}}] - \Delta[\mathrm{P}]$$
$$[\mathrm{B}] = [\bar{\mathrm{B}}] + \Delta[\mathrm{B}] = [\bar{\mathrm{B}}] - \Delta[\mathrm{P}]$$

since $\Delta[\mathrm{P}] = -\Delta[\mathrm{A}] = -\Delta[\mathrm{B}]$ for this example. The values of $[\bar{\mathrm{P}}]$, $[\bar{\mathrm{A}}]$, and $[\bar{\mathrm{B}}]$ refer to the equilibrium concentrations at the final T or P. Therefore,

$$\frac{d[\mathrm{P}]}{dt} = \frac{d([\bar{\mathrm{P}}] + \Delta[\mathrm{P}])}{dt} = \frac{d(\Delta[\mathrm{P}])}{dt}$$

Because $\dfrac{d[\bar{\mathrm{P}}]}{dt} = 0$

$$\begin{aligned}\frac{d(\Delta[\mathrm{P}])}{dt} = \frac{d[\mathrm{P}]}{dt} &= k_1[\mathrm{A}][\mathrm{B}] - k_{-1}[\mathrm{P}]\\ &= k_1([\bar{\mathrm{A}}] - \Delta[\mathrm{P}])([\bar{\mathrm{B}}] - \Delta[\mathrm{P}]) - k_{-1}([\bar{\mathrm{P}}] + \Delta[\mathrm{P}])\\ &= k_1[\bar{\mathrm{A}}][\bar{\mathrm{B}}] - k_{-1}[\bar{\mathrm{P}}] - k_1([\bar{\mathrm{A}}]\Delta[\mathrm{P}] + [\bar{\mathrm{B}}]\Delta[\mathrm{P}] - (\Delta[\mathrm{P}])^2)\\ &\qquad - k_{-1}\Delta[\mathrm{P}]\end{aligned} \tag{8.29}$$

$$-\frac{d(\Delta[\mathrm{P}])}{dt} = \{k_1 \cdot ([\bar{\mathrm{A}}] + [\bar{\mathrm{B}}]) + k_{-1}\}\,\Delta[\mathrm{P}] \tag{8.30}$$

In order to achieve the last equation, we have dropped the term involving $(\Delta[\mathrm{P}])^2$, because it is small for small displacements from equilibrium, and we have used Eq. (8.25) to show that the first two terms of Eq. (8.29) sum to zero. The resulting equation (8.30) is a simple first-order equation, because $\{k_1([\bar{\mathrm{A}}] + [\bar{\mathrm{B}}]) + k_{-1}\}$ is independent of time. If we rewrite Eq. (8.30) as

$$-\frac{d(\Delta[\mathrm{P}])}{dt} = \frac{\Delta[\mathrm{P}]}{\tau}$$

it can be integrated to give

$$\Delta[\mathrm{P}] = \Delta[\mathrm{P}]_0\, e^{-t/\tau} \tag{8.31}$$

where τ is the *relaxation time* for the process. For this example

$$\tau = \frac{1}{k_{-1} + k_1([\bar{\mathrm{A}}] + [\bar{\mathrm{B}}])} \tag{8.32}$$

Results for other examples are presented in Table 8.3.

A very important and at first surprising result of analyses of each of the examples listed in Table 8.3 is that the relaxation kinetics are always simple first order (exponential in time) regardless of the number of molecules involved as reactants or products. This is true because the perturbations produce only small changes in the equilibrium concentrations and we can ignore squared terms. The behavior is illustrated in Fig. 8.5 for the case where the perturbation shifts the

Table 8.3 Relaxation times for reactions involving single steps

Mechanism	Relaxation time*
$\mathrm{A} \underset{k_{-1}}{\overset{k_1}{\rightleftharpoons}} \mathrm{B}$	$\tau = \dfrac{1}{k_1 + k_{-1}}$
$\mathrm{A} + \mathrm{B} \underset{k_{-1}}{\overset{k_1}{\rightleftharpoons}} \mathrm{P}$	$\tau = \dfrac{1}{k_{-1} + k_1 \cdot ([\bar{\mathrm{A}}] + [\bar{\mathrm{B}}])}$
$\mathrm{A} + \mathrm{B} + \mathrm{C} \underset{k_{-1}}{\overset{k_1}{\rightleftharpoons}} \mathrm{P}$	$\tau = \dfrac{1}{k_{-1} + k_1 \cdot ([\bar{\mathrm{A}}][\bar{\mathrm{B}}] + [\bar{\mathrm{B}}][\bar{\mathrm{C}}] + [\bar{\mathrm{A}}][\bar{\mathrm{C}}])}$
$\mathrm{A} + \mathrm{B} \underset{k_{-1}}{\overset{k_1}{\rightleftharpoons}} \mathrm{P} + \mathrm{Q}$	$\tau = \dfrac{1}{k_1 \cdot ([\bar{\mathrm{A}}] + [\bar{\mathrm{B}}]) + k_{-1} \cdot ([\bar{\mathrm{P}}] + [\bar{\mathrm{Q}}])}$
$2\,\mathrm{A} \underset{k_{-1}}{\overset{k_1}{\rightleftharpoons}} \mathrm{A}_2$	$\tau = \dfrac{1}{4k_1[\bar{\mathrm{A}}] + k_{-1}}$

* $[\bar{\mathrm{A}}]$, $[\bar{\mathrm{B}}]$, etc., represent the equilibrium concentrations.

SOURCE: M. Eigen and L. de Maeyer, in *Investigation of Rates and Mechanisms of Reactions*, 3rd ed., Vol. 6, Part II, G. G. Hammes (ed.), Wiley-Interscience, New York, 1974, Chap. 3.

temperatures. The primary step in the photo response of rhodopsin occurs with a quantum yield of about unity. In the eye, the light stimulus requires additional processing before it arrives as a signal at the brain. Nevertheless, an incidence of only a few photons per second is sufficient to give a significant visual sensation.

The visual process is a cyclic one, and the rhodopsin is regenerated in the dark by processes that are incompletely understood. It is likely that the reconstitution of active rhodopsin has a requirement for metabolic energy. There is no evidence that any of the energy of the photon is stored by the rhodopsin photochemistry; light simply serves to trigger or activate an otherwise exergonic ($\Delta G < 0$) reaction. The principal function of the photon is to overcome an activation barrier. This is the most common kind of photochemical process.

Photosynthesis

The process of photosynthesis in plants, algae, and bacteria involves the incorporation of carbon into the various compounds such as carbohydrates, proteins, lipids, nucleic acids, and so on, that make up the material substance and the essential functional components of the organism. The source of carbon for plants and most algae is carbon dioxide; in other organisms, small carbon-containing compounds (acetate, succinate, malate, etc.) are required. The representative reaction for higher plants is

$$CO_2 + H_2O \longrightarrow (CH_2O) + O_2$$

where (CH_2O) represents the fixed carbon of carbohydrate, the major metabolic product. This reaction is endothermic by about +116 kcal $(\text{mol } CO_2)^{-1}$, and the energy required to drive this and the other biosynthetic reactions comes ultimately from sunlight. By contrast with most photochemical reactions and with vision, a portion of the energy of the absorbed photons is retained in photosynthesis in the form of chemical potential of the metabolic products. [For purposes of estimating the efficiency of such processes as photosynthesis, we should compare ΔG for the carbon-fixation process with the free energy change associated with the absorption of radiation. The value of ΔG for carbon fixation is about +118 kcal $(\text{mol } CO_2)^{-1}$, but the free-energy change for the absorption of radiation is more difficult to estimate. A detailed analysis for solar irradiation is given by Ross and Calvin (1967).]

The overall process of photosynthesis can be divided into light-dependent and dark reactions. Light absorbed by chlorophyll pigments serves to split water molecules into molecular oxygen and hydrogen atom equivalents. The latter are transferred as electrons and hydrogen ions along a transport chain of cytochromes, quinones, and iron-, manganese-, and copper-containing proteins to nicotinamide adenine dinucleotide phosphate ($NADP^+$), which becomes reduced. During this process a portion of the energy is stored as the high-energy product

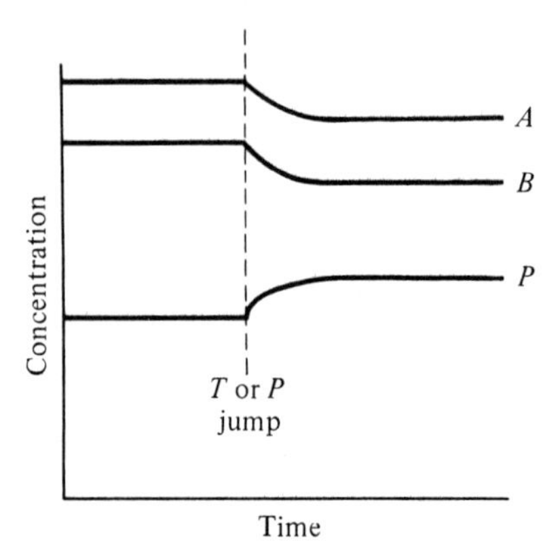

Fig. 8.5 Effect of a temperature or pressure jump on concentrations of reactants or products for a system initially at equilibrium or in a steady state. $A + B \rightarrow P$.

equilibrium in favor of product. The reverse situation is also found, depending on the signs of ΔH^0 and ΔV^0.

An experimental trace is shown in Fig. 8.6 for the dimerization of proflavin, according to the reaction

$$2\,\mathrm{P} \underset{k_{-1}}{\overset{k_1}{\rightleftharpoons}} \mathrm{P}_2 \tag{8.33}$$

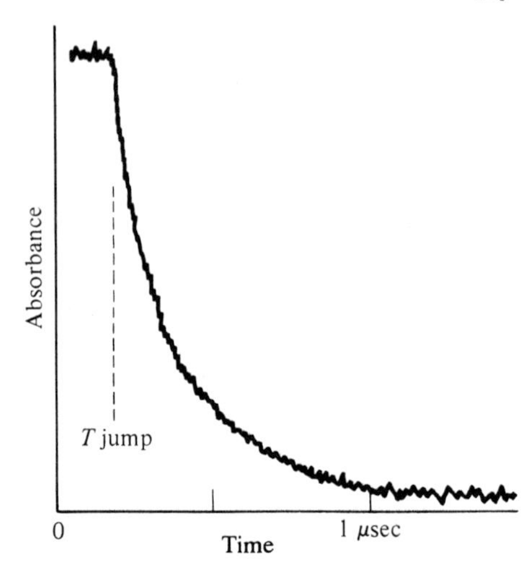

Fig. 8.6 Relaxation of proflavin dimerization. Total concentration $4.64 \times 10^{-3}\,M$, pH 7.8, final temperature 6°C. Absorbance monitored at 455 nm. [Curve from D. H. Turner, G. W. Flynn, S. K. Lundberg, L. D. Faller, and N. Sutin, *Nature 239*, 215 (1972).]

Because of the fast kinetics involved, it was necessary to use a pulsed laser to provide the temperature jump. For dimerizations it is possible to simplify the relaxation expression by writing it in terms of the total concentration $[\mathrm{P}]_t$ of monomers and dimers. Turner et al. (1972) used the expression

$$\frac{1}{\tau^2} = k_{-1}^2 + 8k_1k_{-1}[\mathrm{P}]_t \tag{8.34}$$

The analysis according to Eq. (8.34) does not require a prior knowledge of the equilibrium constant to determine both rate constants. The data should fall on a straight line when plotted according to Eq. (8.34). The plot shown in Fig. 8.7 demonstrates not only that this is observed, but also that the rate constants are independent of pH between 4 and 7.8. From the slope and intercept of the line shown in Fig. 8.7, $k_1 = 8 \times 10^8\,M^{-1}\,\mathrm{s}^{-1}$ and $k_{-1} = 2.0 \times 10^6\,\mathrm{s}^{-1}$ at 25°C were

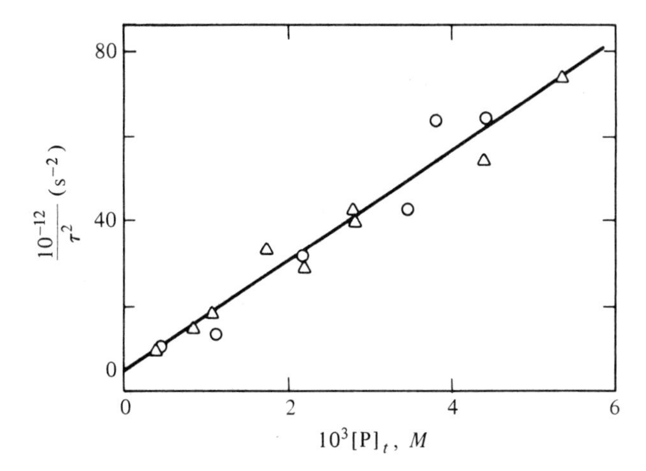

Fig. 8.7 Plot of relaxation data for proflavin dimerization as a function of total proflavin concentration, P_t ○, pH 7.8; △, pH 4.0. Temperature 25°C. [From D. H. Turner, G. W. Flynn, S. K. Lundberg, L. D. Faller, and N. Sutin, *Nature 239*, 215 (1972).]

obtained. The dimerization rate constant corresponds to a process that is almost at the limit placed by the diffusion of the monomeric species in aqueous solution.

Example 8.2 The dimerization of the decanucleotide A_4GCU_4 has been studied by Pörschke et al. (1973). The letters A, G, C, and U represent the bases adenine, guanine, cytosine, and uracil, and this oligonucleotide forms base-paired double-stranded helices, which are models for nucleic acids. The two strands are antiparallel; therefore, the oligonucleotide is self-complementary.

$$2\,A_4GCU_4 \underset{k_{-1}}{\overset{k_1}{\rightleftharpoons}} \begin{matrix} A-A-A-A-G-C-U-U-U-U \\ \vdots\quad\vdots\quad\vdots\quad\vdots\quad\vdots\quad\vdots\quad\vdots\quad\vdots\quad\vdots\quad\vdots \\ U-U-U-U-C-G-A-A-A-A \end{matrix}$$

(a) A temperature jump was applied to a solution containing $7.45 \times 10^{-6}\,M$ oligonucleotide at pH 7.0 and a final temperature of 32.4°C.

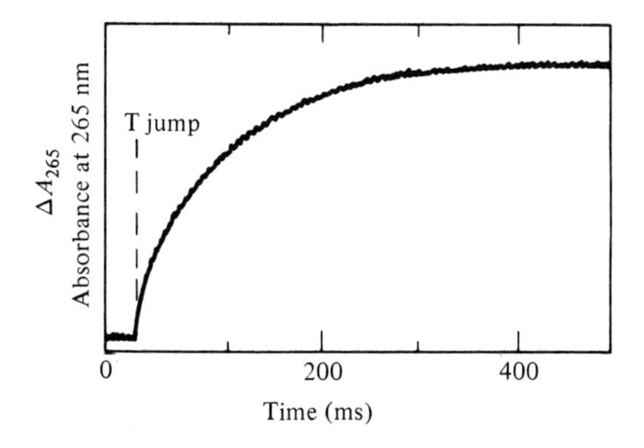

From the response in the uv absorption shown in the figure, the following data can be extracted:

Time (ms)	0	50	100	150	200	250	300	∞
$100 \times \Delta A_{265}$	0	2.0	3.2	3.9	4.28	4.47	4.57	4.70

The difference in absorbance at 265 nm (ΔA_{265}) is directly proportional to the concentration of product at time t minus the concentration at zero time. Use the results to calculate the relaxation time constant under these conditions.

(b) Similar experiments were carried out at 23.3°C and pH 7.0 for a series of concentrations of single-stranded oligomer. The observed relaxation times were as follows:

τ (ms)	455	370	323	244
$[\bar{\text{M}}]$ (μM)	1.63	2.45	3.45	5.90

Determine k_1 and k_{-1} for double-strand (dimer) formation and dissociation, respectively, under these conditions.

Solution

(a) First, test the data for first-order relaxation. The absorbance should approach its value at infinite time (its equilibrium value) exponentially.

$$\frac{\Delta A_\infty - \Delta A_t}{\Delta A_\infty} = e^{-t/\tau}$$

This is equivalent to Eq. (8.31).

Time (ms)	0	50	100	150	200	250	300
$100 \times (\Delta A_\infty - \Delta A_t)$	4.7	2.7	1.5	0.8	0.42	0.23	0.13

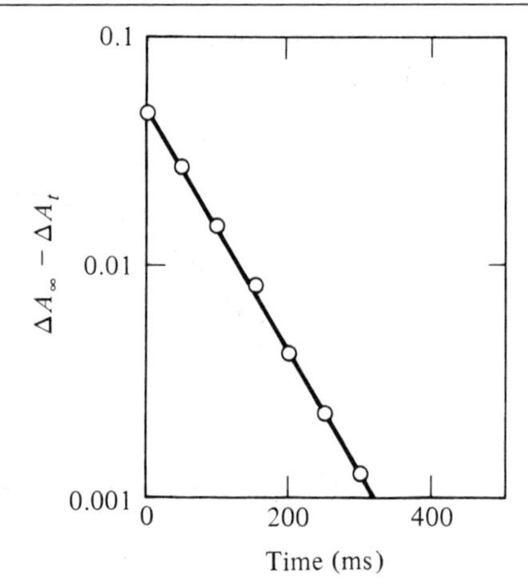

The data do fit a linear first-order plot (use semilog graph paper) with a slope of $5.19\ \mathrm{s}^{-1}$. The relaxation time is

$$\tau = \frac{1}{(2.30)(5.19\ \mathrm{s}^{-1})}$$
$$= 0.0838$$
$$= 84\ \mathrm{ms}$$

(b) We use the appropriate expression from Table 8.3,

$$\tau = \frac{1}{4k_1[\bar{\mathrm{M}}] + k_{-1}}$$

where $[\bar{\mathrm{M}}]$ is the concentration of oligonucleotide. Thus, a plot of $1/\tau$ versus $[\bar{\mathrm{M}}]$ should give a straight line.

$[\bar{\mathrm{M}}]$ (μM)	1.63	2.45	3.45	5.90
$1/\tau$ (s^{-1})	2.20	2.70	3.10	4.10

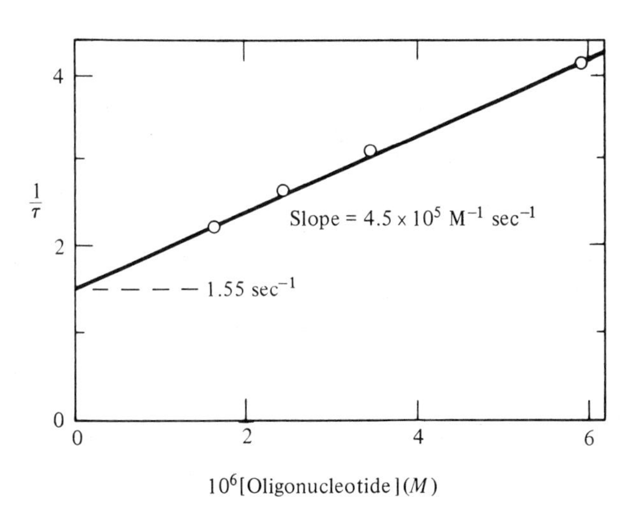

The straight line corresponds to the equation

$$\frac{1}{\tau} = 4k_1[\bar{\mathrm{M}}] + k_{-1}$$

Therefore, from the plot shown,

$$4k_1 = 4.5 \times 10^5\ M^{-1}\ \mathrm{s}^{-1}$$
$$k_1 = 1.12 \times 10^5\ M^{-1}\ \mathrm{s}^{-1}$$
$$k_{-1} = 1.55\ \mathrm{s}^{-1}$$

In this case, the dimerization step is several powers of 10 slower than the diffusion limit.

The presence of intermediates in complex equilibria can produce marked alterations in the relaxation-response curves. Each of the examples in Table 8.3 involves only a single measurable step between reactants and products. In many cases of interest, there are intermediate states, such as enzyme-substrate complexes, that participate in the relaxation process. The mechanism

$$\mathrm{E} + \mathrm{S} \underset{k_{-1}}{\overset{k_1}{\rightleftharpoons}} \mathrm{ES} \underset{k_{-2}}{\overset{k_2}{\rightleftharpoons}} \mathrm{E} + \mathrm{P} \tag{8.35}$$

illustrates a representative situation. At equilibrium (or under steady-state conditions), the concentrations are governed by

$$K_1 = \frac{k_1}{k_{-1}} = \frac{[\mathrm{ES}]}{[\mathrm{E}][\mathrm{S}]} \tag{8.36}$$

and

$$K_2 = \frac{k_2}{k_{-2}} = \frac{[\mathrm{E}][\mathrm{P}]}{[\mathrm{ES}]} \tag{8.37}$$

A perturbation, such as a temperature jump, will influence both K_1 and K_2, and the relaxation process will, in general, be biphasic. The faster processes will occur first, followed by the slower ones.

To pursue the example qualitatively, suppose that a T jump is applied and it shifts the overall equilibrium from S to P. If the rate of step 2 in reaction (8.35) is faster than that of step 1, the concentration curves will appear as shown in Fig. 8.8. The response of the components is clearly biphasic and not simple exponential. Even if only P can be monitored experimentally, it is clear that the relaxation occurs in two stages and that additional kinetic information is revealed.

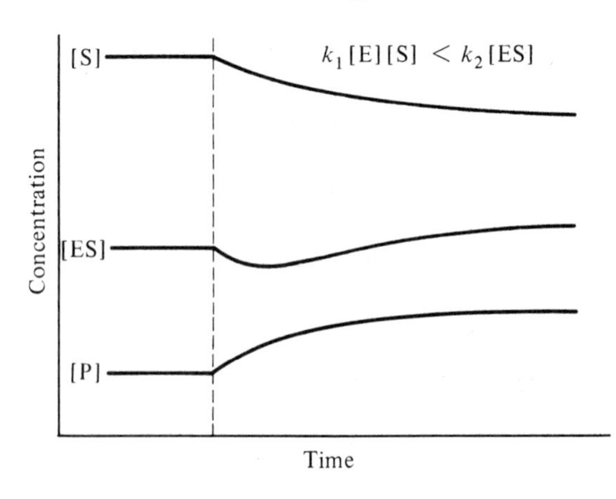

Fig. 8.8 Biphasic transient response to a jump perturbation by a system with a single intermediate, where the formation rate is slower than the dissociation rate.

This is important, because reaction (8.35) has more rate constants to be determined than do those shown in Table 8.3.

For the alternative case where step 1 occurs faster than step 2, the behavior is illustrated in Fig. 8.9. Typically, a sigmoidal or biphasic response is observed, at least when the rates of the two steps are not widely different from one another.

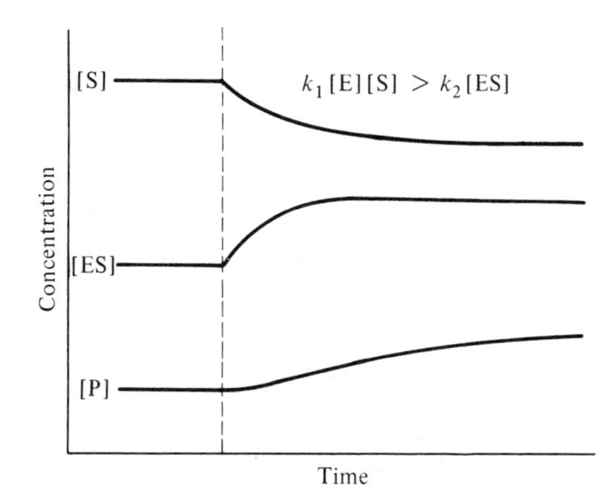

Fig. 8.9 Biphasic transient response to a jump perturbation by a system with a single intermediate, where the formation rate is faster than the dissociation rate.

A representative trace is shown in Fig. 8.10 for the reaction

$$\text{aspartate} + \text{ketoglutarate} \underset{\text{transferase}}{\overset{\text{aspartate amino-}}{\rightleftharpoons}} \text{oxaloacetate} + \text{glutamate}$$

The relaxation process in this experiment begins with glutamate and a pyridoxal form of the enzyme aspartate amino-transferase. The relaxation clearly has several kinetic components, and subsequent detailed studies were required to unravel the complicated mechanism (Fasella and Hammes, 1967; Hammes and Haslam, 1968).

Mathematical treatments can be carried out for the multistep relaxation processes, but they profit from the use of matrix algebra and are more involved

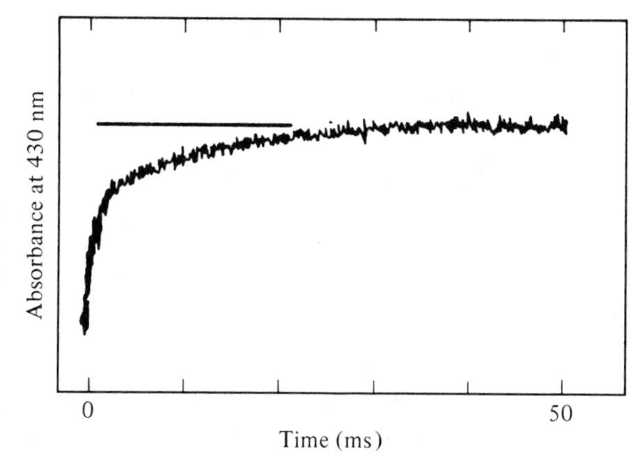

Fig. 8.10 Relaxation in the amino-transferase system beginning with glutamate and pyridoxal enzyme. [From G. G. Hammes and P. Fasella, *J. Amer. Chem. Soc. 84*, 4644 (1962). Copyright by the American Chemical Society.]

than we wish to consider here. An excellent treatment has been given by Eigen and de Maeyer (1974).

MECHANISMS OF ENZYMATIC REACTIONS

We discussed temperature-jump methods in the previous section to emphasize that it is important to try kinetic methods other than the simple steady-state ones. However, now we return to analyze variations of the traditional Michaelis-Menten kinetics.

The process by which an enzyme selects a suitable substrate from the vast array typically present in cytoplasmic fluids is analogous to the process by which a person selects a correctly fitting and styled suit or dress in a clothing store. The substrate selection is harder, because it is as if the clothes are strewn about without regard to size, shape, or gender of the intended wearer, and the selection process occurs mainly through a set of random encounters between the garments and the intended purchaser. But the process is so efficient that many successful fits are provided for each customer each second.

The great specificity and selectivity of enzymatic reactions has been well documented. A corollary consequence of this closeness of fit between enzyme and substrate occurs in the action of inhibitors. Several classes of inhibitors are known, and studies of their behavior made a major early contribution to our knowledge of enzyme mechanisms.

Competitive Inhibition

A molecule that resembles the substrate may be able to occupy the catalytic site because of its similarity in structure, but may be nearly or completely unreactive. By occupying the active site, this molecule acts as a *competitive inhibitor* in preventing normal substrates from being examined and catalyzed. Operationally, competitive inhibitors are those that bind reversibly to the active site. The inhibition can be reversed by (1) diluting the inhibitor, or (2) swamping the system with excess substrate.

Mechanistically, we can write a step

$$\mathrm{E + I \rightleftharpoons EI} \tag{8.38}$$

in addition to the usual Michaelis-Menten formulation. Thus

$$\begin{array}{l} \mathrm{E + S \rightleftharpoons ES \rightleftharpoons E + P} \\ + \\ \mathrm{I} \\ \downarrow\uparrow \\ \mathrm{EI} \end{array} \tag{8.39}$$

where EI is an inactive form of the enzyme. The kinetic behavior in the presence of the competitive inhibitor is seen in Figs. 8.11 and 8.12. Because the equilibria

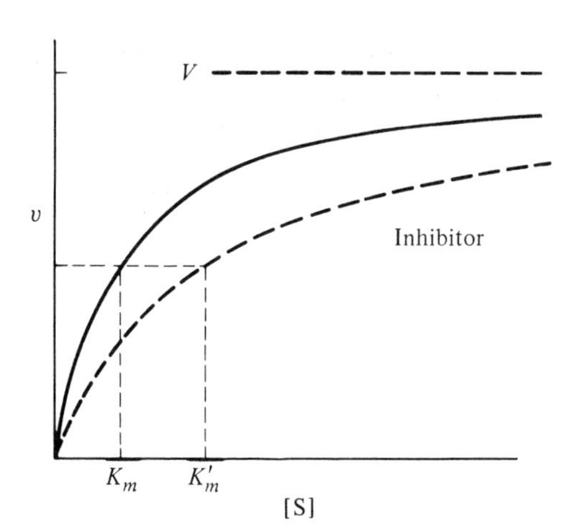

Fig. 8.11 Effect of competitive inhibitor on velocity of substrate reaction: increase of Michaelis constant, but no change in maximum velocity.

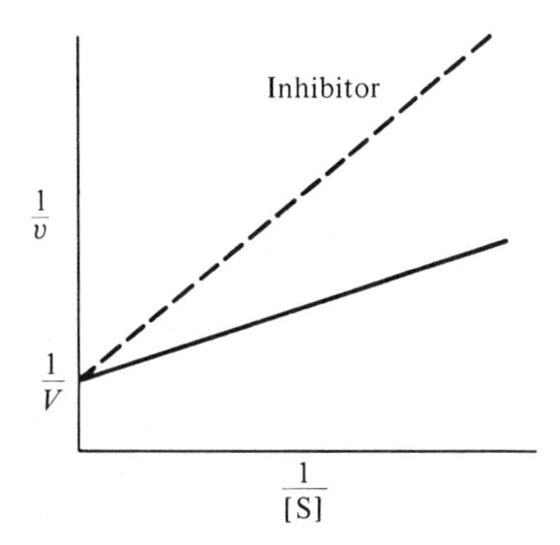

Fig. 8.12 Competitive inhibition viewed by a Lineweaver-Burk plot. The extrapolated maximum velocity, V, is not changed by a competitive inhibitor.

are reversible, the same maximum velocity is reached at sufficiently high substrate concentrations whether the inhibitor is present or not. But the Michaelis constant is different in the presence of the inhibitor; that is, the concentration of substrate required to reach one-half V is greater in the presence of competitive inhibitors than in their absence. It is not difficult to show that the new apparent Michaelis constant, K'_m, is given by

$$K'_m = K_m\left(1 + \frac{[\mathrm{I}]}{K_\mathrm{I}}\right) \tag{8.40}$$

where [I] is the inhibitor concentration and K_I is the dissociation constant for the enzyme-inhibitor complex. We obtained Eq. (8.40) by writing the total enzyme concentration as

$$[\mathrm{E}]_0 = [\mathrm{E}] + [\mathrm{ES}] + [\mathrm{EI}]$$

and using the definition of K_I,

$$K_\mathrm{I} = \frac{[\mathrm{E}][\mathrm{I}]}{[\mathrm{EI}]}$$

A classic example of competitive inhibition occurs in the case of the enzyme succinic dehydrogenase. The normal substrate of this enzyme is succinic acid, and the enzyme catalyzes the oxidation to fumaric acid.

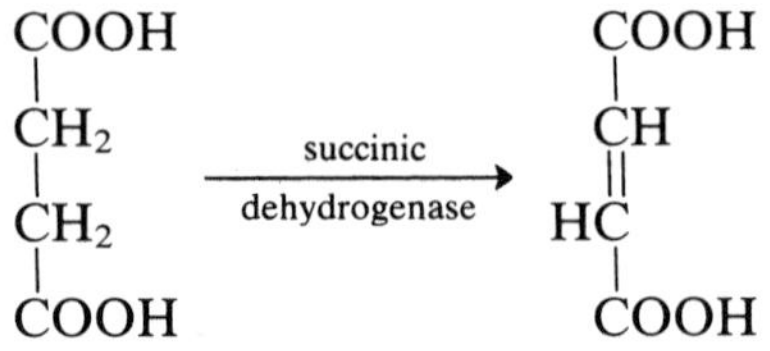

$$\begin{matrix} COOH \\ | \\ CH_2 \\ | \\ CH_2 \\ | \\ COOH \end{matrix} \xrightarrow[\text{dehydrogenase}]{\text{succinic}} \begin{matrix} COOH \\ | \\ CH \\ \| \\ HC \\ | \\ COOH \end{matrix}$$

Succinic acid Fumaric acid

Malonic acid, which has the structure

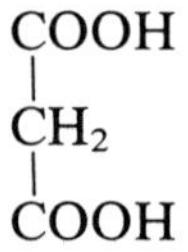

$$\begin{matrix} COOH \\ | \\ CH_2 \\ | \\ COOH \end{matrix}$$

Malonic acid

acts as a competitive inhibitor for succinic dehydrogenase. There is a difference of only one methylene group between the structures of succinic acid and malonic acid; hence the two molecules resemble one another. In this case it is clear that malonic acid cannot act as an alternative substrate, because there is no oxidized form of the molecule analogous to fumaric acid. The value of K_I for inhibition of yeast succinic dehydrogenase by malonate is $1 \times 10^{-5}\,M$, which means that a concentration of $1 \times 10^{-5}\,M$ malonate decreases the apparent affinity of the enzyme for succinate by a factor of 2.

Inhibition by the product of the reaction is another common example of competitive inhibition. Again, a similarity in structure is implicit in the relation between a substrate and its product. Furthermore, inhibition by the product can serve the useful purpose of turning off (or down) an enzyme's action when it has made sufficient product for the biochemical needs of the cell. It is one of the most immediate forms of regulation or metabolic control.

Noncompetitive Inhibition

Several types of inhibition occur that cannot be overcome by large amounts of substrate. These may occur as a consequence of (1) some permanent (irreversible) modification of the active site; (2) reversible binding of the inhibitor to the enzyme, but not at the active site itself; (3) reversible binding to the enzyme-substrate complex.

The simplest form of noncompetitive inhibition occurs when only the value of V is affected, but the affinity of the enzyme for substrate, as measured by $1/K'_m$, is not affected. The kinetic behavior in this case is illustrated in Figs. 8.13

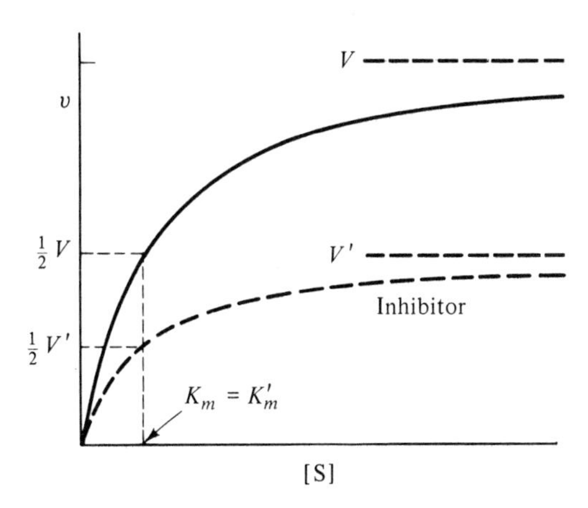

Fig. 8.13 Effect of noncompetitive inhibitor on velocity of substrate reaction: decrease of maximum velocity, but no change in Michaelis constant.

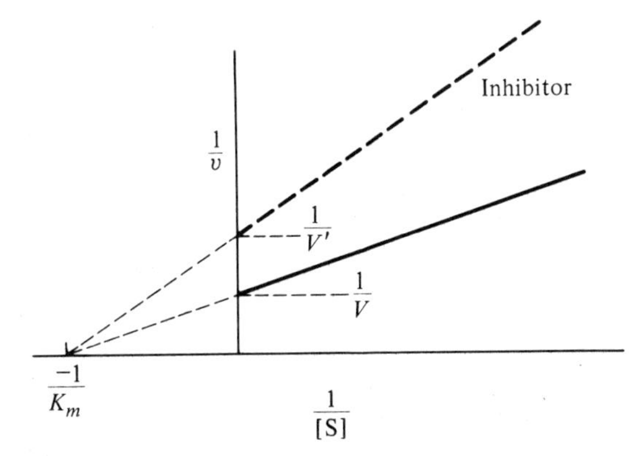

Fig. 8.14 Noncompetitive inhibition viewed by a Lineweaver-Burk plot.

and 8.14. As an example, consider the action of an irreversible modifier, such as the chemical alkylating agent iodoacetamide. This compound reacts with exposed sulfhydryl groups, in particular with cysteine residues of the enzyme protein, to form a covalently modified

$$\text{enzyme—S—CH}_2\underset{\substack{\| \\ \text{O}}}{\text{C}}\text{NH}_2$$

In triose phosphate dehydrogenase and in many other enzymes, the chemically modified enzyme is inactive in its catalytic role. In this case it is clear why the V decreases, because the inhibitor effectively inactivates a portion of the enzyme molecules irreversibly. At the same time, the unreacted enzyme molecules are perfectly normal, in the sense that the K_m value is unaffected by the prior addition of the inhibitor.

Other forms of noncompetitive inhibition occur in which both V and K_m are affected by the inhibitor. A type designated *uncompetitive* results in Lineweaver–Burk plots that are parallel but displaced upward in the presence of inhibitors. In addition, where the enzyme has more than one substrate, the action of an inhibitor must be evaluated with respect to each of the substrates and to the order of addition of components.

An interesting form of inhibition is known as *substrate inhibition.* In this case the velocity of the reaction reaches a maximum and then declines at high substrate concentrations. The consequence in terms of the Lineweaver-Burk plot is shown in Fig. 8.15. The origin of this behavior can be that the substrate binds at a second site on the enzyme, and this site indirectly modifies the main catalytic site so as to render it less effective.

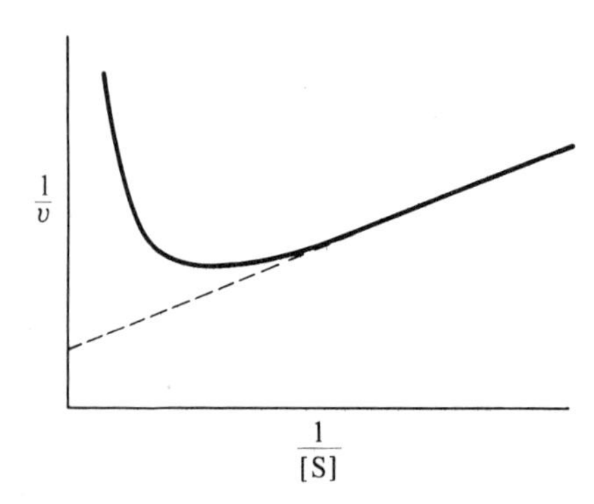

Fig. 8.15 Substrate inhibition viewed by a Lineweaver-Burk plot.

We can interpret the term "inhibitor" in a more general sense. There are many substances that serve to promote a reaction. These may be essential cofactors, such as pyridine nucleotides or ATP, that are really second substrates but operate in cyclic fashion in the overall biochemistry of the cytoplasm. Various ions, such as H^+, OH^-, alkali metals, alkaline earths, and so on, can bind to the enzyme and cause specific and profound changes in the activity of the catalytic site. Transition metals or organic functional prosthetic groups may be involved directly in reactions catalyzed by certain enzymes. The subject of biochemistry possesses a rich and unfolding history of examples of such behavior, and we can only begin to appreciate the variety that nature has incorporated into enzyme function and control.

SUMMARY

Typical Enzyme Kinetics

Velocity of substrate conversion is linear in enzyme concentration:

$$v = k'[\mathrm{E}]$$

Velocity is linear in substrate concentration at low [S]; *first-order reaction:*

$$v = k''[\mathrm{S}] \tag{8.1}$$

Velocity approaches saturation at high [S]; *zero-order reaction*:

$$v = V, \qquad \text{maximum velocity} \tag{8.2}$$

$$\text{turnover number} = \frac{V}{n[E]_0}, \qquad (\text{in s}^{-1}) \tag{8.3}$$

$[E]_0$ = total enzyme concentration, mol ℓ^{-1}
n = number of active centers per molecule

Michaelis-Menten Mechanism

Initial velocity of forward reaction:

$$E + S \underset{k_{-1}}{\overset{k_1}{\rightleftharpoons}} ES$$

$$ES \xrightarrow{k_2} E + P$$

$$v_0 = \frac{V}{1 + K_m/[S]} \tag{8.10}$$

$V = k_2[E]_0$ = maximum velocity, concentration time^{-1}

$[E]_0 = [E] + [ES]$ = total enzyme concentration

$K_m = \dfrac{k_{-1} + k_2}{k_1}$ = Michaelis constant, mol ℓ^{-1}

Lineweaver-Burk:

$$\frac{1}{v_0} = \frac{1}{V} + \frac{K_m}{V} \cdot \frac{1}{[S]} \tag{8.11}$$

Eadie-Hofstee:

$$v_0 = -K_m \cdot \frac{v_0}{[S]} + V \tag{8.13}$$

Michaelis-Menten mechanism, including reverse reaction:

$$E + S \underset{k_{-1}}{\overset{k_1}{\rightleftharpoons}} ES \underset{k_{-2}}{\overset{k_2}{\rightleftharpoons}} E + P$$

Velocity in either direction:

$$v = \frac{V_S K_m^S[S] - V_P K_m^P[P]}{K_m^S K_m^P + K_m^S[S] + K_m^P[P]} \tag{8.23}$$

V_S = maximum velocity for utilization of S
V_P = maximum velocity for utilization of P

$$K_m^S = \frac{k_{-1} + k_2}{k_1}$$

$$K_m^P = \frac{k_{-1} + k_2}{k_{-2}}$$

Reversible inhibition:

$$\begin{array}{l} E + S \underset{k_{-1}}{\overset{k_1}{\rightleftharpoons}} ES \underset{k_{-2}}{\overset{k_2}{\rightleftharpoons}} E + P \\ + \\ I \\ k_{-3} \upharpoonleft\downharpoonright k_3 \\ EI \end{array}$$

$$v_0 = \frac{V_S}{1 + K'_m/[S]}$$

$$K'_m = K_m \cdot (1 + [I]/K_I)$$

$$K_I = \frac{k_{-3}}{k_3} = \frac{[E][I]}{[EI]}$$

Relaxation Kinetics

$$A + B \underset{k_{-1}}{\overset{k_1}{\rightleftharpoons}} P$$

$$\Delta[P] = \Delta[P]_0 \, e^{-(t/\tau)} \tag{8.31}$$

$\Delta[P] = -\Delta[A] = -\Delta[B]$ = small displacement from equilibrium concentration
$\Delta[P]_0$ = displacement at zero time
$1/\tau = k_{-1} + k_1([\bar{A}] + [\bar{B}])$
$[\bar{A}], [\bar{B}]$ = equilibrium concentrations of A, B

See Table 8.3 for other examples.

MATHEMATICS NEEDED FOR CHAPTER 8

Least-squares programs for fitting the best line to a set of points are available in most programmable calculators. To fit a line to a set of points labeled (x_i, y_i)

where each x_i is an abscissa and each y_i is an ordinate, calculate

$$\text{slope} = m = \frac{n \sum_i y_i x_i - \sum_i y_i \sum_i x_i}{D}$$

$$\text{intercept} = b = \frac{\sum_i y_i \sum_i x_i^2 - \sum_i y_i x_i \sum_i x_i}{D}$$

where $D = n \sum_i x_i^2 - \left(\sum_i x_i\right)^2$.

The sums are over the total number, n, of points in the set. The slope and intercept determine the line that minimizes the sum of the squares of the vertical (y_i) distances of each point from the line; this is the best least-squares line. It is always a good idea to plot this line ($y = mx + b$) on the same graph as the points to check that the fit is reasonable. Chapter 20 of Daniels et al., *Experimental Physical Chemistry*, 7th ed. (McGraw-Hill, New York, 1970) discusses the least-squares method plus other methods for treating experimental data.

Four partial derivatives are written in this chapter. We remind the reader that a partial derivative is just like a total derivative except that it explicitly shows what variables are held constant. For example,

$$\left(\frac{\partial \ln K}{\partial P}\right)_T = -\frac{\Delta V^0}{RT}$$

means the change in equilibrium constant with pressure at constant temperature.

REFERENCES

The following books treat enzymes and enzyme kinetics in great detail.

BARMAN, T. E., 1969. *Enzyme Handbook*, Vols. 1 and 2, Springer-Verlag, Berlin.

BENDER, M. L., and L. J. BRUBACKER, 1973. *Catalysis and Enzyme Action*, McGraw-Hill, New York.

BERNHARD, S., 1968. *The Structure and Function of Enzymes*, W. A. Benjamin, Menlo Park, Calif.

BRAY, H. G., and K. WHITE, 1966. *Kinetics and Thermodynamics in Biochemistry*, 2nd ed., Academic Press, New York.

DIXON, M., and E. C. WEBB, 1964. *Enzymes*, 2nd ed., Academic Press, New York.

HAMMES, G. G. (ed.), 1974. *Investigation of Rates and Mechanisms of Reactions: Part II. Investigation of Elementary Reaction Steps in Solution and Very Fast Reactions*, John Wiley, New York.

LAIDLER, K. J., and P. S. BUNTING, 1973. *The Chemical Kinetics of Enzyme Action*, 2nd ed., Oxford University Press, New York.

See also biochemistry texts listed in Chapter 1.

SUGGESTED READINGS

ALBERTY, R. A., and W. H. PEIRCE, 1957. Studies of the Enzyme Fumarase: V. Calculation of Minimum and Maximum Values of Constants for the General Fumarase Mechanism, *J. Amer. Chem. Soc. 79*, 1526.

BRIGGS, G. E., and J. B. S. HALDANE, 1925. A Note on the Kinetics of Enzyme Action, *Biochem. J. 19*, 338.

EIGEN, M., and L. DE MAEYER, 1974. Theoretical Basis of Relaxation Spectrometry, in *Investigation of Rates and Mechanisms of Reactions*, 3rd ed., Vol. 6, Part II, G. G. Hammes (ed.), Wiley-Interscience, New York, pp. 63–146.

FASELLA, P., and G. G. HAMMES, 1967. A Temperature Jump Study of Aspartate Aminotransferase, *Biochemistry 6*, 1798.

HAMMES, G. G., and J. L. HASLAM, 1968. A Kinetic Investigation of the Interaction of α-Methylaspartic Acid with Aspartate Aminotransferase, *Biochemistry 7*, 1519.

LINEWEAVER, H., and D. BURK, 1934. The Determination of Enzyme Dissociation Constants, *J. Amer. Chem. Soc. 56*, 658.

PÖRSCHKE, D., O. C. UHLENBECK, and F. H. MARTIN, 1973. Thermodynamics and Kinetics of the Helix-Coil Transition of Oligomers Containing GC Base Pairs, *Biopolymers 12*, 1313.

TURNER, D. H., G. W. FLYNN, S. K. LUNDBERG, L. D. FALLER, and N. SUTIN, 1972. Dimerization of Proflavin by the Laser Raman Temperature Jump Method, *Nature 239*, 215.

WANG, J. H., 1955. On the Detailed Mechanism of a New Type of Catalase-like Action, *J. Amer. Chem. Soc. 77*, 4715.

WANG, J. H., 1970. Synthetic Biochemical Models, *Accounts Chem. Res. 3*, 90.

PROBLEMS

1. The decarboxylation of a β-keto acid catalyzed by a decarboxylation enzyme can be measured by the rate of CO_2 formation. From the initial rates given in the table, determine the Michaelis-Menten constant for the enzyme and the maximum velocity by a graphical method.

Keto acid concentration [M]	Initial velocity (μmol CO_2/2 min)
2.500	0.588
1.000	0.500
0.714	0.417
0.526	0.370
0.250	0.256

2. The hydration of CO_2,

$$CO_2 + H_2O \rightleftharpoons HCO_3^- + H^+$$

is catalyzed by the enzyme carbonic anhydrase. The steady-state kinetics of the forward (hydration) and reverse (dehydration) reactions at pH 7.1, 0.5°C, and $2 \times 10^{-3}\,M$ phosphate buffer were studied using bovine carbonic anhydrase [H. DeVoe and G. B. Kistiakowsky, *J. Amer. Chem. Soc. 83*, 274 (1961)]. Some typical results for an enzyme concentration of $2.8 \times 10^{-9}\,M$ are:

Hydration		Dehydration	
$\frac{1}{v}$ (M^{-1} s)	$[CO_2]$ (mM)	$\frac{1}{v}$ (M^{-1} s)	$[HCO_3^-]$ (mM)
36×10^3	1.25	95×10^3	2
20×10^3	2.5	45×10^3	5
12×10^3	5	29×10^3	10
6×10^3	20	24×10^3	15

Using graph paper make suitable plots of the data and determine the Michaelis constant, K_m, and the rate constant, k_2, for the decomposition of the enzyme-substrate complex to form product for:

(a) The hydration reaction.

(b) The dehydration reaction.

(c) From your results, calculate the equilibrium constant for the reaction

$$CO_2 + H_2O \rightleftharpoons HCO_3^- + H^+$$

(*Hint*: Start by writing an equation that defines the equilibrium condition in terms of the Michaelis-Menten formulation of this reaction. Note that the kinetics were measured at pH 7.1.)

3. At pH 7 the measured Michaelis constant and maximum velocity for the enzymatic conversion of fumarate to L-malate,

$$\text{fumarate} + H_2O \longrightarrow \text{L-malate}$$

are $4.0 \times 10^{-6}\,M$ and $1.3 \times 10^3\,[E]_0\,s^{-1}$, respectively, where $[E]_0$ is the total molar concentration of the enzyme. The Michaelis constant and maximum velocity for the reverse reaction are $1.0 \times 10^{-5}\,M$ and $800[E]_0\,s^{-1}$, respectively. What is the equilibrium constant for the hydration reaction? (The activity of water is set equal to unity; see Chapter 4.)

4. Consider the simple Michaelis-Menten mechanism for an enzyme catalyzed reaction,

$$E + S \underset{k_{-1}}{\overset{k_1}{\rightleftharpoons}} ES \xrightarrow[k_2]{} E + P$$

The following data were obtained:

k_1, k_{-1} very fast

$k_2 = 100\,s^{-1}$, $K_m = 10^{-4}\,M$ at 280 K

$k_2 = 200\,s^{-1}$, $K_m = 1.5 \times 10^{-4}\,M$ at 300 K

(a) For $[S] = 0.1\,M$ and $[E]_0 = 10^{-5}\,M$, calculate the rate of formation of product at 280 K.

(b) Calculate the activation energy for k_2.
(c) What is the value of the equilbrium constant at 280 K for the formation of the enzyme-substrate complex ES from E and S?
(d) What is the sign and magnitude of the standard thermodynamic enthalpy ΔH^0 for the formation of ES from E and S?

5. Consider the following mechanism for the role of an inhibitor, I, of an enzyme,

$$E + S \underset{k_{-1}}{\overset{k_1}{\rightleftharpoons}} ES \xrightarrow{k_2} E + P$$

$$E + I \underset{k_{-3}}{\overset{k_3}{\rightleftharpoons}} EI$$

The concentrations of [S] and [I] are much larger than the total enzyme concentration $[E]_0$. Derive an expression for the rate of appearance of products.

6.

$$E + F \underset{k_{-1}}{\overset{k_1}{\rightleftharpoons}} EF \underset{k_{-2}}{\overset{k_2}{\rightleftharpoons}} EM \underset{k_{-3}}{\overset{k_3}{\rightleftharpoons}} E + M$$

(a) What rate constant would you need to measure as a function of temperature to determine the $\Delta H^\ddagger$ for formation of EF from E + F? Formation of EM from EF? Formation of EM from E + M?
(b) What rate constants would you need to measure as a function of temperature to determine ΔH^0 for formation of EF from E + F? Formation of EM from E + M?
(c) How could you determine ΔH^0 for formation of M from F using only kinetic experiments?

7. The hydrolysis of sucrose by the enzyme invertase was followed by measuring the initial rate of change in polarimeter (optical rotation) readings, α, at various initial concentrations of sucrose. The reaction is inhibited reversibly by the addition of urea.

$[\text{Sucrose}]_0$ (mol ℓ^{-1})	0.0292	0.0584	0.0876	0.117	0.175	0.234
Initial rate, $\left(\frac{d\alpha}{dt}\right)_0 = v_0$	0.182	0.265	0.311	0.330	0.372	0.371
Initial rate (2 M urea), v_0'	0.083	0.119	0.154	0.167	0.192	0.188

(a) Make a suitable plot of the data in the absence of urea and determine the Michaelis constant K_m for this reaction.
(b) Carry out a suitable analysis of the data in the presence of urea and determine whether urea is a competitive or a noncompetitive inhibitor of the enzyme for this reaction. Justify your answer.

8. The reaction of an enzyme E with a substrate S typically passes through at least two intermediates, ES and EP, and three transition states before reaching the dissociated product P:

$$E + S \longrightarrow [ES]^{\ddagger} \longrightarrow ES \longrightarrow [EZ]^{\ddagger} \longrightarrow EP \longrightarrow [EP]^{\ddagger} \longrightarrow E + P$$

From the following data, prepare a diagram of the energy of reaction of fumarate to malate catalyzed by the enzyme fumarase as a function of the reaction coordinate. (See

Fig. 8.1 for an example of a simpler reaction.) Carefully label all enthalpy or energy differences for which you have information.

	ΔH (cal)	E_a (cal)
Fumarate + $H_2O \rightarrow$ malate	−3,600	
Fumarate + fumarase $\rightarrow$ fumarate-fumarase complex, ES	+4,200	
Fumarate-fumarase complex $\rightarrow$ transition state, $[EZ]^{\ddagger}$		+6,100
Malate + fumarase $\rightarrow$ malate-fumarase complex, EP	−1,200	
Malate-fumarase complex $\rightarrow$ transition state, $[EZ]^{\ddagger}$		+15,000

9. The initial rate of ATP dephosphorylation by the enzyme myosin can be estimated from the amount of phosphate produced in 100 s, $[P_i]_{100}$. Use the following data and graphical methods to evaluate K_m and V for this reaction at 25°C. $[\text{Myosin}]_0 = 0.040$ g/l.

$$\text{ATP} \xrightarrow{\text{myosin}} \text{ADP} + \text{P}$$

$[\text{ATP}]_0$ (μM)	7.1	11	23	40	77	100
$[P_i]_{100}$ (μM)	2.4	3.5	5.3	6.2	6.7	7.1

Explain any curvature that you observe in the plot of the data. Justify the slope that you choose to characterize the reaction.

10. Since it was first discovered in 1968, the enzyme superoxide dismutase, SOD, has been found to be one of the most widely dispersed in biological organisms. In one form or another, its presence has been demonstrated in mammalian tissues (heart, liver, brain, etc.), blood, in invertebrates, plants, algae, and aerobic bacteria. It catalyzes the reaction

$$O_2^- + O_2^- + 2H^+ \xrightarrow{\text{SOD}} O_2 + H_2O_2$$

where O_2^- represents an oxygen molecule with an extra (unpaired) electron. SOD is, therefore, of great importance in detoxifying tissues from the potentially harmful O_2^-. (Several of the proteins had been known since 1933, but their enzymatic activity went unrecognized for 35 years!)

The enzyme has some unusual properties, as illustrated by the following data (for the velocity of the reaction in terms of O_2 formed) taken from a paper by Bannister et al. [*FEBS Letters 32*, 303 (1973)]. The enzyme kinetics is independent of pH in the range 5 to 10.

$[\text{SOD}]_0 = 0.4\ \mu M$. Buffer pH = 9.1.

$\frac{1}{v}$ (s M^{-1})	$\frac{1}{[O_2^-]}$ (M^{-1}
260	13×10^4
175	8.5×10^4
122	6.0×10^4
60	3.0×10^4
10	0.50×10^4

(a) Use graph paper to plot these data according to the Lineweaver-Burk method.
(b) What values do you obtain for V and K_m for this reaction? (Do not be surprised if they are somewhat unusual.)
(c) How can you interpret the results of part (b) in terms of a Michaelis-Menten type of mechanism?
(d) What is the kinetic order of the reaction with respect to O_2^-? Explain your answer.
(e) The reaction is independent of pH (in the region investigated) and first order in enzyme concentration. Write the simplest rate law consistent with the experimental observations. What is the value of the rate constant, including appropriate units?

Klug-Roth, Fridovich, and Rabani [*J. Amer. Chem. Soc. 95*, 2768 (1973)] have proposed a ping-pong-like mechanism for this reaction, involving the steps

$$E + O_2^- \xrightarrow[k_1]{} E^- + O_2$$

$$E^- + O_2^- \xrightarrow[k_2]{2H^+} E + H_2O_2$$

where $k_2 = 2k_1$. (The enzyme contains copper that is oxidized in E and reduced in E^-.)

(f) Show that this mechanism is consistent with the rate data of Bannister et al.
(g) What is the ratio $[E]/[E^-]$ under steady-state conditions?
(h) Given the answer to part (e) for the "observed" rate constant, calculate k_1 and k_2.

11. Alcoholism in humans is a disorder that has serious sociological as well as biochemical consequences. Much research is currently under way to discover the nature of the biochemical consequences to understand better how this disorder can be treated. You can analyze this problem using a variety of kinetic approaches.

Alcohol taken orally is transferred from the gastrointestinal tract (stomach, etc.) to the bloodstream by a first-order process with $t_{1/2} = 4$ min. (These and subsequent figures are subject to individual variations of $\pm 25\%$.) The transport by the bloodstream to various aqueous body fluids is very rapid; thus, the ethanol becomes rapidly distributed throughout the approximately 40 liters of aqueous fluids of an adult human. These fluids behave roughly as a sponge from which the alcohol must be removed.

The removal occurs in the liver, where the alcohol is oxidized in a process that follows zero-order kinetics. A typical value for this rate of removal is about 10 mℓ ethanol h^{-1}, or 4×10^{-3} mol (liter of body fluid)$^{-1}$ h^{-1}. Consumption of about 1 mol (46 g, or 60 mℓ) of ethanol produces a state defined as legally intoxicated. (This amount of alcohol is contained in about 4.5 oz of 80 proof liquor.) At this level, alcohol is rapidly taken up by the body fluids and only slowly removed by the liver.

In the liver, ethanol is oxidized to acetaldehyde in the presence of the enzyme liver alcohol dehydrogenase (LADH). The overall process may be represented by the equation

$$C_2H_5OH\text{ (stomach)} \xrightarrow[\text{transfer}]{k_1} C_2H_5OH\text{ (body fluids)} \xrightarrow[\substack{\text{LADH}\\ \text{liver}}]{k_0} CH_3CHO \downarrow \text{acetate, etc.}$$

(a) On the basis of the data above, calculate at least three points that will enable you to construct a semiquantitative ($\pm 10\%$) sketch of the concentration of ethanol in body fluids as a function of time following the consumption of 60 ml of C_2H_5OH at time zero. Pay particular attention to the rising portion, the maximum level reached, and the decaying portion of the curve.

(b) What is the maximum concentration of ethanol attained in the body fluids after consuming 60 mℓ of C_2H_5OH? Take this to be the threshold level defining legal intoxication and draw a horizontal line at this level. Approximately how many hours are required to reduce the ethanol concentration essentially to zero? This is the recuperation time or hangover period.

(c) On the same graph add a curve showing the time dependence following initial consumption of 120 mℓ of ethanol. (Label the two curves "60 mℓ" and "120 mℓ," respectively.) The concentration in body fluids remains above the "legally intoxicated" level for about how many hours? The hangover period is how long?

(d) Without constructing further plots, estimate the length of time that a person would remain "legally intoxicated" if initially 180 mℓ of ethanol were consumed.

(e) If 60 mℓ of ethanol were consumed initially, what is the maximum amount that this person may drink at subsequent hourly intervals and still avoid exceeding the level of legal intoxication?

(f) It is a popular opinion that a "cocktail party" drinker is less susceptible to intoxication because the consumption of alcohol is stretched over a longer period interspersed with conversation. Using dashes, add a curve to the graph that shows the body fluid concentration for the social drinker who extends the consumption of 120 mℓ of ethanol over a period of 1 hr in 15-min intervals. Justify your plot and comment on the effect of the social drinker's tactic on the intoxication period.

12. The removal of ethanol in the liver involves its oxidation to acetaldehyde by nicotinamide adenine dinucleotide (NAD^+) catalyzed by the enzyme liver alcohol dehydrogenase (LADH). The overall reaction is

$$C_2H_5OH + NAD^+ \xrightleftharpoons{\text{LADH}} CH_3CHO + NADH + H^+$$

The reaction follows a sequential or ordered mechanism wherein NAD^+ must bind to the enzyme before C_2H_5OH binds to form a ternary complex, and CH_3CHO and H^+ dissociate from the ternary complex before the NADH is released.

(a) Write a detailed set of reactions involving intermediate complexes so as to represent the mechanism of this reaction. Indicate each step as being reversible and designate the rate constants (in order) by k_1, k_2, etc., for forward reaction steps and k_{-1}, k_{-2}, etc., for reverse steps.

(b) Give a plausible explanation for the observation that the removal of ethanol from the body obeys zero-order kinetics.

(c) The rate-limiting step for the overall reaction under physiological conditions and in the presence of an intoxicating level of ethanol is the final dissociation of the LADH · NADH complex. The rate constant for this step has been measured to be $3.1\ s^{-1}$. Using appropriate data given in Problem 11, calculate the amount (in μmoles) of the LADH enzyme present in the liver.

(d) Relaxation measurements using the temperature-jump technique were applied to the binding of NAD^+, at two different concentrations, to LADH in the absence of ethanol. The relaxation times measured were

$$\tau = 1.65 \times 10^{-3}\ s, \text{ at } [NAD^+] = 1.0\ mM$$

$$\tau = 7.9 \times 10^{-3}\ s, \text{ at } [NAD^+] = 0.1\ mM$$

where [LADH] ≪ [NAD^+]. Use these relaxation times to calculate the constants k_1 and k_{-1} for the reactions

$$\text{LADH} + \text{NAD}^+ \underset{k_{-1}}{\overset{k_1}{\rightleftharpoons}} [\text{LADH} \cdot \text{NAD}^+]$$

(e) By contrast with ethanol, other simple alcohols, such as methanol (CH_3OH) and ethylene glycol ($HOCH_2CH_2OH$), are highly toxic to humans. Methanol is oxidized to formaldehyde, which reacts irreversibly with proteins (it is a commonly used cross-linking agent, or fixative). Ethylene glycol is not toxic but its oxidation product oxalic acid is. Each of these alcohols is about as good as ethanol as a substrate for LADH. A common antidote for methanol or ethylene glycol poisoning is to administer ethanol in large quantities. Propose a rationale for this therapy.

13. Pyrazole has been proposed as a possible nontoxic inhibitor of LADH-catalyzed ethanol oxidation. Its kinetics was studied by Li and Theorell [*Acta Chem. Scand. 23*, 892 (1969)]. In separate series of experiments, the velocity of the reaction was measured as a function of [C_2H_5OH] or of [NAD^+]. Portions of their results are tabulated below.

Ethanol as Variable Substrate
[LADH] = 4 μg mℓ^{-1},
[NAD^+] = 350 μM, pH 7.4, 23.5°C

	$1/v$ (Relative units)	
$\frac{1}{[C_2H_5OH]}$ (M^{-1})	Control (no pyrazole)	1×10^{-5} M pyrazole
0.125×10^3	0.47	0.59
0.5×10^3	0.55	0.97
1.0×10^3	0.66	1.45
1.5×10^3	0.74	1.91

NAD^+ as Variable Cofactor
[LADH] = 4 μg mℓ^{-1},
[C_2H_5OH] = 5 mM, pH 7.4, 23.5°C

	$1/v$ (Relative units)	
$\frac{1}{[NAD^+]}$ (M^{-1})	Control (no pyrazole)	4×10^{-5} M pyrazole
0.9×10^4	0.59	1.17
1.8×10^4	0.70	1.41
3.0×10^4	0.77	1.59
6.0×10^4	1.05	2.15

(a) Plot these data as Lineweaver-Burk plots on two separate graphs.
(b) What are the Michaelis constants $K_m(C_2H_5OH)$ and $K_m(NAD^+)$, in the absence of inhibitor?
(c) What type of inhibition is exhibited by pyrazole against C_2H_5OH as a substrate?
(d) What type of inhibition is exhibited against NAD^+ as cofactor?
(e) With reference to your answer to Problem 12(a), suggest a mechanism for the action of this inhibitor.
(f) It has been proposed that the damaging effects of ethanol on the liver results from the pronounced decrease in the ratio $[NAD^+]/[NADH]$ caused by the oxidation of ethanol. (NAD^+ is required for several key dehydrogenation steps in the pathway of glucose synthesis. NADH promotes the accumulation of fat in the liver cells.) Assuming that pyrazole proves to be nontoxic and that it is transported to the liver, evaluate its use as an antidote to liver damage in treating chronic alcoholism.

14. The equilibrium

$$I_2 + I^- \underset{k_{-1}}{\overset{k_1}{\rightleftharpoons}} I_3^-$$

has been studied using a laser-induced temperature-jump technique. Relaxation times were measured at various equilibrium concentrations of the reactants at 25°C:

$[\bar{I}^-]$ (mM)	$[\bar{I}_2]$ (mM)	τ (ns)
0.57	0.36	71
1.58	0.24	50
2.39	0.39	39
2.68	0.16	38
3.45	0.14	32

SOURCE: Turner et al., *J. Amer. Chem. Soc. 94*, 1554 (1972).

(a) Calculate k_1 and k_{-1} for this system at 25°C.
(b) Compare your results with the value for the equilibrium constant, $K = 720\ M^{-1}$.
(c) Use simple diffusion theory to estimate a value for k_1. For I^- and I_2 the effective radii are 2.16 and 2.52×10^{-8} cm, respectively, and diffusion coefficients are 2.05 and $2.25 \times 10^{-5}\ cm^2\ s^{-1}$, respectively. Comparing this result with the experimental value from part (a), what do you conclude about the effectiveness of collisions of I_2 and I^- in leading to reaction?

9

Quantum Mechanics

"The ultimate aim of the modern movement in biology is in fact to explain *all* biology in terms of physics and chemistry. . . . Quantum mechanics, together with our empirical knowledge of chemistry, appears to provide us with a 'foundation of certainty' on which to build biology" (Francis Crick, *Of Molecules and Men*, University of Washington Press, Seattle, Wash., 1966).

Quantum mechanics (and statistical mechanics), in principle, provides all the thermodynamics, kinetic, spectroscopic, and so on, properties of molecules. In practice, because of computer limitations, we are severely limited in the types of properties we can calculate, except for the smallest molecules. As computers become better, quantum mechanics will become more useful.

At present, biologists should know at least a minimum of quantum mechanics so they can feel confident in using the many spectroscopic methods. Spectra do not make sense without quantum mechanics. It is also important to know the vocabulary of bonding, orbitals, electron distribution, and charge densities to understand reaction mechanisms. These concepts are based on very simple quantum mechanical descriptions of molecules. The goal of molecular biologists is to explain biological functions in terms of molecular interactions. The goal of quantum biologists, or submolecular biologists, is to explain molecular interactions in terms of electronic motion. In this chapter we shall present the most fundamental aspects of quantum mechanics to provide at least a glimpse of this powerful method.

A good example of the way quantum mechanics is applied in biology is in the understanding of vision. We want to know how the absorption of a photon by rhodopsin in the eye ultimately leads to a signal at the brain that is interpreted as vision. The light-absorbing molecule in rhodopsin is 11-*cis*-retinal. It is isomerized by light to all-*trans*-retinal (see Fig. 9.1). The history of this problem

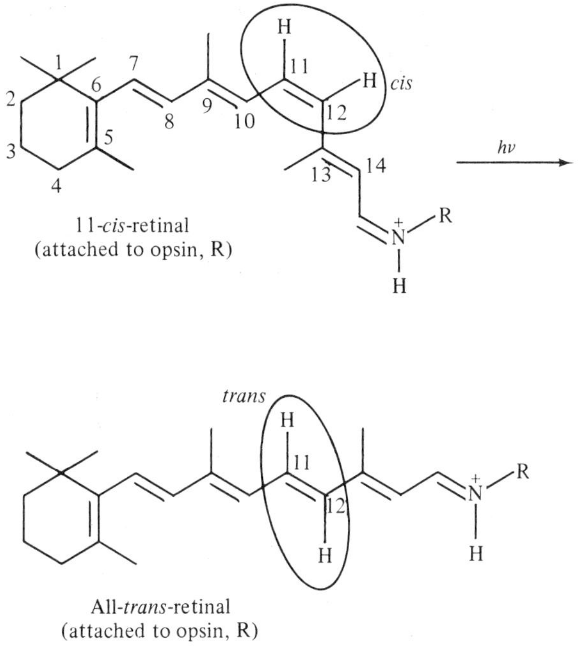

11-*cis*-retinal
(attached to opsin, R)

All-*trans*-retinal
(attached to opsin, R)

Fig. 9.1 Light-driven isomerization of 11-*cis*-retinal attached to opsin by a Schiff's base linkage. This is the first step which leads to vision in all known vertebrates. The hydrogens are not shown except on the 11–12 double bond and the nitrogen atom.

is fairly typical (Wald, 1968; Menger, 1975). First the visual pigment proteins were isolated and characterized; the structures of the active parts were determined. Both organic chemical and spectroscopic techniques were applied. Once the reactants were identified it became important to determine the mechanism of the process. Molecular orbital calculations on retinal (Salem and Bruckmann, 1975; Salem, 1976) showed that the excited electronic state produced by light absorption decreases the double-bond character at the 11–12 double bond. This makes the necessary rotation around this bond in going from the *cis* to the *trans* conformation much easier. Furthermore, the quantum mechanical calculations predicted a very large change in electron distribution for the excited state of 11-*cis*-retinal compared to the ground state. Experimental evidence about the structure of the excited state of retinal has been obtained from spectroscopic measurements such as the effect of an intense electric field on the absorption of light (Mathies and Stryer, 1976). A large shift in charge density was measured; the positive charge is moved from the N^+ toward the six-membered ring. The photon energy has thus been changed to an electrical signal. The hope is that further experimental and theoretical methods based on quantum mechanics will eventually help us to understand the transmission of the signal through the retinal rod membrane, along the optic nerve to the brain.

In this chapter we shall discuss only the simplest quantum mechanical problems, so as to keep the mathematics to a minimum. However, the student should feel confident that the physical chemical ideas are simple; it is only the computation that is difficult. Once the methods are understood, a computer or mathematician can be directed to seek the answer. It is important, then, for the student to learn the vocabulary of quantum mechanics and to learn the main ideas.

Our clearest understanding of the forces that hold molecules together comes from an application of quantum mechanics to the assembly of nuclei and electrons that we call a molecule. Quantum mechanics is successful because there is close correlation between theory and experiment. The best theoretical work makes very precise predictions for such properties as bond energies, bond lengths, electron distributions, and so on, giving excellent agreement with experimentally determined values. Unfortunately, the labor and time involved in such calculations is so great that only relatively small molecules have been solved in detail. For the larger molecules encountered by the biochemist or biologist, it is necessary to make approximations to tackle the problems at all. Among the most useful approximations are those that correspond to our chemical intuition.

1. Molecules can be thought of as collections of atoms or radicals (such as sulfate, ammonium) whose internal structure (nuclei, electrons) need not be known in detail.
2. Electrons may be localized primarily on individual atoms, they may be shared between pairs of atoms, or they may be delocalized over a large portion of the molecule.
3. The motion of electrons is very rapid in comparison with the much heavier nuclei. As a consequence, one may assume that changes in electronic structure or distribution (caused, for example, by photon absorption) will occur with the nuclear arrangement essentially frozen.

Electrons as Waves

To appreciate how quantum mechanics can contribute to our knowledge of molecular structure, we need to go beyond the early Bohr picture of electrons in atoms. That view emphasized the particlelike character of electrons, with their characteristic electronic charge, rest mass, velocity, and angular momentum computed using Newtonian mechanics and classical electromagnetic theory. It introduced the idea of electrons moving in fixed orbits, like planets moving about the sun. Today we know that the fixed-orbit picture is erroneous.

In the 1920s it became clear that for many purposes an electron (and any other particle, for that matter) is better understood as a wave phenomenon. One of the most direct experimental verifications of this wavelike character was the observation of electron diffraction by Davisson and Germer in 1927. They projected a fine beam of electrons onto a single crystal of nickel metal and observed a diffraction pattern of concentric rings around the transmitted beam. Such diffraction patterns were well known from studies of light, sound, x rays, ripples on the surface of a liquid, and other phenomena. From their measurements, Davisson and Germer were able to calculate a wavelength for the electron, and the wavelength λ turned out to be inversely proportional to the momentum (mv) as predicted two years earlier by de Broglie. The *de Broglie relation* is stated as

$$\lambda = \frac{h}{mv} \tag{9.1}$$

where h is Planck's constant, (6.62×10^{-27} erg s), m is the rest mass of the electron (9.11×10^{-28} g), and v is its velocity (cm s^{-1}). The wavelength thus has units of centimeters. Since the velocity is determined by the electric potential used to accelerate the electron, it is easy to change v by changing the accelerating potential and to observe the related change in the spacing of the diffraction pattern. Subsequent experiments showed that other particles, such as protons and neutrons, also exhibit wavelike behavior and fit the de Broglie relation when the appropriate rest masses are used. Since these are much heavier than the electron, their characteristic wavelengths are much shorter for the same velocity. The nuclei of atoms are thus particles with wavelengths that are typically very much shorter than those of the lighter and faster electrons.

Wave Mechanics and Wavefunctions

Once the wave nature of matter was recognized, it became possible for Heisenberg and Schrödinger, independently, to formulate mathematical descriptions of the electron wave motion. The Schrödinger formulation, which is the more familiar of the two, assigns an amplitude ψ to the electron wave; ψ is known as the *wavefunction* of the system. For classical wave motion, the wavefunction is simply the displacement of the system from its equilibrium position; for example, the height of waves on the surface of a lake. For electron wavefunctions there is no simple interpretation of the wavefunction itself, but ψ^2 is proportional to the probability of finding the electron.

In general the wavefunction of a system is a function both of the space coordinates and of time. However, we shall consider only time-independent wavefunctions in this chapter. Once we know the wavefunction for a molecule, we can calculate all the properties of the molecule. To illustrate this, consider a

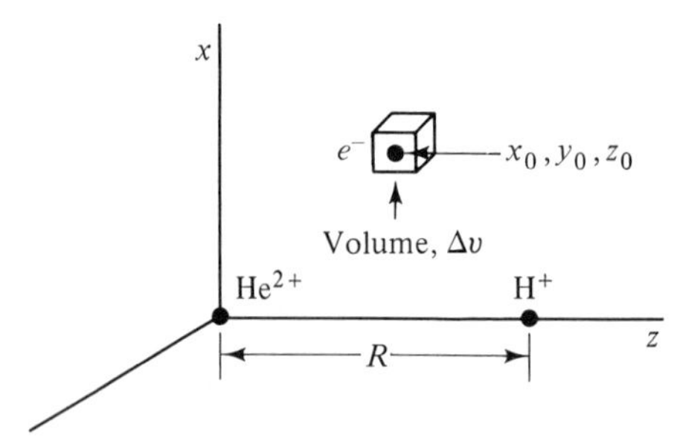

Fig. 9.2 Diatomic molecular HeH^{2+} with a given internuclear distance, R. The probability of finding the electron in a volume, Δv, around point x_0, y_0, z_0 is just the square of the electronic wavefunction multiplied by the volume.

possible molecule, HeH^{2+} (Fig. 9.2). This molecule has one electron, so its electronic wavefunction is a function of the x, y, z coordinate of the electron.

$$\psi = \psi(x, y, z)$$

We consider the He and H nuclei fixed and we are interested in the electron distribution around the nuclei. We will see that for each position of the two nuclei there is a set of time-independent states of the electron characterized by a definite energy, E_n, and wavefunction, ψ_n. The lowest energy and the electron distribution corresponding to it is the ground state. There are excited states corresponding to higher energies and different electron distributions. Because there is only one electron in this molecule, it is fairly easy to obtain $\psi_n(x, y, z)$ for many internuclear distances. That is, we can obtain ψ_n as a function of possible bond lengths. What properties can we now calculate?

The *electron density or distribution* is given by $[\psi_n(x, y, z)]^2$. For a chosen internuclear distance and a particular ground or excited state, the probability of finding the electron in a small volume of size Δv around point x_0, y_0, z_0 is

$$[\psi_n(x_0, y_0, z_0)]^2 \, \Delta v \tag{9.2}$$

The probability of finding the electron in a finite volume of space is obtained by integration over the volume:

$$\int [\psi_n(x, y, z)]^2 \, dv \tag{9.3}$$

We use one integral sign to represent integration over all variables and dv to represent all the coordinates of integration. The probability of finding the electron in all of space is equal to 1. This is called the *normalization condition.* It states that the electron must be found somewhere; the probability of finding it somewhere is unity.

The electron distribution ψ_n^2 directly determines the scattering of x rays by the molecule. The electron distribution, of course, also determines the chemical reactivity of the molecule, but this is harder to quantify.

The *average position* of the electron can be calculated from the wavefunction. As a diatomic molecule is symmetric around the bond, the average position of the electron will be on the z axis, the internuclear axis. The average position for the positive charges of the nuclei is $R/3$, where R is the internuclear distance measured from He. The average position of negative charge (the electron) may change markedly with excitation from the ground to the excited state. This was what was found in retinal.

The *energy* of the electron in each electronic state can also be calculated from the corresponding wavefunction. The energy of the electron is of vital importance because it tells us if the molecule is stable and what its stable bond distance is. We can calculate the electronic energy as a function of internuclear distance. This energy will decrease (become more negative) as the internuclear distance decreases. The repulsion between the positively charged nuclei, which is easily calculated from Coulomb's law for two point charges, will increase as the internuclear distance decreases. If there is a minimum in the total (electron plus nuclear) energy at some internuclear distance, a stable molecule can exist (Fig. 9.3). The energy versus distance can also be calculated for excited electronic

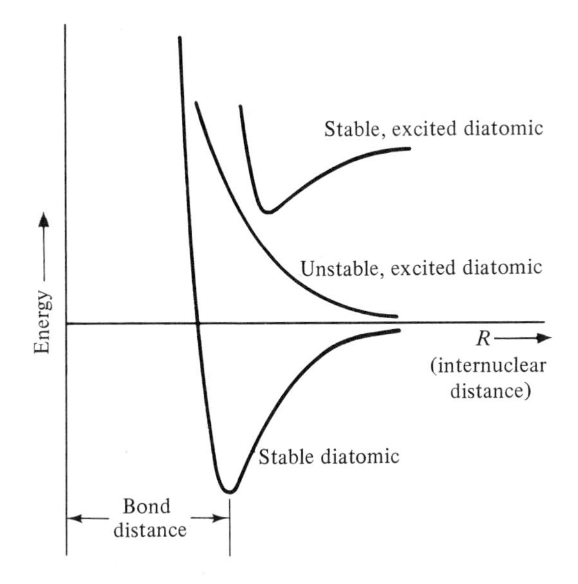

Fig. 9.3 Plot of calculated energy against distance between nuclei, R, in a diatomic molecule.

states. This tells us when excitation (by light, for example) will cause dissociation of the molecule. If dissociation does not occur, how does the bond distance change on electronic excitation? The position of the minimum in energy for the excited electronic state tells us the stable bond distance in the excited state. We can see how electronic excitation in a molecule such as retinal leads to a change in conformation.

Calculation and measurement of energies of molecules in their ground and excited states involves all of spectroscopy; we will save detailed discussion of this for Chapter 10.

We have treated a simple, one-electron, diatomic molecule to make the notation simpler, but the same principles apply to more complicated molecules.

For a two-electron molecule, the electronic wavefunction depends on the positions of both electrons.

$$\psi = \psi(x_1, y_1, z_1, x_2, y_2, z_2)$$

The subscripts refer to electrons 1 and 2. The equations for calculating electron distribution, position, energy, and so on, are the same, but we now must integrate over six coordinates instead of three. It is clear why many-electron molecules quickly begin to tax present-day computers. Adenine (Fig. 9.4) has 70 electrons,

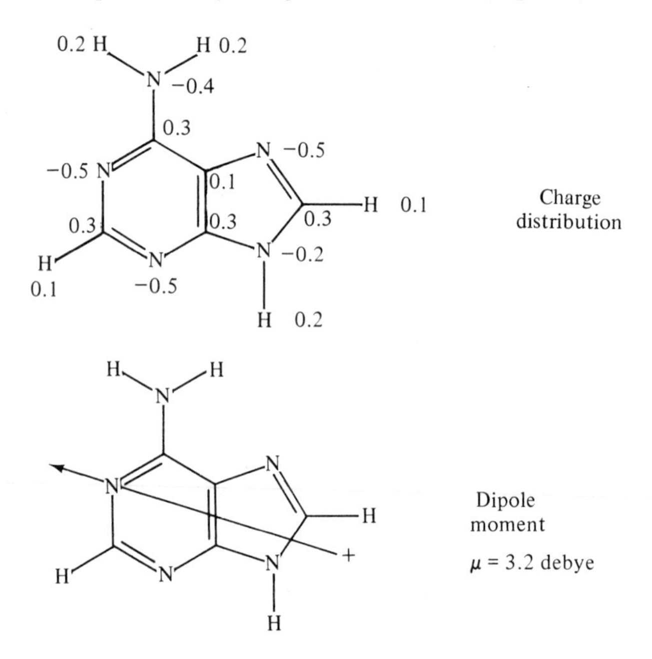

Fig. 9.4 Approximate wavefunctions were used to calculate the charge distribution and dipole moment of adenine. 1 debye is 10^{-18} esu cm. The charge distribution is given as the fraction of an electronic charge at each atom relative to the charge (equal to zero) on the isolated atom. [From B. Pullman and A. Pullman, *Progr. Nucleic Acid Res. 9*, 327 (1969).]

so we have electronic wavefunctions that depend on 210 coordinates. Therefore, at present we need to make approximations to calculate useful energies, bond distances, electron densities, and so on. A useful approximation for many applications is to consider only the π electrons. This approximation is adequate for treating the ultraviolet spectrum of adenine, for example.

Given an approximate wavefunction for the ground state of adenine, what properties could we calculate? The charge distribution can be calculated from Eq. (9.3); an example of such a calculation is given in Fig. 9.4. This gives us some ideas about relative reactivity of various positions on adenine. We cannot make firm predictions because (1) the wavefunctions are very approximate, (2) the solvent is important, and (3) calculations on the possible product species would need to be made. However, the charge densities are consistent with our chemical intuition. The nitrogens are all negative, so positive substituents would be attracted there. Because the three unsubstituted ring nitrogens are more negative than the amino nitrogen, we could predict that protonation occurs first on the ring. This is found experimentally. Predictions of ionization constants for each nitrogen have been made, but their discussion would require more detailed comment than we wish to make.

One consequence of the charge distribution is easily measured and can be used as a check on the calculated wavefunctions. The *dipole moment*, $\boldsymbol{\mu}$, is a measure of the polarity of a molecule. If the center of positive charge from all the nuclei is at the same position as the center of negative charge from all the electrons, the molecule has zero dipole moment; it is nonpolar. Otherwise, it is polar and it has a dipole moment. The dipole moment is calculated by multiplying the distance between the centers of positive and negative charge by the total positive charge or the total number of electrons (we will only discuss dipole moments of uncharged molecules). The dipole moment is a vector, because it has a magnitude (as calculated above) and a direction (from the center of positive charge to the center of negative charge). Symmetry is very helpful in calculating dipole moments. Any molecule with a point, or center, of symmetry has zero dipole moment. This includes H_2, N_2, O_2, CH_4, benzene, naphthalene, and so on. For a molecule with a plane of symmetry (such as adenine), the dipole moment, if it exists, must lie in the plane. To calculate the magnitude and direction of the dipole moment for a molecule such as adenine or retinal, we need to know the locations of all the negative and positive charges in the molecule. The components of the dipole moment vector are

$$\begin{aligned} \boldsymbol{\mu}_x &= \sum_i q_i x_i \\ \boldsymbol{\mu}_y &= \sum_i q_i y_i \\ \boldsymbol{\mu}_z &= \sum_i q_i z_i \end{aligned} \tag{9.4}$$

Each q_i is the charge at position x_i, y_i, z_i. For each nucleus we know its charge and its position, so the sum over nuclei is easy to do. For the electrons we use the electronic wavefunction.

The magnitude of the dipole moment is

$$|\boldsymbol{\mu}| = \sqrt{\boldsymbol{\mu}_x^2 + \boldsymbol{\mu}_y^2 + \boldsymbol{\mu}_z^2} \tag{9.5}$$

The magnitude of the dipole moment can be determined from dielectric measurements on solutions containing the molecule. As we mentioned before, it provides one check on the wavefunctions. The calculated dipole moment for adenine is shown in Fig. 9.4. The dimensions of dipole moment are charge times distance; chemists have traditionally used electrostatic units (esu) times centimeters. The charge on the electron is 4.8×10^{-10} esu. The magnitudes of dipole moments are thus about 10^{-18} esu cm, so a special unit has been defined in honor of Peter Debye, who first explained the difference between polar and nonpolar molecules. One debye, D, equals 10^{-18} esu cm. The calculated adenine dipole moment is 3.2 D, and the measured value in CCl_4 solvent is 3.0 D; this agreement is considered satisfactory for such calculations.

We will solve *simple* quantum mechanical problems in the next section such as a single electron in a box, or the hydrogen atom. However, the student should remember that approximate wavefunctions can be obtained for any molecule. These wavefunctions can then be used to get some idea about where the electrons are in the molecule and how they change on ionization, tautomerization, or excitation. The electron distribution can, in turn, provide understanding of molecular interactions, chemical reactivity, photochemistry, and so on.

SCHRÖDINGER'S EQUATION

Schrödinger's equation is the differential equation whose solutions give the wavefunctions. From the wavefunctions, measurable properties can be calculated. These properties are functions of the positions and momenta of the electrons and nuclei in the molecule. Schrödinger (and Heisenberg) deduced the form of *operators*, which operate on a wavefunction to give a measurable property. We use operators constantly in mathematics. For example, $\ln x$, $\sqrt{x}$, dx, and $3x$ are all examples of operators operating on x. The *position operator* in quantum mechanics is simply the coordinate itself. For example, to find the average position of the electron in the molecule HeH^{2+} (Fig. 9.2), we use†

$$\bar{z} = \int \psi_n z \psi_n \, dv$$

Here z operates on ψ_n simply by multiplying ψ_n.

The momentum operator in quantum mechanics is slightly more complicated. In classical mechanics the momentum is mass times velocity; in the x direction it is $p_x = mv_x$. In quantum mechanics the *momentum operator* in the x direction is

$$p_x = \frac{\hbar}{i}\frac{d}{dx}$$

$$\hbar \equiv \frac{h}{2\pi} = 1.055 \times 10^{-27} \text{ erg s} = 1.055 \times 10^{-34} \text{ J s}$$

$$i \equiv \sqrt{-1}, \text{ imaginary} \qquad i^2 = -1$$

The quantum mechanical momentum operator operates on ψ and involves taking the derivative of ψ with respect to a space coordinate and multiplying by $\hbar/i$.

† We shall consider only real wavefunctions here. In general, the equation must be written as $\int \psi_n^* z \psi_n \, dv$, where ψ_n^* is the complex conjugate of ψ_n. This means wherever the imaginary, i, appears in ψ_n it is replaced by $(-i)$ in ψ_n^*.

The Schrödinger equation is simply an equation describing the conservation of energy for a system written in terms of its wavefunction, ψ. In the operator notation that we have developed, and for a single particle moving in one dimension, this is

$$\begin{array}{ccccc} \text{kinetic energy} & + & \text{potential energy} & = & \text{total energy} \\ T_x\psi & + & U(x)\psi & = & E\psi \\ \dfrac{p_x^2}{2m}\psi & + & U(x)\psi & = & E\psi \\ -\dfrac{\hbar^2}{2m}\dfrac{d^2\psi}{dx^2} & + & U(x)\psi & = & E\psi \\ & & \mathcal{H}\psi & = & E\psi \end{array} \tag{9.6}$$

The operator $\mathcal{H}$, which operates on the wavefunction to give the energy, is called the *Hamiltonian.* T_x is the kinetic energy operator; $U(x)$ is the potential energy operator; E is the total energy operator.

When E is constant, Eq. (9.6) is the time-independent Schrödinger equation in one dimension for a particle of mass m. It is a differential equation whose solutions give the wavefunctions, ψ_n, and energies, E_n, for the particle in its ground and excited states. To emphasize this, we can write the Schrödinger equation as

$$\mathcal{H}\psi_n = E_n\psi_n$$

$$-\frac{\hbar^2}{2m}\frac{d^2\psi_n}{dx^2} + U(x)\psi_n = E_n\psi_n \tag{9.7}$$

The subscript n is the *quantum number* for the wavefunction or *eigenfunction*, ψ_n, and the *energy* or *eigenvalue*, E_n. Before considering applications of the Schrödinger equation to some simple problems, let us first note that we are not restricted to working in one dimension or to using an x, y, z-coordinate system. For example, the Schrödinger equation written for a particle in three coordinates in the x, y, z system is

$$-\frac{\hbar^2}{2m}\left(\frac{\partial^2\psi}{\partial x^2} + \frac{\partial^2\psi}{\partial y^2} + \frac{\partial^2\psi}{\partial z^2}\right) + U(x, y, z)\psi = E_{n_x,n_y,n_z}\psi \tag{9.8}$$

where $\psi = \psi(x, y, z)$. There will be, in general, three quantum numbers, n_x, n_y, n_z, associated with the system. Many problems involving central symmetry, such as a hydrogen atom, are best solved using spherical coordinates, r, θ, ϕ.

Other coordinate systems are also useful. To write the Schrödinger equation in a form independent of the coordinate system, the Laplace operator is introduced. The *Laplace operator*, ∇^2, indicates the operation of taking second derivatives with respect to all coordinates. In Cartesian coordinates, it has the form

$$\nabla^2 = \frac{\partial^2}{\partial x^2} + \frac{\partial^2}{\partial y^2} + \frac{\partial^2}{\partial z^2} \tag{9.9}$$

Using this operator we can write the time-independent Schrödinger equation as

$$-\frac{\hbar^2}{2m}\nabla^2\psi + U\psi = E\psi \tag{9.10}$$

or, even more compactly, as

$$\mathcal{H}\psi = E\psi \tag{9.11}$$

where $\mathcal{H}$ is the Hamiltonian operator

$$\mathcal{H} = -\frac{\hbar^2}{2m}\nabla^2 + U \tag{9.12}$$

representing the sum of kinetic and potential energy. For more than one particle in the system, we sum over all particles:

$$\mathcal{H} = -\frac{\hbar^2}{2}\sum_i \frac{\nabla_i^2}{m_i} + U \tag{9.13}$$

SOLVING WAVE MECHANICAL PROBLEMS

An understanding of the basic elements and procedures of wave mechanics can be of great value in appreciating the contribution of theoretical chemistry to problems of biological interest. It also serves as a basis for understanding the origins of molecular spectroscopy and the kinds of information that can be obtained from spectroscopic studies. We shall choose a few simple examples to illustrate the approach. The examples include a particle in a potential well, an electron in a central force field, a harmonic oscillator, and molecular orbitals constructed as linear combinations of atomic orbitals. We shall then point out what approximations are needed to approach the real molecular situation. The simple examples are more than mere exercises; insights derived from them will be used to describe and interpret important aspects of biomolecular structure and spectra.

First, let us examine the general strategy used in solving real problems in wave mechanics. We shall consider only stationary states of molecules, so that time will not enter explicitly as a variable. If the system does change with time (by the absorption or emission of radiation), we can still describe the initial and final states in this stationary approximation. The dynamics (time course) of the changes can be treated wave mechanically also, but the calculations are mathematically more involved than we need to consider here.

The following outline gives a brief summary of the sequence of operations involved in a typical wave mechanical calculation:

I. Define the problem to be solved.
II. Write the Schrödinger equation for the problem.
III. Solve for the eigenfunctions and for the eigenvalues of energy.
IV. Interpret the solutions and test them against experimentally observable properties.

Outline of Wave Mechanical Procedures

I. Definition of the problem

A. *Write the appropriate potential for the particular problem.* Although the kinetic-energy operator has the same form for all problems, the potential-energy operator distinguishes each problem.

1. Particle in a box: We consider a particle that is contained in a box but is otherwise free. The potential energy is zero inside the box but infinite at the walls and outside the box. For a one-dimensional box,

$$\begin{aligned} U(x) &= 0 \qquad \text{for } 0 < x < a \\ U(x) &= \infty \qquad \text{for } x \leq 0 \quad \text{and} \quad x \geq a \end{aligned} \tag{9.14}$$

Think of a bead on a wire capped at each end. If there is no friction between the bead and wire, the bead is free to move along the wire between the caps.

2. Harmonic oscillator: A harmonic oscillator represents a particle attached to a spring. The potential energy increases whenever the spring is stretched or compressed. This problem is easy to solve and it approximately represents the vibrations of nuclei of a molecule:

$$U(x) = \tfrac{1}{2}kx^2 \qquad k = \text{force constant} \tag{9.15}$$

3. Coulomb's law: This is the dominant potential involving the interactions between nuclei and electrons in molecules; it is the only one we will consider. Other much weaker interactions depend on the spin of the electrons and nuclei. For two charges separated by a distance r, the potential energy is

$$U(r) = -\frac{q_1 q_2}{r} \tag{9.16}$$

For charges q_1 and q_2 in electrostatic units (esu) and r in centimeters, the potential energy is in ergs. This is the most usual form used by chemists and biologists. It is consistent with using m in grams and $\hbar = 1.055 \times 10^{-27}$ erg s in the kinetic-energy operator. In the equations we have used up to now, we did not have to explicitly write different equations for different units. However, the SI equations involving charges have a different form from those using esu units. With SI units the potential energy is

$$U(r) = -\frac{q_1 q_2}{4\pi\varepsilon_0 r} \tag{9.17}$$

Here the charges are in coulombs, r is in meters, ε_0 (the permittivity constant) $= 8.854 \times 10^{-12}$, and the energy is in joules. It is consistent with using m in kilograms and $\hbar = 1.055 \times 10^{-34}$ J s in the kinetic-energy operator.

We can now write the potential-energy operator for any atom or molecule. For a hydrogen atom, in cgs units,

$$U(r) = -\frac{e^2}{r} \tag{9.18}$$

where e = electronic charge and r is the distance between the proton and the electron. For a helium ion, He^+,

$$U(r) = -\frac{2e^2}{r} \tag{9.19}$$

For a helium atom,

$$U(r_1, r_2) = -\frac{2e^2}{r_1} - \frac{2e^2}{r_2} + \frac{e^2}{r_{12}} \tag{9.20}$$

The coordinates r_1 and r_2 refer to electrons 1 and 2 in the helium atom; r_{12} is the absolute (positive) distance between the electrons. The three terms represent, respectively, the electron-nuclear attraction for each electron and the electron-electron repulsion. For a molecule we can easily write the potential energy just by adding terms corresponding to each electron-nucleus attraction, each electron-electron repulsion, and each nucleus-nucleus repulsion.

B. *Establish the boundary conditions for the wavefunction.* The wavefunctions are solutions to the Schrödinger equation, but in addition they must satisfy some auxiliary conditions. Examples of these are:

1. The wavefunction should be single-valued and finite everywhere.
2. The wavefunction of a bound electron, molecule, and so on, should vanish at infinite distance and in any region where the potential is plus infinity.
3. The wavefunction should be continuous and have a continuous first derivative, except at the nucleus of an atom, where the potential energy becomes minus infinity.

II. Writing the Schrödinger equation for the problem

Write down the Schrödinger equation in a suitable coordinate system. Some examples are:

1. Particle in a box (one dimension):

$$-\frac{\hbar^2}{2m}\frac{d^2\psi_n(x)}{dx^2} = E_n\psi_n(x) \qquad \text{for } 0 < x < a$$

2. Harmonic oscillator (one dimension):

$$-\frac{\hbar^2}{2\mu}\frac{d^2\psi_v(x)}{dx^2} + \tfrac{1}{2}kx^2\psi_v(x) = E_v\psi_v(x)$$

3. Hydrogen atom:

$$-\frac{\hbar^2}{2\mu}\nabla^2\psi_{n,l,m}(r,\theta,\phi) - \frac{e^2}{r}\psi_{n,l,m}(r,\theta,\phi) = E_n\psi_{n,l,m}(r,\theta,\phi)$$

We have introduced a new symbol, μ, to represent the *reduced mass.* In the Schrödinger equations for the harmonic oscillator and hydrogen atom we are interested in relative motions of particles. For example, in the hydrogen atom we are solving for the wavefunction of the electron relative to the nucleus; we ignore the motion of the hydrogen atom through space. However, both the electron and the nucleus move relative to the center of mass of the hydrogen atom. Because of this, we must use the reduced mass of the two particles in the Schrödinger equation. The reduced mass, μ, for two particles is

$$\frac{1}{\mu} = \frac{1}{m_1} + \frac{1}{m_2} \tag{9.21}$$

Because the proton is nearly 2000 times more massive than the electron, for a hydrogen atom the reduced mass is nearly equal to the mass of an electron.

III. Solving for the eigenfunctions and for the eigenvalues of energy

Solve the Schrödinger equation using appropriate methods of solving differential equations. Obtain the set of wavefunctions, ψ_n, and eigenvalues, E_n, that satisfy the Schrödinger equation and the boundary conditions. This is usually straightforward, but often exceedingly tedious. We shall adopt the compromise of presenting the solutions that were obtained by the pioneers in this field about 50 years ago. It is relatively simple to convince yourself that they are indeed valid; simply substitute the solutions into the Schrödinger equation and show that they satisfy it exactly.

IV. Interpretation of the wavefunctions

The interpretation of the solutions to the Schrödinger equation involves comparison with available experimental data on the system. A great variety of properties can be tested. Some of the most important of these are:

1. Energies (eigenvalues)—comparison with values obtained from spectroscopy, ionization potentials, electron affinities, and so on.
2. Electron distribution (eigenfunctions)—comparison with atomic and molecular dimensions, bond dissociation energies, dipole moments, probabilities of transition from one state to another, directional character of bonding, intermolecular forces.
3. Systematic changes for different, but related systems—comparison of spectra, ionization potentials, bond lengths, bond energies, force constants, dipole moments, and so on, for atoms or molecules that differ from one another in a simple manner.

PARTICLE IN A BOX

The particle-in-a-box problem is perhaps the simplest of any that can be solved using the wave equation. In this problem the *particle*, which may be an electron, a nucleus, or even a baseball, is considered to move freely ($U = 0$) within a defined region of space, but is prohibited ($U = \infty$) from appearing outside that region. The problem can be solved readily in one, two, or three dimensions and for any shape of rectangular box.

Some biologically relevant applications of this problem include: the behavior of π electrons that are delocalized over large portions of molecules, as in linear polyenes (carotenoids, retinal), planar porphyrins (heme, chlorophyll), and large aromatic hydrocarbons; conduction electrons that may move over extensive regions of biopolymers; and the exchange of protons involved in hydrogen bonding between two nucleophilic atoms, as between the oxygens of adjacent water molecules. Even in cases where the real potential well is not infinite and square, the particle-in-a-box calculation is a useful approximation for obtaining the order of magnitude of the energies involved.

The potential energy for a particle in a one-dimensional box,

$$U = 0 \qquad 0 < x < a$$

$$U = +\infty \qquad \begin{matrix} x \leqslant 0 \\ x \geqslant a \end{matrix}$$

is shown diagrammatically in Fig. 9.5. The solution of the Schrödinger equation outside the box is trivial; because the particle cannot exist there, its wavefunction

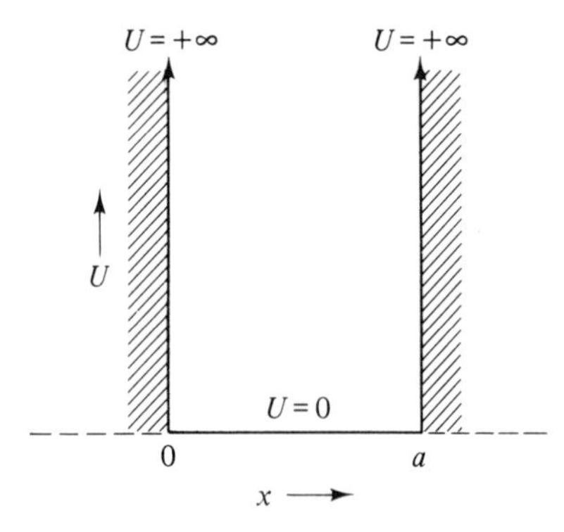

Fig. 9.5 Square potential well of length a.

must be zero everywhere outside the box. Within the box, it must satisfy the one-dimensional Schrödinger equation for $U \equiv 0$:

$$-\frac{\hbar^2}{2m}\frac{d^2\psi_n(x)}{dx^2} = E_n\psi_n(x) \tag{9.22}$$

Rather than present the rigorous solution to this second-order differential equation by conventional methods, which may be unfamiliar to the reader, let us look at the requirements for a wavefunction that solves Eq. (9.22). It must be a function of x such that its second derivative, $d^2\psi/dx^2$, is equal, apart from constant factors, to the original function, ψ. Two such functions are $\sin bx$ and $\cos bx$, since

$$\frac{d^2}{dx^2}(\sin bx) = -b^2 \sin bx$$

and

$$\frac{d^2}{dx^2}(\cos bx) = -b^2 \cos bx$$

In fact, a general solution can be written in the form

$$\psi = A \sin bx + B \cos bx \tag{9.23}$$

We can immediately simplify the solution by applying the continuity condition (I.B.3 in the previous section). We know that the wavefunction must be zero at the boundary of the box (since it is zero everywhere outside); therefore, $\psi(0) = \psi(a) = 0$. When $x = 0$, $\sin bx = 0$, but $\cos bx = 1$. Therefore, we must set $B = 0$ in Eq. (9.23) to satisfy the boundary condition at $x = 0$. Thus

$$\psi = A \sin bx$$

To satisfy the boundary condition at $x = a$, $\sin bx$ must $= 0$ when $x = a$. We know that $\sin n\pi = 0$ for any integer, n; therefore it is easy to see that the function

$$\psi_n = A \sin \frac{n\pi}{a} x \qquad n = 1, 2, 3, 4, \ldots \tag{9.24}$$

is the only one that satisfies the second part of the boundary conditions as well. Here a is the length of the box and n can take on any integral value; it is the quantum number for the problem. We obtain A by using the normalization condition:

$$A^2 \int_0^a \sin^2 \frac{n\pi x}{a} dx = 1$$

Using a table of integrals, we obtain $A = \sqrt{2/a}$. We now have the wavefunction for a particle in a box in its ground ($n = 1$) and excited ($n = 2, 3, 4, \ldots$) states:

$$\psi_n = \sqrt{\frac{2}{a}} \sin \frac{n\pi x}{a} \tag{9.25}$$

We can calculate any property of the particle.

Figure 9.6 shows the behavior of ψ versus x for several values of n. (In each case, the negative of the ψ shown is also a valid solution.) The analogy of a standing wave for the classical vibrating string is worth noting here. The violin

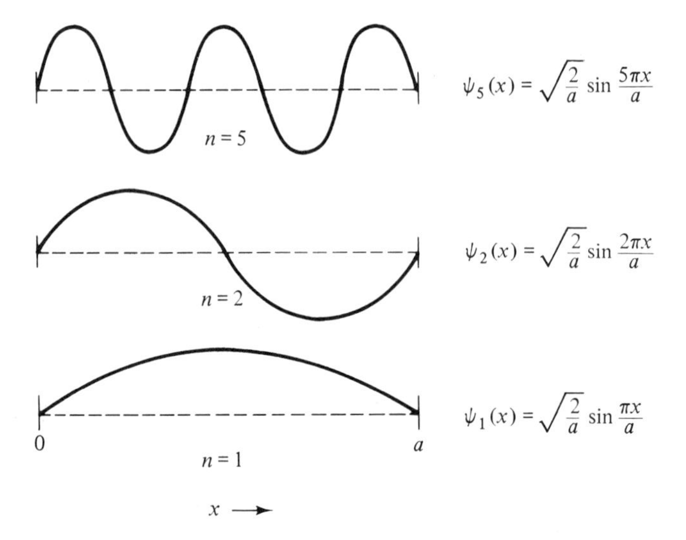

Fig. 9.6 Variation of $\psi_n(x)$ with x for $n = 1, 2,$ and 5. The magnitude of $\psi_n(x)$ is plotted versus x. The extreme values of $\psi_n(x)$ are $\pm\sqrt{2/a}$.

string or a rope fixed at both ends can be set into motion as a standing wave with nodes (points where the string has zero displacement) only at the ends, or with more additional nodes between the ends. These are called the *fundamental mode* ($n = 1$), the *first harmonic* or *overtone* ($n = 2$), the *second overtone* ($n = 3$), and so on. The requirements that the ends are fixed and undergo no displacement are the boundary conditions that limit standing waves to these particular (integral) modes. The properties of the harmonic progression arise naturally from the constraints on the system, just as the quantum numbers arise naturally in wave mechanics from the boundary conditions characteristic of the particular problem. It is this natural occurrence of quantized states that makes wave mechanics a superior approach compared with the arbitrary introduction of quantization as had been done earlier by Bohr.

Let us turn now to the energies associated with the particle-in-a-box problem. Having determined the eigenfunctions [Eq. (9.25)] that constitute the solutions to the Schrödinger equation for this problem, we proceed to substitute them into Eq. (9.22). Thus,

$$-\frac{\hbar^2}{2m}\frac{d^2}{dx^2}\left(A \sin\frac{n\pi x}{a}\right) = E_n\left(A \sin\frac{n\pi x}{a}\right)$$

$$+\frac{\hbar^2}{2m}\frac{n^2\pi^2}{a^2}\left(A \sin\frac{n\pi x}{a}\right) = E_n\left(A \sin\frac{n\pi x}{a}\right) \tag{9.26}$$

from which we can readily extract the eigenvalue solutions

$$E_n = \frac{\hbar^2\pi^2 n^2}{2ma^2} = \frac{h^2 n^2}{8ma^2} \qquad n = 1, 2, 3, \ldots \tag{9.27}$$

Each eigenfunction has an associated energy or eigenvalue, determined in part by the quantum number, n. In fact, the energy values increase with the square of the quantum number, as illustrated in Fig. 9.7. Apart from the fundamental constants that enter into Eq. (9.27), notice the dependence on m and a. The spacing depends inversely on the mass of the particle and inversely on the square of the

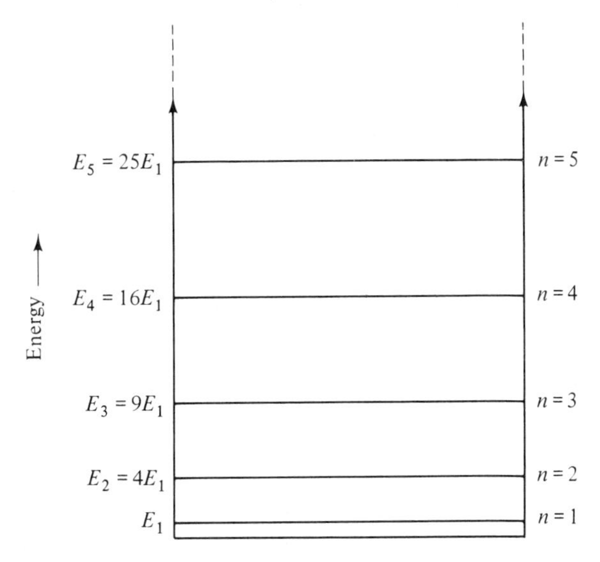

Fig. 9.7 Eigenvalues of energy for a particle in a one-dimensional potential well.

length of the box. A light particle such as an electron will have widely spaced energy levels in comparison with a heavy particle like a nucleus or a baseball. A given particle in a small box will have widely spaced eigenvalues compared with the same particle in a larger box. It is this property that makes quantization important for light particles (electrons) in small regions of space (atoms or molecules). For a baseball in a room or an electron traversing a vacuum tube, the eigenvalues are so closely spaced that they may be considered to provide a continuum of energy. This is just the approach used by classical physics, and it is entirely appropriate for problems of large dimensions. The wave mechanical analysis allows us to determine just when the classical picture breaks down and gives us a method of dealing with truly submiscrosopic problems.

The eigenfunctions and eigenvalues for the particle-in-a-box problem are more than mathematical abstractions. To appreciate their physical significance, we choose to look at the family of carotenoids, which are examples of linear polyenes. A widely distributed example is β-carotene, whose molecular structure is as follows:

CH_3 CH_3 CH_3 H_3C H_3C CH_3 CH_3 CH_3 CH_3 CH_3

It is a symmetrical molecule containing 40 carbon atoms. If the molecule is oxidized, it breaks at the center (dashed line) and forms two molecules of retinal, or vitamin A_1, which is closely related to the pigment of rhodopsin, the protein involved in vision. Many isomers and chemically modified derivatives of β-carotene occur in nature. They play important roles in vision, as sensitizers in photosynthesis and as protective agents against harmful biological oxidations.

β-Carotene and the other carotenoids possess two classes of bonding electrons. For the present, consider one kind, the sigma (σ) electrons, to be responsible for the single bonds and for half of each double bond. These σ electrons are localized in the small regions of space between adjacent atoms. The other type, the pi (π) electrons, contribute the other halves of the double bonds. In systems, like β-carotene or benzene, where double and single bonds alternate in the classical molecular structure, the π electrons are not strongly confined to regions between adjacent atoms. In fact, they are nearly free to roam the full length of the conjugated system. Thus they are confined approximately in a one-dimensional potential well which is about as long as the conjugated system and within which the potential energy is virtually constant. The particle-in-a-box model provides a reasonable approach to the eigenfunctions and eigenvalues for these π electrons. Having adopted this model, we can now proceed to examine its predictions and compare them with physically measurable properties.

β-Carotene has 11 conjugated double bonds and thus has 22 π electrons. We shall assume that the energies and wavefunctions for these 22 electrons can be *very* approximately described by the energies and wavefunctions of a particle in a box. The energy-level pattern shown in Fig. 9.7 is extended and redrawn schematically in Fig. 9.8 to show the lowest 13 energy levels for the particle-in-a-box problem. The energy levels are not properly scaled in Fig. 9.8. The increasing separation of the upper levels is suppressed so that all of them can be shown in the figure. In fact, the energy of the lowest level should be less than 1% of the highest levels shown. As we will discuss later, the electron configuration of the lowest-energy state for the 22 π electrons of β-carotene is obtained by placing the electrons in the levels or orbitals of lowest available energy, such that each orbital has two electrons with opposite spins. As shown in Fig. 9.8, this causes the 11 orbitals to be completely filled and just uses up the 22 π electrons. If we could somehow measure the energy of the electrons in the $n = 11$ level, we could use Eq. (9.27) to calculate a, the length of the potential well, and compare it with the sum of bond lengths for the conjugated system. There is, in fact, no good way to measure directly the energy of the bound electrons in a molecule such as β-carotene. We can, however, relate this model to an *energy difference* that can be measured from the absorption spectrum of the molecule.

When light of the appropriate wavelength is incident on a molecule, it may absorb some of the radiation, resulting in the promotion or excitation of electrons to higher-energy orbitals. The process occurs only if the energy of the incident photons corresponds to the difference between two possible states (eigenstates) of the molecule. Because there are many filled and many empty states available, the absorption spectrum is potentially a rich source of information about the location

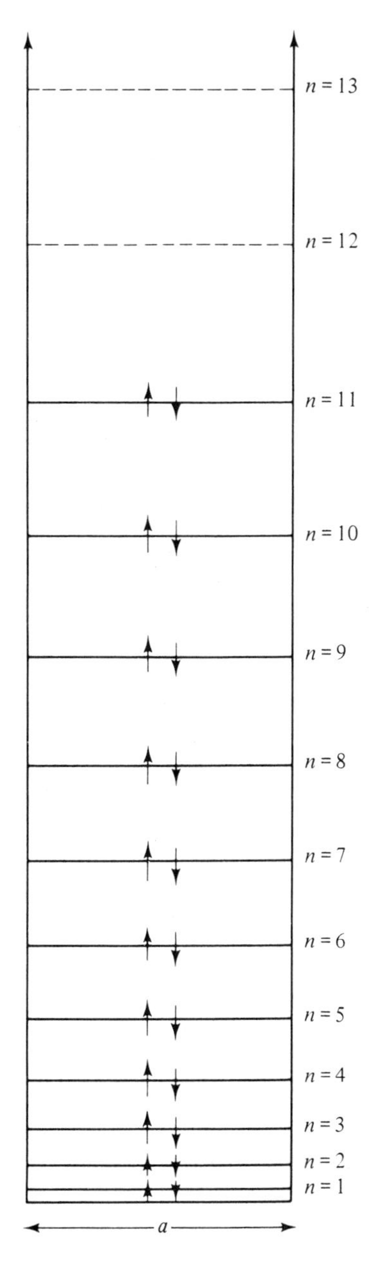

Fig. 9.8 Particle-in-a-box approximation to the π-electron energy levels for β-carotene.

of the energy levels. In practice, the lowest-energy transition, corresponding to the longest-wavelength electronic absorption band, is the easiest to identify and observe. The lowest-energy absorption occurs when an electron in the highest filled energy level is excited to the lowest unfilled energy level. This is shown in Fig. 9.9 for that segment of the energy-level diagram. Thus the energy associated with the photons absorbed in the long-wavelength absorption band is just the

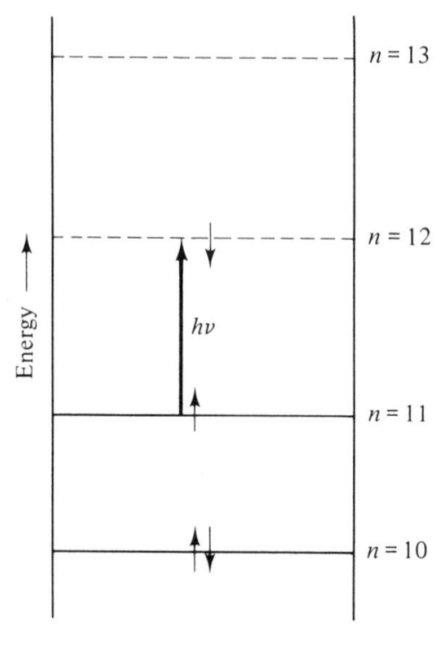

Fig. 9.9 Longest-wavelength electronic transition for the particle-in-a-box approximation to β-carotene.

difference in energy between level $n = 11$ and level $n = 12$. As we shall see, this is just as useful as knowing E_{11} or E_{12} for the purposes of making an estimate of the length a.

Using Eq. (9.27), we can easily derive a general expression for the energy difference between adjacent levels for the particle-in-a-box problem. Let N represent the quantum number of the highest filled level; then $N + 1$ is the quantum number for the lowest unfilled level. The energy difference between these two levels is

$$\Delta E = E_{N+1} - E_N = \frac{h^2}{8ma^2}(N+1)^2 - \frac{h^2}{8ma^2}N^2$$

$$= \frac{h^2}{8ma^2}(N^2 + 2N + 1 - N^2) = \frac{h^2}{8ma^2}(2N+1) \tag{9.28}$$

Since

$$\Delta E = \frac{hc}{\lambda}$$

we obtain the result that

$$\lambda\,(N \rightarrow N+1) = \frac{8mca^2}{h(2N+1)} \tag{9.29}$$

For linear polyenes with different extents of conjugation, both the length a and the number of π electrons ($2N$) will be different. The wavelength increases in proportion to the square of the length a and inversely with the number of π electrons. Since the length of the π system increases *linearly* with the number of π bonds, the long-wavelength absorption band increases in wavelength (shifts toward the red) with increasing extent of conjugation.

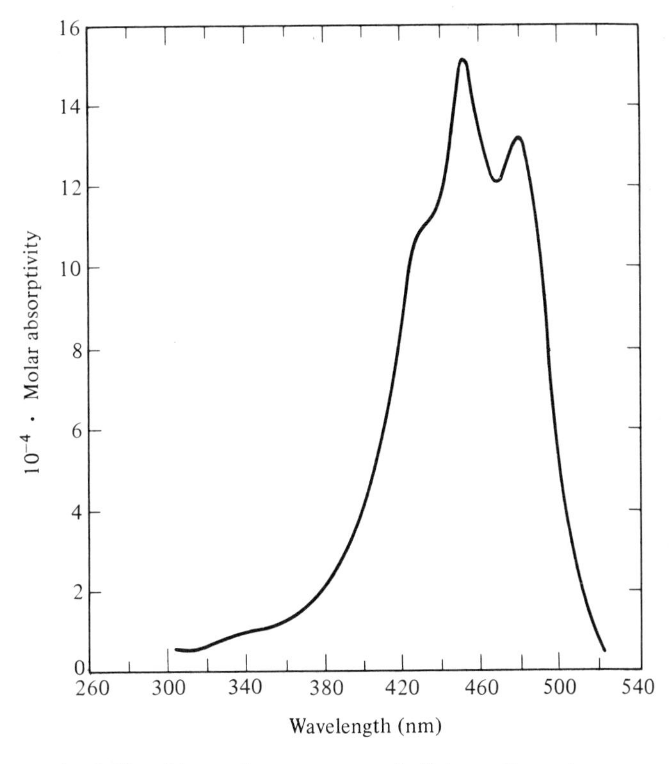

Fig. 9.10 Absorption spectrum of all-trans β-carotene.

The long-wavelength absorption of β-carotene is shown in Fig. 9.10. The structure of the band actually arises from additional vibrational effects and need not concern us here. The long-wavelength maximum is readily identified and occurs at about 480 nm. By substitution into Eq. (9.29) and solving for a, we obtain for $N = 11$,

$$a^2 = \frac{h\lambda(2N+1)}{8mc} = \frac{(6.62 \times 10^{-27} \text{ erg s})(4.8 \times 10^{-5} \text{ cm})(23)}{8(9.11 \times 10^{-28} \text{ g})(3.0 \times 10^{10} \text{ cm s}^{-1})}$$

$$a = 1.83 \times 10^{-7} \text{ cm} = 18.3 \text{ Å}$$

This is an effective length for the one-dimensional free-electron model of carotene.

The actual length of the zigzag chain can also be calculated from the sum of the lengths of the single and double bonds. For such a conjugated system, the length of a single bond is 1.46 Å and the length of a double bond is 1.35 Å. There are 11 double bonds and 10 single bonds for the conjugated chain in β-carotene; therefore, the calculated length is 10×1.46 Å $+ 11 \times 1.35$ Å or 29 Å. The actual length and effective length are different for several reasons. The actual molecular potential does not have a flat bottom and infinitely steep sides. There are also important electron-electron repulsions. Nevertheless, if we compare the spectra of carotenoids with different lengths of conjugation, there is a very good correlation with the form of Eq. (9.29).

Particles in two- and three-dimensional boxes can be solved just like the one-dimensional box. An electron in a two-dimensional box provides a reasonable model for discussing the spectrum of such molecules as benzene, naphthalene, and anthracene. A particle in a three-dimensional box accurately describes a dilute gas in a real container. For a box of sides a, b, c, the wavefunctions are

$$\psi(x, y, z) = \sqrt{\frac{8}{abc}} \sin\frac{n_x \pi x}{a} \sin\frac{n_y \pi y}{b} \sin\frac{n_z \pi z}{c} \tag{9.30}$$

The energies are

$$E = \frac{h^2}{8m}\left(\frac{n_x^2}{a^2} + \frac{n_y^2}{b^2} + \frac{n_z^2}{c^2}\right) \tag{9.31}$$

These energies correspond to the translational energies of an ideal gas. They can be used to derive the ideal gas equation and to predict when quantum effects will cause deviations from classical ideal gas behavior. These quantum effects can become important for He gas at temperatures near absolute zero.

SIMPLE HARMONIC OSCILLATOR

We can gain a clearer understanding of the vibration of molecules by considering the quantum mechanics of a simple harmonic oscillator; the vibration of a diatomic molecule, for example, can be approximated as the motion of two atoms linked by a spring. For simplicity, we consider a particle of mass m attached to one end of a spring, with the other end of the spring attached to a wall (Fig. 9.11). This is a good approximation for a diatomic molecule such as HBr, where one of the atoms is much lighter than the other.

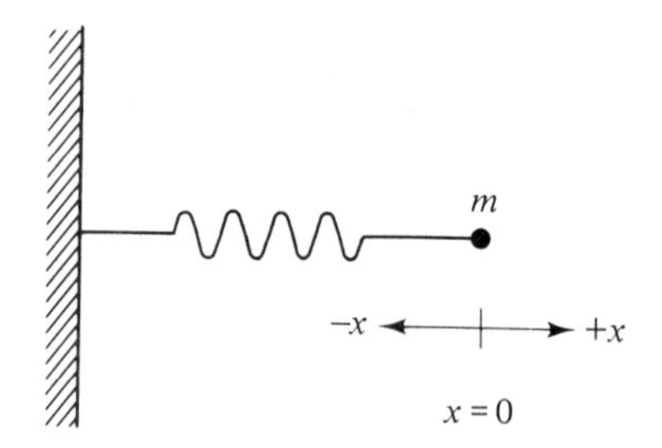

Fig. 9.11 Harmonic oscillator.

We can approximate the vibration of any diatomic molecule by simply substituting the reduced mass, μ, for that of the particle. If the particle is displaced from its equilibrium position $x = 0$ to a position $x = x$, then, according to Hooke's law (Chapter 2), there is a restoring force

$$F = -kx \tag{9.32}$$

where k is the force constant of the spring. The minus sign specifies that the force is a restoring one; it is opposite to the displacement, x.

The change in potential energy due to a change dx in position can be calculated from

$$dU = -F\,dx = kx\,dx$$

Upon integration, we obtain

$$U = \tfrac{1}{2}kx^2 \tag{9.33}$$

if we take $U = 0$ at $x = 0$.

Before we give a quantum mechanical treatment of the problem, let us consider the case from a classical point of view. According to Newton's law, the force exerted on a particle is equal to the mass of the particle, m, times its acceleration. Thus

$$m\frac{d^2x}{dt^2} = -kx \tag{9.34}$$

The solution of this differential equation is

$$x = A \sin 2\pi\nu_0 t \tag{9.35}$$

where A is the amplitude, and the vibration frequency is

$$\nu_0 = \frac{1}{2\pi}\left(\frac{k}{m}\right)^{1/2} \tag{9.36}$$

The particle oscillates periodically from $x = -A$ to $x = A$, with a frequency ν_0 determined by the force constant k and the mass m. The potential energy is

$$\begin{aligned} U &= \tfrac{1}{2}kx^2 \\ &= \tfrac{1}{2}kA^2 \sin^2 2\pi\nu_0 t \end{aligned}$$

and is a maximum at $x = \pm A$. The kinetic energy T of the particle is

$$\begin{aligned} T &= \tfrac{1}{2}m\left(\frac{dx}{dt}\right)^2 \\ &= \tfrac{1}{2}kA^2 \cos^2 2\pi\nu_0 t \end{aligned}$$

By using the identity $\cos^2\theta + \sin^2\theta = 1$,

$$\begin{aligned} T &= \tfrac{1}{2}kA^2(1 - \sin^2 2\pi\nu_0 t) \\ &= \tfrac{1}{2}k(A^2 - x^2) \end{aligned}$$

and is a maximum at $x = 0$ (at which $U = 0$). T becomes 0 at $x = \pm A$ (at which U becomes maximal). The total energy E is equal to $T + U$:

$$\begin{aligned} E &= T + U \\ &= \tfrac{1}{2}kA^2(\cos^2 2\pi\nu_0 t + \sin^2 2\pi\nu_0 t) \\ &= \tfrac{1}{2}kA^2 \end{aligned} \tag{9.37}$$

The energy is a constant determined by the force constant and the amplitude of oscillation.

Treating the problem quantum mechanically, we start with the Schrödinger equation,

$$\mathcal{H}\psi_v = E_v\psi_v$$

$$T\psi_v + U\psi_v = E_v\psi_v$$

We use the subscript v for the vibrational quantum number. The harmonic oscillator Schrödinger equation is thus

$$-\frac{\hbar^2}{2m}\frac{d^2\psi_v}{dx^2} + \tfrac{1}{2}kx^2\psi_v = E_v\psi_v \tag{9.38}$$

This differential equation can be solved by standard methods. Some of the wavefunctions and their corresponding energies are tabulated in Table 9.1. In Fig. 9.12, the wavefunctions, $\psi_v(x)$, and their squares, $\psi_v^2(x)$, which give the probability of finding the particle at x, are plotted for several quantum numbers.

Table 9.1 Some eigenfunctions and eigenvalues for a simple harmonic oscillator*

$\psi_0 = \left(\frac{2a}{\pi}\right)^{1/4} e^{-ax^2}$	x	$E_0 = \frac{1}{2}h\nu_0$
$\psi_1 = \left(\frac{2a}{\pi}\right)^{1/4} 2a^{1/2}x\,e^{-ax^2}$	x	$E_1 = \frac{3}{2}h\nu_0$
$\psi_2 = \left(\frac{2a}{\pi}\right)^{1/4} (4ax^2 - 1)\,e^{-ax^2}$	x	$E_2 = \frac{5}{2}h\nu_0$
$\vdots$		$\vdots$
ψ_v		$E_v = (v + \frac{1}{2})h\nu_0$

* $\nu_0 = \frac{1}{2\pi}\left(\frac{k}{m}\right)^{1/2}, \quad a = \frac{\pi}{h}(km)^{1/2}.$

It is interesting to compare the quantum mechanical results with the classical mechanical results. First, the energy is quantized into energy levels, as is always true in quantum mechanics. The energy of the harmonic oscillator can be only

$$E_v = (v + \tfrac{1}{2})h\nu_0 \tag{9.39}$$

where the quantum number, v, is a positive integer. According to classical mechanics, the energy can be any value depending on the amplitude. Second, classical mechanics allows the oscillator to be at rest and therefore have an energy equal to zero, but according to quantum mechanics, the lowest energy level is $\frac{1}{2}h\nu_0$,

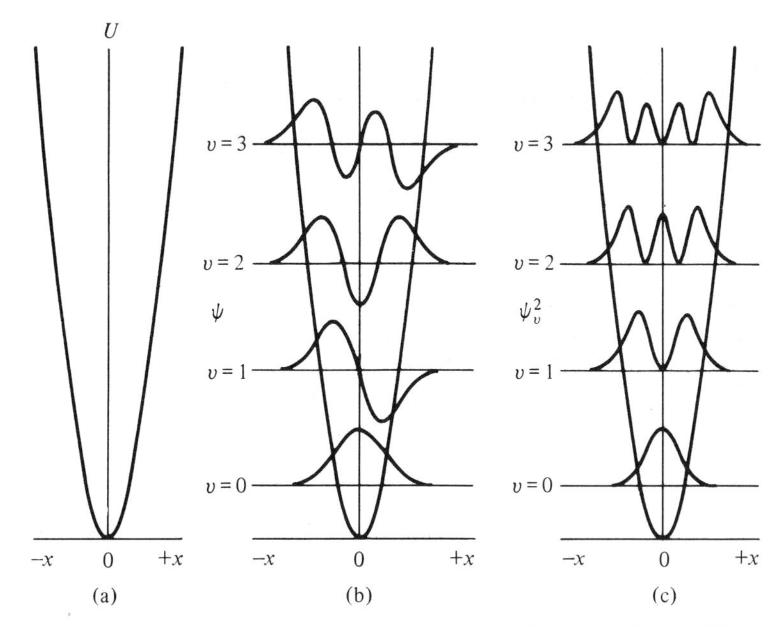

Fig. 9.12 (a) Potential well for a classical harmonic oscillator. (b) Allowed energy levels and wavefunctions for a quantum mechanical harmonic oscillator. $\psi_v(x)$ is plotted versus x; each different wavefunction is arbitrarily displaced vertically. (c) Probability functions for a quantum mechanical harmonic oscillator. The square of each wavefunction $[\psi_v(x)]^2$ is plotted versus x.

called the *zero-point energy*. Both quantum mechanics and classical mechanics give the same fundamental vibration frequency, ν_0, of the oscillator, however. It is also true that when v is large or when (k/m) is small, the quantum mechanics and classical mechanics pictures become very similar.

The quantum mechanical harmonic oscillator is used in interpreting infrared spectra which correspond to molecular vibrations. The difference in energy corresponds to a change in vibrational quantum number $\Delta v = 1$:

$$\Delta E = h\nu_0 \tag{9.40}$$

Therefore, light will be absorbed if its frequency ν is equal to a vibrational frequency of a harmonic oscillator. A harmonic oscillator is a good approximation for the vibration of diatomic molecules and the vibration of bonds in polyatomic molecules. Vibrational frequences are characteristic of a particular type of bond such as C—H, C—C, C=C, and C—N, but they can also be used to determine force constants, k, for the particular vibration. For example, when the force constant for a double-bond vibration decreases (the frequency decreases), it means that the bond has become weaker; the double-bond character has decreased. These differences in bond strength for the same type of bond in different molecules are a useful clue to the distribution of electrons in the molecules. An important application of infrared vibrational spectra is the study of hydrogen bonding. The spectra of O—H and N—H vibrations are very different for the hydrogen-bonded and non-hydrogen-bonded species.

HYDROGEN ATOM

A hydrogen atom consists of two particles, a nucleus of mass m_1 and charge $+e$, and an electron of mass m_2 and charge $-e$. The force acting between the particles is coulombic, and the potential energy U is equal to $-e^2/r$, where r is the distance between the particles. For SI units, $U = -e^2/4\pi\varepsilon_0 r$. The location of the electron relative to the nucleus (actually relative to the center of mass of the hydrogen atom) can be specified by a wavefunction of three coordinates. Because the potential energy depends only on r (it is spherically symmetrical), the wavefunction is also spherically symmetrical. It can be written as a product of two parts; one depends only on r, the other on two angular coordinates. Solving the Schrödinger equation for a hydrogen atom (see the references at the end of the chapter), one obtains the wavefunctions and energy levels.

Three quantum numbers, n, l, and m, characterize the angular part of the wavefunction, while n characterizes the distance dependence. These quantum numbers are called the *principal quantum number*, n; the *angular momentum quantum number*, l; and the *magnetic quantum number*, m. The permissible values of n are positive integers 1, 2, 3, . . .; the permissible values of l for a given n are 0 and positive integers up to $n - 1$; and the permissible values of m for a given l are integers from $-l$ to $+l$. The symbols s, p, d, and f are assigned to functions with $l = 0, 1, 2$, and 3, respectively.

The energy E of the hydrogen atom is quantized, as is the particle in a box, the simple harmonic oscillator, and all other systems. The eigenvalues for E for the H atom are

$$E_n = -\frac{\mu e^4}{2\hbar^2 n^2} \tag{9.41}$$

Note that the energy of a hydrogen atom depends only on the principal quantum number n. All three quantum numbers, n, l, and m, characterize the electron distribution, but the energy depends only on n. With μ in g, $\hbar$ in erg s, and e in esu, energy is obtained in ergs. In SI units, we have for the H atom,

$$E_n = -\frac{\mu e^4}{32\pi^2 \varepsilon_0^2 \hbar^2 n^2} \tag{9.42}$$

Now m is in kilograms, e in coulombs, and $\hbar$ in J s; E_n is in joules. As chemists, spectroscopists, and others each seem to use different energy units in discussing the hydrogen atom, it is convenient to write

$$E_n = -\frac{R}{n^2} \tag{9.43}$$

$$R = 2.179 \times 10^{-11} \text{ erg molecule}^{-1}$$

$$R = 2.179 \times 10^{-18} \text{ J molecule}^{-1}$$

$$R = 313.6 \text{ kcal mol}^{-1}$$

$$R = 13.60 \text{ eV molecule}^{-1}$$

The energy is given relative to the separated proton and electron as the zero of energy; the negative value means that the hydrogen atom is stable. The values of the Rydberg constant, R, given are the ones most used. The first two values correspond to Eqs. (9.41) and (9.42). The value of R in kcal mol^{-1} is useful for comparison with bond energies and heats of chemical reactions. The ionization energy of an atom or molecule is the energy required to remove an electron; it is traditional to quote ionization energies in electron volts. The value of R in eV tells us that the ionization energy of the hydrogen atom is -13.6 eV.

The Schrödinger equations for other single-electron species (He^+, Li^{2+}, Be^{3+}, etc.) can be solved with the potential energy equal to $-Ze^2/r$, where Z is the atomic number. The hydrogen atom itself corresponds to $Z = 1$. Several of the hydrogenlike wavefunctions are tabulated in Table 9.2. The energy of a

Table 9.2 Hydrogenic wavefunctions*

n	l	m	
1	0	0	$\psi(1s) = \frac{1}{\sqrt{\pi}}\left(\frac{Z}{a_0}\right)^{3/2} e^{-Zr/a_0}$
2	0	0	$\psi(2s) = \frac{1}{4\sqrt{2\pi}}\left(\frac{Z}{a_0}\right)^{3/2}\left(2 - \frac{Zr}{a_0}\right) e^{-Zr/2a_0}$
2	1	0	$\psi(2p_z) = \frac{1}{4\sqrt{2\pi}}\left(\frac{Z}{a_0}\right)^{5/2} (z)\, e^{-Zr/2a_0}$
2	1	± 1	$\psi(2p_x) = \frac{1}{4\sqrt{2\pi}}\left(\frac{Z}{a_0}\right)^{5/2} (x)\, e^{-Zr/2a_0}$ $\psi(2p_y) = \frac{1}{4\sqrt{2\pi}}\left(\frac{Z}{a_0}\right)^{5/2} (y)\, e^{-Zr/2a_0}$

*$z = r\cos\theta$
$x = r\sin\theta\cos\phi$
$y = r\sin\theta\sin\phi$
$r^2 = x^2 + y^2 + z^2$

a_0 = Bohr radius $= \frac{\hbar^2}{me^2}$
$= 0.529$ Å $= 5.29 \times 10^{-2}$ nm
Z = charge on nucleus

hydrogenlike orbital is dependent on the quantum number n only, and is

$$E = -\frac{Z^2 \mu e^4}{2\hbar^2 n^2} = -\frac{Z^2 R}{n^2} \tag{9.44}$$

The value of R depends slightly on the mass of the nucleus, because of the reduced mass μ; however, to an accuracy of 0.1%, μ equals the mass of the electron.

ELECTRON DISTRIBUTION

Electron Distribution in a Hydrogen Atom

The square of the wavefunction $\psi(r, \theta, \phi)$ is proportional to the probability of finding the electron at a position r, θ, ϕ, where the origin of the spherical coordinates is at the nucleus (Fig. 9.13). We want to be able to visualize the hydrogen wavefunctions and their squares, because we use hydrogenlike orbitals to represent electron distributions in molecules. Figure 9.14 plots $\psi(1s)$, $\psi(2s)$, and $\psi(3s)$ for a hydrogen atom and $\psi(1s)$ for a helium ion. Each ψ is a maximum at $r = 0$ (the position of the nucleus) and decreases rapidly as the distance from the nucleus increases. The ground state of the hydrogen atom ($n = 1$) has the electron closest to the nucleus. For successive excited states ($n = 2, 3, \ldots$), the electron is spread out over more space away from the nucleus. Note how the $\psi(1s)$ for He^+ with a nuclear charge of +2 is pulled toward the nucleus relative to $\psi(1s)$ for H with a nuclear charge of +1.

A better feeling for the shape of wavefunctions can be obtained from computer plots of the functions as illustrated in Fig. 9.15. The pictures for $\psi(1s)$

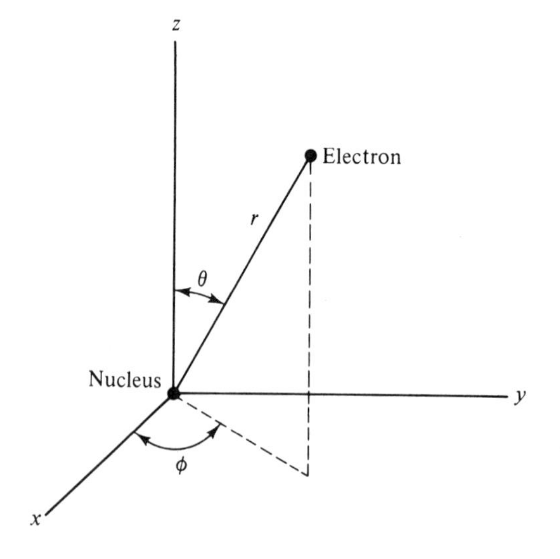

Fig. 9.13 Spherical coordinate system. For an atom the origin should be at the center of mass, but negligible error is made by having it at the nucleus.

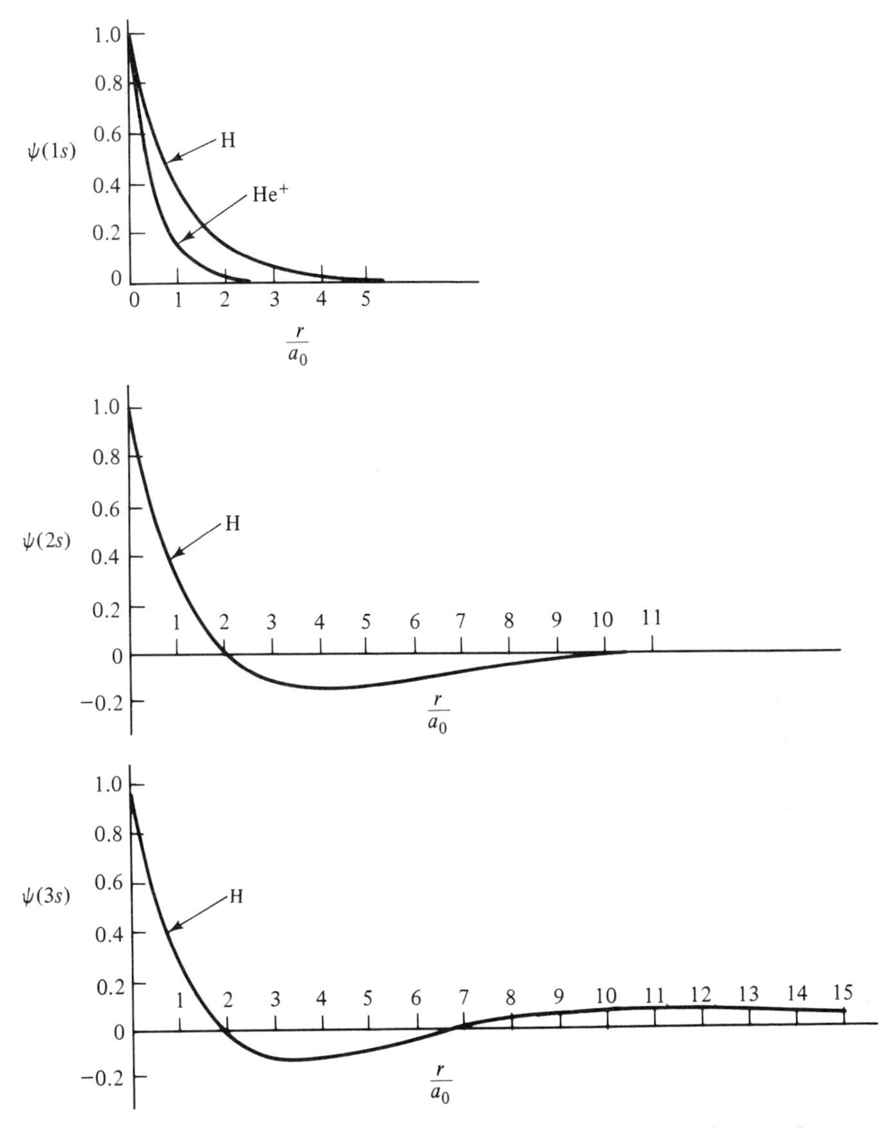

Fig. 9.14 Plots of hydrogenlike wavefunctions versus distance for spherically symmetrical wavefunctions, $\psi(ns)$. The distance scale has units of $a_0 = 0.529$ Å. ψ is set equal to 1 at $r = 0$. ψ can be positive or negative; the number of nodes ($\psi = 0$) is equal to the principal quantum number, n.

and $\psi(2s)$ are just the first two plots in Fig. 9.14 rotated about the origin. The $\psi(2p_x)$ picture shows a wavefunction that is not symmetrical about the origin. $\psi(2p_y)$ and $\psi(2p_z)$ are identical to $\psi(2p_x)$ but are oriented along y and z. The wavefunctions $\psi(2s)$, $\psi(2p_x)$, $\psi(2p_y)$, and $\psi(2p_z)$ all correspond to the same energy for hydrogen atoms; $E_2 = -R/4$.

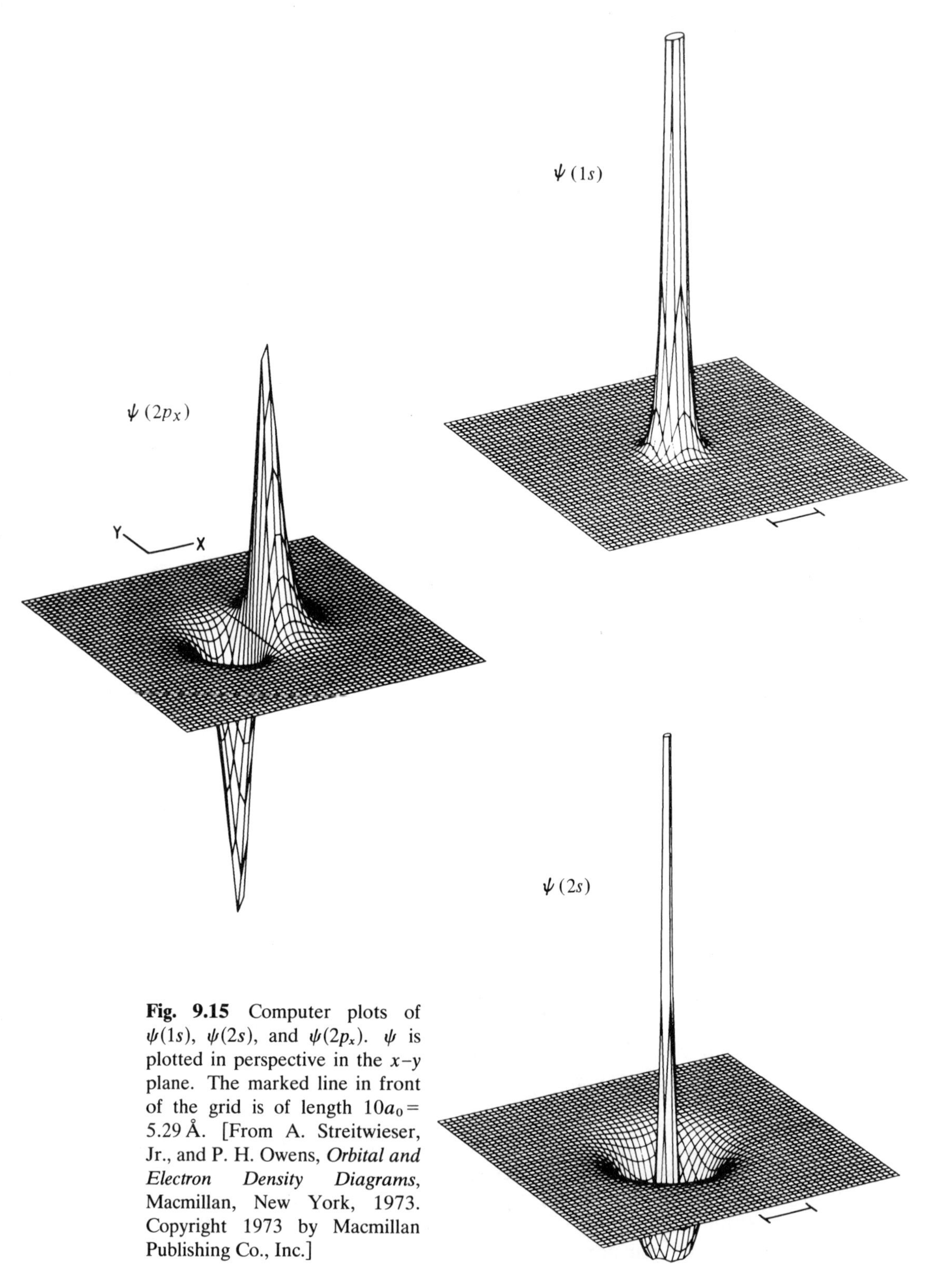

Fig. 9.15 Computer plots of $\psi(1s)$, $\psi(2s)$, and $\psi(2p_x)$. ψ is plotted in perspective in the x–y plane. The marked line in front of the grid is of length $10a_0 = 5.29$ Å. [From A. Streitwieser, Jr., and P. H. Owens, *Orbital and Electron Density Diagrams*, Macmillan, New York, 1973. Copyright 1973 by Macmillan Publishing Co., Inc.]

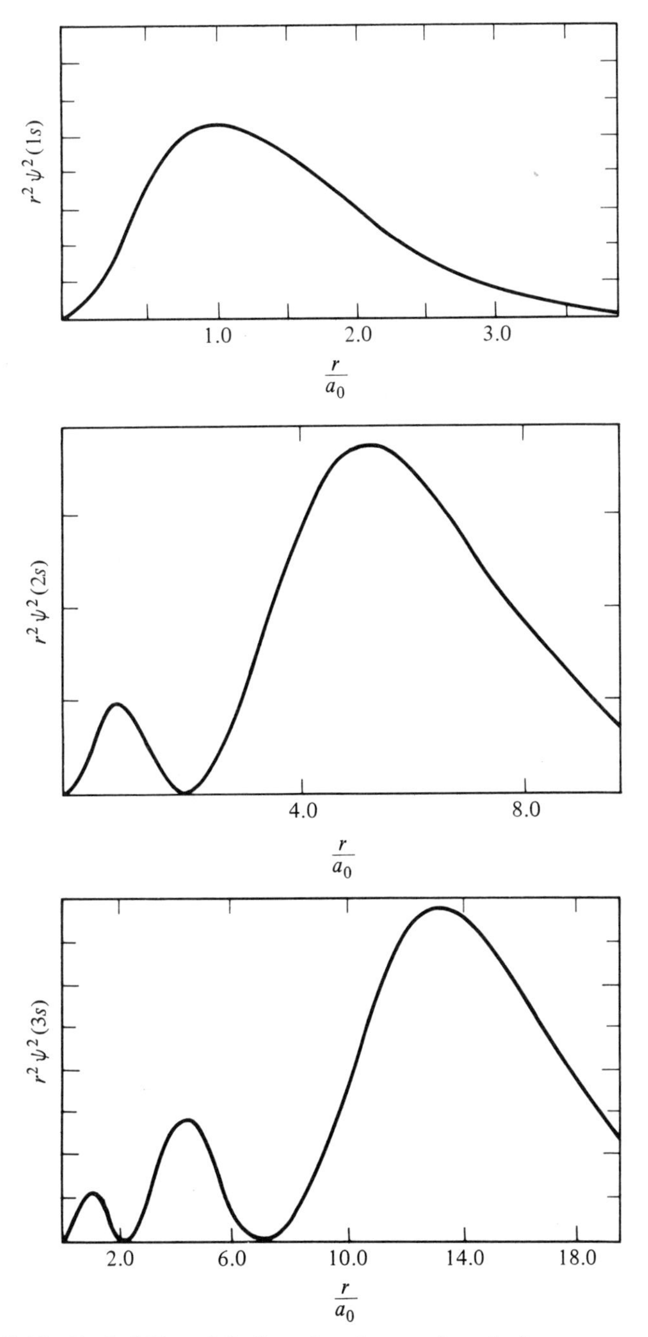

Fig. 9.16 Probability of finding the electron in a hydrogen atom in a spherical shell of radius r around the nucleus. $r^2\psi^2$, which is proportional to this probability, is plotted versus distance in units of $a_0 = 0.529$ Å. Note that the nodes occur at the same location as in Fig. 9.14.

The probability of finding the electron at a distance r from the nucleus is proportional to the volume of a spherical shell of radius r around the nucleus. The volume of a spherical shell is directly proportional to r^2; therefore, the probability of finding the electron at distance r is proportional to $r^2\psi^2$. This is plotted versus r/a_0 in Fig. 9.16 for $\psi(1s)$, $\psi(2s)$, and $\psi(3s)$. The probability of finding the electron is a maximum at $r = a_0$ for the ground state, $\psi(1s)$, of the hydrogen atom. Bohr had deduced a circular orbit at this distance for the electron in a ground-state hydrogen atom; therefore, this distance is called the *Bohr radius*. Quantum mechanics states that the electron can be anywhere relative to the nucleus, but its most probable value is still at a_0. For the excited state $\psi(2s)$ the most probable value of r for the electron is at $5.24a_0$.

Still another way of visualizing an electron distribution in space is shown in Fig. 9.17. Here the enclosed, shaded areas represent the space where there is 90% probability of finding the electron.

So far we have not mentioned the spin of the electron. Experimental results as well as theoretical considerations indicate that an electron has an intrinsic magnetic moment and behaves as if it were a small magnet. In addition to the quantum numbers n, l, and m, a fourth quantum number, m_s, called the *spin*

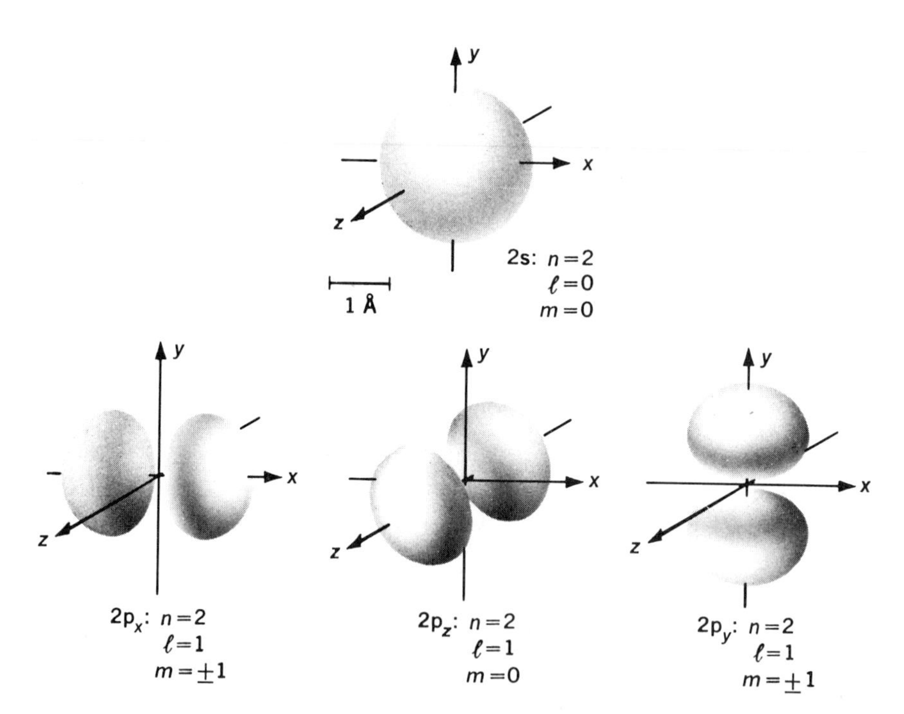

Fig. 9.17 Pictures that show where there is 90% probability of finding the electron in a hydrogen atom in a $(2s)$, $(2p_x)$, $(2p_y)$, and $(2p_z)$ orbital. [From G. C. Pimentel and R. D. Spratley, *Understanding Chemistry*, Holden-Day, San Francisco, 1971, p. 486.]

quantum number, is used to specify the intrinsic magnetic moment. The spin quantum number can either be $+\frac{1}{2}$ or $-\frac{1}{2}$. We have ignored the spin of the electron up to now, because it has an extremely small effect on the energy levels of the hydrogen atom. However, as soon as we consider more than one electron in an atom, the effect of spin becomes important.

Many-electron Atoms

For atoms with more than one electron we can easily write the Schrödinger equation, but we cannot solve it exactly. Consider, for example, the helium atom that has two electrons. The Hamiltonian operator for the two electrons is

$$\mathscr{H} = -\frac{\hbar^2}{2\mu} \cdot (\nabla_1^2 + \nabla_2^2) - \left(\frac{2e^2}{r_1} + \frac{2e^2}{r_2} - \frac{e^2}{r_{12}}\right) \tag{9.45}$$

where μ, the reduced mass, is essentially the mass of an electron, the subscripts 1 and 2 refer to quantities for electrons 1 and 2, respectively. The distance from the nucleus to electron 1 is r_1, the distance to electron 2 is r_2, and r_{12} is the distance between the electrons.

Because of the term containing r_{12}, it is not possible to separate the variables in the Schrödinger equation as we can do for the hydrogen atom. Exact solutions are therefore unavailable. This means we cannot write an exact ψ for He explicitly as we can for a hydrogen atom. However, we can calculate energy levels, electron distributions, and other electronic properties for helium or any atom to a high degree of accuracy. The calculated values agree with experiments within experimental error. Therefore, we are confident in the predicted values for properties that have not yet been measured.

We will discuss many-electron atoms qualitatively; we will not go into details about the approximations used in obtaining wavefunctions and energy levels. We consider a many-electron atom to have hydrogenlike wavefunctions. These orbitals have the shapes shown in Figs. 9.14 through 9.17, but their energies are very different from those of a hydrogen atom. The ordering of the energy levels is $1s$, $2s$, $2p_x = 2p_y = 2p_z$, $3s$, $3p$ (three orbitals), $4s$, $3d$ (five orbitals), and so on. The *Pauli exclusion principle* states that one, or at most two, electrons can be in each orbital; these electrons must have opposite spins. Pauli deduced that no two electrons in an atom can have the same four quantum numbers: n, l, m, and m_s. Two electrons in the same orbital have identical values of n, l, and m; therefore, they must have antiparallel spins ($m_s = +\frac{1}{2}$ for one and $-\frac{1}{2}$ for the other). These simple facts are enough to understand the arrangement of elements in the periodic table. The student is encouraged to review an elementary chemistry book to obtain a more extensive discussion of atomic orbitals and the periodic table. Here we shall just introduce the notation for specifying the electron distribution in a many-electron atom. Figure 9.18 illustrates this notation.

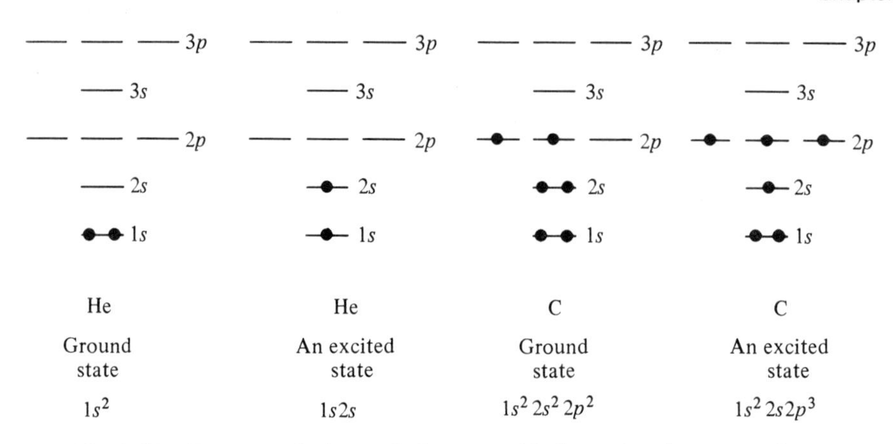

Fig. 9.18 Energy ordering and electron orbital notation for many-electron atoms. At most two electrons are placed in each orbital in accord with the Pauli exclusion principle. The electronic state of the atom is specified by the number of electrons in each orbital. For example, the fluorine ground state is $1s^2 2s^2 2p^5$.

Molecular Orbitals and Linear Combinations of Atomic Orbitals

The electronic wavefunctions of a molecule can be approximated by a set of molecular orbitals (MO's) in which one, or at most two, electrons can be placed. The electronic wavefunctions depend on the internuclear distances, but we can consider the nuclei fixed in determining the electronic wavefunctions and orbitals. Each molecular orbital can be represented as a *linear combination of atomic orbitals* (LCAO). This just means that we can write the molecular orbital as a sum of atomic orbitals on each nucleus. We know what each of the atomic orbitals looks like (Figs. 9.14 to 9.17); therefore we know the shapes of the molecular orbitals. From a chosen number of atomic orbitals we can write the same number of molecular orbitals.

The first step in using and understanding molecular orbitals is to learn the vocabulary. Consider the H_2 molecule with two nuclei labeled A and B. The simplest LCAO we can write is

$$\psi_{H_2} = \frac{1}{\sqrt{2}}[\psi_A(1s) + \psi_B(1s)] \tag{9.46}$$

This says that an MO of H_2 can be approximated by a $1s$ hydrogen atomic orbital (AO) on nucleus A plus a $1s$ hydrogen AO on nucleus B. The $1/\sqrt{2}$ is necessary to normalize the H_2 MO. Because we have chosen to use two AO's, we can write two MO's. The other one is

$$\psi_{H_2} = \frac{1}{\sqrt{2}}[\psi_A(1s) - \psi_B(1s)] \tag{9.47}$$

We can calculate the electron distributions corresponding to these MO's by squaring them and we can calculate their energy levels by using the Hamiltonian operator for a H_2 molecule. These two MO's are examples of sigma, σ, and sigma star, σ^*, MO's.

Equations (9.46) and (9.47) give approximations to the wavefunctions of H_2. To calculate reasonable values for the electronic energies of H_2, much more complicated wavefunctions would be needed. Nevertheless, MO's that are LCAO's are useful concepts to understand electron distributions in molecules qualitatively.

The names of MO's depend on their symmetry properties. A *sigma-bonding orbital* (σ-bonding MO) has cylindrical symmetry around the line joining the nuclei and has no node between the nuclei. A *sigma-star antibonding orbital* (σ^* antibonding MO) has cylindrical symmetry around the line joining the nuclei but has a node between the nuclei. Pictures of a σ and a σ^* MO are shown in Fig. 9.19.

The electron density for a σ bond has cylindrical symmetry around the bond and there is a high probability of finding the electrons between the nuclei. For a σ^* orbital there is still cylindrical symmetry around the bond, but the probability of finding the electrons between the nuclei becomes zero between the nuclei (at the node). The energy calculated for a σ MO is less (more negative) than the sum of the energies for the two AO's, while the energy of a σ^* MO is greater than the sum of the two AO's.

A π *(pi)-bonding molecular orbital* has a node along the line joining the two nuclei. The electron density for a π-bonding MO has a plane of symmetry passing through the bond. A π^* *(pi star)-antibonding molecular orbital* has a node along the line joining the nuclei and also has a node between the two nuclei. Pictures of π and π^* MO's are shown in Fig. 9.20.

A molecule can be thought of as a collection of molecular orbitals. We have discussed σ, σ^*, π, and π^* MO's; there are other MO's with different symmetry which are sometimes important in metal coordination complexes. The distribution of electrons among these orbitals determines the bonding and stability of the molecule. In general, the proper atomic orbitals to be used for the formation of the molecular orbitals are those with comparable energies and the same symmetry with respect to the bond axis. Atomic orbitals of greatly different energies are poor choices for constructing molecular orbitals.

As an example, let us consider 10 MO's which can be formed from combinations of 5 AO's ($1s$, $2s$, $2p_x$, $2p_y$, $2p_z$) on each of two identical nuclei A and B. These 10 MO's can hold up to 20 electrons, so they allow us to discuss the bonding and stability of possible homonuclear diatomic molecules from H_2 to Ne_2. We start adding electrons to the MO's and see what qualitative conclusions we can make about the resulting molecules. Figure 9.21 shows the usual energy-level pattern obtained for the 10 MO's. The actual spacing will depend on which nuclei are involved, and for some molecules $\sigma(2p_x)$ may be above the $\pi(2p_y)$, $\pi(2p_z)$ levels. However, this pattern of levels is a useful one to remember for most molecules, not just diatomics. Figure 9.21 is consistent with the known

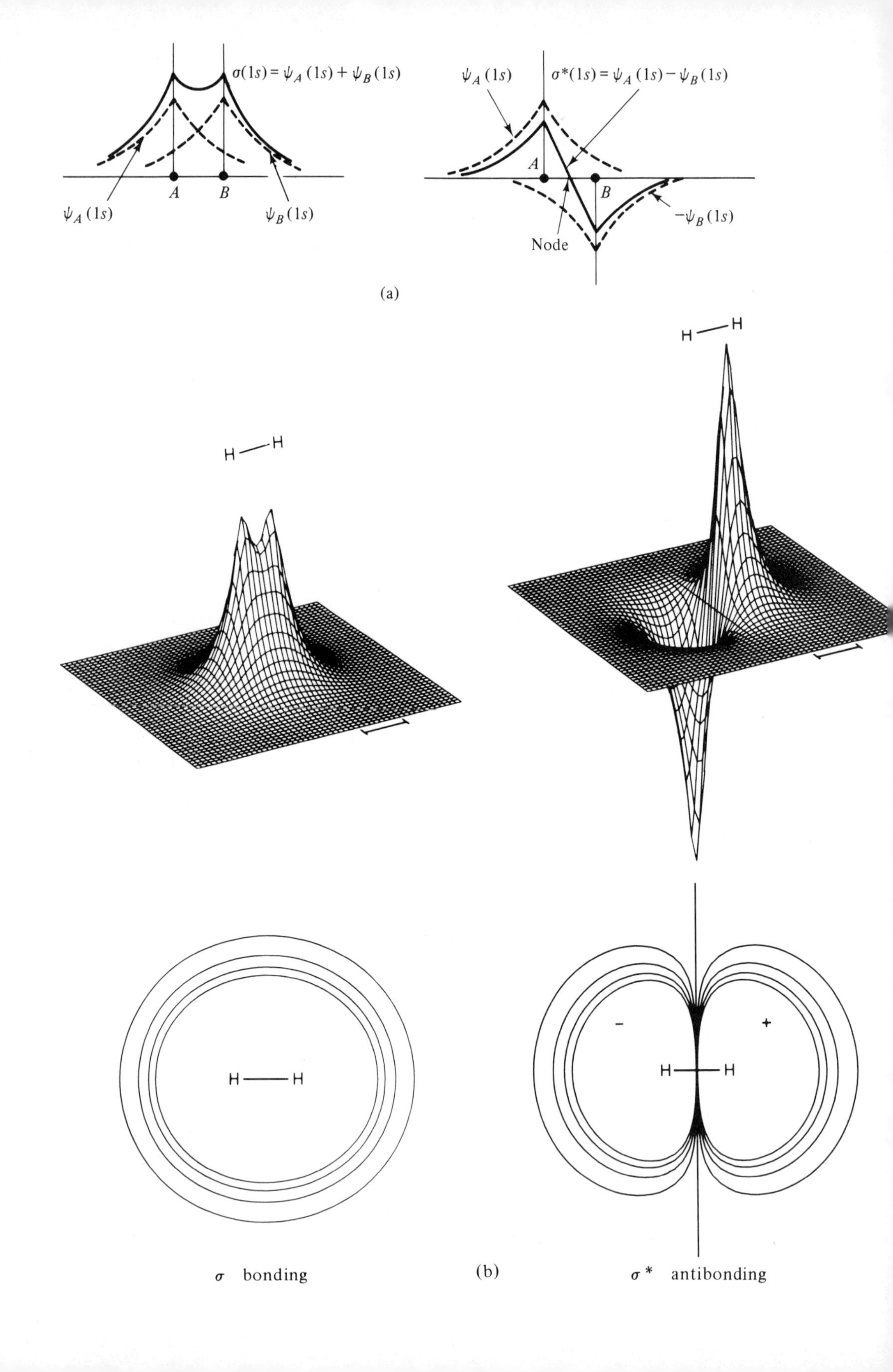

$\sigma(1s) = \psi_A(1s) + \psi_B(1s)$
A
B
$\psi_A(1s)$
$\psi_B(1s)$
$\psi_A(1s)$
$\sigma^*(1s) = \psi_A(1s) - \psi_B(1s)$
A
B
$-\psi_B(1s)$
Node
(a)
H—H
H—H
H—H
-
+
H—H
σ bonding
(b)
σ* antibonding

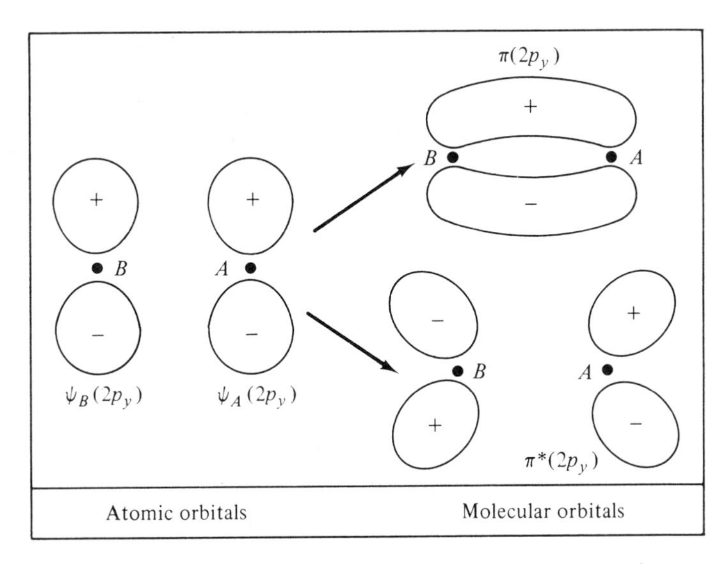

Fig. 9.20 Combination of two $2p_y$ atomic orbitals into a bonding π and antibonding π^* molecular orbital. The drawings represent the angular distributions of the electron densities; the signs + and − specify the signs of the wavefunction.

data on the diatomic molecules of the first row of the period table. He_2, Be_2, and Ne_2 are unstable; H_2, Li_2, B_2, and F_2 form single bonds; C_2 and O_2 form double bonds; and N_2 forms a triple bond. The number of bonds, sometimes called the *bond order*, is one half the number of electrons in bonding MO's minus one half the number of electrons in antibonding orbitals. Each pair of filled bonding and antibonding MO's does not contribute to any bonds. For molecules containing C, N, O, F, and higher atomic-number elements, the $\sigma(1s)$ and $\sigma^*(1s)$ MO's involving these atoms are filled and are not involved in bonding. The $1s$ electrons are then called *core electrons* and are not explicitly considered in the MO's.

Fig. 9.19 (a) Sum $[\sigma(1s)]$ and difference $[\sigma^*(1s)]$ of two $1s$ orbitals. These are the simplest examples of a σ and a σ^* molecular orbital. (b) Computer plots of LCAO approximations to the ground state (σ bonding) and first excited state (σ^* antibonding) for the H_2 molecule. In the middle part of the figure ψ is plotted in perspective in the x–y plane. The marked line in front of the grid is $2a_0 = 1.058$ Å long. The bottom part of the figure shows contour lines on which the value of the wavefunction is constant. The + and − on the σ^* orbital indicate the signs of the wavefunction. [Reprinted with permission from A. Streitwieser, Jr. and P. H. Owens, *Orbital and Electron Density Diagrams*, Macmillan Co. New York, 1973. Copyright 1973 by Macmillan Publishing Co., Inc.]

Figure 9.21 indicates what kind of electronic transitions to expect on excitation by light. For H_2 the longest-wavelength absorption corresponds to a $\sigma \rightarrow \sigma^*$ transition. The excited state with one electron in a σ MO and one in a σ^* MO is unstable, as expected. For O_2, a $\pi \rightarrow \pi^*$ transition occurs, and the double-bond character decreases on excitation. This results in a weaker bond in the excited molecule.

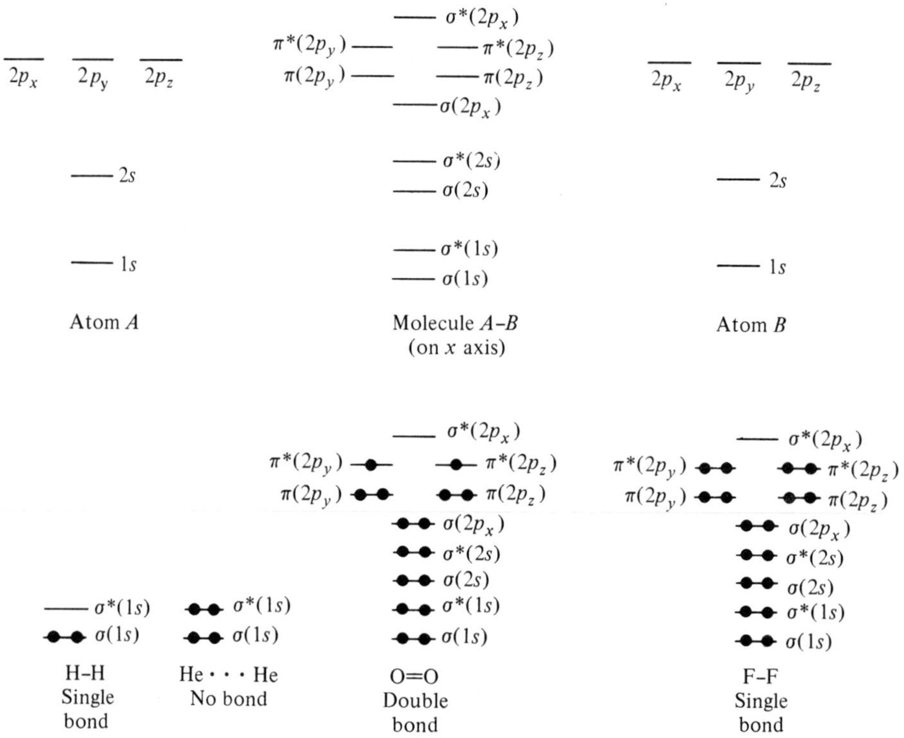

$2p_x$ $2p_y$ $2p_z$

$2s$

$1s$

Atom A

$\sigma^*(2p_x)$
$\pi^*(2p_y)$ $\pi^*(2p_z)$
$\pi(2p_y)$ $\pi(2p_z)$
$\sigma(2p_x)$
$\sigma^*(2s)$
$\sigma(2s)$
$\sigma^*(1s)$
$\sigma(1s)$

Molecule A–B
(on x axis)

$2p_x$ $2p_y$ $2p_z$

$2s$

$1s$

Atom B

$\sigma^*(1s)$
$\sigma(1s)$
H–H
Single
bond

$\sigma^*(1s)$
$\sigma(1s)$
He · · · He
No bond

$\sigma^*(2p_x)$
$\pi^*(2p_y)$ $\pi^*(2p_z)$
$\pi(2p_y)$ $\pi(2p_z)$
$\sigma(2p_x)$
$\sigma^*(2s)$
$\sigma(2s)$
$\sigma^*(1s)$
$\sigma(1s)$
O=O
Double
bond

$\sigma^*(2p_x)$
$\pi^*(2p_y)$ $\pi^*(2p_z)$
$\pi(2p_y)$ $\pi(2p_z)$
$\sigma(2p_x)$
$\sigma^*(2s)$
$\sigma(2s)$
$\sigma^*(1s)$
$\sigma(1s)$
F–F
Single
bond

Fig. 9.21 Typical energy-level pattern for a homonuclear diatomic molecule. The type of bonding and the change in bonding on exciting, ionizing, or capturing an electron can be understood for the first 10 elements.

We must emphasize again that this MO picture is an approximation. However, it can help us remember and systematize the experimental results. There are other reasonable explanations of the bonding in O_2 we could give. One is based on hybridization of AO's before forming MO's. We shall define the various hybrid AO's in the next section, but here we will mention that an alternative description of O_2 involves first forming an *sp* hybrid on each O atom. The $\sigma(2p_x)$ MO in Fig. 9.21 is replaced by a $\sigma(sp)$ and the $\sigma(2s)$ plus $\sigma^*(2s)$ become two *nonbonding orbitals*, $n(sp)$, on each O atom. A *nonbonding MO* is an orbital that is located on a single atom only and is thus not involved in bonding.

Hybridization

In principle, the molecular orbitals of a multinuclear molecule can be obtained from the appropriate combinations of the atomic orbitals of the proper symmetry types. It is easier to visualize a complex molecule if the atomic orbitals of a given atom are combined into a set of *hybrid* orbitals first. These hybrid orbitals are then used to form molecular orbitals with neighboring atoms. A hybrid AO is just a linear combination of AO's on the same atom.

The three types of hybrid AO's that we will consider are sp, sp^2, and sp^3. There are two sp hybrids made from possible combinations of a $2s$ and a $2p$ AO. There are three sp^2 hybrids made from possible combinations of a $2s$ and two $2p$ AO's. There are four sp^3 hybrids from possible combinations of a $2s$ and three $2p$ AO's. All four sp^3 hybrids have the same shape, but they are oriented in different directions in space. The three sp^2 hybrids are identical to each other in shape and similar to the sp^3 hybrids. The two identical sp hybrids are also similar to the sp^2 and sp^3 hybrids. The equations for the hybrids are

$$\psi(sp) = \frac{1}{\sqrt{2}}\psi_A(2s) + \frac{1}{\sqrt{2}}\psi_A(2p)$$

$$\psi(sp^2) = \frac{1}{\sqrt{3}}\psi_A(2s) + \sqrt{\frac{2}{3}}\psi_A(2p) \tag{9.48}$$

$$\psi(sp^3) = \frac{1}{\sqrt{4}}\psi_A(2s) + \sqrt{\frac{3}{4}}\psi_A(2p)$$

Figure 9.22 can represent any one of these hybrid AO's.

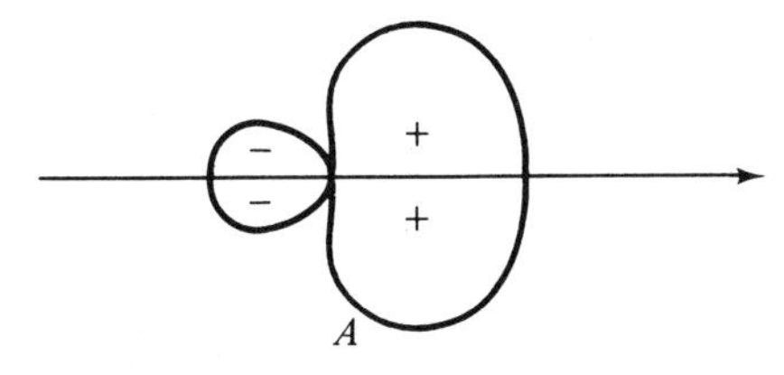

Fig. 9.22 Angular dependence of an sp^3 hybrid orbital; sp and sp^2 orbitals are similar. The arrow indicates the orientation of the hybrid. The plus and minus indicate the sign of the wavefunctions. A bond can be formed by combining this atomic orbital with an atomic orbital on another nucleus in the direction of the arrow.

The important difference among the different hybrids is their orientation in space. The two sp hybrids point in opposite directions. The bond angle between two sp hybrids is 180°. The three sp^2 hybrids point toward the corners of an equilateral triangle centered on the nucleus. The bond angle between two sp^2 hybrids is 120°. The four sp^3 hybrids point toward the corners of a tetrahedron centered on the nucleus. The bond angle between two sp^3 hybrids is 109° 28′. Examples of molecules that can be described in terms of hybrid orbitals are shown in Fig. 9.23.

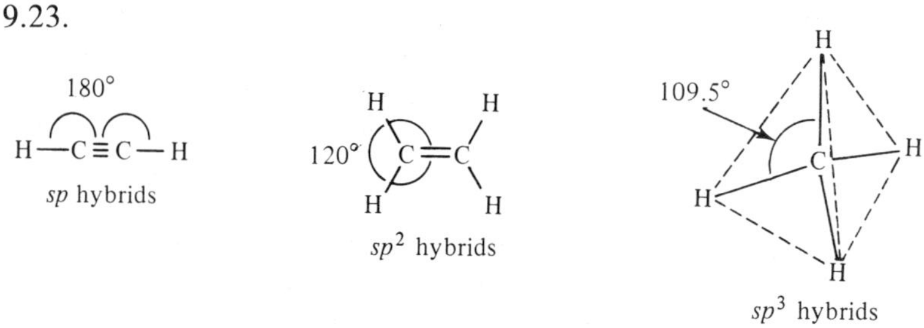

Fig. 9.23 Examples of molecules whose σ bonds can be represented with hybrid atomic orbitals. In acetylene each C—H bond is a linear combination of an *sp* hybrid on C plus a 1*s* orbital on *H*. In ethylene the six σ bonds involve sp^2 hybrids on C. In methane the four σ bonds are combinations of the four sp^3 hybrids on C plus a 1*s* orbital on each H. The π bonds in acetylene and ethylene are sums of 2*p* orbitals.

Delocalized Orbitals

We have previously thought of MO's as two-centered with each orbital localized around two neighboring atoms. Although this approximation is often suitable for σ bonds, it may not be useful for conjugated π bonds. We have already seen, in discussing the free-electron model, that the π electrons in a conjugated system behave as if they are free to move throughout the conjugated system. Let us therefore consider the molecular orbitals of a simple conjugated system, 1,3-butadiene ($CH_2=CH-CH=CH_2$). Butadiene is a planar molecule, with angles between adjacent σ bonds close to 120°. This immediately suggests that sp^2-type hybrid orbitals of the carbon atoms are involved: each carbon atom forms three σ bonds with its three neighboring atoms. If the neighboring atom is a hydrogen, the bonding σ orbital is formed from the 1*s* orbital of the hydrogen and the sp^2 orbital of the carbon. If the neighboring atom is a carbon, the bonding σ orbital is formed from two sp^2 orbitals, one from each carbon. These σ bonds are localized; each includes two nuclei of the molecule. They are illustrated in Fig. 9.24(a).

For each carbon atom, the formation of three sp^2 orbitals from a 2*s* and two 2*p* orbitals leaves it with one 2*p* orbital. If we choose the molecular plane as the *xy* plane, the $2p_z$ orbitals of the carbon atoms are not involved in the formation of the sp^2 orbitals and the σ bonds. The $2p_z$ orbitals form a set of multicentered π orbitals, as illustrated in Fig. 9.24(b). The four $2p_z$ orbitals from the four carbons can form four MO's with π or π^* symmetry. Two turn out to be bonding, π, and two antibonding, π^*. The two bonding ones are filled. The delocalized orbitals mean that writing the double bonds as fixed is not very accurate. It is better to think of the double-bond character also being partly on the central C—C bond.

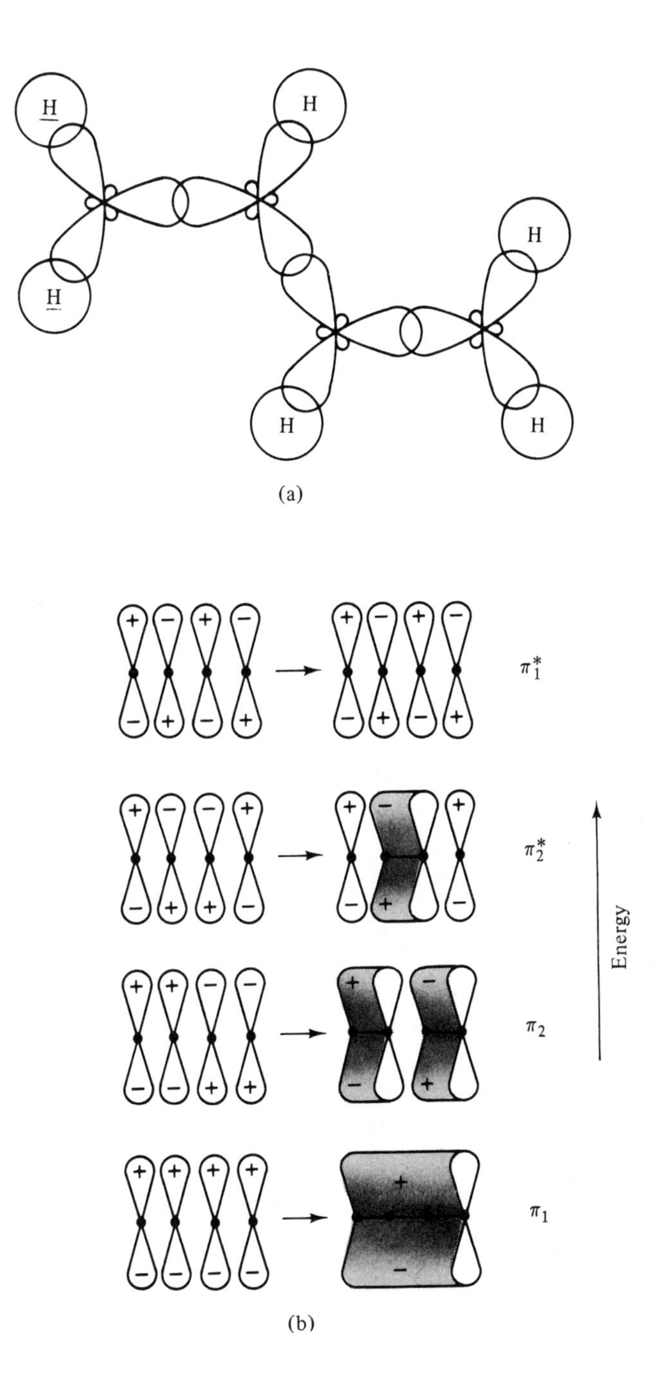

Fig. 9.24 (a) The σ bonds in butadiene $CH_2{=}CH{-}CH{=}CH_2$. (b) The π molecular orbitals in butadiene. The four π electrons occupy the two bonding orbitals, π_1 and π_2. The dots (•) represent the positions of the carbon atoms.

Molecular Structure and Molecular Orbitals

The language of molecular orbitals is used to help us systematize and remember facts of chemical structure. It presents us with a framework that simplifies the logical ordering of the data. The two main ideas we want to emphasize are:

1. The electron distribution in a molecule can be represented by a set of molecular orbitals and corresponding energy levels. These MO's are filled by adding electrons to the lowest energy levels with at most two electrons in each level. The properties of the molecule depend on which MO's are occupied and which excitations of electrons to unfilled MO's can occur.
2. Each molecular orbital can be thought of as a sum of hydrogenlike atomic orbitals on different nuclei (except for nonbonding MO's which are localized on a single nucleus).

Geometry and stereochemistry

It is straightforward to correlate bond angles, bond lengths, and orbitals. We have mentioned the correlation of sp^3, sp^2, and sp hybrids with 109.5°, 120°, and 180° bond angles. We also know that two unhybridized p orbitals on the same atom will form bonds at 90°. We can now use measured bond angles to help us decide on a consistent set of MO's and we can use assumed MO's to predict bond angles. Similarly, bond types and bond distances can be correlated with bonding orbitals. A single bond (a sigma bond) is longer than a double bond (a sigma bond plus a pi bond), which is longer than a triple bond (a sigma bond plus two pi bonds). Rotation about a single bond is easy, whereas rotation about a double bond requires enough energy to break the π bond. Figure 9.25 shows some measured bond angles and bond lengths. The expected trends are seen. The C≡C bond is 20% shorter than the C—C bond; the benzene C⋯C bond is intermediate in length between ethane and ethylene. The H—C—H bonds in ethane are tetrahedral. In ethylene the H—C—H is less than the 120° trigonal angle expected, but in benzene the C—C—H angles are precisely 120°.

The structure of formamide is informative. The H—N—H angle of 119° implies that the bonding around the N is trigonal, not tetrahedral as in NH_3. The O—C—N angle is also trigonal, as expected. The C—N bond distance is 1.343 Å, which is shorter than a C—N single bond (1.47 Å) such as is found in methyl amine. These data indicate that formamide is planar and that the C—N bond is partially a double bond. The π MO's assumed for formamide would therefore be delocalized over three nuclei (C, N, O). The peptide bond is similar to formamide in that it is planar with delocalized π MO's.

In butadiene the delocalized π MO's are consistent with the fact that the two terminal double bonds in CH_2=CH—CH=CH_2 are slightly longer than a C=C double bond in a nonconjugated system: 1.35 Å in the former versus 1.33 Å in the latter. On the other hand, the central C—C bond of the molecule is considerably shorter than a typical C—C single bond: 1.46 Å versus 1.54 Å.

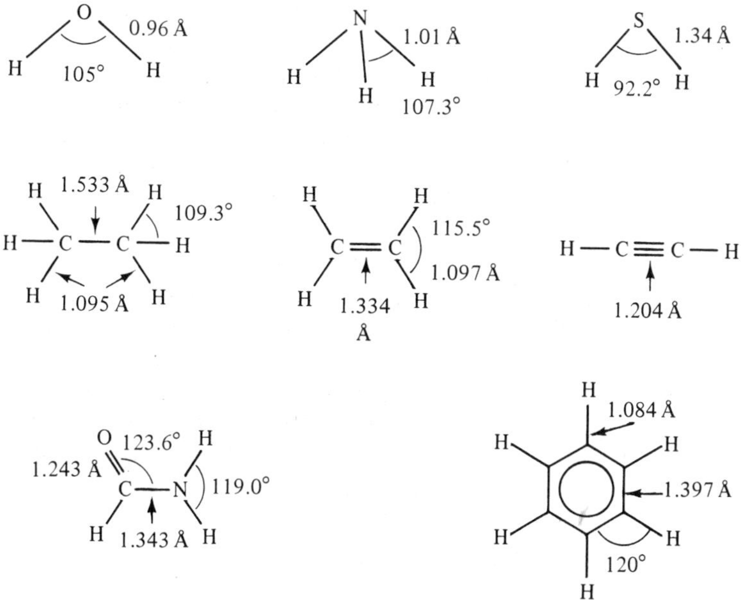

Fig. 9.25 Molecular geometry. Bond angles are given in degrees, and bond distances are in angstroms (1 Å = 0.1 nm).

Also, while rotation around a regular single bond has a low energy barrier, the π-bond character of the central C—C link prohibits rotation around this bond.

A very important application of molecular orbital methods is in the understanding of chemical reactions and reaction mechanisms. The Woodward-Hoffmann rules (Hoffmann and Woodward, 1968) use the symmetry characteristics of molecular orbitals to rationalize and predict the stereochemical course of concerted organic reactions. An example is the conversion of cyclobutene to its isomer butadiene. An isotopically labeled reactant can form either of two products:

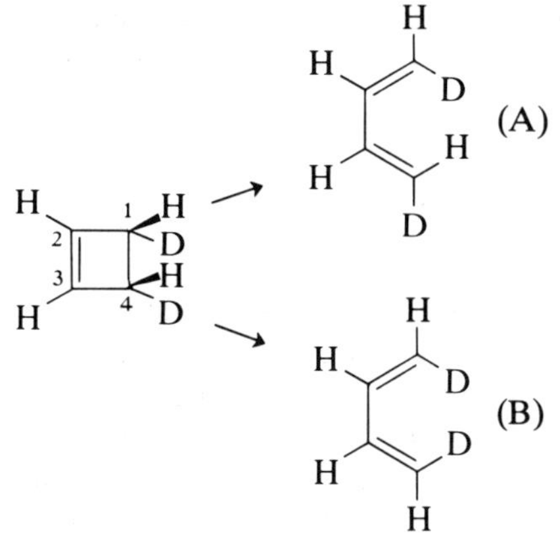

The important MO's are the π and π^* of the $C_2—C_3$ double bond, the σ and σ^* of the $C_1—C_4$ single bond of cyclobutene, and the four π, π^* MO's of butadiene. If we can find a path that converts occupied bonding MO's in the reactant to occupied bonding MO's in the product, the reaction should occur easily. Hoffmann and Woodward conclude that ground-state cyclobutene should form product (A) while excited-state cyclobutene should form product (B). They thus predict, and find, that thermal reaction gives (A), but photochemical reaction gives (B).

The catalytic efficiency and stereochemical selectivity of enzymes can be related to the ability of the enzyme to constrain reactants and products to particular orientations. This specific orientation will greatly enhance certain reactions and reaction paths over the more random collisions produced by thermal motion. The literature in this field is large and it often contains molecular orbital language. (See for example, Kirsch, 1973, Bruice, 1976.) We hope that the MO fundamentals presented here will help the biochemistry student to evaluate and understand the often contradictory papers on this subject.

Charge distributions and dipole moments

An LCAO-MO description of a molecule characterizes the electron distribution in the molecule. When we write, for example, a π MO for formamide, it will have the form

$$\psi(\pi) = C_N\phi_N(2p_z) + C_C\phi_C(2p_z) + C_O\phi_O(2p_z)$$

where N, C, and O refer to the nitrogen, carbon, and oxygen nuclei in formamide, $\phi_N(2p_z)$, etc., are corresponding $2p_z$ AO's on each nucleus and C_N, C_C, and C_O are numerical coefficients. Each electron that occupies that MO is distributed according to the square of the wavefunction. Therefore a π electron in that MO will be distributed as follows: C_N^2 probability on N, C_C^2 probability on C, and C_O^2 probability on O. The sum of the probabilities is, of course, equal to 1. The total electronic charge on each nucleus is just the sum of charges contributed by each occupied MO. In this approximation the net charge on each nucleus is the nuclear charge minus the electronic charge calculated from the MO's. The dipole moment in this approximation is given by Eq. (9.4), with each charge being the net charge on each nucleus.

Hydrogen Bonds

The hydrogen bond is a weak but important bond which occurs between two atoms that have high net negative charges, and one of which (the donor) has a hydrogen atom attached. In biological systems the most important hydrogen bonds are O—H···O, O—H···N, N—H···O, and N—H···N. These bonds help determine protein and nucleic acid structure and, of course, help establish the vital properties of water. The experimental criterion which establishes that a hydrogen bond exists is that the X···Y distance in X—H···Y is about 2.8 to 3.0 Å. Without a hydrogen bond, that distance is greater than 3.0 Å. A quantum mechanical description of a hydrogen bond involves the electrostatic attraction of the positively charged hydrogen nucleus of the donor to a negatively charged nonbonding orbital on the acceptor. This attraction of one proton to two nuclei can provide an efficient path for moving charges; the proton can move from one nucleus to another.

$$\mathrm{O{-}H\cdots O} \longrightarrow \mathrm{O\cdots H{-}O}$$

The very high mobility of H^+ in water is explained by this mechanism. A charge relay system of this type has also been postulated to explain the catalytic mechanism in chymotrypsin (Blow, 1976).

SUMMARY

Schrödinger's Equation

Schrödinger's time-independent equation:

$$\mathcal{H}\psi = E\psi \tag{9.6}$$

$\mathcal{H} \equiv$ Hamiltonian operator

$$\equiv -\frac{h^2}{8\pi^2}\sum_i \frac{1}{m_i}\cdot\left(\frac{\partial^2}{\partial x_i^2} + \frac{\partial^2}{\partial y_i^2} + \frac{\partial^2}{\partial z_i^2}\right) + U(x_i, y_i, z_i)$$

h = Planck's constant
$\sum_i$ = sum over all particles in system
m_i = mass of each particle i in system
x_i, y_i, z_i = coordinates of each particle i in system
$U(x_i, y_i, z_i)$ = potential-energy operator for system
ψ = eigenfunction or wavefunction, a function of coordinates of all particles in system; there will be an infinite number of wavefunctions corresponding to different states of the system and usually each wavefunction will correspond to a different energy of the system
E = energy of system

The value of any measurable property for a system characterized by wavefunction ψ can be calculated from the appropriate operator:

$$\text{value} = \int \psi^* \text{ operator } \psi\, dv$$

ψ = normalized wavefunction of system. ψ^* = its complex conjugate. and $\int \psi^*\psi\, dv = 1$; for real wavefunctions, $\psi^* = \psi$

Some Useful Operators

Energy E:

Operator for Cartesian coordinates for one particle:

$$\mathcal{H} = -\frac{h^2}{8\pi^2 m}\left(\frac{\partial^2}{\partial x^2} + \frac{\partial^2}{\partial y^2} + \frac{\partial^2}{\partial z^2}\right) + U$$

Position: x, y, z:
Momentum: p_x, p_y, p_z*:*

The linear momentum components along x, y, z are p_x, p_y, p_z. The classical definition of momentum is mass · velocity. Units are, for example, g cm s^{-1} or kg m s^{-1}.

Operators in Cartesian coordinates:

$$p_x = \frac{h}{2\pi i}\frac{\partial}{\partial x} \qquad p_y = \frac{h}{2\pi i}\frac{\partial}{\partial y} \qquad p_z = \frac{h}{2\pi i}\frac{\partial}{\partial z}$$

Systems Whose Schrödinger Equation Can Be Solved Exactly

Particle in a box: potential energy for a one-dimensional box is $U(x) = 0$ *in box and* $U(x) = \infty$ *outside box*:

$$-\frac{h^2}{8\pi^2 m}\frac{d^2\psi}{dx^2} = E\psi \tag{9.22}$$

$$\frac{h^2}{8\pi^2} = \frac{\hbar^2}{2} = 5.561 \times 10^{-55}\ \text{erg}^2\ \text{s}^2 = 5.561 \times 10^{-69}\ \text{J}^2\ \text{s}^2$$

m = mass of particle
ψ = wavefunction of particle in box
E = energy of particle in box

Normalized wavefunctions for a particle in a one-dimensional box of length a:

$$\psi_n = \sqrt{\frac{2}{a}}\sin\frac{n\pi x}{a} \tag{9.25}$$

ψ_n = normalized wavefunction of particle in box in quantum state n
n = quantum number, a positive integer
a = length of box

Energies for a particle in a one-dimensional box of length a are proportional to n^2*:*

$$E_n = \frac{n^2h^2}{8ma^2} \tag{9.27}$$

E_n = energy of particle in quantum state n; each energy corresponds to a unique quantum state; that is, no energy value is degenerate

n = quantum number, a positive integer
$h^2/8 = 5.4883 \times 10^{-54}$ erg^2 s^2 = 5.4883×10^{-68} J s
m = mass of particle
a = length of box
1 kcal mol^{-1} = 6.946×10^{-14} erg particle^{-1}

Normalized wavefunctions for a particle in a two-dimensional box:

$$\psi_{n_x n_y} = \frac{2}{\sqrt{ab}} \sin \frac{n_x \pi x}{a} \sin \frac{n_y \pi y}{b}$$

$\psi_{n_x n_y}$ = normalized wave function of particle in box of quantum state specified by quantum numbers n_x and n_y
n_x, n_y = quantum numbers, positive integers
a = length of box in x direction
b = length of box in y direction

Energies for a particle in a three-dimensional box:

$$E_{n_x n_y n_z} = \left(\frac{n_x^2}{a^2} + \frac{n_y^2}{b^2} + \frac{n_z^2}{c^2}\right)\frac{h^2}{8m} \qquad (9.31)$$

Schrödinger equation for a one-dimensional harmonic oscillator:

$$-\frac{h^2}{8\pi^2 m}\frac{d^2\psi}{dx^2} + \frac{k}{2}x^2\psi = E\psi \qquad (9.38)$$

ψ = wavefunction for a particle on x axis whose potential energy is directly proportional to its square of distance from origin
E = energy of harmonic oscillator
k = force constant of harmonic oscillator, mass time^{-2}
h = Planck's constant
m = mass of particle

Energies for a harmonic oscillator are equally spaced:

$$E_v = (v + \tfrac{1}{2})h\nu_0 \qquad (9.39)$$

$$\nu_0 = \frac{1}{2\pi}\sqrt{\frac{k}{m}}$$

$E_0 = h\nu_0/2$ = zero-point energy
v = quantum number for a harmonic oscillator = 0, or 1, or 2, etc.
h = Planck's constant
ν_0 = fundamental vibration frequency of oscillator
k = force constant of harmonic oscillator, mass time^{-2}
m = mass of particle

Energies for an electron in a hydrogen atom are inversely proportional to n^2*:*

$$E_n = -\frac{2\pi^2\mu e^4}{(4\pi\varepsilon_0)^2h^2n^2} \qquad \text{SI units} \tag{9.42}$$

$$E_n = -\frac{2\pi^2\mu e^4}{h^2n^2} \qquad \text{cgs-esu units} \tag{9.41}$$

$$\mu = \text{reduced mass of electron and proton} = \frac{m_p m_e}{m_p + m_e}$$

m_p = mass of proton
m_e = mass of electron
e = elementary charge
h = Planck's constant
ε_0 = permittivity of vacuum
n = principal quantum number

For SI units, $2\pi^2\mu e^4/(4\pi\varepsilon_0)^2h^2 = 2.179 \times 10^{-18}$ J; $m_e = 9.110 \times 10^{-31}$ kg; $e = 1.602 \times 10^{-19}$ C; $h = 6.626 \times 10^{-34}$ J s; $\varepsilon_0 = 8.8542 \times 10^{-12}$ C^2 N^{-1} m^{-2}.

For cgs-esu units, $2\pi^2\mu e^4/h^2 = 2.179 \times 10^{-11}$ erg; $m_e = 9.110 \times 10^{-28}$ g; $e = 4.803 \times 10^{-10}$ esu; $h = 6.626 \times 10^{-27}$ erg s.

$$E_n = \frac{-2.179 \times 10^{-18}}{n^2} \qquad \text{J atom}^{-1}$$

$$E_n = \frac{-2.179 \times 10^{-11}}{n^2} \qquad \text{erg atom}^{-1}$$

$$E_n = \frac{-313.6}{n^2} \qquad \text{kcal mol}^{-1}$$

$$E_n = \frac{-13.60}{n^2} \qquad \text{eV atom}^{-1}$$

MATHEMATICS NEEDED FOR CHAPTER 9

In this chapter for the first time we use a few trigonometric identities.

$$\begin{aligned}
\cos n\pi &= 1 && \text{for } n = 0 \text{ or an even integer} \\
&= -1 && \text{for } n = \text{an odd integer} \\
\sin n\pi &= 0 && \text{for } n = \text{any integer} \\
\sin(\theta + 2\pi) &= \sin\theta && \\
\cos(\theta + 2\pi) &= \cos\theta && \\
\sin^2\theta + \cos^2\theta &= 1 && \text{for any value of } \theta
\end{aligned}$$

We discuss differential equations, but the reader is not expected to solve any. The reader should be able to use calculus to verify the solutions, however. Remember that

$$\frac{d}{d\theta}(\sin a\theta) = a \cos a\theta$$

$$\frac{d}{d\theta}(\cos a\theta) = -a \sin a\theta$$

$$\frac{d}{dx}(e^{ax}) = ae^{ax}$$

$$\frac{d}{dx}(e^{ax2}) = 2axe^{ax2}$$

Integrals of functions other than powers of x are used in solving some of the problems.

$$\int xe^{ax2}\, dx = \frac{e^{ax2}}{2a} + C$$

$$\int \sin^2 ax\, dx = \frac{x}{2} - \frac{\sin 2ax}{4a} + C$$

$$C = \text{a constant}$$

The complex conjugate of a function is obtained by changing i (the imaginary) to $-i$ in the function.

REFERENCES

Good discussions of atomic and molecular orbitals can be found in the following introductory chemistry books and organic chemistry books.

DICKERSON, R. E., H. B. GRAY, and G. P. HAIGHT, JR., 1974. *Chemical Principles*, 2nd ed., W. A. Benjamin, Menlo Park, Calif.

HEARST, J. E., and J. B. IFFT, 1976. *Contemporary Chemistry*, W. H. Freeman, San Francisco.

MAHAN, B. H., 1975. *University Chemistry*, 3rd ed., Addison-Wesley, Reading, Mass.

MORRISON, R. E., and R. N. BOYD, 1974. *Organic Chemistry*, 3rd ed., Allyn and Bacon, Boston.

PIMENTEL, G. C., and R. D. SPRATLEY, 1971. *Understanding Chemistry*, Holden-Day, San Francisco.

STREITWIESER, A., JR., and C. H. HEATHCOCK, 1976. *Introduction to Organic Chemistry*, Macmillan, New York.

For other useful books, see

HALLIDAY, D., and R. RESNICK, 1970. *Fundamentals of Physics*, John Wiley, New York.

HANNA, M. W., 1969. *Quantum Mechanics in Chemistry*, 2nd ed., W. A. Benjamin, Menlo Park, Calif.

PAULING, L., 1960. *The Nature of the Chemical Bond*, 3rd ed., Cornell University Press, Ithaca, N.Y.

PAULING, L., and E. B. WILSON, JR., 1935. *Introduction to Quantum Mechanics*, McGraw-Hill, New York.

STREITWIESER, A., JR., and P. H. OWENS, 1973. *Orbital and Electron Density Diagrams*, Macmillan, New York.

SUGGESTED READINGS

BLOW, D. M., 1976. Structure and Mechanism of Chymotrypsin, *Accounts Chem. Res. 9*, 145.

BRUICE, T. C., 1976. Some Pertinent Aspects of Mechanism as Determined with Small Molecules, *Ann. Rev. Biochem. 45*, 331.

HOFFMANN, R., and R. B. WOODWARD, 1968. The Conservation of Orbital Symmetry, *Accounts Chem. Res. 1*, 17.

KIRSCH, J., 1973. Mechanism of Enzyme Action, *Ann. Rev. Biochem. 42*, 205.

MATHIES, R., and L. STRYER, 1976. Retinal Has a Highly Dipolar Vertically Excited Singlet State: Implications for Vision, *Proc. Natl. Acad. Sci. U.S. 73*, 2169.

MENGER, E. L. (ed.), 1975. Chemistry of Vision, Special Issue, *Accounts Chem. Res. 8*, 81.

SALEM, L., 1976. Theory of Photochemical Reactions, *Science 191*, 822.

SALEM, L., and P. BRUCKMANN, 1975. Conversion of a Photon to an Electrical Signal by Sudden Polarization in the N-retinylidene Visual Chromophore, *Nature 258*, 526.

WALD, G., 1968. The Molecular Basis of Visual Excitation, *Nature 219*, 800.

PROBLEMS

1. Calculate the de Broglie wavelengths of the following:
(a) Electrons accelerated by 80,000 V.
(b) Neutrons with an energy of RT per mole, at a temperature of 300 K.
(c) Protons accelerated by 80,000 V.

2. Consider a one-dimensional box of length l. For parts (a), (b), and (c), calculate the energy of the system in the ground state and the wavelength (in Å) of the first transition (longest wavelength).
(a) $l = 10$ Å, and the box contains one electron.

(b) $l = 20$ Å, and the box contains one electron.
(c) $l = 10$ Å, and the box contains two electrons (assume no interaction between the electrons).
(d) Write the complete Hamiltonian for the system described in part (c).
(e) What will happen to the answers in part (c) if the potential of interaction between the electrons is included in this Hamiltonian?

3. The exact ground-state energy for a particle in a box is $0.125h^2/ma^2$; the exact wavefunction is $(\sqrt{2/a})\sin(n\pi x/a)$. Consider an approximate wavefunction $= x \cdot (x - a)$.
(a) Show that the approximate wavefunction fits the boundary conditions for a particle in a box.
(b) Given this approximate wavefunction, $\psi(x) = x \cdot (x - a)$, the approximate energy E' of the system can be calculated from

$$E' = \int_0^a \psi(x)H\psi(x)\,dx \Big/ \int_0^a \psi^2(x)\,dx$$

This is called the *variational method.* Calculate the approximate energy E'.
(c) What is the percent error for the approximate energy?
(d) According to the variational theorem, the more closely the wavefunction used approximates the correct wavefunction, the lower the energy will be calculated according to part (b). The converse is also true. Propose another approximate wavefunction which could be used to calculate an approximate energy. How would you find out if it was a better approximation?

4. The π electrons of metal porphyrins, such as the iron-heme of hemoglobin or the magnesium-porphyrin of chlorophyll, can be visualized using a simple model of free electrons in a two-dimensional box.
(a) Obtain the energy levels of a free electron in a two-dimensional square box of length a. [*Hint*: Extrapolate from Eqs. (9.27) and (9.31).]
(b) Sketch an energy-level diagram for this problem. Set $E_0 = h^2/8ma^2$ and label the energy of each level in units of E_0.
(c) For a porphyrin like hemin that contains 26π electrons, indicate the electron population of the filled or partly filled π orbitals of the ground state of the molecule. (Note that orbitals that are degenerate in energy, $E_{12} = E_{21}$, can still hold two electrons each.)
(d) The porphyrin structure measures about 1 nm on a side ($a = 1$ nm). Calculate the longest-wavelength absorption band position for this molecule. (Experimentally these bands occur at about 600 nm.)

5. If G is a measurable parameter and g is the quantum mechanical operator corresponding to this parameter, the expected value of G, $\langle G \rangle$, can be calculated from

$$\langle G \rangle = \int \psi^* g \psi\, dv \Big/ \int \psi^* \psi\, dv$$

where the integration extends over all space important to the problem and ψ^* is the complex conjugate of ψ (if ψ has no imaginary part, $\psi^* = \psi$).
Calculate the expected position $\langle x \rangle$ and momentum $\langle p \rangle$ for a harmonic oscillator in its ground state.

6. For a diatomic species A–B, the vibration of the molecule can be approximately treated by considering the molecule as a harmonic oscillator.

(a) Calculate the reduced mass for HF.

(b) Take $k = 9.7 \times 10^5$ dyn cm^{-1}; plot the potential energy versus internuclear distance near the equilibrium distance of 0.92 Å.

(c) Sketch in the same diagram horizontal lines representing the allowed vibrational energy levels.

7. A physicist wishes to determine whether the covalently bonded species $(HeH)^{2+}$ could be expected to exist in an ionized gas.

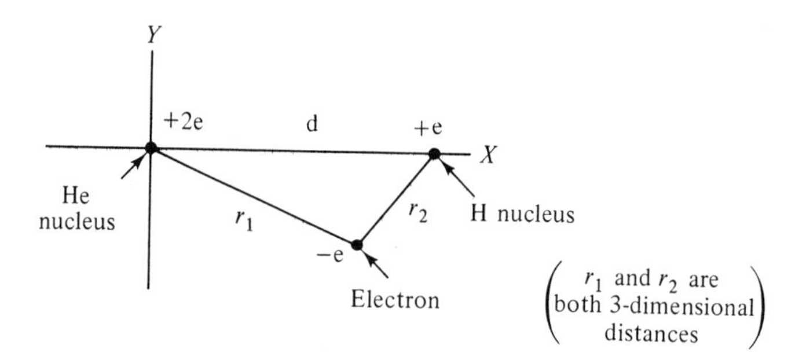

(a) Assuming that the nuclei remain separated by a fixed distance d (the nuclei do not move) and that the electron is free to move in *three dimensions*, write the Hamiltonian operator for this molecule.

(b) Write out fully the equation that must be solved exactly to determine the various wavefunctions and allowed energies of $(HeH)^{2+}$.

(c) Suppose that you have obtained all the exact solutions to the equation in part (b). How can you tell whether the covalently bonded species $(HeH)^{2+}$ is energetically stable?

8. The ionization potential of an electron in an atom or molecule is the energy required to remove the electron. Calculate the ionization potential for a 1s electron in He^+. Give your answer in units of eV and kcal mol^{-1}.

9. Consider the diatomic molecules NO, NO^-, and NO^+.

(a) Give MO diagrams for each of these molecules similar to those in Fig. 9.21.

(b) Rank these species in increasing order of bond energies and in increasing order of bond lengths.

10. In a protein, a basic structural unit is the amide residue:

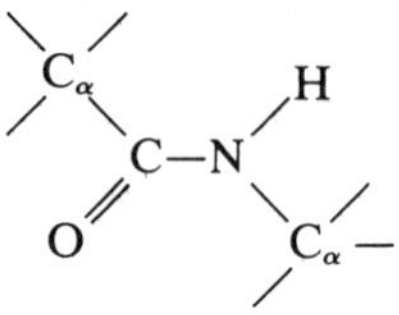

where C_α represents a carbon at the α position to the carbonyl group. A key clue to the solution of the three-dimensional structures of proteins was the realization that the six atoms shown lie in a plane. By constructing the molecular orbitals around the atoms O, C, and N, explain why the planar structure is expected. What are the expected bond angles between adjacent bonds?

11. The fundamental vibration frequency of gaseous $^{14}N^{16}O$ is 1904 cm^{-1}.
(a) Calculate the force constant, using the simple harmonic oscillator model.
(b) Calculate the fundamental vibration frequency of gaseous $^{15}N^{16}O$.
(c) When $^{14}N^{16}O$ is bound to hemoglobin A, an absorption band at 1615 cm^{-1} is observed and is believed to correspond to the bound NO species. Assuming that when an NO binds to hemoglobin, the oxygen is so anchored that its effective mass becomes infinite, estimate the vibration frequency of bound $^{14}N^{16}O$ from the vibration frequency of gaseous NO.
(d) Using another model, that binding of NO to hemoglobin does not change the reduced mass of the NO vibrator, calculate the force constant of bound NO.
(e) Estimate the vibration of $^{15}N^{16}O$ bound to hemoglobin using, respectively, the assumptions made in parts (c) and (d).

12. The C=C stretching vibrational frequency of ethylene can be treated as a harmonic oscillator.
(a) Calculate the ratio of the fundamental frequency for ethylene to completely deuterated ethylene.
(b) Putting different substituents on the ethylene can make the C=C bond longer or shorter. For a shorter C=C bond, will the vibration frequency increase or decrease relative to ethylene? State why you think so.
(c) If the fundamental vibration frequency for the ethylene double bond is 2000 cm^{-1}, what is the wavelength in nm for the first harmonic vibration frequency?

13. Consider a particle in a one-dimensional box.
(a) For a box of length 1 nm, what is the probability of finding the particle within 0.01 nm of the center of the box for the lowest energy level?
(b) Answer part (a) for the first excited state.
(c) The longest-wavelength transition for a particle in a box (not the above box) is 200 nm. What is the wavelength if the mass of the particle is doubled? What is the wavelength if the charge of the particle is doubled? What is the wavelength if the length of the box is doubled?

14. Ammonia NH_3 is known to be a pyramidal molecule. Let us compare it with aniline.

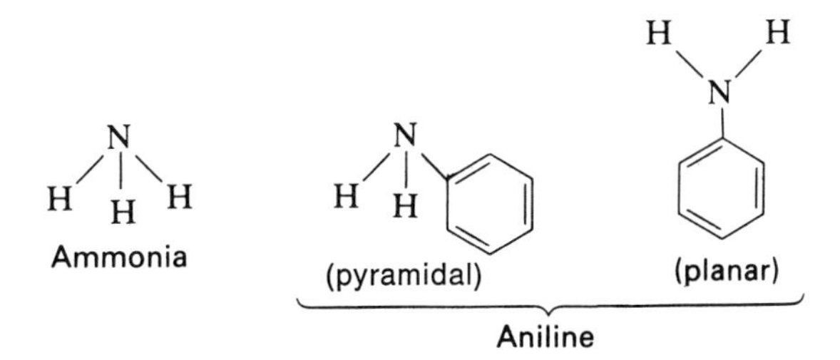

(a) For each of the three structures shown, answer the following questions:
(1) What hybrid orbital can best describe the bonding around the nitrogen atom?
(2) How many nonbonding electrons are there on nitrogen?
(3) How many π electrons are there in each structure?
(b) Which structure would you predict for aniline? Why?

15. An approximate MO calculation gives the following π-electron charges on an amide group:

	Ground-state charges	Excited-state charges
O	$-0.33e$	$+0.08e$
C	$+0.18e$	$-0.16e$
N	$+0.15e$	$+0.08e$

$e = 4.8 \times 10^{-10}$ esu. Use the bond lengths given in Fig. 9.25 and 120° for the O—C—N bond angle.

(a) Calculate the components of the π dipole moment parallel and perpendicular to the C—N bond for the ground and excited state. Use debyes for units.

(b) Calculate the magnitudes of the π dipole moment for the ground state, for the excited state, and for the change in dipole.

(c) Plot the magnitudes and directions to scale.

10

Spectroscopy

How do we learn about molecules when most of them are too small to be seen even with the most powerful microscopes? There is no simple answer to this question, but a large amount of useful information has come from various spectroscopic investigations. The purpose of this chapter is to acquaint the reader with several important and representative methods in spectroscopy and to show how they can be of value in some typical biological situations.

Traditionally, spectroscopy referred to the interaction of light with matter. Now it includes electromagnetic radiation that lies outside the visible region of the spectrum as well. Figure 10.1 shows the wavelength regions of the electromagnetic spectrum that are useful in spectroscopy. Note that the visible spectrum is a very small part of this wide range. We can characterize the electromagnetic spectrum in terms of wavelength, frequency, or energy. Different types of spectroscopists prefer different units. In the visible and near-ultraviolet region, wavelengths in nanometers are used: 1 nm ≡ 1 mμ (millimicron) ≡ 10 Å (angstrom) ≡ 10^{-7} cm. In the microwave and radiowave region, frequencies are usually used.

$$\nu = \frac{c}{\lambda} \tag{10.1}$$

where ν = frequency, Hz (s^{-1})
c = speed of light in vacuum = 3×10^{10} cm s^{-1}
λ = wavelength in vacuum, cm

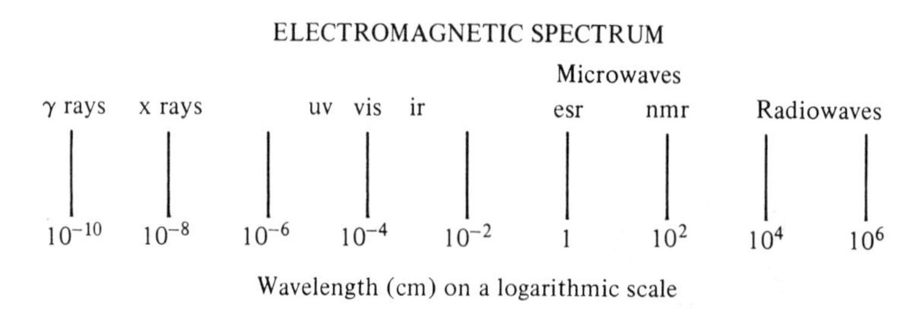

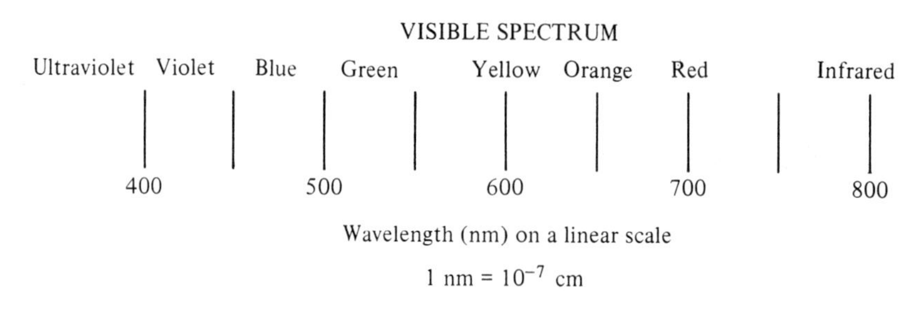

Fig. 10.1 Wavelength regions of the electromagnetic spectrum.

Infrared spectroscopists plot their spectra versus wavenumbers, $\bar{\nu}$:

$$\bar{\nu} = \frac{1}{\lambda} \tag{10.2}$$

where $\bar{\nu}$ = "frequency," cm^{-1}
λ = wavelength in vacuum, cm

Units of energy are popular for spectra in the far-ultraviolet, x-ray, and γ-ray regions. Planck's equation, $\Delta\varepsilon = h\nu$, is used to convert frequencies to energies. For example, 200-nm ultraviolet light corresponds to about 6 eV or 143 kcal mol^{-1}. Interaction of radiation with matter may result in the absorption of incident radiation, emission of fluorescence or phosphorescence, scattering into new directions, rotation of the plane of polarization, or other changes. Each of these interactions can give useful information about the nature of the sample in which they occur. Before discussing spectroscopy in detail, let us look at a few familiar examples.

Color

One of the most obvious properties of a substance is its color. The green color of leaves is due to chlorophyll absorption, the orange of a carrot or tomato arises from carotenes, and the red of blood results from hemoglobin. The color is characteristic of the spectrum (in the visible region) of light transmitted by the substance when white light (or sunlight) shines through it, or when light is reflected from it. The quantitative measure of color that we use is the absorption spectrum or reflectance spectrum. This is a plot of the variation with wavelength or frequency of the absorption of radiation by the substance. Figure 10.2 compares the visible spectra of green chlorophyll, orange β-carotene, and red oxyhemoglobin. Note that the spectra of the three different molecules have both similarities and differences; the spectra are much richer in information than the colors would indicate.

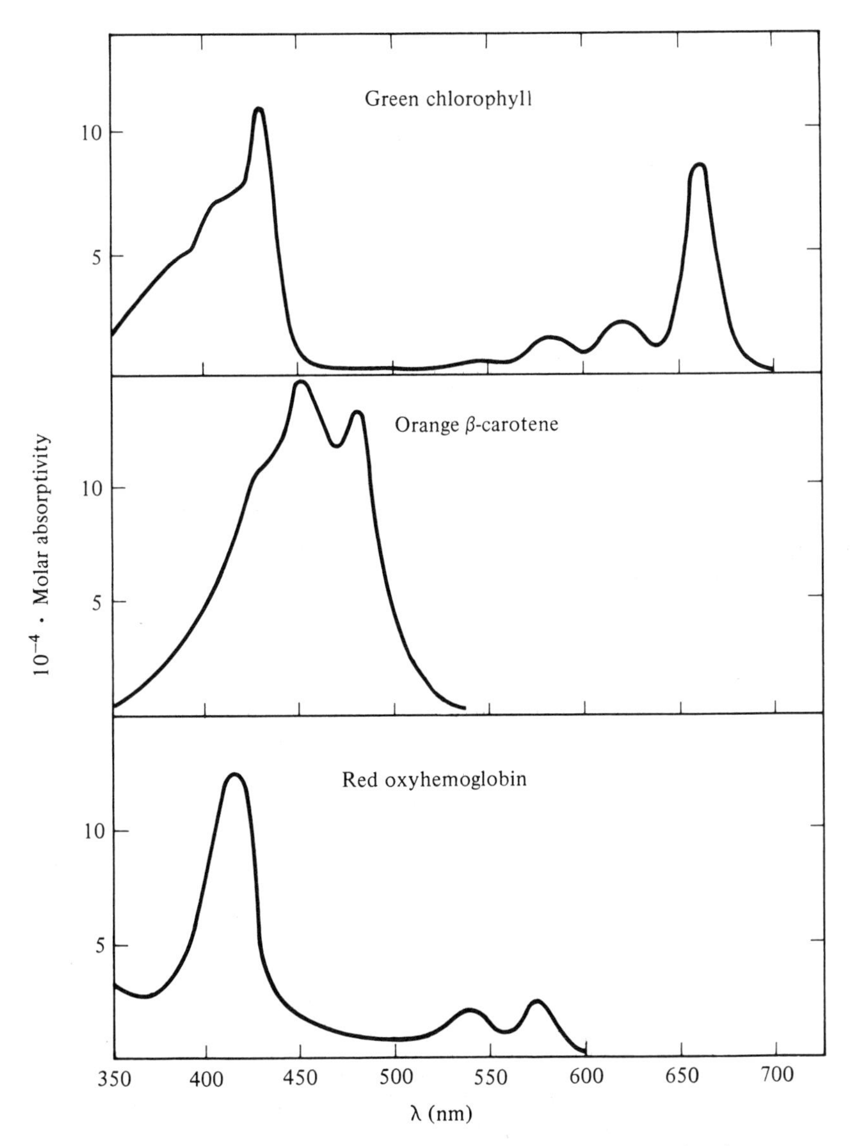

Fig. 10.2 Comparison of the visible spectra of three molecules.

Refractive index

The familiar bent appearance of a straight stick partly immersed in a tub of water results from the difference in refractive index of light in air and in water. The refractive index, n, is simply a measure of the ratio of the velocity of light in vacuum, c, to the velocity in the medium, v. The velocity of light in vacuum is approximately 3×10^8 m s^{-1} (3×10^{10} cm s^{-1}); this corresponds to about 1 ft ns^{-1}:

$$n = \frac{c}{v} \tag{10.3}$$

The refractive index of air for visible light is about 1.00028 at 15°C and 1 atm pressure, so the velocity of light in air is very close to that in vacuum. The refractive index at a wavelength of 589 nm for some common substances is given in Table 10.1. The reason that we are able to see objects that are not colored is because the changes of refractive index at the boundaries result in bending of the light rays directed at our eyes. Techniques such as phase contrast have been used in microscopy to increase the contrast between different regions of biological organelles. The appearance of detailed internal structure in such organelles occurs because their refractive index differs in the different regions. Because refractive index is an intrinsic property of a material that is related to its composition, a careful measurement in a refractometer can be used to characterize the substance.

Table 10.1 Refractive indexes for some common substances at 589 nm

Substance	Conditions	Refractive index, n_{589}
Air	1 atm, 15°C	1.0002765
Water	Liquid, 15°C	1.33341
Ethanol	Liquid, 20°C	1.3611
Carbon tetrachloride	Liquid, 20°C	1.4601
Hexane	Liquid, 20°C	1.37506
Quartz	Fused	1.45843
Protein		1.51–1.54
Sucrose	Crystal, 20°C	1.5376
Lipids		1.40–1.44

Even more information can be gained by examination of the refractive index using polarized light. Objects such as single crystals or biological cells frequently exhibit different refractive indices depending on the direction of polarization of the light relative to the fixed object. This anisotropy of refractive index is known as *birefringence* (literally, double refraction). It is an indication of ordering of molecules within the sample and can give useful information regarding intermolecular associations in cell membranes, nerve fibers, muscle, and the like.

Crystals of some substances, such as quartz, fluorite (CaF_2), and gypsum ($CaSO_4 \cdot 2\,H_2O$), exhibit birefringence that is related to the distinctive crystal axes.

ABSORPTION AND EMISSION OF RADIATION

Light that is incident on a colored sample is partly absorbed by it. This means that at certain wavelengths there is less intensity coming out the other side. Careful examination shows that there is no breakdown in the law of conservation of energy, however. The result of the absorption may appear as *heat* producing a temperature rise in the sample, *luminescence* in which a photon of the same or lower energy is emitted, *chemistry* that incorporates energy into altered bonding structures, or a combination of these. Nevertheless, the total energy change of the processes is always exactly equal to the energy of the photon that was absorbed.

There are three common ways of describing the absorption of radiation. These are (1) a kinetic picture of the probability of a photon being absorbed, (2) a description involving classical electrical oscillators, and (3) a quantum mechanical description of the behavior of a molecule with time in the presence of an oscillating electromagnetic field (light). Each of these provides useful insights, and the literature contains extensive references to each of them. We will look at them in a qualitative way, present a few important equations that can be derived, and show how the different descriptions can be related to one another.

The interactions leading to absorption of light are essentially electromagnetic in origin. The oscillating electromagnetic field associated with the incoming photon generates a force on the charged particles in a molecule. If the electromagnetic force results in a change in the arrangement of the electrons in a molecule, we say that a transition to a new electronic state has occurred. The new state will be of higher energy if radiation is absorbed, and the process can be illustrated using an energy-level diagram such as that shown in Fig. 10.3. The absorbed photon results in the excitation of the molecule from its normal or ground state, G, to a higher energy or excited electronic state, E. The excited electronic state has a rearranged electron distribution. It can sometimes be approximately described as a promotion of one of the ground-state electrons to an orbital of higher energy. This completes the absorption process, which usually occurs very rapidly (within 10^{-15} s), but the excited state produced is usually not stable for very long. Typically, within 10^{-8} s it has disappeared by pathways involving fluorescence emission, the dissipation of heat, or the initiation of photochemistry. Fluorescence and heat conversion result in the return of the molecule to the ground state, and the only net effect is the conversion of the energy of the absorbed photon into heat and (perhaps) a photon of lower energy. Figure 10.4 shows the energy of a molecule (versus time), which has absorbed a photon and emitted some of the energy as fluorescence.

At ordinary intensities the relaxation of the system back to the ground state is so rapid that the light produces very little change in the population of molecules

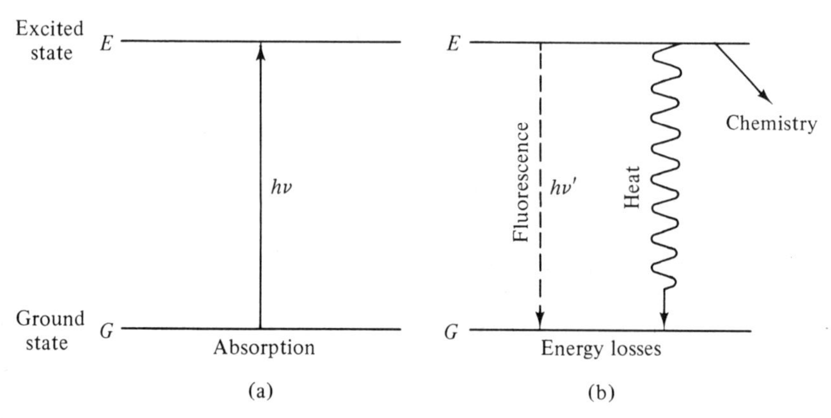

Fig. 10.3 Energy-level diagram: (a) the absorption of a photon of energy $h\nu$, and (b) three important processes by which the excitation energy is subsequently released or converted.

in the ground state. At the very high intensities encountered in flash-lamp or laser experiments, it is possible to excite nearly all of the molecules out of the ground state simultaneously, but this occurs only under unusual conditions in the laboratory.

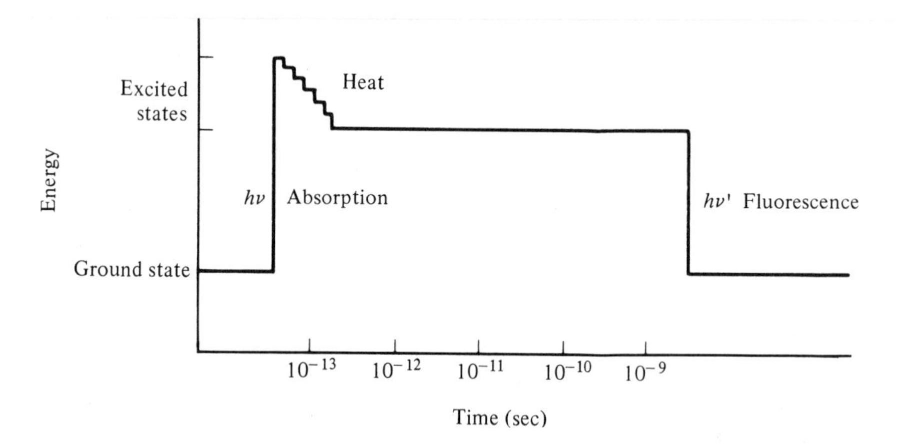

Fig. 10.4 The energy versus time for a single molecule in the presence of light. The molecule absorbs a photon and is excited to a vibrational level of an excited electronic state. The excess vibrational energy is rapidly lost as heat, and the molecule reaches the relatively stable lowest vibrational level of the excited electronic state. After a certain time the molecule emits a photon and returns to its ground state. Each molecule will have a different excited state lifetime, but the collection of molecules will have a characteristic relaxation time or half-life.

Radiation-induced Transitions

Einstein considered the dynamics of an atom or molecule placed in a radiation field using a rather simple model. Consider any two states of a molecule, such as the excited (E) state and ground (G) state of Fig. 10.3. The states will differ in energy by an amount $E_E - E_G = h\nu$. If the energy separation is large (compared with kT), virtually all of the molecules will be in the ground state in the dark at ordinary temperatures. In the presence of a radiation field, this population distribution will be slightly disturbed. We can characterize the radiation field by a radiation energy density $\rho(\nu)$, which is the amount of energy near frequency ν and per unit volume of the sample. The dimensions of $\rho(\nu)$ are energy volume^{-1}. It is related to the more familiar intensity of illumination of a surface $I(\nu)$, in energy area^{-1} time^{-1}, by the equation

$$I(\nu) = \frac{c}{n}\rho(\nu) \tag{10.4}$$

where c is the velocity of light and n is the refractive index of the medium. Some representative values of approximate illumination intensities and radiation-energy densities are given in Table 10.2. Values are also given in terms of quanta or photons using the Planck equation.

Table 10.2 Radiation densities and illumination intensities values for polychromatic sources are given per nanometer of spectral bandwidth

Source of phenomenon	λ	Illumination intensity, $I(\lambda)$*		Radiation density, $\rho(\lambda)$*	
	(nm)	μW cm^{-2}	quanta s^{-1} cm^{-2}	erg cm^{-3}	quanta cm^{-3}
Threshold of completely dark-adapted human eye	507	4×10^{-9}	1×10^{4}	1.3×10^{-18}	3×10^{-7}
Firefly luminescence	562	5×10^{-2}	15×10^{10}	1.7×10^{-11}	5
Full moon at surface of earth	500	0.3	1×10^{12}	1×10^{-10}	30
Bright summer sunlight	500	1×10^{5}	2×10^{17}	3×10^{-5}	1×10^{7}
Ruby laser, 1 kJ, 1 ms duration	694.3	1×10^{12}	4×10^{24}	300	1.2×10^{14}

* $I(\lambda)$ and $\rho(\lambda)$ are given for a wavelength range of 1 nm around the λ quoted.

As a consequence of the introduction of the radiation field to an absorbing sample, three processes are induced:

1. *Absorption.* The radiation induces transitions from the ground state G to the excited state E. The rate of this absorption process will depend on the number of molecules in state G, on the radiation density, and on the absorption coefficient of the molecule.

2. *Stimulated emission.* Once molecules are present in the excited state E, the radiation field can stimulate transitions down to the lower state, giving emission of radiation.
3. *Spontaneous emission.* Molecules put into the excited state E by the radiation field are substantially out of equilibrium. Spontaneous relaxation processes restore the system to equilibrium. These are spontaneous in the sense that they do not depend on the radiation density.

Classical Oscillators

The classical theory of light absorption considers matter to consist of an array of charges that can be set into motion by the oscillating electromagnetic field associated with light. The electric-dipole oscillators set in motion by the field have certain characteristic, or natural, frequencies, ν_i, that depend on the particular substance. When the frequency of the radiation is near the oscillator frequency, absorption will occur, and the intensity of the radiation will be decreased upon passing through the substance. The refractive index of the substance also undergoes large changes, called *dispersion anomalies*, in the same spectral region. The intensity of the interaction is known as the *oscillator strength*, f_i, and it can be thought of as characterizing the number of electrons per molecule that oscillate with the characteristic frequency, ν_i.

Although the classical picture has been replaced by the much more informative quantum mechanical description of the absorption of radiation, there are classical holdovers in the current literature. For example, a transition that is fully allowed quantum mechanically is said to have an *oscillator strength* of 1.0. Operationally, the oscillator strength, f, is related to the intensity of the absorption, that is, to the area under an absorption band plotted versus frequency:

$$f = \frac{2303 \text{ cm}}{\pi N_0 e^2 n} \int \varepsilon(\nu)\, d\nu \tag{10.5}$$

where ε is the molar absorptivity, c is the velocity of light, m the mass of the electron and e its charge, n the refractive index of the medium, and N_0 is Avogardro's number. The integration is carried out over the frequency range associated with the absorption band. Oscillator strengths can be observed from magnitudes of unity down to very small values ($<10^{-4}$). Measured values greater than unity, as for the blue bands of hemoglobin or chlorophyll, usually indicate the overlap of two or more electronic transitions.

Quantum Mechanical Description

The most satisfactory and complete description of the absorption of radiation by matter is based on time-dependent wave mechanics. It analyzes what happens to the eigenfunctions of molecules in the presence of an external oscillating electromagnetic field. Because it requires a high level of mathematical knowledge to follow the development of time-dependent wave mechanics, we will simply cite some of the useful results obtained from the detailed treatment.

Because the interaction is basically electromagnetic, the operator that conveniently describes it is the dipole-moment operator, $\boldsymbol{\mu} = e\mathbf{r}$. This operator is just the position operator, $\mathbf{r} = \mathbf{x} + \mathbf{y} + \mathbf{z}$, multiplied by the electronic charge. For example, the *permanent dipole moment* of a molecule in its ground state is obtained by using the ground-state wavefunction of the molecule to calculate the average position of negative charge. When the centers of positive and negative charge coincide, the permanent dipole moment is zero.

A transition from one state to another occurs when the radiation field connects the two states. In wave mechanics, the means for making this connection is described by the *transition dipole moment*

$$\boldsymbol{\mu}_{0A} = \int \psi_0 \boldsymbol{\mu} \psi_A \, dv \qquad (10.6)$$

where ψ_0, ψ_A = wavefunctions for the ground and excited state, respectively
dv = volume element; integration is over all space

The transition dipole moment will be nonzero whenever the symmetry of the ground and excited state differ. A hydrogen atom will always have zero permanent dipole moment, but if ψ_0 is a $1s$ orbital and ψ_A is a $2p$ orbital, the transition dipole does not equal zero. Similarly, ethylene has zero permanent dipole moment, but if ψ_0 is a π MO and ψ_A is a π^* MO, then $\boldsymbol{\mu}_{\pi\pi^*}$ is not zero. The transition moment has both magnitude and direction; the direction is characterized by the vector components: $\langle \boldsymbol{\mu}_x \rangle_{0A}$, $\langle \boldsymbol{\mu}_y \rangle_{0A}$, and $\langle \boldsymbol{\mu}_z \rangle_{0A}$. Most electronic transitions are polarized, which means that the three vector components are not all equal. In the examples above, the hydrogen-atom transition from s to p_x is polarized along x. The ethylene $\pi \rightarrow \pi^*$ transition is polarized along the C=C double bond. The magnitude of the transition is characterized by its absolute-value squared, which is called the *dipole strength*, D_{0A}:

$$D_{0A} = |\boldsymbol{\mu}_{0A}|^2 \qquad (10.7)$$

This is a scalar quantity rather than a vector. It has the units of a dipole moment squared. Dipole moments are frequently expressed in *debyes*, D, where 1 D = 10^{-18} esu cm, and the units of the dipole strength are then debyes^2.

The relations among dipole strength, oscillator strength, and spectra are given in Table 10.3.

Beer-Lambert Law

Consider a sample of an absorbing substance (liquid solution, solid, or gas) placed between two parallel windows that will transmit the light, as shown in Fig. 10.5. (Although the discussion involves "light," these principles apply equally well to radiation in the uv, ir, or other spectral regions.) Suppose that light of intensity I_0

Table 10.3 Spectroscopic relations*

1. Dipole strength ⟷ transition dipole moment

$$D_{0A} = |\boldsymbol{\mu}_{0A}|^2$$

2. Dipole strength ⟷ absorption spectrum

$$D_{0A} = \frac{3\hbar \cdot 2303c}{4\pi^2 N_0 n} \int \frac{\varepsilon(\nu)}{\nu} d\nu$$

$$= \frac{9.185 \times 10^{-39}}{n} \int \frac{\varepsilon(\nu)}{\nu} d\nu \qquad (\text{esu cm})^2$$

3. Oscillator strength ⟷ absorption spectrum

$$f_{0A} = \frac{2303 \text{ cm}}{\pi N_0 e^2 n} \int \varepsilon(\nu)\, d\nu$$

$$= \frac{1.441 \times 10^{-19}}{n} \int \varepsilon(\nu)\, d\nu \qquad \text{unitless}$$

* Frequency of light, ν, in s^{-1}; N_0 = Avogardro's number; n = refractive index; m, e = mass, charge of electron; $\hbar = h/2\pi$, c = speed of light.

is incident from the left, propagates along the x direction, and exits from the right, with decreased intensity I_t. At any point x within the sample, it has intensity I, which will decrease smoothly from left to right. (We ignore for the moment the discontinuous intensity decrease, usually about 10%, which results from reflections at the windows.)

If the sample is homogeneous, the *fractional* decrease in the light intensity is the same across a small interval dx, regardless of the value of x. The fractional decrease for a solution depends linearly on the concentration of the absorbing

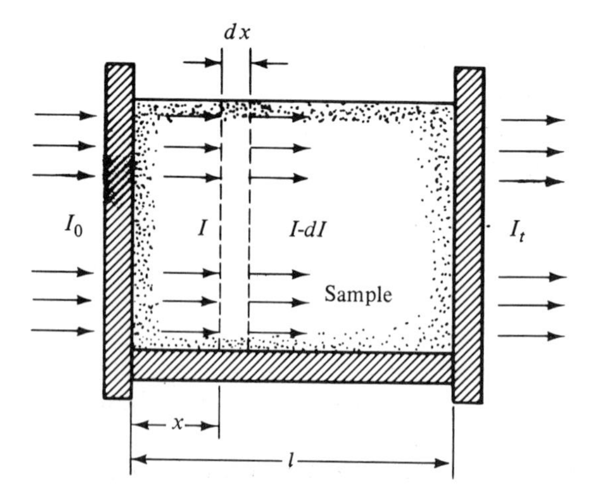

Fig. 10.5 Transmission of light by an absorbing sample.

substance. In mathematical form, this can be written

$$-\frac{dI}{I} = \alpha c\, dx \tag{10.8}$$

where dI/I is the fractional change in light intensity, c is the concentration of absorber, and α is a constant of proportionality. Equation (10.8) is the differential form of the Beer-Lambert law. It is straightforward to integrate it between limits I_0 at $x = 0$ and I_t at $x = l$, where l is the optical path length. Since neither α nor c depends on x, we can write

$$-\int_{I_0}^{I_t} \frac{dI}{I} = \alpha c \int_0^l dx \tag{10.9}$$

$$\ln \frac{I_0}{I_t} = \alpha c l \qquad I_t = I_0\, e^{-\alpha c l} \tag{10.10}$$

For measurements made with cuvets (optical sample cells) of different path lengths, the transmitted intensity, I_t, decreases exponentially with increasing path length (Fig. 10.6). Alternatively, for a single-cuvet path length, the transmitted

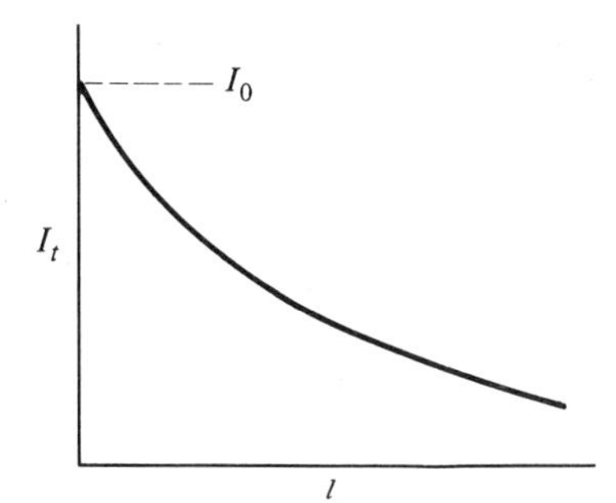

Fig. 10.6 Dependence of transmitted light intensity on path length.

intensity decreases exponentially with increasing concentration of an absorbing solute. A more common formulation utilizes base 10 rather than natural logarithms,

$$A = \log \frac{I_0}{I_t} = \varepsilon c l \tag{10.11}$$

where A is the *absorbance* or *optical density*, and $\varepsilon = \alpha/2.303$ is the molar absorptivity (or molar extinction coefficient), with units $\ell\ \text{mol}^{-1}\ \text{cm}^{-1}$, when the concentration, c, is in $\text{mol}\ \ell^{-1}$. Equation (10.11) is the integrated form of the Beer-Lambert law and shows that the absorbance is linearly related to concentration (or path length). The relation between absorbance and transmission ($T = I_t/I_0$) is exactly analogous to that between pH and (H^+):

$$A = -\log T$$

Because absorption depends strongly on wavelength for nearly all compounds, we must specify the wavelength at which measurement is made. This is usually done using a subscript λ, indicating the particular wavelength, as A_λ or ε_λ. For a single substance at a specified wavelength, ε_λ is a constant, characteristic of the absorbing sample, and is *independent* of both c and l. The wavelength dependence of ε_λ (or of A_λ) is known as the *absorption spectrum* of the compound.

It was mentioned earlier that a small but significant portion of light is lost by reflection at the cuvet windows. Corrections for this effect, as well as for any absorption by the solvent, are usually made by substituting solvent for the solution in the same cuvet and making a second measurement. The transmitted intensity of this second measurement is then used as I_0 in the Beer-Lambert expressions. Alternatively, a double-beam method may be used, where the solution (sample) and solvent (reference) are placed in matched cuvets and their transmissions are measured simultaneously.

Deviations from the Beer-Lambert law can occur for a variety of reasons. Problems arising from insufficiently monochromatic radiation, uncollimated light beams, or imprecise sample geometry are minimized by the design of most modern spectrophotometers. Certain properties of individual samples require investigation to determine whether corrections need to be applied. Beer-Lambert law deviations can arise from inhomogeneous samples, light scattering by the sample, dimerization or other aggregation at high concentrations, or changes in equilibria involving dissociable absorbing solutes. The most common consequence is that the measured absorbance does not increase linearly with increasing concentration or path length. In many cases this behavior can be used to obtain useful information about the sample. It is essential to understand the origin of such effects in order for spectrophotometry to be a useful tool for experimental investigations.

Quantitative determinations using Beer's law

One of the most widely used applications of spectroscopy is for the quantitative determination of the concentrations of substances in solution. Noting that both the absorbance, A, and the molar absorptivity, ε, depend on wavelength, we now write

$$A_\lambda = \varepsilon_\lambda c l \tag{10.12}$$

A plot of A_λ versus λ (or A_ν versus ν) is the absorption spectrum of the solution. For a single substance, the absorbance A_λ at any wavelength is directly proportional to concentration c for a fixed path length, l. Experimentally, the absorbance is determined by taking the log of the ratio of the incident to the transmitted light intensities.

The absorbance of a solution of more than one independent species is additive. For two components, M and N,

$$\begin{aligned} A_\lambda &= A_\lambda^{\mathrm{M}} + A_\lambda^{\mathrm{N}} = \varepsilon_\lambda^{\mathrm{M}} l[\mathrm{M}] + \varepsilon_\lambda^{\mathrm{N}} l[\mathrm{N}] \\ &= (\varepsilon_\lambda^{\mathrm{M}}[\mathrm{M}] + \varepsilon_\lambda^{\mathrm{N}}[\mathrm{N}])l \end{aligned} \tag{10.13}$$

If measurements are made at two (or more) wavelengths where the ratios of extinction coefficients differ, the resulting equations,

$$A_1 = (\varepsilon_1^M[M] + \varepsilon_1^N[N])l$$
$$A_2 = (\varepsilon_2^M[M] + \varepsilon_2^N[N])l \qquad (10.14)$$

can be solved for the concentrations of the absorbing solutes,

$$[M] = \frac{1}{l}\frac{\varepsilon_2^N A_1 - \varepsilon_1^N A_2}{\varepsilon_1^M\varepsilon_2^N - \varepsilon_2^M\varepsilon_1^N}$$
$$[N] = \frac{1}{l}\frac{\varepsilon_1^M A_2 - \varepsilon_2^M A_1}{\varepsilon_1^M\varepsilon_2^N - \varepsilon_2^M\varepsilon_1^N} \qquad (10.15)$$

These relations are widely used in the spectrophotometric analysis of mixtures of absorbing species.

A wavelength at which two or more components have the same extinction coefficient is known as an *isosbestic*. [$\varepsilon_\lambda^M = \varepsilon_\lambda^N = \varepsilon_{iso}$]. At this wavelength, the absorbance can be used to determine the total concentration of the two components.

$$\lambda = \text{isosbestic:} \quad A_{iso} = \varepsilon_{iso}l[M] + \varepsilon_{iso}l[N] = \varepsilon_{iso}l \cdot ([M] + [N]) \qquad (10.16)$$

Measurements at an isosbestic wavelength plus one other wavelength where the extinction coefficients differ for the two components provide a particularly simple solution to the Beer-Lambert law equations:

$$[M] + [N] = \frac{A_{iso}}{\varepsilon_{iso}l}$$
$$\frac{[M]}{[N]} = \frac{\varepsilon_1^N A_{iso} - \varepsilon_{iso}A_1}{\varepsilon_{iso}A_1 - \varepsilon_1^M A_{iso}} \qquad (10.17)$$

Example 10.1 Solutions containing the amino acids tryptophan and tyrosine can be analyzed under alkaline conditions (0.1 *M* KOH) from their different uv spectra. The extinction coefficients under these conditions at 240 and 280 nm are

$\varepsilon_{240}^{Tyr} = 11{,}300\ M^{-1}\ cm^{-1}$	$\varepsilon_{240}^{Trp} = 1960\ M^{-1}\ cm^{-1}$
$\varepsilon_{280}^{Tyr} = 1500\ M^{-1}\ cm^{-1}$	$\varepsilon_{280}^{Trp} = 5380\ M^{-1}\ cm^{-1}$

A 10-mg sample of the protein glucagon is hydrolyzed to its constituent amino acids and diluted to 100 mℓ in 0.1 *M* KOH. The absorbance of this solution (1-cm path) was 0.717 at 240 nm and 0.239 at 280 nm. Estimate the content of tryptophan and tyrosine in μmol g^{-1} of protein.

Solution Neither of the wavelengths is an isosbestic for these amino acids, so we use Eq. (10.15):

$$[\text{Tyr}] = \frac{(5380)(0.717) - (1960)(0.239)}{(11{,}300)(5380) - (1500)(1960)}$$

$$= \frac{3.39 \times 10^3}{57.9 \times 10^6}$$

$$= 5.85 \times 10^{-5}\ M$$

$$[\text{Trp}] = \frac{(11{,}300)(0.239) - (1500)(0.717)}{57.9 \times 10^6}$$

$$= 2.81 \times 10^{-5}\ M$$

Since 10 mg of protein was hydrolyzed and diluted to 100 mℓ of solution, the estimated contents are:

585 μmol of Tyr g^{-1} of protein

281 μmol of Trp g^{-1} of protein

When only two absorbing compounds are present in a solution, one or more isosbestics are frequently encountered if we examine the entire wavelength range of the uv, visible, and ir regions. The isosbestics do not *necessarily* occur, however, for the molar absorptivity of one compound may be less than that of the other in every accessible wavelength region. An easy way to spot isosbestics is to superimpose plots of ε versus λ for the two compounds. Wherever the curves cross, there is an isosbestic. An example is shown in Fig. 10.7 for cytochrome *c*.

The additivity of absorbance can be generalized to any number of components. Thus, for *n* components,

$$A = A_1 + A_2 + A_3 + \cdots + A_n = (\varepsilon_1 c_1 + \varepsilon_2 c_2 + \varepsilon_3 c_3 + \cdots + \varepsilon_n c_n)l \quad (10.18)$$

To determine the *n* concentrations by absorption methods, we need to make measurements at a minimum of *n* wavelengths, each characterized by a unique set of ε's. This may be impractical when *n* is large. Even so, a single component can sometimes be measured accurately in a complex mixture if there is a wavelength where its absorbance is much greater than that of any of the other components. For example, hemoglobin in red blood cells can be measured spectrophotometrically at 541 or 577 nm (Fig. 10.1), because the other cell constituents (proteins, lipids, carbohydrates, salts, water and so on) do not absorb significantly at those wavelengths.

In multicomponent solutions isosbestic points almost never occur. The probability that three or more compounds have identical molar absorbances at any wavelength is exceedingly small. The probability of two or more isosbestic points for the spectra of the set of compounds is so small as to be completely negligible. This property is useful for diagnostic purposes and results in the

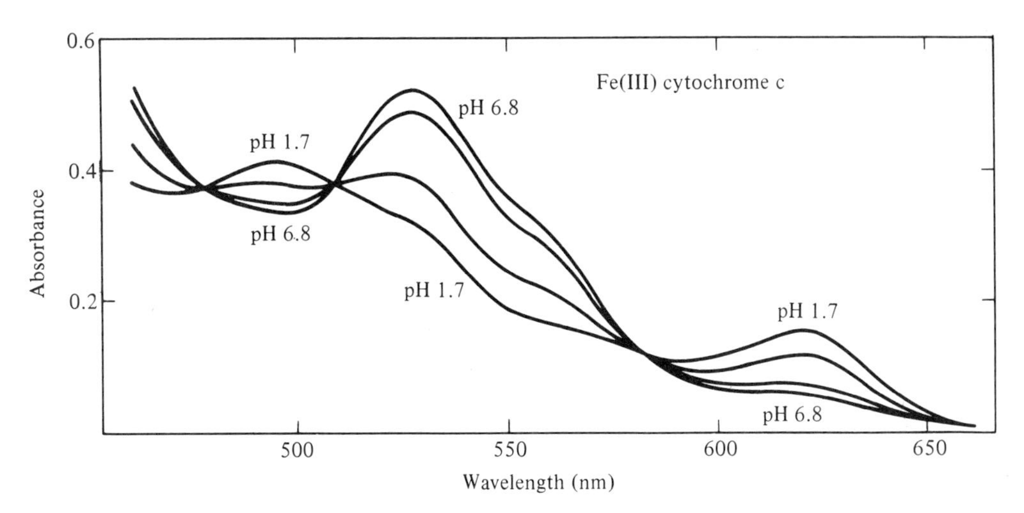

Fig. 10.7 Absorption spectra of Fe(III) cytochrome c (5.3×10^{-5} *M*) at various pH values from 6.8 to 1.7. Isosbestic points occur at 477, 509, and 584 nm. (E. Yang, unpublished spectra.)

following generalization: *The occurrence of two or more isosbestics in the spectra of a series of solutions of the same total concentration demonstrates the presence of two and only two components absorbing in that spectral region.*

It is possible, in principle, for this generalization to be violated, but the chances are vanishingly small. Note that the rule does not necessarily apply to components that do not absorb at all in the wavelength region investigated. It also does not apply if two chemically distinct components have identical absorption spectra (such as ADP and ATP). In this case, the entire spectrum is a set of isosbestics for these two components alone.

Two common examples of the usefulness of isosbestics are the study of equilibria involving absorbing compounds and investigations of reactions involving absorbing reactants and products. The presence of isosbestics is used as evidence that there are no intermediate species of significant concentration between the reactants and products.

Exercise Consider the reaction of an absorbing reactant M to give an absorbing product P by the reaction

$$M \xrightarrow{k_1} N \xrightarrow{k_2} P \qquad k_1 \sim k_2$$

Show that no isosbestic points would be expected for this system, even if the intermediate N has no measurable absorbance in the same spectral region.

PROTEINS AND NUCLEIC ACIDS: ULTRAVIOLET ABSORPTION SPECTRA

Most proteins and all nucleic acids are colorless in the visible region of the spectrum; however, they do exhibit absorption in the near-ultraviolet region. These spectroscopic absorptions contain information relevant to the composition of these complex molecules, as well as about the conformation or three-dimensional structure of the molecules in solution. Figure 10.8 shows the uv absorption of serum albumin, a representative protein. There is a distinctive absorption maximum at around 280 nm and a much stronger maximum at 190 to 210 nm. The 280-nm band is assigned to $\pi \rightarrow \pi^*$ transitions in the aromatic amino acids, the 200-nm band to $\pi \rightarrow \pi^*$ transitions in the amide group. These features are sufficiently characteristic that they are often used in the preliminary identification of proteins in biological materials. The spectra of nucleic acids in the same region (Fig. 10.9) show somewhat different characteristics. An absorption maximum occurs at 260 nm, and a shoulder is seen near 200 nm on a background rising to shorter wavelengths. Again these features are assigned to $\pi \rightarrow \pi^*$ transitions and are characteristic of all nucleic acids. The observations are sufficiently general that the ratio of absorbances A_{260}/A_{280} has been used to determine quantitatively the ratio of nucleic acid to protein in a mixture of the two.

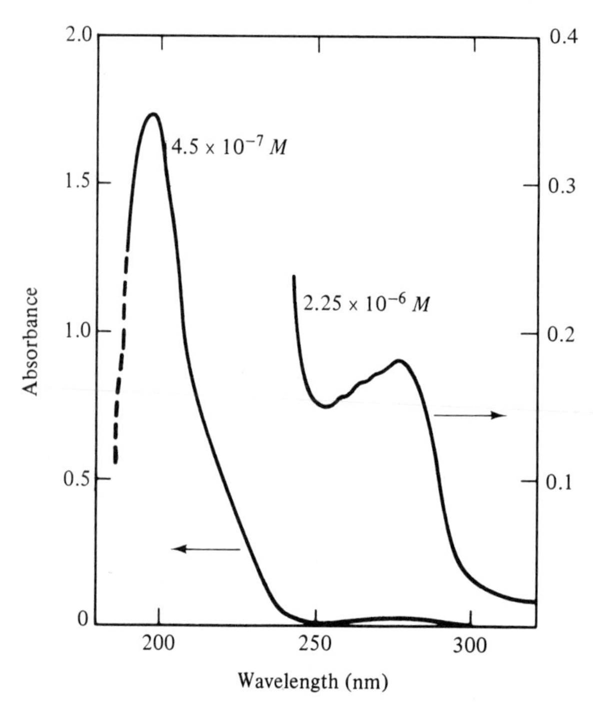

Fig. 10.8 Ultraviolet absorption spectrum of bovine serum albumin. Solution in 10^{-3} M phosphate buffer pH 7.0, 1.0-cm path. The wavelength region above 240 nm was measured at a higher concentration and on an expanded absorbance scale (right) so that the weaker absorption bands in that region would be visible. (E. Yang, unpublished results.)

Proteins are natural polyamino acids. That is, they are polymers in which an assortment of the 20 natural amino acids are connected by amide linkages in a

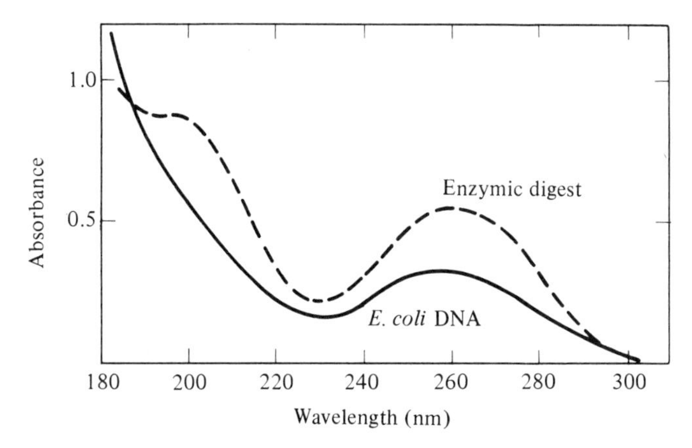

Fig. 10.9 Ultraviolet absorption spectrum of DNA from *E. coli* in the native form at 25°C (solid curve) and as an enzymic digest of nucleotides (dashed curve). [From D. Voet, W. B. Gratzer, R. A. Cox, and P. Doty, *Biopolymers 1*, 193 (1963).]

long chain. Each protein has a characteristic sequence, but there are thousands of different proteins in nature and accordingly many different compositions and sequences. Nucleic acids are also linear polymers, but the monomer unit is the nucleotide, consisting of an heterocyclic base attached to a sugar phosphate. Only four different bases commonly occur, but there are many different ways in which they can be combined and arranged in the linear sequence. The chain involves covalent links between the sugars (ribose for RNA, deoxyribose for DNA) and the phosphate linking groups. To understand the uv absorption of proteins or nucleic acids, we need to examine the various contributions to the spectra. We shall consider the following important factors:

1. Absorption spectra of individual monomers.
2. Contribution of the polymer backbone.
3. Secondary and tertiary structure, including helix formation.
4. Hydrogen bonding and solvent effects.

Amino Acid Spectra

The uv spectra of various amino acids are summarized in Figs. 10.10 and 10.11. Several distinctive categories of spectra appear. The aromatic amino acids tryptophan, tyrosine, and phenylalanine are the only ones absorbing significantly at wavelengths longer than 230 nm. In particular, at the "characteristic" protein wavelength of 280 nm, absorbance reflects only the presence of tryptophan and tyrosine in a protein and cannot be used for quantitative purposes without supplementary analysis. For example, the absorbance in a 1-cm cell at 280 nm for 1% solutions of proteins in water varies from 3.1 for NAD nucleosidase from pig

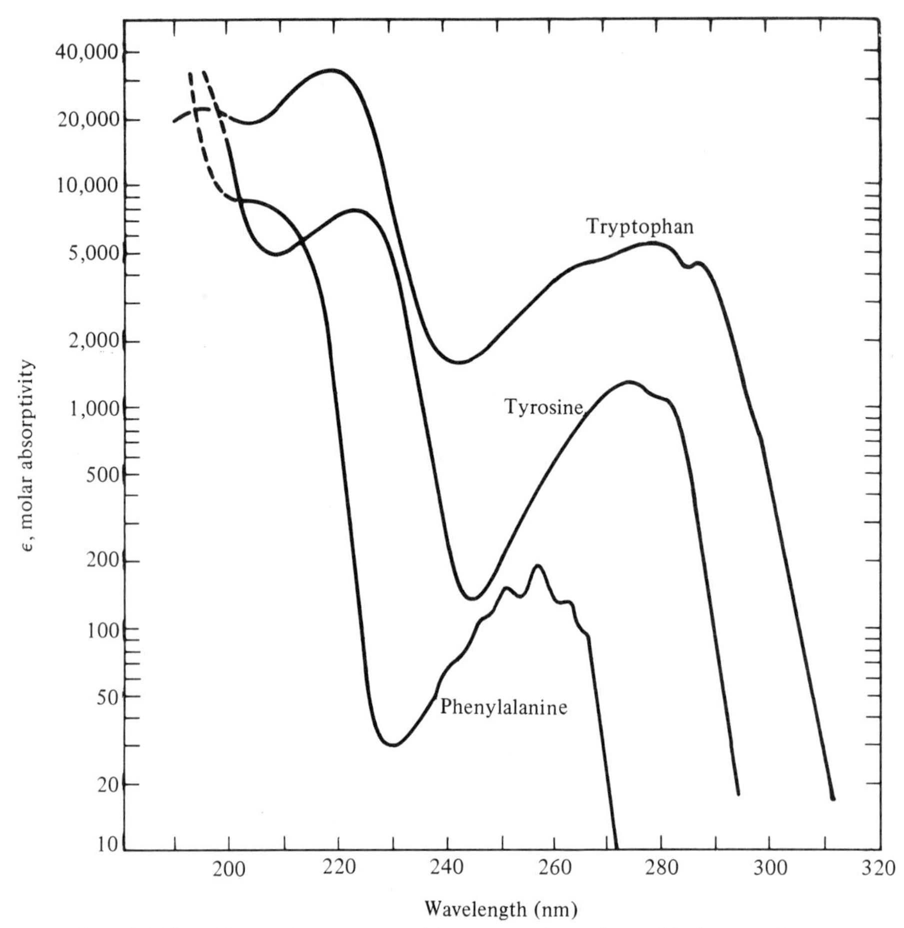

Fig. 10.10 Absorption spectra of the aromatic amino acids (tryptophan, tyrosine, phenylalanine) at pH 6. [From S. Malik, cited by D. B. Wetlaufer in *Advan. Protein Chem. 17*, 303 (1962).]

to 27 for egg white lysozyme. Between 200 and 230 nm, the aromatic amino acids (especially tryptophan) have absorption and so do histidine, cysteine, and methionine. Between 185 and 200 nm, only phenylalanine and tyrosine have distinct maxima, but the curves of all the other amino acids rise sharply to shorter wavelengths as illustrated by alanine. Extensive tables of amino acid absorption data are given in the *Handbook of Biochemistry and Molecular Biology*, 3rd ed. (CRC Press, Inc., Cleveland, Ohio, 1976).

Polypeptide Spectra

The contribution of the amide linkages to the absorption spectra can be seen by comparing the spectrum of lysine hydrochloride in Fig. 10.11 with that of poly-L-lysine hydrochloride in the random-coil form (Fig. 10.12). The broad

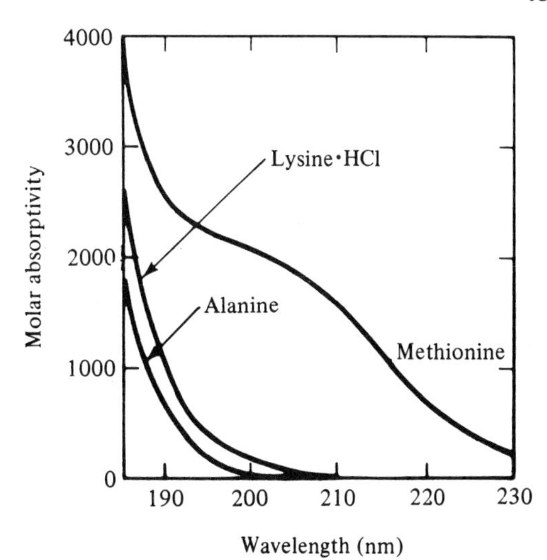

Fig. 10.11 Ultraviolet absorption spectra of three α-amino acids in aqueous solution at pH 5. (Data from *Handbook of Biochemistry and Molecular Biology*, 3rd ed., CRC Press, Inc., Cleveland, Ohio, 1976.)

absorption centered at 192 nm ($\varepsilon_{192} = 7100\, M^{-1}\, \text{cm}^{-1}$) is characteristic of the amide linkage in poly-L-lysine and increases the absorbance in this region by eight-fold over that of the free amino acid. All proteins have contributions to the absorption spectra in the region around 190 nm (180 to 200 nm) from the polypeptide backbone; however, they are accompanied by absorption contributions from certain of the side chains, especially the aromatic ones.

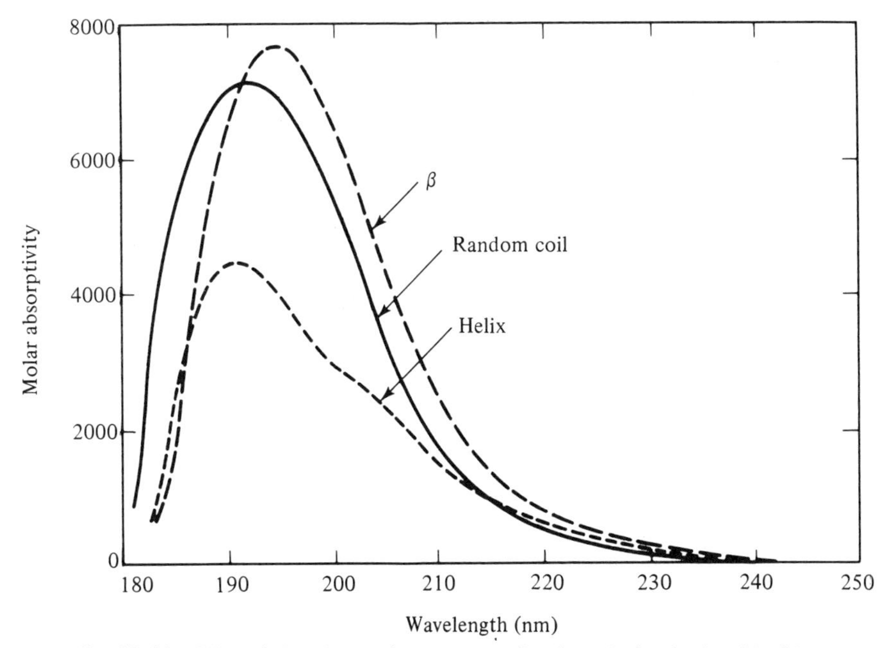

Fig. 10.12 Ultraviolet absorption spectra of poly-L-lysine hydrochloride in aqueous solution: random coil, pH 6.0, 25°C; helix, pH 10.8, 25°C; β form, pH 10.8, 52°C. [From K. Rosenheck and P: Doty, *Proc. Natl. Acad. Sci. U.S. 47*, 1775 (1961).]

Secondary Structure

The influence of conformation of a protein is sensitively detected by uv spectroscopy. Figure 10.12 shows the effect on the poly-L-lysine spectrum resulting from raising the pH to induce helix formation (by reducing the net positive charge on the lysine side chains) and by raising the temperature, which converts the polypeptide to the β-sheet structure. Proteins are more complicated in their native secondary structure and they usually contain simultaneously several elements of secondary structure, at different locations along the peptide chain. The process of denaturation destroys much of this secondary structure, and changes in the absorption spectra occur as a consequence. The absorption features associated with helix, β sheet, or random coil are sufficiently distinctive to be used for diagnostic purposes for the native protein (Rosenheck and Doty, 1961). In this respect, circular dichroism spectra (see later in this chapter) can be more informative, however.

Origin of Spectroscopic Changes

It is meaningful to analyze the various factors that contribute to the spectroscopic changes that occur when a protein undergoes a change in secondary structure. Ultimately, most of the effects are electrical in origin, because the spectra are sensitive to charge effects on the ground and excited electronic states. A list of such influences includes (1) changes in local charge distribution; (2) changes in dielectric constant; (3) changes in bonding interactions, such as hydrogen bonds; and (4) changes in dynamic (or resonance) coupling between different parts of the molecule.

Some of these can be modeled theoretically and quantitative predictions can be made. The theoretical analysis is generally rather complex and the agreement between predictions and experiment ranges from excellent to fair (Amos and Burrows, 1973). Here, we will look at a few representative examples.

The uv or visible spectrum of a substance depends markedly upon such factors as its state (gas, liquid, solid), the solvent, temperature, extent of aggregation or dissociation, and specific molecular complexes. Figure 10.13 illustrates this behavior for the molecule anisole.

$CH_3O-C_6H_5$

and is representative of other aromatic molecules. The spectrum of the vapor (bottom) is typically the "best resolved," in the sense that the absorption-band envelope contains a number of sharp spikes or peaks. These are characteristic of particular vibrations associated with the molecular structure and are superimposed on the underlying electronic transition. The energy of the different vibrations is added to the energy of the electronic transition when both the electronic and vibrational quantum numbers change as the result of absorption of a photon. This is called a *vibronic transition*. The energy of the electronic

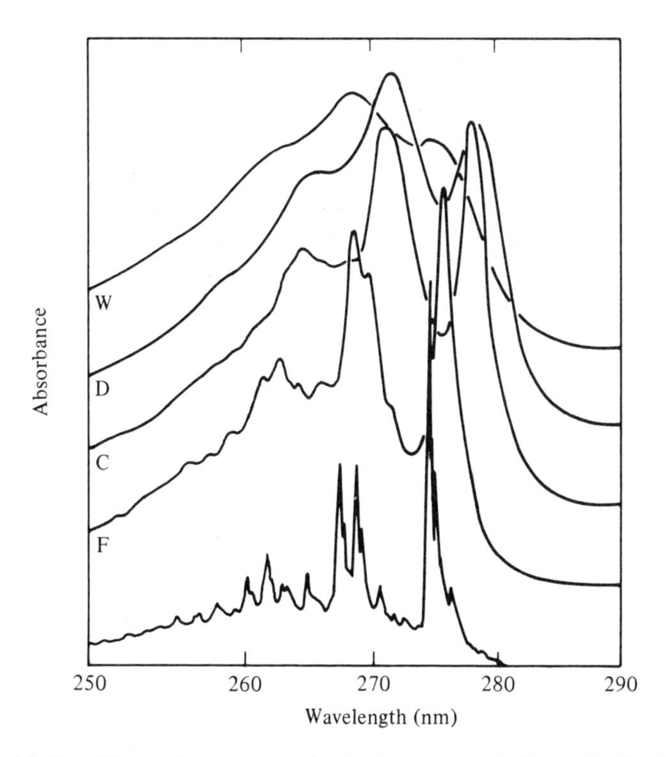

Fig. 10.13 Absorption spectra of anisole vapor and of anisole dissolved in perfluorooctane, F; in cyclohexane, C; in dioxane, D; and in water W. All at 30°C. The spectra of anisole solutions are displaced vertically to decrease overlap. [From G. L. Tilley, Ph.D. thesis, Purdue University, 1967; cited by M. Laskowski, Jr., in *Spectroscopic Approaches to Biomolecular Conformation*, D. W. Urry (ed.), Americam Medical Association, Chicago, 1970.]

transition itself (without added vibrational energy) is determined from the low-frequency or long-wavelength edge of the vibronic spectrum. A detailed vibrational analysis may be needed to locate this energy precisely. Because there are many ways in which large molecules can vibrate (many *normal modes* of vibration), the envelope of an electronic transition of a molecule in the gas phase is often very complex. The reason that the vibronic structure is so sharp is that most of the molecules in the gas phase are relatively isolated from one another; only a small fraction are in the process of undergoing a collision during the very brief time (about 10^{-15} s) that is required for the photon to be absorbed. Thus, nearly all of the gas-phase molecules are in essentially the same environment and not interacting with one another.

In the liquid state, either pure liquid or in solution, the environment of the molecules is quite different. Close contacts exist between molecules in liquids, and essentially all of the molecules are in the process of undergoing collisions. The forces between molecules are strong enough to perturb the molecular energy levels significantly and, together with the much greater variety of environments present at the instants of photon absorption, this leads to a broadening and shifting of the spectra. The effects depend on the polarity and hydrogen-bonding ability of the solvent, as can be seen in the upper spectra in Fig. 10.13.

The local environment of the amino acids in a native protein depends sensitively on the electric properties of the nearby peptide chain and associated solvent. Careful measurements of the absorption spectrum of a protein near 280 nm show small but distinct differences from the spectrum of the constituent aromatic amino acids. These differences reflect the sum of the effects of the local environments on the aromatic amino acids. Changes of the same magnitude are encountered upon denaturation of the protein.

Nucleic Acids

By contrast with the amino acid units of proteins, the nucleotides that make up the nucleic acid polymers have rather similar absorption spectra. The aromatic bases that are attached to the ribose- or deoxyribose-phosphates all have absorption maxima near 260 nm (Table 10.4). The free base, the nucleoside (the base attached to the sugar), the nucleotide (the base attached to the sugar phosphate), and the denatured polynucleotide all have very similar absorption spectra in this region. For example, the wavelength of maximum absorption is at 260.5 nm for adenine ($\varepsilon = 13.4 \times 10^3$), at 260 nm for adenosine ($\varepsilon = 14.9 \times 10^3$), at 259 nm for adenosine-5′-phosphate ($\varepsilon = 15.4 \times 10^3$), and at 257 nm for the tetranucleotide pApApApA. The latter, however, exhibits a lower absorbance per nucleotide in aqueous solution, where $\varepsilon_{259} = 11.3 \times 10^3\ M^{-1}\ \text{cm}^{-1}$. The ratio of the absorbance of the hydrolyzed compound to that of the intact compound at the same wavelength is known as the *hyperchromicity ratio* of the polynucleotide. For

Table 10.4 Absorption properties of nucleotides*

	λ_{max} (nm)	$10^{-3}\ \varepsilon_{max}$ (M^{-1} cm^{-1})
Ribonucleotides		
Adenosine-5′-phosphate	259	15.4
Cytidine-5′-phosphate	271	9.1
Guanosine-5′-phosphate	252	13.7
Uridine-5′-phosphate	262	10.0
Deoxyribonucleotides		
Deoxyadenosine-5′-phosphate	—	15.3
Deoxycytidine-5′-phosphate	271	9.3
Deoxyguanosine-5′-phosphate	—	—
Thymidine-5′-phosphate	267	10.2

* Wavelengths of maxima and molar absorptivities of nucleotides at pH 7.
SOURCE: *Handbook of Biochemistry and Molecular Biology*, 3rd ed., CRC Press, Inc., Cleveland, Ohio, 1976.

the tetranucleotide, it is $15.4 \times 10^3/11.3 \times 10^3 = 1.36$ at 250 nm. Hyperchromicities range from 1.0 to 1.4 (see Fig. 10.9), depending on the identity of the nucleotide, the identities of its nearest neighbors in the polynucleotide, and the conformation.

The *hypochromicity* (decreased absorptivity) of the polynucleotides or nucleic acids relative to the nucleotides results primarily from interactions between adjacent bases in the stacked arrangement of the helical polymer. The change upon denaturation to the random-coil form of the polymer or upon hydrolysis to the mononucleotides is easily measured (typically 30 to 40% change) and is often used to follow the kinetics or thermodynamics of the denaturation process. The origin of the hypochromism is again basically electromagnetic in nature. It is somewhat more complex than the solvent effects discussed previously, however, because it involves interactions between the electric-dipole transition moments of the individual bases with those of their neighbors. Thus, it depends not only on the intrinsic transition moments of each base, which differ both in magnitude and direction for the chemically distinct bases, but also on the relative orientations of the interacting bases. It is the latter dependence that produces the hyperchromicity when the bases become unstacked upon denaturation to the random coil form.

Rhodopsin: A Chromophoric Protein

Many proteins include groups other than the common amino acids. Often, but not always, these groups are covalently linked to the polypeptide chain. Some examples and the nature of the attached group are: glycoproteins (sugars), hemeproteins (iron porphyrins), flavoproteins (flavin), and rhodopsin (retinal; vitamin A). In the last three cases the group is a *chromophore*, and contributes to the absorption spectrum in the visible or near-uv regions. We shall examine some features of the visual pigment protein rhodopsin, which not only absorbs radiation in the visible region of the spectrum but also undergoes a photochemical transformation that triggers the visual stimulus.

Rhodopsins in nature occur widely in both vertebrates and invertebrates. Recently, a form of rhodopsin, called "bacteriorhodopsin," has been found in the outer membranes of certain halophilic (salt-loving) bacteria. Rhodopsins consist of a colorless protein (or glycoprotein), opsin, to which the chromophore retinal is covalently attached. In mammals the isomer 11-*cis*-retinal is attached to the ε-amino group of a lysine side chain of the opsin by a protonated Schiff's base linkage:

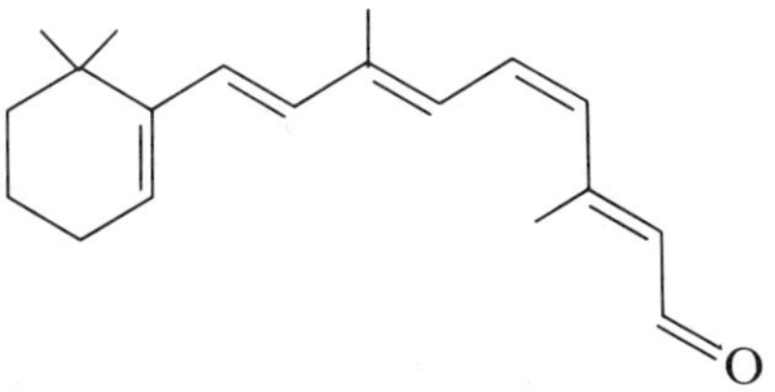

O

11-*cis*-Retinal

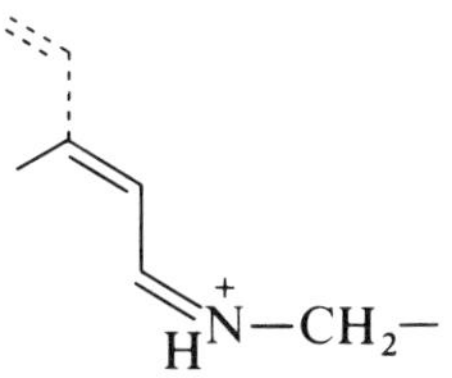

$\overset{+}{\mathrm{HN}}-CH_2-$

Protonated Schiff's base

In other organisms the chromophore appears to be the 9-cis or 13-cis isomer or the molecule 3-dehydroretinal. Rhodopsins have intense, broad absorption bands with maxima lying between 440 and 565 nm, depending on the specific chromophore and its environment. Cow rhodopsin absorbs at 500 nm (Fig. 10.14) and is a bright red-orange color; animals that are deficient in pigments (many albinos) show this red-orange in the pupil.

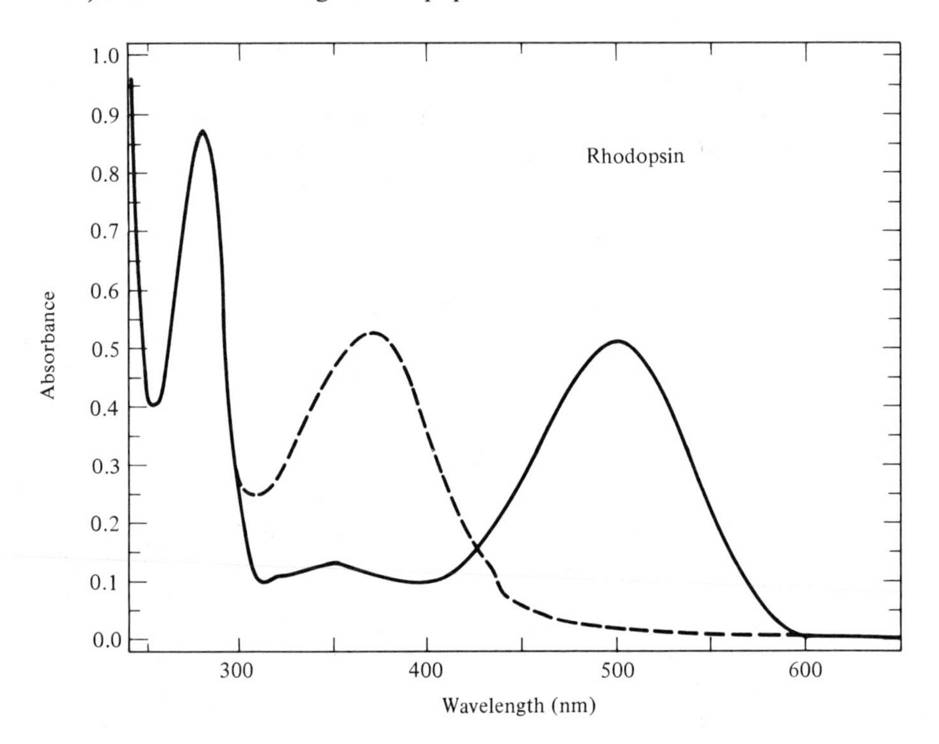

Fig. 10.14 Absorption spectrum of cow rhodopsin in the uv and visible. Purified rhodopsin was measured as isolated from dark-adapted cow retinas (solid curve) and again following its conversion by illumination (dashed curve). [From M. L. Applebury, D. M. Zuckerman, A. A. Lamola, and T. M. Jovin, *Biochemistry 13*, 3448 (1974). Copyright by the American Chemical Society.]

Retinal is an example of the linear polyenes that we have examined in Chapter 9. The six conjugated double bonds of the aldehyde contain π-electrons that can be approximately treated by the particle-in-a-box model. The long-wavelength absorption band (a $\pi \rightarrow \pi^*$ transition) of isolated retinal in a variety of organic solvents occurs between 366 and 377 nm, which is the expected region for polyenes of this length. In fact, the spectrum of rhodopsin after it has been photoconverted is very similar to that of retinal in an organic solvent. The origin of the shift of this band to 500-nm or even longer wavelengths in unilluminated rhodopsin presents an interesting spectroscopic problem that is not solved at present.

FLUORESCENCE

There is a certain fascination in observing the irridescent glow of many minerals under ultraviolet illumination. Because our eyes are not sensitive in the uv, we do not detect the exciting illumination directly, but only sense the visible fluorescence emitted by the sample. Many biological substances emit characteristic fluorescence. Chlorophyll from leaves emits red fluorescence; proteins fluoresce in the ultraviolet primarily from tryptophan residues; reduced pyridine nucleotides, flavoproteins, and the Y base of transfer RNA also exhibit characteristic fluorescence emission. In other cases, interesting and useful information can be obtained from fluorescent species that are introduced to biochemical molecules or systems. Fluorescent "labels" have been prepared by equilibrium binding or covalent attachment of fluorescent chromophores to enzymes, antibodies, membranes and so on. Modified fluorescent substrate analogs, such as ε(etheno)-ATP are useful as probes of enzyme active sites. In some cases natural chromophores that are not themselves particularly fluorescent, such as retinal in rhodopsin, can serve as quenchers of the fluorescence of added probe molecules.

Bioluminescence occurs in a wide variety of biological organisms. Although fireflies are among the most spectacular of these species, by far the greatest number of bioluminescent species occur in the ocean. The luminescence is produced biochemically and usually does not require prior illumination. It serves a variety of purposes, including visual assistance in minimum-light environments, communication for social or sexual purposes, as camouflage, and for repelling predators. The full role of light emission is only beginning to be appreciated.

In many respects fluorescence or luminescence spectroscopy is even richer than absorption in the variety of information and the range of sensitivity that it can provide. As with absorption, changes in the shape or position of the fluorescence spectrum reflect the environment in which the fluorescing molecule (*fluorophore*) exists. In addition, the intensity of fluorescence or the fluorescence yield (per photon absorbed) can vary over a large range. In some cases changes of 100-fold or more can be produced. These changes in fluorescence yield are accompanied by changes in the fluorescence lifetime that can be measured directly. The polarization and depolarization of fluorescence reflect the relative immobilization of the fluorophore, the extent to which the excitation is transferred to other molecules, and relative conformational changes among the different parts of a complex fluorophore. Quenching of fluorescence by added molecules provides evidence of geometrical relations, of the accessibility of the surface of organelles, and of the presence of intermediates such as triplet states or free radicals. In carefully designed experiments fluorescent properties can be used as an accurate meter stick to measure distances at the molecular level.

Excited-state Properties

The origin of fluorescence can be seen by reference to Fig. 10.3. One of the principal paths by which the energy of electronic excited states may be released is by the emission of radiation. Kinetically, the process is commonly of first order and is characterized by a rate law of the form

$$F \propto -\frac{d[\mathrm{M}^*]}{dt} = k_d[\mathrm{M}^*] \tag{10.19}$$

where F = intensity of fluorescence
$[\mathrm{M}^*]$ = concentration of the excited electronic state that undergoes fluorescence.

In accordance with this rate law, the decay is commonly exponential with time, described by a *fluorescence decay time*, τ, given by

$$\tau = \frac{1}{k_d} \tag{10.20}$$

Measurements of the decay of fluorescence of simple molecules in dilute solution usually give $\log F$ versus t plots that are linear. An example is shown for anthracene in Fig. 10.15.

If fluorescence is the only decay path for the excited state, then the decay rate constant (designated k_f) is the reciprocal of the *natural fluorescence lifetime*,

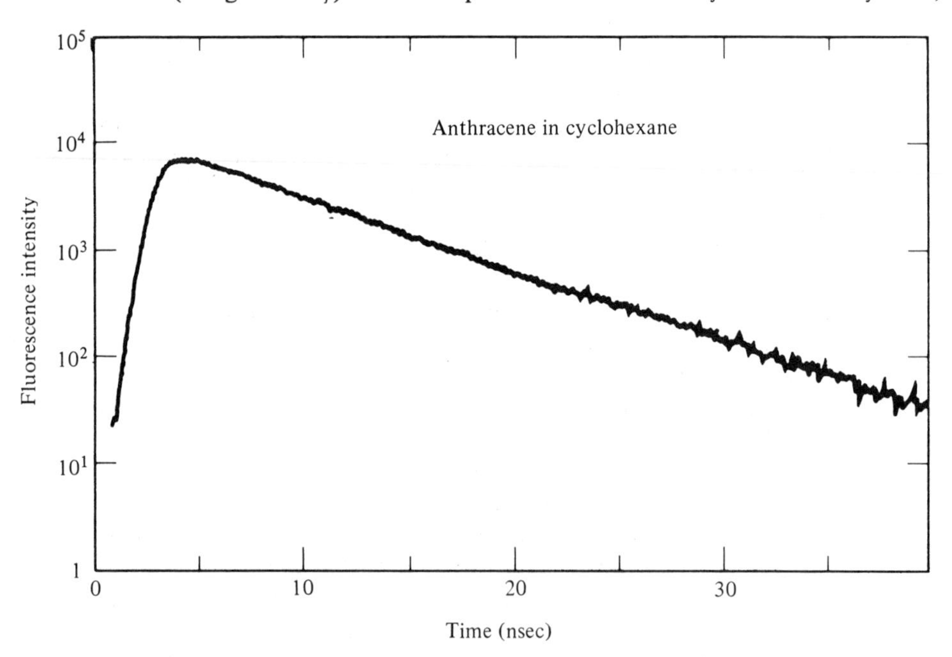

Fig. 10.15 Fluorescence decay of anthracene in cyclohexane (1.7×10^{-3} *M*), excited by a light flash of 1.4-nsec duration (half-width) and at approximately 360 nm. Emission measured at 450 nm. Logarithmic decay curve, obtained using single-photon counting method. [From P. R. Hartig, K. Sauer, C. C. Lo and B. Leskovar, *Rev. Sci. Instr. 47*, 1122 (1976).]

τ_0. Thus

$$k_f = \frac{1}{\tau_0} \tag{10.21}$$

In most cases there are significant nonradiative processes competing for the decay of the excited state. These include thermal deactivation, photochemistry, and quenching by other molecules, Q. The overall rate of decay of the excited state is, therefore, the sum of the rates of all of these processes:

$$-\frac{d[\mathrm{M}^*]}{dt} = k_f[\mathrm{M}^*] + k_t[\mathrm{M}^*] + k_p[\mathrm{M}^*] + k_Q[\mathrm{M}^*][\mathrm{Q}]$$

$$= k_d[\mathrm{M}^*] \tag{10.22}$$

where k_f, k_t, k_p, and k_Q = rate constants for fluorescence, thermal deactivation, photochemistry, and quenching, respectively, and

$$k_d = k_f + k_t + k_p + k_Q[\mathrm{Q}] \tag{10.23}$$

The observed lifetime of fluorescence will then be

$$\tau = \frac{1}{k_f + k_t + k_p + k_Q[\mathrm{Q}]} \tag{10.24}$$

The *quantum yield* of fluorescence, ϕ_f, is the fraction of absorbed photons that lead to fluoresence; it is the number of photons fluoresced divided by the number of photons absorbed. Obviously, the quantum yield is less than or equal to 1.

$$\phi_f = \frac{\text{number of photons fluoresced}}{\text{number of photons absorbed}} \tag{10.25}$$

The quantum yield can also be considered as the ratio of the rate of fluorescence to the rate of absorbance. But the rate of absorbance must equal (in the steady state) the rate of decay of the excited state. Therefore,

$$\phi_f = \frac{k_f[\mathrm{M}^*]}{k_d[\mathrm{M}^*]} \tag{10.26}$$

But, using Eqs. (10.20) and (10.21),

$$\phi_f = \frac{\tau}{\tau_0} \tag{10.27}$$

Equation (10.27) provides a relation between the quantum yield of fluorescence and the fluorescence lifetime. Table 10.5 gives a summary of fluorescence quantum yields for a variety of fluorophores commonly encountered in biological studies.

Fluorescence almost always occurs only from the lowest excited (singlet) state of the molecule. We might otherwise expect to see fluorescence from each of the excited states reached by progressively greater frequency (photon energy) of the exciting radiation absorbed by the fluorophore, but this is never the case.

Table 10.5 Fluorescence quantum yields and radiative lifetimes

Compound	Medium	τ, ns	ϕ	Reference*
Fluorescein	0.1 M NaOH	4.62	0.93	a
Quinine sulfate	0.5 M H_2SO_4	19.4	0.54	a
9-Aminoacridine	Ethanol	15.15	0.99	a
Phenylalanine	H_2O	6.4	0.004	b
Tyrosine	H_2O	3.2	0.14	b
Tryptophan	H_2O	3.0	0.13	b
Cytidine	H_2O, pH 7	—	0.03	c
Adenylic acid (AMP)	H_2O, pH 1	—	0.004	c
Etheno-AMP	H_2O, pH 6.8	23.8	1.00	d
Chlorophyll a	Diethyl ether	5.0	0.32	e
Chlorophyll b	Diethyl ether	—	0.12	e
Chloroplasts	H_2O	0.35–1.9	0.03–0.08	f
Riboflavin	H_2O, pH 7	4.2	0.26	g
DANSYL sulfonamide†	H_2O	3.9	0.55	h
DANSYL sulfonamide + carbonic anhydrase	H_2O	22.1	0.84	h
DANSYL sulfonamide + bovine serum albumin	H_2O	22.0	0.64	h

* (a) W. R. Ware and B. A. Baldwin, *J. Chem. Phys. 40,* 1703 (1964); (b) R. F. Chen, *Anal. Letters 1,* 35 (1967); (c) S. Udenfriend, *Fluorescence Assay in Biology and Medicine,* Vol. II, Academic Press, New York, 1969; (d) R. D. Spencer et al., *Eur. J. Biochem. 45,* 425 (1974); (e) G. Weber and F. W. J. Teale, *Trans. Faraday Soc. 53,* 646 (1957); (f) A. Müller, R. Lumry, and M. S. Walker, *Photochem. Photobiol. 9,* 113 (1969); (g) R. F. Chen, G. G. Vurek, and N. Alexander, *Science 156,* 949 (1967); (h) R. F. Chen and J. C. Karnohan, *J. Biol. Chem. 242,* 5813 (1967).

† DANSYL sulfonamide is 1-dimethylaminonaphthalene-5-sulfonamide.

(The molecule azulene is an example of one of the rare exceptions; azulene fluorescence comes predominantly from the second excited singlet state.) The reason that only the lowest state normally emits radiation is that the processes of internal conversion of the higher states (thermal deactivation from higher electronic states to the lowest excited state) are exceedingly rapid. This is illustrated for bacteriochlorophyll in Fig. 10.16, where the absorption and fluorescence spectra are plotted in the vertical direction (turned 90° from the usual orientation) to correspond to the energy-level diagram. Internal conversion from the lowest excited state to the ground state also occurs. It is one of the important sources of thermal deactivation, $k_t[M^*]$, that compete with fluorescence. The rate is often slower for this step, however, partly because of the greater energy separation between the ground state and the first excited electronic state compared with the energy differences among the excited states.

Fluorescence Quenching

A decrease in fluorescence intensity or quantum yield occurs by a variety of mechanisms. These include collisional processes with specific quenching

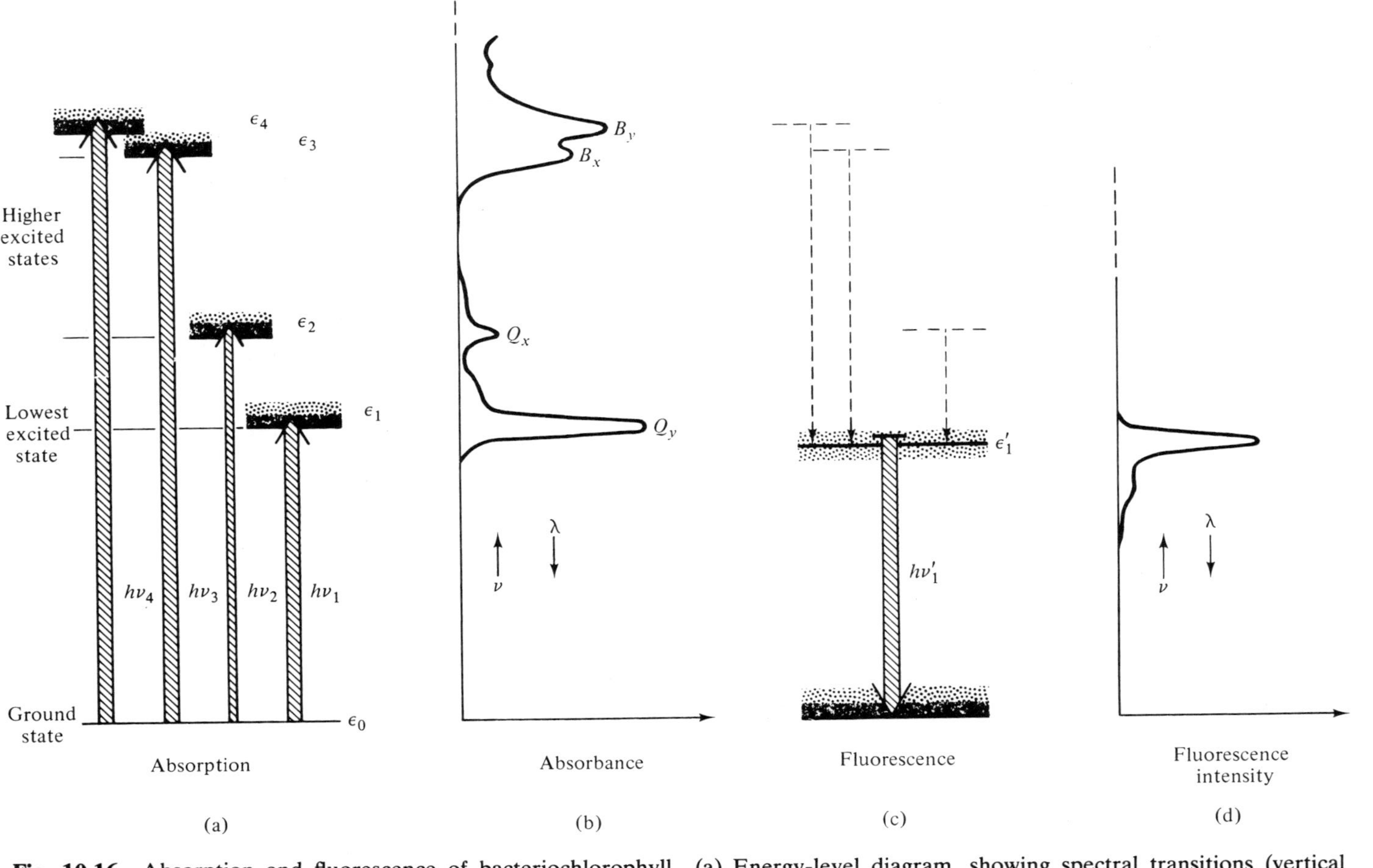

Fig. 10.16 Absorption and fluorescence of bacteriochlorophyll. (a) Energy-level diagram, showing spectral transitions (vertical arrows). The energy levels are broadened (shading) by vibrational sublevels that are not usually resolved in solution spectra. (b) Absorption spectrum corresponding to energy levels of part (a). This spectrum is turned 90° from the usual orientation to show the relation to the energy levels. (c) Radiationless relaxation (dashed arrows) and fluorescence (shaded arrow). (d) Fluorescence emission spectrum corresponding to part (c). Note the red shift (shorter shaded arrow) of the fluorescence compared with the corresponding Q_Y absorption illustrated in parts (a) and (b). (From K. Sauer, in *Bioenergetics of Photosynthesis*, Govindjee (ed.), Academic Press, New York, 1975, pp. 115–181.)

molecules, excitation transfer to nonfluorescent species, complex formation or aggregation that forms nonfluorescent species (concentration quenching), and radiative migration leading to self-absorption. There are important biological applications or consequences of each of these quenching mechanisms.

Quenching of fluorescence by added substances or by impurities can occur by a collisional process. Because it is the excited state of the fluorescent molecule that must undergo collisonal quenching, the encounters must occur frequently and the quenching process must be efficient. Molecular oxygen is one of the most widely encountered quenchers. This is because O_2 is a triplet species (two unpaired electrons) in its ground electronic state. Most excited fluorophores emit from a singlet state (no unpaired electrons), and O_2 quenches the fluorescence by means of the reaction

$$M^* \text{ (singlet)} + O_2 \text{ (triplet)} \longrightarrow M^* \text{ (triplet)} + O_2 \text{ (singlet)}$$

Normally, M^* (singlet) is fluorescent and M^* (triplet) is not.

Collisional quenching is a bimolecular process kinetically; however, the excited oxygen molecules quickly return to the ground triplet state upon subsequent collisions or interactions with the solvent. As a consequence, they are not consumed in the process, and collisional quenching obeys pseudo-first-order kinetics. For a generalized quencher molecule, Q, the relevant equations are

$$M + h\nu \longrightarrow M^* \quad \text{excitation}$$

$$M^* \xrightarrow{k_f} M + h\nu' \quad \text{fluorescence}$$

$$M^* + Q \xrightarrow{k_Q} M + Q^* \quad \text{quenching}$$

In the absence of quenchers,

$$\phi_f^0 = \frac{k_f}{k_f + k_t + k_p} \tag{10.28}$$

In the presence of a quencher at concentration [Q],

$$\phi_f = \frac{k_f}{k_f + k_t + k_p + k_Q[\mathrm{Q}]} \tag{10.29}$$

Therefore, we obtain a result known as the Stern-Volmer relation

$$\frac{\phi_f^0}{\phi_f} = 1 + \frac{k_Q[\mathrm{Q}]}{k_f + k_t + k_p} = 1 + K[\mathrm{Q}] \tag{10.30}$$

We can also write the equation in terms of the lifetime, τ', in the absence of quencher.

$$\frac{\phi_f^0}{\phi_f} - 1 = k_Q\tau'[\mathrm{Q}] \tag{10.31}$$

where

$$\tau' = \frac{1}{k_f + k_t + k_p} \tag{10.32}$$

Because the intensity of fluorescence, F, is proportional to the quantum field, ϕ_f, a plot of F^0/F versus [Q] will give a straight line with slope $k_Q\tau'$. Thus, the longer the lifetime of the excited state, the greater is the probability of quenching.

The consequences of concentration quenching can be quite dramatic. The fluorescence of benzene in oxygen-free solutions occurs with a lifetime of 29 ns; in a solution in equilibrium with O_2 at 1 atm pressure, this lifetime is decreased fivefold, to 5.7 ns. Chlorophyll a is strongly fluorescent in dilute solutions ($\phi_f \cong 0.3$), but the fluorescence intensity is quenched essentially to zero with added quinones or carotenoids. (In this case, the mechanism may involve complex formation between chlorophyll and the quencher molecule.) It is clear from these examples that great care must be taken to remove all extraneous quenching species in the determination of the intrinsic properties of fluorescing molecules.

Concentration quenching may occur as a consequence of aggregation, dissociation, or other changes in the fluorophore itself. *Excimers* (excited dimers) may form because of greater interactions of the excited-state species. In each case there will result a concentration-dependent quantum yield of fluorescence. For example, if nonfluorescent excimers form according to

$$\text{M}^* + \text{M} \xrightarrow{k_e} [\text{M} \cdot \text{M}]^*$$

then

$$\phi_f = \frac{k_f}{k_f + k_t + k_p + k_e[\text{M}]} \tag{10.33}$$

Chlorophyll a at concentrations of about 10^{-2} exhibits concentration quenching in most solvents. Depending on the particular medium, the aggregated chlorophylls may be completely nonfluorescent or they may have a weak but distinctive fluorescence of their own. In photosynthetic membranes the chlorophyll concentrations locally are typically 0.05 to 0.1 *M* and fluorescence yields are only about one tenth of the monomer value.

Excitation transfer processes, which will be discussed next, provide additional paths for fluorescence quenching.

Excitation Transfer

Some of the most valuable applications of fluorescence to biochemical systems involve the transfer of excitation from one chromophore to another. Because such transfer processes depend strongly on the distance between the chromophores and on their relative orientations, experiments can be designed to obtain useful information concerning macromolecular geometry. Excitation transfer has been treated theoretically by a number of authors (see Knox, 1975), and several of the most important relations have been verified quantitatively in carefully designed model experiments.

We shall consider the transfer of excitation from an excited donor molecule, D*, to an acceptor, A, which then fluoresces. The efficiency of excitation can be measured by the quenching of fluorescence of the donor in the presence of acceptor. Both the quantum yield and lifetime of the donor will be decreased. Alternatively, the sensitized fluorescence of the acceptor can be measured. The mechanism of transfer depends on the distance between donor and acceptor. In the range from about 1 to 10 nm, *resonance energy transfer* (*Förster transfer*) occurs. For each donor-acceptor pair, the efficiency of transfer depends on r^{-6}, where r is the distance between them. The energy transfer efficiency is

$$Eff = \frac{r_0^6}{r_0^6 + r^6} \tag{10.34}$$

where Eff = efficiency of transfer ($0 \leqslant Eff \leqslant 1$)
r_0 = characteristic distance for the donor-acceptor pair; it is the distance for which $Eff = 0.5$
r = distance between donor and acceptor

The value of r_0 depends on the amount of overlap between the fluorescence spectrum of the donor and the absorption spectrum of the acceptor. It also depends on the angular orientation between donor and acceptor. Once r_0 is known, r can be obtained from a measured efficiency. The efficiency can be obtained from the change of fluorescent lifetime of the donor caused by the presence of the acceptor:

$$Eff = \frac{\tau_D - \tau_{D-A}}{\tau_D} \tag{10.35}$$

where τ_D = fluorescence lifetime of the donor
τ_{D-A} = fluorescence lifetime of the donor in the presence of the acceptor

Molecular Rulers

Extensive experimental tests of resonance or Förster transfer have been carried out. Förster (1949) verified the expected concentration dependence in a study of the quenching of fluoresence of trypaflavin in methanol by the dye rhodamine B. In this case a value $r_0 = 5.8$ nm was obtained, which shows that transfer can occur over distances several times larger than the actual molecular dimensions. The dependence on the inverse sixth power of distance has been tested in an elegant series of experiments by Latt et al. (1965) and by Stryer and Haugland (1967). In these studies two chromophores are attached covalently to a rigid molecular framework. Then excitation transfer is measured for a series of synthetic species where the distance between the donor and acceptor molecule is different. Stryer and Haugland examined transfer from a naphthyl group at one end to a dansyl group at the other end of oligoprolines with 1 to 12 monomer units in the rigid chain. The results showed excellent agreement with the r^{-6} dependence from 1.2 to 4.6 nm separation, as shown in Fig. 10.17.

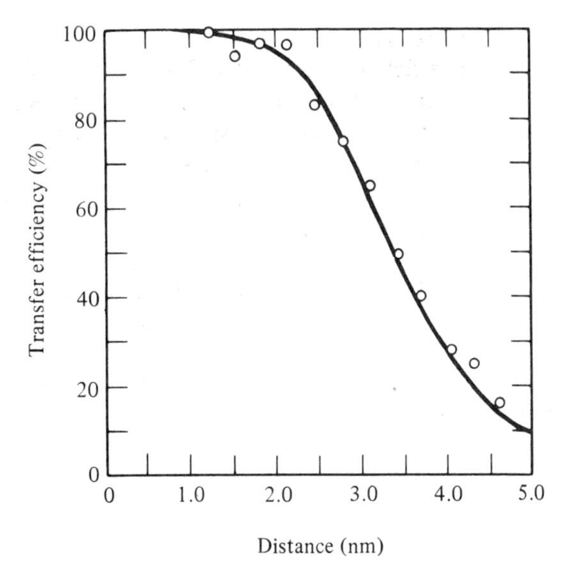

Fig. 10.17 Efficiency of energy transfer as a function of distance in dansyl-(L-prolyl)$_n$-α-naphthyl, $n = 1$ to 12. Energy transfer is 50% efficient at 3.46 nm. Solid line corresponds to r^{-6} dependence. [From L. Stryer and R. P. Haugland, *Proc. Natl. Acad. Sci. U.S. 58*, 719 (1967).]

An important determinant of excitation transfer is the spectral overlap between the donor emission and the acceptor absorption spectrum. This overlap was varied fortyfold by solvent effects for a modified steroid by Haugland et al. (1969), and the transfer rate varied almost in parallel. These extensive tests have confirmed the validity of the Förster resonance transfer mechanism, at least over distances in the range 1 to 10 nm.

Applications of excitation transfer to biochemical questions are rapidly increasing in number. One of the first attempts to measure distances in this way was carried out by Beardsley and Cantor (1970), who labeled the aminoacyl attachment site of a transfer RNA from yeast. The particular transfer RNA studied (specific for phenylalanine attachment) has an unusual fluorescent base, the Y base, located next to the anticodon region. Using several excitation transfer acceptor labels, they were unable to demonstrate significant quenching of the Y-base fluorescence and concluded that the anticodon and the aminoacyl attachment site are at least 4.0 nm from one another. In the subsequent structure determination, the distance was found to be 7.0 nm in the crystal.

Phosphorescence

An excited singlet state (with all electrons paired) can become an excited triplet state (with two unpaired electrons). The triplet state is of lower energy than the singlet, but the probability of a singlet-triplet transition is usually small. It is a forbidden transition. The singlet-triplet conversion is catalyzed by certain molecules, such as O_2, which is a triplet in the ground state, and by species with a high atomic number, such as I^-. Once a molecule is in an excited triplet state, it can emit light and return to its ground state. *The emission of light from an excited triplet state is phosphorescence*. Quantum yields and lifetimes for phosphorescence are defined just as for fluorescence. Experimentally, phosphorescence is usually distinguished from fluorescence by its lifetime. Phosphorescence lifetimes occur in the millisecond and above range while fluorescence lifetimes are in the microsecond and below range.

At room temperature DNA does not phosphoresce; the triplet state is quickly quenched. However, in frozen solution DNA shows a characteristic phosphorescence due to thymidylic acid. The triplet states of the other nucleotides are either quenched or they transfer excitation to the thymidylic nucleotides, which then emit the phosphorescence.

OPTICAL ROTATORY DISPERSION AND CIRCULAR DICHROISM

A property of most biological molecules is molecular asymmetry or *chirality*; such molecules are not identical to their mirror images. The simplest examples result from the occurrence of asymmetric carbon atoms in these molecules. A carbon that is tetrahedrally bonded to four different atoms or groups can exist in two different structures that are mirror images of one another, as shown in Fig. 10.18.

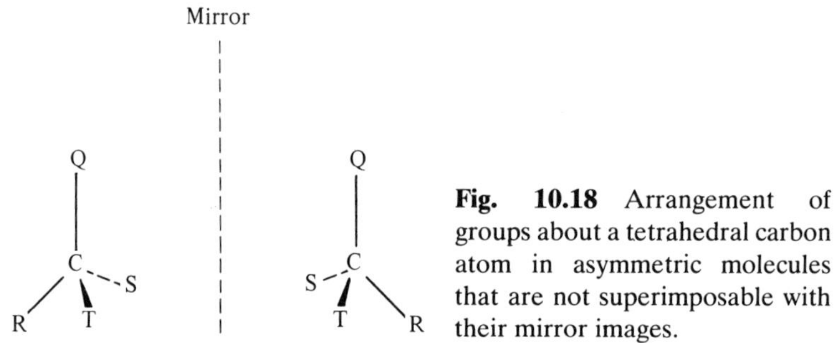

Fig. 10.18 Arrangement of groups about a tetrahedral carbon atom in asymmetric molecules that are not superimposable with their mirror images.

They are frequently referred to as left- and right-handed forms. The rules for characterizing such molecules are described in organic chemistry texts.

Chiral molecules have distinctive properties that bear emphasis. First, consider some common examples of such "handedness" that are more familiar. Screws or nuts and bolts can be cut with a right-hand or a left-hand thread. Even if the heads of the bolts are cut off, the threaded shanks are still distinguishable left from right on the basis of the handedness of the thread cutting. This emphasizes the fact that there is no necessary spatial orientation dependence in making the handedness distinction. A tumbled assortment of left- and right-handed bolt shanks could be sorted by using a test nut. Once the shanks are separated into different bags, they will have different interactions with nuts of a given handedness. In much the same way, chiral molecules in solution and with random, constantly changing orientations will have a different interaction with light that is polarized in a chiral way (circularly polarized light).

These examples emphasize also that chirality of materials does not necessarily depend on the presence of asymmetric carbon atoms, although that is commonly the origin for small molecules. In the case of nucleic acids, an important source of chirality is whether the helical polynucleotide winds in a left- or a right-handed sense. Nucleic acids, and proteins in the α-helical conformation, wind in the right-handed sense. The origin of this handedness preference lies in the fact that the chiral monomer units found in them are all of one type: the L (levo or left) amino acid isomer for proteins and the D (dextro) sugar for nucleic acids. The overall structure of biologically active proteins (enzymes, antibodies) depends on additional determinants of tertiary structure that are not so easily characterized. These are very important, however, for they serve to make the detailed surface structure of an enzyme or antibody nonsuperimposable with its mirror image. As a consequence, the active sites of these molecules are able to distinguish between mirror-image substrate molecules, much as left-handed nuts will interact only with left-handed bolts. Enzymes that are involved with the synthesis of chiral molecules may make a single chiral isomer from a substrate that is not itself asymmetric. It is clear from these few examples that the property of chirality is of fundamental and wide-reaching importance in biology.

Polarized Light

As mentioned above, materials that are asymmetric in structure also exhibit chirality in their optical properties. The two most commonly studied are *optical rotation*, which depends on the difference in refractive index of a substance for left- and right-circularly polarized light, and *circular dichroism* (CD), which is the difference in absorption of the two circular polarization components. The spectrum of optical rotation is known as *optical rotatory dispersion* (ORD).

To understand the phenomena of optical activity, we need to appreciate the characteristics of circularly polarized light. Electromagnetic radiaton can be described by an electric field vector that oscillates with a characteristic frequency or wavelength. It moves through a medium with a velocity $v = c/n$, where c is the velocity of light in a vacuum and n is the refractive index of the material. For *unpolarized* light, the electric vector may oscillate in any direction perpendicular to the direction of propagation. For a large number of photons in an unpolarized beam, all directions are equally represented. The electric vectors can be pictured as radiating spokes on a many-spoked wheel, as shown in Fig. 10.19 (top). For *plane-polarized* light, the electric vector oscillates in a single plane that includes the propagation direction. A pictorial description of vertically plane-polarized light is shown in Fig. 10.19 (middle). For an observer moving at the photon's velocity, the electric vector appears to be oscillating back and forth along a line. For this reason, plane-polarized light is sometimes referred to as being linearly polarized. *Circularly polarized* light propagates so that the tip of its electric vector sweeps out a helix (Fig. 10.19, bottom). To the observer moving with the photons, the electric vector appears to be moving in a circle, like the hands of a

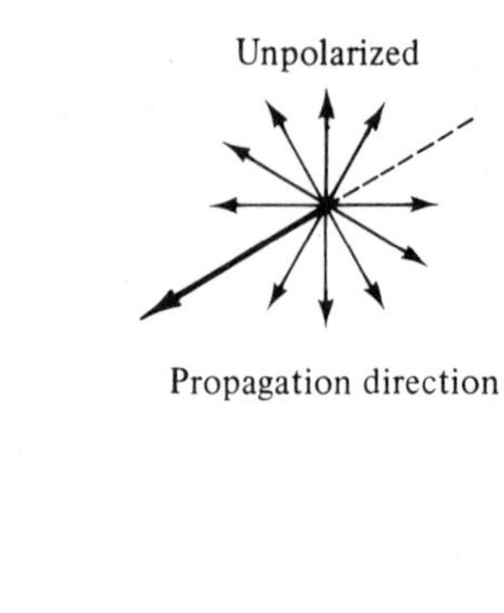

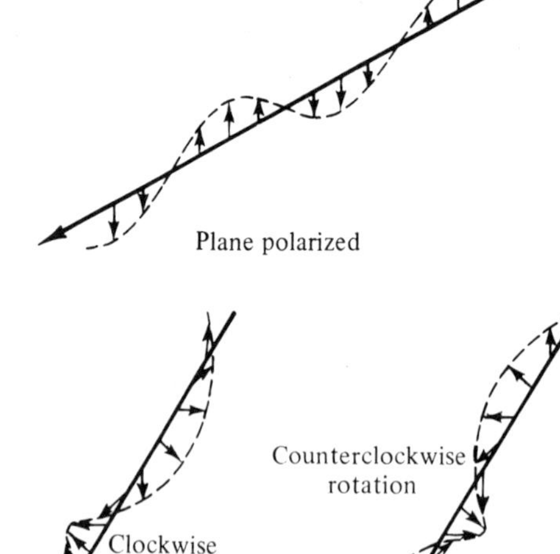

Fig. 10.19 Change in electric vector orientation during propagation of light that is unpolarized (top), plane-polarized (middle), or circularly polarized (bottom).

clock. The convention is that for left-circularly polarized light, the electric vector moves counterclockwise as the light comes toward the observer; for right-circularly polarized light, it moves clockwise from the point of view of the observer. Elliptically polarized light, which we shall not consider in detail here, propagates such that its electric vector sweeps out an ellipse, as seen by the observer moving with the light. Elliptically polarized light is either right- or left-handed, and it may be thought of as intermediate between circularly and plane-polarized. In fact, optical retardation plates will progressively produce plane, elliptical, circular, elliptical, plane, elliptical, and so on, polarizations as the thickness of the retarder is increased.

It is useful to consider plane-polarized light as the resultant of two (hypothetical) circular polarized components of opposite handedness but propagating exactly in phase with one another. The two opposite circularly polarized components will add to give a plane-polarized resultant. This can be seen in Fig. 10.19 (bottom) by adding vectorially the right- and left-circularly polarized electric vectors at corresponding points along the propagation vector. The resultant electric vector at each point will have only a vertical component; the horizontal components will always cancel exactly. For this example the resultant is vertically plane-polarized.

We have described previously materials that are (linearly) birefringent toward plane-polarized light. These materials must be geometrically anisotropic. Birefringence occurs because the refractive index, and hence the light propagation velocity, is different for polarization planes oriented in different directions on the incident surface. Substances that are optically active (chiral) exhibit *circular birefringence,* where circular birefringence is $n_L - n_R$, the difference between the refractive index for left-circularly polarized and the refractive index for right-circularly polarized light. In other words, the velocities of propagation for left- and right-circular polarized light are different. There is an important difference between circular birefringence and birefringence toward plane-polarized light. A homogeneous chiral sample, such as a solution of an optically active solute, is geometrically isotropic, in the sense that its optical properties are identical viewed from any direction. The circular birefringence results from an intrinsic property of the material that persists even though the orientations of the molecules are random. Chiral molecules also show *circular dichroism* ($A_L - A_R$), a difference in absorbance for left- and right-circularly polarized light.

A quantum mechanical derivation shows how circular birefringence or dichroism is related to electronic wavefunctions of the ground and excited states of the molecule. An optically active transition must have an electric-dipole transition moment and a magnetic-dipole transition moment that are not perpendicular to each other. This can occur only for molecules that are not mirror images. A qualitative understanding of the electronic motion in these molecules is that the electrons do not move in a line or in a circle; instead, they move in a helical path.

Optical Rotation

Optical rotation by chiral samples results from and is a measure of their circular birefringence. The term "optical rotation" comes from the usual experimental measurement procedure. If plane-polarized light is propagated through a transparent chiral sample, the emerging beam will also be plane-polarized, but its plane of polarization (still including the direction of propagation) will be rotated by an angle ϕ with respect to the direction of the polarization of the incident light. If the rotation is clockwise as seen by the observer, ϕ is positive; counterclockwise rotation is assigned a negative value of ϕ (Fig. 10.20). The origin of optical rotation can be rationalized by considering the incident plane-polarized light to be made up of two opposite circularly polarized, but in-phase, components, as mentioned above. Because the chiral sample is circularly birefringent, the two circular-polarized component electric vectors will propagate through the sample

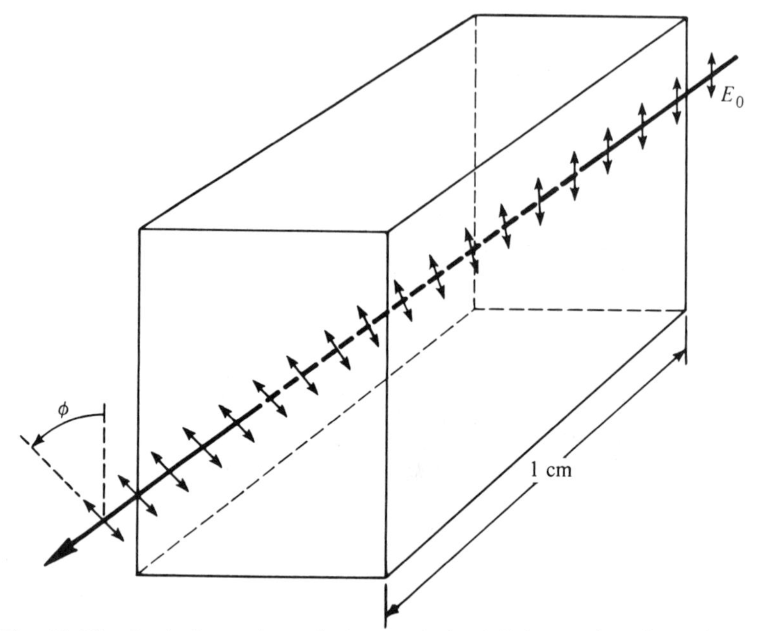

Fig. 10.20 Optical rotation of plane-polarized light passing through a chiral sample. For the example shown the sign of the rotation angle, ϕ, is negative.

with different velocities. As a consequence, one of them becomes advanced in phase with respect to the other. The resultant, which remains plane-polarized, becomes rotated progressively upon passage through the sample. When it emerges from the other side, it has undergone a rotation that can be shown to be given by

$$\text{rotation (rad cm}^{-1}) = \phi = \frac{\pi}{\lambda}(n_L - n_R) \tag{10.36}$$

where λ is the wavelength of the light in vacuum. Note that ϕ is given per centimeter of pathlength in the sample. The actual angle of rotation clearly increases linearly with the path length through the sample. The rotation is measured by placing a polarizing element (suitable crystal or Polaroid sheet) called an *analyzer* in the emergent beam. By turning the analyzer so that it is crossed (perpendicular) to the direction of polarization of the emergent beam, the intensity is extinguished and the angle of rotation can be measured to better than 1 millidegree. To appreciate how sensitively circular birefringence can be measured, consider Eq. (10.36) for a measurement made at $\lambda = 314$ nm ($= \pi \times 10^{-5}$ cm). A rotation of 1 rad cm^{-1} (1 rad = 57.3 degrees) corresponds to $n_L - n_R = 10^{-5}$.

Circular Dichroism

Materials that exhibit (linear) dichroism toward plane-polarized light do so because they have different absorbances as a function of the orientation of the sample with respect to the polarization direction of the incident light. Polaroid sheets are examples of dichroic materials. As with birefringence, linear dichroism results from geometric anisotropy in the sample.

Circular dichroism, analogously, results from a differential absorption of left- and right-circularly polarized light by a sample that exhibits molecular asymmetry. A simple expression of the circular dichroism is given by

$$\Delta A = A_L - A_R \tag{10.37}$$

where A_L and A_R are the absorbances of the sample for pure left- and right-circularly polarized light, respectively. As with circular birefringence, circular dichroism may be either positive or negative. Circular dichroism occurs only in a region of the spectrum where the sample absorbs, whereas circular birefringence occurs in all wavelength regions of an opticaly active substance. This latter property is of obvious advantage for transparent substances such as sugars, in which the lowest energy electronic transitions occur in the far ultraviolet.

Referring again to Fig. 10.20, the passage of plane-polarized light, E_0, through a circularly dichroic sample produces not only a phase shift due to the circular birefringence (which occurs in the absorption region as well) but also a differential decrease of the amplitudes (the electric vector magnitudes) of the left- and right-circularly polarized components. The emerging beam, E, is found to be elliptically polarized (Fig. 10.21) as a consequence. A detailed analysis shows that the ellipticity is given by

$$\theta \text{ (rad cm}^{-1}) = \frac{2.303(A_L - A_R)}{4l} \tag{10.38}$$

where l is the path length through the sample.

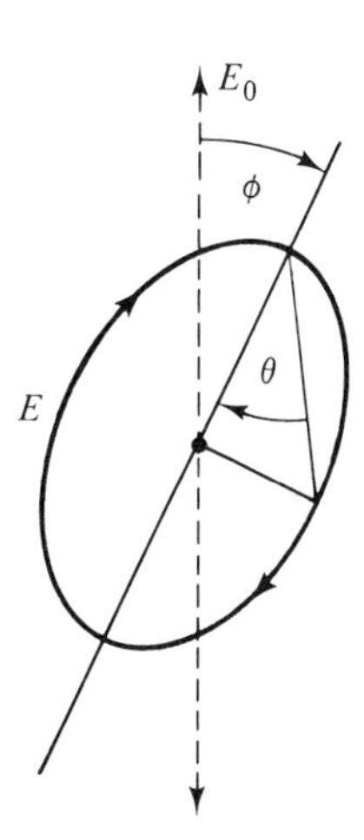

Fig. 10.21 Elliptically polarized light emerging toward the observer from a circularly dichroic sample. The sign convention is that ϕ is positive for clockwise rotation of the major axis and θ is positive for right-elliptically polarized light, as shown. The rotation angle is also called α, the ellipticity angle is also called ψ.

Experimental measurements of optical rotation and ellipticity are burdened with a history of measurements in cells of length 10 cm (path length d, in decimeters) and the use of different symbols for the same quantity. The equations summarized in Table 10.6 represent a consensus of current usage.

Table 10.6 Summary of experimental parameters for optical rotation and circular dichroism*

Optical rotation

$$\text{Specific rotation} = [\alpha] = \frac{\alpha}{dc} \quad (10.39)$$

$$\text{Molar rotation} = [\phi] = \frac{M[\alpha]}{100} = \frac{100\alpha}{lm} \quad (10.40)$$

Circular dichroism

$$\text{Specific ellipticity} = [\psi] = \frac{\psi}{dc} \quad (10.41)$$

$$\text{Molar ellipticity } [\theta] = \frac{M[\psi]}{100} = \frac{100\psi}{lm} \quad (10.42)$$

$$\text{Circular dichroism} = \Delta\varepsilon = \varepsilon_L - \varepsilon_R = \frac{A_L - A_R}{lm} \quad (10.43)$$

$$[\theta] = 3298(\varepsilon_L - \varepsilon_R) \quad (10.44)$$

* α = rotation angle, degrees
ψ = ellipticity, degrees
ε = molar absorptivity
M = molecular weight
d = path length, dm
l = path length, cm
c = concentration, g cm^{-3}
m = concentration, mol ℓ^{-1}

CIRCULAR DICHROISM OF NUCLEIC ACIDS AND PROTEINS

The optical activity of a nucleic acid or protein is the sum of the individual contributions from the monomeric units and the contribution from their interactions in the polymeric arrangement. By comparing the circular dichroism (CD) of the native polymer to that of its monomeric units, their separate contributions can be determined experimentally and subtracted from the optical activity of the intact polymer. The difference then gives the contribution from the interactions in the native polymer conformation. This is illustrated for DNA and RNA in Fig. 10.22. The dramatic differences between the solid and dashed curves demonstrate that the principal contribution to the CD of the nucleic acids arises in the conformational structure of the polymer. This property is of great diagnostic value and indicates the usefulness of CD measurements in nucleic acid research. A similar situation occurs for proteins, with the added advantage that several forms of secondary structure are commonly present in native proteins. These elements can be characterized as predominantly random-coil, α helix, or β chain (pleated sheet). By measuring homopolypeptides under conditions where the conformation is uniform throughout the polymer, Gratzer and Cowburn (1969) prepared plots of ORD and CD spectra in the peptide region that show the range of values encountered for polypeptides of different amino acids (Fig. 10.23). The

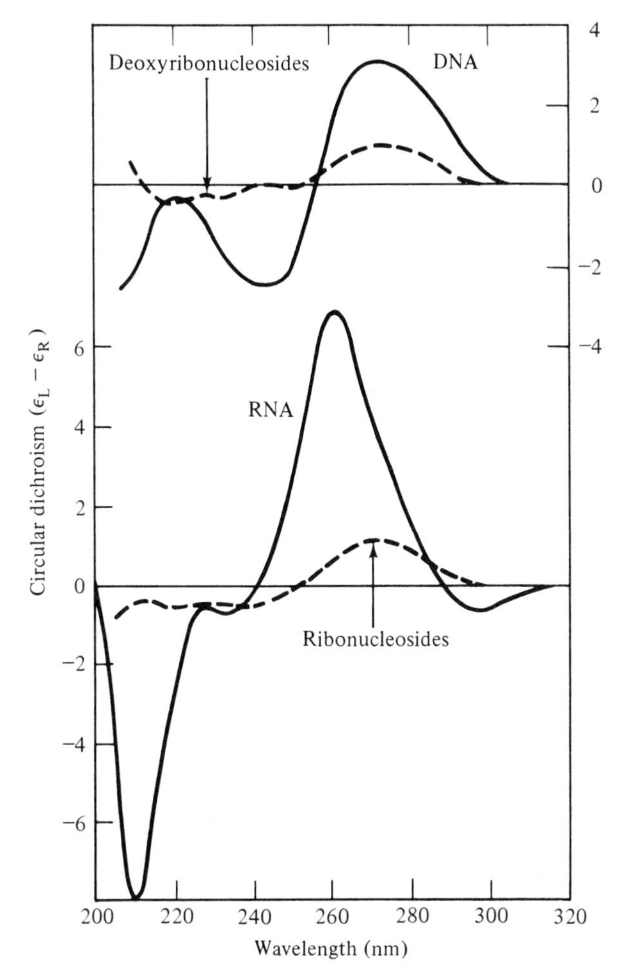

Fig. 10.22 Circular dichroism of double-stranded DNA and RNA compared with their component mononucleosides. *M. Lysodeikticus* DNA data from F. Allen et al., *Biopolymers 11*, 853 (1972). Rice dwarf virus RNA data from T. Samejima et al., *J. Mol. Biol. 34*, 39 (1968). Nucleoside spectra calculated from the base composition (72% G+C for the DNA; 44% G+C for the RNA) and CD data of C. R. Cantor et al., *Biopolymers 9*, 1059, 1079 (1970). [From V. A. Bloomfield, D. M. Crothers, and I. Tinoco, Jr., *Physical Chemistry of Nucleic Acids*, Harper & Row, New York, 1974, p. 134.]

ranges are relatively small compared with the differences among the CD spectra of the different conformations. The poorest reliability is expected for random conformations, especially when they occur in the segments of proteins. These include connecting links between helical or β-sheet structures, bends or loops at the exterior surface of the protein, flexible ends of the peptide chains, and interactions with prosthetic groups such as hemes or flavins incorporated in the structure.

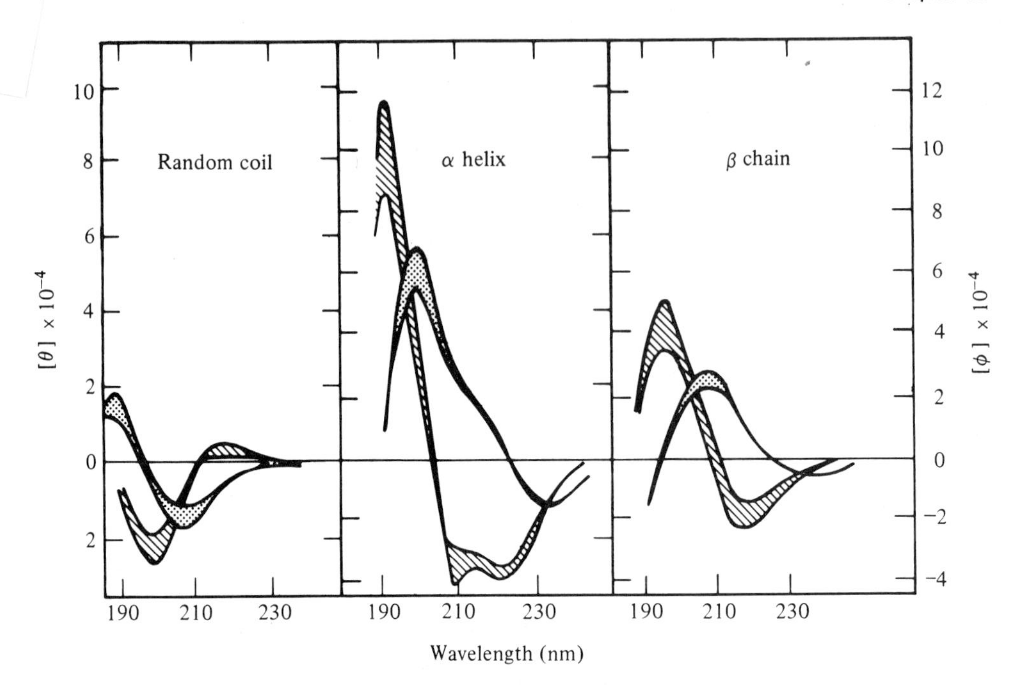

Fig. 10.23. CD (crosshatched) and ORD (dotted) of homopolypeptides in the random coil, α-helical, and β conformations. The shaded areas indicate the range of values. [From W. B. Gratzer and D. A. Cowburn, *Nature 222*, 426 (1969).]

Induced Circular Dichroism of Chromophores

The CD spectra of chromophores in biochemical systems can indicate interactions with molecules in the immediate vicinity. A nice illustration of this characteristic is the phenomenon of induced CD. If a symmetric dye molecule or other absorbing solute is dissolved in a transparent chiral solvent, such as amyl alcohol or menthol, then circular dichroism is observed in the region of the absorption band of the solute. This induced CD is not present if the same chromophore is dissolved in nonchiral solvents. The induced CD results from an interaction of the transition moment of the chromophore with the asymmetric molecules in its environment (Hayward and Totty, 1971).

Certain dye molecules, such as acridine, proflavin, and ethidium, exhibit strong binding to biopolymers, such as proteins, nucleic acids, or polysaccharides in aqueous solutions. Such interactions can lead to induced CD in the absorption region of the bound dye. Many biological molecules of interest contain native chromophores that exhibit induced CD. Examples in which the separated chromophore itself is nonchiral include hemoglobin, myoglobin, cytochromes, hemocyanin, rhodopsin, ferredoxins, transferrin, copper-proteins, vitamin B_{12}, and flavoproteins. The absorption and CD of frog rhodopsin are shown in Fig. 10.24. This example is particularly illustrative because, upon bleaching of the rhodopsin by light, both the absorption and the CD (bottom curve in Fig. 10.24) in the visible region disappear. The isolated chromophore, retinal, does not exhibit any measurable circular dichroism, as expected for this symmetric molecule.

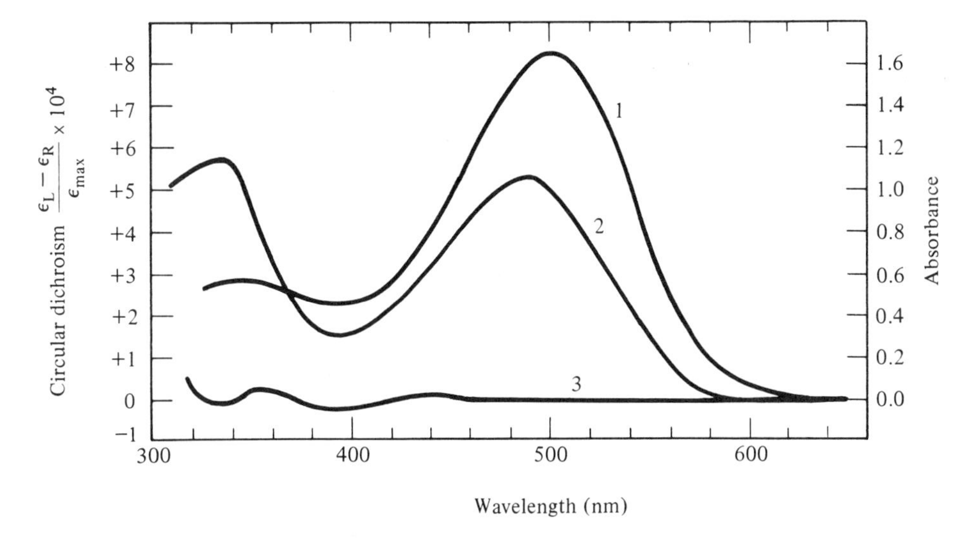

Fig. 10.24. Circular dichroism of rhodopsin from *Rana pipiens*. Curve 1: absorbance curve of rhodopsin in 2% digitonin, pH = 6.7. Curve 2: circular dichroism of same solution expressed in terms of a rhodopsin with unit absorbance at the maximum; 10-mm pathlength. Curve 3: Same solution after exposure to tungsten light of 60 W until all purple had faded. [From F. Crescitelli, W. F. H. M. Mommaerts, and T. I. Shaw, *Proc. Natl. Acad. Sci. U.S. 56*, 1729 (1966).]

NUCLEAR MAGNETIC RESONANCE

We have been discussing electronic spectroscopy in which the energy levels correspond to different arrangements of electrons in the molecule. We shall now describe a spectroscopic method in which the energy levels correspond to different orientations of the magnetic moment of a nucleus in a magnetic field.

Some nuclei have magnetic dipole moments that will orient in a magnetic field. The possible energies of the nuclear magnetic moment in the magnetic field are of course quantized. Therefore, characteristic frequencies of electromagnetic radiation are absorbed. The frequencies depend on the nucleus, the strength of the magnetic field, and, most important to us, the chemical and magnetic environment of the nucleus. Nuclei with an even number of protons and an even number of neutrons have zero nuclear magnetic moments; no nuclear magnetic spectra are possible. These nuclei include many isotopes that commonly occur in nature: $^{12}C_6$, $^{16}O_8$, and $^{32}S_{16}$. (The atomic number is the subscript; the mass number is the superscript.) Other nuclei have magnetic moments and their nuclear magnetic resonance (nmr) can be studied. The simplest nuclei to consider have a spin of $\frac{1}{2}$; these include $^{1}H_1$, $^{13}C_6$, $^{15}N_7$, $^{19}F_9$, and $^{31}P_{15}$. (The electron also has a spin of $\frac{1}{2}$ and a magnetic moment that allows study of electron spin resonance in a magnetic field.) Nuclei with spins greater than $\frac{1}{2}$, such as $^{14}N_7$, have more complicated spectra, which we will not discuss. The most extensively studied nucleus has been the proton. Recently, $^{13}C_6$ nmr has become important, despite the fact that the natural abundance of the ^{13}C isotope is only about 1%.

The reason that the magnetic spectrum of nuclei of spin $\frac{1}{2}$ are easy to interpret is that the nucleus can have only two orientations in the magnetic field: with the field or against the field. The magnetic quantum number, which specifies the orientation, can have only the values of $\pm\frac{1}{2}$. Therefore, an isolated nucleus of spin $\frac{1}{2}$ in a magnetic field will have a single absorption frequency, as shown in Fig. 10.25. The absorption frequency, ν_0, is

$$h\nu_0 = 2\mu_m H_0 \tag{10.45}$$

where μ_m is the component of the nuclear magnetic moment along the direction of the magnetic field H_0. It is characterisic of each different nucleus.

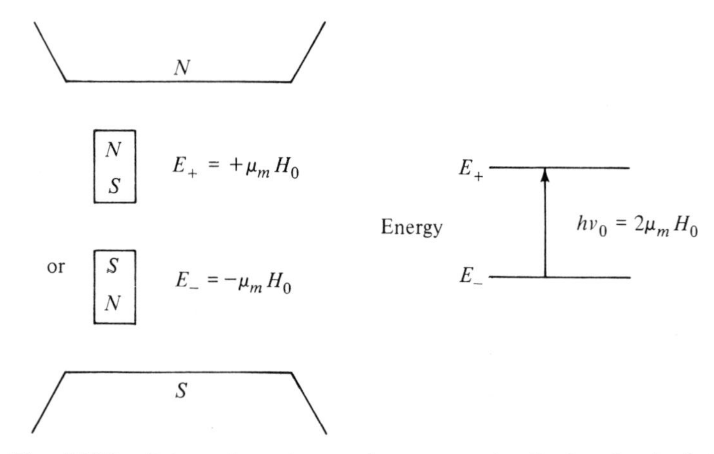

Fig. 10.25 Orientation of a nuclear magnetic dipole of spin $\frac{1}{2}$ in a magnetic field. The resonance condition for absorption of radiation ($h\nu = 2\mu_m H_0$) shows that the nmr spectrum will depend both on the applied magnetic field and the properties of the nucleus.

Chemical Shifts and Spin-Spin Splitting

The fact that makes nmr useful is that the nuclei in molecules are not isolated. They interact with the surrounding electrons and with the other nuclei in the molecule. The applied magnetic field induces a magnetic moment in the electrons

usually field flips protons or nuclear spins

which is in the opposite direction to the applied field. The electrons thus shield the nucleus from the magnetic field; the shielding is small but measurable. The amount of shielding is characterized by a unitless shielding contant, σ. The resonance condition is now

$$h\nu = 2\mu_m H_0(1 - \sigma) \tag{10.46}$$

Each magnetically distinct nucleus in a molecule will have a different amount of shielding and thus a different value of σ. An nmr spectrum can be measured either by keeping the magnetic field H_0 constant and varying the frequency of the radiation, or the frequency can be held constant and the magnetic field varied. It is impractical to measure σ directly, so a relative shielding is measured. A standard molecule (such as tetramethylsilane) is introduced into the nmr spectrometer, and resonance frequencies (or magnetic fields) are measured relative to it. A chemical shift can be measured as a change in frequency for the nucleus relative to the standard. The shift is quoted in hertz (Hz) as upfield (toward a higher magnetic field) or downfield (toward a lower magnetic field) relative to the standard. The chemical shift in Hz is directly proportional to the applied magnetic field; therefore, the spectrometer frequency must also be given. To avoid an instrument-dependent parameter, a unitless chemical shift, δ, has been defined as

$$\delta\ (\text{ppm}) \equiv \frac{\nu\ (\text{sample, Hz}) - \nu\ (\text{reference, Hz})}{\nu\ (\text{spectrometer, MHz})} \tag{10.47}$$

The spectrometer frequency is an instrument characteristic which depends on the magnitude of the magnetic field and the nucleus of interest. There is some disagreement in the literature, but the main convention is to have δ *positive* for *downfield* shifts in frequency. The chemical shift, δ, is a unitless number given in parts per million (ppm). It is defined to be independent of applied magnetic field and characteristic of the molecule. The chemical shift of a nucleus depends on many factors, but the electron density at the nucleus is often dominant. A high electron density causes a large shielding and means that the applied magnetic field must be increased to obtain resonance. This upfield shift means a decrease in magnitude of δ. Conversely, a low electron density at a nucleus means a downfield shift and an increase in δ. For example, the chemical shifts of the methyl protons (underlined below) in CH_3X relative to tetramethylsilane become larger as X becomes a better electron-withdrawing group: $\delta(C\underline{H_3}{-}CH_3) = 1$, $\delta(C\underline{H_3}{-}C_6H_5) \approx 2$, $\delta(C\underline{H_3}{-}OH) \approx 4$. A proton attached directly to an electronegative atom such as in a carboxyl group has a very low electron density (which is why it is acidic); it can have a $\delta(COO\underline{H}) \approx 10$. The range of 0 to 10 ppm nearly covers the range of δ for protons. The chemical shift for protons ranges from 0 to 600 Hz for a 60-MHz spectrometer and up to 3600 Hz for a 360-MHz spectrometer.

The chemical shift is a measure of nuclear-electron interactions. The *spin-spin splitting*, J, is a measure of the interaction of two or more nuclei, transmitted by the electrons. Only nonequivalent nuclei (nuclei with different chemical shifts) cause spin-spin splitting. The two possible orientations of a spin $\frac{1}{2}$ nucleus in a magnetic field can split the energy levels of neighboring nuclei. This means that the absorption line of a set of equivalent nuclei is split into a multiplet. The frequency separation between the lines of the multiplet is the spin-spin splitting, J, in Hz. If the spin-spin splitting is less than one tenth the frequency difference due to the chemical shifts, simple first-order theory, as illustrated in Fig. 10.26, can be applied. We must note that J in Hz (unlike δ in Hz) is independent of the applied magnetic field. The values of J for protons range from 0 to about 20 Hz. If we are measuring proton nmr in a hydrocarbon or carbohydrate, there are no effects from the carbon or oxygen nuclei, because they have no magnetic moment (except for negligible amounts of ^{13}C and ^{17}O). Naturally occurring nitrogen (^{14}N) has a spin of 1 and tends to broaden neighboring proton lines rather than split them. Consequently, we usually only need to consider proton-proton splittings. Figure 10.26 gives several examples of spin-spin splittings. Each set of n magnetically equivalent* protons, such as the three on a methyl group, split neighboring protons into a multiplet of $(n + 1)$ lines. The number of lines and their relative intensities can be easily derived by counting the number of ways of arranging the nuclear spins in a magnetic field. One proton creates a doublet of intensity 1 : 1. Two equivalent protons create a triplet of intensity 1 : 2 : 1. Three equivalent protons create a quadruplet of intensity 1 : 3 : 3 : 1. The student should easily be able to obtain the result that n protons create $(n + 1)$ lines with relative intensities corresponding to the coefficients of a binomial expansion $(1 + x)^n$.

For protons on adjacent atoms, the spin-spin coupling constant, J_{AB}, is of order of magnitude 10 Hz, but its exact value is very characteristic of the configuration and conformation. For example, the protons on a carbon-carbon double bond have a $J(cis) \cong 12$ Hz and a $J(trans) \cong 19$ Hz. For protons on a carbon-carbon single bond.

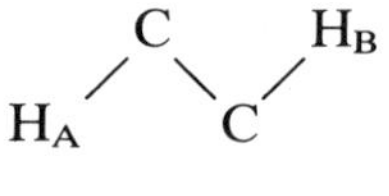

```
      C       HB
    /   \    /
  HA      C
```

the coupling constant is related to the dihedral angle between the protons. The dihedral angle is defined as the angle between the

```
      C
    /   \
  HA      C
```

plane and the

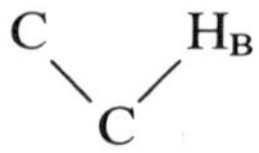

```
  C       HB
    \    /
      C
```

plane. When this angle is 90°, the spin-spin coupling constant is near zero; when the dihedral angle is 0° or 180°, the coupling constant is near 10 Hz. Protons separated by four bonds (H_A—C—C—C—H_B) often produce negligible spin-spin

* The protons could be chemically distinct, but still magnetically equivalent.

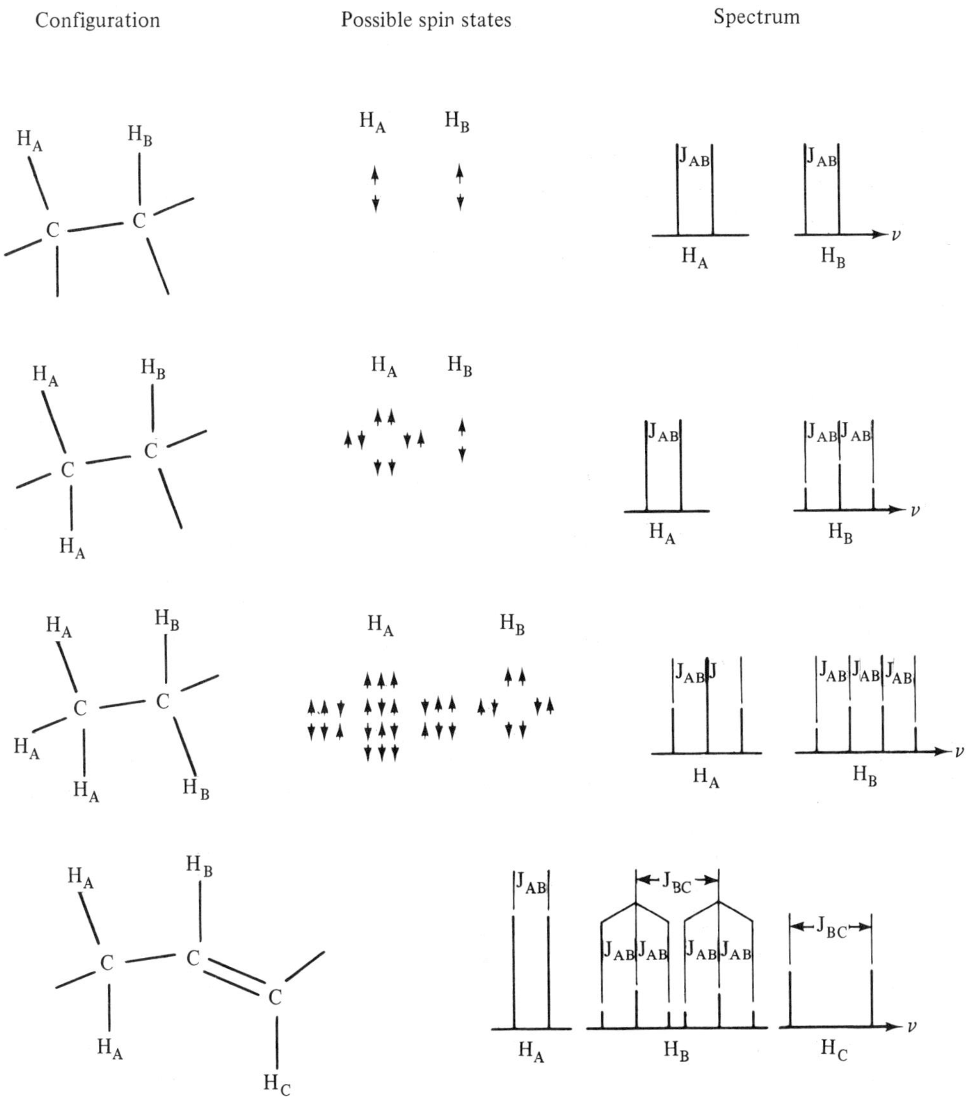

Fig. 10.26 Spin-spin splittings. A group of n equivalent protons will have $(n+1)$ possible spin states in a magnetic field. This will give rise to a multiplet of $(n+1)$ lines in adjacent protons. Note that the magnitude of the splitting of A by B equals that of B by A. The central position of the multiplet depends on the chemical shift. The total intensity of each multiplet is a measure of the number of equivalent protons that are split. In the example at the bottom of the figure proton B is split by two equivalent protons and a different, single proton. This gives rise to a pair of triplets.

splitting. In aromatic or conjugated systems however, a splitting of up to 1 or 2 Hz may be seen.

Spin-spin decoupling is a useful method for assigning protons in a complicated spectrum. If one proton (A) has been assigned, a proton (B) which is coupled to it can be determined. Intense radiation is applied of a frequency which proton A absorbs. This saturates A and decouples it from the other protons; the splitting of B by A disappears in the spectrum. This proves that proton B was indeed coupled to A.

Nuclear magnetic resonance is a very powerful method for determining the structure and conformation of molecules in solution. The chemical shifts are characteristic, the spin-spin splittings indicate neighboring nuclei, and the area under each peak gives the number of equivalent nuclei.

There are many other capabilities and applications of nmr which we cannot discuss here. We will mention only two other measurable parameters obtainable from magnetic resonance experiments. These are the *spin-lattice relaxation time*, T_1, and the *spin-spin relaxation time*, T_2. The time T_1 characterizes the time necessary for the nuclear spins to return to their equilibrium distribution after the orienting magnetic field has been removed. It is analogous to a fluorescence decay time, τ. The spin-spin relaxation time T_2 is a measure of energy exchange among nuclear spins. It can be measured from the line width of a nuclear magnetic absorption band. These two characteristic times can be used to study chemical kinetics and rotational and conformational motion of molecules.

Figure 10.27 shows idealized *proton magnetic resonance* (pmr) spectra for some simple molecules to illustrate typical magnitudes of the chemical shifts. Spectra can be easily measured in D_2O, where only the nonexchangeable protons remain. The protons on O, N, S, and some very acid carbons are quickly exchanged for deuterons. Measurements in H_2O are more difficult because of the tremendous absorption of water protons. However, these measurements allow a distinction between hydrogen-bonded and non-hydrogen-bonded N—H groups in nucleic acids and proteins. Thus pmr studies in H_2O are very good measures of secondary structure in proteins and nucleic acids (Bolton et al., 1976). For organic molecules, completely deuterated solvents are commonly used.

Figure 10.28 shows the measured spectrum of a mutagenic molecule, ethidium. The spectrum is complicated, but all of the peaks could be assigned to protons. The H4 and H7 protons are the only ones without protons on neighboring carbons, so they will only show 1- or 2-Hz splittings. The H7 proton was assigned to the highest upfield peak ($\delta = 6.544$ ppm) because of its magnetic shielding by the phenyl ring. This leaves $\delta = 7.407$ as H4. The phenyl can rotate relative to the phenanthridene group; therefore, the two ortho (Ho, Ho′) and the two meta (Hm, Hm′) protons are equivalent. The three similar protons on the phenyl (Hp, Hm, Hm′) are not resolved in this spectrum and are assigned to the $\delta = 7.81$ region (three protons in area) by analogy with other substituted benzene rings. The two ortho protons are assigned to the doublet with area corresponding to two protons at $\delta = 7.486$ ppm. The assignments of H1, H2, H9, and H10 are also done by analogy and previous experience. Proof of the

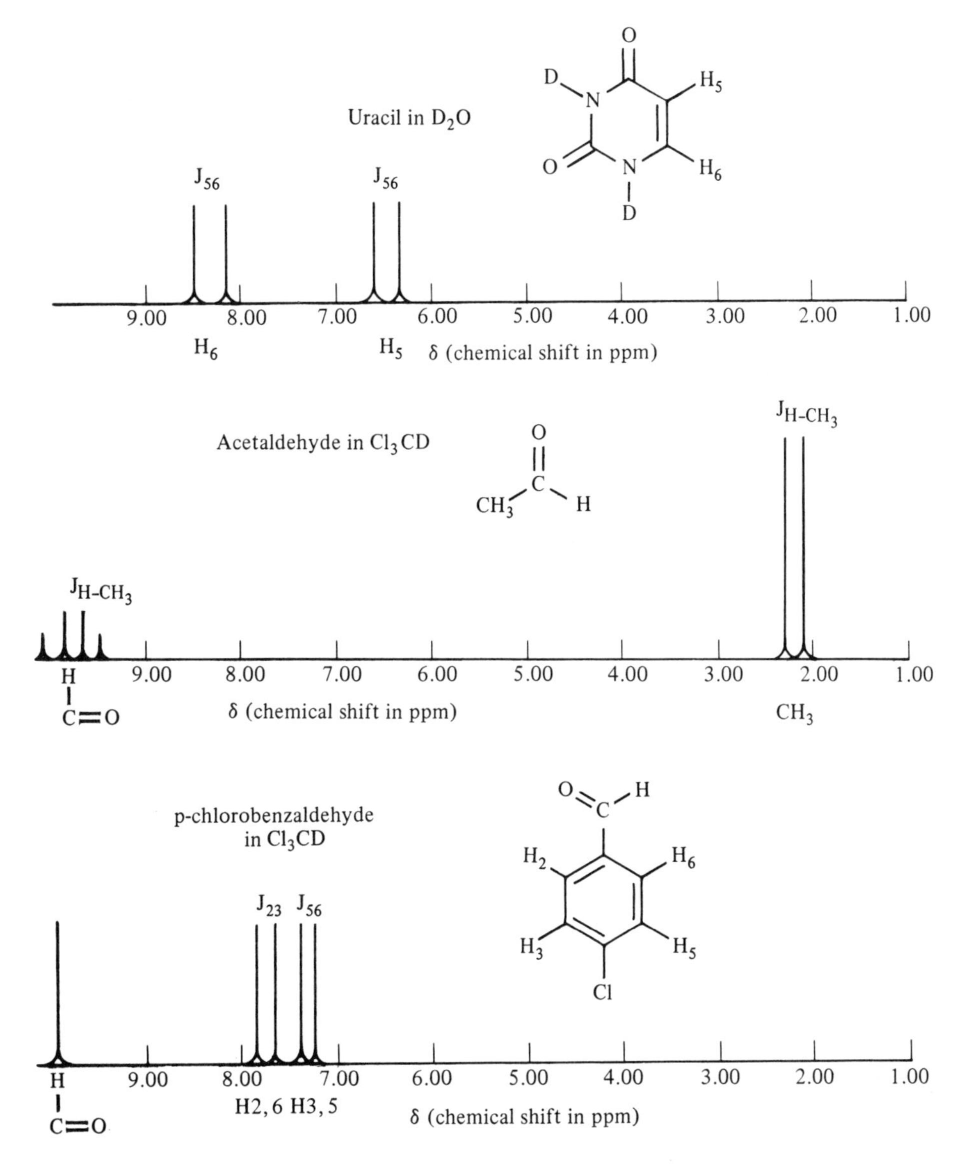

Fig. 10.27 Idealized sketches of proton magnetic resonance spectra for three compounds. Chemical shifts are given in parts per million (ppm) relative to tetramethylsilane. The magnetic field increases to the right. Note that aldehyde protons appear near 10 ppm, phenyl protons between 7 and 8 ppm, and methyl protons near 2 ppm. This progression from aldehyde to methyl corresponds to increasing electron density at the proton. This provides increasing shielding at the proton and requires a higher magnetic field to achieve resonance. The spin-spin splittings are not drawn to scale.

Chemical shift, δ, in ppm
downfield from
trimethylsilylpropionate-d_4

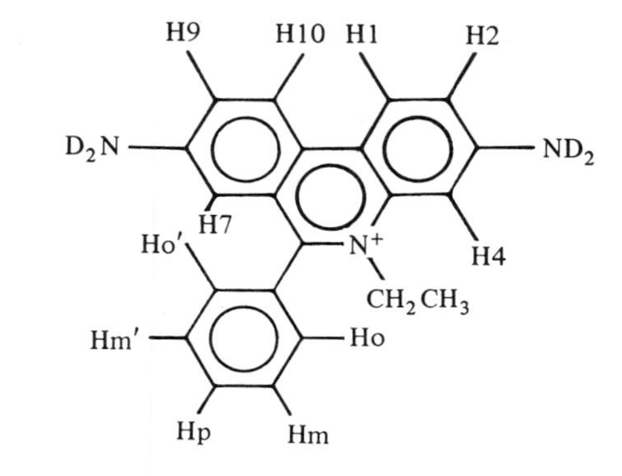

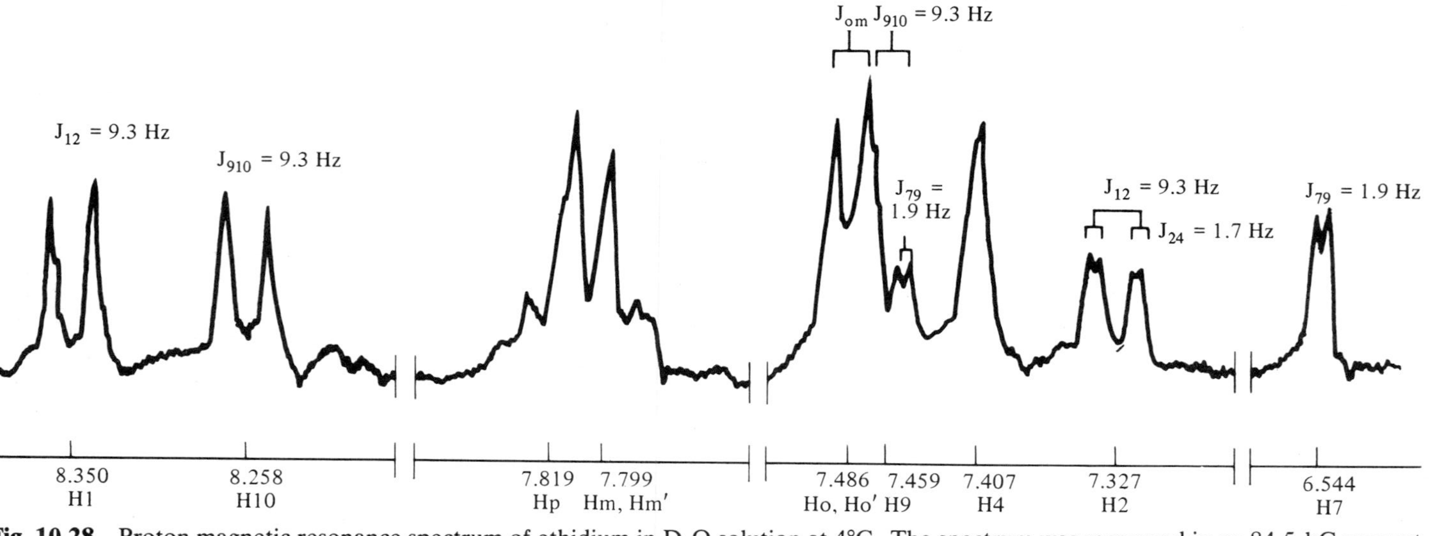

Fig. 10.28 Proton magnetic resonance spectrum of ethidium in D_2O solution at 4°C. The spectrum was measured in an 84.5-kG magnet at 360 MHz. The original spectrum was plotted with a scale of 11.654 Hz cm^{-1}. As 1 Hz is $\frac{1}{360}$ ppm, the chemical shifts can easily be measured to ±0.003 ppm. The ethyl protons are further upfield (smaller δ) and are not shown on this spectrum. (Data provided by Che-Hung Lee.)

assignments could be obtained by study of selectively deuterated derivatives. Spin-spin splittings and spin-spin decoupling are also helpful. The 9-10 and 1-2 spin-spin splittings are about 9 Hz, characteristic of a three-bond separation in an aromatic system. The aromatic four-bond spin-spin splitting of about 2 Hz is clearly seen for the 2-4 and 7-9 interactions.

This molecule binds strongly to DNA, and various spectroscopic and hydrodynamic methods have shown that it intercalates into the DNA. That is, the flat aromatic rings fit in between two adjacent base pairs in the DNA. PMR studies on ethidium bound to DNA fragments are consistent with this interpretation. Furthermore, the relative shifts of the different protons give very detailed information about the structure of the complex. The H2 and H9 protons have the largest shifts on binding to DNA. The H4, H7 and H1, H10 show similar, but smaller, shifts. When ethidium is bound to DNA, Ho and Ho′ are no longer equivalent; the binding immobilizes the phenyl group. There is no appreciable effect on the Hm, Hm′, and Hp. Similar pmr studies of the nucleic acid part of the complex give further useful information.

Molecules that intercalate into DNA can cause frame-shift mutations. These are mutations in which base pairs are added to, or deleted from, the replicated DNA at this site. NMR (and other spectroscopic) studies of mutagen-DNA complexes should help in understanding how frame-shift mutations occur.

ADDITIONAL SPECTROSCOPIC METHODS

The variety of spectroscopic tools that have been applied to biochemical and biological problems is enormous. We cannot hope to cover them all in sufficient detail to introduce them properly here. Instead, we shall list a few representative books, reviews, and surveys and suggest that the reader read them for further enlightenment.

Magnetic circular dichroism (MCD) can be induced in the absorption regions of any chromophoric molecule, whether it possesses intrinsic chirality or not. This property, and the related magnetic optical rotation first discovered by Michael Faraday in 1845, results from an asymmetric coupling between the applied external magnetic field and the light-driven transition moments. It measures very different properties from the normal CD and has been of considerable value in interpreting the spectra and electronic properties of heme proteins. For a review, see Stephens (1974).

Infrared spectroscopy uses long-wavelength (low-energy) radiation to excite vibrations, rotations, and related nuclear motions of functional groups in molecules. These can be interpreted in terms of changing interactions between these groups and their surroundings (e.g., the presence of hydrogen bonding). Polarized infrared spectra (ir dichroism) can give even more specific information about the orientation of such groups in macroscopic samples such as viruses or membranes.

Recently, resonance raman spectroscopy has been used to obtain vibrational information about certain chromophoric species. Raman spectroscopy is a form of inelastic light scattering in which the scattered photon differs in energy from the incident photon by an amount equal to the energy of a molecular vibration. Like ORD, raman scattering can be induced in both transparent and absorbing regions of the spectrum, and it is commonly excited by visible or uv light. For exitation within absorption bands, an enhancement or resonance raman effect greatly increases the signals from vibrations associated with the chromophore. Thus, the technique becomes applicable to chromophores that are relatively dilute. Resonance raman has been applied to heme proteins and to rhodopsin. For a review, consult Spiro (1974).

Electron paramagnetic resonance (epr) can be applied to free radicals or other molecular species that possess unpaired electrons (transition metal complexes, triplet states). Although its range of applicability is necessarily narrower than that of nmr, it can exhibit great sensitivity and is particularly valuable in detecting intermediate or transitory species where electron rearrangements are occurring. It has been applied to heme proteins, ferredoxins, copper-protein complexes, and so on (Swartz et al., 1972). Furthermore, it has been developed as a probe technique. Paramagnetic ions, such as Mn^{2+} or Eu^{3+}, can be used to characterize the nature of ion binding and its role in the catalytic sites of many enzymes. Stable free radicals, such as nitroxide derivatives, have been incorporated into the fatty acid chains of lipids (Hubbell and McConnell, 1969). When these are incorporated into artificial or natural membranes, their epr spectra give detailed information about the flexibility of the lipid chains and the rotational and translational mobility within the lipid phase.

Many new and promising spectroscopies are appearing over the horizon. These include studies of the absorption spectra of x rays, x-ray fluorescence, x-ray absorption fine structure, photoelectron spectroscopy (stimulated by x rays, uv photons, or electron bombardment), circular polarized raman spectroscopy, fluorescence detected circular dichroism and circular polarization of fluorescence, two-photon (laser) spectroscopy, coherent anti-Stokes raman spectroscopy, picosecond and subpicosecond spectroscopy stimulated by mode-locked laser pulses, solid-state nmr, spin-polarization measurements, microwave-induced double resonance, electron-nuclear double resonance, and more. While some of these may prove to be of only limited use to biochemists and biologists, their contribution will nonetheless be valuable. It is certain that many such techniques lie just below the horizon.

SUMMARY

Absorption and Emission

Oscillator strength:

$$f = \frac{2303\ \text{cm}}{\pi N_0 e^2 n} \int \varepsilon(\nu)\, d\nu \qquad (10.5)$$

c = velocity of light
m = mass of electron
N_0 = Avogadro's number
e = electronic charge
n = refractive index
ε = molar absorptivity

Transition dipole moment:

$$\boldsymbol{\mu}_{0A} = \int \psi_0 \boldsymbol{\mu} \psi_A d\nu \quad (10.6)$$

$$\boldsymbol{\mu} = e \sum_i \mathbf{r}_i$$

Dipole strength:

$$D_{0A} = |\boldsymbol{\mu}_{0A}|^2 \quad (10.7)$$

Transformation relations:

See Table 10.3.

Beer-Lambert law:

$$A = \varepsilon c l \quad (10.11)$$

$A = \log \dfrac{I_0}{I_t}$ = absorbance

ε = molar absorptivity or molar extinction coefficient, $M^{-1}\ \mathrm{cm}^{-1}$
c = concentration, M^{-1}
l = path length, cm

Fluorescence rate constant:

$$k_f = \frac{1}{\tau_0} \quad (10.21)$$

k_f = first-order rate constant for fluorescence
τ_0 = natural lifetime of fluorescence

Fluorescence quantum yield:

$$\phi_f = \frac{\text{number of photons fluoresced}}{\text{number of photons absorbed}} = \frac{\tau}{\tau_0} \quad (10.25), (10.27)$$

Excitation Transfer

Energy-transfer efficiency:

$$Eff = \frac{r_0^6}{r_0^6 + r^6} \tag{10.34}$$

Eff= fractional efficiency of transfer
r_0 = characteristic distance for the donor-acceptor pair
r = distance between donor and acceptor

$$Eff = \frac{\tau_D - \tau_{D-A}}{\tau_D} \tag{10.35}$$

τ_D = fluorescence lifetime of donor
τ_{D-A} = fluorescence lifetime of donor in the presence of acceptor

Optical Rotatory Dispersion and Circular Dichroism

$$\text{Circular birefringence} = n_L - n_R$$

$$\text{Optical rotation} = \phi = \frac{\pi}{\lambda}(n_L - n_R), \text{ rad cm}^{-1} \tag{10.36}$$

n_L and n_R are refractive indices for left and right circularly polarized light, respectively, at wavelength λ.

$$\text{Circular dichroism} = \Delta A = A_L - A_R$$

$$\text{Ellipticity} = \theta = \frac{2.303(A_L - A_R)}{4l}, \text{ rad cm}^{-1} \tag{10.38}$$

A_L and A_R are absorbances for left- and right-circularly polaried light, respectively; l is the path length in cm.

Experimental parameters:

See Table 10.6.

REFERENCES

JAFFE, H. H., and M. ORCHIN, 1962. *Theory and Applications of Ultraviolet Spectroscopy*, John Wiley, New York.

JAMES, T. L., 1975. *Nuclear Magnetic Resonance in Biochemistry*, Academic Press, New York.

SELECTED READINGS

AMOS, A. T., and B. L. BURROWS, 1973. Solvent-Shift Effects on Electronic Spectra and Excited-State Dipole Moments and Polarizabilities, *Advan. Quantum Chem. 7*, 289–313.

BEARDSLEY, K., and C. R. CANTOR, 1970. Studies of Transfer RNA Tertiary Structure by Singlet-Singlet Energy Transfer, *Proc. Natl. Acad. Sci. U.S. 65*, 39–46.

BOLTON, P. H., C. R. JONES, O. BASTEDO-LERNER, K. L. WONG, and D. R. KEARNS, 1976. Quantitative Determination of the Number of Secondary and Tertiary Structure Base Pairs in Transfer RNA in Solution, *Biochemistry 15*, 4370–4377.

FÖRSTER, TH., 1949. Experimental and Theoretical Investigation of the Intermolecular Transfer of Electronic Excitation Energy, *Z. Naturforsch. 4*a, 321–327.

GRATZER, W. B., and D. A. COWBURN, 1969. Optical Activity of Biopolymers, *Nature 222*, 426–431.

HAUGLAND, R. P., J. YGUERABIDE, and L. STRYER, 1969. Dependence of the Kinetics of Singlet-Singlet Energy Transfer on Spectral Overlap, *Proc. Natl. Acad. Sci. U.S. 63*, 23–30.

HAYWARD, L. D., and R. N. TOTTY, 1971. Optical Activity of Symmetric Compounds in Chiral Media: I. Induced Circular Dichroism of Unbound Substrates, *Can. J. Chem. 49*, 624–631.

HUBBELL, W. L., and H. M. MCCONNELL, 1969. Orientation and Motion of Amphiphilic Spin Labels in Membranes, *Proc. Natl. Acad. Sci. U.S. 64*, 20–27.

KNOX, R. S., 1975. Excitation Energy Transfer and Migration: Theoretical Considerations, in *Bioenergetics of Photosynthesis*, Govindjee (ed.), Academic Press, New York, pp. 183–221.

LATT, S. A., H. T. CHEUNG, and E. R. BLOUT, 1965. Energy Transfer: A System with Relatively Fixed Donor-Acceptor Separation, *J. Amer. Chem. Soc. 87*, 995–1003.

ROSENHECK, K., and P. DOTY, 1961. The Far Ultraviolet Absorption Spectra of Polypeptide and Protein Solutions and Their Dependence on Conformation, *Proc. Natl. Acad. Sci. U.S. 47*, 1775–1785.

SPIRO, T. G., 1974. Raman Spectra of Biological Materials, in *Chemical and Biochemical Applications of Lasers*, Vol. 1, C. B. Moore (ed.), Academic Press, New York, pp. 29–70.

STEPHENS, P. J., 1974. Magnetic Circular Dichroism, *Ann. Rev. Phys. Chem. 25*, 201–232.

SUSSMAN, R., and W. B. GRATZER, 1962. Cited by D. B. Wetlaufer in Ultraviolet Spectra of Proteins and Amino Acids, *Advan. Protein Chem. 17*, 303–390.

SWARTZ, H. M., J. R. BOLTON, and D. C. BORG (eds.), 1972. *Biological Applications of Electron Spin Resonance*, Academic Press, New York, pp. 213–264.

PROBLEMS

1. The longest-wavelength absorption band of chlorophyll a peaks *in vivo* at a wavelength of about 680 nm.

(a) For photons with a wavelength of 680 nm, calculate the energy in ergs photon^{-1}, in electron volts, and in kcal einstein^{-1}.

(b) CO_2 fixation in photosynthesis can be represented by the reaction

$$CO_2 + H_2O \longrightarrow (CH_2O) + O_2$$

where (CH_2O) represents $\frac{1}{6}$ of a carbohydrate molecule such as glucose. The enthalpy change for this endothermic reaction is $\Delta H = +116$ kcal mol^{-1} of CO_2 fixed. What is the minimum number of einsteins of radiation that need to be absorbed to provide the energy needed to fix 1 mol of CO_2 via photosynthesis?

(c) Experimentally, many values have been determined for the number of photons required to fix one CO_2 molecule. The presently accepted "best values" lie between 8 and 9 photons per CO_2 molecule. What is the photochemical quantum yield, based on your answer to part (b), for this process?

2. Using the data given in Table 10.2 for a 1-kJ ruby laser pulse of 1-ms duration, estimate the absorbance of safety goggles needed to protect the operator's eyes against damage from a potential accident. Base your estimate on the information in Problem 23(b) of Chapter 7.

3. The mechanism of sun tanning is initiated by the absorption of the 0.2% of solar radiation energy lying in the range 2900 to 3132 Å. The earth's atmosphere cuts off radiation <2900 Å and the burning efficiency drops from 100% at 2967 Å to 2% at 3132 Å. The *average* efficiency across the 2900 to 3132 Å band is 50%. The total incident radiation from sunlight at the surface of the earth corresponds to 500 cal cm^{-2} day^{-1}.

(a) Calculate the energy equivalent (in cal cm^{-2} s^{-1}) of the radiation involved in sunburning. (Assume that a typical day length is 12 h.)

(b) Many commercial ointments contain *o*- or *p*-amino benzoates:

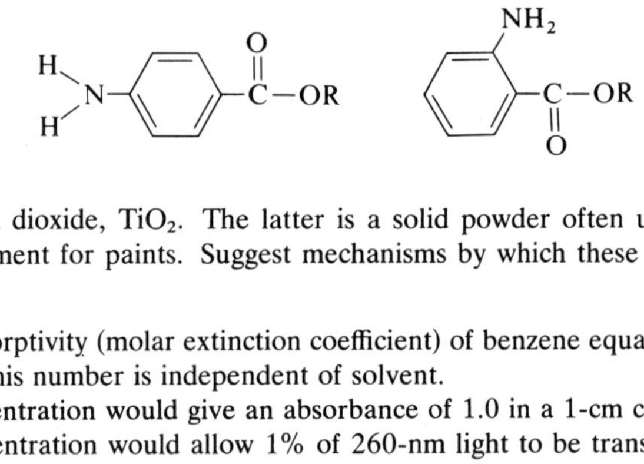

$H_2N{-}C_6H_4{-}C(=O){-}OR$ $\qquad$ $C_6H_4(NH_2){-}C(=O){-}OR$

or titanium dioxide, TiO_2. The latter is a solid powder often used as a "super-white" pigment for paints. Suggest mechanisms by which these substances might work.

4. The molar absorptivity (molar extinction coefficient) of benzene equals 100 at 260 nm. Assume that this number is independent of solvent.

(a) What concentration would give an absorbance of 1.0 in a 1-cm cell at 260 nm?

(b) What concentration would allow 1% of 260-nm light to be transmitted through a 1-cm cell?

(c) If the density of liquid benzene is 0.8 g cm^{-3}, what thickness of benzene would give an absorbance of 1.0 at 260 nm?

5. Indicator dyes are often used to measure the hydrogen-ion concentration. From the measured spectra of an indicator in solution, the ratio of acid to base species and the pH can be calculated:

$$HIn \rightleftharpoons In^- + H^+$$

The molar extinction coefficients are:

	ε (300 nm)	ε (400 nm)
HIn	10,000	2,000
In^-	500	4,000

(a) Calculate the ratio of concentrations In^-/HIn for a solution of the indicator which has A (300 nm) = 1.2 and A (400 nm) = 7 in a 1 cm cell.
(b) What is the pH if the pK of the indicator is 5.0?

6. The content of chlorophylls a and b in leaves or green algae is determined by extracting the pigments into acetone : water (80 : 20 v/v) solution. The specific absorptivities of the pigments in this solvent system reported by Mackinney [*J. Biol. Chem. 140*, 315 (1941)] at 645 and 663 nm are as follows:

Wavelength (nm)	Specific absorptivity (ℓ g^{-1} cm^{-1})	
	Chl a	Chl b
645	16.75	45.60
663	82.04	9.27

Derive expressions, with numerical coefficients evaluated, that give the concentrations in $\mu g\, m\ell^{-1}$ of chlorophyll a, of chlorophyll b, and of total chlorophylls (a + b) for specified values of A_{645} and A_{663} measured using a 1-cm path length.

7. A sample of RNA is hydrolyzed in base and the resulting mixture of ribonucleotides is passed through a chromatographic column. Cytidylic and uridylic acids are obtained in separate peaks, but adenylic and guanylic acids elute together from the column. The fraction containing the mixed adenylic and guanylic acids is adjusted to pH 7 and determined to have absorbances of 0.305 at 280 nm and 0.655 at 250 nm in a 1-cm cell. The molar absorptivities at pH 7 are as follows:

	ε_{280} (M^{-1} cm^{-1})	ε_{250} (M^{-1} cm^{-1})
Adenylic acid	2,300	12,300
Guanylic acid	9,300	15,750

Calculate the mole ratio of adenine to guanine in the RNA.

8. An indicator is a dye whose spectrum changes with pH. Consider the following data for the absorption spectrum of an indicator (pK 4) in its ionized and un-ionized form.

$$InH^+ \rightleftharpoons In + H^+$$

	Molar absorptivity ε (M^{-1} cm^{-1})	
λ (nm)	InH^+	In
400	10,000	0
420	15,000	2,000
440	8,000	8,000
460	0	12,000
480	0	3,000

The absorbance of the indicator solution is measured in a 1-cm cell and found to be:

λ	400	420	440	460	480
A	0.250	0.395	0.400	0.300	0.075

(a) Calculate the pH of the solution.
(b) Calculate the absorbance of the solution at 440 nm and pH 6.37 for the same total indicator concentration.
(c) If you want to measure the quantum yield of fluorescence, of In and InH^+ independently, what wavelength would you choose for exciting InH^+? For exciting In? Why?

9. Use the table of data to answer the following questions; $\varepsilon(\lambda) \equiv$ molar absorptivity at wavelength λ; $\varepsilon_L - \varepsilon_R(\lambda) \equiv$ molar circular dichroism at wavelength λ.

	Compound A	Compound B
ε (3000 Å)	$10^4\ M^{-1}$ cm^{-1}	0
ε (4000 Å)	$10^3\ M^{-1}$ cm^{-1}	$10^5\ M^{-1}$ cm^{-1}
Fluorescence quantum yield	0.4	0
Phosphorescence quantum yield	0	0.1
Maximum λ of fluorescence	5000 Å	—
Maximum λ of phosphorescence	—	6000 Å
$\varepsilon_L - \varepsilon_R$ (3000 Å)	0	0
$\varepsilon_L - \varepsilon_R$ (4000 Å)	0	10

(a) Four milliliters of a 0.1 M solution of compound A is put into a 1-cm cell. What is the absorbance at 3000 Å?
(b) What is the absorbance of a solution containing 0.1 M of A and 0.1 M of B in a 1-cm cell at 3000 Å?
(c) Light of wavelength 3000 Å and intensity 10^6 photons cm^{-2} s^{-1} is incident on a cell of compound B. What fraction of the light is emitted as phosphorescence?
(d) The circular dichroism of a solution containing 0.1 M A and 0.1M B is measured in a 2-cm cell. What is the value of $A_L - A_R$ at 4000 Å?

10. For the following questions, a single sentence or phrase should be sufficient as an answer.

Glycine   Phenylalanine   Adenine

(a) Which, if any, of these molecules would you expect to rotate the plane of linearly polarized incident on them? Why?
(b) Which, if any, of these molecules would you expect to be circularly dichroic? Why?
(c) Which, if any, of these molecules could in principle emit fluorescence when appropriately excited? Why?
(d) If a molecule emitted light on excitation, what experiment would you do to tell if it was fluorescence or phosphorescence?

11. A fluorescent dansyl group has a lifetime of 21.0 ns when attached to a protein. When a naphthyl group is attached to the amino group of the terminal amino acid in the protein, the dansyl lifetime becomes 15.0 ns. Use the data in Fig. 10.17 to calculate the distance between the dansyl and the naphthyl group.

12. Cytidylic acid (molecular weight = 323.2) in water at pH 7 has the following optical properties at 280 nm: $\varepsilon = 8000\ M^{-1}\ \text{cm}^{-1}$; $\varepsilon_L - \varepsilon_R = 3\ M^{-1}\ \text{cm}^{-1}$; $[\phi] = 7500$ deg $M^{-1}\ \text{cm}^{-1}$.
(a) Calculate the specific rotation, $[\alpha]$, and the molar ellipticity, $[\theta]$.
(b) For $10^{-4}\ M$ cytidylic acid in a 1-cm cell, calculate the absorbance, A, the circular dichroism $(A_L - A_R)$, and the rotation angle in degrees at 280 nm.
(c) Calculate the rotation in rad cm^{-1}, the circular birefringence $(n_L - n_R)$, and the ellipticity in rad cm^{-1} for the solution in part (b).

13. The fluorescence lifetime of benzene in cyclohexane is 29 ns when air is completely removed. In the presence of 0.0072 M dissolved O_2, the measured fluorescence lifetime is 5.7 ns because of quenching. Calculate the rate constant k_Q for the quenching reaction. If the relative fluorescence intensity is 100 for pure benzene, what is the relative fluorescence intensity of benzene containing 0.0072 M dissolved O_2?

14. The proton magnetic resonance of selectively deuterated methyl ethyl ether was measured in a 100-MHz spectrometer to give the following spectrum:

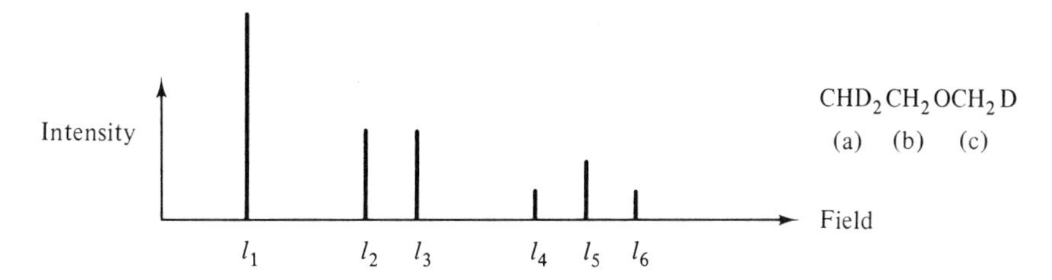

(a) Assign each of the six lines to protons (a), (b), or (c).
(b) If the separation between lines l_1 and l_5 is 200 Hz and is due to a chemical shift, what will the separation be in a 220-MHz spectrometer?
(c) If the spin-spin splitting of proton (a) is 5 Hz, what is the spin-spin splitting of proton (b)? What will the spin-spin splitting be in a 220-MHz spectrometer?
(d) Sketch the proton magnetic resonance spectrum for $CD_3CH_2OCD_3$.

15. Deduce the structure of the following compounds from their pmr spectra. Designate which protons go with which pmr lines. Chemical shifts are given in ppm relative to water.

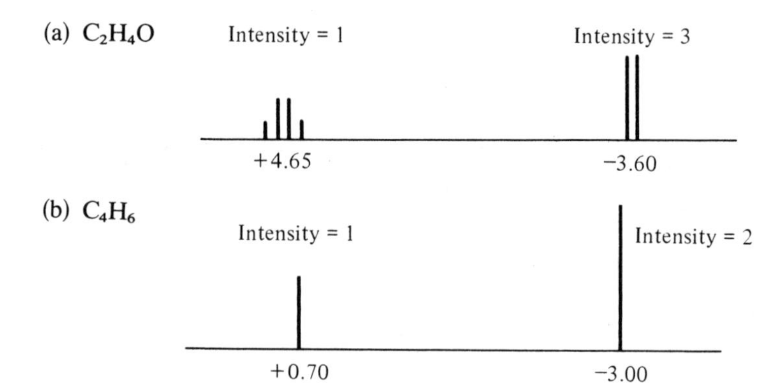

11

Molecular Distributions and Statistical Thermodynamics

A unique aspect of biological systems is that many important species are polymers of very high molecular weight. For example, the enzyme lysozyme, which is a relatively small protein, has a molecular weight of 14,600 and contains 129 amino acid residues. A typical DNA molecule of a bacterium has a molecular weight of the order of 10^9 and contains several million nucleotides. Macromolecules like these participate in every aspect of the activities of a cell. In fact, the interactions among such macromolecules and between the macromolecules and small molecules define the life process itself. To understand the properties of such macromolecules better, certain statistical concepts are helpful.

Consider the DNA from the bacterial virus T4. When purified, it is a double helix with a molecular weight of $\sim 10^8$. In solution, this long, threadlike molecule can assume many conformations: some with the molecule in an extended form, others with the molecule in a tight coil. Some properties of the molecule, such as how fast it can diffuse or sediment in a solution, depend on the *average* dimension of the molecule. Others depend upon the *range* of possible conformations of the molecule. For example, if the DNA solution is stirred vigorously, some of the molecules will break into smaller pieces. Molecules in the extended form are more easily broken. Both the average dimension and the fraction of molecules in an extended conformation are relevant to this process. Information bearing on the distribution of molecular conformations can be obtained by statistical methods.

If the DNA solution is heated, at a certain temperature the double helix begins to "melt": Base pairing is disrupted, regions of single-stranded loops form, and, if the temperature is high enough, the two strands of the double helix dissociate completely. A schematic illustration of the melting phenomenon is shown in Fig. 11.1. The transition from the helix to the dissociated strands occurs

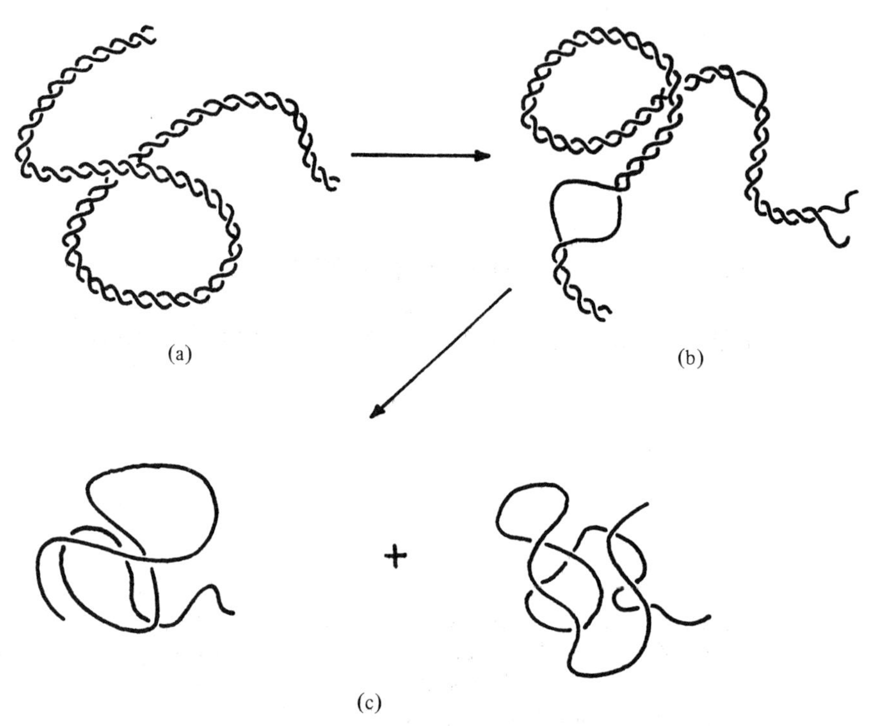

Fig. 11.1 Schematic illustration of the thermal "melting" of a double-stranded DNA (a) to its complementary single strands (c). At an intermediate temperature, single-stranded loops and/or ends can form, as illustrated in (b). See the text for a more detailed discussion.

in a temperature range of only a few degrees; hence, the term "melting" is used to describe this transition. Statistical analysis provides an explanation for the sharp transition. Furthermore, some of the thermodynamic parameters for the transition can be evaluated by a statistical analysis. These parameters are helpful in understanding the replication of the DNA, because the replication of the DNA molecules involves the dissociation of the two parent strands and the concurrent formation of new double strands for each of the two daughter DNA molecules.

Either inside or outside the cell, the DNA interacts with a variety of small and not so small molecules. Among the small molecules the antibiotic actinomycin binds to many sites of the DNA and can block transcription of the DNA into RNA. Many heterocyclic dye molecules and polycyclic aromatic hydrocarbons, such as 9-aminoacridine and benz(a)pyrene can bind to the DNA and cause mutations. Such molecules are often carcinogenic. The modes of binding of small molecules to DNA can be understood by statistical analyses of the binding experiments. Among the larger molecules which interact with the DNA, some bind only to a few specific sites; others bind with much less discrimination. The RNA polymerase of the host bacterium of T4, for example, initiates RNA synthesis (transcription) from only a few sites on the DNA. The product of the T4 gene 32 (the gene 32 protein), which is involved in DNA replication and recombination, is much less site-specific. It does, however, bind preferentially to single-stranded DNA in a highly "cooperative" fashion. If a certain amount of gene 32 protein is mixed with an excess of single-stranded DNA, some DNA molecules will be completely covered with the protein while other DNA molecules are bare. Statistical analysis again provides some insight into such divergent patterns of interactions.

The usefulness of statistical concepts is not limited to large molecules. These concepts provide a connection between the molecular-dynamical picture of matter and the conventional thermodynamic laws for bulk matter that we explored at the beginning of this book. Intuitively, it seems reasonable to expect that the laws governing the thermodynamic properties of matter should be related to the laws of mechanics governing its microscopic parts. Indeed, thermodynamic properties of a macroscopic system can be obtained from the molecular parameters of its microscopic constituents. In the first part of this chapter, the usefulness of statistical concepts will be illustrated by a number of examples. In the second part some relations between macroscopic and microscopic properties will be discussed.

BINDING OF SMALL MOLECULES BY A POLYMER

For the interaction between two simple species, A and B, we found in Chapter 4 that the system at equilibrium can be described by an equilibrium constant K, and that the standard free energy change ΔG^0 is related to the equilibrium constant K by the equation $\Delta G^0 = -RT \ln K$. As an example, for the combination of A and B to form AB,

$$A + B \rightleftharpoons AB$$

the equilibrium constant and the standard free energy change are simply

$$K = \frac{[AB]}{[A][B]} \tag{11.1}$$

and

$$\Delta G^0 = -RT \ln K \tag{11.2}$$

In many cases of biological interest, one of the two interacting species is a complex macromolecule and the other a relatively small molecule. Usually, the macromolecule has more than one site where it can bind the small molecule. As a result, many different molecular complexes can form, depending upon which sites of the macromolecule are occupied by the small molecules. For example, a hemoglobin molecule has four sites for oxygen. A high-molecular-weight DNA has numerous sites for the binding of actinomycin. In such cases an expression such as Eq. (11.1) is not very descriptive in comparison with the richness of detail of the different species at the molecular level. One could write an equilibrium constant for each possible species, but the numbers often are very large and such cases can be treated more easily by the use of statistical methods. In the sections below we shall treat a number of examples.

Identical-and-Independent-Sites Model

Suppose that a polymer molecule P has a number of identical and independent sites for binding a smaller molecule, A. If the sites are *identical* (each has the same affinity for A) and *independent* (the occupation of one site has no effect on the binding to another site), a very simple description of this system can be obtained. Examples for which the identical-and-independent-sites model is applicable are the binding of DNA polymerase molecules to a single-stranded DNA, the binding of the drug ethidium to a DNA double helix, and the binding of substrates to certain enzymes or proteins containing several identical subunits per molecule.

As a simple example, let us consider a polymer molecule P with four identical and independent sites. This polymer molecule might be a tetrameric protein, with each one of the four subunits having one site for the binding of a substrate A. Suppose that the four sites are distinguishable and can be labeled 1, 2, 3, and 4. Since the sites are identical, the equilibrium constants of the following reactions are identical:

$$\mathrm{P + A \rightleftharpoons PA^{(1)}} \qquad K = \frac{[\mathrm{PA^{(1)}}]}{[\mathrm{P}][\mathrm{A}]}$$

$$\mathrm{P + A \rightleftharpoons PA^{(2)}} \qquad K = \frac{[\mathrm{PA^{(2)}}]}{[\mathrm{P}][\mathrm{A}]}$$

$$\mathrm{P + A \rightleftharpoons PA^{(3)}} \qquad K = \frac{[\mathrm{PA^{(3)}}]}{[\mathrm{P}][\mathrm{A}]}$$

$$\mathrm{P + A \rightleftharpoons PA^{(4)}} \qquad K = \frac{[\mathrm{PA^{(4)}}]}{[\mathrm{P}][\mathrm{A}]}$$

The superscript (j) in each reaction specifies which site of the polymer is involved.

The total concentration of species with one A molecule per P molecule, [PA], is

$$[\mathrm{PA}] = [\mathrm{PA}^{(1)}] + [\mathrm{PA}^{(2)}] + [\mathrm{PA}^{(3)}] + [\mathrm{PA}^{(4)}]$$
$$= 4K[\mathrm{P}][\mathrm{A}]$$

In other words, for the reaction

$$\mathrm{P} + \mathrm{A} \overset{K_1}{\rightleftharpoons} \mathrm{PA} \qquad \text{(any site)}$$

$$K_1 = \frac{[\mathrm{PA}]}{[\mathrm{P}][\mathrm{A}]} = 4K \tag{11.3}$$

The factor 4 in Eq. (11.3) can also be thought of as a result of the fact that there are four ways for the formation of a PA (by binding of an A to any one of the 4 sites of P), but only one way for any PA to dissociate into P and A.

For the reaction

$$\mathrm{PA}\ \text{(any site)} + \mathrm{A} \overset{K_2}{\rightleftharpoons} \mathrm{PA}_2 \qquad \text{(any two sites)}$$

$$K_2 = \frac{[\mathrm{PA}_2]}{[\mathrm{PA}][\mathrm{A}]} = \left(\frac{4-1}{2}\right)K = \frac{3}{2}K \tag{11.4}$$

There are (4 − 1), or three, ways of picking one of the three remaining sites in PA for the formation of a PA_2, and there are two ways for the dissociation of a PA_2 into PA and A. Similarly, the equilibrium constants K_3 and K_4 are

$$K_3 = \left(\frac{4-2}{3}\right)K = \frac{2}{3}K \tag{11.5}$$

and

$$K_4 = \left(\frac{4-3}{4}\right)K = \frac{1}{4}K \tag{11.6}$$

The concentrations of the various species are:

$$[\mathrm{PA}] = 4K[\mathrm{P}][\mathrm{A}] = 4[\mathrm{P}](K[\mathrm{A}])$$

$$[\mathrm{PA}_2] = \frac{3}{2}K[\mathrm{PA}][\mathrm{A}] = \left(\frac{4\cdot 3}{2}\right)K^2[\mathrm{P}][\mathrm{A}]^2 = 6[\mathrm{P}](K[\mathrm{A}])^2$$

$$[\mathrm{PA}_3] = \frac{2}{3}K[\mathrm{PA}_2][\mathrm{A}] = \left(\frac{4\cdot 3\cdot 2}{3\cdot 2}\right)K^3[\mathrm{P}][\mathrm{A}]^3 = 4[\mathrm{P}](K[\mathrm{A}])^2 \tag{11.7}$$

$$[\mathrm{PA}_4] = \frac{1}{4}K[\mathrm{PA}_3][\mathrm{A}] = K^4[\mathrm{P}][\mathrm{A}]^4 = [\mathrm{P}](K[\mathrm{A}])^4$$

Let $K[\mathrm{A}] = S$. The total concentration of the polymer is

$$[\mathrm{P}]_T = [\mathrm{P}] + [\mathrm{PA}] + [\mathrm{PA}_2] + [\mathrm{PA}_3] + [\mathrm{PA}_4]$$
$$= [\mathrm{P}](1 + 4S + 6S^2 + 4S^3 + S^4)$$
$$= [\mathrm{P}](1 + S)^4 \tag{11.8}$$

The total concentration of A molecules that are bound to the polymer molecules is

$$\begin{aligned}[A]_{bound} &= [PA] + 2[PA_2] + 3[PA_3] + 4[PA_4] \\ &= [P](4S + 12S^2 + 12S^3 + 4S^4) \\ &= 4[P]S(1 + 3S + 3S^2 + S^3) \\ &= 4[P]S(1 + S)^3 \end{aligned} \quad (11.9)$$

The number of bound A molecules per polymer molecule is ν, as defined in Eq. (5.8):

$$\nu = \frac{[A]_{bound}}{[P]_{total}} = \frac{4[P]S(1 + S)^3}{[P](1 + S)^4} = \frac{4S}{1 + S} \quad (11.10)$$

It can be shown that, in general, if there are N identical and independent sites,

$$\begin{aligned}\nu &= \frac{NS}{1 + S} \\ &= \frac{NK[A]}{1 + K[A]}\end{aligned} \quad (11.11)$$

From Eq. (11.11), it follows that

$$\nu(1 + K[A]) = NK[A]$$

$$\nu = (N - \nu)K[A]$$

and

$$\frac{\nu}{[A]} = K(N - \nu) \quad (11.12a)$$

or

$$\frac{\nu}{[A]} = KN - K\nu \quad (11.12b)$$

When N is large, it is usually neither experimentally possible nor desirable to determine the concentrations of the individual species [P], [PA], $[PA_2]$, ... The parameter ν is most frequently used to express the extent of binding. If $\nu/[A]$ is plotted versus ν, according to Eq. (11.12a) or (11.12b), a straight line results. The slope of the line is $-K$, where K is the equilibrium constant for binding to one site. The intercept of the line with the ν axis is N, where N is the total number of sites per polymer molecule. Such a plot is called a *Scatchard plot.*

Example 11.1 The formyltetrahydrofolate synthetases can utilize the energy of hydrolysis of ATP (to ADP and phosphate) for the formation of a carbon-nitrogen bond between (l)-tetrahydrofolate and formate. Binding of the Mg complex of ATP to such an enzyme from *C. cylindrosporum* has been measured; the data are plotted in Scatchard form in Fig. 11.2.

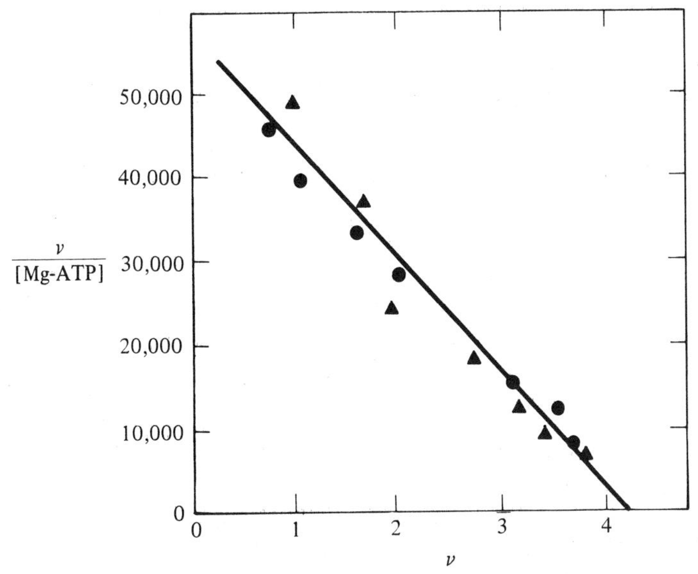

Fig. 11.2 Binding of Mg-ATP by tetrahydrofolate synthetase from *C. cylindrosporum*. The circles and triangles represent two sets of data obtained by two different techniques. Measurements were carried out at 23°C and a pH of 8.0. [Data from N. P. Curthoys and J. C. Rabinowitz, *J. Biol. Chem. 246*, 6942 (1971).]

The data fit the identical-and-independent-sites model rather well. The slope of the plot gives an equilibrium constant K of $1.37 \times 10^4 \, M^{-1}$. The intercept gives $N = 4.2$. Since the enzyme is known to have four identical subunits (each has a molecular weight of 60,000), the binding results are consistent with the notion that there is one Mg-ATP binding site per subunit.

Example 11.2 In the discussion leading to Eq. (11.12), we have used the notation of Chapter 5 and defined the quantities ν and N as follows:

ν = number of bound A *per polymer molecule*

N = number of binding sites *per polymer molecule*

A polymer is made of monomer units. A DNA molecule is a polynucleotide consisting of nucleotides as its monomer building blocks. Sometimes it is more convenient to define the binding parameters on a *per monomer unit basis* as follows:

r = number of bound A *per monomer unit* of polymer

n = number of binding sites *per monomer unit* of the polymer

If the average degree of polymerization or the average number of monomer units per polymer molecule is $\bar{P}$, we obtain readily

$$\nu = r\bar{P} \tag{11.13}$$

and

$$N = n\bar{P} \tag{11.14}$$

Substituting these relations into Eq. (11.12),

$$\frac{r\bar{P}}{[\mathrm{A}]} = K\bar{P}(n - r)$$

or

$$\frac{r}{[\mathrm{A}]} = K(n - r) \tag{11.15}$$

Thus, the Scatchard plot can also be presented with r and n as the parameters.

In Fig. 11.3, data for the binding of the trypanocide drug ethidium (ethidium is sometimes used to treat animals infected by parasitic trypanosomes) to a double-stranded DNA are plotted in the Scatchard fashion for two temperatures. Note that the data are represented reasonably well by the identical-and-independent-sites model. The r intercept gives $n \approx 0.23$, or approximately one binding site per 4 nucleotides. Note also that the line

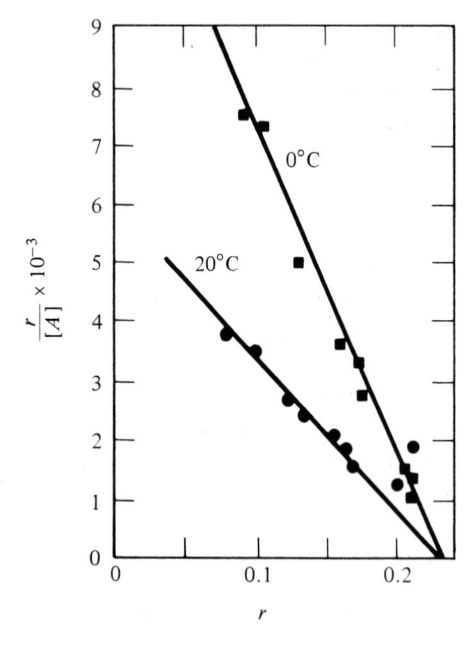

Fig. 11.3 Scatchard plot of the binding of ethidium to DNA in a medium containing 3 M CsCl, 0.01 M Na_3 EDTA.

is steeper, indicating a greater binding constant, at the lower temperature. This temperature dependence of the binding constant yields a negative ΔH^0 (calculated to be ~ -6 kcal mol^{-1}) for the binding of the drug. The fairly strong interaction ($K = 2.66 \times 10^4 \, M^{-1}$ and $5.60 \times 10^4 \, M^{-1}$ at 20°C and 0°C, respectively) between the drug and DNA is evidence supporting the notion that the target of the drug *in vivo* is the DNA of the trypanosome.

Langmuir Adsorption Isotherm

A binding curve at a constant temperature is called an *isotherm.* In the independent-sites model, since the occupation of one site has no effect on binding to other sites, the spatial distribution of the sites has no effect on the final binding equation (11.12).

Equation (11.12) [or (11.15)] can also be changed into the following form:

$$\frac{\nu/N}{1 - \nu/N} = K[\mathrm{A}]$$

Since ν is the number of sites occupied per polymer molecule and N is the total number of sites per polymer molecule, ν/N is simply the fraction of sites occupied, which is designated as θ. We obtain, then,

$$\frac{\theta}{1 - \theta} = K[\mathrm{A}] \tag{11.16}$$

Equation (11.16) was first derived by Langmuir in 1916 for the adsorption of a gas on a solid surface.

Nearest-neighbor Interactions, Three Sites

In the previous section, the many binding sites on a polymer are assumed to be independent or noninteracting. Frequently, however, the binding of a small molecule to one site affects the binding of small molecules to other sites. The interactions between the sites greatly increase the complexity of the problem, and analytical solutions of the binding isotherms can only be obtained for special cases. In this section we consider a special case, with N identical sites arranged in a linear array. In other words, the identical sites form a one-dimensional lattice. This model was first used by Ising and is called the *Ising model.* To simplify matters we shall first consider only nearest-neighbor interactions. Longer-range interactions will be discussed in a later section.

The N identical sites in a linear array can be expressed by an N-digit number of 0's and 1's such as 00111000 . . . 1011100011, in which the symbol 0 represents a free site and the symbol 1 represents an occupied site. The number written represents a linear polymer with N identical sites, such that the first two sites are unoccupied, the next three sites are occupied, and so on. Let K be the equilibrium constant for the binding of an A molecule to a site with no occupied nearest neighbors, and τK be the equilibrium constant for the binding of an A molecule to a site with one adjacent site occupied. The parameter τ accounts for the interaction between the two adjacent occupied sites. With these definitions the equilibrium constant for any reaction can be written quickly. A few examples follow for a simple case with $N = 3$:

Reaction	Equilibrium constant
$000 + A \rightleftharpoons 001$	K
$001 + A \rightleftharpoons 101$	K
$001 + A \rightleftharpoons 011$	τK
$101 \rightleftharpoons 110$	τ
$101 + A \rightleftharpoons 111$	$\tau^2 K$

The ratio of concentrations of [001] to [000], the relative concentration of [001], is

$$\frac{[001]}{[000]} = K[\mathrm{A}]$$

Similarly, the relative concentrations of [101], [011], and [111] are $(K[\mathrm{A}])^2$, $\tau(K[\mathrm{A}])^2$, and $\tau^2(K[\mathrm{A}])^3$, respectively. We shall refer to these relative concentrations as the *statistical weights* of the species. The free species [000] has a statistical weight of one by convention. Defining the product $K[\mathrm{A}]$ as S, the statistical weight of each state for the case $N = 3$ is given in Table 11.1.

Note that the statistical weight of each state can be written as $\tau^i S^j$, where j is the number of 1's (occupied sites) in the state and i is the number of 1's following 1 (number of nearest-neighbor interactions).

Table 11.1 Statistical weights of species resulting from binding to a trimer

Number of sites occupied	Species	Statistical weight
0	000	1
1	001	S
	010	S
	100	S
2	101	S^2
	011	τS^2
	110	τS^2
3	111	$\tau^2 S^3$

Because the statistical weight of each species is the relative concentration of that species, the mole fraction of any species X_i is the statistical weight of the species, ω_i, divided by the sum of the statistical weights of all species:

$$X_i = \frac{\omega_i}{\sum_i \omega_i} \tag{11.17}$$

Since $\sum_i \omega_i$ appears in many calculations, we define a function

$$Z = \sum_i \omega_i \tag{11.18}$$

For our present case, summing the statistical weights listed in Table 11.1 gives

$$Z = 1 + 3S + S^2 + 2\tau S^2 + \tau^2 S^3 \tag{11.19}$$

The average number of bound A per P is

$$\begin{aligned} \nu &= \frac{1[PA] + 2[PA_2] + 3[PA_3]}{[P] + [PA] + [PA_2] + [PA_3]} \\ &= \frac{3S + 2(S^2 + 2\tau S^2) + 3\tau^2 S^3}{1 + 3S + S^2 + 2\tau S^2 + \tau^2 S^3} \\ &= \frac{3S + 2(S^2 + 2\tau S^2) + 3\tau^2 S^3}{Z} \end{aligned} \tag{11.20}$$

Equation (11.20) can be written in a simple form. Because

$$Z = 1 + 3S + S^2 + 2\tau S^2 + \tau^2 S^3$$

$$\left(\frac{\partial Z}{\partial S}\right)_\tau = 3 + 2S + 4\tau S + 3\tau^2 S^2$$

and

$$S\left(\frac{\partial Z}{\partial S}\right)_\tau = 3S + 2(S^2 + 2\tau S^2) + 3\tau^2 S^3 \tag{11.21}$$

The right side of Eq. (11.21) is identical to the numerator in Eq. (11.20). Thus,

$$\begin{aligned} \nu &= S\left(\frac{\partial Z}{\partial S}\right)_\tau \Big/ Z \\ &= \left[\frac{\partial Z/Z}{\partial S/S}\right]_\tau \\ &= \left(\frac{\partial \ln Z}{\partial \ln S}\right)_\tau \end{aligned} \tag{11.22}$$

Although Eq. (11.22) is derived for a special case, it can be shown that the equation is generally applicable.

Exercise Show that for the case of four identical and independent sites discussed previously, Eq. (11.22) gives the same result as Eq. (11.10).

Cooperative Binding, Anticooperative Binding, and Excluded-site Binding

It is informative to examine the shape of a plot of the fraction of binding sites occupied, $\theta(\theta \equiv \nu/N)$ as a function of S. With $N = 3$, θ is plotted against S for the case $\tau = 1$ and $\tau = 10^3$ in Fig. 11.4. The case $\tau = 1$ is, of course, equivalent

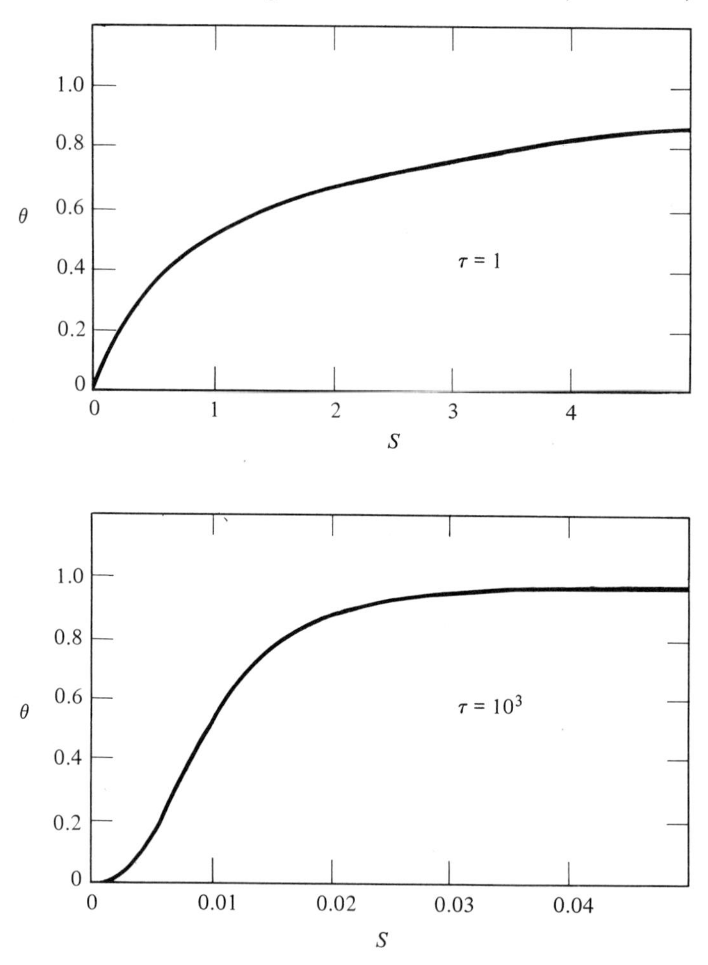

Fig. 11.4 Fraction of sites occupied, θ, as a function of the substrate concentration times the binding constant. The curves shown are for the cases of no interaction between occupied sites ($\tau = 1$), and for positive interaction between adjacent sites with a value of $\tau = 10^3$.

to the case when the sites are independent. For such a case, θ increases gradually with S and approaches 1 asymptotically. The case $\tau = 10^3$, or in general for $\tau > 1$, is termed *cooperative binding*, because the binding of A to one site makes it easier to bind another A to an adjacent site. Note that the shape of the θ versus S plot is sigmoidal (s-shaped). As S increases, θ increases slowly at first, showing an upward curvature, and then increases more rapidly. At still higher values of θ, a decrease occurs in the slope of the θ versus S plot with increasing S, showing a downward curvature. Such a sigmoidal plot of ν versus S is indicative of cooperative binding.

In Fig. 11.5, θ is plotted against log S for the same data shown in Fig. 11.4. Note that the shapes of the curves are dependent on whether the plot is *versus S* or log S. Several qualitative effects of cooperative binding can be seen in these

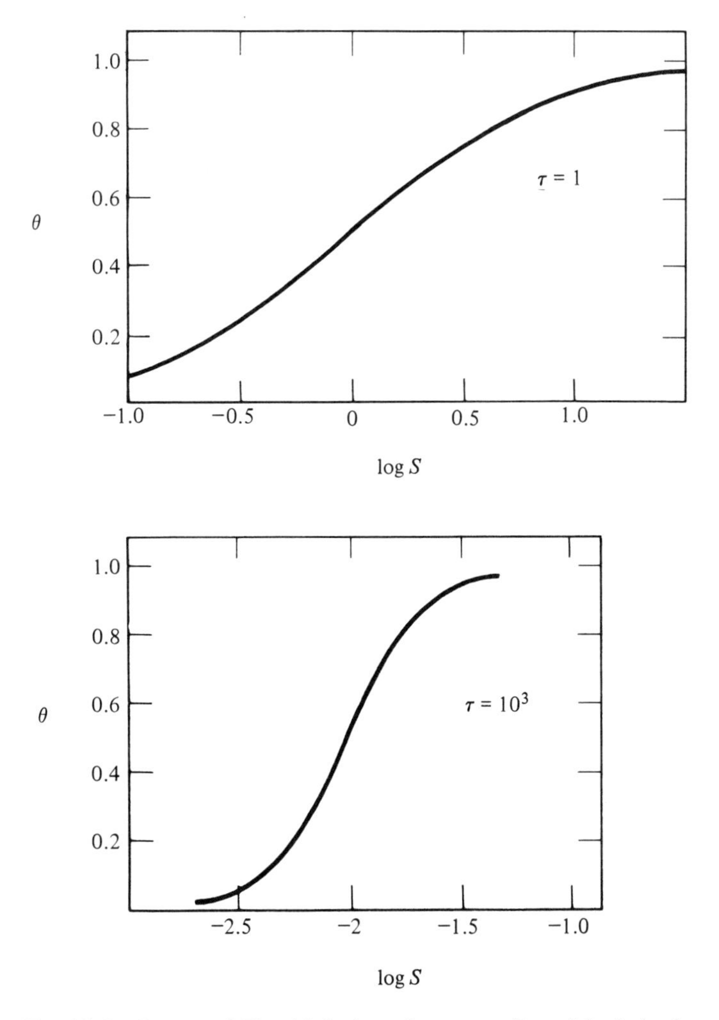

Fig. 11.5 Curves of Fig. 11.4 plotted on a semilogarithmic basis.

figures. For the independent-sites case ($\tau = 1$), θ changes from 0.25 (25% of the sites occupied) to 0.75 (75% of the sites occupied) over 1 log S unit, or a tenfold change in S. For the case with $\tau = 10^3$, the same change in θ occurs over 0.38 log S unit, or a 2.4-fold change in S. Thus the binding occurs more abruptly in the latter case.

For the independent-sites model, the shape of the θ versus S (or θ versus $\log S$) curve is not affected by the number of sites N per polymer molecule, since N is absent in the binding equation (11.16). For cooperative binding the shape of the θ versus S (or θ versus $\log S$) curve is dependent on N. As N increases, the number of terms in Eqs. (11.18) to (11.21) increases. For the same value of the interaction parameter τ, the difference in S for a given change in θ is less for larger N. Thus the abruptness of the cooperative binding is even more extreme. If N is very large, it can be shown that θ changes from essentially zero to essentially 1 in the range $(1 - 2/\sqrt{\tau}) < \tau S < (1 + 2/\sqrt{\tau})$.

Another important difference between independent-sites binding (noncooperative binding) and cooperative binding is the relative amounts of the various species. For our case with $N = 3$, let us examine the situation when $\nu = \frac{3}{2}$ (when half of the sites are occupied). The mole fractions of the species with 0, 1, 2, and 3 sites occupied are $1/Z$, $3S/Z$, $(S^2 + 2\tau S^2)/Z$, and $\tau^2 S^3/Z$, respectively. These quantities are calculated for the case $\tau = 1$ ($\nu = \frac{3}{2}$ at $S = 1$) and $\tau = 10^3$ ($\nu = \frac{3}{2}$ at $S = 9.75 \times 10^{-3}$), and are plotted in Fig. 11.6.

The difference between the two distributions is evident. In the case $\tau = 1$, the most abundant species center around $\nu = \frac{3}{2}$. In the case $\tau = 10^3$, however, the predominant species are the one with all sites free (47 mol %) and the one with all sites occupied (43.6 mol %). Such an uneven distribution is a consequence of the cooperative nature of binding. Since the occupation of one site makes the binding to adjacent sites more favorable, the bound A molecules tend to cluster. When τ is very large, only the species with all sites occupied and the species with all sites free are of importance. This is the *all-or-none limit.*

In the all-or-none limit, for our case of $N = 3$, we can simply consider the reaction as

$$P + 3A \longrightarrow PA_3$$

since none of the other species, PA and PA_2, is present in significant amounts. The equilibrium constant for the reaction is

$$K = \frac{[PA_3]}{[P][A]^3}$$

and

$$\nu = \frac{3[PA_3]}{[P] + [PA_3]} = \frac{3K[A]^3}{1 + K[A]^3} \tag{11.23}$$

Because $\nu/3 = \theta$, the fraction of sites occupied, it is straightforward to show that

$$\frac{\theta}{1 - \theta} = K[A]^3 \tag{11.24}$$

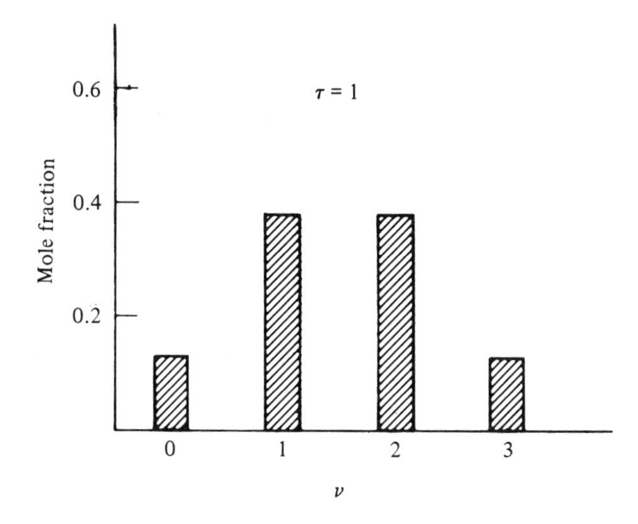

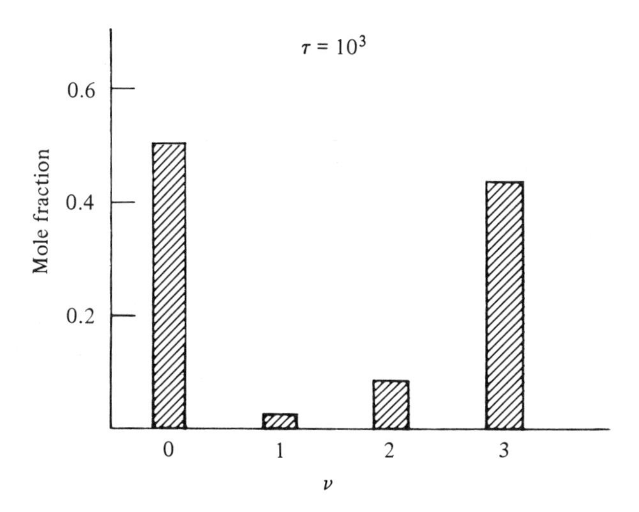

Fig. 11.6 Distribution of species at half saturation for $N = 3$. Independent sites ($\tau = 1$); cooperative binding ($\tau = 10^3$). Note that the cooperative binding leads to an all-or-none type binding.

or

$$\frac{d \log [\theta/(1 - \theta)]}{d \log [\mathrm{A}]} = 3$$

In general, if there are N sites per polymer molecule, the all-or-none model yields

$$\frac{d \log [\theta/(1 - \theta)]}{d \log [\mathrm{A}]} = N \tag{11.25}$$

It is useful sometimes to plot $\log[\theta/(1-\theta)]$ as a function of $\log[A]$. If the slope is 1, the sites are independent. If the slope is greater than 1, there is some cooperativity. If the slope is equal to N, the binding can be approximated by the all-or-none model.

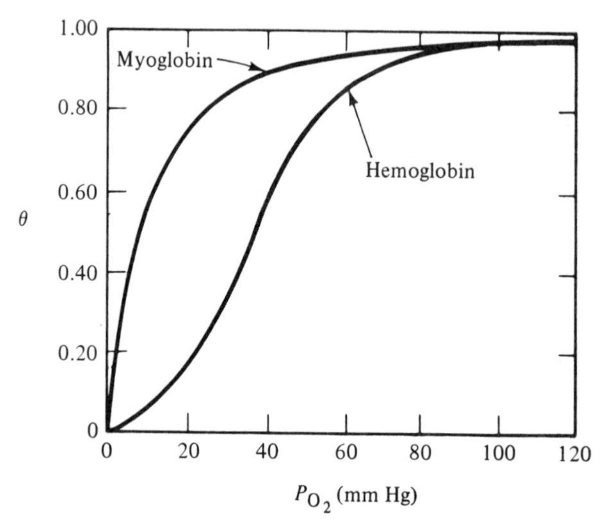

Fig. 11.7 Fraction of hemes of myoglobin or hemoglobin occupied by oxygen as a function of the pressure of oxygen. Myoglobin has only one heme group and therefore cannot show cooperativity. Hemoglobin has four hemes; the binding is cooperative.

Example 11.3 In Fig. 11.7, θ is plotted as a function of the partial pressure of O_2, P_{O_2}, for myoglobin and for hemoglobin. Myoglobin is an oxygen storage protein with one oxygen-binding heme group per molecule. The binding curve for this protein has the same shape as the curve in Fig. 11.4 for the independent-sites model ($\tau = 1$). The O_2-carrying protein of the blood, hemoglobin, has four heme groups per molecule. The binding curve for this protein is sigmoidal, indicating cooperative binding. If $\log[\theta/(1-\theta)]$ is plotted against $\log P_{O_2}$, the slope is 2.8 for hemoglobin, which is greater than 1 but less than 4.

The case $\tau < 1$ is also of interest. This occurs when binding to one site reduces the affinity of adjacent sites (*anticooperative* or *interfering binding*). In the limit $\tau = 0$, the occupation of one site excludes binding to adjacent sites, and the case is referred to as the *excluded-site model.*

N Identical Sites in a Linear Array with Nearest-neighbor Interactions

In the preceding section, we discussed quantitatively the binding of small molecules to a polymer molecule with three identical sites in a linear array. A number of important concepts came out of the discussion. These include cooperative and anticooperative binding, the all-or-none limit, and the excluded-site model. The transition from a polymer with $N = 3$ to one with a large number of monomer units requires no new physical concepts, but the mathematics is more involved. This is described in detail in the references given at the end of the chapter. Instead of repeating the lengthy derivations here, we shall examine some illustrative examples.

Example 11.4 Adenosine molecules can form hydrogen-bonded complexes with the bases of polyuridylic acid. In Fig. 11.8, the fraction of sites

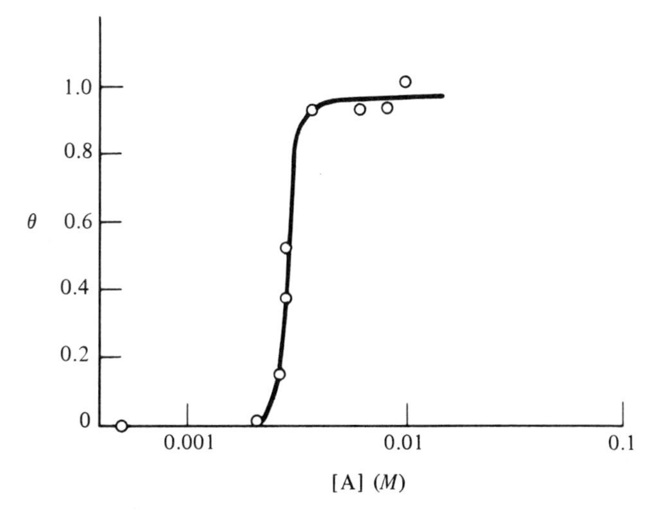

Fig. 11.8 Binding of adenosine, A, to polyuridylic. The fraction of sites occupied, θ, is plotted against the concentration of free adenosine, [A]. [Data from W. M. Huang and P. O. P. Ts'o, *J. Mol. Biol. 16*, 523 (1966).]

occupied in polyuridylic acid is shown as a function of the concentration of free adenosine. The shape of the curve clearly indicates cooperative binding. The interaction parameter τ is calculated to be $\sim 10^4$. The cooperativity is interpreted as a result of the favorable "stacking" interactions between adjacent adenosine molecules.

Example 11.5 The bacteriophage T4 gene 32 protein is involved in both DNA recombination and DNA replication. If the amount of the protein is insufficient to saturate the amount of the single-stranded DNA present, the distribution of the protein molecules among the DNA molecules is far from random. Some DNA molecules are nearly saturated with the protein, while others are essentially free of bound protein molecules, as shown in the electron micrograph in Fig. 11.9.

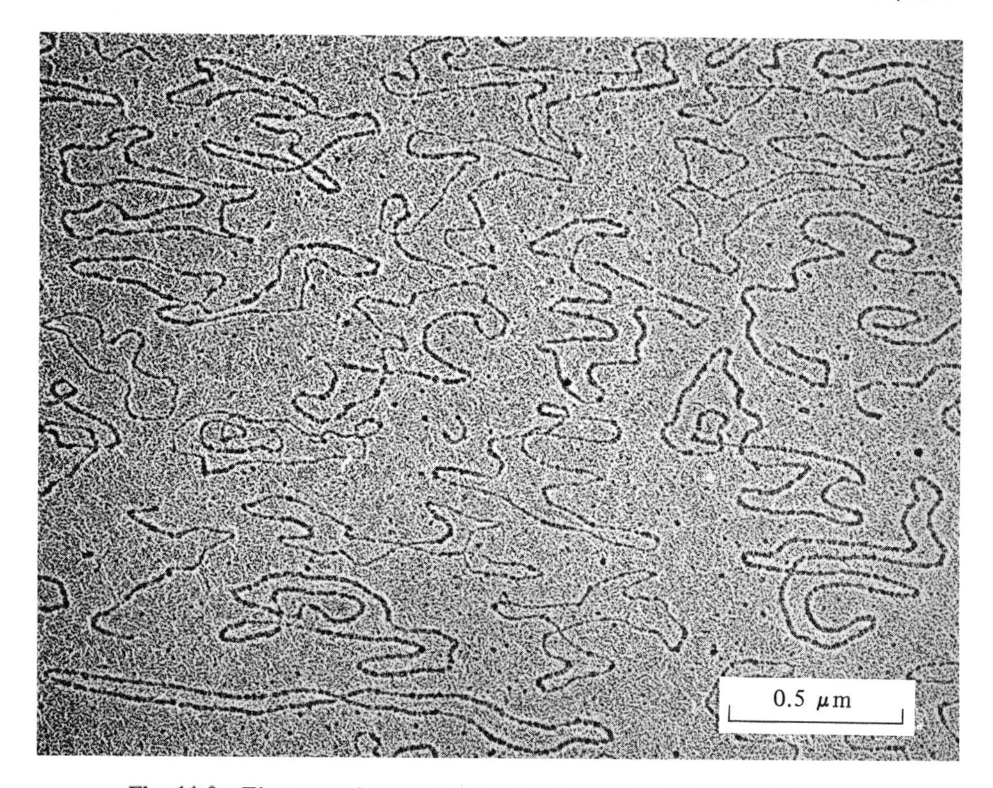

Fig. 11.9 Electron micrograph, showing loops of single-stranded circular DNA molecules. Look carefully and note that some segments of the DNA appear dark and thick, while others are light and thin. The thick segments show where the DNA gene 32 protein forms complexes with the DNA. The protein distribution is far from uniform along the DNA. You can find some DNA molecules that are completely thick and others that consist mostly of thin segments. [From H. Delius, N. J. Mantell, and B. Alberts, *J. Mol. Biol. 67,* 341 (1972).]

Example 11.6 An actinomycin molecule consists of a heterocyclic ring phenoxazone, with two cyclic pentapeptides attached to the ring. It binds to a double-stranded DNA by inserting the phenoxazone ring between two adjacent base pairs and tucking the peptides into the narrow groove of the DNA helix. One guanine-cytosine (G · C) base pair must be present at each binding site. Data for the binding of actinomycin to calf thymus DNA are shown in Fig. 11.10.

THE RANDOM WALK

In the preceding sections, we have discussed the binding of small molecules to a polymer with many sites. We shall now discuss an important problem in statistics:

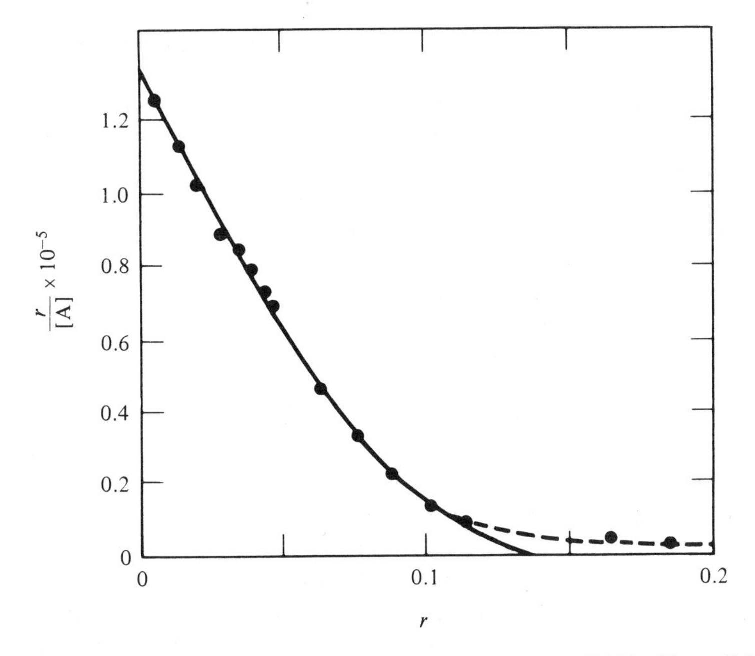

Fig. 11.10 Binding of actinomycin C_3 to calf thymus DNA. The solid curve shown, which fits the data well for $r \leq 0.1$, is obtained by using the excluded-site model. The parameters for the curve are $K = 1.6 \times 10^5$ and that two bound actinomycin molecules must be separated by at least six base pairs. It has also been assumed that every G · C is a site of binding and that the G + C pairs (41% of the base pairs are G · C in calf thymus DNA) are randomly distributed in the DNA. [Data from W. Müller and D. M. Crothers, *J. Mol. Biol. 35*, 251 (1968).)

the random-walk problem or the problem of random flight. The simplest example of such a problem involves someone walking along a sidewalk but unable to decide at each step whether to go forward or back. The direction of each step is random. This is a *one-dimensional random walk.* The *two-dimensional random walk* is analogous to a dazed football player who may take steps randomly in any direction on the field. The problem is also easily generalized to three (or more) dimensions.

If we suppose that the pace of each step is l, what is the probability, or chance, that after N steps the person is at a distance kl from the starting point? (The parameter k may be assigned any value between zero and N.) Intuitively, we may conclude that, if the steps are truly random in direction, the chance of $k = 0$ (finishing exactly at the starting point) or for $k = N$ (all steps in the same direction) will be small in comparison with the probability of having an intermediate value of k. We shall see that this intuitive picture is correct.

While the formulation of the random-walk problem may seem rather abstract and to bear little relation to physical problems, we will soon see that this is not so. A number of phenomena of interest, such as molecular diffusion and the average dimension of a long polymer molecule, can be treated as "random-walk problems" in three dimensions.

To introduce an analysis of the random-walk problem, it is best to start with the one-dimensional version. If we use the symbol 1 for a step forward and the symbol 0 for a step back, the record of a walk of N steps is represented by an N-digit number, such as 11000101011 . . . 0011. Let p be the probability of a step forward and q be the probability of a step back. When we said that the direction of each step was "random," we implied that $p = q$. In general, though, p does not have to be the same as q.

Before going further, let us elaborate on what we mean by "the probability of a step forward being p." Imagine that N similar fleas (N is a very large number) are hopping randomly forward or backward. These N fleas form an ensemble. After each has taken one hop, m are found to have moved forward and the rest, $N - m$, to have moved backward. The fraction of fleas that moved forward, m/N, is the probability p. Similarly, $(N - m)/N = q$. Obviously, $p + q = 1$, a consequence of the fact that we allow only two possibilities: either a hop forward or a hop back. This imaginary experiment can be performed in a different way. Instead of having N fleas take one hop each, we can require one flea to take N hops. We expect that, if there are no time-dependent changes, the fraction of steps forward is p. *It is a basic postulate of statistical mechanics that the time average of a certain property of a system is the same as the ensemble average at any instant.*

Now we come back to the one-dimensional random-walk problem. For a given sequence of N steps represented by 11000101011 . . . 0011, the probability for this sequence to occur is $ppqqqpqpqpp \ldots qqpp = p^m q^{N-m}$, where m is the number of 1's. There are many ways that we can take N steps with m of them forward. For example, if $N = 4$ and $m = 2$, the possible ways are 0011, 0110, 1100, 1010, 0101, and 1001. Several additional examples are presented in Table 11.2.

In Table 11.2 we can see that the sum of the probabilities of all possible outcomes of an N-step walk can be represented in the form $(q + p)^N$. Since $q + p = 1$, this sum is equal to 1, but we need to know how the sum depends on q, p, and N. The sum can be expanded by the binomial theorem:

$$(q + p)^N = q^N + Nq^{N-1}p + \frac{N \cdot (N - 1)}{2!} q^{N-2}p^2 + \cdots + \frac{N!}{m!(N - m)!} q^{N-m}p^m + \cdots + p^N \tag{11.26}$$

Remember that

$$N! = N \cdot (N - 1) \cdot (N - 2) \cdots (3) \cdot (2) \cdot (1)$$

Each term in the expansion gives the probability of the corresponding outcome. The term q^N, for example, is the probability that all N steps are

Table 11.2 Some random-walk examples

Total number of steps	Record of random walk	Probability	Sum of probabilities of all possibilities
1	0	q	$q + p$
	1	p	
2	00	q^2	$q^2 + 2qp + p^2$
	01, 10	$2qp$	$= (q + p)^2$
	11	p^2	
3	000	q^3	$q^3 + 3q^2p + 3qp^2 + p^3$
	001, 010, 100	$3q^2p$	$= (q + p)^3$
	011, 110, 101	$3qp^2$	
	111	p^3	
N	0000...	q^N	$(q + p)^N$
	...	...	
	...	...	
	1111...	p^N	

backward. The term

$$W(m) = \frac{N!}{m!(N - m)!} q^{N-m} p^m \tag{11.27a}$$

gives the probability that m steps are forward and $N - m$ steps are backward. The coefficient

$$\frac{N!}{m!(N - m)!} \tag{11.27b}$$

is the number of ways one can take N steps with m of them forward. Since $p + q = 1$, the numerical value of $(q + p)^N$ is of course always 1.

Calculation of Some Mean Values for the Random-walk Problem

The important equation is Eq. (11.27a). Knowing the probability of a random walk of N steps with m steps forward, we can now calculate certain averages of interest.

Mean displacement

First, let us calculate the average of the displacement. If the pace of each step is l, the net displacement for m steps forward is $[m - (N - m)]l$ or $(2m - N)l$. So if we determine $\bar{m}$, the average value of m, the *mean displacement* forward is $(2\bar{m} - N)l$.

The average value of m is the sum over each value of m times the probability of taking those m steps.

$$\bar{m} = \sum_{m=0}^{N} mW(m) \tag{11.28a}$$

$$\bar{m} = 0 \cdot q^N + 1 \cdot Nq^{N-1}p + 2 \cdot \frac{N \cdot (N-1)}{2!} q^{N-2}p^2 + \cdots$$

$$+ m \cdot \frac{N!}{m!(N-m)!} q^{N-m}p^m + \cdots + Np^N \tag{11.28b}$$

where Eq. (11.28b) is obtained by substituting Eq. (11.27) in Eq. (11.28a). The terms in Eq. (11.28b) look like those in Eq. (11.26), except that each term is multiplied by the exponent of p in the term. This suggests that Eq. (11.28b) might be related to the derivative of Eq. (11.26). We therefore let

$$Z = (q + p)^N = q^N + Nq^{N-1}p + \frac{N \cdot (N-1)}{2!} q^{N-2}p^2 + \cdots$$

$$+ \frac{N!}{m!(N-m)!} q^{N-m}p^m + \cdots + p^N \tag{11.26}$$

$$\frac{\partial Z}{\partial p} = Nq^{N-1} + 2 \cdot \frac{N \cdot (N-1)}{2!} q^{N-2}p + \cdots$$

$$+ m \cdot \frac{N!}{m!(N-m)!} q^{N-m}p^{m-1} + \cdots + Np^{N-1} \tag{11.29}$$

$$p\left(\frac{\partial Z}{\partial p}\right) = Nq^{N-1}p + 2 \cdot \frac{N \cdot (N-1)}{2!} q^{N-2}p^2 + \cdots$$

$$+ m \cdot \frac{N!}{m!(N-m)!} q^{N-m}p^m + \cdots + Np^N \tag{11.30}$$

Comparing Eqs. (11.30) and (11.28b), we see that

$$\bar{m} = p\left(\frac{\partial Z}{\partial p}\right) \tag{11.31}$$

But

$$Z = (q + p)^N$$

Therefore,

$$\left(\frac{\partial Z}{\partial p}\right) = N(q + p)^{N-1}$$

and

$$\bar{m} = p\left(\frac{\partial Z}{\partial p}\right) = Np(q + p)^{N-1} \tag{11.32}$$

Equation (11.32) is true for any values of p and q. For the random (unbiased) walk, $p = q = 1/2$, and

$$\bar{m} = \frac{N}{2} \tag{11.33}$$

and the net displacement forward is

$$(2\bar{m} - N)l = 0 \tag{11.34}$$

This seems completely reasonable. If the probability of going forward is the same as that of going backward, on the average half of the total number of steps taken will be forward. After taking N steps, sometimes one might be a few paces in front of the origin, sometimes a few paces behind the origin. The average or mean displacement comes out to be zero.

Mean-square displacement

After taking a larger and larger number of steps, N, at random, it becomes increasingly unlikely that we shall end up exactly where we started. The *mean-square displacement* is a measure of how far from the origin we can expect to be, on the average. The importance of the difference between the average distance from the origin and the average net displacement forward is that in calculating the average distance we do not distinguish whether we finally end up in front of or behind the origin. For example, if after N steps person A ends up three steps forward of the origin and person B ends up three steps behind the origin, the average of their net displacements is zero, but the average of their distances from the origin is clearly not zero, it is three steps.

One way to characterize the average distance from the origin is to calculate the *mean-square displacement.* In a single journey of N steps, if m steps are forward, the net displacement forward is

$$d = (2m - N)l \tag{11.35}$$

and the square of the displacement is

$$d^2 = (2m - N)^2 l^2 \tag{11.36}$$

Whether $m > N/2$ (a net displacement forward) or $m < N/2$ (a net displacement backward), d^2 is positive.

The mean value of $(2m - N)^2$ can be calculated as follows, where a bar above a quantity designates the average of that quantity:

$$\begin{aligned} \overline{(2m - N)^2} &= 4\overline{m^2} - 4\overline{mN} + \overline{N^2} \\ &= 4\overline{m^2} - 4N\bar{m} + N^2 \end{aligned} \tag{11.37}$$

But $\bar{m} = N/2$ for a random walk; thus

$$\overline{(2m - N)^2} = 4\overline{m^2} - 2N^2 + N^2 = 4\overline{m^2} - N^2 \tag{11.38}$$

To calculate $\overline{m^2}$, we start with its definition by analogy with Eq. (11.28),

$$\overline{m^2} = \sum_{m=0}^{N} m^2 W(m)$$

$$= 0 \cdot q^N + 1 \cdot Nq^{N-1}p + 2^2 \cdot \frac{N \cdot (N-1)}{2!} q^{N-2}p^2 + \cdots$$

$$+ m^2 \cdot \frac{N!}{m!(N-m)!} q^{N-m}p^m + \cdots + N^2p^N \tag{11.39}$$

If we differentiate Eq. (11.30) with respect to p, we obtain

$$\frac{\partial}{\partial p}\left[p\left(\frac{\partial Z}{\partial p}\right)\right] = Nq^{N-1} + 2^2 \cdot \frac{N \cdot (N-1)}{2!} q^{N-2}p + \cdots$$

$$+ m^2 \cdot \frac{N!}{m!(N-m)!} q^{N-m}p^{m-1} + \cdots$$

$$+ N^2p^{N-1} \tag{11.40}$$

Comparing Eqs. (11.39) and (11.40), we note that multiplying the right side of (11.40) by p gives the right side of (11.39). Therefore,

$$\overline{m^2} = p\left\{\frac{\partial}{\partial p}\left[p\left(\frac{\partial Z}{\partial p}\right)\right]\right\} \tag{11.41}$$

From Eq. (11.32),

$$p\left(\frac{\partial Z}{\partial p}\right) = Np(p+q)^{N-1}$$

Therefore,

$$\frac{\partial}{\partial p}\left[p\left(\frac{\partial Z}{\partial p}\right)\right] = N\left[p\frac{\partial (p+q)^{N-1}}{\partial p} + (p+q)^{N-1}\right]$$

$$= N[p(N-1)(p+q)^{N-2} + (p+q)^{N-1}]$$

and

$$p\frac{\partial}{\partial p}\left[p\left(\frac{\partial Z}{\partial p}\right)\right] = Np[p(N-1)(p+q)^{N-2} + (p+q)^{N-1}] \tag{11.42}$$

For $p = q = 1/2$ and $p + q = 1$,

$$\overline{m^2} = p\frac{\partial}{\partial p}\left[p\left(\frac{\partial Z}{\partial p}\right)\right] = \frac{N}{2}\left(\frac{N-1}{2} + 1\right) = \frac{N(N+1)}{4} \tag{11.43}$$

$$\overline{(2m-N)^2} = 4\overline{m^2} - N^2 \tag{11.38}$$

and, after all these complicated manipulations, we obtain the strikingly simple result that

$$\overline{(2m - N)^2} = N(N + 1) - N^2$$
$$= N \qquad (11.44)$$

Therefore, the mean-square displacement $\overline{d^2}$ is

$$\overline{d^2} = Nl^2 \qquad (11.45)$$

The root-mean-square displacement $\overline{(d^2)}^{1/2}$ is

$$\overline{(d^2)}^{1/2} = N^{1/2}l \qquad (11.46)$$

Equation (11.46) says that after N random steps, on the average one is $\sqrt{N}$ paces from where one started. Although we have derived this equation for a one-dimensional random walk, it can be shown that it holds for a random walk in two- or three-dimensional space as well. We presented the derivations for the random-walk displacements in detail to illustrate how average properties are obtained from statistical weights and sums over states.

Diffusion

Molecules, either in the gas phase or the liquid phase, are always undergoing many collisions. Let us assume that the average distance traveled by a molecule between two successive collisions (the mean free path) is l. Whenever a molecule has a collision, its direction changes, depending upon the direction of the molecule hitting it. Starting from a certain position at time zero, the total number, N, of collisions a given molecule has is proportional to the time t. There is a close relation between the diffusional process and the random walk. The fact that the individual "steps" in the diffusion process are not all of the same length turns out to be unimportant as long as the measurements are made over distances that are large compared with the mean free path. For a given spatial distribution of molecules (concentration gradient) at time zero, the spatial distribution of the molecules at time t can be obtained by considering each molecule as a random walker. The average value of the square of the displacement for the diffusional process based on Eq. (11.45) is expected to be proportional to N and, therefore, to the elapsed time. The proportionality constant is a measure of how fast the molecule diffuses and is directly related to the diffusion coefficient D, as was seen in Chapter 6.

Using the random-walk model, Einstein was able to derive the diffusion equation from the molecular point of view.

Average Dimension of a Linear Polymer

A flexible polymer can assume many conformations which differ little in energy. Consider a polypeptide,

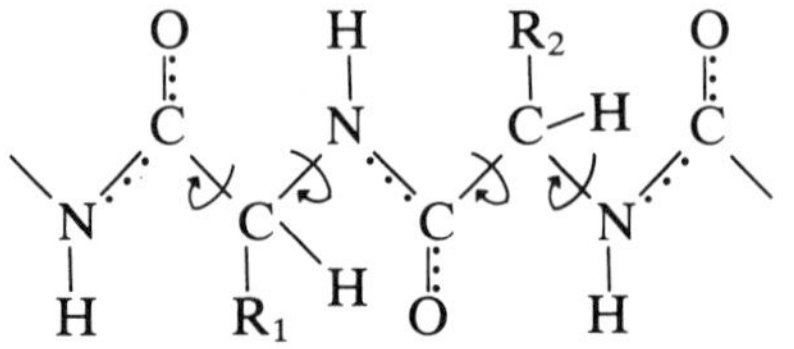

for example. Along the backbone of the polymer, rotation around the amide bonds N⋯C has a high-energy barrier because of the partial double-bond character of these bonds. Rotation around the other bonds is relatively free, however, resulting in many polymer conformations of similar stability. Some of these conformations may be highly extended, while others may be much more compact. Thus, in discussing the dimensions of such a polymer, it is necessary to specify *average* dimensions.

A quantity frequently used to express the average dimension of a flexible polymer is the root-mean-square end-to-end distance, $(\overline{h^2})^{1/2}$. If a polymer molecule is highly extended, its end-to-end distance is much greater than for the same molecule in a compactly coiled form. A simple and useful model for the evaluation of the average end-to-end distance is the random-coil or Gaussian model. In this model the linear polymer is considered to be made of N segments each of length l, linked by $(N - 1)$ universal joints (unrestricted bending) so that the angle between any pair of adjacent segments can take any value with equal probability.

The reader may recognize immediately that the random-coil model is exactly the same as the random-walk problem we have discussed. From Eq. (11.45), the mean-square end-to-end distance is $\overline{h^2} = \overline{d^2} = Nl^2$. For a given type of polymer, $\overline{h^2}$ is proportional to N and therefore proportional to the molecular weight of the polymer. We can also write

$$\overline{h^2} = Nl^2 = Nl \cdot l = Ll \qquad (11.47)$$

where $L \equiv Nl$ is the contour length of the molecule.

Real polymer molecules are, of course, made of monomers linked by chemical bonds rather than segments linked by universal joints. Let us consider polyethylene.

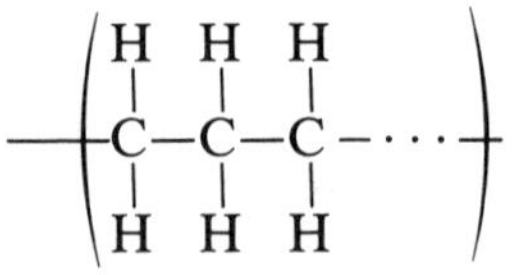

For simplicity, we consider only the carbon backbone chain

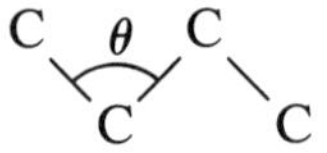

with a bond angle θ between two adjacent bonds and bond length b for each C—C bond. If θ could assume any value, one could imagine a universal joint at each carbon atom, and the mean-square end-to-end distance $\overline{h^2} = Nl^2 = Nb^2$, where N is the number of carbon atoms per polymer chain. We know, however, that the bond angle θ cannot assume any value. Rather, the most stable conformation for

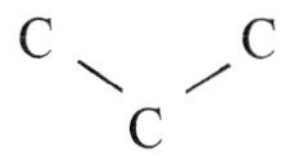

bonds has an angle θ close to the tetrahedral angle of 109°. The restriction in θ makes the molecule a little less flexible. If the random-coil model is used for polyethylene, we expect that $l > b$, the effective distance, l, between carbon atoms will be greater than the bond length, b. In fact, it can be shown that $l = \sqrt{2}b$ for polyethylene.

For a polymer such as a double-stranded DNA, the effective segment length, l, of an equivalent random coil cannot be related directly to the bond lengths. Hydrodynamic measurements give $l \approx 10^3$ Å for DNA. Thus for a flexible molecule such as polyethylene, the segment length is of the order of the bond length [$l = \sqrt{2}b \cong 2.2$ Å]; for a stiff molecule such as a double-stranded DNA, the effective length is several orders of magnitude larger than a bond length.

A quantity closely related to mean-square end-to-end distance is the mean-square radius, which is defined by

$$\overline{R^2} = \frac{\sum_i m_i r_i^2}{\sum_i m_i} \qquad (11.48)$$

for any collection of mass elements, where m_i is the mass of the ith element and r_i is the distance of this element from the center of mass of the collection.

It can be shown that for an open-ended random coil,

$$\overline{R^2} = \frac{Nl^2}{6} \qquad (11.49a)$$

and for a circular random coil (formed by linking the two ends of the open-ended chain),

$$\overline{R^2} = \frac{Nl^2}{12} \qquad (11.49b)$$

It was first shown by Debye that $\overline{R^2}$ can be measured experimentally by light scattering.

Note that the random-coil model predicts that $(\overline{h^2})^{1/2}$ or $(\overline{R^2})^{1/2} \equiv R_g$ is proportional to $\sqrt{N}$, or the square root of the molecular weight. This relation can be tested experimentally. If it is found to be true, the polymer is said to behave as a random coil, or a Gaussian chain.

For real polymers, in general, there are interactions between adjacent segments, and they will have preferred orientations instead of equally probable orientations, as required by the random-coil model. At a given temperature the random-coil model is expected to be applicable only in solvents ("θ solvents") in which the interactions between polymer segments are balanced by solvent-polymer interactions. Otherwise, the interactions between the polymer segments will cause deviations from the random-coil model.

HELIX-COIL TRANSITIONS

Helix-coil Transition in a Polypeptide

In a polypeptide chain the partial double-bond character of the N⋯C and C⋯O bonds for each amide residue in the chain requires the group of six atoms

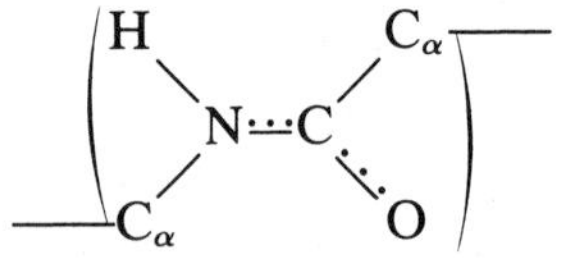

to lie in a plane. It was first suggested by Pauling and his coworkers in 1951 that the chain of planar groups can be rotated around the α-carbons (C atoms between the CO group of one amino acid and the NH group of the next) into a helix such that a hydrogen bond is formed between each CO group and the fourth preceding NH group. The locations of the hydrogen and oxygen atoms involved in hydrogen bonding are illustrated in Fig. 11.11, and a projection of the resulting helix, called

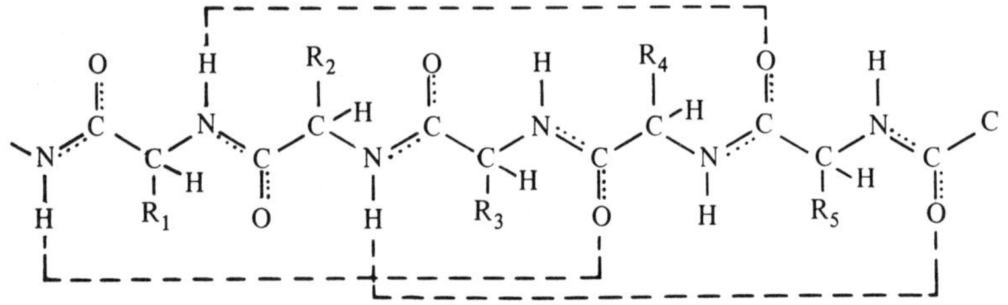

Fig. 11.11 Atoms linked together by hydrogen bonds when a polypeptide chain is wound into an α helix.

an α *helix*, is shown in Fig. 11.12. The polypeptide backbone of the α helix follows closely along the contour described by a helical spring made from a single strand of wire. The existence of the α-helix structure in many proteins has now been confirmed, mainly by x-ray diffraction studies of proteins. For example, in the enzyme adenyl kinase from porcine muscle, over one half of the amide residues are in the α or distorted α-helix conformation.

For a number of synthetic polypeptides, the entire chain of the molecule can assume the α-helix structure, depending upon the solvent, temperature, pH, and so on. In the non-hydrogen-bonded form, the polypeptide chain is flexible and

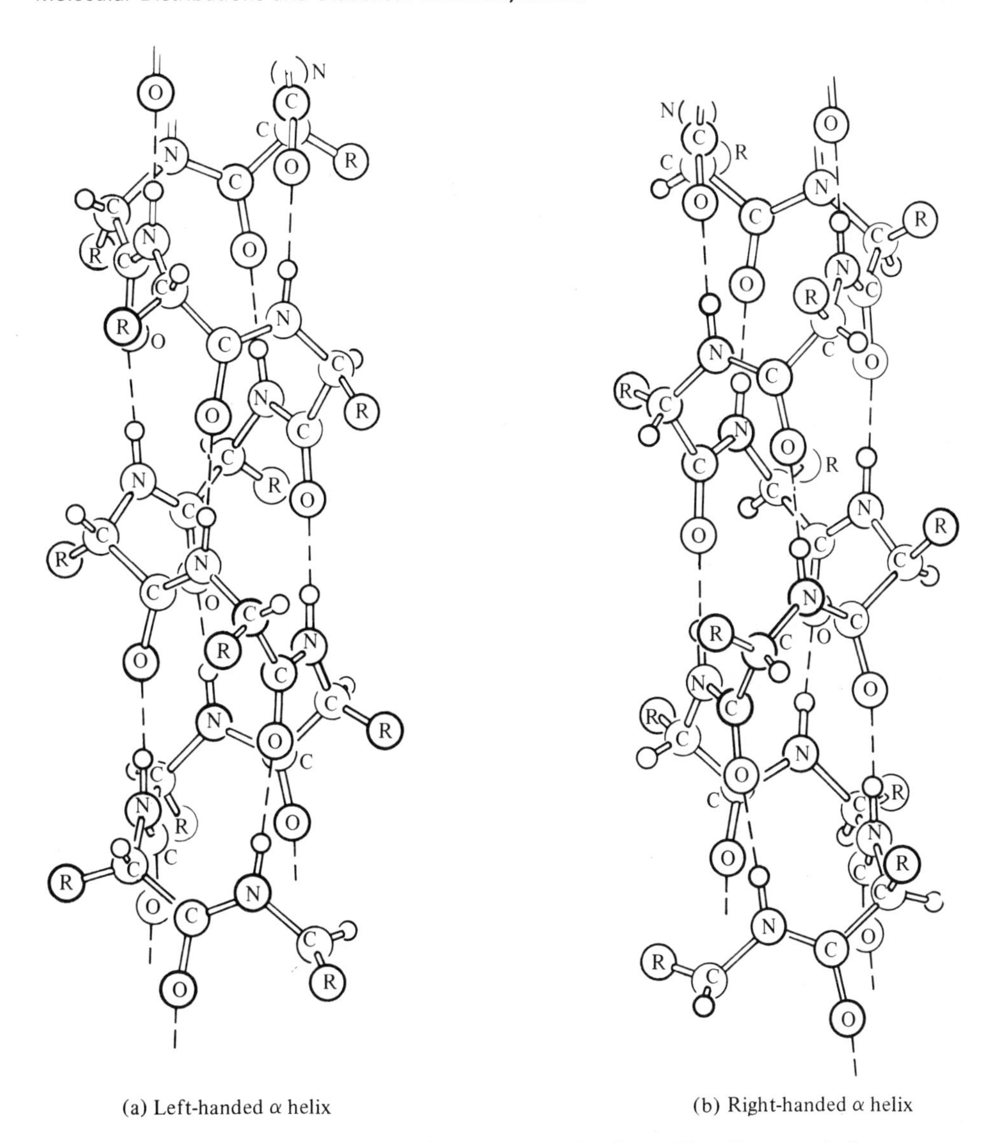

(a) Left-handed α helix (b) Right-handed α helix

Fig. 11.12 The α helix. (From Linus Pauling, *The Nature of the Chemical Bond*, Cornell University Press, Ithaca, N.Y., 1960, p. 500. Copyright © 1939 and 1940, third edition © 1960 by Cornell University. Used by permissions of Cornell University Press.)

can assume many conformations as discussed previously. Such a chain is frequently said to be in the "coiled" form. The transition from the coiled form to the helical form (or vice versa) is referred to as a *coil-helix transition.* For polypeptides of high molecular weight, if one of the parameters (such as temperature) is changed, the transition from the coiled form to the α-helix form (or vice versa) generally occurs within a narrow range.

We can understand the basic features of the helix-coil transition using a formulation that is similar to our discussion of the binding of small molecules by a polymer. The conformation of the chain can be specified by the states of the carbonyl oxygen atoms. We use the symbol 1 to represent a carbonyl oxygen if it is hydrogen-bonded and the symbol 0 if it is not. Whenever an oxygen is in a bonded state, it is to be understood that it is hydrogen-bonded to the fourth preceding NH group. (It follows, then, that the first three oxygen atoms of the chain are necessarily unbonded.) It is further assumed that the sense of the helix is unique (either left-handed or right-handed, but not both). In this way, the conformation of a chain of N residues is represented by an N-digit number, with the first three digits always zeros.

Now consider the transition from a state such as 00000000 . . . to a state 00000010 . . . , which represents the transition of the seventh residue from the nonbonded state to the α-helix configuration. In the nonbonded state, as shown earlier, there are two bonds in each residue around which free rotation may occur. In a helical form, however, the chain assumes a much more rigid conformation. Since the oxygen atom of the seventh residue is H-bonded to the fourth preceding NH group, the transition of the seventh residue from the coiled state to the helical state means that the fifth and sixth residues must also be ordered into the helical conformation (even though their oxygen atoms are not H-bonded). In other words, the formation of the first helical element involves the ordering of three residues. The transition of the state 00000010 . . . to the state 00000011 . . . , however, requires the ordering of only one more residue, the eighth, since the sixth and seventh residues are already ordered. Intuitively, then, we would expect that the formation of the first helical element is more difficult than the formation of the next adjacent helical element.

Based on this discussion, we can assign statistical weights according to the following rules:

1. For each 0, the statistical weight is taken as unity.
2. For each 1 after a 1, the statistical weight is s.
3. For each 1 after a 0, the statistical weight is σs.

Note that s is the equilibrium constant for the reaction

$$00000010\ldots \longrightarrow 00000011\ldots$$

or the addition of a helical element to the end of a helix; σs is the equilibrium constant for the reaction

$$00000000\ldots \longrightarrow 00000010\ldots$$

or the *initiation* of a helical element. We expect σ to be less than 1, because initiation of an α helix is more difficult than propagation of the helix.

The reader should recognize the similarity between the present case and the previously discussed binding problem with nearest-neighbor interactions. The similarity can be made more apparent if the parameters used in the previous discussion on binding, τ and S, are transformed into σ and s by the relation $\tau = 1/\sigma$ and $S = \sigma s$.

There is one difference between the two cases, however. For binding, we consider only nearest-neighbor interactions. For an α-helix formation, the very nature of the α-helix structure dictates more than nearest-neighbor influence: For example, if the seventh residue goes into a helical conformation, both the fifth and the sixth residues are also brought into the helix conformation. The consequence of this is that the conformations of three adjacent residues are closely related, and conformations with two 1's separated by no more than two 0's are expected to be rare. To see the last point more clearly, let us consider a reaction

$$000001111\ldots \longrightarrow 000001001\ldots$$

as an example. For 000001111 . . . , the fourth to the ninth residues are in the α-helix structure, because the sixth residue is bonded to the third residue. In 000001001 . . . , the fourth to ninth residues are still constrained in the α-helix structure, since we have specified that the sixth and the ninth residues are in the α-helix form. Thus for this reaction, although two hydrogen bonds are broken, there is no gain in freedom for any of the residues. Therefore, the reaction is not favored. This leads to the fourth rule for the assignment of statistical weights:

4. For two 1's separated by no more than two 0's, the statistical weight is zero.

To assign a statistical weight of zero is of course equivalent to saying that such conformations are not permitted. This is a bit of overkill, but the rule is a good approximation for our purpose.

With these four rules, we can assign the statistical weight for any conformation. The statistical weight for the conformation 0001111000100011100 is $\sigma^3 s^9$, for example.

Once the rules for the statistical weights have been formulated, the helix-coil transition problem can be handled by mathematical techniques similar to the ones we have discussed for the binding of small molecules to sites with nearest-neighbor interactions. The function Z can be formulated and, in a way similar to Eq. (11.22), $(\partial \ln Z/\partial \ln s)_\sigma$ gives the average number of monomer units in the helical conformation per polymer molecule. We shall not give the mathematical details here. Certain features of the coil-helix transition will be discussed in the following examples.

Example 11.7 Results of the classical treatment of Zimm et al. (1959) for the coil-helix transition of poly-γ-benzyl-L-glutamate are shown in Fig. 11.13. The curves shown were obtained by the statistical analysis we outlined. The value of σ used to calculate the curves is 2×10^{-4}. As we have discussed, the parameter s represents the equilibrium constant for the change of a coil element at the end of a helical segment into a helical segment:

$$\ldots 111110 \ldots \overset{s}{\rightleftharpoons} \ldots 111111 \ldots$$

If the polypeptide molecule is very long, the free-energy difference between a monomer unit in the coil form and in the helix form is zero at the

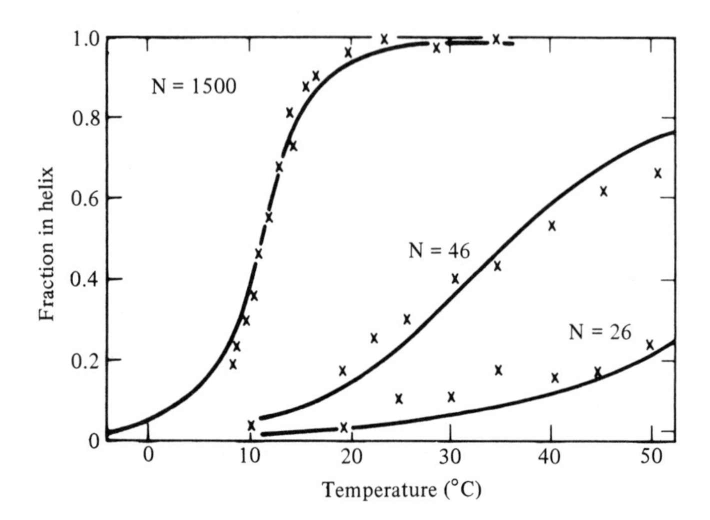

Fig. 11.13 Temperature dependence of the fraction of monomer units in the helix form for poly-γ-benzyl-L-glutamate (in a 7:3 mixture of dichloracetic acid and 1,2-dichloroethane as the solvent) with polymer lengths of 26, 46, and 1500 monomer units. In this solvent the helix is favored at high temperatures. [Data from B. H. Zimm and J. K. Bragg, *J. Chem. Phys. 31*, 526 (1959); B. H. Zimm, P. Doty, and K. Iso, *Proc. Natl. Acad. Sci. U.S. 45*, 1601 (1959).]

temperature corresponding to the midpoint of the transitions (T_m, the melting temperature) shown in Fig. 11.13. Thus s is equal to 1 at T_m. At any other temperature, s can be calculated from

$$\ln \frac{s_2}{s_1} = -\frac{\Delta H^0}{R}\left(\frac{1}{T_2} - \frac{1}{T_1}\right) \tag{4.43}$$

where ΔH^0 is the enthalpy change for the reaction shown above.

When N is large, the transition is very sharp. The polypeptide changes from 80% in the coil form to 80% in the helix form in a temperature range of ~7°. In this temperature range the change in s is rather small, from 0.97 to 1.03. The cooperative transition is the result of a small σ. If we consider the formation of a helix from a coil, a small σ means that to start a helical region in the middle of a coil region is difficult, but once a helical region is initiated, it is much easier to extend it. This can be seen from the statistical weights of the species for the following transition:

	...0000000...	$\rightleftharpoons$	...0000100...	$\rightleftharpoons$	...0000110...
Statistical weight:	1		σs		σs^2

For the first reaction, the equilibrium constant is $\sigma s/1$ or σs. For the second reaction, the equilibrium constant is $\sigma s^2/\sigma s$ or s, which is much greater than σs.

Conversely, if we consider the formation of a coil from a helix, to start a coil region in the middle of a helix is difficult, as follows:

$$\ldots 0001111111 \ldots \rightleftharpoons \ldots 0001100011 \ldots \rightleftharpoons \ldots 0001100001 \ldots$$

Statistical weight: σs^7 $\qquad$ $\sigma^2 s^4$ $\qquad$ $\sigma^2 s^3$

The equilibrium constant for the first reaction is $\sigma^2 s^4/\sigma s^7$ or σ/s^3. The equilibrium constant for the second reaction is $\sigma^2 s^3/\sigma^2 s^4$ or $1/s$. Since in the region of interest, $s \approx 1$ and $\sigma \ll 1$, to initiate a coil region in the middle of a helix is more difficult. A consequence of this is that, for short polypeptides, coil regions are expected to be at the end of the molecules. For long polypeptides there are so many sites in the middle of the molecules that, in the transition zone, coil regions are expected to be present in the middle of the molecules as well.

Because of the difficulty of initiation and the ease of propagation, an entire cluster of monomers tends to go from one state to the other in a cooperative manner. Such cooperative behavior can also be observed in a collection of more complex elements. Many examples can be found in human societies as well.

The sharpness of the transitions shown in Fig. 11.13 differs for different N, although all curves can be fitted by the same parameters. For the shorter polypeptides, the transition is broader. This is similar to the case of the binding of small molecules to interacting sites discussed previously.

Example 11.8 The unfolding of ribonuclease A by heating. As shown in Fig. 11.14, heating ribonuclease A (in 0.04 *M* glycine buffer, pH 3.15) causes the unfolding of the protein. The unfolding was monitored by the change in light absorption. The transition from the folded form to the unfolded form occurs in a narrow temperature range. The unfolding of a protein is more complex than the helix-coil transition of a simple polypeptide, however. The folded protein has structural features other than the α helix. In addition, factors such as disulfide bridges, ionic bonds, and interactions between the nonbonded groups can contribute to the stability of the folded form. However, since most of these factors contribute significantly only if the protein is in a correctly folded form, the unfolded to folded, or folded to unfolded, transition is highly cooperative. In fact, the all-or-none model has been used in such studies until very recently, when kinetic studies indicated that there are more than two states involved.

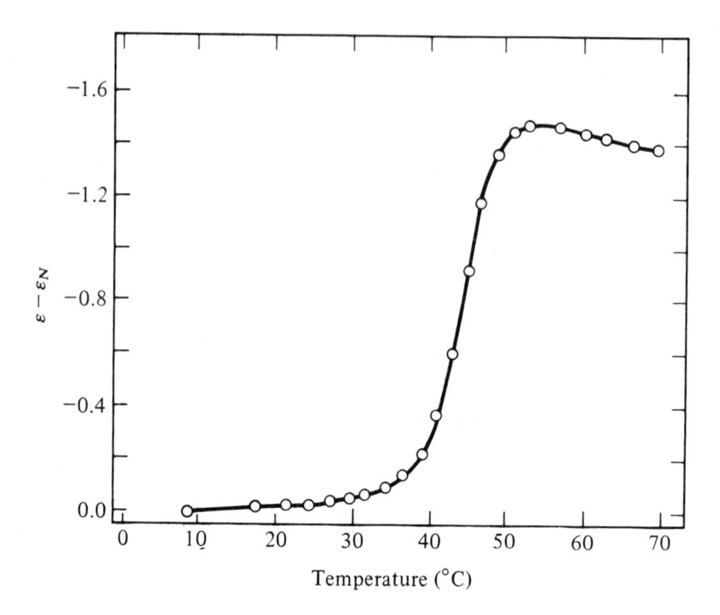

Fig. 11.14 Temperature dependence of $\varepsilon - \varepsilon_N$ of ribonuclease A. ε is the molar absorptivity of the protein at 287-nm wavelength; ε_N is the same quantity when the protein is in its native (folded) form. [Data from J. Brandts and L. Hunt, *J. Amer. Chem. Soc. 89*, 4826 (1967).]

Helix-coil Transition in a Double-stranded Nucleic Acid

In the preceding discussion on the helix-coil transition in an α-helix, we have seen that the difficulty in initiating the formation of a helical segment ($\sigma \ll 1$) is the essential reason for the cooperative transition. In a double-stranded nucleic acid, a somewhat similar situation exists. Double-stranded DNA is held together by the interactions between complementary bases in two single strands of DNA as illustrated in Fig. 11.15. Let us represent a base pair (in the hydrogen-bonded form) by the symbol 1, and a pair of bases in the nonbonded form by the symbol 0. Let s be the equilibrium constant for the formation of an additional bonded pair at the end of a helical segment:

$$11110000\ldots \underset{}{\overset{s}{\rightleftharpoons}} 11111000\ldots$$

If we compare the reaction above with the reaction

$$11110000\ldots \underset{}{\overset{\sigma s}{\rightleftharpoons}} 11110100\ldots$$

we note that there is an important difference. In the first reaction, the newly formed base pair is directly "stacked" on the base pair at the end of the original helix. For polynucleotides, either DNA or RNA, the stacking of base pairs is

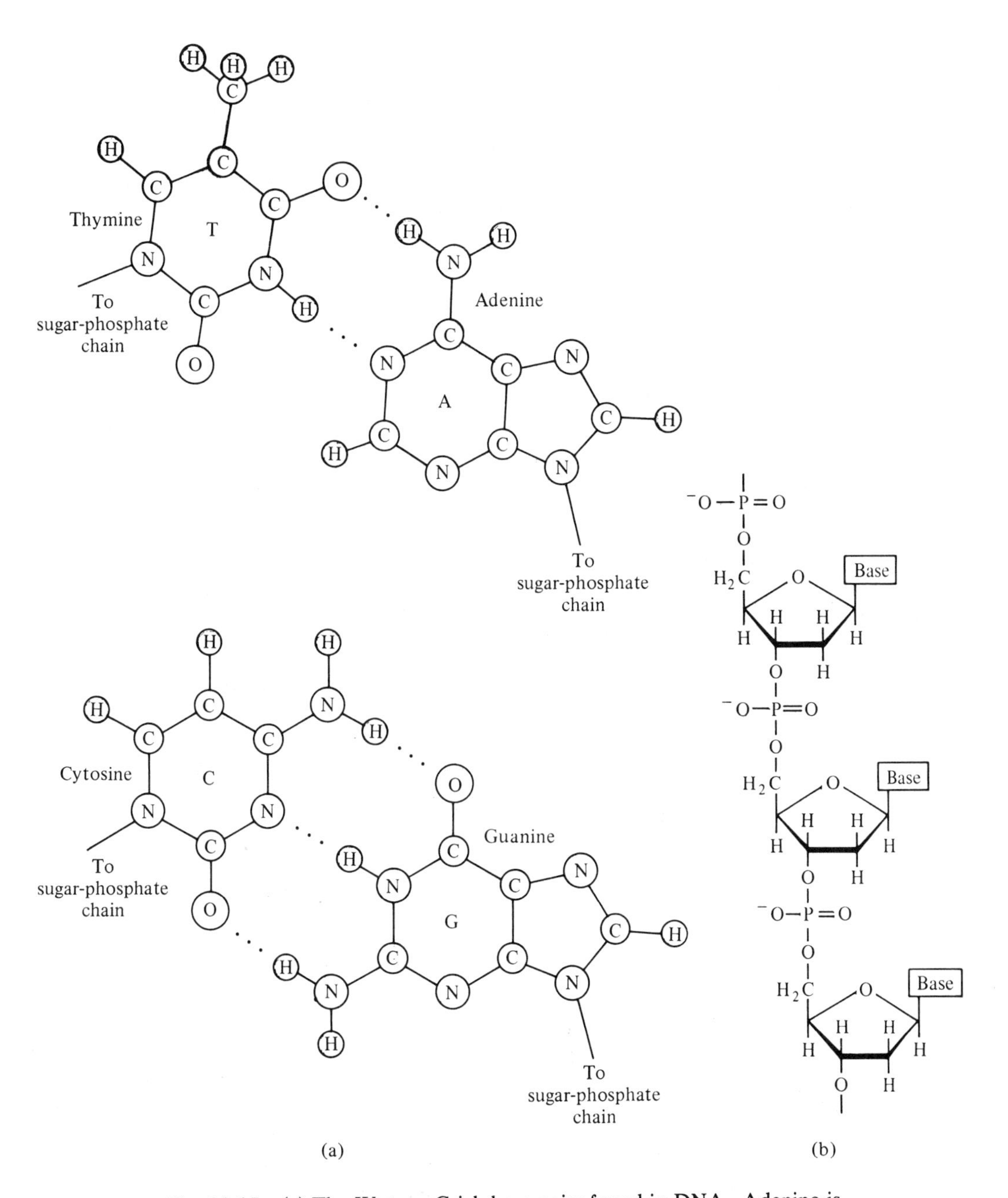

Fig. 11.15 (a) The Watson-Crick base pairs found in DNA. Adenine is complementary to thymine (A · T) and guanine is complementary to cytosine (G · C). In RNA uracil substitutes for thymine (A · U). (b) The sugar-phosphate chain in a single strand of deoxyribonucleic acid, DNA. The sugar-phosphate chain in RNA differs only in that the sugar is ribose instead of deoxyribose as shown above.

thermodynamically favored. In the second reaction, there is little stacking interaction between the newly formed pair and the last pair in the preceding helical segment. If the equilibrium constant is σs for the second reaction, we expect that $\sigma \ll 1$ if stacking interaction is thermodynamically favored. Thus, the helix-coil transition in a double-stranded nucleic acid is similar to that in a polypeptide.

There are several new features in the nucleic acid case, however. First, let us consider the two reactions

$$11110000\ldots \rightleftharpoons 11110100\ldots$$

and

$$1111\underbrace{0000\ldots00}_{j\text{ zeros}}00\ldots \rightleftharpoons 1111\underbrace{0000\ldots00}_{j\text{ zeros}}10\ldots$$

If j is very large, it means that the pair of bases which are to form the additional base pair are on the average far apart in the nonbonded form. From Eq. (11.46), the root-mean-square average distance between this pair of nonbonded bases is proportional to $\sqrt{j}$. It is therefore expected that the larger j is, the more difficult it is to bring the two together to form a base pair. Thus, unlike the case for a polypeptide, σ for a nucleic acid is not a constant but a function of j. The equilibrium constant for the first reaction is $\sigma_1 s$ and that for the second is $\sigma_j s$. This complicates the situation quite a bit. To compare this case with the binding of small molecules by a polymer, the dependence of σ on j is equivalent to the situation where we have to consider not only the nearest-neighbor interactions ($j = 0$) but also the next-nearest-neighbor interaction ($j = 1$), the next..., and so on.

Second, the formation of the very first base pair between two complementary chains means that the two chains, which are free to move about in solution independently of each other, must be brought together. After the formation of the first base pair one of the two chains can still be considered as free to be anywhere, but the second chain is constrained by the pairing to move with the first chain. Thus for the reaction

$$\underset{\text{all zeros}}{000\ldots000} \longrightarrow \underset{\text{all zeros except one}}{0000100\ldots000}$$

the equilibrium constant is taken as ξs, with ξ expected to be much less than 1. The parameter ξ is also expected to be dependent on total concentration. The lower the concentration, the harder it is to bring two chains together, and the smaller is ξ.

Third, there are two major types of base pairs in a nucleic acid. For a DNA these are A · T pairs and G · C pairs; for an RNA they are A · U pairs and G · C pairs. Since the stabilities of the two kinds of base pairs are different, in general, we should use the parameters s_A and s_G for these types rather than a single s. Strictly speaking, the stacking interaction between two adjacent pairs is dependent on what kinds of pairs they are, and therefore s_A and s_G are further dependent on at least the nearest-neighbor base sequence.

To include all these considerations in a theoretical analysis of the helix-coil transition is beyond the scope of this text. Nevertheless, from our analyses of the problem a number of features of the helix-coil transition in a nucleic acid can be understood:

1. When N is large, the transition from the completely helical form to the completely coiled form in a given solvent occurs within a narrow range of temperature. This cooperative process is frequently referred to as the *melting* of a nucleic acid. The temperature corresponding to the midpoint of the transition is designated as T_m, the *melting temperature.* The cooperativity of the transition is primarily a result of the stacking interactions between the adjacent base pairs, which make the initiation of a helical segment as well as the disruption of a base pair inside a helical segment difficult.
2. The enthalpy change for the reaction $11110000\ldots \overset{s}{\rightleftharpoons} 11111000\ldots$ is negative. In most of the solvents $s_G > s_A$; thus T_m for a nucleic acid rich in G + C is higher than that for a nucleic acid rich in A + T (or A + U). If the base composition of a nucleic acid is intramolecularly heterogeneous, that is, some segments of each molecule are richer in A + T (or A + U) than the other portions, the melting profile is broadened, with the melting of the regions richer in A + T (or A + U) preceding the melting of the (G + C)-rich regions. Because of the cooperative nature of the transition, segments heterogeneous in base composition must be sufficiently long to show independent melting profiles.
3. For a large molecule, the parameter ξ contributes negligibly to $\ln Z$. Since ξ is the only concentration-dependent term in Z, T_m for a large molecule is expected to be independent of the concentration. For short helixes this is not true. T_m should be lower at lower concentration. Similarly, at the same total concentration of nucleotides, T_m for a high-molecular-weight nucleic acid is expected to be higher than that of a low-molecular-weight nucleic acid of the same base composition. This dependence is expected to be evident only in the molecular-weight range where ξ contributes significantly to $\ln Z$.
4. As discussed for the helix-coil transition in a polypeptide, if N is small, the coiled regions are expected to be at the ends of the molecules. This is frequently referred to as "melting from the ends."

Example 11.9 The transition temperatures for the helix-coil transitions of DNAs correlate well with the base composition of the DNA. Plots of a large number of such measurements for two sets of experimental conditions are shown in Fig. 11.16. Analogously, for a particular DNA molecule the range of temperatures (transition width) over which melting occurs is related to the uniformity of base composition along major segements of the chain. Synthetic DNA's with regular repeating sequences will have sharp melting transitions.

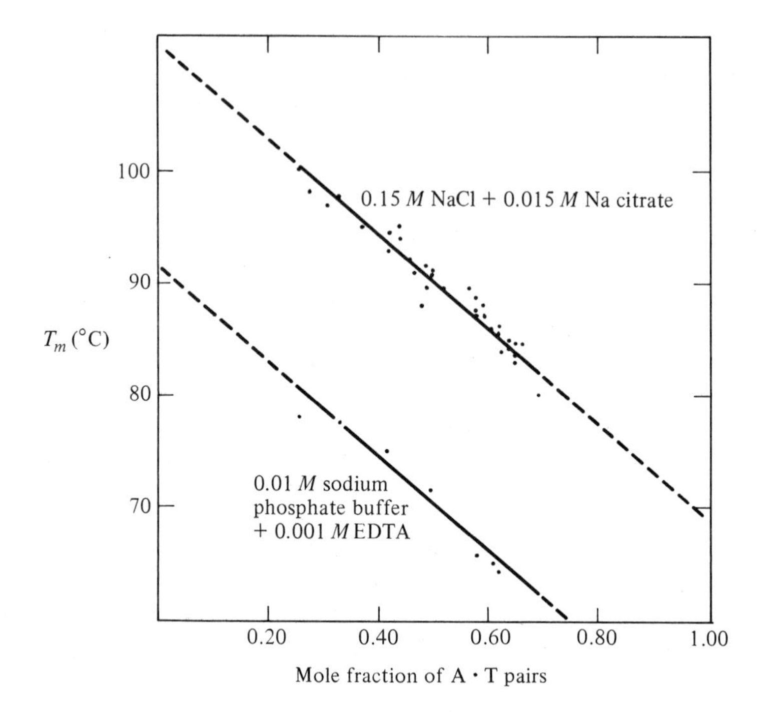

Fig. 11.16 Melting temperatures of double-stranded DNA's are plotted against the mole fraction of A · T pairs in the DNA's. In a given ionic medium, T_m decreases with increasing A · T content, indicating that $s_A < s_G$. Note also that for a given base composition, T_m is lower in a medium with lower Na^+ concentration. This is because the repulsion between the negatively charged phosphate groups on the two complementary strands is reduced by the positively charged counterions (Na^+ in this case). Increasing the Na^+ concentration increases the stability of the double helix and, therefore, increases its T_m. [Data from J. Marmur and P. Doty, *J. Mol. Biol. 5*, 109 (1962).]

Example 11.10 The DNA of coliphage 186 has a molecular weight of 20×10^6. The base composition of the DNA is not homogeneous along the molecule, and segments rich in AT pairs can be denatured before the denaturation of segments rich in G + C pairs. The consequences of such inhomogeneous denaturation are seen in Fig. 11.17.

Example 11.11 The effect on T_m of shearing T2 DNA to reduce its length is seen in Fig. 11.18. Shorter fragments have lower T_m values and broader melting transitions.

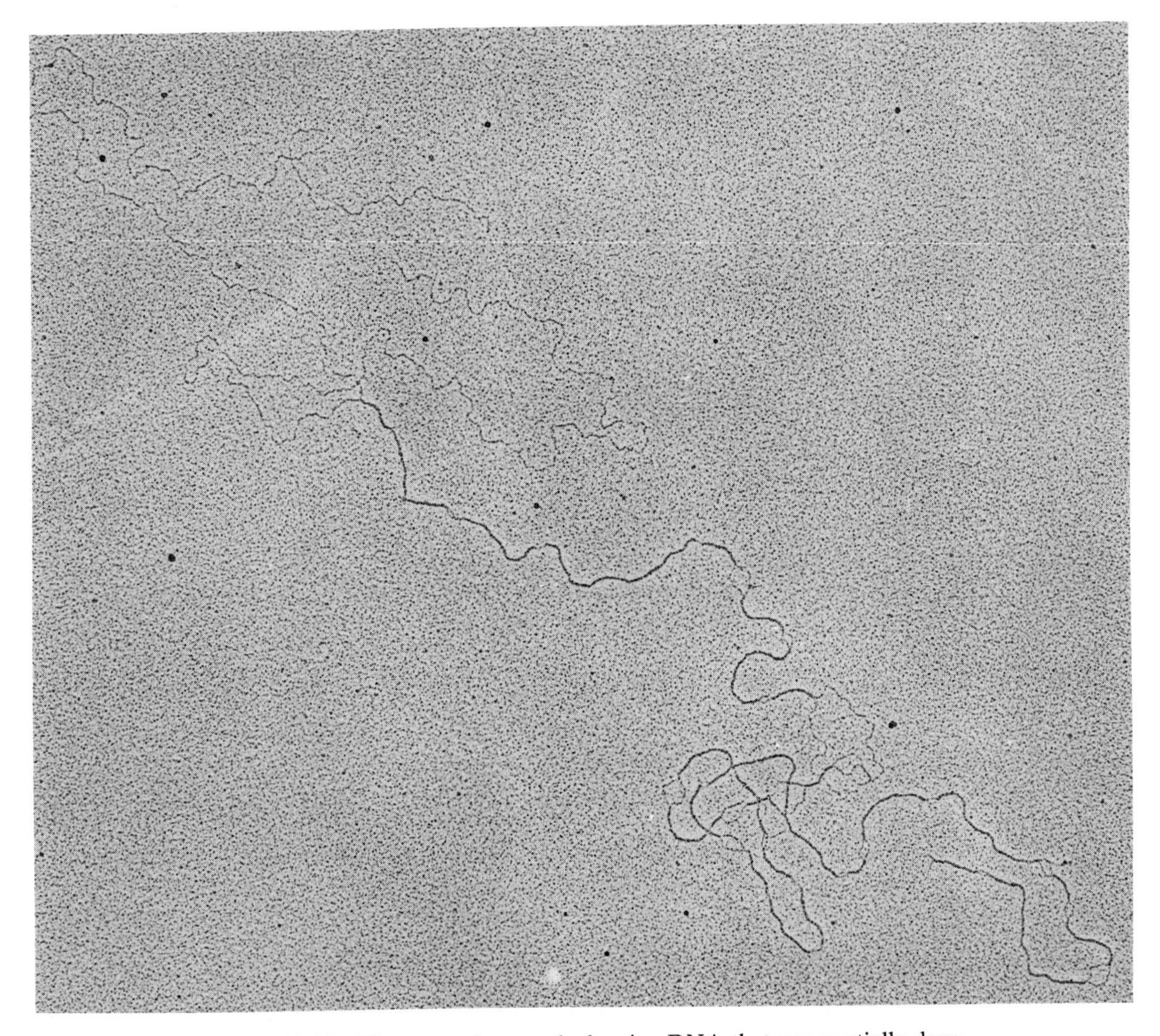

Fig. 11.17 Electron micrograph showing DNA that was partially denatured by exposure to a high pH (11.20) for 10 min, then processed for viewing. Denatured regions shown as thinner and kinkier lines in the micrograph. (Courtesy of R. Inman.)

Example 11.12 The difference in T_m for short helices of identical base composition but different sequence shows that s is dependent not only on whether a base pair is A · U or G · C but also on the *sequence* of the base pairs (Fig. 11.19). For large double-stranded DNA's and RNA's, the parameter s_A is therefore the *average* s of an A · T pair. Similarly, s_G is the *average* s of a G · C pair.

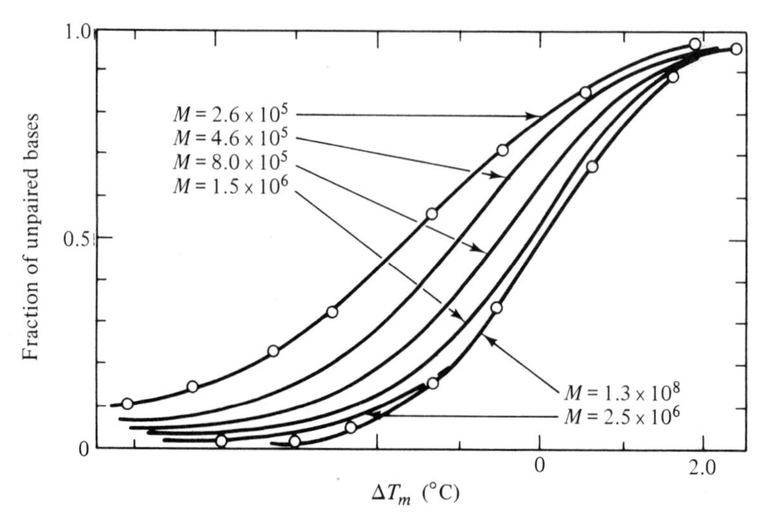

Fig. 11.18 Fraction of unpaired bases plotted against ΔT_m, the difference between the melting temperature of a DNA of molecular weight M and that of a very large DNA. For $M > 2.5 \times 10^6$, ΔT_m is very small. For DNA's with molecular weights less than 2.5×10^6, ΔT_m is clearly dependent on the size of the DNA. [Data from D. M. Crothers, N. R. Kallenbach, and B. H. Zimm, *J. Mol. Biol. 11*, 802 (1965).]

STATISTICAL THERMODYNAMICS

In previous sections we have illustrated the usefulness of statistical concepts in treating problems of biological interest. We stated at the beginning of this chapter that statistical methods also provide a bridge linking the thermodynamic properties of a macroscopic system and the molecular parameters of its microscopic constituents. In the sections below we shall discuss certain aspects of statistical thermodynamics, which treat an equilibrium system by statistical methods.

To illustrate the basic concepts in simple terms, our *quantitative* discussions will be limited to a system of noninteracting particles (that is, an ideal gas). The purpose of these discussions is to provide a molecular basis for the thermodynamic laws. In the section on the statistical mechanical concept of entropy, some qualitative examples are provided for more complex systems.

Statistical Mechanical Internal Energy

In Chapter 2 we discussed the internal energy for a macroscopic system and the relation between the energy, E, the heat absorbed by the system, q, and the work done on the system, w. A *macroscopic* system, such as 1 ℓ of air, 1 drop of water, or 1 carat of diamond, contains many *microscopic* particles. One liter of air at room temperature and atmospheric pressure contains approximately 5×10^{21} molecules of O_2, approximately four times as many molecules of N_2, and many

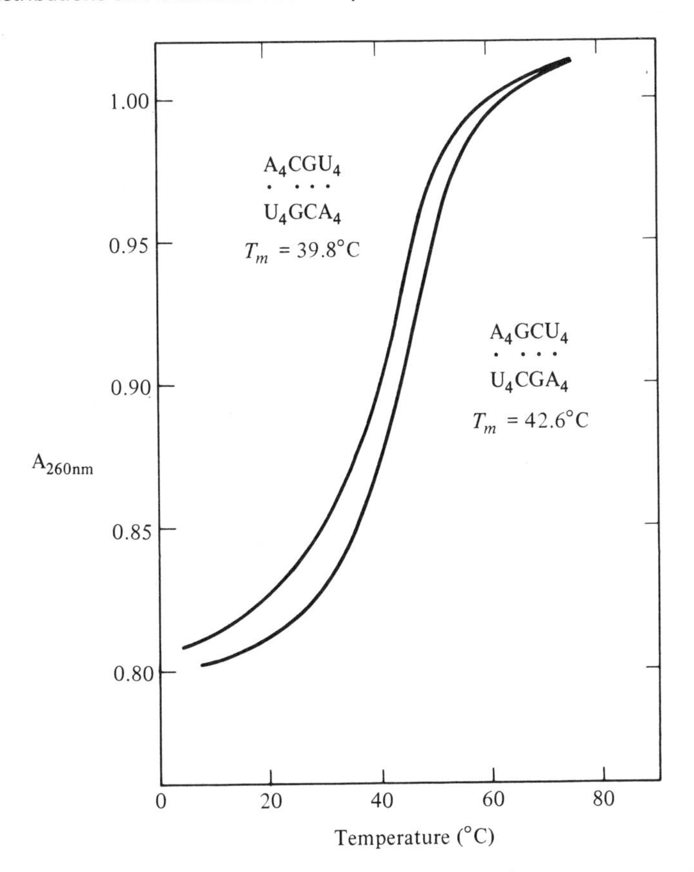

Fig. 11.19 Helix-coil transition curves for two short double-stranded RNA helices. $A_{260\text{nm}}$ is the absorbance at a wavelength of 260 nm, which was measured to monitor the helix-coil transition. The two helices have identical base composition and had the same concentration in media of identical salt concentrations. [Data from Olke Uhlenbeck.]

other molecules. A tiny drop of water 1 μm (10^{-4} cm) in diameter contains about 2×10^{10} water molecules, and 1 carat (200 mg) of diamond contains about 10^{22} carbon atoms. In Chapter 9 we discussed the *energy levels* for a single molecule. For very simple systems (such as ideal gases) the energy levels can be calculated from quantum mechanics. For more complex systems, the energy levels can be obtained in principle, although the actual computation can be very difficult and in many cases is not possible. As an example, we can readily show that for a monatomic gas in a container of dimensions a, b, and c in the directions x, y, and z, respectively, the energy levels of a molecule of mass m are

$$E(n_x, n_y, n_z) = \frac{h^2}{8m}\left(\frac{n_x^2}{a^2} + \frac{n_y^2}{b^2} + \frac{n_z^2}{c^2}\right) \tag{9.31}$$

where n_x, n_y, and n_z are integers.

At any particular time, each gas molecule will have an energy E_i given by Eq. (9.31). We can describe the energy distribution of the monatomic gas by specifying the number of molecules $N_1, N_2, \ldots, N_i, \ldots$ in each possible energy level $E_1, E_2, \ldots, E_i, \ldots$. Because the molecules of an ideal gas do not interact with one another, the total energy E of the system is simply the sum of the energies of the molecules:

$$E = N_1E_1 + N_2E_2 + N_3E_3 + \cdots + N_iE_i + \cdots$$

$$= \sum_i N_iE_i \tag{11.50}$$

The set of numbers $N_1, N_2, \ldots, N_i, \ldots$ are called *occupation numbers.*

Work

Consider a gas in a container with a movable wall (a piston). Let the *external* force in the direction of movement of the piston be F_x. If we move the piston by a distance dx in the direction of the external force, the work done on the gas by the surroundings is

$$dw = F_x\,dx \tag{11.51}$$

The total force in the x direction exerted by all molecules is

$$\sum_i N_iF_{ix} \tag{11.52}$$

where F_{ix} is the force in the x direction exerted by a molecule in the ith energy level.

If the change dx is carried out *reversibly*, the external force is balanced by the forces exerted by the molecules:

$$F_x = \sum_i N_iF_{ix} \tag{11.53}$$

The work is done by a reversible process

$$dw_{\text{rev}} = F_x\,dx = \sum_i N_iF_{ix}\,dx \tag{11.54}$$

For any molecule of the ideal gas, the force F_{ix} is related to the change in its energy, E_i, due to the change of the dimension of the container by dl. This is expressed by the equation

$$F_{ix} = -\frac{dE_i}{dl} \tag{11.55}$$

Substituting Eq. (11.55) into Eq. (11.54),

$$dw_{\text{rev}} = \sum_i N_i \cdot \left(-\frac{dE_i}{dl}\right) dx$$

similar to our explanation for the equation $dV = -A\,dx$ in Chapter 2; $dx = -dl$, and

$$dw_{\text{rev}} = \sum_i N_i\,dE_i \tag{11.56}$$

Equation (11.56) states that the work done on the system for a reversible process is related to the change in the energy levels due to the change in the dimension of the container.

This relation can be illustrated for a particle in a one-dimensional box of length a. The energy levels are

$$E_i = \frac{h^2 n_x^2}{8ma^2} \tag{9.27}$$

and the force is therefore

$$F_{ix} = -\frac{dE_i}{dl} = -\frac{dE_i}{da} = \frac{n_x^2 h^2}{4ma^3} \tag{11.57}$$

In the classical picture, the particle is thought to be moving back and forth in the box with an average velocity u_x such that the average kinetic energy is equal to the energy E_i given by Eq. (9.27):

$$\text{Kinetic energy} = \tfrac{1}{2}mu_x^2 = \frac{h^2}{8m}\frac{n_x^2}{a^2} \tag{11.58}$$

The momentum change per collision with a wall is $2mu_x$. The number of collisions between the particle and one of the walls is $n_x/2a$ per unit time. From Newton's law the total momentum change with the wall per unit time is equal to the force exerted by the particle on the wall.

$$F_x = 2mu_x \cdot \frac{u_x}{2a} = \frac{mu_x^2}{a} = \frac{n_x^2 h^2}{4ma^3}$$

This is the same as the result given by Eq. (11.57).

Heat

Starting with (11.50), we can write the differential expression for the change of E with change of E_i and N_i.

$$dE = \sum_i N_i\,dE_i + \sum_i E_i\,dN_i \tag{11.59}$$

Combining Eqs. (11.56) and (11.59), we obtain

$$dE = dw_{\text{rev}} + \sum_i E_i\,dN_i \tag{11.60}$$

but from the first law of thermodynamics, Eq. (2.19),

$$dE - dw_{\text{rev}} = dq_{\text{rev}}$$

Therefore,

$$dq_{\text{rev}} = \sum_i E_i \, dN_i \tag{11.61}$$

Equation (11.61) states that the heat absorbed by the system undergoing a reversible change is related to changes in the number of particles in the various energy levels. This immediately suggests that dq_{rev} is the part of the total energy that is related to the *distribution* of the particles among the various energy levels.

Distributions

Let us again take the energy levels of a monatomic gas molecule, Eq. (9.31), as an example. Consider now the number of ways we can distribute N_1 particles in the first energy level with energy E_1 and a degeneracy of g_1. The degeneracy, g_i, represents the number of distinguishable states that have the same energy, E_i. A simple example of degeneracy is given by Eq. (9.31). The energy for a particle in a cube (a = b = c) is degenerate whenever all three quantum numbers are not identical. For example, there are three states ($n_x = 1$, $n_y = 1$, $n_z = 2$; $n_x = 1$, $n_y = 2$, $n_z = 1$; $n_x = 2$, $n_y = 1$, $n_z = 1$) which all have energy $E = 6h^2/8ma^2$. The distribution of particles among states is the same problem as the number of ways of distributing N_1 objects into g_1 boxes. Before we can proceed, two questions must be asked: (1) Are the N_1 objects distinguishable or indistinguishable? (2) Is there any rule that specifies how many objects can be placed in each box? (As we will see in the next section, nature has set certain rules with regard to the distribution of particles.)

Example 11.13 To illustrate the importance of these questions, let us consider a simple example with $N_1 = 2$ and $g_1 = 3$.

Case 1. The objects are distinguishable, and there is no restriction on the number of objects in each box. All possible ways of distributing the objects are:

A	(1)	(2)	—
B	(1)	—	(2)
C	—	(1)	(2)
D	(1)(2)	—	—
E	—	(1)(2)	—
F	—	—	(1)(2)
G	(2)	(1)	—
H	(2)	—	(1)
I	—	(2)	(1)

The horizontal lines after each letter represent the three degenerate states with energy E_1, and 1 and 2 are the distinguishable objects. The total number of different ways of distributing the objects is 9.

Case 2. The objects are indistinguishable and there is no restriction on the number of objects in each box. If the objects are identical so that they cannot be distinguished, we can no longer label them 1 and 2 as we did in case 1. Therefore, the possible ways are:

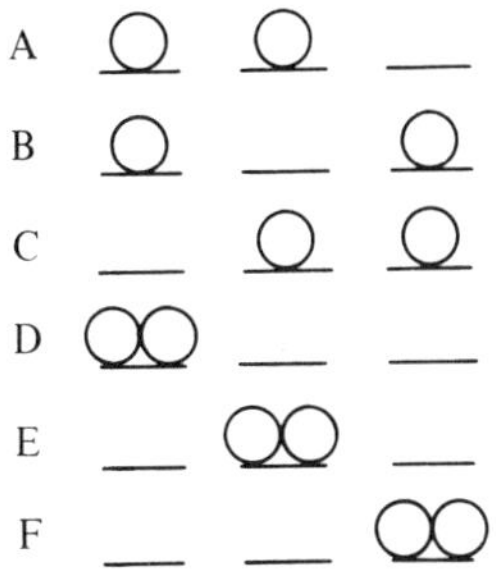

A
B
C
D
E
F

The total number of ways is 6.

Case 3. The objects are indistinguishable, and no more than one object can be in each box. The possible ways are:

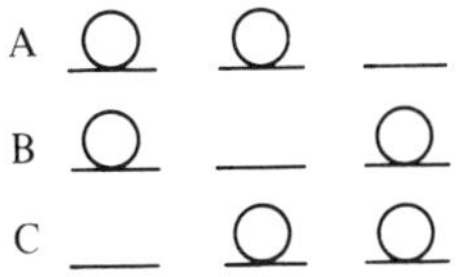

A
B
C

The total number of ways is 3.

You may wish now to confirm that, if the objects are distinguishable and no more than one object can be in each box, the total number of ways is 6.

The number of ways of distributing N_1 objects into g_1 boxes is clearly dependent on the conditions that we specify. Statistical treatments based on the conditions specified in cases 1, 2, and 3 are called *Boltzmann statistics*, *Bose-Einstein statistics*, and *Fermi-Dirac statistics*, respectively. The general solutions for distributing N_1 objects into g_1 boxes according to the three kinds of statistics are tabulated in Table 11.3. The expression for Boltzmann statistics is easiest to understand. Each unique object can be placed in any of the g_1 boxes. Therefore there are g_1 ways to place one object, g_1^2 ways for two objects, and $g_1^{N_1}$ ways for N_1 objects. The expression for Fermi-Dirac statistics is just the number of ways of choosing a subgroup of N_1 objects from a larger group of g_1 objects. One example is the number of ways of taking m steps forward out of a total of N steps. Equation (11.27b) which gives the answer is identical to the expression in Table 11.3 for Fermi-Dirac statistics. The expression for Bose-Einstein statistics is similar except we are now choosing a subgroup of N_1 objects from a larger group of $(g_1 + N_1 - 1)$ objects. Working out some simple examples with g_1 or N_1 equal to a small number will quickly convince you of the correctness of the table.

Table 11.3 Number of ways t_1 for distributing N_1 objects into g_1 boxes

Statistics	Objects: Distin-guishable	Objects: Indistin-guishable	Objects/box: No limit	Objects/box: No more than 1	Number of ways t_1
Boltzmann	√		√		$g_1^{N_1}$
Fermi-Dirac		√		√	$\dfrac{g_1!}{(g_1 - N_1)!N_1!}$
Bose-Einstein		√	√		$\dfrac{(g_1 + N_1 - 1)!}{(g_1 - 1)!N_1!}$

We can now calculate the *total number of ways t* of distributing a total of $N = \sum_i N_i = N_1 + N_2 + \cdots + N_i + \cdots$ objects such that there are N_1 objects in the group of g_1 boxes, N_2 objects in the group of g_2 boxes, and so on. (The different subscripts can now be used to designate different energy levels, if we wish.) For the Fermi-Dirac or the Bose-Einstein case, t is just the product $t_1 t_2 \ldots t_i \ldots$ or $\prod_i t_i$, since for each of the t_1 distributions, there are t_2 ways of distributing N_2 objects into g_2 boxes, t_3 ways of distributing N_3 objects into g_3 boxes, and so on. Thus

$$t\text{ (Fermi-Dirac)} = \prod_i \frac{g_i!}{(g_i - N_i)!\, N_i!} \tag{11.62}$$

and

$$t\text{ (Bose-Einstein)} = \prod_i \frac{(g_i + N_i - 1)!}{(g_i - 1)!\, N_i!} \tag{11.63}$$

For the Boltzmann case, besides the product $\prod_i t_i$ there is an additional factor: Since the objects are distinguishable, there are many ways of grouping N objects into subgroups $N_1, N_2, \ldots$. For example, if we want to group five objects A, B, C, D, and E into two subgroups containing two and three objects, respectively, there are 10 ways:

AB CDE
AC BDE
AD BCE
AE BCD
BC ADE
BD ACE
BE ACD
CD ABE
CE ABD
DE ABC

In general, there are $N!/N_1!N_2!\cdots N_i!\cdots$ ways of grouping N distinguishable objects into subgroups $N_1, N_2, \ldots, N_i, \ldots$. Thus for the Boltzmann case,

$$t\,(\text{Boltzmann}) = \frac{N!}{\prod_i N_i!}\prod_i g_i^{N_i}$$

$$= N!\prod_i \frac{g_i^{N_i}}{N_i!} \qquad (11.64)$$

Quantum Mechanical Rules Governing a Distribution

For molecules, atoms, or subatomic particles, nature has set certain rules governing their distributions into the various quantum states. Particles with half-integral spin, such as electrons, protons, and ^{3}He nuclei, are subject to the *Pauli exclusion principle*; that is, no two particles can be in the same quantum state. Fermi-Dirac statistics therefore apply to these particles. Particles with integral spin, such as photons, deuterons, and ^{4}He nuclei, are not subject to the Pauli exclusion principle. Therefore, Bose-Einstein statistics represent their distributions. Boltzmann statistics was developed before the quantum mechanical rules were realized. No natural particles obey Boltzmann statistics strictly.

It can be shown (Problem 9), however, that if the system is sufficiently dilute, such that $N_i \ll g_i$,

$$t\,(\text{Fermi-Dirac}) \approx t\,(\text{Bose-Einstein}) \approx \frac{t\,(\text{Boltzmann})}{N!}$$

Defining

$$t\,(\text{corrected Boltzmann}) = \frac{t\,(\text{Boltzmann})}{N!}$$

$$= \prod_i \frac{g_i^{N_i}}{N_i!} \qquad (11.65)$$

we can use the "corrected Boltzmann" enumeration for dilute systems such as gases at temperatures that are not too low. Because the mathematics for the corrected Boltzmann statistics is simpler, we shall use it to gain some insight into the statistical mechanical nature of entropy. (For low-molecular-weight gases at low temperatures the method does not apply, because $N_i \geq g_i$.)

Most Probable Distribution

For a given thermodynamic state there is a large number of possible combinations of the quantum states of the particles. For example, if we have an ideal gas that consists of N particles in a container of volume V, there are many ways of distributing the N particles among the various energy levels (which are dependent on V) such that the total energy is E.

For a macroscopic system, N is enormously large, and so is the number of energy states and their degeneracies. Thus the possible quantum states for a given thermodynamic state is an extremely large number.

Consider a hypothetical case of five equally spaced energy levels with energies E_1, $2E_1$, $3E_1$, $4E_1$, and $5E_1$, each level with a degeneracy of 2, and four particles that obey the Boltzmann statistics. A number of distributions, all having the same total energy $E = 12E_1$, and the respective number of states are as follows:

Distribution $(N_1, N_2, N_3, N_4, N_5)$	Number of states t (Boltzmann)
(1, 0, 2, 0, 1)	192
(1, 1, 0, 1, 1)	384
(0, 1, 2, 1, 0)	192
(0, 0, 4, 0, 0)	16
(0, 2, 0, 2, 0)	96
(1, 0, 1, 2, 0)	192
. . . . .	. . .

Note that for certain distributions the number of states is larger than for others. If all particles are in the same energy level, the number of quantum states is small. If the particles are distributed among as many energy levels as possible, the number of quantum states is large. We shall take the point of view that for the macroscopic system, the properties of the system at equilibrium are represented, to a very high degree of approximation, by the properties of *the most probable* distribution.

Partition Function: The Most Probable Distribution for an Isolated System

To find the most probable distribution, one must find the set of occupation numbers $N_1, N_2, \ldots, N_i, \ldots$ for which t is a maximum. It happens that mathematically it is easier to treat the logarithm of t. If t is a maximum, the function $\ln t$ is also a maximum. Thus, to maximize t, we can maximize $\ln t$ by setting

$$\left(\frac{\partial \ln t}{\partial N_1}\right)_{N_2, N_3, \ldots} = 0$$

$$\left(\frac{\partial \ln t}{\partial N_2}\right)_{N_1, N_3, \ldots} = 0, \text{ etc.}$$

In other words, $\ln t$ can be maximized by setting

$$\left(\frac{\partial \ln t}{\partial N_i}\right)_{N_j \neq N_i} = 0 \qquad \text{(for all } i\text{)} \tag{11.66}$$

However, if we have chosen an *isolated system*, its energy is a constant, E. For most cases of interest the total number of particles at equilibrium in an isolated system is also a constant, N. These two constraints,

$$\sum_i N_i E_i = E \tag{11.50}$$

$$\sum_i N_i = N \tag{11.67}$$

must be obeyed in maximizing t. The mathematical procedure for solving Eqs. (11.66), (11.50), and (11.67) involves the use of the Lagrange undetermined multipliers. We shall omit the mathematics and give the final results. If t (corrected Boltzmann) is used, the result is an expression for the *Boltzmann distribution*

$$\frac{N_i}{N} = g_i\, e^{-E_i/kT} \bigg/ \sum_i g_i\, e^{-E_i/kT} \tag{11.68a}$$

where k, the gas constant R divided by Avogadro's number, is the Boltzmann constant. The function $\sum_i g_i\, e^{-E_i/kT}$, which sums over all the energy levels of the system, is called the *partition function*, Z. Thus,

$$\frac{N_i}{N} = g_i\, e^{-E_i/kT} \Big/ Z \tag{11.68b}$$

$$Z = \sum_i g_i\, e^{-E_i/kT} \tag{11.69}$$

The ratio of the number of molecules in two energy levels is

$$\frac{N_i}{N_j} = \frac{g_i}{g_j}\, e^{-(E_i - E_j)/kT} \tag{11.70}$$

Note that there will tend to be more molecules in the lower energy states unless the degeneracy increases rapidly as the energy increases.

Entropy and the Most Probable Distribution

In Eq. (11.61), we showed that dq_{rev} is related to a change in the *distribution* of the particles. Let us therefore examine the change in $\ln t$ due to changes in the N_i's:

$$d \ln t = \sum_i \left(\frac{\partial \ln t}{\partial N_i}\right)_{N_j \neq N_i} dN_i \tag{11.71}$$

Taking

$$t = \prod_i \frac{g_i^{N_i}}{N_i!} \tag{11.65}$$

$$\ln t = \sum_i (N_i \ln g_i - \ln N_i!)$$

$$\left(\frac{\partial \ln t}{\partial N_i}\right)_{N_j \neq N_i} = \ln g_i - \frac{d \ln N_i!}{dN_i} \tag{11.72}$$

(All other terms contain $N_j \neq N_i$, and therefore their partial derivatives with respect to N_i are zero.)

If N_i is large, *Stirling's approximation*,

$$\ln N_i! = N_i \ln N_i - N_i \tag{11.73}$$

can be used, and

$$\frac{d \ln N_i!}{dN_i} = \frac{d}{dN_i}(N_i \ln N_i - N_i) = 1 + \ln N_i - 1$$

$$= \ln N_i \tag{11.74}$$

Thus

$$\left(\frac{\partial \ln t}{\partial N_i}\right)_{N_j \neq N_i} = \ln g_i - \ln N_i = \ln \frac{g_i}{N_i} \tag{11.75}$$

But if the system is at equilibrium, according to Eq. (11.68b),

$$\ln \frac{g_i}{N_i} = \ln\left(\frac{Z}{N}\right) + \frac{E_i}{kT}$$

Therefore,

$$d \ln t = \sum_i \left(\frac{\partial \ln t}{\partial N_i}\right)_{N_j \neq N_i} dN_i$$

$$= \sum_i \left[\ln\left(\frac{Z}{N}\right) dN_i + \frac{E_i}{kT} dN_i\right]$$

$$= \left(\ln \frac{Z}{N}\right) \sum_i dN_i + \frac{1}{kT} \sum_i E_i \, dN_i \tag{11.76}$$

But

$$\sum_i dN_i = d \sum_i N_i = 0 \tag{11.77}$$

since the total number of particles is constant. Furthermore,

$$\sum_i E_i \, dN_i = dq_{\text{rev}} \tag{11.61}$$

Substituting these relations into Eq. (11.76), we obtain

$$d \ln t = \frac{dq_{\text{rev}}}{kT}$$

But in Chapter 3 we defined the change of entropy, $dS = dq_{\text{rev}}/T$. Therefore,

$$dS = kd \ln t \tag{11.78}$$

where S is the thermodynamic entropy. *Note that in the above equation t is the maximized distribution or the most probable distribution.*

Equation (11.78) is a very important result. It provides a microscopic interpretation of entropy. It is the quantitative statement of the relation between entropy and disorder or randomness that was discussed in Chapter 3.

Statistical Mechanical Entropy

We now define a function

$$S_{\text{sm}} = k \ln t$$

as the statistical mechanical entropy. The corrected Boltzmann statistics gives, from Eq. (11.78),

$$dS_{\text{sm}} = dS$$

Therefore, the statistical mechanical entropy cannot differ from the thermodynamic entropy by more than an additive constant.

As we have discussed, at ordinary temperatures t is very large. When the absolute temperature T approaches zero, however, the energy approaches a minimum, and all particles will be in the lowest energy levels available to them. Thus, t is either unity or a very small number compared with the values of t at ordinary temperatures. In other words, as T approaches zero, S_{sm} is either rigorously zero or vanishingly small compared with S_{sm} at ordinary temperatures. From the third law of thermodynamics, S is also zero as T approaches zero for a perfectly ordered crystalline substance. Therefore, S_{sm} and S must be identical:

$$S = S_{\text{sm}} = k \ln t \tag{11.79}$$

We can see that from the point of view of statistical mechanics, the third law of thermodynamics is a natural consequence of the occupation of the lowest available energy levels by the particles as T approaches zero.

One statement of the second law of thermodynamics is that for an isolated system, the equilibrium state is the one for which the entropy is a maximum. From the statistical mechanical point of view, the equilibrium state of an isolated system is one that represents the most probable distribution and has the maximum randomness.

Examples of Entropy and Probability

The relation

$$S = k \ln t$$

is not only important in providing a microscopic interpretation of entropy, but it also contains useful qualitative insights for many problems. In this section we shall consider a number of qualitative and quantitative examples.

Example 11.14 The entropy changes for the melting of ice and the evaporation of water are as follows:

	ΔS^0 (cal deg^{-1} mol^{-1})
$H_2O(s, 0°C) \rightarrow H_2O(l, 0°C)$	+ 5.25
$H_2O(l, 100°C) \rightarrow H_2O(g, 100°C)$	+26.1

Qualitatively, the solid is the most ordered, the liquid the next, and the gas the least ordered. The positive ΔS values for the transitions are in agreement with this qualitative interpretation.

Example 11.15 The ΔS^0 values for a few reactions in the gas phase at 25°C are as follows:

	ΔS^0 (cal deg^{-1} mol^{-1})
$2\,CO(g) + O_2(g) \rightarrow 2\,CO_2(g)$	−41.48
$2\,H_2(g) + O_2(g) \rightarrow 2\,H_2O(g)$	−21.2
1-Butene(g) + $H_2(g)$ → n-butane(g)	−30.1
Ethylene(g) + $H_2(g)$ → ethane(g)	−28.8
$H_2(g) + Cl_2(g) \rightarrow 2\,HCl(g)$	+ 4.78

Note that for each of the first four reactions, there is a decrease in the total number of molecules in proceeding from reactants to products. There is also a large decrease in entropy. The last reaction listed in the table has no net change in the number of molecules. Note that the magnitude of ΔS^0 for this reaction is small and positive.

Example 11.16 For reactions in solution, the interpretation of entropy changes is much more complex, owing to the interactions between the solvent molecules and the solute molecules. A few examples based on data in Table A.5 are as follows:

	ΔS^0 (cal deg^{-1} mol^{-1})
$CO_3^{2-}(aq) + H^+(aq) \rightarrow HCO_3^-(aq)$	+35.4
$NH_4^+(aq) + CO_3^{2-}(aq) \rightarrow NH_3(aq) + HCO_3^-(aq)$	+34.7
$OH^-(aq) + H^+(aq) \rightarrow H_2O(l)$	+19.2
$NH_4^+(aq) + HCO_3^-(aq) \rightarrow CO_2(aq) + H_2O(l) + NH_3(aq)$	+22.3
$NH_3(aq) + H^+(aq) \rightarrow NH_4^+(aq)$	+ 0.7
$HC_2O_4^-(aq) + OH^-(aq) \rightarrow C_2O_4^{2-}(aq) + H_2O(l)$	− 5.3
$CH_4(aq) \rightarrow CH_4(CCl_4)$	+18

For the first four reactions, neutralization of charges occurs. Since a charged species tends to orient the water molecules around it, charge neutralization results in disorientation of some of the solvent molecules and an increase in entropy. The next two reactions involve transfer of charge but no neutralization. The entropy changes are small as a consequence. The last reaction involves the transfer of a nonpolar methane (CH_4) molecule from a polar solvent (H_2O) to a nonpolar solvent (CCl_4). The large positive ΔS^0 is primarily the result of the ordering of water molecules around a nonpolar molecule. Removal of the nonpolar molecule results in the randomization of the water molecules and hence a positive ΔS. (See discussion of Table 3.2, Chapter 3.) This kind of effect is important for many reactions in aqueous media, such as the binding of a nonpolar substrate to the nonpolar region of an enzyme, or the folding of a protein.

Example 11.17 Consider two chambers, each of volume V, connected through a stopcock as shown in Fig. 11.20. Initially, the left chamber contains n moles of a gas at pressure P_0 and temperature T, the stopcock is closed, and the right chamber has been evacuated. At a certain time the stopcock is opened, and the gas expands adiabatically into a vacuum. You may readily show from the discussions in Chapters 2 and 3 that for this process

$$q = 0 \quad \text{and} \quad w = 0$$

and therefore $\Delta E = 0$, $\Delta T = 0$, and $P_{\text{final}} = P_0/2$. To calculate ΔS from

$$\Delta S = \frac{q_{\text{rev}}}{T}$$

we must carry out a reversible process which arrives at the same final conditions: temperature T and pressure $P_0/2$. This can be done by an isothermal, reversible expansion:

$$\Delta E = 0$$

$$q_{\text{rev}} = w_{\text{rev}} = nRT \ln \frac{2V}{V} = nRT \ln 2$$

and

$$\Delta S = nR \ln 2$$

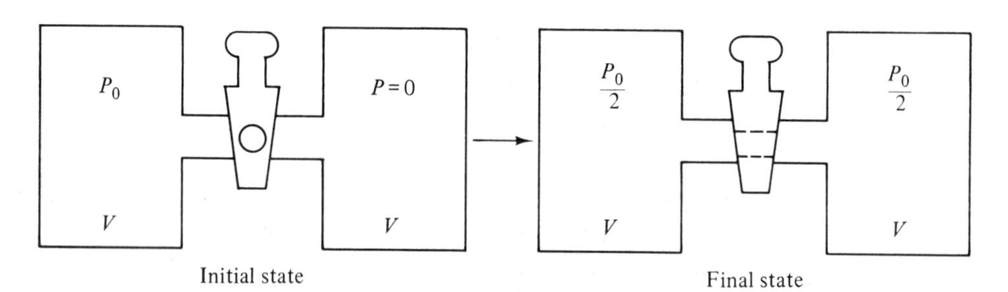

Fig. 11.20 Gas, initially all contained in a volume V at pressure P_0, expands adiabatically to volume $2V$ and pressure $P_0/2$ spontaneously, with a consequent increase in entropy.

We shall now see that Eq. (11.79) gives the same result. For the process under consideration,

$$\Delta S = S_2 - S_1 = k \ln t_2 - k \ln t_1$$

$$= k \ln \frac{t_2}{t_1}$$

where the subscripts 1 and 2 refer to the initial and final states, respectively.

The ratio of the number of ways that the molecules can be arranged, t_2/t_1, is equivalent to the ratio of the probabilities. Because there are more ways to distribute the gas molecules in a volume $2V$ than in a volume V, the final state is more probable than the initial state. The ratio can be evaluated as follows. Imagine that we place the molecules one by one randomly into either of the two sides. The chance that the first molecule will be in the left side is $\frac{1}{2}$. The chance that both the first and the second molecules will be in the left side is $(\frac{1}{2})(\frac{1}{2}) = (\frac{1}{2})^2$. The chance that all nN_0 molecules are in the left side is $(\frac{1}{2})^{nN_0}$. Thus $t_1/t_2 = (\frac{1}{2})^{nN_0}$ or $t_2/t_1 = 2^{nN_0}$. Therefore,

$$\Delta S = k \ln 2^{nN_0} = nN_0 k \ln 2 = nR \ln 2$$

The problem can also be looked at in a different way. Imagine that the volume V is divided into y cubicles of equal volume. Initially, the nN_0

molecules are to be distributed among the y cubicles. Because the molecules of an ideal gas occupy an insignificant volume themselves, we assume that there is no limit to how many molecules can occupy each cubicle. The number of ways of placing one molecule in y cubicles is y. The number of ways t_1 of placing nN_0 molecules in y cubicles is $t_1 = y^{nN_0}$. For the final state there are $2y$ cubicles; therefore, $t_2 = (2y)^{nN_0}$. It follows then that

$$\Delta S = k \ln \frac{t_2}{t_1} = k \ln 2^{nN_0} = nR \ln 2$$

Example 11.18 Suppose that we have n_D moles of an ideal gas D and n_E moles of an ideal gas E, at the same temperature and pressure and separated initially by a partition, as shown in Fig. 11.21. If the barrier is withdrawn,

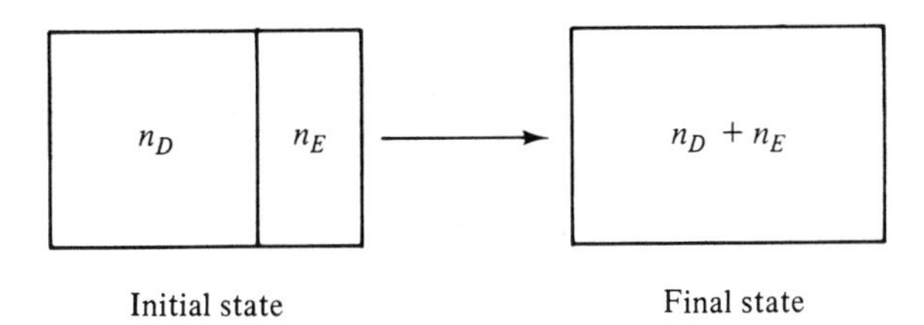

Fig. 11.21 Two different gases, initially separated by a barrier, will mix spontaneously, and the entropy of the system will increase, when the barrier is removed.

mixing of the two gases occurs spontaneously. Let the volumes occupied initially by D and E be V_D and V_E, respectively. Since the gases are at the same temperature and pressure,

$$\frac{V_D}{V_E} = \frac{n_D}{n_E}$$

To calculate ΔS for this process, we can divide V_D into y_D and V_E into y_E cubicles of equal size. The number of ways of placing $n_D N_0$ molecules of D in y_D cubicles is $y_D^{n_D N_0}$, and the number of ways of placing $n_E N_0$ molecules of E in y_E cubicles is $y_E^{n_E N_0}$. Initially, with the barrier present, the total number of ways t_1 is, therefore,

$$t_1 = y_D^{n_D N_0} \cdot y_E^{n_E N_0}$$

If the barrier is removed, there are $y_D + y_E$ cubicles for the gases, and

$$t_2 = (y_D + y_E)^{n_D N_0} \cdot (y_D + y_E)^{n_E N_0}$$

Thus

$$\frac{t_1}{t_2} = \left(\frac{y_D}{y_D + y_E}\right)^{n_D N_0} \cdot \left(\frac{y_E}{y_D + y_E}\right)^{n_E N_0}$$

But

$$\frac{y_D}{y_E} = \frac{V_D}{V_E} = \frac{n_D}{n_E}$$

so

$$\frac{y_D}{y_D + y_E} = \frac{n_D}{n_D + n_E} = X_D$$

and

$$\frac{y_E}{y_D + y_E} = \frac{n_E}{n_D + n_E} = X_E$$

where X_D and X_E are the mole fractions of D and E, respectively, in the mixture. Thus

$$\frac{t_2}{t_1} = X_D^{-n_D N_0} \cdot X_E^{-n_E N_0}$$

and

$$\begin{aligned} \Delta S &= k \ln \frac{t_2}{t_1} \\ &= -(n_D k N_0 \ln X_D + n_E k N_0 \ln X_E) \\ &= -(n_D R \ln X_D + n_E R \ln X_E) \end{aligned}$$

This equation can be generalized to give the ideal entropy of mixing for any number of components

$$\Delta S = -R \sum_i n_i \ln X_i \tag{11.80}$$

Example 11.19 Certain DNA molecules, such as those from phage λ, contain single-stranded ends of complementary base sequences which enable the molecules to circularize, as illustrated diagrammatically in Fig. 11.22.

Let the length of the double-stranded portion of the DNA be L. Since the ends are very short, L is essentially the length of the whole molecule. For a flexible molecule of linear DNA, we have seen that the mean-square

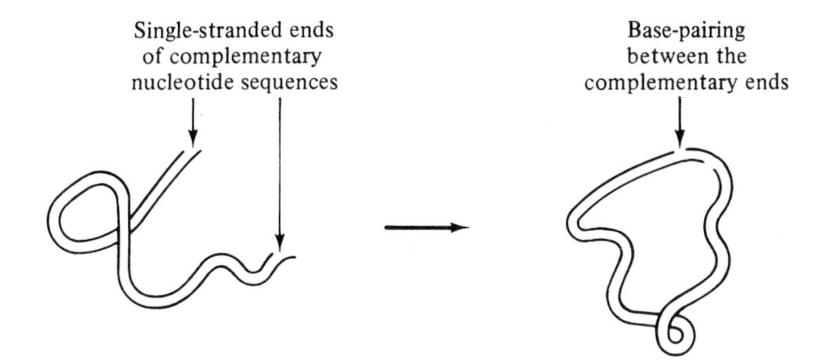

Fig. 11.22 Cyclization of a DNA molecule with "cohesive" ends. Cyclization involves a loss of entropy proportional to the logarithm of the length of the DNA.

end-to-end distance is

$$\overline{h^2} = Ll \tag{11.47}$$

where l is the effective segment length.

Let us choose one end of the linear DNA molecules as the origin. The other end of the DNA will be found on the average in a sphere of volume $V = (4\pi/3)(\overline{h^2})^{3/2}$, whose center is the origin.

In the circular form, the two ends are constrained to be in a much smaller volume, v_i. Therefore, the circularization of each DNA molecule is associated with an unfavorable (negative) entropy term:

$$\begin{aligned}\Delta S_c &= k \ln \frac{v_i}{V} \\ &= k \ln v_i - k \ln \left(\frac{4\pi}{3}\right) - \frac{3}{2} k \ln \overline{h^2} \\ &= \left[k \ln v_i - k \ln \left(\frac{4\pi}{3}\right) - \frac{3k}{2} \ln l\right] - \frac{3k}{2} \ln L\end{aligned}$$

For a group of DNA molecules with the same cohesive ends but of different lengths, the sum of the terms in the brackets is a constant, and ΔS_c is more negative for larger L. Thus the longer such a DNA molecule is, the less favorable is ring formation. This has been observed experimentally, as shown in Fig. 11.23. The curve shown is the function predicted by the ΔS_c term. Note that at the low end of the molecular-weight scale, the discrepancy between the experimental results and the theoretical prediction is large. For such short DNA molecules, the equation $\overline{h^2} = Ll$ (derived from the random-walk formula) is no longer applicable, because the stiffness of the double-stranded DNA makes it difficult for short segments to circularize.

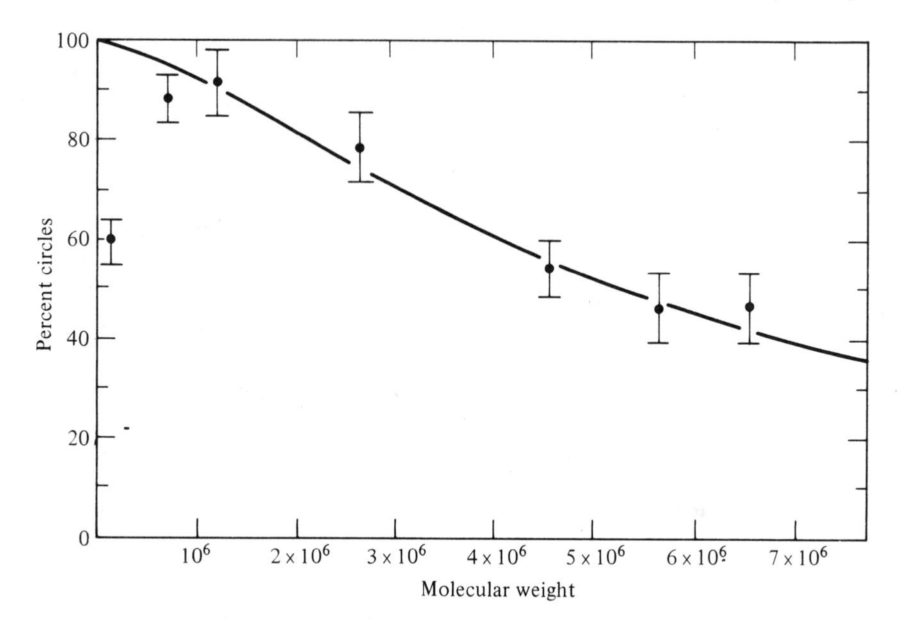

Fig. 11.23 Percentage of DNA molecules in the circular form as a function of the molecular weight of the DNA. The DNA samples were obtained by treating an *E. coli* F-factor DNA with a restriction enzyme Eco RI, which introduces two staggered breaks at sites with base sequence $\begin{array}{l}\text{-G}^{\downarrow}\text{AATTC-}\\ \text{-CTTAA}_{\uparrow}\text{G-}\end{array}$, yielding DNA molecules with cohesive ends of sequences AATT-. [Data from J. E. Mertz and R. W. Davis, *Proc. Natl. Acad. Sci. U.S. 69*, 3370 (1972).]

Partition Function: Applications

In our treatment leading to Eq. (11.68), we have chosen an isolated system. The partition function

$$Z = \sum_i g_i e^{-E_i/kT} \tag{11.69}$$

sums over all the energy states that a molecule can occupy and is therefore called the *molecular partition function.*

For simple molecules the various energy levels can be obtained by quantum mechanical calculations or by spectroscopic measurements, and the partition functions can be obtained. It can be shown that, if the partitition function of a system is known, all the thermodynamic properties of the system can be calculated. For a system of N noninteracting particles, the pressure P, internal energy

E, and entropy S are related to the partition function Z by the relations

$$P = NkT\left(\frac{\partial \ln Z}{\partial V}\right)_T \tag{11.81}$$

$$E = NkT^2\left(\frac{\partial \ln Z}{\partial T}\right)_V \tag{11.82}$$

$$S = kN \ln \frac{Z}{N} + \frac{E}{T} + kN \tag{11.83}$$

Another important property of the partition function can be better seen if we elect to sum over all quantum states rather than energy levels. That is, the g_i degenerate states are summed up individually:

$$Z = \sum_{\substack{\text{all}\\ \text{quantum}\\ \text{states}}} e^{-E/kT} \tag{11.84}$$

To a high degree of approximation the energy of a molecule in a particular state is the simple sum of various types of energy, such as translational energy, E_{tr}; rotational energy, E_{rot}; vibrational energy, E_{vib}; electronic energy, E_{el}; and so on. If

$$E = E_{\text{tr}} + E_{\text{rot}} + E_{\text{vib}} + E_{\text{el}} + \cdots$$

it follows immediately that

$$\begin{aligned} Z &= (\sum e^{-E_{\text{tr}}/kT})(\sum e^{-E_{\text{rot}}/kT})(\sum e^{-E_{\text{vib}}/kT})(\sum e^{-E_{\text{el}}/kT})\cdots \\ &= Z_{\text{tr}}Z_{\text{rot}}Z_{\text{vib}}Z_{\text{el}} \cdots \end{aligned} \tag{11.85}$$

In other words, if the energy can be expressed as a sum of terms, the partition function can be partitioned into corresponding terms, the product of which gives the total partition function.

Systems other than an isolated system can also be treated by statistical mechanics. Because systems of chemical or biological interest are seldom isolated systems, it is more useful to obtain the partition functions for closed systems or open systems. It is beyond the scope of this book to treat such problems, however.

For a system consisting of complex molecules, it is not possible to obtain the partition function rigorously. Consider the example of a polypeptide molecule. Each amide residue in the chain is made of atoms, and each atom has its translational, etc., energy levels. Since the atoms interact, the energy levels of one are strongly affected by many others. It is obviously an impossible task to obtain the complete molecular partition function. On the other hand, if we are interested only in the helix-coil transition, there is no need to know the fine detail of the energy levels. All we really need are the *relative* contributions to the partition function of a residue in the helix and the coil states of the molecule. These relative contributions are what we have termed "statistical weights" in a number of discussions in the first part of this chapter. In fact, several of the functions that we denoted by Z are examples of partition functions. Therefore, in the first part of this chapter we have already given a number of examples involving the applications of statistical mechanics to complex systems of biological interest.

SUMMARY

Binding of Small Molecules by a Polymer

Polymer molecule with N identical and independent sites for the binding of A, a small molecule:

$$\frac{\nu}{[A]} = K(N - \nu) \tag{11.12}$$

or

$$\frac{r}{[A]} = K(n - r) \tag{11.15}$$

or

$$\frac{\theta}{1 - \theta} = K[A] \tag{11.16}$$

$\nu \equiv$ number of bound A molecules per polymer molecule
$r \equiv$ number of bound A molecules per monomer unit of the polymer molecule
$\theta \equiv$ fraction of sites occupied
$N \equiv$ number of sites per polymer molecule
$n \equiv$ number of sites per monomer unit of the polymer molecule
$K \equiv$ intrinsic binding constant

Polymer with N identical sites in a linear array with nearest-neighbor interactions:

$$\nu = \left(\frac{\partial \ln Z}{\partial \ln S}\right)_{\tau} \tag{11.22}$$

Cooperative binding:

If $\tau > 1$, the binding of an A molecule to one site makes it easier to bind another A to an adjacent site. θ versus A plot is sigmoidal in shape. If τ is very large, the binding approaches the all-or-none limit, and the predominant species have $\theta = 0$ and $\theta = 1$. In this limit,

$$\frac{d \log [\theta/(1 - \theta)]}{d \log [A]} = N \tag{11.25}$$

Anticooperative binding:

If $\tau < 1$, the binding of an A molecule to one site makes it more difficult to bind another A to an adjacent site. In the limit $\tau = 0$, binding to one site *excludes* binding to adjacent sites.

Random Walk and Related Topics

$$\overline{h^2} = Nl^2 \qquad (11.47)$$

$$= Ll \qquad (11.47)$$

$\overline{h^2}$ = mean-square end-to end distance
N = number of "segments" of a polymer molecule
l = length of a segment; l is a measure of the stiffness of the polymer molecule
L = contour length of a polymer

$$\overline{R^2} = \frac{Nl^2}{6} \quad \text{(open-ended random coil)} \qquad (11.49a)$$

$\overline{R^2}$ = mean-square radius

$$\overline{R^2} = \frac{Nl^2}{12} \quad \text{(circular random coil)} \qquad (11.49b)$$

Helix-coil Transitions

Simple polypeptides:

The α helix is characterized by hydrogen bonds between CO groups and their respective fourth preceding NH groups in a polypeptide chain. The *initiation* of a helical element in a coiled region is more difficult than the addition of a helical element to the end of a helix. For polypeptides of high molecular weight, helix-coil transitions can be very sharp. Statistical methods have been used successfully to treat the helix-coil transition.

Proteins:

Native proteins are in a highly ordered or folded structure. Unfolding of a protein involves the disruption of hydrogen bonds, disulfide bridges, ionic bonds, and interactions between the nonbonded groups. The transition from a folded to an unfolded form (and vice versa) is usually highly cooperative and approaches the all-or-none limit.

Double-stranded nucleic acids:

Owing to favorable interactions ("stacking interactions") between neighboring base pairs in a double helix, the formation of an additional base pair at the end of a helical region is favored compared with the formation of a base pair in the middle of a coiled region. This leads to a cooperative transition between the helical and coiled forms. The basic features of the transition, such as the sharpness of the transition, the temperature dependence, the effect of the molecular weight of the polynucleotide, and the effect of base composition, can be understood from statistical mechanical considerations of the problem.

Statistical Thermodynamics of an Ideal Gas

The most probable distribution:

The properties of a system at equilibrium is represented, to a high degree of approximation, by the properties of the most probable distribution,

$$\frac{N_i}{N} = g_i\, e^{-E_i/kT}/Z \qquad (11.68b)$$

$$Z = \sum_i g_i\, e^{-E_i/kT} \qquad (11.69)$$

$$\frac{N_i}{N_j} = \frac{g_i}{g_j}\, e^{-(E_i-E_j)/kT} \qquad (11.70)$$

$N_i \equiv$ number of molecules in an energy level E_i
$N_j \equiv$ number of molecules in an energy level E_j
$N \equiv$ total number of molecules
$g_i, g_j \equiv$ degeneracy, the number of states with energy E_i, E_j
$k \equiv$ Boltzmann constant
$\equiv R/N_0$ (R = gas constant and N_0 = Avogadro's number)
$Z \equiv$ molecular partition function, which sums up all the energy states of the molecule; for simple molecules Z can be calculated by quantum mechanics

Entropy:

$$S = k \ln t \qquad (11.79)$$

$t \equiv$ number of ways of distributing the molecules; the larger the t, the more probable the state is, and the larger is the entropy; Eq. (11.79) provides an interpretation of the second and third laws of thermodynamics

Molecular partition function and thermodynamic functions:

All thermodynamic functions, such as P, E, and S, can be obtained from the partition function Z. If the energy of a molecule can be expressed as a sum of terms (translational energy, rotational energy, vibrational energy, electronic energy, etc.), the partition function can be factored (partitioned) into corresponding terms, the product of which gives the total partition function.

MATHEMATICS NEEDED FOR CHAPTER 11

We use permutations and combinations in calculating distributions of particles among states. The number of *permutations* of n objects is

$$P_n = n! = n \cdot (n-1) \cdot (n-2) \ldots \cdot 3 \cdot 2 \cdot 1$$

This is the number of ways of arranging n distinguishable objects in n boxes; only one object per box is allowed. A simple example is the $3! = 6$ permutations of the three letters a, b, c.

A combination of objects is a group of objects without respect to order. The three letters (a, b, c) form one combination. The number of *combinations* of n objects taken r at a time is

$${}_nC_r = \frac{n!}{r!(n-r)!}$$

This is the number of ways of arranging N indistinguishable particles in g boxes; only one object per box is allowed. Note that this is the expression for the distribution of Fermi-Dirac particles with $n = g$ and $r = N$. The n (or g) boxes are divided into combinations of r (or N) boxes containing particles and combinations of $n - r$ (or $g - N$) boxes without particles. A simple example is the number of ways of arranging 3 pennies into groups of 2 heads and 1 tail:

$${}_3C_2 = \frac{3!}{2!1!} = 3$$

Stirling's approximation for factorials is useful for large N

$$\ln N! = N \ln N - N$$

If a group of particles can be arranged in t_1 ways and independently they can also be arranged in t_2 ways, the total number of arrangements is $t_1 \cdot t_2$. Similarly, if the probability of an event occurring is p_1 and if the probability of an independent event occurring is p_2, then the joint probability of the two independent events occurring is $p_1 \cdot p_2$. For n independent arrangements or n independent probabilities, the expressions are

$$t = \prod_{i=1}^{n} t_i = t_1 \cdot t_2 \cdot t_3 \ldots t_n$$

$$p = \prod_{i=1}^{n} p_i = p_1 \cdot p_2 \cdot p_3 \ldots p_n$$

REFERENCES

Textbooks on Statistical Mechanics

DAVIDSON, N., 1962. *Statistical Mechanics*, McGraw-Hill, New York.

HILL, T. L., 1960. *An Introduction to Statistical Thermodynamics*, Addison-Wesley, Reading, Mass.

NASH, L. K., 1970. *Introduction to Statistical Thermodynamics*, Addison-Wesley, Reading, Mass.

REIF, F., 1965. *Fundamentals of Statistical and Thermal Physics*, McGraw-Hill, New York.

For a useful book that treats some of the material of this chapter in more detail, see

TANFORD, C., 1961. *Physical Chemistry of Macromolecules*, John Wiley, New York.

The relation between entropy and probability is discussed entertainingly in

BENT, H. A., 1965. *The Second Law*, Oxford University Press, New York.

PROBLEMS

1. For a dicarboxylic acid $HO_2C—R—CO_2H$, where the two $—CO_2H$ groups are far apart, using statistical methods show that the ratio of the acid dissociation constants K_1/K_2 is expected to be ~4. (The acid dissociation constants K_1 and K_2 are the equilibrium constants for the reactions

$$HO_2C—R—CO_2H \rightleftharpoons {}^-O_2C—R—CO_2H + H^+$$

and

$$^-O_2C—R—CO_2H \rightleftharpoons {}^-O_2C—R—CO_2^- + H^+$$

respectively.)

2. The binding of a ligand A to a macromolecule was studied by equilibrium dialysis. In each measurement, a $1 \times 10^{-6}\,M$ solution of the macromolecule was dialyzed against an excess amount of a solution containing A. After equilibrium was reached, the total concentrations of A on each side of the dialysis membrane were measured. The following data were obtained:

Total concentration of A (M)

The side without macromolecules	The side with $1 \times 10^{-6}\,M$ macromolecules
0.51×10^{-5}	0.67×10^{-5}
1.02×10^{-5}	1.28×10^{-5}
2.01×10^{-5}	2.34×10^{-5}
5.22×10^{-5}	5.62×10^{-5}
10.5×10^{-5}	10.91×10^{-5}
20.0×10^{-5}	20.47×10^{-5}

From these data, calculate ν as a function of [A]. Make a Scatchard plot, and evaluate the intrinsic equilibrium constant K and the total number of sites per macromolecule, if the independent-and-identical-sites model appears to be applicable.

3. Plot $\nu/[A]$ versus ν for the following cases:
(a) There are a total of 10 identical and independent sites per polymer, and the intrinsic binding constant is $K = 5 \times 10^5$.
(b) There are a total of 10 independent sites per polymer. Nine of the 10 sites are identical, with $K = 5 \times 10^5$. The binding constant for the tenth site is 5×10^6. For more than one type of site, the Scatchard equation can be generalized to

$$\frac{\nu}{[A]} = \sum_i \frac{N_i K_i}{1 + K_i[A]}$$

[Note that the deviation from linearity is small for part (b). Because experimental error in such measurements is usually appreciable, one should be cautious in concluding from a linear Scatchard plot that the sites are identical.]

4. A certain macromolecule P has four identical sites in a linear array for the binding of a small molecule A. Prepare a table listing all species with half of the sites occupied (species with two occupied sites and two unoccupied sites) and their statistical weights for the following cases:
(a) The sites are independent.
(b) Nearest-neighbor interactions are present.
(c) Binding to one site excludes the binding to a site immediately adjacent to it.
(d) Give also, for part (b), an expression for the concentration ratio $[PA_4]/[PA_2]$.

5. A DNA has short single-stranded ends which can join either intramolecularly to form a ring or intermolecularly to form aggregates. Discuss briefly and concisely under what condition you expect ring formation will predominate and under what condition you expect intermolecular aggregation will predominate.

6. An oligopeptide has seven amide linkages (or eight amino acid residues including the terminal carboxyl group). With the rules for the statistical weights we have discussed for the formation of an α helix, obtain the function Z, which is the sum of the statistical weights of all species. Also obtain an expression for ν, the average number of helical residues per molecule, as a function of σ and s.

7. An RNA oligonucleotide has the sequence $A_6C_7U_6$. It can form a hairpin loop held together by a maximum of 6 A · U base pairs.

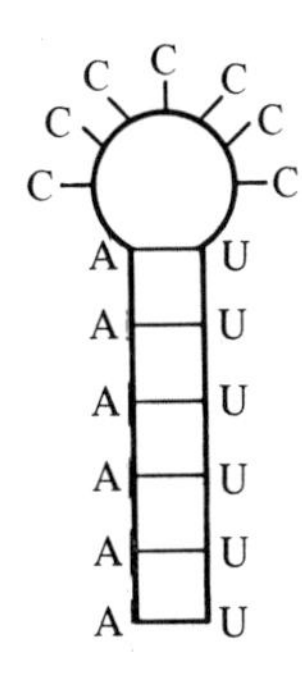

Assume that the helical region can melt only from either end. Use the notation s = equilibrium constant for adding a base pair to a helical region and $\sigma_j s$ = equilibrium constant for initiating the first base pair forming a loop. The subscript $j = 7, 9, 11, 13, 15, 17$ characterizes the number of unpaired bases in the loop.

(a) Calculate the statistical weight of each species which can be present.

(b) Calculate the mole fraction of each species which can be present.

(c) Assume that the molar absorptivity of each species depends only upon the number of base pairs formed: ε_0 = absorptivity per mole of mononucleotide for species with no base pairs; ε_1 = absorptivity of all species with 1 base pair; ε_2 = etc. Write an expression for the absorbance in a 1-cm cell of a solution of $A_6C_7U_6$ in terms of s, σ_j, ε_j, and c = concentration of $A_6C_7U_6$ in moles of nucleotides per liter.

8. The statistical effective segment length, l, of a DNA molecule is 100 nm. Calculate the mean square end-to-end distance and the contour length for a bacterial DNA with 10^7 base pairs. The distance between base pairs is 0.34 nm.

9. Show that if a system is sufficiently dilute such that $N_i \ll g_i$.

$$t\text{ (Fermi-Dirac)} \approx t\text{ (Bose-Einstein)} \approx \frac{t\text{ (Boltzmann)}}{N!}$$

{*Hint*: $t\text{ (Fermi-Dirac)} = \prod_i [g_i!/N_i!(g_i - N_i)!]$

$$= \prod_i \frac{g_i(g_i - 1)(g_i - 2)\cdots(g_i - N_i + 1)}{N_i!}$$

Since $g_i \gg N_i$,

$$g_i \approx g_i - 1 \approx g_i - 2 \cdots \approx g_i - N_i + 1$$

Therefore, $g_i(g_i - 1)\cdots(g_i - N_i + 1)$ is the product of N_i terms, each of which is approximately equal to g_i. Thus

$$t\text{ (Fermi-Dirac)} \approx \prod_i [(g_i^{N_i}/N_i!)] = \frac{t\text{ (Boltzmann)}}{N!}$$

t (Bose-Einstein) $\approx t$ (Boltzmann)/$N!$ can be similarly shown.}

10. Calculate the degeneracy for each of the first five levels for a particle in a three-dimensional cubic box.

11. (a) How many three-letter words can be made from twenty-six letters? A word is any sequence of three letters from AAA to ZZZ.

(b) How many different basketball teams of five players can be chosen from a group of 100 people?

(c) How many different proteins containing 100 amino acids can be made from the 20 commonly occurring amino acids?

12. Consider two systems: one consists of an electron in a cubic box of 1 nm on a side; the other consists of a He atom in a cubic box of 1 cm on a side. For each system, calculate the ratio of the probabilities (N_2/N_1) of finding the particle in the first two energy levels at (a) 10 K, (b) 1000 K, (c) 10,000 K.

13. A proton in a magnetic field can be in only one of two energy levels. For an energy spacing corresponding to a transition frequency of 100 megacycles, use the Boltzmann distribution to calculate the absolute temperature when:

(a) The proton is in the lower energy state.
(b) The proton is in the upper energy state.
(c) The probability of finding the proton in either level is equal.
(d) The probability of finding the proton in the lower level is 1.000015 times the probability of finding it in the upper level. (Remember that $e^x = 1 + x + \cdots$ when x is small.)

14. What is the change of entropy for the following reactions?

(a) One hundred pennies are changed from all heads to all tails.
(b) One hundred pennies are changed from all heads to 50 heads plus 50 tails.
(c) One mole of heads is mixed with one mole of tails to give two moles of half heads and half tails.

12

X-ray Diffraction and Related Topics

When one looks at photographs of the lunar landscape under the sun, one is struck by the sharp contrast between the bright lunar surface and the dark lunar sky. We see things around us because they scatter light. The daytime sky on earth is bright because the gas molecules and dust particles in the atmosphere scatter light. Because the moon is devoid of an atmosphere, darkness fills the space above the lunar horizon.

If macroscopic objects can be seen by their scattering of light, it seems that we should also be able to see molecules, and the detailed atomic structures within the molecules, by light scattering. There are several different aspects about seeing an object, large or small: contrast, sensitivity, and resolution. *Contrast* is dependent on the amount of light scattered by the object and the amount of light scattered by other things around it. We have difficulty seeing a white fox in a snow-covered field because the object and the background scatter about the same amount of light. *Sensitivity* is dependent on the absolute amount of light scattered by the object. A dark-adapted human eye can detect a pulse of visible light containing some 50 photons, and devices of even higher sensitivity have been constructed. *Resolution* is a measure of the spatial separation of two sources. For small enough separation the two no longer appear separated. The average human eye, for example, has an angular resolution of about 1 minute. Suppose that we write the number 11 on a wall with the 1's about 1 mm apart. At a distance of about 3.5 m from the wall it will be difficult for us to tell whether the number is an

11 or a fuzzy 1. At this distance the two 1's subtend an angle of 1×10^{-3} m/3.5 m $\cong 3 \times 10^{-4}$ rad $\cong$ 1 min at the eye.

Because of diffraction, even with the help of optically perfect lenses, the minimal resolvable separation cannot be much less than the wavelength of light. Since the wavelength of light is several thousand angstroms and atomic separation in a molecule is of the order of several angstroms, to see molecular details other types of radiation of much lower wavelengths must be employed.

X RAYS

In 1895, W. C. Roentgen accidentally discovered that when a beam of fast-moving electrons struck a solid surface, a new radiation was emitted. The new rays, which Roentgen called *x rays,* were found to cause certain minerals to fluoresce, to expose covered photographic plates, and to be transmitted through matter. One of the first experiments Roentgen did was to show that different kinds of matter were transparent to x rays to different degrees. The bones and the flesh of the experimenter's hand could clearly be distinguished when the hand was placed in front of a screen made of a material that fluoresced when x-rayed. Three months after Roentgen's discovery, x rays were put to use in a surgical ward. Eighteen years after the discovery, the first crystal structure (that of NaCl) was solved by x-ray diffraction. Today x-ray diffraction is one of the most powerful techniques in studying biological structures.

Emission of X Rays

Figure 12.1 illustrates the intensity distribution of the x rays as a function of wavelength, when a target (silver in this case) is bombarded with electrons accelerated by a certain potential (40 kV). The distribution has the appearance of a broad spectrum with four sharp spikes superimposed on it. The broad spectrum has a precipitous drop on the short-wavelength side but tails gradually into the long-wavelength side.

When electrons strike the target atoms, some are deflected by the field of the atoms with a loss of energy. This energy loss is accompanied by the emission of photons with energies corresponding to the energy differences of the incoming and the scattered electrons. Since the maximum energy of the photons emitted cannot exceed the energy of the incoming electrons, there is a sharp short-wavelength limit of the spectrum at a given accelerating voltage. The spikes in the emission spectrum are due to discrete energy levels of the orbital electrons of the target atoms. If the energy of an incoming electron is sufficiently high, it can eject an electron from an inner orbital. A photon of a characteristic wavelength is emitted when an electron in an outer orbital falls into the inner orbital to fill the vacancy left by the ejected electron. Thus x rays are photons (electromagnetic radiation) with wavelengths in the range from a few tenths to a few angstroms.

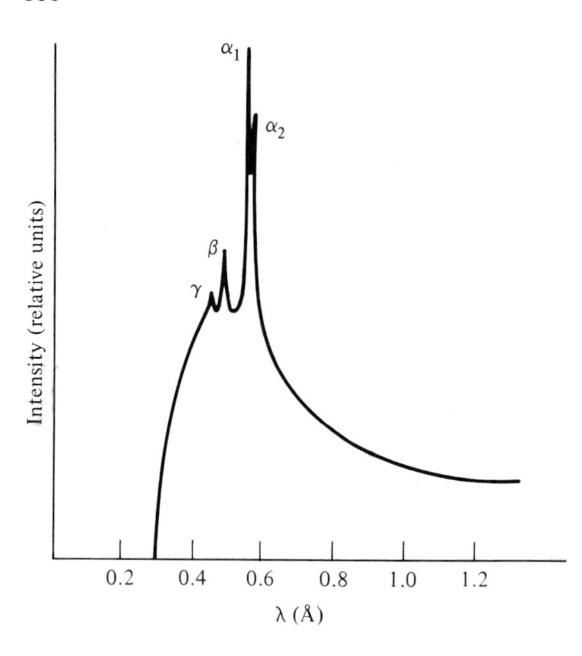

Fig. 12.1 Relative intensity of x rays produced as a function of wavelength when a silver target is bombarded with electrons accelerated by a 40-kV potential.

Image Formation

Since the wavelengths of x rays are of the same order of magnitude as interatomic distances in molecules, it seems that we should be able to construct a powerful x-ray camera, take a photograph, and see the fine structure. Let us analyze first how an optical image is formed. When an object is placed under a parallel beam of light, light is scattered by the object in all directions. Because of the wave nature of light, the scattered light in any particular direction is characterized by an amplitude term and a phase term. The pattern of the scattered waves is called the *diffraction pattern* of the object. In the presence of a lens, part of the diffraction pattern is intercepted by the lens and is refocused to give an image in the image plane. Unfortunately, no lens is available to refocus scattered x rays. Nevertheless, it should be clear that all the information that can be deduced from the image of an object must be present in its diffraction pattern, as a lens cannot provide any information not present in the diffraction pattern. The determination of molecular structures by x-ray diffraction therefore depends on two aspects: obtaining the diffraction patterns and determination of the structures.

Scattering of X Rays

X-ray photons are scattered by electrons in matter. The scattering can be *coherent*, in which the wavelengths of the incident and the scattered radiation are the same, or *incoherent*, in which the wavelengths differ. If inner orbital electrons are ejected by the incident radiation and secondary radiation is emitted when electrons in outer orbitals fall back to fill the vacancies, the emitted secondary radiation is referred to as fluorescence "scattering." In the discussions below we will consider only coherent scattering.

Figure 12.2 illustrates the scattering of a wave front by two identical point scatterers i and j. A reference plane, AB, perpendicular to the incident radiation,

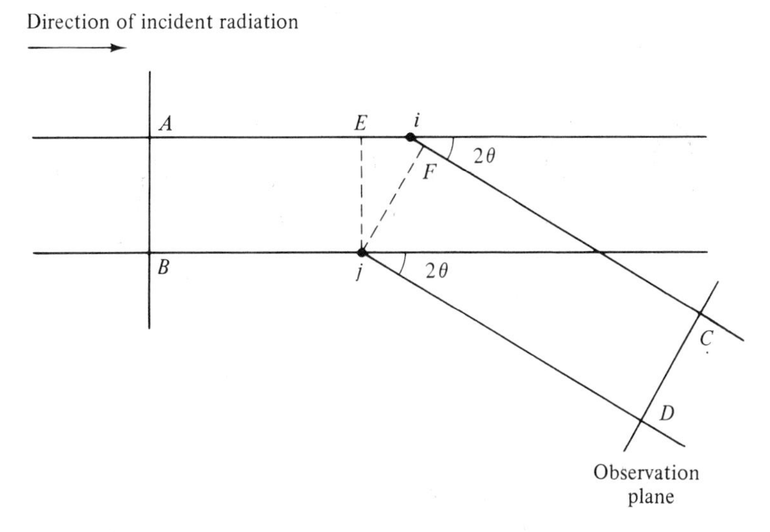

Fig. 12.2 Scattering of a wave front by two identical point scatterers i and j. The incident rays are in phase at plane AB. Rays scattered at an angle 2θ with respect to the incident rays are in phase at the plane of detection CD only if the path difference, $\overline{AiC}-\overline{BjD}$ (which is equal to $\overline{Ei}+\overline{iF}$, as shown), is an integral multiple of the wavelength of the radiation.

is drawn where all of the incident radiation is in phase. The plane CD represents the observation plane where the scattered radiation is measured. Because of the difference in path lengths between AiC and BjD, waves that are in phase at AB are not necessarily in phase at CD. A special case is when the angle 2θ between the incident radiation and the direction of observation is 0. In this case the difference in path lengths is zero and the scattered waves at CD are necessarily in phase: the amplitude of the resultant radiation scattered by the two scatterers is twice that scattered by one scatterer [Figure 12.3(a)]. Such total reinforcement also occurs when the difference in path lengths is an integral multiple of the wavelength. In all other cases there will be less than total reinforcement, and in the extreme case total destructive interference can occur. Two examples are illustrated in Figure 12.3(b) and (c).

For a single atom, total reinforcement of scattering by all the electrons in the atom occurs only at zero scattering angle. Figure 12.4 depicts the dependence on $\sin(\theta/\lambda)$ of the atomic scattering factor f_0, which is the ratio of the amplitude

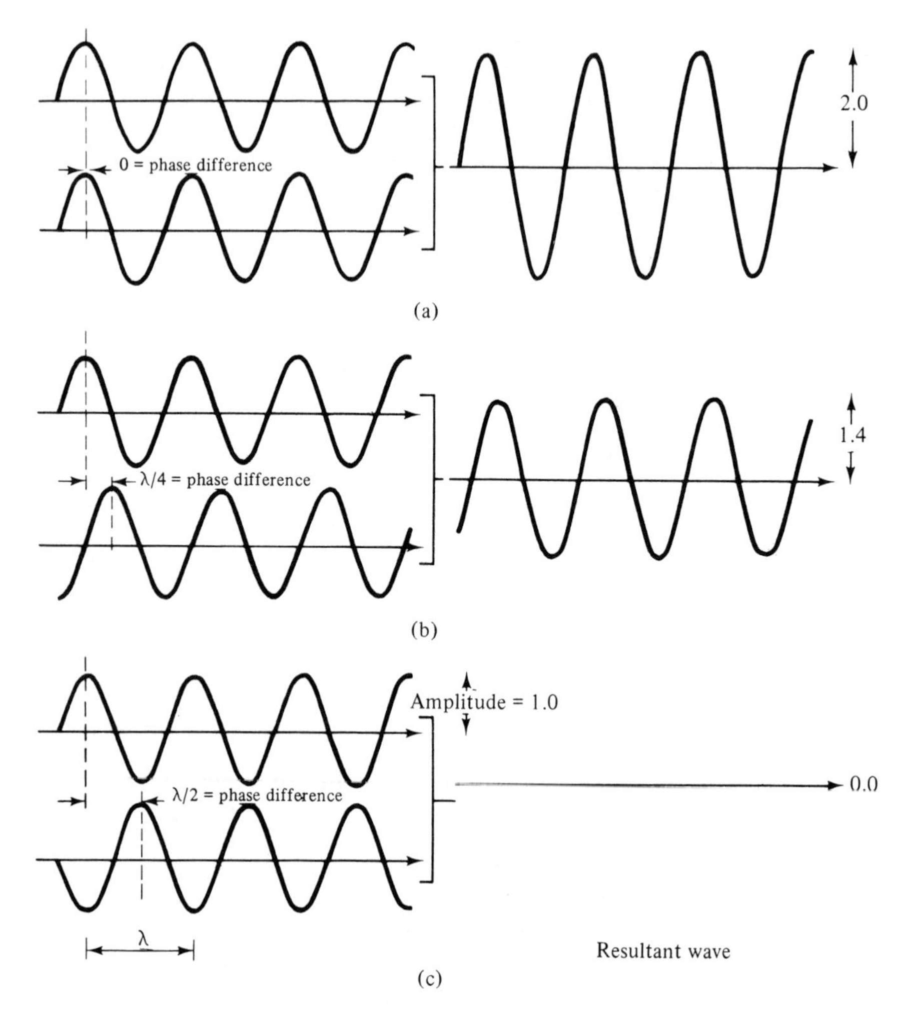

Fig. 12.3 Summing up of two waves: (a) waves completely in phase; (b) waves out of phase by one-quarter wavelength (partial reinforcement occurs); (c) waves out of phase by one-half wavelength (total destructive interference occurs). (From J. P. Glusker and K. N. Trueblood, *Crystal Structure Analysis: A Primer*, Oxford University Press, New York, 1972, p. 19, Fig. 5.)

scattered by the atom to that by an electron. At zero scattering angle f_0 is equal to the number of electrons in the atom. Because of the spatial distribution of the electrons, f_0 decreases at larger $\sin \theta/\lambda$, owing to interference. In a crystal, the atomic scattering factor is reduced further because of thermally induced vibrations of the atom. This temperature effect can be corrected by multiplying f_0 by a temperature-dependent factor. The symbol f is used for the corrected scattering factor.

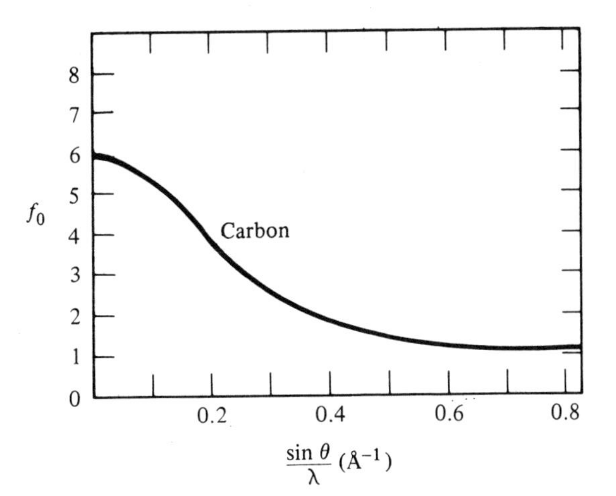

Fig. 12.4 Dependence of atomic scattering f_0 on $\sin\theta/\lambda$. The angle between the incident and scattered rays is 2θ. The curve shown is for carbon. Note that when θ approaches zero, the value of f_0 approaches the number of electrons per atom.

Laue's Equations for the Diffraction of X Rays by a Crystal

Let us consider first the diffraction of x rays by a linear array of identical *point* scatterers which are equally spaced, with a distance a between each adjacent pair. As illustrated in Figure 12.5, we let α_0 be the angle the parallel incident rays

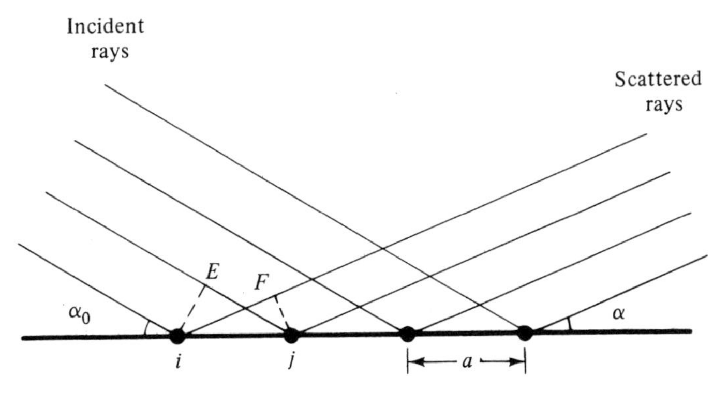

Fig. 12.5 Scattering of incident rays by a row of point scatterers of spacing a. The path difference of rays scattered by any two adjacent scatterers i and j is $(\overline{iF} - \overline{jE})$, which is equal to $a\cos\alpha - a\cos\alpha_0$ or $a(\cos\alpha - \cos\alpha_0)$.

make with the array. The incident rays are scattered in all directions. In a given direction α, the difference in path of rays scattered by two adjacent scatters is

$$\text{path difference} = a(\cos\alpha - \cos\alpha_0)$$

as illustrated in Figure 12.5. If this path length is an integral multiple of the wavelength, then *all* the rays scattered by the points will reinforce. This condition is expressed by

$$a(\cos\alpha - \cos\alpha_0) = h\lambda \tag{12.1}$$

where h is an integer. For a fixed value of a and a fixed glancing angle α_0, there are several values of α at which constructive interference occurs, corresponding to $h = 0, 1, 2, \ldots$. For all other values of α, rays scattered from any two scatterers will be somewhat out of phase. The resultant wave from a large number of scattered rays with random phases has zero amplitude. In summary, then, when parallel monochromatic x rays strike a row of equidistant point scatterers at a glancing angle α_0, the scattered radiation will concentrate in a number of cones, each cone with a single α given by Eq. (12.1). A special case with $\alpha_0 = 90°$ is illustrated in Fig. 12.6. We might interject at this point that incoherently scattered rays, because of changes in λ, will be more or less randomly distributed spatially. They contribute to the background intensity but are not important for our discussions.

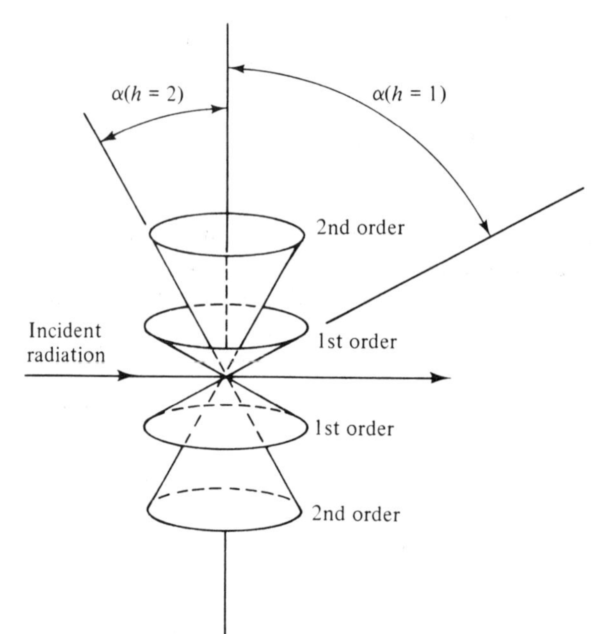

Fig. 12.6 Equidistant point scatterers arranged in the vertical direction at the apexes of the cones. With an incident angle $\alpha_0 = 90°$, Eq. (12.1) reduces to $a\cos\alpha = h\lambda$, or $\alpha = \cos^{-1}(h\lambda/a)$. Cones with $h = 1$ (first order) and $h = 2$ (second order) are shown. Note that $\cos\alpha = \cos(-\alpha)$; thus in this particular case, for each order there are two cones 180° apart. The zero-order diffraction ($h = 0$) is not shown.

We can now generalize our treatment to the case of a three-dimensional lattice. Figure 12.7 illustrates a simple lattice formed by three sets of equidistant planes. The points of intersection of these planes form the space lattice. The characteristic interplanar distances are a, b, and c, as indicated. We can also consider this simple lattice as generated by three sets of fundamental translations along three nonplanar axes a, b, and c. For the most general case, the interaxial angles are unequal, and so are a, b, and c. The lattice is called *triclinic*. If the interaxial angles are all 90° and $a \neq b \neq c$, the lattice is called *orthorhombic*; if the interaxial angles are 90° and $a = b = c$, the lattice is *cubic*. There are other combinations that we need not mention here.

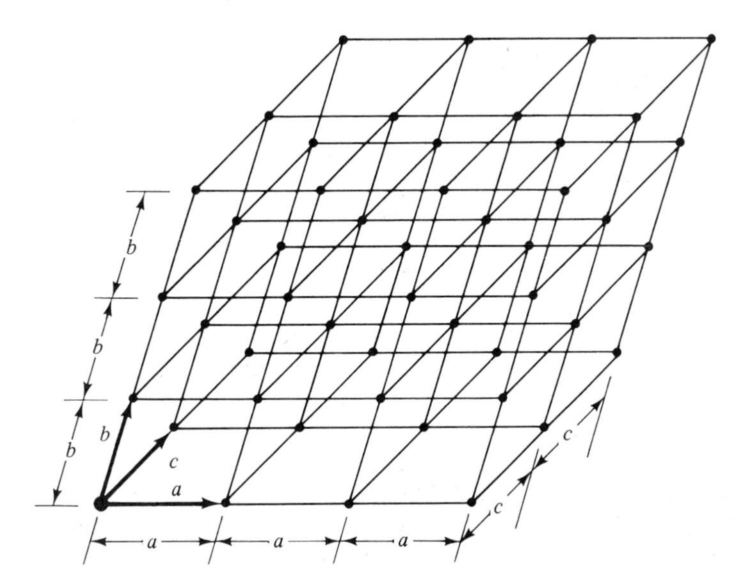

Fig. 12.7 Lattice of points. The lattice axes are depicted at the lower left corner.

If we have an identical scatterer located at each lattice point in a triclinic lattice, it is easy to generalize Eq. (12.1), that total reinforcement occurs if the following *three* equations are satisfied:

$$a\,(\cos\alpha - \cos\alpha_0) = h\lambda \tag{12.1}$$

$$b\,(\cos\beta - \cos\beta_0) = k\lambda \tag{12.2}$$

$$c\,(\cos\gamma - \cos\gamma_0) = l\lambda \tag{12.3}$$

These equations were first derived by M. von Laue and are referred to as *Laue's equations.* The angles α_0, β_0, and γ_0 and α, β, and γ are the angles the incident and diffracted beams make with the three axes, respectively.

Measuring the Diffraction Pattern

If one measures the diffraction pattern of a crystal by using a *monochromatic* beam of x rays and placing a photographic film behind the crystal (or in front of the crystal, in which case forward scattering will be measured), one might be surprised to find that aside from a very intense spot due to the incident beam, little characteristic scattering above the diffuse background intensity could be seen. The reason can be found by a careful examination of the three Laue's equations. For a given set of values of a, b, c, α_0, β_0, γ_0, and λ, the angles α, β, and γ must satisfy the three equations. In addition, α, β, and γ must satisfy a fourth equation. This is easy to see for an orthorhombic lattice, for which the three axes are mutually perpendicular. In this system if a line is perpendicular to two of the axes, it must be parallel to the third one. In other words, we cannot independently specify α, β, and γ. It can be shown that for an orthorhombic system $\cos^2\alpha + \cos^2\beta + \cos^2\gamma \equiv 1$. Now if three variables α, β, and γ must satisfy four equations, there can be no general solution. Looking at the situation in a different way, by analogy with Fig. 12.6, for a three-dimensional crystal there will be three conical surfaces extending from the crystal at the origin and forming angles α, β, and γ with the axes. The direction of constructive interference is given by a line common to all three conical surfaces. Since three conical surfaces generally do not have a common intersecting line, there will be no scattered rays of significant intensity.

This discussion shows that to measure the diffraction pattern, a special arrangement must be made. We have several choices:

1. Instead of using a monochromatic incident beam, we can use "white" x rays, which contain a broad distribution of wavelengths. For certain values of λ, the four equations can be satisfied. This option is of historical interest, as when Laue and his coworkers did their first experiment on x-ray diffraction by a crystal, little was known about x rays, and the x-ray source they used gave a broad spectrum of radiation. Now monochromatic radiation is almost always used in x-ray diffraction studies. We shall explain in later sections how to obtain monochromatic x rays.

2. If monochromatic radiation is used, characteristic diffraction rays can be observed if the orientation of the crystal relative to the incident beam is systematically changed. This can be achieved by rotating or oscillating a crystal about one axis of the crystal at a time, or by changing the position of the detector (a photographic film or an ionization or solid-state detector). Sometimes the rotation or the oscillation of the crystal is coupled to the movement of detector so as to obtain a diffraction pattern which is easier to interpret. A particular mode of coupling can be achieved mechanically, or a computer can be programmed to control the movements.

3. If monochromatic radiation is incident upon a powder of little crystals randomly oriented, a fraction of the crystals will have the correct orientation with respect to the incident beam to give characteristic diffractions.

Physical Meaning of the Indices (*h, k, l*) in Laue's Equations

So far we have only mentioned that the parameters h, k, and l are integers. The physical meaning of h, k, and l became clear primarily through the work of W. L. Bragg.

For a three-dimensional lattice, many parallel planes can be drawn through the lattice points. For simplicity, a two-dimensional lattice, and several groups of parallel lines going through the lattice points, are illustrated in Fig. 12.8. Each group of parallel lines can be indexed in the following way: pick any line such as the heavy line in the figure. The intercepts this line makes with the a axis and b axis are $3a$ and $6b$, respectively. The reciprocals of the coefficients are $(\frac{1}{3}, \frac{1}{6})$. Now

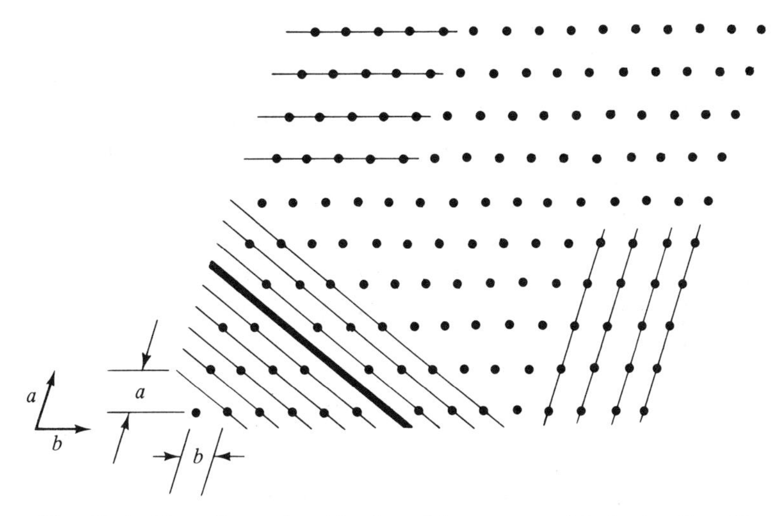

Fig. 12.8 Two-dimensional lattice with some possible lattice lines illustrated.

multiply these reciprocals by the smallest number that will give integral indices. In this example, the smallest integer is 6, and the indices obtained are (2, 1). These indices are called the *Miller indices.* The steps are summarized below:

	Axis	
Steps	a	b
1. Intercepts	3	6
2. Reciprocals	$\frac{1}{3}$	$\frac{1}{6}$
3. Miller indices	2	1

Exercise Pick a few arbitrary lattice lines parallel to the one we discussed above. Convince yourself that these parallel lines all have the same Miller indices.

The distance d_{hk} between any two adjacent parallel lines is determined by their Miller indices. For example, if the a axis and b axis are perpendicular to each other, it is easy to show that

$$d_{hk} = \frac{1}{\sqrt{\frac{h^2}{a^2} + \frac{k^2}{b^2}}}$$

If the interaxial angle is different from 90°, the expression for d_{hk} is more cumbersome unless vector-algebra notations are used. We shall not concern ourselves with the details here.

Each group of parallel lines of indices (h, k) shown in Fig. 12.8 is characterized by two parameters: their direction and the distance d_{hk}. Both parameters can be represented by a single point in the following way: (1) pick an arbitrary point O as our origin; (2) draw a line perpendicular to the parallel lines (h, k) from the origin; and (3) terminate the line drawn at a point P_{hk} such that the length $OP_{hk} \equiv \sigma_{hk} \equiv 1/d_{hk}$.

Exercise From a common origin O, obtain graphically P_{hk} for each of the three groups of parallel lines shown in Fig. 12.8. You may choose a convenient scale factor K in the actual drawing. That is, let

$$\sigma_{hk} = \frac{K}{d_{hk}}$$

By similar procedures, we can obtain all other points corresponding to all other possible parallel lines through the lattice points. All the P_{hk}'s form a new two-dimensional lattice. In other words, we can transform a two-dimensional lattice into a two-dimensional *reciprocal lattice.* We call the new lattice a "reciprocal lattice" because the distance from its origin O to any point P_{hk} is the reciprocal of the distance d_{hk} in the original lattice.

The situation is not much different for a three-dimensional lattice. Any group of parallel planes can be represented by three Miller indices (h, k, l). The interplanar spacing d_{hkl} can be calculated from the indices (for an orthorhombic lattice $d = 1/[(h/a)^2 + (k/b)^2 + (l/c)^2]^{1/2}$). A reciprocal lattice point P_{hkl} can be obtained by picking an arbitrary origin O, drawing a line perpendicular to the planes (h, k, l), and terminating the line at P_{hkl} such that the length $OP_{hkl} \equiv \sigma_{hkl} \equiv 1/d_{hkl}$. Each group of parallel crystal planes is represented by a single point in a reciprocal lattice.

With the preceding discussions on the reciprocal lattice, we can now illustrate the geometric relations given by Laue's equations. The incident x rays can be thought of as scattered by the scatterers in the various groups of crystal planes. Suppose that a crystal C is located at the center of a sphere of radius $1/\lambda$, where λ is the wavelength of the x rays. This sphere is called the Ewald sphere. The incident beam passes through C and leaves the sphere at O, as depicted in Fig. 12.9. It can be shown that Laue's equations (with integers, h, k, and l) are satisfied if, using O as our origin of the reciprocal lattice, the point P_{hkl} is located on the Ewald sphere. Put in a slightly different way, for a group of crystal planes (h, k, l) located at C, reinforcement of the rays scattered by scatterers in this group of planes can occur only if the corresponding point P_{hkl} in the reciprocal lattice is located on the Ewald sphere.

We shall not give a mathematical proof of the interpretation above. The proof is cumbersome and tedious without using vector algebra. For students who had an introductory course in vector algebra, proof can be found in a number of references on x-ray diffraction.

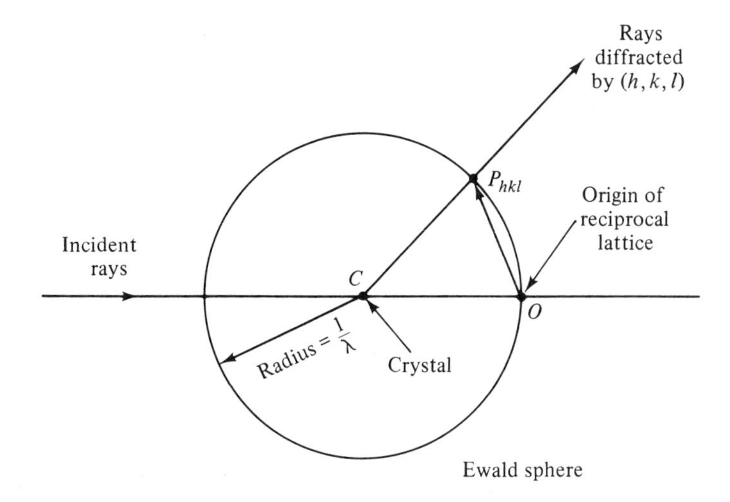

Fig. 12.9. Ewald sphere of radius $1/\lambda$. Reinforcement of rays scattered by a group of planes of Miller indices (h, k, l) occurs in the direction CP_{hkl} if P_{hkl} is on the Ewald sphere.

Bragg's Interpretation

A simpler physical picture relating the directions of incident and scattered rays by a crystal plane was noted by Bragg. For a family of parallel crystal planes (h, k, l) with spacing d_{hkl}, a monochromatic incident beam of wavelength λ will *appear* to be reflected by the planes if the incident angle θ is

$$\theta = \sin^{-1}\frac{n\lambda}{2d_{hkl}} \tag{12.4}$$

where n is an integer 0, 1, 2, ... To state it in different words, if the incident rays make an angle θ with a family of crystal planes (h, k, l) such that

$$2d_{hkl}\sin\theta = n\lambda \tag{12.5}$$

reinforcement of the scattered rays occurs in a direction that also makes an angle θ with the planes. Equation (12.4) or (12.5) is called the *Bragg law.*

The Bragg law and Laue's equations can be shown to be equivalent. The case $n = 0$ is not of much interest. We shall illustrate the case $n = 1$ graphically; Fig. 12.10 shows a cross section of the Ewald sphere shown in Fig. 12.9. The following relations result directly from plane geometry, since OA is the diameter and C is the center of the circle:

$$\angle OP_{hkl}A = 90°$$

$$\angle OCP_{hkl} = 2\,\angle OAP_{hkl}$$

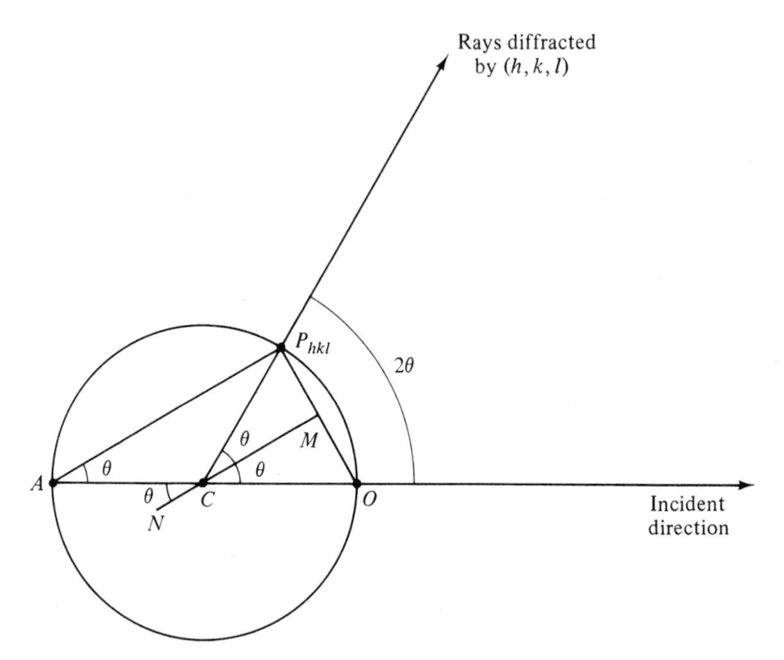

Fig. 12.10 Section through the Ewald sphere. The Bragg law is satisfied when the reciprocal lattice point P_{hkl} is on the Ewald sphere.

Furthermore, if CM is perpendicular to OP_{hkl},

$$\angle OCM = \angle P_{hkl}CM$$

Also,

$$\angle NCA = \angle OCM$$

Let $\angle OAP_{hkl} = \theta$; then $\angle OCP_{hkl} = 2\theta$, and $\angle OCM = \angle P_{hkl}CM = \angle NCA = \theta$. These angles are indicated in Fig. 12.10.

From our discussion on how the reciprocal lattice is constructed, it should be clear that the planes (h, k, l) are perpendicular to OP_{hkl}. Therefore, the direction of the crystal planes (h, k, l) is given by the straight line NCM shown in Fig. 12.10. It follows, then, that:

1. Both the incident beam (in the direction AC) and the constructively diffracted rays (in the direction CP_{hkl}) make an angle θ with the parallel crystal planes (h, k, l).
2. Since $\angle OP_{hkl}A = 90°$, $\overline{OP_{hkl}} = \overline{AO} \sin\theta$. But $\overline{AO} = 2 \cdot (1/\lambda)$ and $\overline{OP_{hkl}} = 1/d_{hkl}$. Thus

$$\frac{1}{d_{hkl}} = 2 \cdot \frac{1}{\lambda} \sin\theta$$

or

$$2d_{hkl} \sin\theta = \lambda$$

This is the Bragg law for $n = 1$.

For higher values of n, we can write the Bragg law in the form

$$2 \cdot \frac{d_{hkl}}{n} \sin \theta = \lambda \tag{12.6}$$

The quantity (d_{hkl}/n) can be viewed as the spacing between a family of planes with Miller indices (nh, nk, nl). The higher-order Bragg diffractions can be viewed as coming from these planes.

Monochromatization of X Rays by Bragg Reflection

From the discussions of the preceding section, it is apparent that a beam of nonmonochromatic x rays will be dispersed by a crystal. Rays of different wavelengths are diffracted with different angles at the crystal. Monochromatic radiation can therefore be obtained by selecting a particular diffracted beam. Various crystal monochromators based on this principle have been designed, because structural determination by x-ray diffraction almost always uses monochromatic rays. A second way of obtaining monochromatic rays is to use x-ray filters. This will be described later.

Intensity of Diffraction

So far we have considered the diffraction of x rays by a lattice of identical *point* scatterers. Laue's equations, or the equivalent Bragg equation, provide the relation between the diffraction pattern and the lattice parameters. For a point scatterer the intensity of the scattered ray is independent of the scattering angle. If we have a real atom instead of a point scatterer, the intensity of the scattered ray is a function of $\sin \theta/\lambda$, as we have discussed already (see Fig. 12.4). Since constructive interference occurs only at $\sin \theta/\lambda = n/2d_{hkl}$ (Eq. 12.6), the atomic scattering factor for a group of planes of indices (h, k, l) is independent of wavelength. The intensities of rays scattered by different groups of planes vary, of course, because of differences in $\sin \theta/\lambda$ values.

The situation is more complex when the lattice contains several kinds of atoms. Suppose that we have a crystal of diatomic molecules AB as illustrated in Fig. 12.11. Consider the rows of A atoms first. The reinforcement of scattered rays occurs at angles

$$\theta_A = \sin^{-1} \frac{n\lambda}{2d}$$

where d is the spacing between the rows. For the rows of B atoms, the reinforcement of scattered rays also occurs at angles

$$\theta_B = \sin^{-1} \frac{n\lambda}{2d} = \theta_A$$

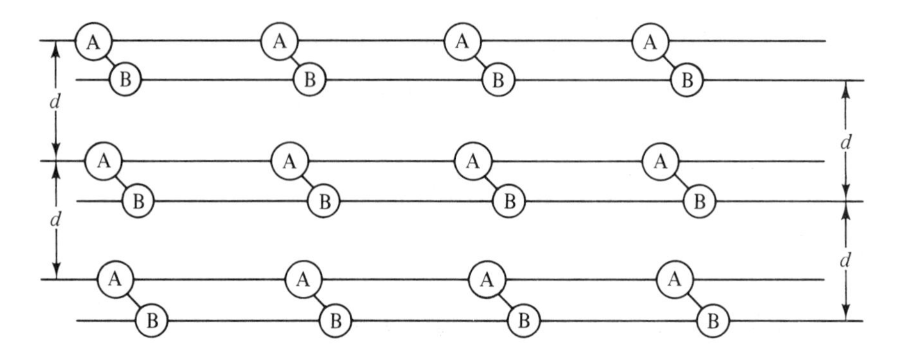

Fig. 12.11 Three rows of diatomic molecules AB in a certain crystal. The distance d between adjacent rows of A atoms is necessarily the same as the distance d between rows of B atoms. The distance between adjacent rows of A and B atoms is not the same as d.

as the spacing between the rows of B atoms must be the same as the spacing between the rows of A atoms in the crystal. The spacing between a row of A atoms and a row of B atoms is different from d, however. This means that rays scattered by rows of A and rows of B atoms are not in phase at the angle $\theta_A = \theta_B$. This results in partial interference, and therefore the total intensity at the angle $\theta_A = \theta_B$ is less than the sum of intensities due to independent scattering by A atoms and by B atoms. To put it a slightly different way, we can consider the crystal as being made of two interpenetrating lattices for the A atoms and B atoms, respectively. The two lattices have identical lattice parameters. At a given Bragg angle, the intensity of rays diffracted by the lattice of A atoms is dependent on f_A, the atomic structure factor for A; the intensity of rays diffracted by the lattice of B atoms is dependent on f_B, the atomic structure factor for B. In addition to its dependence on f_A and f_B, the total intensity of rays diffracted by the two interpenetrating lattices is also dependent of the relative *phases* of rays diffracted by the two lattices.

For more complex structures, it is more convenient to view the intensity problem in terms of the unit-cell concept described below.

Unit Cell

Any crystal can be considered as formed by placing a basic structural unit on every point of a lattice. (The crystal so formed is, of course, a perfect one. We shall not be concerned here with crystal imperfections.) An example is shown in Fig. 12.12.

We can consider the lattice as made of *unit cells*. The translations of the unit cell along the lattice axes generate the lattice. Figure 12.13 illustrates a two-dimensional lattice and several different unit cells that might be chosen. Usually the one with the highest symmetry and the lowest volume is chosen. For our

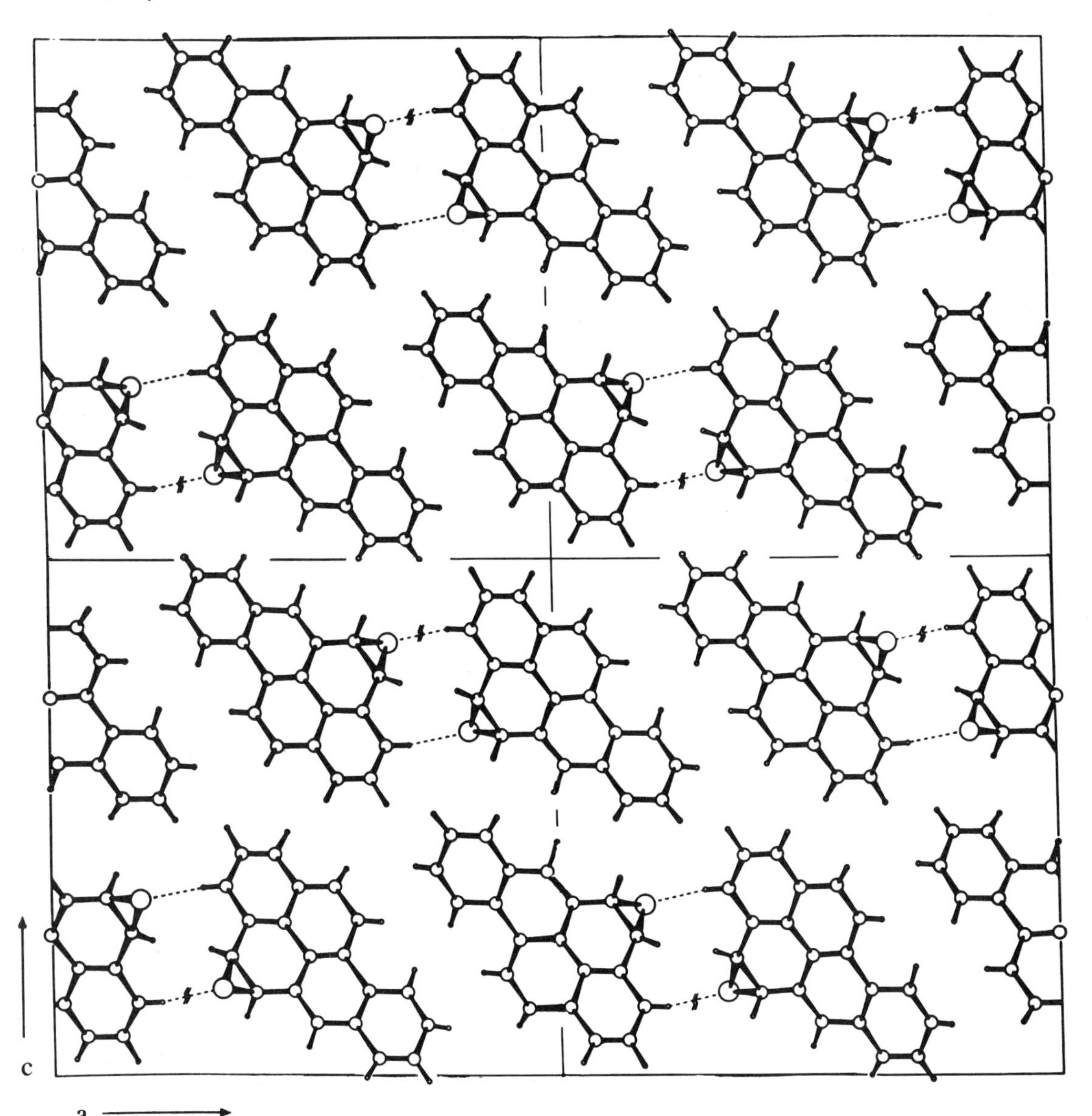

Fig. 12.12 Crystal structure of benz[*a*]pyrene-4,5-oxide. The molecules form planar layers. Four unit cells are shown. [Courtesy of Dr. Jenny P. Glusker.]

two-dimensional lattice, a unit cell is specified by two axes **a** and **b**. Let the dimensions of a unit cell be a and b in these directions; then the position of any atom in the unit cell is given by coordinates (X, Y), where X and Y are given in fractions of the unit-cell dimensions. For example, an atom at the center of a unit cell has coordinates $(\frac{1}{2}, \frac{1}{2})$. For a three-dimensional lattice, a unit cell is specified

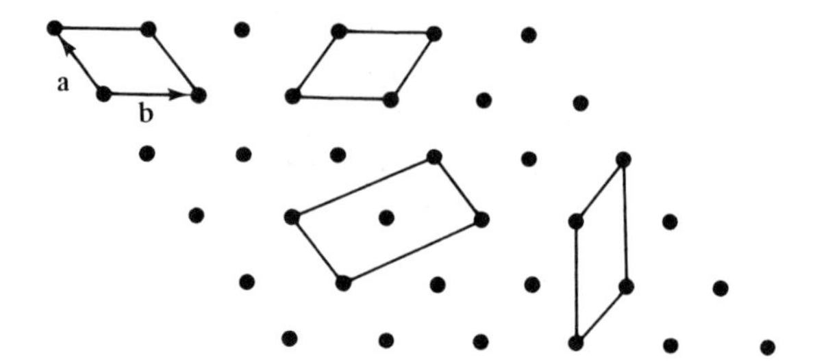

Fig. 12.13 Two-dimensional lattice of points and some of the unit cells that can be chosen. Each unit cell is defined by two axes, **a** and **b**, of length a and b, respectively, which are depicted for the unit cell at the upper left corner. For a three-dimensional lattice, a unit cell is defined by three axes, **a**, **b**, and **c**. Translation of a unit cell along its axes generates the lattice.

by three axes **a**, **b**, and **c** and the position of an atom in the unit cell will be given by (X, Y, Z). If we determine the unit-cell dimensions and the coordinates of its atomic contents, we have all the information of the crystal structure.

Summing Waves

Coming back to the intensity problem, our basic strategy is to sum the waves scattered by all atoms in a unit cell, at the same time keeping in mind the requirement of constructive interference of scattering from different parts of the crystal, which is expressed by the Bragg equation or Laue's equations.

Let us first consider how to add waves. We could do it graphically, as we did in Fig. 12.3. But this is tedious and impractical. There are several more convenient ways.

A particular scattered wave can be expressed by a sine (or a cosine) function:

$$y_1 = a_1 \sin (2\pi\nu_0 t + \alpha_1) \tag{12.7}$$

where a_1 is the amplitude or maximum displacement, y_1 the displacement at any time t, ν_0 the frequency, and α_1 the phase angle relative to an arbitrary origin. Let $2\pi\nu_0 t \equiv \phi$. We write

$$y_1 = a_1 \sin (\phi + \alpha_1) \tag{12.8}$$

This function is illustrated in Fig. 12.14. This wave can be generated by the y projection of a rotating vector $\mathbf{a}_1$ of length a_1 [Fig. 12.14(b)]. At $t = 0$ (therefore, $\phi = 0$), the vector $\mathbf{a}_1$ makes an angle α_1 with the horizontal axis. Its y projection is $a_1 \sin \alpha_1$. The vector rotates counterclockwise at an angular velocity of $2\pi\nu_0$. After a time t, its angle with the horizontal axis becomes $2\pi\nu_0 t + \alpha_1$ or $\phi + \alpha_1$, and its y axis projection is just $y_1 = a_1 \sin (\phi + \alpha_1)$. Thus the y projection of this rotating vector always gives the displacement of the wave at any time.

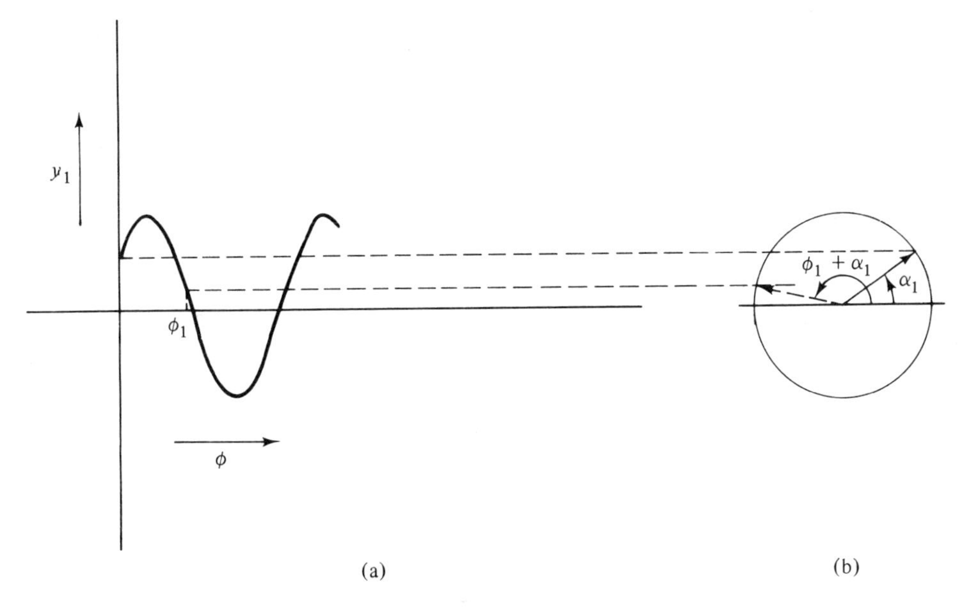

Fig. 12.14 (a) Sine wave represented by the equation $y_1 = a_1 \sin(\phi + \alpha_1)$. (b) The sine function can be considered as generated by the y projection of a rotating vector of length a_1. The positions of the vector at $\phi = 0$ and $\phi = \phi_1$, and the corresponding points on the sine curve are illustrated.

A second scattered wave of the same frequency but different amplitude and phase as the first wave can be represented by

$$y_2 = a_2 \sin(\phi + \alpha_2) \tag{12.9}$$

This wave can be similarly considered as generated by the y projection of a vector $\mathbf{a}_2$ of length a_2, which initially makes an angle α_2 with the horizontal axis and rotates at the same angular velocity $2\pi\nu_0$ as the first vector.

For the sum of the two waves, the displacement y_r at any time t is the sum of the displacement of the individual waves y_1 and y_2, as we have done in Fig. 12.3:

$$y_r = y_1 + y_2 \tag{12.10}$$

As illustrated in Fig. 12.15, this sum is just the y projection of the vector $\mathbf{a}_r$, which is the *vectorial sum* of $\mathbf{a}_1$ and $\mathbf{a}_2$. Readers who have not worked with vectors should not be intimidated at this point. If from an origin O we draw OA in the direction of $\mathbf{a}_1$ (OA makes an angle α_1 with the horizontal axis) with length $\overline{OA} = a_1$, then draw AB in the direction of $\mathbf{a}_2$ (AB makes an angle α_2 with the horizontal axis) with length $\overline{AB} = a_2$, the resultant vector $\mathbf{a}_r$ is just OB.

Therefore, the sum of the waves can be considered as generated from the vector $\mathbf{a}_r$ rotating at the same angular velocity as $\mathbf{a}_1$ and $\mathbf{a}_2$. The phase of the resulting wave is most easily seen at $t = 0$ (Fig. 12.15). The y projection of $\mathbf{a}_r$ is

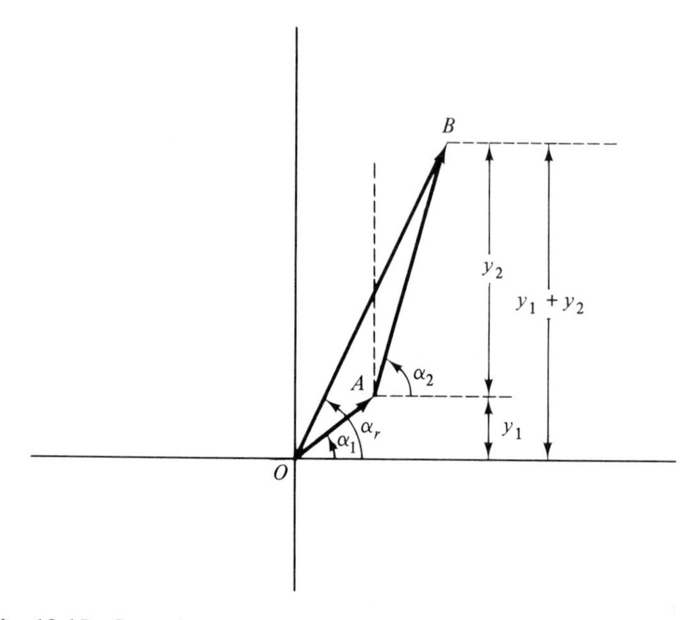

Fig. 12.15 Summing up of two waves. The wave $y_1 = a_1 \sin(\phi + \alpha_1)$ is generated by the y projection of the rotating vector OA. The wave $y_2 = a_2 \sin(\phi + \alpha_2)$ is generated by the y projection of the rotating vector AB. The sum of the two waves, $y_1 + y_2$, is generated by the y projection of the resultant vector OB.

$y_r = y_1 + y_2 = a_1 \sin \alpha_1 + a_2 \sin \alpha_2$. Its horizontal projection is $a_r \cos \alpha_r$ or $a_1 \cos \alpha_1 + a_2 \cos \alpha_2$. Thus

$$\tan \alpha_r = \frac{a_r \sin \alpha_r}{a_r \cos \alpha_r} = \frac{a_1 \sin \alpha_1 + a_2 \sin \alpha_2}{a_1 \cos \alpha_1 + a_2 \cos \alpha_2} \tag{12.11}$$

or

$$\alpha_r = \tan^{-1} \frac{a_1 \sin \alpha_1 + a_2 \sin \alpha_2}{a_1 \cos \alpha_1 + a_2 \cos \alpha_2} \tag{12.12}$$

The discussion above can be generalized to the sum of n coherently scattered rays represented by n rotating vectors. The resultant wave is represented by the vectorial sum of these n vectors. The phase of the resultant wave is

$$\alpha_r = \tan^{-1} \left(\sum_{j=1}^{n} a_j \sin \alpha_j \Big/ \sum_{j=1}^{n} a_j \cos \alpha_j \right) \tag{12.13}$$

where $\sum_{j=1}^{n} a_j \sin \alpha_j$ is the vertical component of the resultant vector and $\sum_{j=1}^{n} a_j \cos \alpha_j$ is the horizontal component of the resultant vector.

The magnitude or length of the resultant vector is, from the Pythagorean theorem,

$$|\mathbf{a}_r| = [(a_r \sin \alpha_r)^2 + (a_r \cos \alpha_r)^2]^{1/2}$$

$$= \left[\left(\sum_{j=1}^{n} a_j \sin \alpha_j \right)^2 + \left(\sum_{j=1}^{n} a_j \cos \alpha_j \right)^2 \right]^{1/2} \tag{12.14}$$

Note that the amplitude of the wave, or the maximum y-axis projection of α_r, is just $|\mathbf{a}_r|$.

We shall use the vectorial summation method later in our discussion of the phasing problem. But for an analytical formulation of problems, it is more convenient to use the complex-number representation. Each wave is represented in complex notation by

$$a_j e^{i\alpha_j}$$

where $i = \sqrt{-1}$. From the identity

$$e^{i\alpha_j} \equiv \cos \alpha_j + i \sin \alpha_j \tag{12.15}$$

we have, for the resultant of n waves,

$$\sum_{j=1}^{n} a_j \cos \alpha_j + i \sum_{j=1}^{n} \sin \alpha_j$$

The magnitude of any complex number $A + Bi$ is $(A^2 + B^2)^{1/2}$. The magnitude or length of $\mathbf{a}_r$ is thus

$$|\mathbf{a}_r| = \left[\left(\sum_{j=1}^{n} a_j \cos \alpha_j\right)^2 + \left(\sum_{j=1}^{n} a_j \sin \alpha_j\right)^2\right]^{1/2}$$

We notice immediately that this result would be identical to our result obtained using the vector notations, if we take the imaginary component of the complex number to be the vertical component in our vector system and the real component of the complex number as the horizontal component in our vector system.

Summing Scattered Rays

At a given Bragg diffraction angle, $\theta = \sin^{-1}(\lambda/2d_{hkl})$, corresponding to diffraction by planes of indices (h, k, l), the rays scattered by an atom j can be expressed by $f_j e^{-i\alpha_j}$, where f_j is the atomic scattering factor and α_j is the phase. The scattering by one unit cell containing n atoms is the sum

$$F(h, k, l) = \sum_{j=1}^{n} f_j e^{-i\alpha_j} \tag{12.16}$$

which is called the *structure factor* of the planes (h, k, l). We can also express the structure factor in the form

$$F(h, k, l) = A(h, k, l) + iB(h, k, l) \tag{12.17}$$

According to our previous discussions,

$$A(h, k, l) = \sum_{j=1}^{n} f_j \cos \alpha_j \tag{12.18}$$

$$B(h, k, l) = \sum_{j=1}^{n} f_j \sin \alpha_j \tag{12.19}$$

The magnitude of $F(h, k, l)$ is

$$|F(h, k, l)| = (A^2 + B^2)^{1/2} \tag{12.20}$$

and the phase of $F(h, k, l)$ is

$$\alpha(h, k, l) = \tan^{-1}\frac{B}{A} \tag{12.21}$$

It can be shown that the phase α_j of the scattered rays by an atom j at X_j, Y_j, Z_j, in the direction of a Bragg diffraction angle, is $2\pi(hX_j + kY_j + lZ_j)$. The atomic coordinates (X_j, Y_j, Z_j) are expressed in units of the unit-cell dimensions, as we have defined previously. Substituting into Eqs. (12.18) and (12.19) gives the components of the structure factor as

$$A(h, k, l) = \sum_{j=1}^{n} f_j \cos[2\pi(hX_j + kY_j + lZ_j)] \tag{12.22a}$$

$$B(h, k, l) = \sum_{j=1}^{n} f_j \sin[2\pi(hX_j + kY_j + lZ_j)] \tag{12.22b}$$

We can now summarize the intensity of rays scattered by a crystal as follows. Corresponding to a group of parallel planes of Miller indices (h, k, l), there is a point P_{hkl} in the reciprocal lattice. If this reciprocal lattice point is on the Ewald sphere (Fig. 12.9), constructive interference occurs. The amplitude of the resultant scattering by the contents of a unit cell is given by the magnitude of the structure factor, $|F(h, k, l)|$, which is equal to $(A^2 + B^2)^{1/2}$, with A and B given by Eqs. (12.22a) and (12.22b). The intensity of this resultant is equal to the square of the amplitude, or $A^2 + B^2$. Since A and B can be obtained if the atoms and their coordinates in the unit cell are known [Eqs. (12.21) and (12.22)], we can obtain the relative intensity of rays scattered by any point of the reciprocal lattice (or by any group of planes in the real lattice). In other words, if we know the structure, we can deduce the diffraction pattern. Actually a few minor corrections are yet to be made, such as corrections for the absorption of x rays, thermal vibrations, and so on. But we have learned the essentials of how a diffraction pattern is formed.

Phase Problem

What we have discussed so far is what x-ray diffraction pattern will result from a given crystal. We are usually concerned with the reverse problem of figuring out the structures from the diffraction patterns. Knowing what structure will produce what diffraction pattern is helpful if we have certain structural models in mind. If the model was not too far off, adjustments of the model could be made to make the two agree. But if the three-dimensional structure of a large molecule is very complex, the chance of success by trial and error is nil; we must try to figure out how to deduce the structure from the diffraction pattern.

It would have been trivial to go backward from diffraction pattern to structure were it not for the fact that detectors for x rays can only measure the intensity (proportional to the square of the amplitude), not the phase. Therefore, we must have a way of obtaining the phases in order to derive the information lost in recording the diffraction pattern. Several phasing methods are available for unit cells containing relatively few atoms. One method of particular importance is the *Patterson synthesis.* Patterson defined a function $P(UVW)$ which can be calculated from a diffraction pattern without the phase information:

$$P(UVW) = \frac{1}{V_C} \sum_h \sum_k \sum_l |F|^2 \cos\left[2\pi(hU + kV + lW)\right] \tag{12.23}$$

where V_C is the volume of the unit cell and UVW are coordinates in fractions of unit-cell dimensions. The triple summation means that the quantity is to be summed over all values of h, k, and l. It turns out that the Patterson function is large when two electron-dense regions are separated by a vector with components UVW. Vectors between heavy atoms usually show up clearly as maxima in the Patterson function. A detailed discussion of the function is beyond the scope of this text.

For large molecules, the most powerful method for phasing is isomorphous substitution. Suppose that we have a crystal of a protein and a crystal of a heavy metal derivative of the protein. The two crystals are isomorphous if they have essentially the same unit-cell dimensions and atomic arrangements: the main difference being the substitution of a heavy atom in the derivative, with little distortion of the protein structure. The heavy atom is usually weakly bound to a side chain, or a heavy metal ion can substitute for a lighter metal ion [Hg(II) for Zn(II)]. When a pair of isomorphous crystals are available, their *difference* Patterson map usually permits the location of the heavy atom in the unit cell, and therefore the phase of scattering contribution of the heavy atom can be calculated. Suppose that $F(h, k, l)$, $F'(h, k, l)$, and $f(h, k, l)$ are the structure factors of the protein, the derivative, and the heavy atom, respectively, for a certain reciprocal lattice point (h, k, l). The magnitudes of F and F' are both known from the intensities. If the heavy atom scatters much more than the light atom it replaces, then to a good approximation F' should be the vectorial sum of F and f. We can therefore proceed to obtain F and F' as illustrated in Fig. 12.16. The vector **AB** is drawn with $\overline{AB} = |f|$ and the angle α equal to the phase angle of f. A circle with radius equal to $|F|$ and centered at A is first drawn. Any point X on this circle satisfies the relations $\overline{XA} = |F|$ and $\mathbf{XA} + \mathbf{AB} = \mathbf{XB}$. We can then draw a second circle with radius equal to $|F'|$ and centered at B. Any point Y on the second circle satisfies the relations $\overline{YB} = |F'|$ and $\mathbf{YA} + \mathbf{AB} = \mathbf{YB}$. Clearly, the points C and C' at which the two circles intersect give us the solutions for F and F'. $\overline{CA} = F$ and $\overline{CB} = F'$ are one set of solutions, and $\overline{C'A} = F$ and $\overline{C'B} = F'$ are another set of solutions. This ambiguity can be removed if we have more than one isomorphous derivative. Isomorphous substitution methods have played crucial roles in solving the crystal structures of proteins.

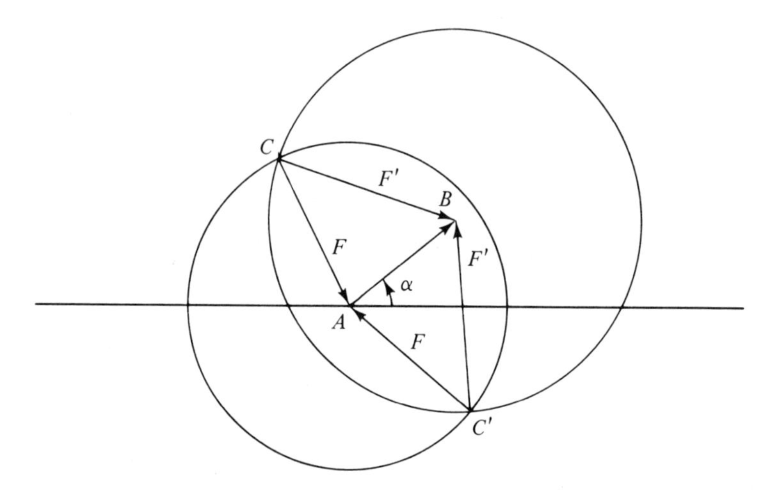

Fig. 12.16 Phase-angle determination by isomorphous replacement. See the text for an explanation.

Symmetry Considerations

Structural determination by x-ray diffraction is made easier by symmetry considerations. If we know the features of our left hands, the features of the right hands can be deduced, since to a good approximation the left hand and the right hand are symmetrical with respect to a mirror plane. Similarly, if there are certain symmetry features of the atomic contents of a unit cell, it is only necessary to consider the asymmetric portion. As symmetry features of a crystal lattice are fully reflected in the diffraction pattern of the crystal, symmetry considerations are very helpful in interpreting diffraction patterns.

Figures 12.17 to 12.19 illustrate several types of symmetry operations. Rotational symmetries are illustrated in Fig. 12.17. For example, the two 7's

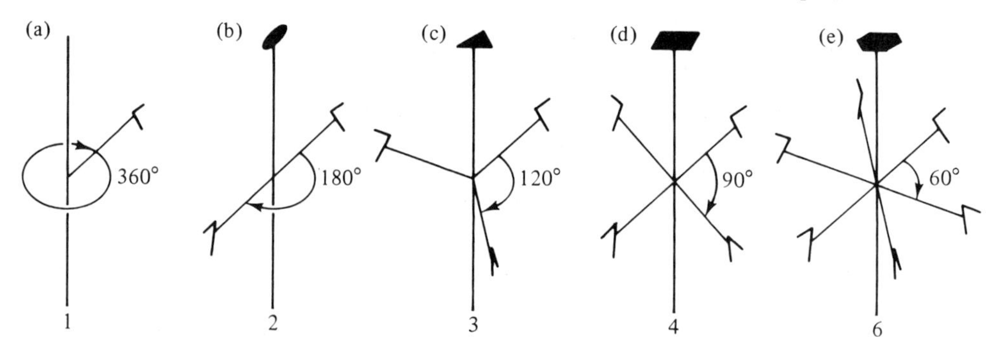

Fig. 12.17 Rotational symmetry axes. It can be shown that the five axes shown are the only ones allowed in crystals. [From L. V. Azároff, *Elements of X-Ray Crystallography*, McGraw-Hill, New York, 1968, p. 4, Fig. 1.5.]

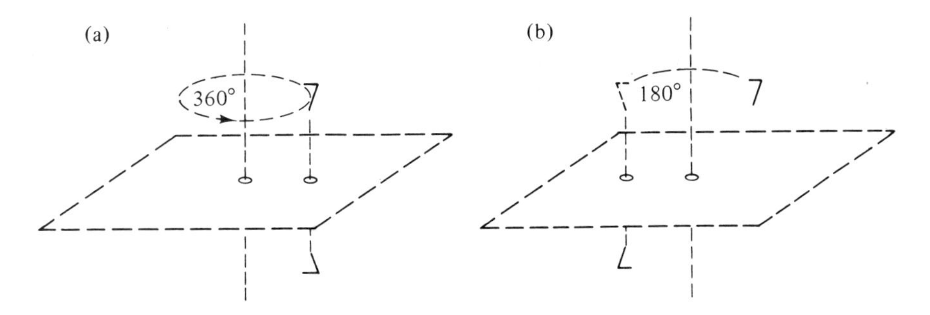

Fig. 12.18 One-fold (left) and two-fold (right) rotoreflection axis. [From L. V. Azároff, *Elements of X-Ray Crystallography*, McGraw-Hill, New York, 1968, p. 5, Fig. 1.6.].

shown in Fig. 12.17(b) are related by a 180° rotation around the symmetry axis shown. We say that the two 7's are related by a twofold rotation axis. Similarly, the 7's in Fig. 12.17(a), (c), (d), and (e) are related by a onefold, threefold, fourfold, and sixfold rotation axis, respectively.

Rotoreflection symmetries are illustrated in Fig. 12.18. The two 7's in Fig. 12.18(b), for example, are related by a twofold rotatory-reflection axis. As illustrated in the figure, if one 7 is first rotated around the twofold axis (to give the dotted 7), then reflected through the mirror plane, it will be superimposed on the other 7. Note that a onefold rotoreflection axis is the same as a mirror plane.

The symmetry center or inversion center is illustrated in Fig. 12.19. Extending a line joining any point of one of the 7's and the inversion center will meet the corresponding point of the other 7 equidistant from the inversion center. Similar to a rotoreflection axis, we can also have rotoinversion axes.

The types of symmetry operations discussed above—rotation, rotoreflection, and inversion—are symmetry operations around a point and are called *point symmetries.* For any isolated molecule or object, point symmetries are sufficient to describe all of its symmetry properties. The combination of point symmetry operations that can be carried out for a given object is called a *point group.* Mathematically, it can be shown that for a three-dimensional crystal there are 32 and only 32 different point groups.

In addition to point symmetries, a three-dimensional lattice also possesses *translational symmetry*, since moving a lattice along one of the axes by any

Fig. 12.19 Inversion center. [From L. V. Azároff, *Elements of X-Ray Crystallography*, McGraw-Hill, New York, 1968, p. 6, Fig. 1.9.]

integral number of lattice spacings gives a lattice that coincides with the original one. Two important kinds of symmetry arise by a combination of translation and point-symmetry operations.

Figure 12.20 illustrates the combination of translation and rotation to give a screw axis. The pair of 7's (at e and g) are related by a rotation of $\phi = 360°/n$

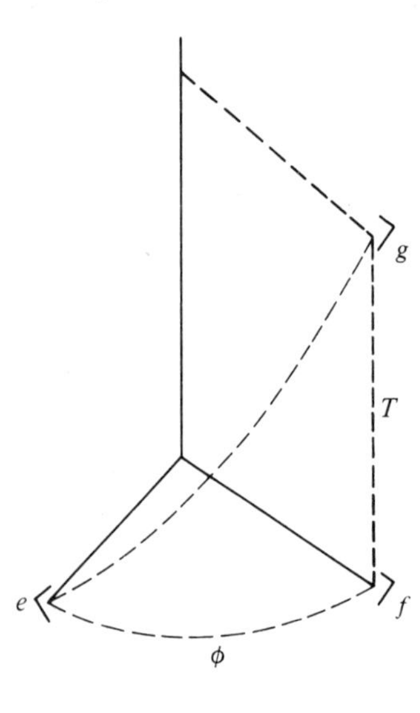

Fig. 12.20 Screw axis. A rotation from e to f followed by a translation from f to g is equivalent to a screw motion from e to g. [From L. V. Azároff, *Elements of X-Ray Crystallography*, McGraw-Hill, New York, 1968, p. 59, Fig. 4.3.]

followed by a translation T. If a lattice possesses such a screw axis, note that after n such operations, the total translation must be an integral multiple of the lattice spacing t in the direction of the translation. Thus

$$nT = mt \tag{12.24}$$

or

$$T = \frac{m}{n}t$$

where m is an integer. Screw axes are denoted by n_m. A 3_2 screw axis, for example, means that the symmetry involves a rotation around a threefold axis and a translation of $\frac{2}{3}$ times the lattice period t.

Another type of symmetry is the *glide plane*, as illustrated in Fig. 12.21. The 7's are related by a translation T followed by a reflection through the plane.

For three-dimensional lattices, there are 230 and only 230 symmetry groups when screw axes and glide planes are included in the symmetry operations in addition to point-symmetry operations. These are the 230 *space*-symmetry groups.

We have mentioned that the diffraction pattern of a crystal depends on its symmetry. Crystallographers have calculated, for each of the 230 space groups, which of the diffraction spots (h, k, l) are absent. A wealth of information is tabulated in *International Tables for X-Ray Crystallography* (Volume I, N. F. M. Henry and K. Lonsdale, eds., Kynoch Press, Birmingham, U.K., 1969.) From

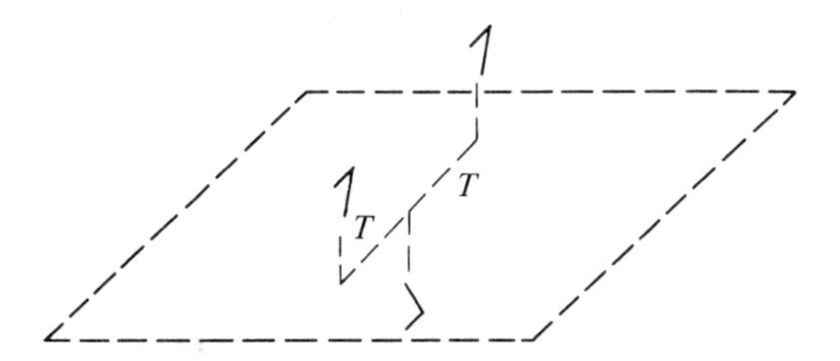

Fig. 12.21 Glide plane or a glide reflection. The glides are shown by dashed lines. The 7's above and below the plane are related by a reflection followed by a translation of one half the regular lattice translation (the distance between the two upper 7's). [From L. V. Azároff, *Elements of X-Ray Crystallography*, McGraw-Hill, New York, 1968, p. 58, Fig. 4.2.]

such information the space group of a crystal can be readily determined from intensity measurement of relatively few spots of the diffraction pattern. When this is done, what remains is the determination of the atomic arrangement of the asymmetric portion of the unit cell, by the methods we have already discussed.

Scattering of X Rays by Noncrystalline Materials

Because of the long-range three-dimensional order in a single crystal, detailed atomic arrangement can be deduced from x-ray diffraction studies of the crystal. X-ray scattering by noncrystalline materials also yields structural information, although in general much less detail is obtainable.

For long, threadlike macromolecules such as DNA, it is often possible to obtain fibers of a bundle of molecules with their long axes parallel or nearly parallel to the fiber axis. The orientations of the molecules in a fiber about the fiber axis is random. When such a fiber is placed in an x-ray beam, any long-range regular feature of the fiber will stand out in the diffraction pattern; the disordered features of the fiber will contribute only to the general background scattering. For example, for a DNA fiber with the molecules in the B-form conformation, since the helix axes of the DNA double helices are all parallel to the fiber axis, the strongest long-range regularity is the helical repeat of 34 Å per helical turn and 3.4 Å per base pair. These features are reflected in the diffraction pattern shown in Fig. 12.22.

For solutions of macromolecules, several types of information can be obtained by x-ray scattering measurements. First, as mentioned earlier, at zero scattering angle constructive interference always occurs for rays scattered by different parts of a scatterer, and the intensity is dependent on the total number of electrons. Therefore, when extrapolated to zero scattering angle, and after subtracting out scattering due to the solvent, the scattered intensity of a solution of macromolecules is related to the number of electrons per macromolecule. Since we can easily determine the elemental composition of the macromolecule,

(a)

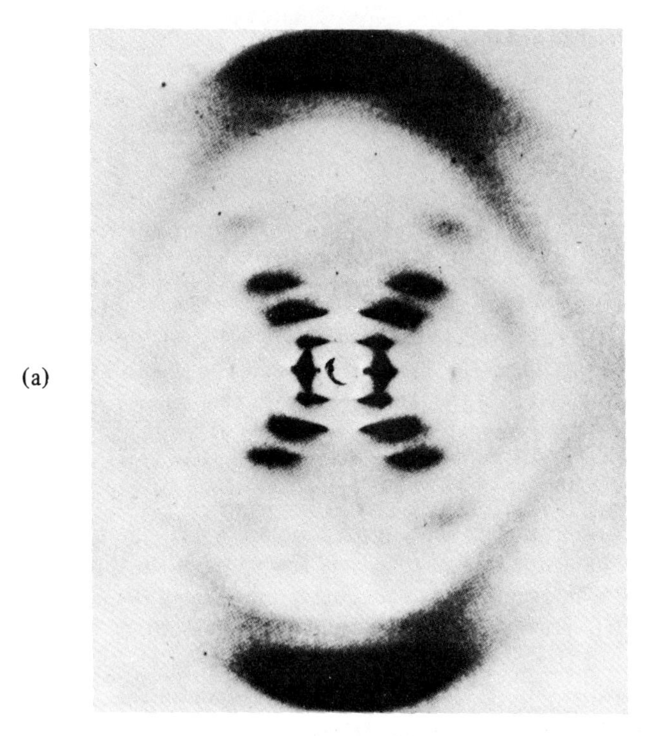

Meridian

(b)

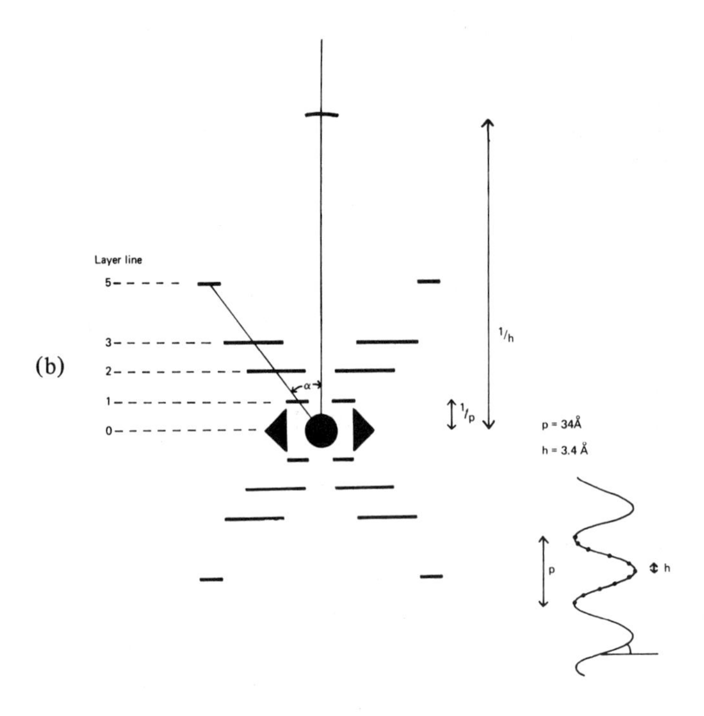

we can calculate the number of electrons per unit mass, and therefore obtain from the corrected zero-angle scattering intensity the molecular weight of the macromolecule. Second, although the macromolecules are randomly oriented in a solution so that little information on the atomic arrangement of the molecule can be deduced from the diffraction pattern, the angular dependence of the scattering intensity does provide us with information on the size of the macromolecule. It can be shown that at very low angles (of the order of milliradians),

$$\ln i(\alpha) = \ln i(0) - \frac{16\pi^4 R_G^2}{3\lambda^2} \sin^2 \frac{\alpha}{2} \tag{12.25}$$

where α is the angle between the incident and the scattered beams, $i(\alpha)$ is the scattering intensity at α, λ is the wavelength of the x rays, and R_G^2 is the mean-square radius (Chapter 11) of the macromolecule. Thus if the experimentally measured $\ln i(\alpha)$ is plotted as a funtion of $\sin^2 (\alpha/2)$, in the low-angle range a straight line results. The intercept gives $\ln i(0)$, from which the molecular weight of the macromolecule can be calculated as we have discussed, and the slope gives $-16\pi^4 R_G^2/3\lambda^2$, from which we can calculate the mean-square radius. The angular dependence of scattering intensity at larger angles is also dependent on the general shape of the macromolecule. We shall not discuss this aspect in this book.

Absorption of X Rays and X-ray Filters

Because of coherent, incoherent, and fluorescence scattering, the intensity of a beam of x rays is reduced after passing through any material. Similar to Beer's law behavior which we discussed in Chapter 10, the absorbance for monochromatic x rays of a given wavelength is proportional to the thickness of the sample and to the number of electrons per unit volume. For a given material, the absorbance is dependent on the energy (wavelength) of the x rays. Figure 12.23 illustrates this dependence. There are several abrupt rises in the curve going from left to right: three in a region a little over 1000 eV and another around 9000 eV. Similar to the sharp spikes shown in Fig. 12.1, these absorption edges are due to the ejection of orbital electrons in the L and K shells, respectively, and are labeled in the figure as L and K edges. The energy at which an edge absorption occurs is determined by the energy of the orbital electron. The value depends on the particular substance.

Fig. 12.22 (a) X-ray diffraction pattern of a NaDNA fiber in the B-helical conformation. Note the characteristic cross and the very strong meridian spots. (b) Line drawings of the pattern above, illustrating the relation between the helix dimensions (lower right) and the layer lines and meridian spots. [From J. P. Glusker and K. N. Trueblood, *Crystal Structure Analysis: A Primer*, Oxford University Press, New York, 1972, p. 137, Fig. 39(b).]

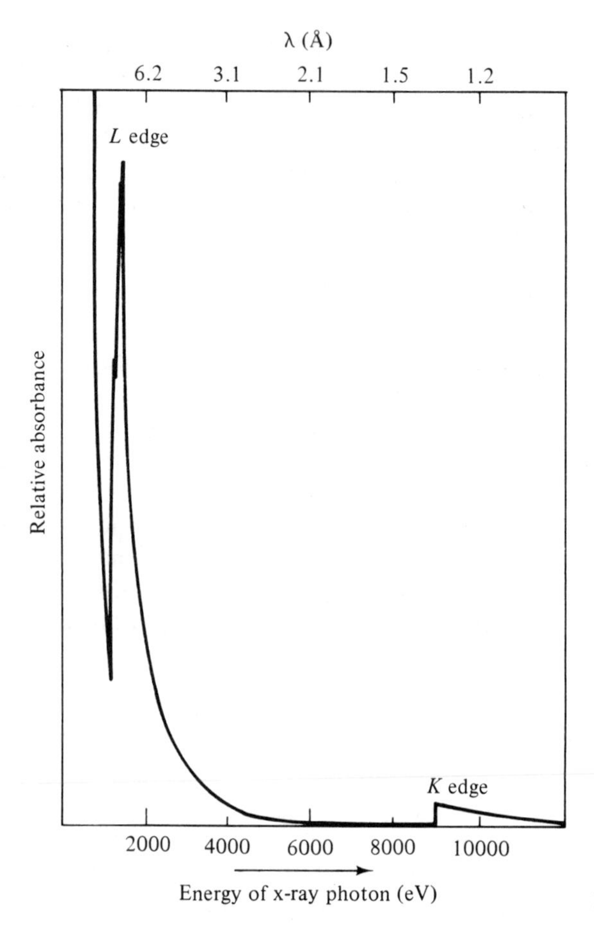

Fig. 12.23 Dependence of the absorption of x rays by Cu on the energy (wavelength) of the x-ray photon.

An important application of the sawtooth-shaped K edge is the selection of essentially monochromatic radiation by the use of two filters with slightly different K edges. As illustrated in Fig. 12.24, we can choose a certain filter A, which absorbs most of the radiation with energy slightly higher than the sharp Cu K_α emission line, and a certain filter B, which absorbs most of the radiation with energy slightly lower than the K_α line. The thicknesses of the two filters can be adjusted so that they absorb about equally in the higher-energy region. If we first measure the diffraction pattern with filter A between the incident beam and the crystal, then measure the diffraction pattern with filter B replacing filter A, the *difference* in the two patterns is the pattern due to the monochromatic K_α line.

Extended Fine Structure of Edge Absorption

In Fig. 12.23, we depicted an edge absorption (the K edge, for example) as a sharp rise followed by a smooth gradual drop. Careful experimental measurements showed, however, that the drop is not smooth. For a given material there

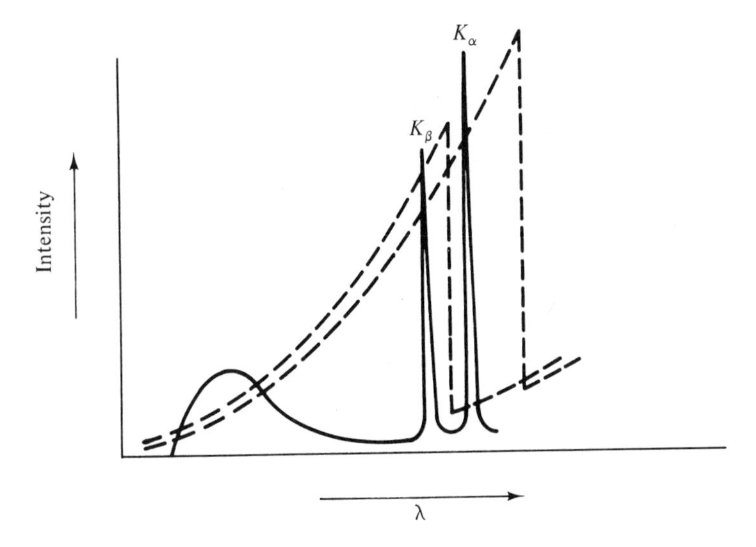

Fig. 12.24 Intensity distribution of x rays from a Cu target (solid line) and absorption curves for two filters with different absorption edges (dashed sawtooth-shaped lines).

are characteristic wiggles on the shorter-wavelength side of the absorption edge, as illustrated in Fig. 12.25. Qualitatively speaking, such wiggles appear because the probability of absorption of an x-ray photon by an orbital electron depends on the initial and final energy states of the electron. The initial energy state of the electron (in the K shell, for example) is, of course, quantized (Chapter 9). If we have an isolated atom, the final energy state of the ejected electron is not quantized, because it can possess different kinetic energies. Indeed, for such an isolated atom no wiggles would appear in the absorption curve. If the atom that absorbs the x-ray photon is not isolated but is surrounded by neighboring atoms, the ejected electron is back-scattered by the neighboring atoms. This situation is reminiscent of the particle-in-a-box problem (Chapter 9). We recall that the

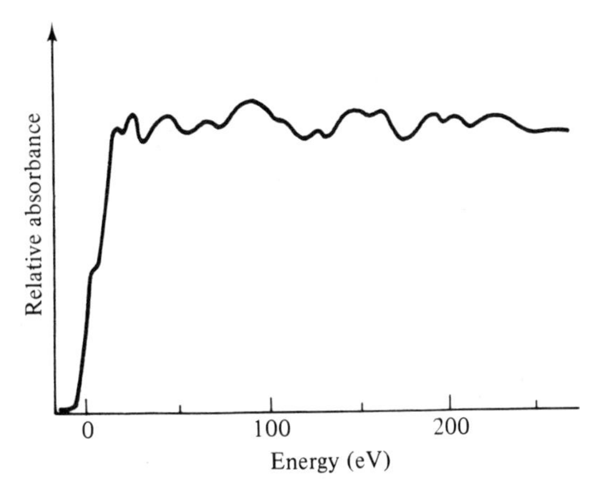

Fig. 12.25 Extended fine structure at Cu K edge. The energy scale is measured from the sharp K edge. [From L. V. Azároff, *Elements of X-Ray Crystallography*, McGraw-Hill, New York, 1968, p. 128, Fig. 6.35.]

energy of a particle in a box is quantized, with the energy levels dependent on the dimensions of the box. Similarly, in the presence of the neighboring atoms, the energy of the electron ejected by an x-ray photon is also quantized, with energy levels dependent on the distances between the atom at which absorption occurs and its neighboring atoms. This quantization of the final energy state of the ejected electron results in the characteristic wiggles in the fine structure of the edge absorption curve. By analyzing the edge absorption fine structure, it is possible to deduce the neighboring atomic arrangement around the absorbing atom.

Since the K absorption edge occurs at different wavelengths for different elements, it is possible to select a particular wavelength so that the atomic arrangement around a particular atom can be deduced. For example, if an enzyme contains a heavy metal ion, the wavelength of the incident x ray can be chosen to correspond to the absorption edge of the heavy metal. Analysis of the fine structure of the absorption profile can then yield information on the atomic environment of the heavy metal. It should be noted that structural analysis by edge absorption fine structure does not require crystalline material. It is the close neighbors of the absorbing atom that back scatter the ejected electron; therefore, long-range order is not necessary for the phenomenon.

X Rays from Synchrotron Radiation

An intense source of x rays is obtainable from the storage rings of electrons associated with an electron accelerator like a synchrotron. In these storage rings the circulating electrons produce electromagnetic radiation over a broad range, including the x-ray region.

The intensity of x rays from a synchrotron storage ring can be tens of thousands times higher than a conventional source. This high intensity permits the use of crystal monochromators which permit the selection of intense monochromatic radiation of the desired wavelength. In cases where high intensity is desirable, the synchrotron radiation is advantageous. For example, if diffraction patterns can be obtained in a very short time because of the high beam intensity, time-dependent structural changes can be studied. Another important application is to obtain the fine structure of edge absorption discussed in the preceding section. Here one must be able to tune the wavelength of the x-ray source to cover the regions of interest for a particular element, such as a heavy metal in an enzyme. Especially in biological systems, where the concentrations of such absorbing atoms are very low, the high intensity of x rays is advantageous. The application of synchrotron radiation to structural studies of biological systems is still in its infancy but appears promising.

ELECTRON DIFFRACTION

According to de Broglie's equation [Eq. (9.1)], a beam of electrons should exhibit wavelike properties. If an electron initially at rest at a point where the electrostatic potential is zero is accelerated by an electric field to a point where the potential is V, its kinetic energy at that point is

$$\frac{p^2}{2m_e} = eV \tag{12.26}$$

where p is the momentum, m_e the mass of an electron, and e the charge of an electron. The corresponding wavelength is, therefore,

$$\lambda = \frac{h}{\sqrt{2m_e eV}} \tag{12.27}$$

Actually, if the electrons are moving at a velocity not negligible compared with the velocity of light c, a relativistic correction is needed:

$$\lambda = \frac{h}{\sqrt{2m_e eV_{\text{corrected}}}} \tag{12.28}$$

where

$$V_{\text{corrected}} = V \cdot \left(1 + \frac{eV}{2m_e c^2}\right) \tag{12.29}$$

It is easy to show that if V is of the order of 10^4 volts, the wavelength is of the same order of magnitude as x rays. Therefore, electron diffraction can also be used for structural determinations of crystals. Electrons are scattered much more strongly by atoms than are x-ray photons. This is a blessing in some ways and a nuisance in others. Because of the strong scattering, a much smaller crystal can be used, and higher diffracted intensity means that less time is needed in data acquisition. Especially for low-energy electrons (V of the order of a few hundred volts or less), scattering is due to the first few layers of the atoms at the surface, and therefore much information can be obtained about the *surface* structures. But also because of the strong scattering, the path for the beam of electrons must be evacuated; otherwise, the beam will be scattered by the gas molecules. This creates problems for many biological samples, which frequently contain water and cannot be kept in vacuum. Also, effects due to interaction between the diffracted rays with the incident rays, which are not important for x-ray diffraction, are much more pronounced in electron diffraction. They make the interpretation of the electron diffraction patterns more difficult.

NEUTRON DIFFRACTION

In Chapter 6, we gave the translational kinetic energy of a gas as

$$U_{\text{tr}} = \tfrac{3}{2}RT$$

This equation is not limited to molecular or atomic particles; it applies to neutrons as well. If high-velocity neutrons from an atomic reactor are slowed down by cōllisions with molecules (D_2O is usually used) at around room temperature, it can be readily shown that their energy corresponds to a wavelength of the order of 1 Å. Such thermal neutrons can be used in diffraction studies of crystals and in low-angle-scattering studies of solutions, similar to the use of x rays.

There are several important differences. X rays are scattered primarily by electrons; neutrons are scattered by nuclei. Since the size of the nucleus is much smaller compared with the wavelength of the thermal neutrons, the scattering factor f_0 of an atom is not much affected by the value of $\sin\theta/\lambda$. (Thermal motions of atoms in a crystal, however, do cause the scattering intensity to decrease at larger scattering angles, and it is necessary to correct f_0 for this.) The scattering factor f_0 of an atom for neutrons is not related to the atomic number or mass in any simple way. Some values are tabulated in Table 12.1 for elements

Table 12.1 Relative neutron scattering factors

Element*	Relative scattering factor
Hydrogen(^{1}H)	−0.37
Deuterium(^{2}H)	0.65
Carbon(^{12}C)	0.66
Oxygen(^{16}O)	0.58
Nitrogen(^{14}N)	0.94
Phosphorus(^{31}P)	0.53
Sulfur(^{32}S)	0.31
Potassium	0.35
Sodium(^{23}Na)	0.35
Magnesium	0.52
Calcium(^{40}Ca)	0.49
Calcium(^{44}Ca)	0.18
Iron	0.96
Iron(^{54}Fe)	0.42
Iron(^{56}Fe)	1.01
Iron(^{57}Fe)	0.23
Cobalt(^{59}Co)	0.25
Nickel(^{58}Ni)	1.44
Nickel(^{60}Ni)	0.30
Nickel(^{62}Ni)	−0.87
Copper	0.79
Zinc	0.59
Bromine	0.67
Iodine	0.52

* Whenever the specific nucleus is not indicated, the value given is the average according to the natural abundance of the element.

SOURCE: Values taken from G. E. Bacon, *Applications of Neutron Diffraction in Chemistry*, Pergamon Press, New York, 1963.

frequently present in biological substances. The negative sign for nuclei ^1H and ^{62}Ni means that scattering by these nuclei is 180° out of phase with that by others.

Several important features of neutron scattering can be deduced by examining the data in Table 12.1. The scattering of neutrons by hydrogen is quite high, unlike the scattering of x rays, for which hydrogen is the weakest scatterer. Therefore, the positions of hydrogens can be more readily deduced from the neutron scattering data of a crystal. The scattering of neutrons by deuterium is quite different from that by hydrogen. Thus deuterium can be used to substitute hydrogen to give isomorphous crystals. For small-angle-scattering studies, neutron diffraction has one important advantage. It can be calculated that the relative scattering factors for H_2O and D_2O (deuterium oxide) are −0.0056 and +0.064, respectively. Most of the biological substances, such as nucleic acids, proteins, and lipids, have scattering factors between these values. Therefore, by using mixtures of H_2O and D_2O as the solvent, the background scattering of the solvent can be made to be the same as one of the components. Only components that scatter differently will stand out and contribute to the scattering *contrast.* Thus if we are studying a protein–nucleic acid complex, we can in essence look at either the protein or the nucleic acid component in the complex by selecting the appropriate H_2O–D_2O mixtures. Similarly, the contrast between membrane and cytoplasm can be varied from positive to negative by increasing the D_2O content of the solvent.

A disadvantage of neutron scattering studies is that there are very few high-flux neutron sources available. If the flux is low, a larger crystal is needed to give a diffraction pattern in a reasonable time. Therefore, neutron-scattering studies are under way at only a few places.

ELECTRON MICROSCOPY

In our discussion of x-ray diffraction, we commented on the fact that the lack of x-ray lenses makes it necessary to deduce the structure mathematically from the diffraction pattern. A beam of electrons, however, can be focused by electrostatic or magnetic lenses. Therefore a beam of electrons, after interacting with a specimen, can be focused to give an image of the specimen on a fluorescent screen or a photographic plate. Because of the short wavelength attainable for a beam of electrons, the resolution of an electron microscope can be made several hundredfold higher than that of a light microscope. As a consequence, electron microscopy has developed into one of the most powerful techniques for studying macromolecules, assemblies of macromolecules, cellular organizations, and so on. One might expect that since the wavelength of electrons can easily be made much less than atomic dimensions, it should be possible to see detailed atomic arrangement within a molecule by electron microscopy. This is difficult, if not impossible, to achieve. The reasons will become clearer after brief discussions on resolution, contrast, and specimen damage by electron bombardment.

Resolution

For an ideal lens, a luminous point source should be imaged into a point in the image plane. In practice, lenses for focusing rays of electrons cannot be constructed to achieve such ideality. A point source will give instead a circle of radius Δr_i at the image plane. If M is the magnification factor, the point source will have an apparent radius $\Delta r = \Delta r_i/M$. This lens aberration is called *spherical aberration.* It has been shown that for small values of α, Δr is proportional to α^3:

$$\Delta r = C_s\alpha^3 \tag{12.30}$$

where α (in radians) is the angle a point source extends at the lens aperture (see Fig. 12.26) and C_s is a proportionality constant. The value of C_s for a good

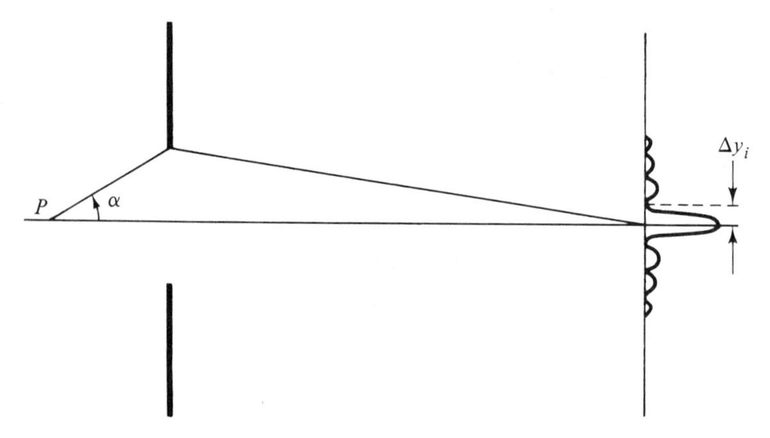

Fig. 12.26 Diffraction by a lens aperture. An axial point source P will give a diffraction pattern at the image plane as depicted. The central bright spot of radius Δy_i at the plane is called the *Airy disk*, after G. Airy.

electron microscope is about 2 mm. If we wish to make $\Delta r \approx 1$ Å, we must use a small aperture of several milliradians. There are other lens aberrations, but they are less important for an electron microscope.

It appears that we can make spherical aberration as small as possible by making α very small. We must keep in mind, however, that when an aperture is very small, diffraction effects become important. This is illustrated in Fig. 12.26. A point source will give the diffraction pattern illustrated. At the image plane, there is a central spot surrounded by concentric circular maxima of decreasing brightness. The distance from the lens axis to the position of the first minimum, Δy_i, can be taken as the radius of the central spot. Thus, because of diffraction, a point will appear to have an apparent radius $\Delta y = \Delta y_i/M$, with M being the magnification as before. For small apertures, it can be shown that

$$\Delta y \approx \frac{0.6\lambda}{\alpha} \tag{12.31}$$

The smaller the aperture, or the larger the wavelength, the greater will be the diffraction effect on resolution.

As a compromise, α can be chosen so that the magnitudes of diffraction error and spherical aberration are about the same. Equating Eqs. (12.30) and (12.31), we obtain

$$\alpha \approx \left(\frac{0.6\lambda}{C_s}\right)^{1/4} \tag{12.32}$$

and

$$\Delta r = \Delta y \approx \frac{0.6\lambda}{(0.6\lambda/C_s)^{1/4}} \approx 0.7C_s^{1/4}\lambda^{3/4} \tag{12.33}$$

At the present time, the point resolution (the smallest separation between two points that can be resolved) of a good electron microscope is about 3 Å. For crystalline materials with periodic spacings, spacing close to 1 Å can be resolved.

Contrast

At the beginning of this chapter, we mentioned that contrast is an important factor in seeing an object. In order to see a macromolecular species by electron microscopy, molecules are usually mounted on a supporting film, such as a carbon film (typically about 100 Å thick, although much thinner films can be used). Contrast between a macromolecule and the background is due to a number of phenomena. One important factor is that some of the incident electrons are scattered by the specimen and will not reach the image plane, either because of the aperture stop or because of lens aberration. A strong scatterer will therefore appear as a dark spot in the image. Diffraction of the incident rays by a crystalline material or at the edges of dense particles, and interference between incident and scattered rays, also contribute to contrast. Detailed discussions can be found in books on electron microscopy. Contrast is frequently much more of a problem than resolution in electron microscopy.

Several methods have been developed to enhance the contrast. The macromolecules can be "shadowed" by impinging evaporated metal obliquely onto them. More metal will be deposited on the sides of the macromolecules than on the supporting film. The macromolecules can also be preferentially "stained" with an electron-dense stain (positive staining), or the supporting film can be stained preferentially with an electron-dense stain (negative staining). Some examples are shown in Fig. 12.27. Contrast is also improved by using a dark-field arrangement: either the incident electron beam is tilted, or the aperture is moved off center, so that only the scattered electrons will reach the image.

Radiation Damage

We have mentioned the problem of keeping a biological specimen in an evacuated chamber. Procedures used in sample preparation for viewing in the microscope

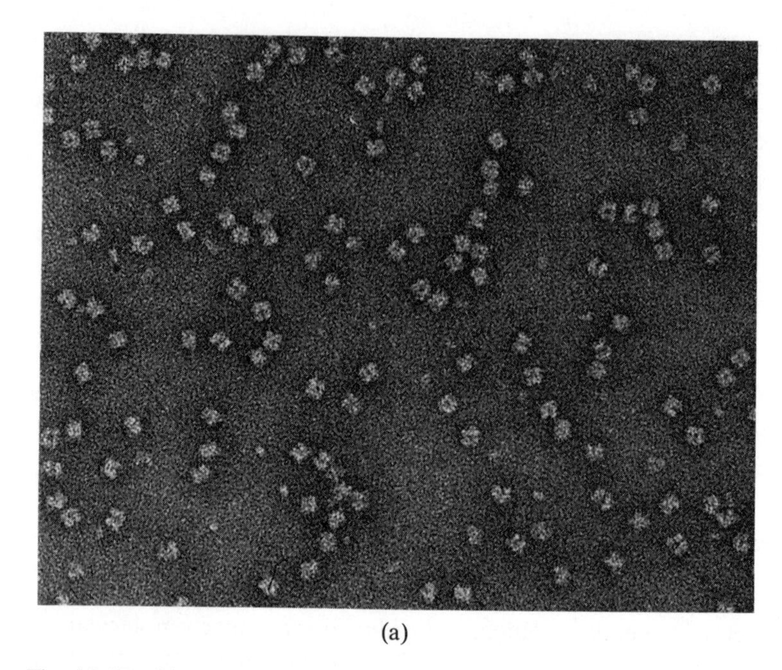

(a)

Fig. 12.27 (a) Electron micrograph of the enzyme δ-aminolevulinic acid dehydratase from bovine liver. The enzyme molecules were negatively contrasted with sodium phosphotungstate. The magnification factor is 250,000. The particles have the appearance of four discrete lobes arrayed at the corners of a square. Sedimentation and gel electrophoresis studies of the enzyme indicate that it is an octamer. A model consistent with all these observations is one in which the subunits of the octameric enzyme are arranged at the corners of a cube. [For details, see W. H. Wu, D. Shemin, K. E. Richards, and R. C. Williams, *Proc. Natl. Acad. Sci. USA*, *71*, 1767 (1974). Photomicrograph courtesy of the Virus Laboratory, University of California at Berkeley.].

(b) An electron micrograph of a part of a "heteroduplex" DNA. When two strands of complementary base sequences associate by base pairing, a duplex or double-stranded molecule is formed (see the section on DNA helix-coil transition in Chapter 11). If the two strands are not completely complementary, the noncomplementary regions will not pair, and will show as single-stranded loops. Such a molecule is called a heteroduplex. For the particular heteroduplex shown, there is a single-stranded loop. One side of the loop contains two tyrosine transfer RNA genes of the bacterium *Escherichia coli*. To determine the positions of these transfer RNA genes, molecules of an iron-containing protein ferritin are linked chemically to the ends of the transfer RNA molecules. The ferritin-carrying transfer RNA and the heteroduplex DNA are annealed to allow the binding of the transfer RNA. The ferritin molecules are essentially opaque to the electron beam and therefore show as dark dots on the micrograph. To see the rest of the molecule, positive staining with uranyl acetate and shadowing with platinum were employed. [For details, see M. Wu and N. Davidson, *J. Mol. Biol.* *78*, 1 (1973). Micrograph courtesy of N. Davidson.]

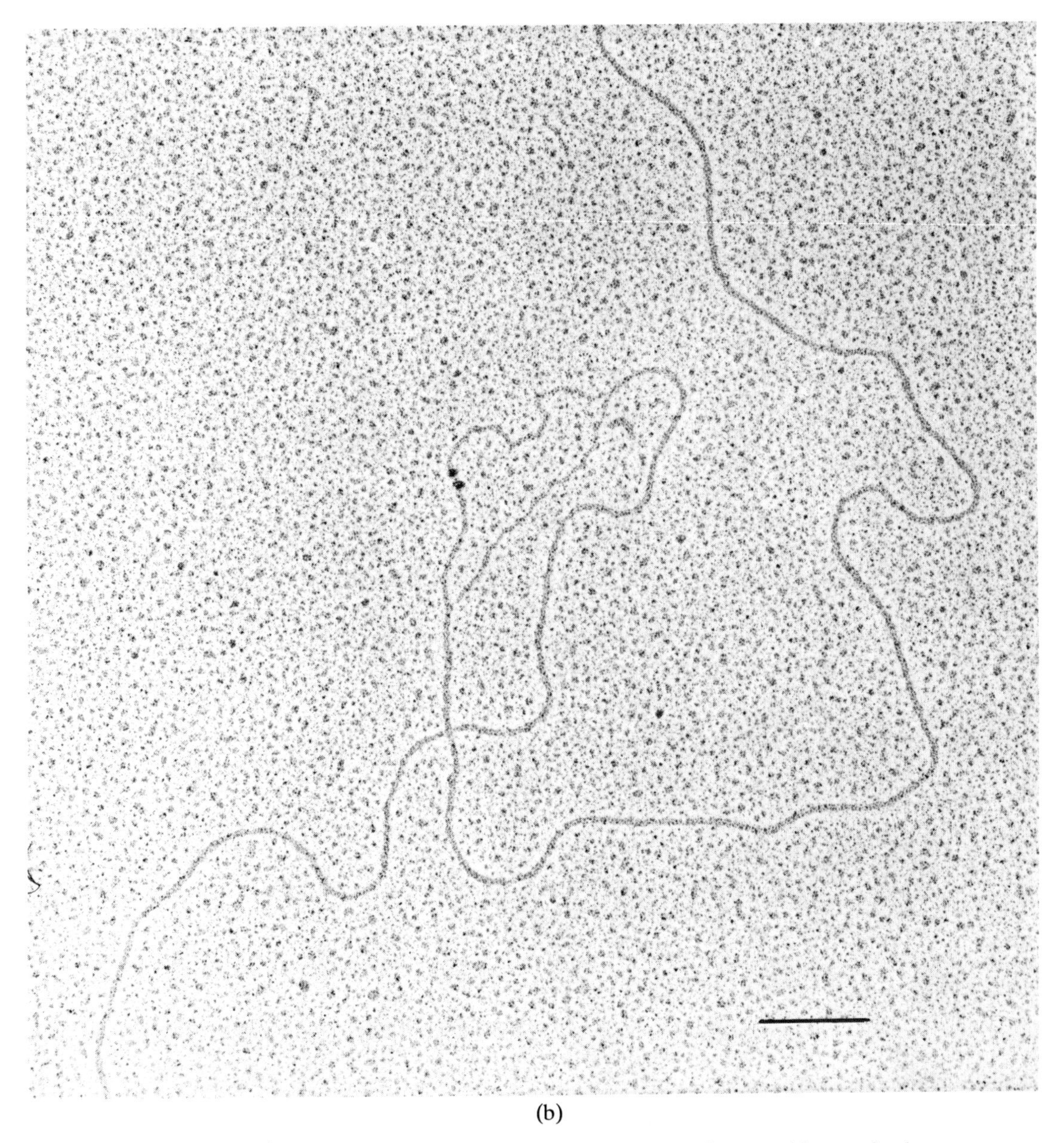

(b)

might also cause deformation of the species. Another problem of electron microscopy is radiation damage. Energy absorbed by a macromolecule from the incident beam causes heating of the molecule, breakage of chemical bonds, and rearrangement of atoms. If the specimen is stained, movement of the stain may also occur. Since the final image of the molecule on a photographic film is the composite picture of the molecule during the period when it is subject to electron bombardment, the image might be quite different from the original structure.

Radiation damage is a serious problem when high-resolution results are attempted on individual molecules.

Transmission and Scanning Electron Microscopes

In a transmission electron microscope, electrons from a source (usually a hot tungsten filament) are first accelerated by an electric field to give the required wavelength. After passing through the specimen, the electrons are detected by a fluorescent screen for viewing or by a photographic plate or videotape for recording. Lenses are used to guide the paths of the electrons. In a scanning electron microscope, electrons from a source, after acceleration, are first focused by a lens to give a very small spot. The specimen is then scanned by moving the focused spot across it point by point and line by line. This scanning is achieved by the use of scanning coils, in much the same way as in a television camera. A suitable detector analyzes either the transmitted electrons, or back-scattered and secondary electrons ejected from the sample by the incident electrons. The signal detected can be displayed on an oscilloscope (television tube), point by point and line by line, in synchronization with the movement of the focused spot.

It was pointed out by A. Crewe that the electron optics for the two types of microscopes are actually rather similar, as illustrated in Fig. 12.28. Transmission electron microscopes have had a longer history of development, and commercial

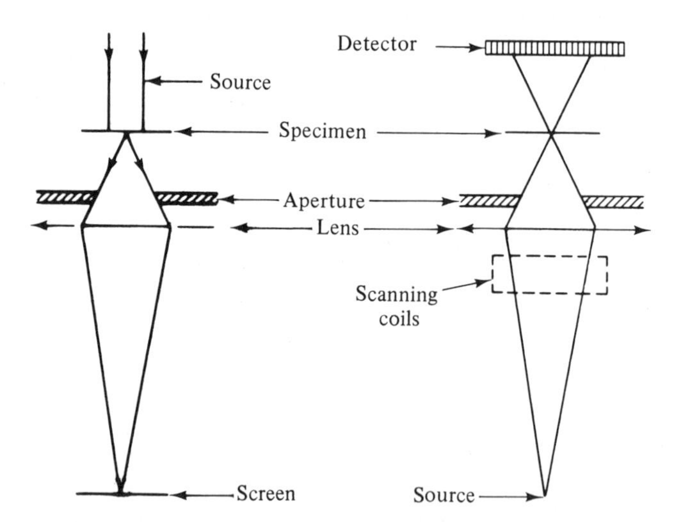

Fig. 12.28 Schematic diagram showing the similarity between a conventional microscope (on the left) and a scanning microscope (on the right). To make the comparison more obvious, the scanning microscope diagram has been drawn so that the electrons move from the bottom of the diagram to the top. [From A. V. Crewe and J. Wall, *J. Mol. Biol. 48*, 377 (1970).]

instruments capable of a point resolution of several angstroms are available. Before 1970, the resolution of scanning electron microscopes was about 100 Å, limited primarily by the size of the focused spot. The scanning microscope has several advantages, however, in certain applications. When secondary electrons are detected, the contrast can be made very high for the surface features of a specimen by coating it with a thin layer of gold. Surface topography can be vividly revealed, as illustrated in Fig. 12.29. One can also use a detector to measure the energy of the x rays emitted at each point of an uncoated specimen when irradiated by the electrons. Since the energy of the x rays emitted is

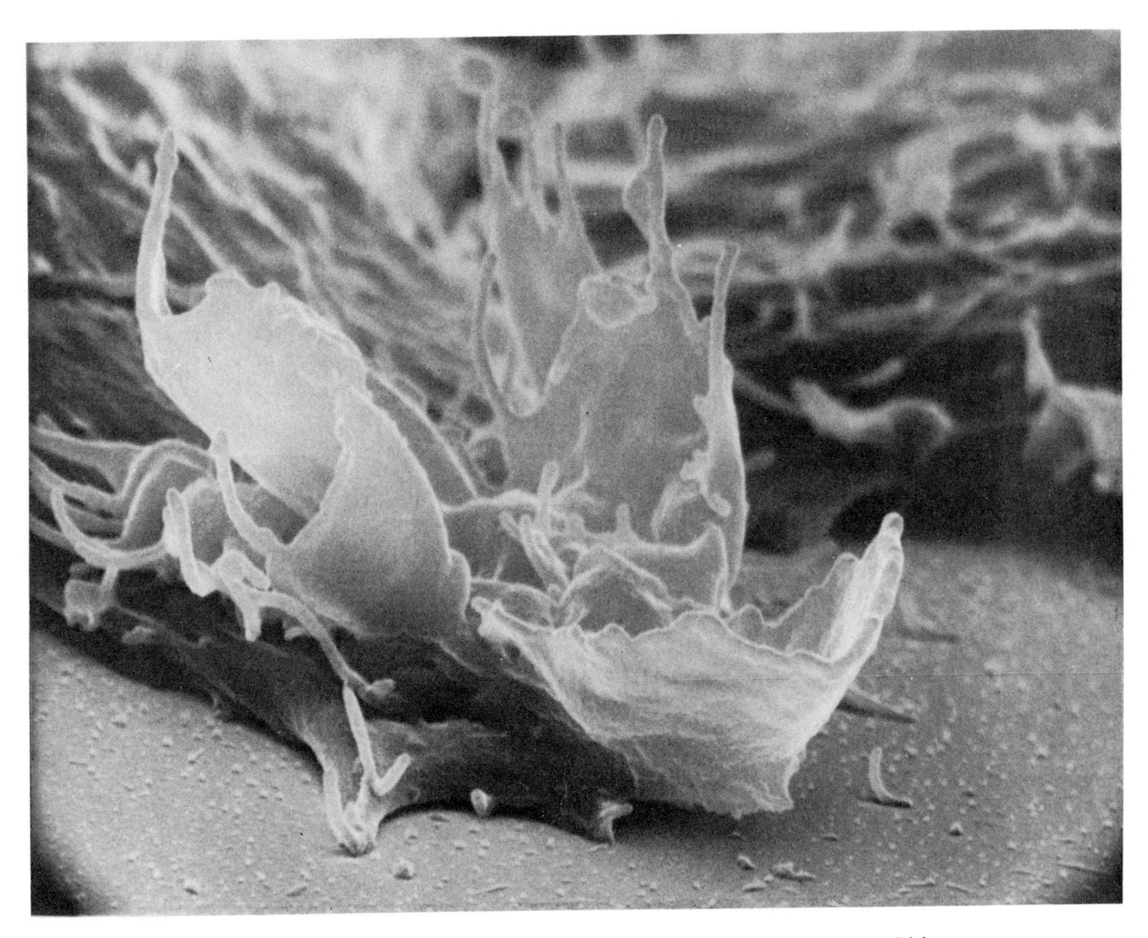

Fig. 12.29 An electron photomicrograph of a cultured hamster kidney cell adhered to a glass surface. Cells grown on glass microscope slide coverslips were "fixed" in glutaraldehyde, dried, uniformly coated with gold, and examined in a scanning electron microscope. The magnification is about 14,000 times. Photograph courtesy of J. P. Revel. [More details about this micrograph can be found in J. P. Revel, P. Hoch, and D. Ho, *Exptl. Cell Research 84*, 207 (1974).]

characteristic of the elements irradiated, a scanning microscope in this mode can be used as a microanalyzer for the elemental composition of the specimen.

A scanning electron microscope with a much improved resolution (about 5 Å) was designed by Crewe around 1970. Instead of using a heated filament as the source, electrons emitted from a cold tungsten tip under the influence of a high electric field (field emission) were used. This greatly reduces the size of the focused spot. Contrast is achieved by analyzing both the coherently and incoherently scattered electrons. Excellent micrographs of biological materials have been obtained with this type of high-resolution electron microscope (see Fig. 12.30).

Image Enhancement and Reconstruction

If a specimen is composed of a lattice of subunits, the periodicity of the structure should be present in its image recorded on an electron micrograph. We have already discussed direct studies of periodic structures by diffraction methods. Powerful mathematical methods have been developed to analyze diffraction patterns of periodic structures. Diffraction techniques can also be applied to study the *image* recorded on a micrograph. As the image has been magnified many times from the original molecular assembly, its diffraction pattern must be studied with radiation of a much longer wavelength. The wavelength of visible light is convenient for studying the diffraction pattern of the image. Certain structural features not readily discernible from the original electron micrograph image can be better characterized from the optical diffraction of the image. Undesirable features of the image, such as noise (which has no periodicity), can be filtered out. The filtered diffraction pattern can be reconstructed if necessary to give an improved image by the use of an appropriate lens. Diffraction, filtration, and reconstruction of the image can also be done mathematically. The intensity of the image is sampled at a grid of points covering the image and digitized. The other operations are then carried out mathematically by the use of a computer.

Since the image recorded on an electron micrograph is a two-dimensional superposition of different layers in a three-dimensional structure, it should be possible to deduce the three-dimensional structure from several images obtained with different viewing angles. Such three-dimensional image reconstruction is facilitated if the specimen is composed of symmetrically arranged subunits. The image of a structure composed of helically arranged subunits, for example,

---

Fig. 12.30 Electron micrographs of a segment of a polyuridylic acid molecule stained by an osmium compound. A high resolution scanning transmission electron microscope was used in obtaining these micrographs. (a) Micrograph prior to noise filtering. (b) Micrograph after noise filtering (see the section on image enhancement and reconstruction). Individual osmium atoms attached to the uridine bases can be clearly seen. [Micrographs courtesy of M. Beer. For detailed discussions see M. Cole, W. Wiggins, and M. Beer *J. Mol. Biol.*, *117*, 387 (1978).]

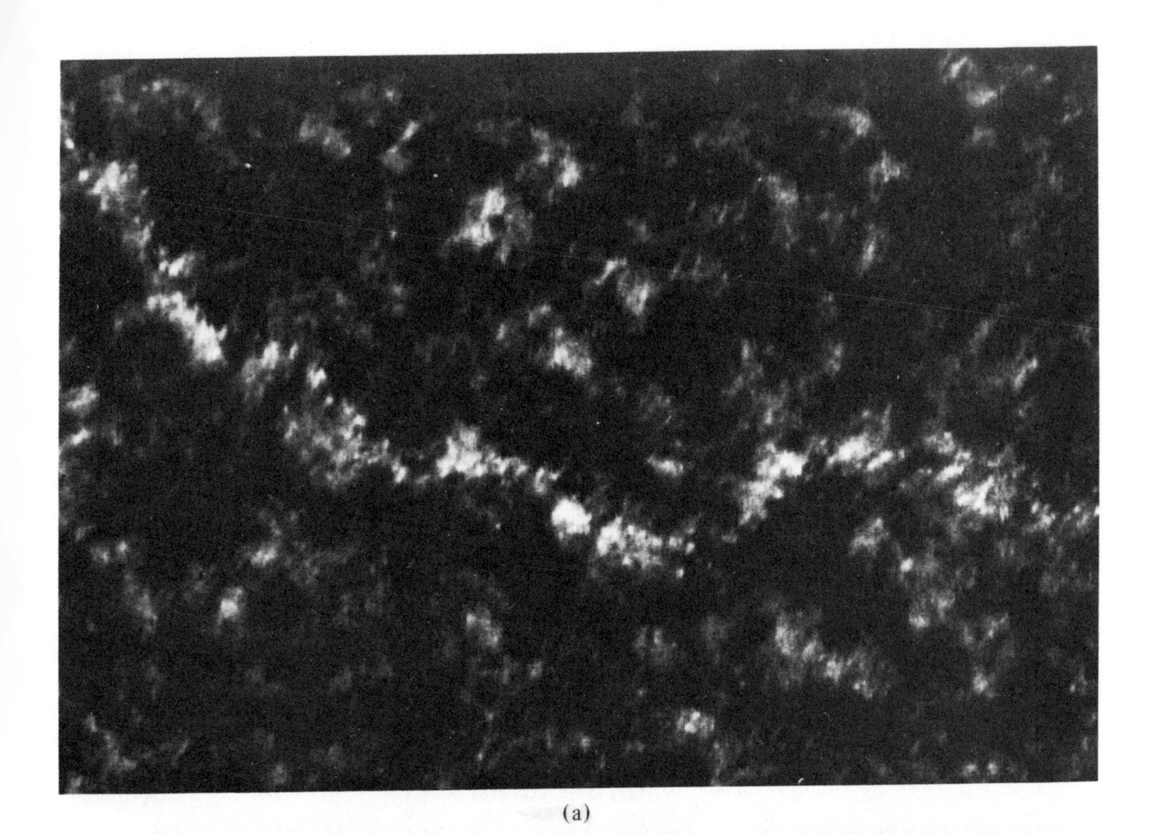

(a)

(b)

contains many different views of the subunit, and therefore one image is sufficient for deducing the three-dimensional structure of a subunit. Three-dimensional image reconstruction has been done in recent years for a number of ordered structures, and resolutions in the range 4 to 20 Å have been achieved.

SUMMARY

X-ray Diffraction

X rays are photons (electromagnetic radiation) with wavelengths in the range from a few tenths of 1 Å to several angstroms. X rays are usually produced by bombarding a metal target with electrons.

Laue's equations:

For a crystal made of identical scatterers at the lattice points, the conditions for diffraction are

$$a\,(\cos\alpha - \cos\alpha_0) = h\lambda \tag{12.1}$$

$$b\,(\cos\beta - \cos\beta_0) = k\lambda \tag{12.2}$$

$$c\,(\cos\gamma - \cos\gamma_0) = l\lambda \tag{12.3}$$

λ = wavelength of x rays
a, b, c = lattice spacings along the axes **a**, **b**, and **c**, respectively
$\alpha_0, \beta_0, \gamma_0$ = angles the incident beam makes with the three axes **a**, **b**, and **c**
α, β, γ = angles the diffracted beam makes with the three axes **a**, **b**, and **c**
h, k, l = integers

Miller indices:

Any plane of Miller indices (h, k, l) is parallel to one that makes intercepts a/h, b/k, and c/l with the lattice axes **a**, **b**, and **c**. For an orthorhombic lattice, the interplanar spacing for planes of Miller indices (h, k, l) is

$$d_{hkl} = \left[\left(\frac{h}{a}\right)^2 + \left(\frac{k}{b}\right)^2 + \left(\frac{l}{c}\right)^2\right]^{-1/2}$$

Reciprocal lattice:

A family of parallel planes of Miller indices (h, k, l) can be represented by a point P_{hkl} in the reciprocal lattice. The vector OP_{hkl}, where O is the origin of the reciprocal lattice, is perpendicular to the planes and the length OP_{hkl} equals the reciprocal of interplanar spacing.

Bragg equation:

For a family of crystal planes of Miller indices (h, k, l) and an interplanar spacing d_{hkl}, the incident x-ray beam of wavelength λ appears to be reflected by the planes when the incident angle θ with respect to the planes satisfies the equation

$$\theta = \sin^{-1}\frac{n\lambda}{2d_{hkl}} \qquad n = 0, 1, 2, \ldots \tag{12.4}$$

The Bragg law is equivalent to Laue's equations. We can also write

$$\theta = \sin^{-1}\frac{\lambda}{2(d_{hkl}/n)} \tag{12.6}$$

d_{hkl}/n can be considered as the interplanar spacing of planes of Miller indices (nh, nk, nl).

Unit cell:

The unit cell is the basic unit of a crystal structure. Translation of the unit cell along the three lattice axes generates the crystal.

Structure factor:

At a Bragg angle $\theta = \sin^{-1}(\lambda/2d_{hkl})$ corresponding to diffraction by planes of indices (h, k, l), the rays scattered by a unit cell can be represented by a complex number

$$F(h, k, l) = A(h, k, l) + iB(h, k, l) \tag{12.17}$$

$$A(h, k, l) = \sum_{j=1}^{n} f_j \cos[2\pi(hX_j + kY_j + lZ_j)] \tag{12.22a}$$

$$B(h, k, l) = \sum_{j=1}^{n} f_j \sin[2\pi(hX_j + kY_j + lZ_j)] \tag{12.22b}$$

f_j = atomic scattering factor of atom j
X_j, Y_j, Z_j = coordinates of atom j, units of the unit-cell dimensions
n = total number of atoms in the unit cell

$F(h, k, l)$ is the structure factor. The intensity of the diffracted rays by planes of indices (h, k, l) is proportional to $A^2 + B^2$. The phase of F is

$$\alpha(h, k, l) = \tan^{-1}\frac{B}{A} \tag{12.21}$$

Phase problem:

A critical step in obtaining the structure of the crystal from its diffraction pattern is to determine the phases of the structure factors. An important method for phase determination of macromolecules is the isomorphous replacement method.

Symmetry:

Structural determinations by x-ray diffraction are much facilitated by symmetry considerations. Symmetry features of a crystal lattice include rotation, rotoreflection, inversion, screw axes, and glide planes.

Scattering by noncrystalline materials:

Some molecular parameters, such as molecular weight, mean-square radius, and neighboring atomic arrangement around an atom from which an electron has been ejected, can be deduced for noncrystalline materials. Structures of fibers can also be studied by x-ray diffraction.

Neutron Diffraction

The de Broglie wavelength of thermal neutrons is of the order of interatomic distances in crystals. Neutrons are scattered by nuclei; x-ray photons are scattered primarily by electrons.

The neutron scattering factor of an atom is not related to the atomic number or mass in any simple way. The scattering of neutrons by hydrogen is quite high, making neutron diffraction a valuable tool for determining thc positions of the hydrogens in a crystal.

The isotopes hydrogen and deuterium scatter neutrons quite differently. This fact is useful in structural studies of macromolecules by neutron diffraction.

Electron Microscopy

The de Broglie wavelength of electrons accelerated by a voltage V is

$$\lambda = \frac{h}{\sqrt{2m_e e V_{\text{corrected}}}} \tag{12.28}$$

$$V_{\text{corrected}} = V \cdot \left(1 + \frac{eV}{2m_e c^2}\right) \tag{12.29}$$

c = speed of light in vacuum
m_e = rest mass of an electron
e = charge of an electron
h = Planck's constant

Electrostatic or magnetic lenses can be used to control the paths of a beam of electrons. Electron microscopes with a point resolution of a few angstroms have been constructed.

In addition to resolution, contrast and specimen damage by the electron beam are also important considerations in electron microscopy.

REFERENCES

AZÁROFF, L. V., 1968. *Elements of X-Ray Crystallography*, McGraw-Hill, New York.

BACON, G. E., 1962. *Neutron Diffraction*, 2nd ed., Oxford University Press, New York.

GLUSKER, J. P., and K. N. TRUEBLOOD, 1972. *Crystal Structure Analysis: A Primer*, Oxford University Press, New York.

HALL, C. E., 1966. *Introduction to Electron Microscopy*, 2nd ed., McGraw-Hill, New York.

HOLMES, K. C., and D. M. BLOW, 1966. *The Use of X-Ray Diffraction in the Study of Protein and Nucleic Acid Structure*, Interscience Publishers, New York. [Reprinted from *Methods of Biochemical Analysis*, D. Glick (ed.), Vol. 13.]

RYMER, T. B., 1970. *Electron Diffraction*, Eyre Methuen, London.

WILSON, H. R., 1966. *Diffraction of X-Rays by Proteins, Nucleic Acids and Viruses*, Edward Arnold, London.

SUGGESTED READINGS

ARNOTT, S., 1970. "The Geometry of Nucleic Acids," in J. A. V. Butler and D. Noble (eds.), *Progress in Biophysics and Molecular Biology*, vol. 21, Pergamon Press, Oxford.

CREWE, A. V., and J. WALL, 1970. A Scanning Microscope with 5 Å Resolution, *J. Mol. Biol. 48*, 375.

DICKERSON, R. E., and I. GEIS, 1969. *The Structure and Action of Proteins*, Harper & Row, New York.

STERN, E. A., 1976. The Analysis of Materials by X-Ray Absorption, *Sci. Amer. 234*, 46.

PROBLEMS

1. The energy of an electron upon acceleration by a potential of V volts is eV, where e is the charge. In units of electron volts, the energy is just V electron volts.

(a) Calculate the energy in ergs of an electron accelerated by a potential of 40 kV. ($1 \text{ eV} = 1.602 \times 10^{-12}$ erg.)

(b) When the electron strikes a target atom, the maximum energy of the photon emitted corresponds to the conversion of all the energy of the incoming electron to that of the photon. Calculate the wavelength of such photons.

(c) Show that in general,

$$\lambda_s = \frac{12{,}390}{V} \quad \text{Å}$$

where λ_s is the shortest wavelength.

2. (a) For Cu metal, the K absorption edge is at $\lambda = 1.380$ Å (Fig. 12.23). For Ag metal, the K absorption edge is at $\lambda = 0.4858$ Å. Calculate the energy required for the ejection of an electron from the K shell of Cu and that from the K shell of Ag.

(b) The sharp x-ray emission lines α_1, α_2, β, and γ shown in Fig. 12.1 result from the ejection of K-shell electrons of the Ag target atoms by the bombarding electrons. From the K-edge wavelength for Ag given in part (a), calculate the minimum accelerating voltage below which the sharp spikes will not be seen.

(c) The wavelengths of the four emission lines shown in Fig. 12.1 are 1.54433 Å (α_2), 1.54051 Å (α_1), 1.39217 Å (β) (which is actually a doublet at high resolution; the value given is the average of the doublet), and 1.38102 Å (γ). These lines result from the emission of x-ray photons when the electrons in the L-shell energy levels fall into the K shell vacated by the ejected electrons. Calculate the L-shell energy levels corresponding to these emission lines.

3. (a) For the arrangement shown in Fig. 12.6, calculate α for the first- and the second-order diffraction cones if the wavelength of the incident x rays is 1.54 Å and the spacing between the equidistant scattering centers is 4 Å.

(b) A cylindrically shaped film, with its axis coincident with the row of scatterers, is used to record the diffracted rays. It is easy to see that the diffraction cones will intersect the film to give layer lines. Calculate the distance between the layer lines formed by the first and the second diffraction cones. The radius of the cylindrical film is 3 cm and all other parameters are the same as in part (a).

4. For $\lambda = 1.54$-Å x rays, obtain the first-order ($n = 1$) Bragg angles for interplanar spacings of 5 Å and 10 Å.

5. From Eq. (12.6) it can be seen that for a fixed λ, the interplanar spacing (d_{hkl}/n) is inversely proportional to $\sin\theta$. In other words scattering corresponding to the smallest spacing occurs at the maximum value of $\sin\theta$ (when $\theta = 90°$). If $\lambda = 1.54$-Å x rays are used, what is the theoretical limit of resolution? (That is, what is the minimal spacing observable?)

6. (a) The lattice of NaCl is cubic with a unit-cell dimension of 5.64 Å. Each unit cell contains 4 Na^+ and 4 Cl^- ions. The density of NaCl is 2.163 g cm^{-3}. Calculate Avogadro's number.

(b) The dimensions of the unit cell of crystalline lysozyme are $a = 79.1$ Å, $b = 79.1$ Å, and $c = 37.9$ Å. The interaxial angles are all 90°. Each unit cell contains eight lysozyme molecules. The density of crystalline lysozyme is 1.242 g cm^{-3}, and chemical analysis showed that 64.4% (by weight) is lysozyme, the rest being water and salt. From these data, calculate the molecular weight of lysozyme.

7. (a) Show that for an orthorhombic lattice, the interplanar spacing d_{hkl} between planes of Miller indices (h, k, l) is

$$d_{hkl} = \frac{1}{\sqrt{\dfrac{h^2}{a^2} + \dfrac{k^2}{b^2} + \dfrac{l^2}{c^2}}}$$

where a, b, and c are the lattice spacings.

(b) Show that for a cubic lattice, $d_{100} : d_{110} : d_{111} = 1 : 0.707 : 0.578$.

8. (a) A cubic unit cell contains one atom at the origin. Calculate the relative intensities of the diffractions by the 100, 200, and 110 planes.

(b) Do the same calculations for a unit cell containing two identical atoms, one at the origin and one at $X = Y = Z = \frac{1}{2}$.

9. Calculate the de Broglie wavelengths of electrons accelerated by a voltage of:
(a) 10 kV.
(b) 100 kV.

10. Calculate the de Broglie wavelengths of thermal neutrons at 300 K.

Appendix

Table A.1 Useful physical constants

		SI units	cgs-esu units
Gas constant	R	$8.3143\ J\ K^{-1}\ mol^{-1}$	8.3143×10^{7} erg $deg^{-1}\ mol^{-1}$ 1.987 cal $deg^{-1}\ mol^{-1}$ 0.08205 ℓ atm $deg^{-1}\ mol^{-1}$
Avogadro's number	N_0	$6.0222 \times 10^{23}\ mol^{-1}$	6.0222×10^{23} molecules mol^{-1}
Boltzmann constant	k	$1.3806 \times 10^{-23}\ J\ K^{-1}$	1.3806×10^{-16} erg deg^{-1}
Faraday constant	F	$9.6487 \times 10^{4}\ C\ mol^{-1}$	9.6487×10^{4} coulombs mol^{-1}
Speed of light	c	$2.9979 \times 10^{8}\ m\ s^{-1}$	$2.9979 \times 10^{10}\ cm\ s^{-1}$
Planck constant	h	6.6262×10^{-34} J s	6.6262×10^{-27} erg s
Elementary charge	e	1.6022×10^{-19} C	4.8030×10^{-10} esu
Electron mass	m_e	9.1096×10^{-31} kg	9.1096×10^{-28} g
Proton mass	m_p	1.6726×10^{-27} kg	1.6726×10^{-24} g
Standard gravity	—	$9.8066\ m\ s^{-2}$	$980.66\ cm\ s^{-2}$
Permittivity of vacuum	ε_0	$8.8542 \times 10^{-12}\ C^2\ N^{-1}\ m^{-2}$	

C = coulomb   kg = kilogram
g = gram   m = meter
J = joule   N = newton
K = degrees Kelvin   s = second

Table A.2 Definition of prefixes

mega ≡ M	means multiply by 10^6
kilo ≡ k	means multiply by 10^3
centi ≡ c	means multiply by 10^{-2}
milli ≡ m	means multiply by 10^{-3}
micro ≡ μ	means multiply by 10^{-6}
nano ≡ n	means multiply by 10^{-9}
pico ≡ p	means multiply by 10^{-12}

Table A.3 Energy conversion factors

To convert from:	To:	Multiply by:
calories	ergs	4.184×10^7
calories	joules	4.184
calories	kilowatt-hours	1.162×10^{-6}
electron volts	kcal mol^{-1}	23.060
electron volts	ergs	1.602×10^{-12}
ergs	calories	2.389×10^{-8}
ergs	joules	10^{-7}
ergs molecule^{-1}	joules mol^{-1}	6.024×10^{16}
ergs molecule^{-1}	kcal mol^{-1}	1.440×10^{13}
joules	calories	0.2389
joules	ergs	10^7
joules	kilowatt-hours	2.778×10^{-7}
joules mol^{-1}	ergs molecule^{-1}	1.660×10^{-17}
joules mol^{-1}	kcal mol^{-1}	2.389×10^{-4}
kilowatt-hours	calories	8.604×10^5
kilowatt-hours	joules	3.600×10^6
kcal mol^{-1}	electron volts	0.04336
kcal mol^{-1}	ergs molecule^{-1}	6.944×10^{-14}
kcal mol^{-1}	joules mol^{-1}	4.184×10^3

Table A.4 Miscellaneous conversions and abbreviations

Length

1 angstrom = 10^{-8} cm = 0.1 nm
1 inch = 2.54 cm
1 foot = 30.48 cm
1 mile = 5280 ft = 1609 m

Weight

1 pound = 453.6 g

Energy

1 erg = 1 g cm^2 s^{-2}
1 joule = 1 kg m^2 s^{-2}

Force

1 dyne = 1 g cm s^{-2} = 10^{-5} N

Pressure

1 atmosphere = 760 mm Hg (torr) = 14.70 lb in^{-2}
= 1.013×10^6 dyn cm^{-2}
= 1.013×10^5 newton m^{-2}
= 1.033×10^4 kg-force m^{-2}

Power

1 watt = 1 J s^{-1} = 1 VA

Volume

1 gallon = 3.785 ℓ

Table A.4 Miscellaneous conversions and abbreviations (*cont.*)

Å = angstrom	P = poise
A = ampere	rad = radian
cal = calorie	V = volt
dyn = dyne	W = watt
hertz (Hz) = cycles per second	y = year
h = hour	

Table A.5 Inorganic compounds*

	ΔH_f^0 (kcal mol^{-1})	S^0 (cal mol^{-1} deg^{-1})	ΔG_f^0 (kcal mol^{-1})
$Ag(s)$	0	10.17	0
$Ag^+(aq)$†	25.31	17.67	18.43
$AgCl(s)$	−30.37	23.00	−26.25
C(*s*, graphite)	0	1.37	0
C(*s*, diamond)	0.45	0.58	0.68
$Ca(s)$	0	9.95	0
$CaCO_3(s)$	−288.45	22.2	−269.78
$Cl_2(g)$	0	53.29	0
$Cl^-(aq)$	−40.02	13.17	−31.35
$CO(g)$	−26.42	47.22	−32.78
$CO_2(g)$	−94.05	51.07	−94.26
$CO_2(aq)$	−98.69	29.0	−92.31
$HCO_3^-(aq)$	−165.18	22.7	−140.31
$CO_3^{2-}(aq)$	−161.63	−12.7	−126.22
$Fe(s)$	0	6.49	0
$Fe_2O_3(s)$	−196.5	21.5	−177.1
$H_2(g)$	0	31.21	0
$H_2O(g)$	−57.80	45.11	−54.64
$H_2O(l)$	−68.32	16.72	−56.69
$H^+(aq)$	0	0	0
$OH^-(aq)$	−54.96	−2.519	−37.60
$H_2O_2\ (aq)$	−45.68	32.5	−31.47
$H_2S(g)$	−4.82	49.15	−7.89
$N_2(g)$	0	45.77	0
$NH_3(g)$	−10.98	46.05	−3.98
$NH_3(aq)$	−19.32	26.3	−6.37
$NH_4^+(aq)$	−31.74	26.97	−19.00
$NO(g)$	21.60	50.34	20.72
$NO_2(g)$	8.09	57.47	12.39
$NO_3^-(aq)$	−49.37	35.0	−26.41
$Na^+(aq)$	−57.28	14.4	−62.59
$NaCl(s)$	−98.23	17.3	−91.78
$NaCl(aq)$	−97.30	27.6	−93.94

Table A.5 Inorganic compounds (*cont.*)

	ΔH_f^0 (kcal mol^{-1})	S^0 (cal mol^{-1} deg^{-1})	ΔG_f^0 (kcal mol^{-1})
$NaOH(s)$	−101.99	12.5	−90.1
$O_2(g)$	0	49.00	0
$O_3(g)$	34.0	56.8	39.06
S(rhombic)	0	7.62	0
$SO_2(g)$	−70.94	59.30	−71.74
$SO_3(g)$	−94.45	61.24	−88.52

* Standard thermodynamic values at 25°C (298 K) and 1 atm pressure. Values for ions refer to an aqueous solution at unit activity on the molarity scale. Standard enthalpy of formation, ΔH_f^0, third-law entropies, S^0, and standard Gibbs free energy of formation, ΔG_f^0, are given.

† The standard state for all ions and for species labeled (*aq*) is that of a solute on the molarity scale.

SOURCE: Data from *CODATA Bulletin No. 10*, 1973, and F. Daniels and R. A. Alberty, *Physical Chemistry*, 3rd ed., John Wiley, New York, 1966.

Table A.6 Hydrocarbons*

	ΔH_f^0 (kcal mol^{-1})	S^0 (cal mol^{-1} deg^{-1})	ΔG_f^0 (kcal mol^{-1})
Acetylene, $C_2H_2(g)$	54.19	48.00	50.00
Benzene, $C_6H_6(g)$	19.82	64.34	30.99
Benzene, $C_6H_6(l)$	11.72	41.41	29.72
n-Butane, $C_4H_{10}(g)$	−30.15	74.12	−4.10
Cyclohexane, $C_6H_{12}(g)$	−29.43	71.28	7.59
Ethane, $C_2H_6(g)$	−20.24	54.85	−7.87
Ethylene, $C_2H_4(g)$	12.50	52.45	16.28
n-Heptane, $C_7H_{16}(g)$	−44.88	102.27	1.91
n-Hexane, $C_6H_{14}(g)$	−39.96	92.83	−0.06
Isobutane, $C_4H_{10}(g)$	−32.15	70.42	−4.99
Methane, $CH_4(g)$	−17.89	44.52	−12.15
Naphthalene, $C_{10}H_8(g)$	36.08	80.22	53.44
n-Octane, $C_8H_{18}(g)$	−49.82	111.55	3.92
n-Pentane, $C_5H_{12}(g)$	−35.00	83.40	−2.00
Propane, $C_3H_8(g)$	−24.82	64.51	−5.61
Propylene, $C_3H_6(g)$	4.88	63.80	14.99

* Standard thermodynamic values at 25°C (298 K) and 1 atm pressure. Standard enthalpy of formation, ΔH_f^0, third-law entropies, S^0, and standard Gibbs free energy of formation, ΔG_f^0, are given.

SOURCE: Data from D. R. Stull, E. F. Westrum, Jr., and G. C. Sinke, *The Chemical Thermodynamics of Organic Compounds*, John Wiley, New York, 1969.

Table A.7 Organic compounds*

	ΔH_f^0 (kcal mol^{-1})	S^0 (cal mol^{-1} deg^{-1})	ΔG_f^0 (kcal mol^{-1})	ΔG_f^0 (1 *M* activity, *aq*) (kcal mol^{-1})
Acetaldehyde $CH_3CHO(g)$	−39.76	63.15	−31.86	−33.28
Acetate$^-$(*aq*)	—	—	—	−88.99
Acetic acid $CH_3CO_2H(l)$	−115.7	38.20	−93.06	−94.79
Acetone $CH_3COCH_3(l)$	−59.3	47.9	−37.14	−38.48
Adenine $C_5H_5N_5(s)$	22.94	36.1	71.58	—
L-Alanine $CH_3CHNH_2COOH(s)$	−134.5	30.88	−88.49	−88.84
L-Alanylglycine $C_5H_{10}N_2O_3(s)$	−197.52	46.62	−127.30	—
L-Aspartate^{+--}(*aq*)	—	—	—	−166.99
Aspartic acid $C_4H_7NO_4(s)$	−232.64	40.66	−174.53	−172.08
Butyric acid $C_3H_7COOH(s)$	−127.6	54.1	−90.27	—
Creatine $C_4H_9N_3O_2(s)$	−128.39	45.3	−63.32	—
L-Cysteine $HSCH_2CHNH_2COOH(s)$	−127.6	40.6	−82.21	−81.34
L-Cystine $C_6H_{12}N_2O_4S_2(s)$	−251.4	67.06	−165.71	−161.16
Ethanol $C_2H_5OH(l)$	−66.20	38.40	−41.62	−43.24
Formaldehyde $CH_2O(g)$	−27.70	52.29	−26.27	−31.2
Formamide $HCONH_2(g)$	−44.5	59.38	−33.71	—
Formic acid $HCOOH(l)$	−101.52	30.82	−86.39	—
Fumarate$^-$(*aq*)	—	—	—	−144.41
Fumaric acid *trans*-$(=CHCOOH)_2(s)$	−193.85	39.7	−156.23	−154.41
α-D-Galactose $C_6H_{12}O_6(s)$	−307.21	49.1	−219.75	−220.98
α-D-Glucose $C_6H_{12}O_6(s)$	−304.6	50.7	−217.62	−219.28
L-Glutamate^{+--}(*aq*)	—	—	—	−165.87
L-Glutamic acid $C_5H_9NO_4(s)$	−241.32	44.98	−174.78	−172.73
Glycerol $HOCH_2CHOHCH_2OH(l)$	−159.8	48.87	−114.02	−116.76
Glycine $H_2CNH_2COOH(s)$	−128.4	24.74	−90.27	−90.8
Glycylglycine $C_4H_8N_2O_3(s)$	−178.12	45.4	−117.25	—

Table A.7 Organic compounds (*cont.*)

	ΔH_f^0 (kcal mol^{-1})	S^0 (cal mol^{-1} deg^{-1})	ΔG_f^0 (kcal mol^{-1})	ΔG_f^0 (1 *M* activity, *aq*) (kcal mol^{-1})
Guanine $C_5H_5N_5O(s)$	−43.96	38.3	11.33	—
L-Isoleucine $C_6H_{13}NO_2(s)$	−152.5	49.71	−82.97	—
Lactate$^-$(*aq*)	—	—	—	−123.76
L-Lactic acid $CH_3CHOHCOOH(s)$	−165.89	34.00	−124.98	—
β-Lactose $C_{12}H_{22}O_{11}(s)$	−534.59	92.3	−374.52	−375.22
L-Leucine $C_6H_{13}NO_2(s)$	−154.6	50.62	−85.34	−84.39
Maleic acid *cis*$(=CHCOOH)_2(s)$	−188.96	38.1	−150.86	—
Methanol $CH_3OH(l)$	−57.02	30.3	−39.73	−41.88
L-Methionine $C_5H_{11}NO_2S(s)$	−181.3	55.32	−120.88	—
Oxalic acid $(-COOH)_2(s)$	−198.36	28.70	−167.58	—
L-Phenylalanine $C_9H_{11}NO_2(s)$	−111.6	51.06	−50.56	—
Pyruvate$^-$(*aq*)	—	—	—	−113.44
Pyruvic acid $CH_3COCOOH(l)$	−139.7	42.9	−110.75	—
L-Serine $HOCH_2CHNH_2COOH(s)$	−173.6	35.65	−121.70	—
Succinate^{2-}(*aq*)	—	—	—	−164.97
Succinic acid $(-CH_2COOH)_2(s)$	−224.88	42.0	−178.64	−178.35
Sucrose $C_{12}H_{22}O_{11}(s)$	−531.1	86.1	−369.18	−370.88
L-Tryptophan $C_{11}H_{12}N_2O(s)$	−99.2	60.00	−28.54	—
L-Tyrosine $C_9H_{11}NO_3(s)$	−160.5	51.15	−92.18	−88.63
Urea $NH_2CONH_2(s)$	−79.63	25.00	−47.12	−48.72
L-Valine $C_5H_{11}NO_2(s)$	−147.7	42.75	−85.80	—

* Standard thermodynamic values at 25°C (298 K) and 1 atm pressure. Values for ions refer to an aqueous solution at unit activity on the molarity scale. Standard enthalpy of formation, ΔH_f^0, third-law entropies, S^0, and standard Gibbs free energy of formation, ΔG_f^0, are given.

SOURCE: Data from D. R. Stull, E. F. Westrum, Jr., and G. C. Sinke, *The Chemical Thermodynamics of Organic Compounds*, John Wiley, New York, 1969, and from J. T. Edsall and J. Wyman, *Biophysical Chemistry*, Vol. 1, Academic Press, New York, 1958.

Table A.8 Atomic weights of the elements*

Element	Symbol	Atomic number	Atomic weight
Actinium	Ac	89	(227)
Aluminum	Al	13	26.98
Americium	Am	95	(243)
Antimony	Sb	51	121.75
Argon	Ar	18	39.948
Arsenic	As	33	74.92
Astatine	At	85	(210)
Barium	Ba	56	137.34
Berkelium	Bk	97	(249)
Beryllium	Be	4	9.012
Bismuth	Bi	83	208.98
Boron	B	5	10.81
Bromine	Br	35	79.909
Cadmium	Cd	48	112.40
Calcium	Ca	20	40.08
Californium	Cf	98	(251)
Carbon	C	6	12.011
Cerium	Ce	58	140.12
Cesium	Cs	55	132.91
Chlorine	Cl	17	35.453
Chromium	Cr	24	52.00
Cobalt	Co	27	58.93
Copper	Cu	29	63.54
Curium	Cm	96	(247)
Dysprosium	Dy	66	162.50
Einsteinium	Es	99	(254)
Erbium	Er	68	167.26
Europium	Eu	63	151.96
Fermium	Fm	100	(253)
Fluorine	F	9	19.00
Francium	Fr	87	(223)
Gadolinium	Gd	64	157.25
Gallium	Ga	31	69.72
Germanium	Ge	32	72.59
Gold	Au	79	196.97
Hafnium	Hf	72	178.49
Helium	He	2	4.003
Holmium	Ho	67	164.93
Hydrogen	H	1	1.0080
Indium	In	49	114.82
Iodine	I	53	126.90
Iridium	Ir	77	192.2
Iron	Fe	26	55.85
Krypton	Kr	36	83.80
Lanthanum	La	57	138.91
Lawrencium	Lw	103	(257)
Lead	Pb	82	207.19
Lithium	Li	3	6.939
Lutetium	Lu	71	174.97
Magnesium	Mg	12	24.312
Manganese	Mn	25	54.94
Mendelevium	Md	101	(256)
Mercury	Hg	80	200.59
Molybdenum	Mo	42	95.94
Neodymium	Nd	60	144.24
Neon	Ne	10	20.183
Neptunium	Np	93	(237)
Nickel	Ni	28	58.71
Niobium	Nb	41	92.91
Nitrogen	N	7	14.007
Nobelium	No	102	(253)
Osmium	Os	76	190.2
Oxygen	O	8	15.9994
Palladium	Pd	46	106.4
Phosphorus	P	15	30.974
Platinum	Pt	78	195.09
Plutonium	Pu	94	(242)
Polonium	Po	84	(210)
Potassium	K	19	39.102
Praseodymium	Pr	59	140.91
Promethium	Pm	61	(147)
Protactinium	Pa	91	(231)
Radium	Ra	88	(226)
Radon	Rn	86	(222)
Rhenium	Re	75	186.23
Rhodium	Rh	45	102.91
Rubidium	Rb	37	85.47
Ruthenium	Ru	44	101.1
Samarium	Sm	62	150.35
Scandium	Sc	21	44.96
Selenium	Se	34	78.96
Silicon	Si	14	28.09
Silver	Ag	47	107.870
Sodium	Na	11	22.9898
Strontium	Sr	38	87.62
Sulfur	S	16	32.064
Tantalum	Ta	73	180.95
Technetium	Tc	43	(99)
Tellurium	Te	52	127.60
Terbium	Tb	65	158.92
Thallium	Tl	81	204.37
Thorium	Th	90	232.04
Thulium	Tm	69	168.93
Tin	Sn	50	118.69
Titanium	Ti	22	47.90
Tungsten	W	74	183.85

Table A.8 Atomic weights of the elements (*cont.*)

Element	Symbol	Atomic number	Atomic weight	Element	Symbol	Atomic number	Atomic weight
Uranium	U	92	238.03	Yttrium	Y	39	88.91
Vanadium	V	23	50.94	Zinc	Zn	30	65.37
Xenon	Xe	54	131.30	Zirconium	Zr	40	91.22
Ytterbium	Yb	70	173.04				

* Based on mass of ^{12}C at 12.000.... The ratio of these weights to those on the older chemical scale (in which oxygen of natural isotopic composition was assigned a mass of 16.0000...) is 1.000050. (Values in parentheses represent the most stable known isotopes.)

Table A.9 Biochemical compounds

Amino acids found in proteins: $^{+}H_3N-C(R)(H)-C(=O)O^-$

R **groups:**

Glycine	H—
Alanine	CH_3—
Valine	$(CH_3)_2CH$—
Leucine	$(CH_3)_2CHCH_2$—
Isoleucine	$CH_3CH_2(CH_3)CH$—
Phenylalanine	C_6H_5—CH_2— (phenyl ring)
Tyrosine	HO—C_6H_4—CH_2— (phenol ring)
Tryptophan	indole ring (with N)—CH_2—
Threonine	$HO(CH_3)CH$—
Methionine	$CH_3SCH_2CH_2$—
Cysteine	$HSCH_2$—
Proline (amino acid)	H_2^+—N—C(H)—C(=O)—O^-, with N—CH_2—CH_2—CH_2—C ring

Table A.9 Biochemical compounds (*cont.*)

R **groups** (*cont.*)			
Aspartic acid	$^{-}O(O=)CCH_2-$	Lysine	$H_2NCH_2CH_2CH_2CH_2-$
Glutamic acid	$^{-}O(O=)CCH_2CH_2-$	Histidine	N, N, H (imidazole ring) $-CH_2-$
Asparagine	$NH_2(O=)CCH_2-$	Arginine	$H_2N\underset{\|}{C}NHCH_2CH_2CH_2-$ NH
Glutamine	$NH_2(O=)CCH_2CH_2-$	Serine	$HOCH_2-$

A polypeptide chain:

$$^{+}H_3N\overset{R_1}{\overset{|}{C}}H\overset{O}{\overset{\|}{C}}NH-\left[\overset{R_n}{\overset{|}{C}}H\overset{O}{\overset{\|}{C}}NH\right]_n-\overset{R_2}{\overset{|}{C}}HCO_2^-$$

The amino terminal residue is R_1; the carboxyl terminal is R_2.

Components of DNA: Nucleotides

O, CH_3, NH, H, N, O, O, $^{-}O-P-O$, O, O^-, OH

Thymidylic acid, T
(Base is thymine.)

NH_2, N, N, H–, N, N, H, O, $^{-}O-P-O$, O, O^-, OH

Deoxyadenylic acid, A
(Base is adenine.)

NH_2, H, N, H, N, O, O, $^{-}O-P-O$, O, O^-, OH

Deoxycytidylic acid, C
(Base is cytosine.)

O, N, NH, H–, N, N, NH_2, O, $^{-}O-P-O$, O, O^-, OH

Deoxyguanylic acid, G
(Base is guanine.)

Table A.9 Biochemical compounds (*cont.*)

Components of RNA: Nucleotides (*cont.*)

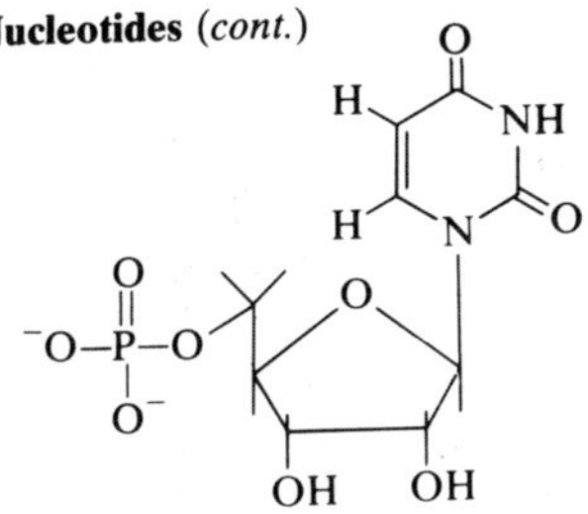

Uridylic acid, U
(Base is uracil.)

Cytidylic, adenylic and guanylic acid are similar to uridylic acid, but with cytosine, adenine, and guanine replacing uracil.

A polynucleotide chain, a single strand of DNA. (A single strand of RNA would have a hydroxyl group at each 2′ position.)

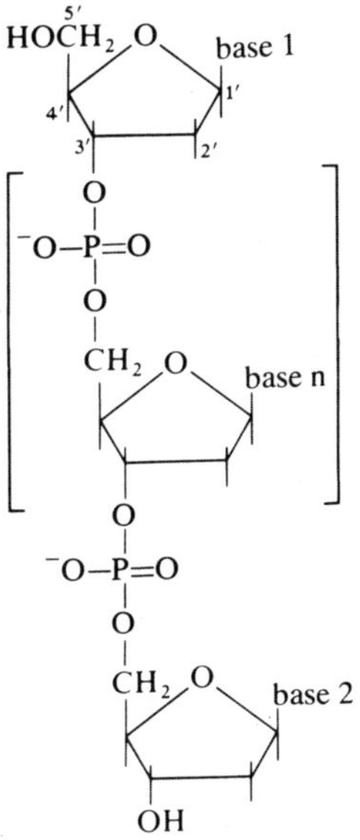

Base 1 is at the 5′ terminal end of the chain; base 2 is at the 3′ terminal end.

Miscellaneous:

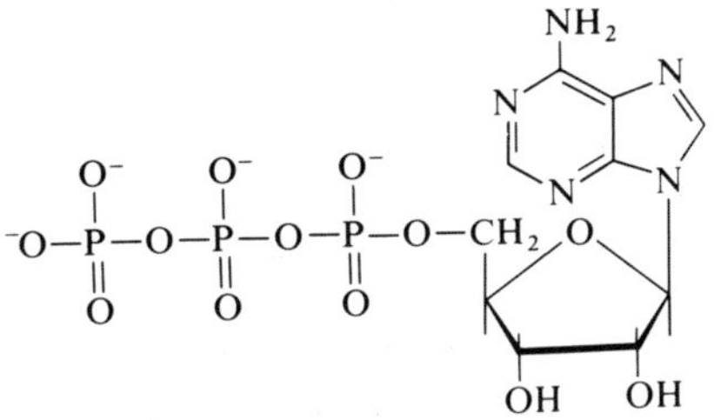

Adenosine triphosphate (ATP)

Table A.9 Biochemical compounds (*cont.*)

Miscellaneous (*cont.*)

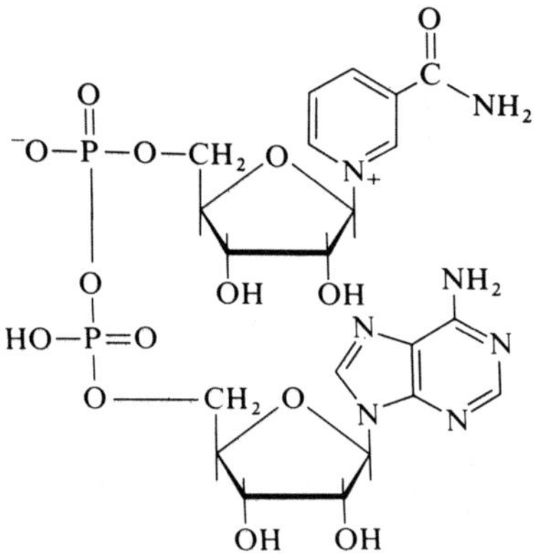

Nicotinamide adenine dinucleotide (NAD^+)

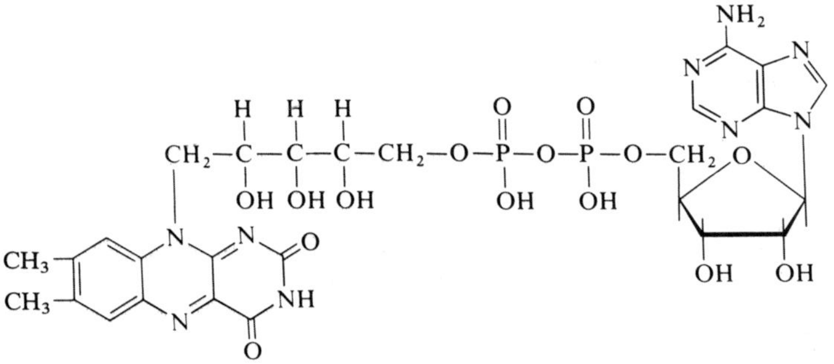

Flavin adenine dinucleotide (FAD)

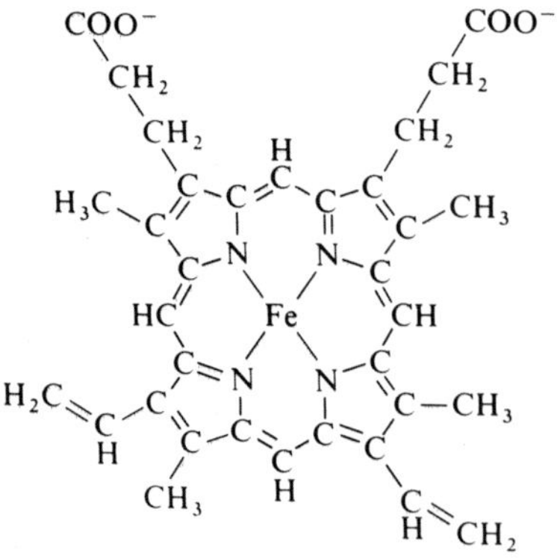

Heme, Fe II-protophorphyrin IX
(Hematin has Fe III.)

Table A.9 Biochemical compounds (*cont.*)

Miscellaneous (*cont.*)

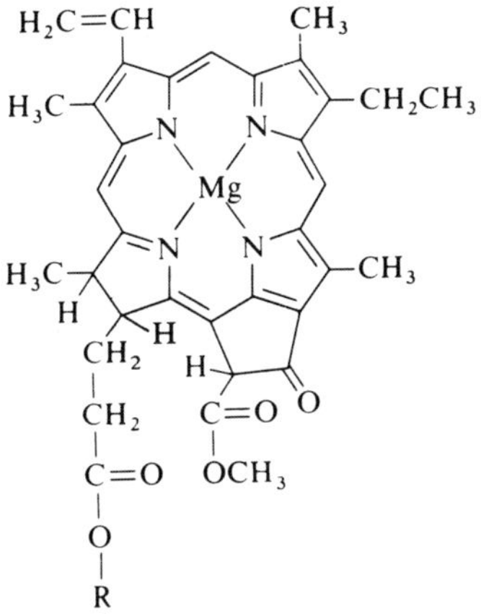

Chlorophyll a (R = phytyl, $C_{20}H_{39}-$)

Answers to Selected Problems

Chapter 2

3. (b) +48 cal **4.** (a) 10 kcal **5.** (a) −597 cal **6.** (a) $q = 0$, $w < 0$, $\Delta E < 0$, $\Delta H < 0$ **7.** (a) $\Delta H = q = 9720$ cal, $w = -741$ cal, $\Delta E = 8979$ cal **8.** (a) −48.44 cal

9. (a) $q = w = \Delta H = \Delta E = 0$. The expansion and contraction of the solids are negligible. **10.** 1 deg hr^{-1} **13.** 40°C **14.** (a) $q = -15.9$ kcal

15. 8.8×10^{-4} M **16.** (a) $\Delta E = \Delta H = 0$

17. (c) $\Delta H^0_{773} = \Delta H^0_{298} + 475 \cdot \Delta C^0_P$

$\Delta C^0_P = (\frac{1}{2}) \cdot \bar{C}^0_P(O_2, g) + \bar{C}^0_P(CH_4, g) - \bar{C}^0_P(CH_3OH, l)$ **20.** (a) −34.2 kcal

Chapter 3

1. (a) $q_1 = 133$ kcal, $q_3 = -33$ kcal. 133 kcal of heat is absorbed at the hot temperature and 33 kcal of heat is evolved at the cold temperature. The sum of these heats (+100 kcal) is converted to work done on the surroundings (w = −100 kcal).

2. (a) $\Delta H^0 = -34.2$ kcal, $\Delta S^0 = -19.5$ cal deg^{-1}, $\Delta G^0 = -28.4$ kcal **3.** (a) ΔG^0 is large and positive for this reaction. Do not buy catalyst. **4.** (b) $\Delta S > 0$, $\Delta G = < 0$

6. (a) −1.436 kcal; (b) −1.436 kcal; (c) −5.26 cal deg^{-1}; (d) 0; (e) −1436 cal evolved; (f) −0.04 cal done by system **8.** (c) 0.347 cal deg^{-1}

9. (c) Do not do it. Refrigerators only transfer heat from inside the refrigerator to outside. If you turn the refrigerator off and open the door you can cool the room slightly. (e) Wrong. The water will freeze, but the temperature will rise to give an increase in entropy.
11. (a) decrease; (b) zero; (c) decrease **13.** (c) Assume that T and P are constant. $q < 0$, $w > 0$, $\Delta T = 0$, $\Delta E < 0$, $\Delta H < 0$, $\Delta S = ?$, $\Delta G < 0$

Chapter 4

1. (a) $K = 77$, $\Delta G^{0\prime}_{298} = -2.57$ kcal mol^{-1}; (b) $\Delta G^{0\prime}_{298} = 4.83$ kcal mol^{-1}, $K = 2.9 \times 10^{-4}$
3. (c) 1.2×10^{-5} **4.** (b) 4.2×10^{-3} atm **5.** (a) 0.22 **7.** (a) -821 cal; (b) 0; (c) -821 cal; (d) -1642 cal; (e) 12.6 kcal; (f) 45.0 cal deg^{-1} **10.** 5×10^{-49} atm. Do not worry.
11. (a) 49.3 kcal. Heat is absorbed. (b) 153 cal deg^{-1}. (c) The increase of entropy on forming the rigid helix suggests solvent is being released from the coil. The dominant effect seems to be breaking of solvent-coil bonds.
13. (b) $H^+ = 10^{-7}$, $OH^- = 10^{-7}$, $Na^+ = 0.138$, $H_2PO_4^- = 0.062$, $HPO_4^{2-} = 0.038$, $PO_4^{3-} = 8.4 \times 10^{-8}$, $H_3PO_4 = 8.0 \times 10^{-7}$
16. (a) $+0.363$ V **18.** (c) -34.84 kcal **19.** (a) 5×10^{-8}; (b) -0.0178

Chapter 5

1. (a) 141 g mol^{-1}; (b) -0.66°C; (c) Solvent flows from a solution of low osmotic pressure to one of high osmotic pressure. (d) The solution of higher osmotic pressure will have the lower vapor pressure. **3.** (a) 2.2 kcal **7.** (a) 29.89 torr; (b) 100.1°C; (c) -0.37°C; (d) 4.9 atm; (e) 51.0m; (f) 0 atm
11. (a) -76°C. (b) 23.2 cal mole^{-1} deg^{-1}. Hydrogen bonding and association in liquid decreases entropy of liquid. (c) 30.65 kcal. Pure liquid is standard state. (d) NH_4Cl dissociates completely in NH_3. Boiling point corresponds to a 2 molal solution.
14. (a) 17.52 torr; (b) 1015 torr; (c) 0.999; (d) 1548 torr **16.** 3.55 atm
21. (a) 0.84 atm **22.** (a) -0.062; (b) 3300 cal

Chapter 6

1. (a) 1.84×10^5 cm sec^{-1}; (b) 813 cal mol^{-1}; (c) 2.7×10^{19} molecule cm^{-3}; (d) 1.34×10^{-5} cm; (e) 1.26×10^{10} sec^{-1}; (f) 1.76×10^{29} cm^{-3} sec^{-1}; (g) 1.5 cm sec^{-1}
3. (b) 6.7×10^{-8} g sec^{-1} **4.** a = 130 Å, b = 8.4 Å

7. $$\frac{s_2}{s_1} = \left(\frac{M_2}{M_1}\right)^{2/3}, \frac{D_2}{D_1} = \left(\frac{M_1}{M_2}\right)^{1/3}, \frac{[\eta_2]}{[\eta_1]} = 1.$$

12. ring/linear = 0.328

Chapter 7

1. (a) I_2 is zero order; ketone is first order; H^+ is first order **3.** (e) $A + A \rightarrow A_2$ (slow), $A_2 + B \rightarrow C + D + A$ (fast) **5.** (a) 6.25%; (b) 14.3%; (c) 0%; (d) 11% **9.** (a) zero order; (b) 1×10^{-3} M s^{-1}; (c) $-dA/dt = k$; (d) $D \rightarrow D'$ (slow), $D' + A \rightarrow AD'$ (fast), $AD' + B \rightarrow C + D$ (fast) **11.** 12.7 kcal mol^{-1} **14.** (a) second order; (b) $A + A \rightarrow P$; (c) 5 M^{-1} min^{-1}; (d) 67 min **16.** (c) 26.0°C

17. (a) I: $\dfrac{dO_2}{dt} = 3k(O_3)^2$, II: $\dfrac{dO_2}{dt} = 2\left(\dfrac{k_2 k_1}{k_{-1}}\right)\dfrac{(O_3)^2}{(O_2)}$;

(c) No. For mechanism I, $\Delta H^{\ddagger}$ must be > 0; for mechanism II, $\Delta H^{\ddagger}$ must be > 25.6 kcal. **20.** (b) 40% **21.** (a) 4.8 hr **22.** 0.12 **23.** (c) 4 mwatt

Chapter 8

1. $K_m = 0.44\ M$, $V = 0.35\ \mu$m CO_2 min^{-1} **2.** (c) $K'_{eq} = 3.75$ at pH 7.1, $K_{eq} = 3.0 \times 10^{-7}$

5. $\dfrac{d(\text{P})}{dt} = \dfrac{k_2 \cdot (E_0)}{1 + \dfrac{K_m}{(\text{S})}\left(1 + \dfrac{(\text{I})}{K_i}\right)}$, $K_i = \dfrac{k_{-3}}{k_3}$

7. (a) 44 mM; (b) noncompetitive **10.** (b) $\dfrac{1}{V}$ and $\dfrac{1}{K_m} = 0$; (c) $k_2 \gg (k_1 + k_{-1})$; (d) first order; (e) 1.25×10^9 M^{-1} s^{-1}; (g) 2; (h) $k_1 = 1.8 \times 10^9$ M^{-1} sec^{-1}, $k_2 = 3.6 \times 10^9$ M^{-1} sec^{-1} **13.** (b) $K_m(C_2H_5OH) = 0.44$ mM, $K_m(NAD^+) = 18\ \mu$M; (c) competitive; (d) noncompetitive **14.** (a) $k_1 = 6.4 \times 10^9$ M^{-1} sec^{-1}, $k_{-1} = 8{,}2 \times 10^6$ sec^{-1}

Chapter 9

2. (a) $E = 6.02 \times 10^{-13}$ erg, $\lambda = 1100$ nm;
(b) $E = 1.51 \times 10^{-13}$ erg, $\lambda = 4400$ nm; (c) $E = 12.04 \times 10^{-13}$ erg, $\lambda = 1100$ nm

3. (b) $\dfrac{5h^2}{4\pi^2 ma^2}$ **4.** (d) 660 nm

6. (c) $E_0 = 4.12 \times 10^{-13}$ erg, $E_1 = 12.36 \times 10^{-13}$ erg, etc.

9. (b) Bond energies: $NO^- < NO < NO^+$
Bond lengths: $NO^+ < NO < NO^-$

11. (a) 1.597×10^6 dyn cm^{-1}; (b) 1870 cm^{-1}; (c) 1390 cm^{-1}; (d) 1.148×10^6 dyn cm^{-1}; (e) $\nu(c) = 1343$ cm^{-1}, $\nu(d) = 1586$ cm^{-1} **13.** (a) 0.020; (b) zero; (c) 400 nm, 200 nm, 800 nm

15. (a) Ground state: $\mu_{\parallel} = 1.95$ D, $\mu_{\perp} = 1.70$ D,
Excited state: $\mu_{\parallel} = 0.28$ D, $\mu_{\perp} = 0.41$ D

Chapter 10

1. (a) 2.923×10^{-12} erg photon^{-1} = 1.82 eV = 42.0 kcal mol^{-1}; (b) 2.76; (c) 30–35%
4. (a) 0.01 M; (b) 0.02 M; (c) 9.8×10^{-4} cm
6. c(Chl a, μg ml^{-1}) = 12.80 A_{663} − 2.585 A_{645}
c(Chl b, μg ml^{-1}) = 22.88 A_{645} − 4.67 A_{663}
c(Chl total, μg ml^{-1}) = 8.13 A_{663} + 20.30 A_{645}
8. (a) 4.0; (b) 0.400 **12.** (a) $[\alpha]=2320$ deg dm^{-1} cm^{-3} g^{-1}, $[\theta]=9894$ deg 1 cm^{-1} mol^{-1};
(b) $A=0.8$, $A_L-A_R=3\times10^{-4}$, $\alpha=0.0075°$;
(c) $\alpha=1.31\times10^{-4}$ radians cm^{-1}, $n_L-n_R=1.16\times10^{-9}$,
$\psi=1.73\times10^{-4}$ radians cm^{-1}
15. (a) CH_3CHO;
(b)

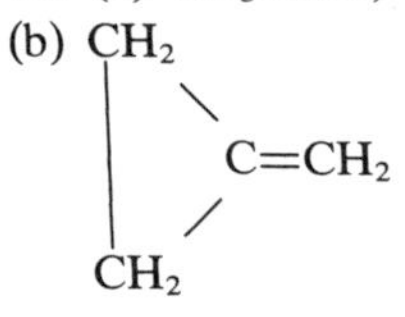

Chapter 11

2. $N=5$, $K=1\times10^5$ **4.** (d) $\dfrac{\tau^3 s^2}{3(1+\tau)}$

7. (c)
$$A=\left(\frac{c}{Z}\right)[\varepsilon_0+\varepsilon_1 s\cdot(\sigma_7+\sigma_9+\sigma_{11}+\sigma_{13}+\sigma_{15}+\sigma_{17})$$
$$+\varepsilon_2 s^2\cdot(\sigma_7+\sigma_9+\sigma_{11}+\sigma_{13}+\sigma_{15})$$
$$+\varepsilon_3 s^3\cdot(\sigma_7+\sigma_9+\sigma_{11}+\sigma_{13})$$
$$+\varepsilon_4 s^4\cdot(\sigma_7+\sigma_9+\sigma_{11})+\varepsilon_5 s^5\cdot(\sigma_7+\sigma_9)+\varepsilon_6 s^6\cdot(\sigma_7)$$

Z = terms in brackets with all ε_i set equal to 1.
11. (a) $26^3=17{,}576$ **12.** (a) Electron: 0, He: 3.0
13. (a) $T\to+0$; (b) $T\to-0$; (c) $T\to\pm\infty$.

Chapter 12

1. (a) 6.408×10^{-8} erg; (b) 0.31 Å;

(c)
$$\lambda_s=\frac{ch}{V}\cdot 1 \text{ electron volt}=\frac{ch}{V}\cdot 1.602\times10^{-12}\text{ erg}$$
$$=\left(\frac{1.24\times10^{-4}}{V}\right)\text{cm}=\left(\frac{12{,}400}{V}\right)\text{Å, with } V \text{ in volts}$$

3. (a) 67.4° and 39.6° respectively; (b) 2.38 cm **5.** 0.77 Å **6.** (a) 6.025×10^{23};
(b) 14,300 **8.** (a) 1:1:1

Index